CRC Handbook of Geophysical Exploration at Sea

2nd Edition Hydrocarbons

Editor

Richard A. Geyer, Ph.D.
Adjunct Professor
Offshore Technology Research Center
Texas A&M University
College Station, Texas

CRC Press
Boca Raton Ann Arbor London

Library of Congress Cataloging-in-Publication Data

Handbook of geophysical exploration at sea / [edited by] Richard A.
Geyer. -- 2nd ed.
p. cm.
Rev. ed. of: CRC handbook of geophysical exploration at sea.
c1983.
Includes bibliographical references and index.
Contents: v. 1. Hydrocarbons.
ISBN 0-8493-4252-X
1. Ocean bottom--Handbooks, manuals, etc. 2. Prospecting-
-Geophysical methods--Handbooks, manuals, etc. I. Geyer, Richard A.
II. CRC handbook of geophysical exploration at sea.
GC87.H35 1991
551.46'083--dc20 91-17946
CIP

This book represents information obtained from authentic and highly regarded sources. Reprinted material is quoted with permission, and sources are indicated. A wide variety of references are listed. Every reasonable effort has been made to give reliable data and information, but the authors and the publisher cannot assume responsibility for the validity of all materials or for the consequences of their use.

Direct all inquiries to CRC Press, Inc., 2000 Corporate Blvd., N.W., Boca Raton, Florida 33431.

International Standard Book Number 0-8493-4252-X

Library of Congress Card Number 91-17946

Printed in the United States of America 1 2 3 4 5 6 7 8 9 0

PREFACE

The percentage composition of the subject matter of this revised edition is approximately the same as that for the first, but the number of chapters has increased significantly. This necessitated the publication of a companion volume. In addition to dealing with various geophysical exploration methods, it adds and emphasizes the use of radiometric methods to explore for hard minerals found in sediments on and beneath the ocean floor. The chapters describing the various methods have been added because of their extensive usefulness in finding hard minerals of economic value, as well as in the search for hydrocarbons. The former also qualify for inclusion in this volume because these too are basically remote sensing methods; and can therefore be included in the generally accepted definition of the basic subject of Geophysics.

There is a great diversity of pertinent scientific areas comprising the broad subject of "Geophysical Exploration at Sea", but these can be classified into three major categories, namely: (a) potential field methods, as represented for example by gravity, magnetic, electrical, and telluric current methods; (b) those based on elastic wave theory, as it applies to the generation, transmission, reception and display of acoustic and seismic waves, as they propagate in the sea and into unconsolidated and consolidated bottom sediments; and (c) radiometric methods.

Each depends for its success on the ability to measure certain basic physical properties of the media under investigation, such as density for gravity, susceptibility for magnetics, resistivity/conductivity for electrical and telluric currents, and velocity and acoustic impedance properties for seismic methods. Actually, it is the differences in these properties between the geological target and the surrounding material that determine success in solving specific pure and applied scientific problems being studied on or beneath the bottom of the sea.

A problem is never completely solved unless the location in which the data are being measured is known to the degree of accuracy required for the particular problem being studied. Therefore, it is imperative that accurate navigational methods are available, especially in the open ocean where landmarks are absent. Thus, chapters are included describing the latest methods available to provide accurate navigational information at sea.

Two of the basic categories, namely, potential field and seismic acoustic methods, can be further subdivided, for purposes of discussion, into two subcategories: (1) the design of specialized instrumentation to obtain data in the marine environment; and (2) the development of theoretical and applied techniques to be used successfully to definitively interpret results of geophysical exploration at sea. These can be used to improve our understanding of various marine phenomena and processes, as they apply to structure, composition and geologic history of the subsurface unconsolidated and consolidated sediments and formations; as well as to explore for diversified resources of economic value that lie beneath it. It should also be emphasized that much of the information thus obtained can also be applied to help solve comparable scientific problems pertaining to the remaining 21% of solid earth that can be studied on land.

Participation was again solicited from members of the academic, industrial, and governmental communities to provide information describing the results of their most recent basic and applied research and operational methods in marine geophysics. Each in its own right has important information which, when shared, becomes mutually advantageous. The response from these three important sectors was again most gratifying and has resulted in achieving the objectives of this revised edition.

Information in the form of case histories was also included because these provide examples of the integrated approach to the solution of special problems indigenous to geographic areas and marine geologic provinces. Seismic-acoustic and geoacoustic methods have also been emphasized because they have dominated and will continue to dominate both pure and applied marine geophysical activities. These also have in general exhibited a greater versatility and degree of resolution in solving a wide range of geophysical and geologic problems, on both a regional and local scale, than potential methods.

The category of seismic-acoustics found in the first edition has been subdivided into the more definitive areas of geoacoustics and geotechnology. The latter is directed specifically toward the solution of problems involving geophysical hazards associated with selecting offshore drilling locations for hydrocarbons and for pipeline route surveys, whereas the term geoacoustics is reserved primarily for remote sensing studies to better understand: (a) the physical properties of sea floor sediments; (b) transmission of sound in the ocean; and (c) structural and tectonic characteristics of specific areas.

The design of specialized instruments and instrumentation systems has also been emphasized, because of the special environmental hazards encountered in the sea, as well as the broader and more efficient operational opportunities provided by conducting surveys in this medium as compared with similar ones on land. For example, continuous seismic surveying on land cannot be conducted because the instrument cables cannot be moved continuously as they can at sea. The same applies to electrical and telluric current surveys and to a lesser extent to gravity and magnetic surveys. For magnetic surveys, however, these can be made continuously from airplanes or helicopters flying over land.

Gravity surveys originally could not be made continuously on land unless confined to roads to obtain the most accurate results for oil exploration. However, since the publication of the first edition tremendous progress has been made in measuring gravity in the air using either fixed wing aircraft or helicopters, as well as from satellite observations. These can also be used to gather gravity data for marine surveys. All result in obtaining gravity data quickly, efficiently, and with sufficient accuracy for hydrocarbon exploration. Numerous aspects of these advances together with case histories are described in a number of chapters which also include discussions of theory, history, and instrumentation.

Geophysical research and operational activities are conducted at sea by citizens of many nations. Therefore, efforts were made to obtain participation from representatives of as many different countries and parts of the world's oceans as possible. For example, there are chapters written by scientists from Canada, Denmark, Sweden, and the United Kingdom, as well as from the United States. Similar chapters are included describing research and operational methods conducted in the Atlantic and Pacific Oceans, both in shallow (continental shelf) as well as in deep water, and in the Gulf of Mexico and Caribbean Sea.

Each of the individuals' sections of the volumes are designed to include a broad overview of the major operational and research geophysical radiometric methods used to obtain accurate information and to definitively interpret the results of exploration surveys at sea, and to provide an insight into the special needs for specific instrumentation to accurately yield the necessary data while at the same time operating in the hostile marine environment in terms of pressure and corrosion effects and rapid water movement from waves and currents. Case histories were again included together with regional studies as a means of presenting an integrated analysis of the solution of specific questions.

More emphasis has been placed on review-type chapters in the revised edition, because they provide a better perspective on the development and applications of each method in terms of increased accuracy and greater diversification of uses. The methods included in review chapters include gravity, resistivity, seismic acoustics and radiometrics. In addition, specific chapters are provided in each section describing the results of recent diversified research and operational techniques that can be applied to markedly improve data-gathering as well as interpretation and display procedures. These results can then help solve an ever-increasingly broader spectrum of theoretical and applied problems. This in turn leads to an improved understanding of our geological knowledge of the earth beneath the sea—knowledge that can be used to explore for the much-needed resources contained therein.

Richard A. Geyer, Ph.D.
Adjunct Professor
Offshore Technology Research Center
Texas A&M University
College Station, Texas

THE EDITOR

Dr. Richard A. Geyer has had a diversified career in geophysics and oceanography in both the industrial and academic sectors. He received his B.S. in Geology in 1937 from New York University and his Ph.D. in Geophysics from Princeton University in 1951. Beginning at the Woods Hole Oceanographic Institute (1943-1946), he continued in oceanography as head of the Oceanographic Section of Humble Oil Co., Technical Director of Oceanography at Texas Instruments, and finally as head of the Department of Oceanography at Texas A&M University (1966-1978). His activities in geophysics include periods as manager of the Gravity and Magnetic Departments of Geophysical Service, Inc., and adjunct professor of Marine Geology & Geophysics at the University of Houston. He is presently consulting in Exploration Geophysics and since September 1989 has served as adjunct professor of the Offshore Technology Research Center of Texas A&M University.

He has received national and international recognition for his efforts in these two fields by giving invited papers and was vice chairman of a presidentially appointed Committee on Marine Science, Engineering and Resources (1967-1969). He was named an honorary member of the Society of Exploration Geophysicists in 1981 and of the Dallas Geophysical Society in 1980, of which he was president in 1964-1965.

Dr. Geyer also has had extensive experience as an editor of several books on marine pollution and the use of submersibles in oceanography and ocean engineering. He edited the journal *Geophysics* from 1950-1952 and has published 30 papers in geophysics and oceanography. In addition, he was co-director of a NASA-sponsored project (1961-1962) to determine what geological and geophysical experiments should be performed by astronauts upon reaching the moon.

CONTRIBUTORS

Richard H. Bennett, Ph.D.
Senior Research Oceanographer
Seafloor Geosciences Division
Naval Oceanographic and Atmospheric Research Laboratory
Stennis Space Center, MS

Kevin B. Briggs, M.S.
Oceanographer
Geoacoustics/Geotechniques Branch
Seafloor Geosciences Division
Naval Oceanographic and Atmospheric Research Laboratory
Stennis Space Center, MS

Frank S. Carnaggio
Senior Electronic Technician
Geoacoustics/Geotechniques Branch
Seafloor Geosciences Division
Naval Oceanographic and Atmospheric Research Laboratory
Stennis Space Center, MS

Bruce Christian, M.S.
Network Software Connectivity Specialist
Communication Networks and Computer Services
Phillips Petroleum Co.
Bartlesville, OK

Peter Fleischer, Ph.D.
Geologist
Geoacoustics/Geotechniques Branch
Seafloor Geosciences Division
Naval Oceanographic and Atmospheric Research Laboratory
Stennis Space Center, MS

Richard A. Geyer, Ph.D.
Adjunct Professor
Offshore Technology Research Center of Texas A&M University
College Station, TX

William R. Gumert, M.A.
Chief Geophysicist
Aerogravity Division
Carson Services, Inc.
Perkasie, PA

Mark W. Johns, Ph.D.
Senior Geologist, Program Manager
PTI Environmental Services
Bellevue, WA

Woo-Yeol Jung, Ph.D.
Geophysicist
Naval Research Laboratory
Washington, D.C.

Douglas N. Lambert
Research Oceanographer
Geoacoustics/Geotechniques Branch
Seafloor Geosciences Division
Naval Oceanographic and Atmospheric Research Laboratory
Stennis Space Center, MS

Clyde Lee
President
Sytech Corp.
Houston, TX

Huon Li, Ph.D.
Senior Research Oceanographer (Retired)
Geoacoustics/Geotechniques Branch
Seafloor Geosciences Division
Naval Oceanographic and Atmospheric Research Laboratory
Stennis Space Center, MS

Paul K. Matthias, M.S.
Polaris Technology
Narragansett, RI

Philip D. Rabinowitz, Ph.D.
Professor
Department of Oceanography
Texas A&M University
College Station, TX

Charles R. Rein, Ph.D.
(Deceased)
Senior Mechanical Engineer
Ocean Technology Division
Naval Oceanographic and Atmospheric Research Laboratory
Stennis Space Center, MS

Michael D. Richardson, Ph.D.
Supervising Oceanographer
Geoacoustics/Geotechniques Branch
Seafloor Geosciences Division
Naval Oceanographic and Atmospheric Research Laboratory

William B. Sawyer
Geologist
Geoacoustics/Geotechniques Branch
Seafloor Geosciences Division
Naval Oceanographic and Atmospheric Research Laboratory
Stennis Space Center, MS

Byron R. Selman
Manager, Offshore Products
O.R.E. International, Inc.
Falmouth, MA

Alan Mark Swiercz
Exploration Geophysicist
Shell Offshore Inc.
New Orleans, LA

Elliott Taylor, Ph.D.
Coastal Science and Engineering Center
Woodward-Clyde Consultants
Seattle, WA

Mary N. Tinker, M.S.
Reservoir Research
Arco Oil & Gas Co.
Plano, TX

Samuel G. Tooma
Supervisory Physical Scientist
Seafloor Geosciences Division
Naval Oceanographic and Atmospheric Research Laboratory
Stennis Space Center, MS

Peter K. Trabant, Ph.D.
Consultant Oceanographer
Houston, TX

Herbert D. Valliant, M.Sc.
ZLS Corporation
Austin, TX

Donald J. Walter
Oceanographer
Geoacoustics/Geotechniques Branch
Seafloor Geosciences Division
Naval Oceanographic and Atmospheric Research Laboratory
Stennis Space Center, MS

Dawn J. Wright, M.S.
Department of Geography and Marine Science Institute
University of California at Santa Barbara
Santa Barbara, CA

David C. Young
Mechanical Engineering Technician
Geoacoustics/Geotechniques Branch
Seafloor Geosciences Division
Naval Oceanographic and Atmospheric Research Laboratory
Stennis Space Center, MS

TABLE OF CONTENTS

Chapter 1

UTILIZATION OF SATELLITE GEOIDAL ANOMALIES FOR COMPUTER ANALYSIS OF GEOPHYSICAL MODELS

Bruce E. Christian

TABLE OF CONTENTS

I. INTRODUCTION

Using radar altimeter data from the Seasat satellite, a geophysical model has been developed that successfully accounts for the anomalous geoid measurements observed over the Tonga Arc-Tonga Trench region. The model indicates an increase in density in the oceanic lithosphere and subducting slab that may be due to aging and compression of these tectonic features. The anomalous geoid appears to be sensitive to the modeling parameters. The effect of removing part of the model is evident in the anomalous geoid profile.

An isostatic and non-isostatic model for the compensation of the ocean bathymetry was analyzed. Airy isostatic compensation was found to over-compensate the Tonga Arc region and under-compensate the Tonga Trench. The non-isostatic model completely compensated the ocean bathymetry, but suggested an oceanic crust thickness of 18 to 24 km, with a compensation depth of 80 km. The pressure at depths of 9, 31, 59, and 80 km seems to indicate regions of maximum shear stress where earthquakes have been observed.

A. THEORY

Gravitational field theory is frequently used to investigate the nature of the Earth's interior. Even though it is well recognized that inversion of gravity anomalies to obtain density distribution leads to nonunique solutions, information is obtainable from the analysis. In utilizing gravity data in the past a common problem has been the lack of data covering large areas; however, global satellite geoid measurements now permit the analysis of large scale geological structures, which often are of a linear extent on the order of thousands of kilometers or more. In particular geoidal anomalies have been used in the study of trench-island arc systems and in the mechanical modeling of these dynamic features (Jones, 1982; Chapman and Talwani, 1982; Chapman, 1979; Leitao and McGoogan, 1976; Watts, 1979).

In this study we investigate the geoid anomaly associated with the Tonga Trench-Arc system. This tectonic feature as well as the bathymetry will be incorporated into the models and a study will be done to determine what effects the various components contribute to the anomalous geoid field. This study will attempt to determine models of the density field constrained by the observed geoid, to determine the corresponding nonhydrostatic stress field at various depths and finally to tentatively correlate the seismic activity in the area being investigated. An isostatic and a nonisostatic model will be used for determining the nonhydrostatic stress field and both models will be compared with the satellite geoid measurements as a constraint.

B. ISOSTASY

In studying the Earth's interior it is evident that some sort of ''compensation'' of the surface features is taking place, as evidenced by the distribution of the gravity anomalies. This compensation is governed by the principle of isostasy. The state of isostasy is usually defined as a compensation of surface loads by subsurface density anomalies distributed above some constant depth. The criteria for defining this distribution may be based on the assumption of constant pressure at a given depth, such as suggested by Airy (1855) and Pratt (1855); on a state of mechanical equilibrium being attained, as in the regional compensation models of Vening Meinesz (1931); or on a condition that strain energy is minimized as suggested by Jefferys (1970).

1. Airy Isostasy

The Airy hypothesis of isostasy (Figure 1) can be illustrated by picturing the Earth's crust as floating in a fluid magma layer that is denser than the crust. The assumptions used for the Airy system of isostasy are:

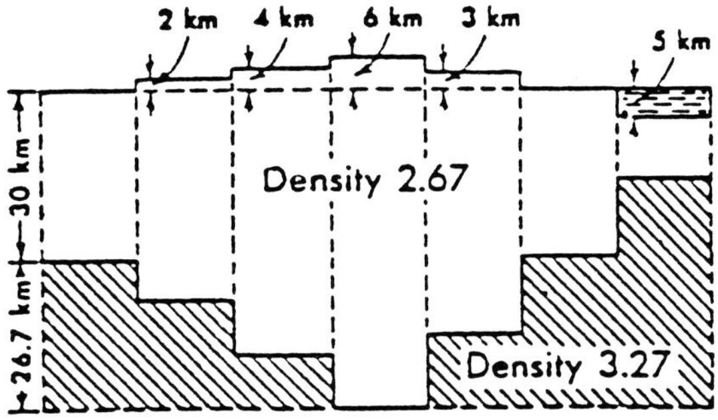

FIGURE 1. Airy isostatic system. (From Heiskanen and Meinesz, *The Earth and its Gravity Field,* McGraw-Hill, New York, 1958. With permission.)

1. The isostatic compensation is complete
2. The compensation is local, i.e., the compensating masses are directly under the topography
3. The density of the Earth's crust has a constant value, typically 2.67 g/cm^3. In this study a value of 2.85 g/cm^3 was used (Carlson and Raskin, in press)
4. At a certain depth the pressure is constant

Airy compensation implies that the Earth's crust is thicker under mountains than under lowlands and that the mass deficiency of ocean basins is compensated by antiroots of a heavy underlayer.

2. Pratt Isostasy

The Pratt model consists in the hypothesis that under the mountains the density of the Earth's crust is smaller, and when the ocean is deeper the density of the crust below is higher (Figure 2). The assumptions used for the Pratt isostatic system are:

1. The isostatic compensation is complete, that is, the density under mountains is smaller than beneath land at low elevations.
2. The compensating masses are located directly underneath the mountains and reach to the depth of compensation, where equilibrium prevails.
3. The depth of compensation is everywhere equal when measured from the geoid.
4. The density ρ' of the compensation, corresponding to a topographic elevation H, and a density of the topography satisfy the equation

$$\rho' = (H/D)\,\rho \tag{1}$$

3. Regional Isostasy

The Vening Meinesz hypothesis which assumes regional compensation is a modified version of the Airy hypothesis (Figure 3). In this theory the topography is a load on an unbroken crust or, in ocean areas, a mass taken away from the crust. The crust is assumed to behave like an elastic plate, strong enough to resist the shear stresses caused by the load floating on the denser substratum. The amount of downbending at each point is used to determine the effect of the isostatic compensation present at the computation point. Horizontal spreading of the compensation occurs in such a way that the compensating density diminishes from the center of the area toward the periphery.

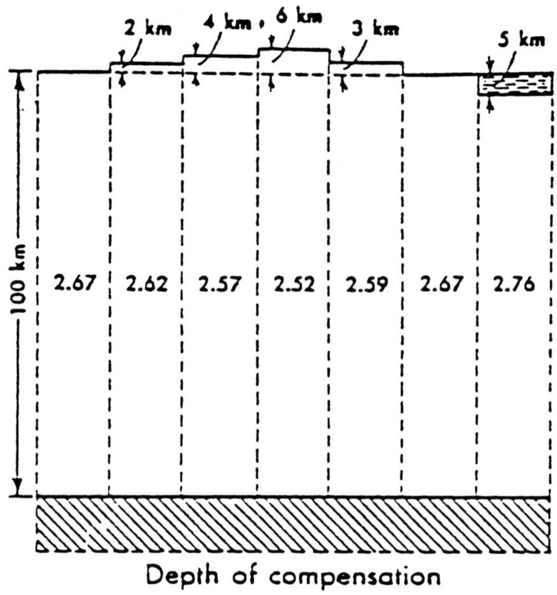

FIGURE 2. Pratt isostasy. (From Heiskanen and Meinesz, *The Earth and its Gravity Field,* McGraw-Hill, New York, 1958. With permission.)

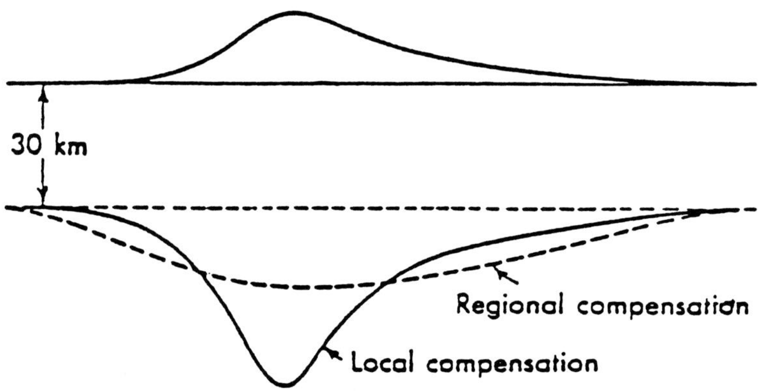

FIGURE 3. Comparison of Meinesz regional isostasy with Airy local isostasy. (From Heiskanen and Meinesz, *The Earth and its Gravity Field,* McGraw-Hill, New York, 1958. With permission.)

C. GEOID

The geoid as generally defined is the surface of the oceans, after some slight idealizations, that form an equipotential surface. This mean sea-level surface is considered to be extended continuously through the continents. This definition has proven highly useful and is considered the fundamental surface of physical geodesy.

Mathematically, an equipotential surface (Figure 4) is described by

$$W(x,y,z) = W_0 = \text{constant} \tag{2}$$

and is also known as a level surface.

Differentiating the gravity potential $W = W(x,y,z)$ yields

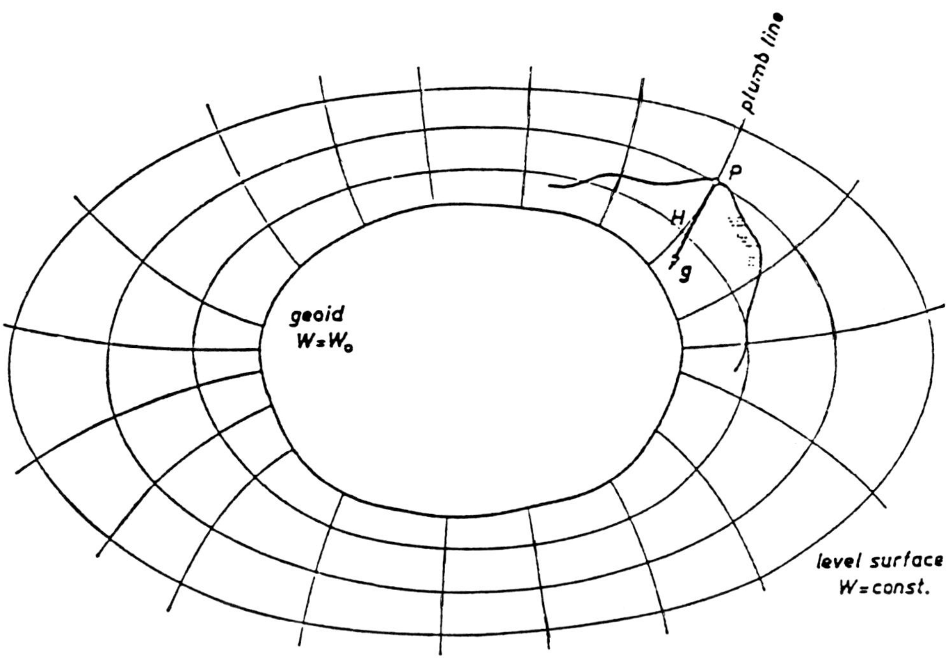

FIGURE 4. Level surfaces and the geoid. (From Heiskanen and Moritz, *Physical Geodosn,* W. H. Freeman Co., San Francisco, 1967. With permission.)

$$dW = \nabla W \cdot dr = g \cdot dr \tag{3}$$

where dr = (dx,dy,dz). If the vector dr is taken along the equipotential surface $W = W_0$, then the potential remains constant and dW = 0, and

$$g \cdot dr = 0 \tag{4}$$

This then implies that the gravity vector is normal to the equipotential surface (geoid) passing through the same point.

D. THE GRAVITY FIELD OF THE EARTH

The quantity called gravity is the result of gravitational force due to mass attraction and centrifugal force. The potential of gravity, W, is the sum of potentials of the gravitational force, U, and the centrifugal force, Ω:

$$W = W(x,y,z) = U + \Omega = \int\int\int k(\rho/l)dV + .5\omega(x^2 + y^2) \tag{5}$$

where

$$l = [(x - \xi)^2 + (y - \eta)^2 + (z - \gamma)^2]^{1/2} \tag{6}$$

and the integration is over the volume of the earth, $k = 6.67 \times 10^{-8}$ in cgs units, $\omega^2 = 0.5317484 \times 10^{-8}$ (rad/sec)^2, and ρ is the density. The vector quantity g (gravity) is then

$$g = \nabla W \qquad (\partial W/\partial x,\ \partial W/\partial y,\ \partial W/\partial z) \tag{7}$$

The scalar components of gravity arc then

$$g_x = \partial W/\partial x = -k\int\int\int((x - \xi)/l^3)\rho + \omega^2 x \tag{8}$$

$$g_y = \partial W/\partial y = -k\int\int\int((y - \eta)/l^3)\rho + \omega^2 y \tag{9}$$

$$g_z = \partial W/\partial z = -k\int\int\int((z - \gamma)/l^3)\rho \tag{10}$$

where ζ, η, and τ define the coordinates of the mass element. The magnitude g is called gravity. It has the dimensions of acceleration and is measured in "Gals". One Gal is equal to one cm/sec^2.

The mass contribution of the gravitational potential is mathematically described by

$$U = k\int\int\int (dM/l)dV \tag{11}$$

where the integration is carried over the whole Earth. In terms of spherical harmonics the gravitational potential is given by

$$U = (Mk/r)\sigma(a_1/r)\Sigma\Sigma P_j^m(\cos\theta)X[C_{jm}\cos mu + S_{jm}\sin mu] \tag{12}$$

where the first summation is from j equal zero to infinity and the second summation is from m equal zero to j.

The International Gravity Formula in closed form, measured in Gals is

$$\gamma_Q = 978.03189[(1 + 0.00193167\sin^2\phi)/(1 - 0.0066946090\sin^2\phi)^{1/2}] \tag{13}$$

which was computed from the following values adopted by the IAU 1964 assembly (Caputo, 1967, page 82)

- $a_1 = 6.378160 \times 10^8$
- $c_{20} = 1.08270 \times 10^{-3}$
- $MG = 3.98603 \times 10^{20}$ in cgs units.

E. THE ANOMALOUS GRAVITY FIELD AND GEOIDAL UNDULATIONS

Defining the disturbing potential T as the difference between the actual gravity potential W and the normal gravity potential U yields

$$T(x,y,z) = W(x,y,z) - U(x,y,z) \tag{14}$$

Comparing the geoid $W(x,y,z) = W_0$ with a reference ellipsoid $U(x,y,z) = W_0$ (Figure 5) gives the geoidal undulation (N) defined as the distance PQ. The gravity anomaly is then the difference in magnitude

$$\Delta g = g_P - \gamma_Q \tag{15}$$

which is a scalar quantity.

Now

$$U_P = U_Q - \gamma N \tag{16}$$

$$U_P = U_Q - T \tag{17}$$

then

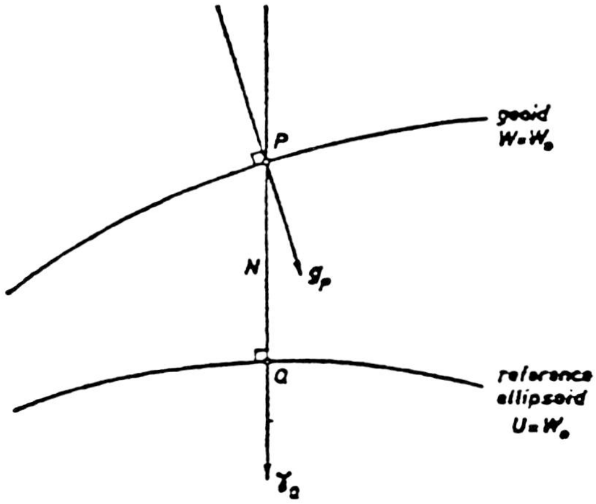

FIGURE 5. Geoid and reference ellipsoid. (From Heiskanen and Moritz, *Physical Geodosn,* W. H. Freeman Co., San Francisco, 1967. With permission.)

$$T = \gamma N \tag{18}$$

which is Brun's formula relating the geoidal undulation to the disturbing potential. Since

$$\gamma_P = \gamma_Q + N(d\gamma/dn) \tag{19}$$

and

$$g_P = \gamma_P - dT/dn \tag{20}$$

then

$$\Delta g = N(d\gamma/dn) - dT/dn \tag{21}$$

Introducing Brun's formula into Equation 21 then yields

$$\Delta g = (T/\gamma)(d\gamma/dn) - (dT/dn) = -\gamma\, d/dn(T/\gamma) \tag{22}$$

which gives the gravimetric anomaly as a function of the disturbing potential and the normal gravity field.

As a spherical approximation

$$\begin{aligned} \gamma &= kM/r^2,\ \partial\gamma/\partial n = \partial\gamma/\partial r = -2kM/r^3 \\ \gamma^{-1}\ \partial\gamma/\partial n &= -2/r \\ dT/dn &= \partial T/\partial r \end{aligned} \tag{23}$$

On the mean surface of the Earth then,

$$\gamma^{-1}\ (d\gamma/dn) = -(2/R) \tag{24}$$

where R is the Earth's mean equivolumetric radius. This then yields the fundamental equation of geodesy

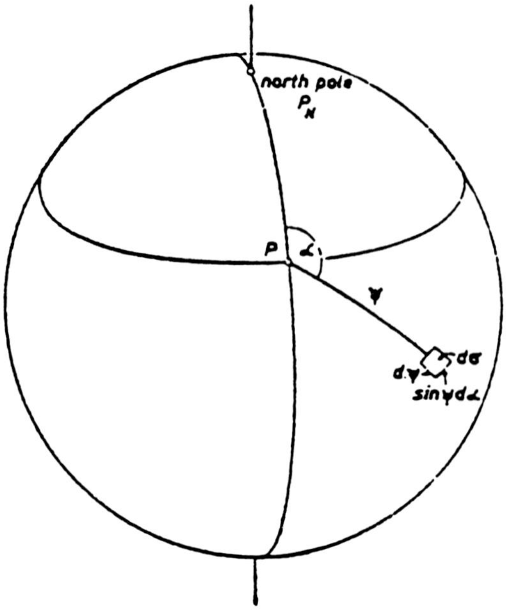

FIGURE 6. Coordinate system for integration of Stokes function. (From Heiskanen and Moritz, *Physical Geodosn,* W. H. Freeman Co., San Francisco, 1967. With permission.)

$$\Delta g = -(\partial T/\partial r) - 2T/a \tag{25}$$

This equation can be solved giving (see Appendix A)

$$N = R/(4\pi\gamma) \int\int \Delta g S(\psi) d\sigma \tag{26}$$

which is Stokes' formula. $S(\psi)$ is known as Stokes' function and is given by

$$\begin{aligned} S(\psi) = {} & \sin(\psi/2)^{-1} - 6 \sin\psi/2 + 1 - 5 \cos\psi \\ & - 3 \cos\psi \ln(\sin\psi/2 + \sin^2\psi/2) \end{aligned} \tag{27}$$

A more useful form of Stokes' formula is

$$N = (R/\gamma)\int \Delta g_{ave}(\psi) F(\psi) d\psi \tag{28}$$

where the integration (Figure 6) is from ψ equal zero to π and

$$F(\psi) = 1/2\ S(\psi)\sin\psi \tag{29}$$

$$\Delta g_{ave} = (1/2\pi)\int \Delta g(\psi\alpha) d\alpha \tag{30}$$

where a goes from zero to 2π. This final form of Stokes' formula now allows the geoidal undulation to be computed from the gravity anomaly due to an anomalous mass distribution (see Appendix B).

II. DESCRIPTION OF DATA

A. SATELLITE ACQUISITION SYSTEM

With the advent of radar altimeters in orbiting satellites, beginning with SKYLAB in 1973 (Leitao and McGoogan, 1976), GEOS-3 in 1975 (Stanley, 1979) and SEASAT in 1978 (Lame and Born, 1982), the geoid height in oceanic regions can be determined to within a meter. This allowed a data set of geoid anomalies covering most of the world's ocean to become available to the scientific community. The geoid height or anomaly is the distance between the geoid and the ellipsoid computed along the normal to the ellipsoid.

On June 27, 1978, the SEASAT satellite was launched. The altimeter experiment contained on the satellite had a range precision of $\pm$ 10 cm allowing determination of the ocean surface to better than a meter. The SEASAT data provides a geoid anomaly with respect to a best fit reference ellipsoid based on the 1° by 1° GEM 10B gravimetric geoid.

A global daset of geoidal anomalies have been computed for the GEOS-3 data by Brace and M. Schultz (1980). While the vertical resolution is adequate for most modeling studies of the subsurface structure the spatial resolution is not. Since the GEOS-3 data provided has a value at every degree, this yields a spatial resolution of approximately 200 km. Unfortunately this distance is much greater than the dimensions of many geological features. Using the more recent SEASAT satellite data it is possible to obtain an average measurement of the geoid every 7 km, thereby permitting increased spatial resolution.

To realize the full potential of the altimeter measurements, a number of corrections to the measurement must be provided with an accuracy compatible with the 10 cm measurement precision. These corrections can be separated into (1) instrument corrections, (2) propagation medium corrections, (3) a geoid model, (4) effects of temporal variations in the ocean surface (solid earth tides, ocean tides, barometric pressure, etc.), and (5) the spacecraft orbit.

As shown in Figure 7, the altitude measurement h can be expressed as

$$h^* = h + h_{sg} + h_i + h_a + h_s + h_g + h_t + h_o + \epsilon \tag{31}$$

where

- h^* is the distance from the satellite center of mass to the reference ellipsoid at the altimeter subsatellite point as determined from SEASAT orbit computation
- h is the satellite height above the ocean surface measured by the altimeter
- h_i is the total of instrument delays
- h_{sg} is the effect of the spacecraft geometry
- h_a is the sum of atmospheric path length corrections
- h_s is the correction due to surface and radar pulse interaction
- h_g is the geoid height
- h_t is the correction for solid earth and ocean tides
- h_o is the surface topography due to ocean circulation phenomena
- ϵ is a random noise measurement.

B. SEASAT GEOPHYSICAL DATA RECORDS

SEASAT geophysical data records (GDR) contain satellite data that have been processed through the sensor and/or geophysical algorithms (NASA publication 622-97, 1980). Each tape consists of sensor and/or geophysical records and text records for one of the four sensors contained on the SEASAT satellite. The processing was done by the Algorithm Development Facility at the Jet Propulsion Laboratory in Pasadena, CA. GDR text records contain bookkeeping and general information while the GDR data records consist of both sensor and/or geophysical records.

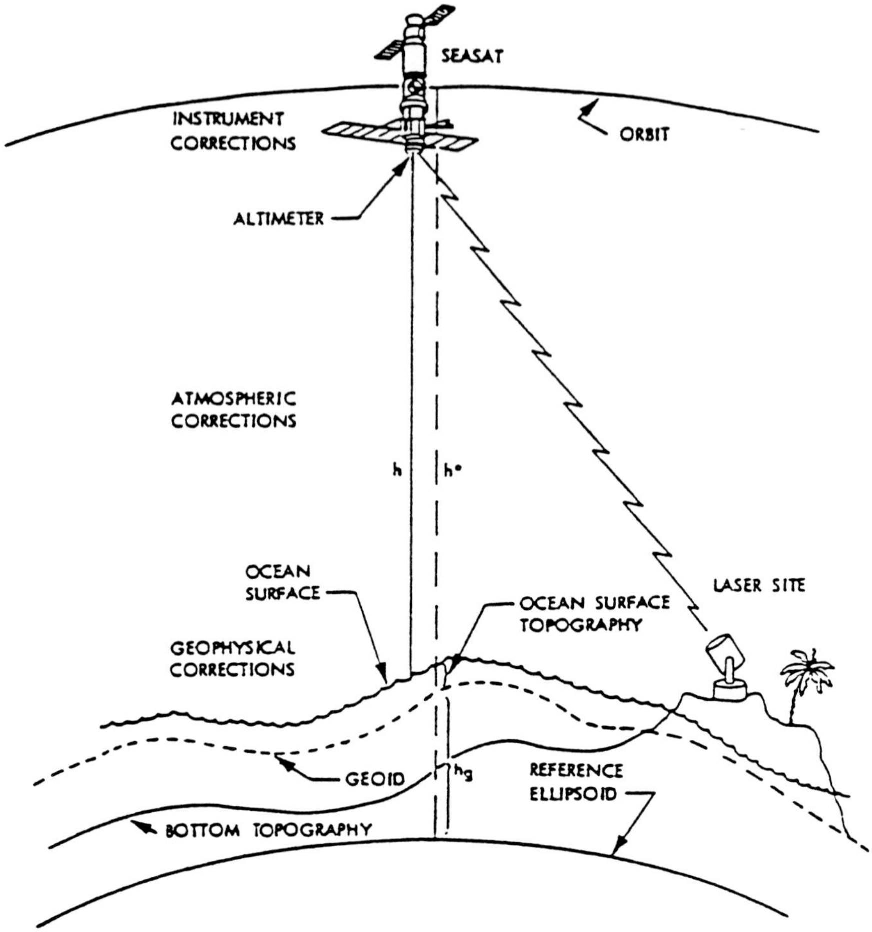

FIGURE 7. SEASAT data collection, modeling and tracking system. (From Tapley et al., *J. Geophys. Res.*, 87, 3179, 1982. With permission.)

The GDR file is read in records of 8-bit bytes with the maximum record being 8064 bytes. Each file contains one or more text records followed by one or more integer valued data records. The data on these records are represented as unsigned integers. The high order byte is not a sign bit, but a significant data bit. These files contain EBCDIC, ASCII, and unsigned integer data. While the individual records will be of one type only, the files will contain records of all three data types.

To utilize these tapes efficiently each individual record was accessed and the first two bytes of information was read and stored in a temporary file. The first integer byte contains the record type, which should be ten. The second integer byte contains the data type which should be one. These numbers are the code numbers that indicate this particular record contains altimeter data. The temporary file was then read concurrently with the tape using the temporary file to determine records of interest. If the temporary file indicated the record was a proper data record all 8064 bytes were read into a one byte array.

In using these data records on Digital Equipment Corp. VAX-780 computer, multi-byte quantities need to be reversed. This was done by equivalencing the one byte data array with a four byte integer array and then reversing the four byte integer number. These integer numbers were then scaled by a scale factor supplied on the tape. The scaling procedure used for the geoid values was to subtract 15,000 then multiply by 0.001. The latitude and longitude were scaled by 10^{-6}.

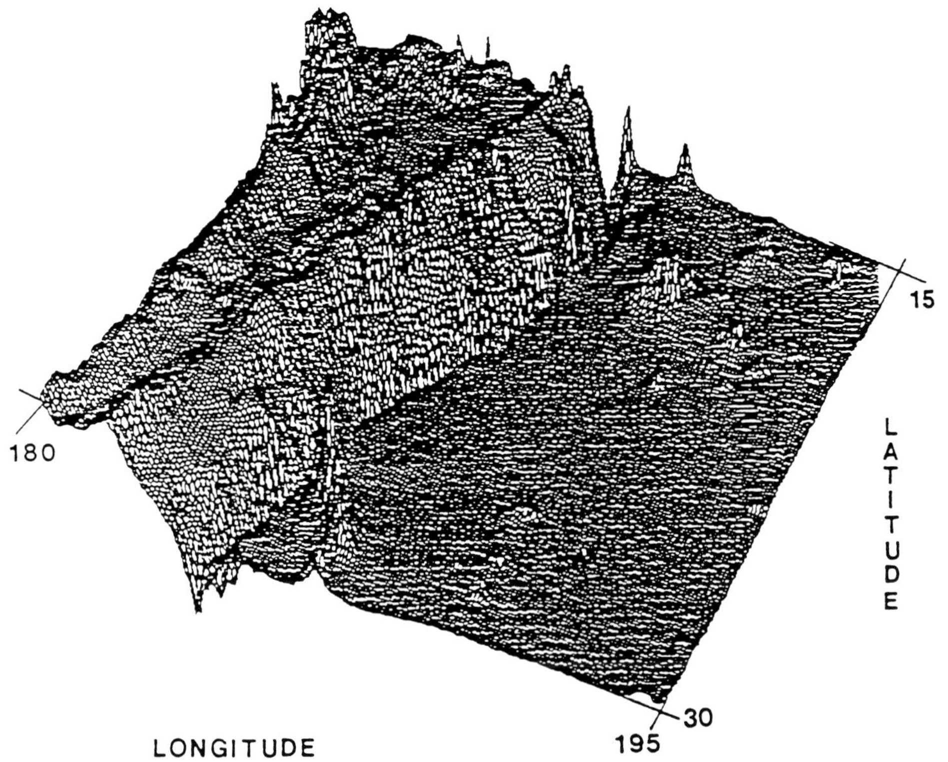

FIGURE 8. Three dimensional plot of area of investigation.

C. SYNBAP BATHYMETRY DATA

An additional data set used was the SYNBAPS global gridded bathymetric data. This data set consists of digital bathymetric data interpolated into a 5-min by 5-min latitude/longitude grid. The gridded data base was interpolated with digitizations of bathymetric maps available up to 1981. Depths are given in uncorrected meters based on a 1500 m/sec sound speed.

Each data record consists of 233 card images containing 3721 depth values in 16I5 format. The 16I5 format represents 16 integer numbers with five bytes per number. The data records are preceded by a record header which contains the colatitude and longitude of the first data point in the data records. Since each data record represents a 5°-by-5° area, then by determining the position of the bathymetry data in the record the latitude and longitude can be assigned to it. All of this data is contained on one tape which is subdivided into 16 files representing a 45°-by-45° area.

A 3D plot of the Tonga Trench, Tonga Arc, and the South Pacific Ocean utilizing this data is shown in Figure 8.

III. AREA OF INVESTIGATION

The area of investigation in this study is the Tonga Trench, the associated lithosphere 1000 km seaward of the trench, and the back-arc region associated with the Tonga Trench. Geographically the Tonga Trench is closely associated with the Kermadec Trench. However, geophysically and geologically they exhibit different characteristics. Current studies of the northern part of the Tonga region suggest a complicated subduction closure system. A detailed investigation (Hanus and Vanek, 1979) of the morphology of the Wadati-Benioff zone based

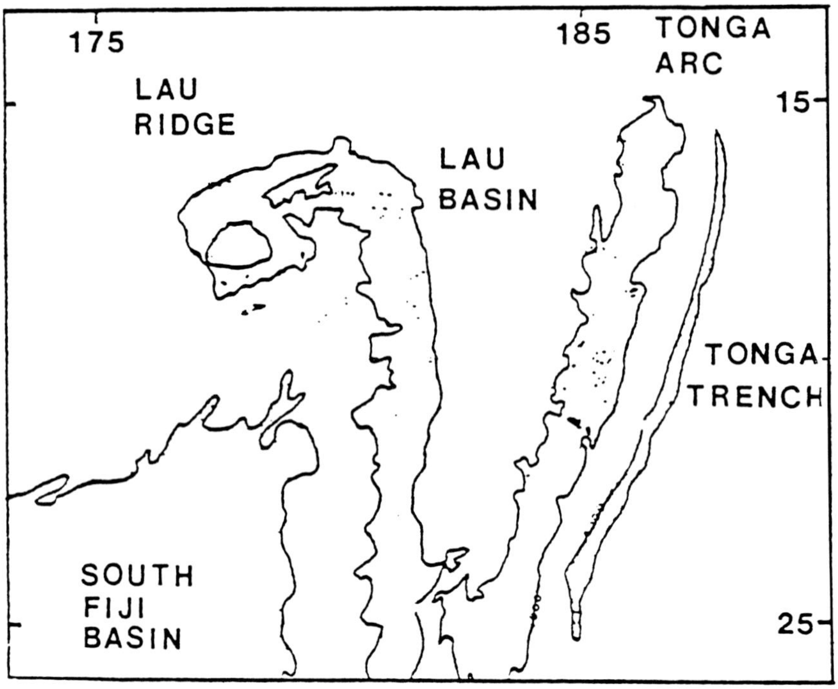

FIGURE 9. Area of investigation.

on the geometry of earthquake distributions suggests that the deep structure of the northern part of the Tonga region is the result of three different subduction systems. The three subduction systems in this area are Tonga, Lau, and Horne along with two deep seated fracture zones (Niuafo'ou, Peggy). The Tonga system represents the active part of the recent subduction of the Pacific plate, and the Lau and Horne systems represent paleoplates that may have become activated by collision with the Tonga system. The seismically active portions of both fracture zones reach to a depth of at least 400 km.

The Tonga island and ridge system forms a north to northeast trending linear system (Figure 9). Active volcanoes are found on the western side of the Tonga Ridge approximately 150 to 200 km west of the Tonga Trench. The Tonga Islands are mainly composed of limestone with some Eocene lavas and Micoene tuffs exposed.

Extending westward from the Tonga Ridge is the Lau Basin. The upper mantle beneath the Lau Basin is highly attenuating and typified by very low P-wave velocities. The eastern boundary of the attenuating zone is near the line of active volcanism and appears to extend to the seismic zone. The Lau Basin is also characterized by relatively high heat flow, thin sedimentary cover, and is underlain by oceanic crust capped by ocean floor type basalts (Sclater et al., 1972), Karig (1971) has interpreted the Lau Basin as a young (Pliocene) and probably still active marginal basin formed by extensional rifting behind the Tonga Arc.

A. MORPHOLOGICAL AND GEOPHYSICAL CHARACTERISTICS

A well defined seismic zone dips toward the west from the trench system beneath the Tonga-Kermadec Ridge (Sykes et al., 1969). This zone appears to be associated with a slab characterized by high seismic velocities. The upper boundary of the slab approximately coincides with the seismic zone and shows a distinct velocity continuity. The seismicity in this area extends to depths greater than 600 km (Figure 10). Strong compressional stresses within and parallel to the strike of the slab have been shown to exist by Strobach (1973). These stresses appear to be due to the folding of the slab as a consequence of the linearity

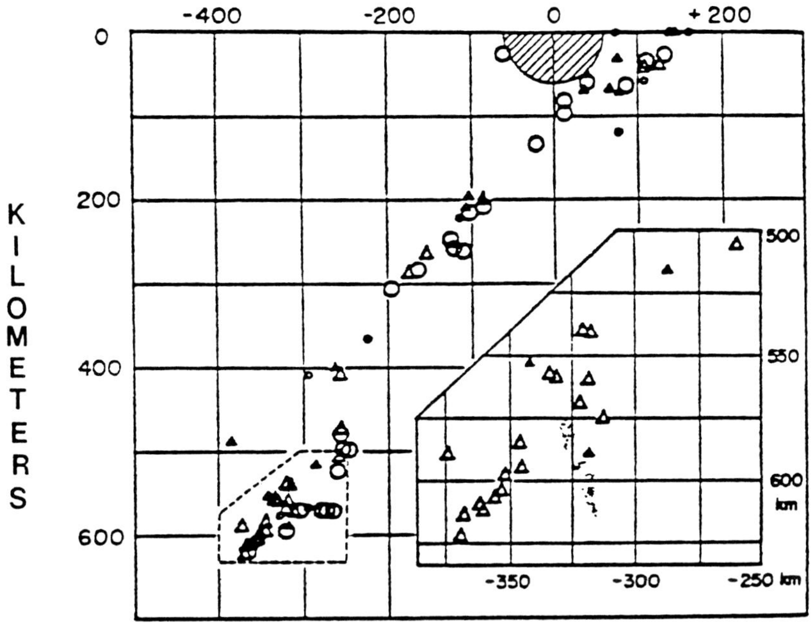

FIGURE 10. Earthquake hypocenters, distance from Tonga in kilometers.

of the Tonga Arc. The dip of the Benioff zone is approximately 43 to 45 degrees while the descent rate of the slab is in the range of 91 to 110 mm per year (Sclater et al., 1972; Le Pichon, 1968).

IV. ANALYSIS AND INTERPRETATION

An analysis by Wainwright (1983) showed that the isostatic geoid over trench regions can be considered constant, thereby allowing the assumption of two dimensionality. This is valid as long as the analysis is confined to areas away from the edge of the trenches, which are usually quite complex in terms of geological and tectonic features, but this complexity appears to diminish within 200 km from the edge of the subduction zone. Therefore the models used in this study are two dimensional.

In this analysis the models being examined will be compared to the observed SEASAT data shown in Figure 11. The characteristics typical of the SEASAT data in this area are a maximum difference of about 30 m and a smooth profile that exhibits no relative high frequency undulations. Parallel profiles a few kilometers away show a variation of only 1 to 2 m with respect to the SEASAT data shown in Figure 11.

A. ISOSTATIC MODEL

This study begins with a test of how well Airy isostasy compensates for the negative load of the bathymetry. To accomplish this analysis the amount of compensating mass needs to be determined. This is done by assuming constant pressure at a fixed depth; below this depth it was assumed compensation was complete. Assuming constant pressure then implies the total mass above any arbitrary point, at the depth of compensation, is constant. Utilizing this fact with a three layer model that uses the crust density as a reference density yields the following equation,

$$(S + C + R)\rho_c = S\rho_s + C\rho_c + R\rho_r \tag{32}$$

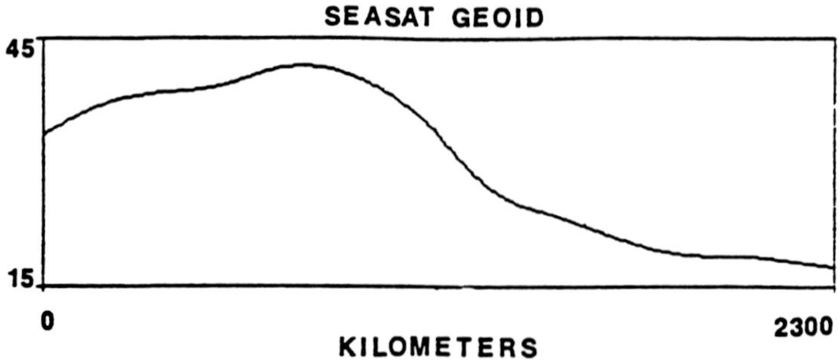

FIGURE 11. SEASAT data profile used in analysis of anomalous density models.

Solving for R, the length of the compensating root, in terms of S the bathymetric depth and C the crustal thickness yields,

$$R = S(\rho_c - \rho_s)/(\rho_r - \rho_c) \tag{33}$$

where ρ_c is the density of the crust, ρ_s is the density of seawater, and ρ_r is the density of the compensating root.

Implementing equation 33 with a density for seawater of 1.03 gm/cm^3, a density for the oceanic crust of 2.85 gm/cm^3, and a density for the compensating root of 3.35 gm/cm^3 results in the mass anomaly distribution shown Figures 12 and 13. In Figure 12 a compensation depth of 45 km is assumed. This depth is used because it put the compensating root within 3 to 4 km from the bottom of the ocean, beneath the trench axis. This then implies 45 km is the minimum (within 2 to 3 km) compensation depth, since any shallower depth would cause the compensating root to intersect the ocean floor. Airy isotasy when normally used is not considered for depths greater than 50 to 60 km, however, Figure 13 shows the results for a compensation depth of 80 km.

A second model was then studied using for seawater a density of 1.03 gm/cm^3, a density for the oceanic crust of 2.67 gm/cm^3, and a density for the compensating root of 3.50 gm/cm^3. The minimum compensation depth for these choices of densities is approximately 30 km (Figure 14) and yields an oceanic crust of about 11 km. Figures 15 and 16 show the results for compensation depths of 50 and 80 km using the same densities as before, giving an oceanic crust of 40 and 60 km, respectively. In the second model, since the anomalous mass density is larger, the vertical thickness of the compensating root is smaller.

All five of these figures have the same functional form but have different valued maxima and minima. Figure 14 (30 km compensation depth) yielded the smallest difference between the maximum and minimum (80 milligals), whereas Figure 13 (80 km compensation depth) was the largest (480 milligals). It was initially hoped that these compensation models would yield smoother gravity/geoid profiles, however, they obviously do not. Instead, by assuming Airy isostasy this tends to introduce high frequency anomalies into the geoid which are not present in the Seasat geoid. Specifically the assumption of Airy isostasy undercompensates the trench and overcompensates the arc and seamount.

B. NON-ISOSTATIC MODEL

Figure 17 depicts a non-isostatic model for the oceanic crust across the Tonga region with its associated gravity and geoid profile. This model has three layers with the density for seawater of 1.03 gm/cm^3, a density for the crust of 2.85 gm/cm^3, and a density for the root of 3.35 gm/cm^3. It does not assume isostasy in a particular form, but does assume that the bathymetry is compensated in some manner, such that the associated geoid profile is as

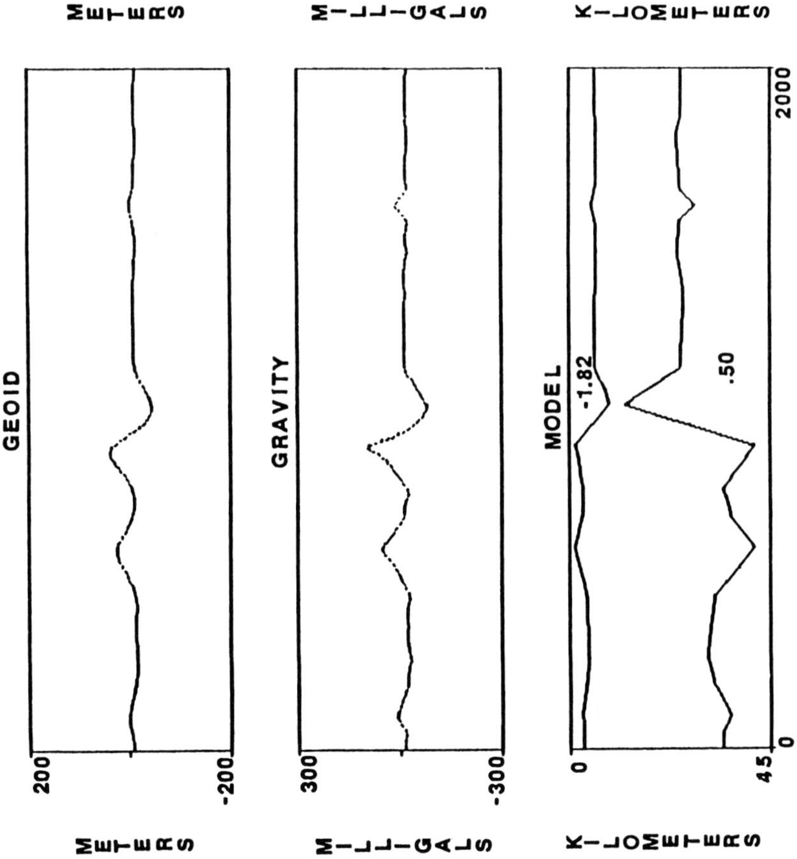

FIGURE 12. Airy isostasy, root density 3.35, compensation depth 45 km.

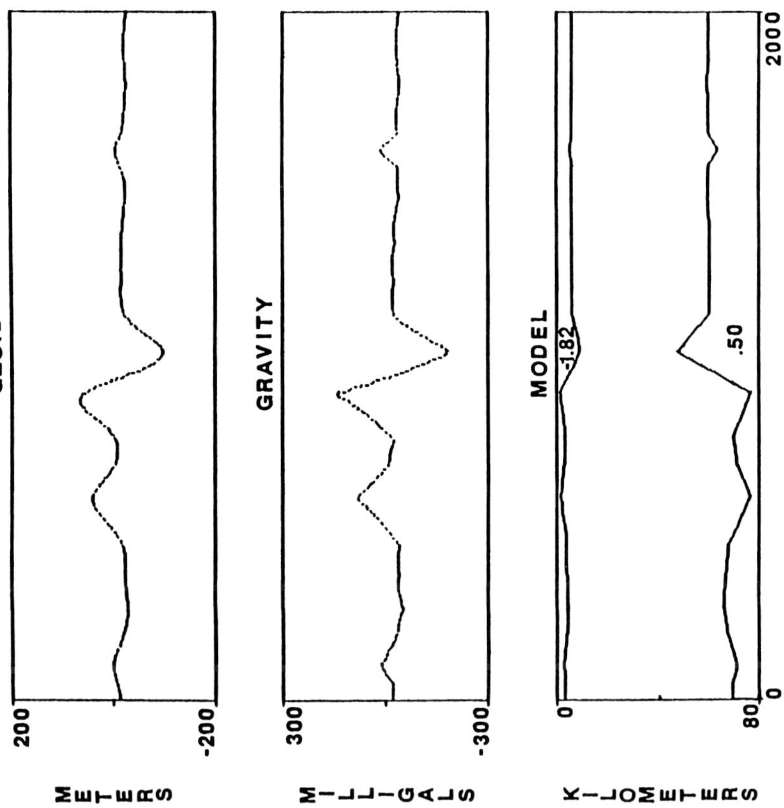

FIGURE 13. Airy isostasy, root density 3.35, compensation depth 80 km.

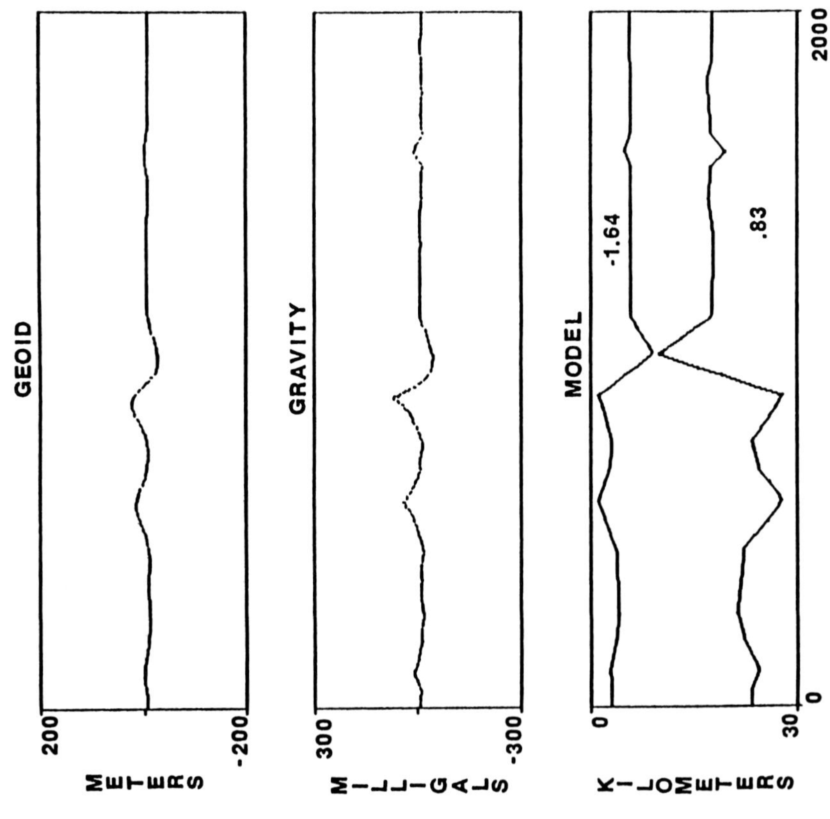

FIGURE 14. Airy isostasy, root density 3.50, compensation depth 30 km.

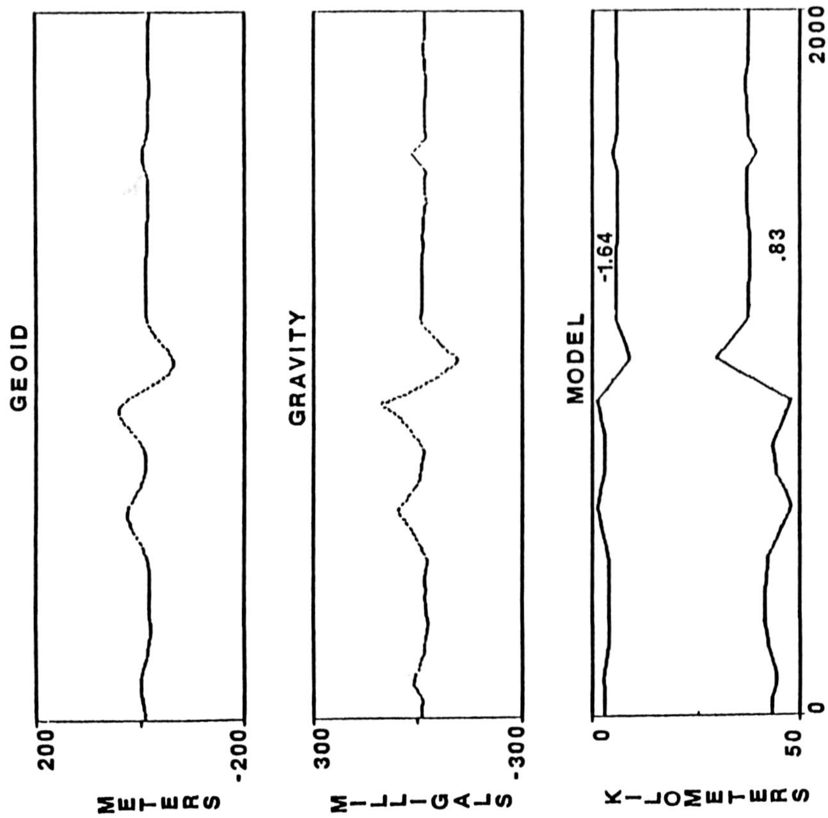

FIGURE 15. Airy isostasy, root density 3.50, compensation depth 50 km.

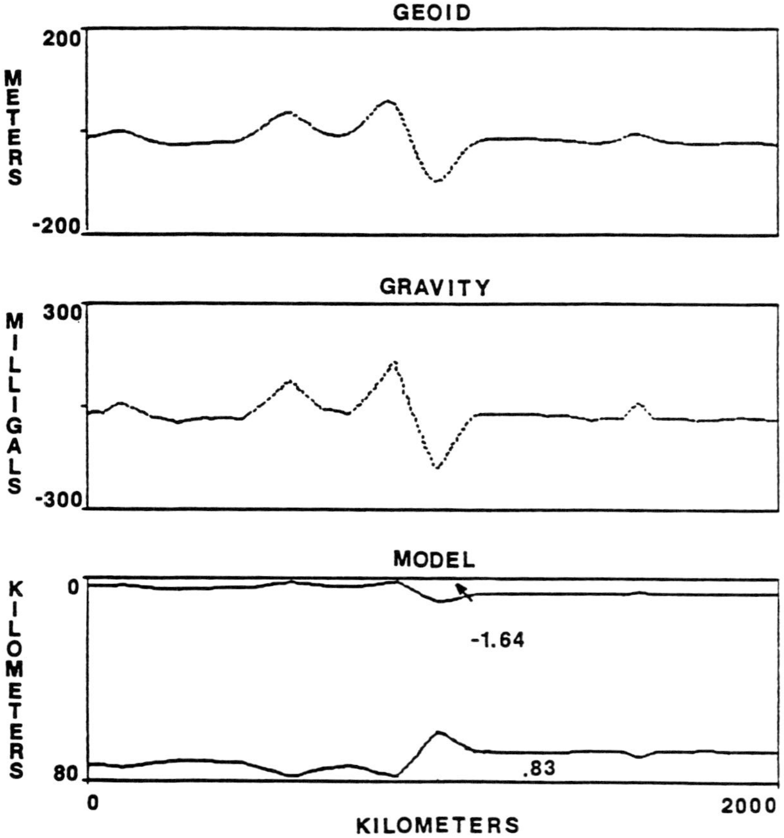

FIGURE 16. Airy isostasy, root density 3.50, compensation depth 80 km.

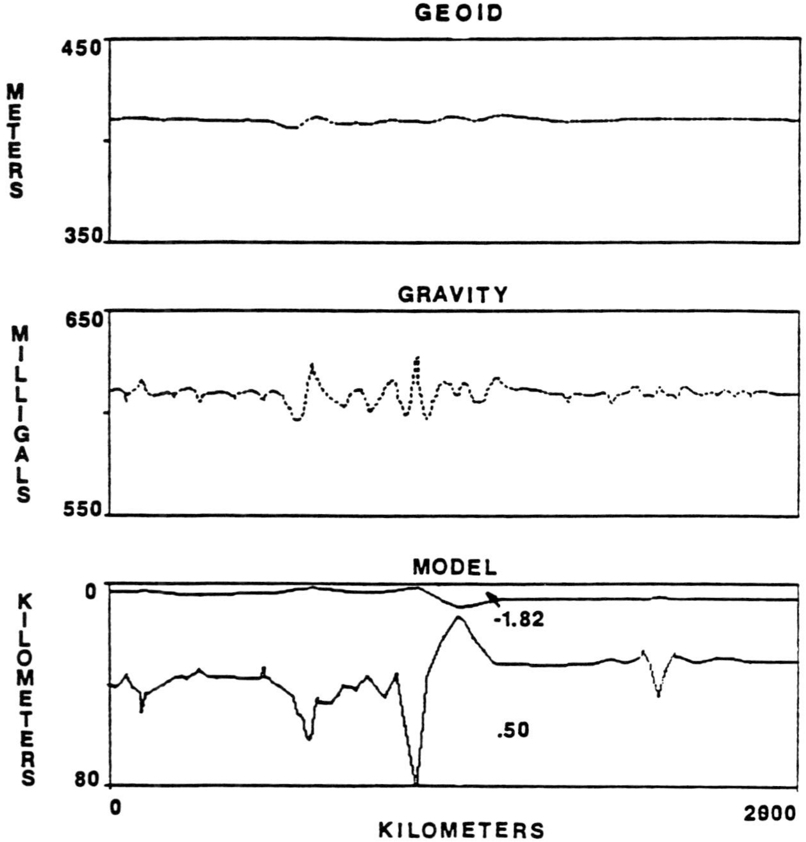

FIGURE 17. Non-isostatic model, seawater density 1.03, crust density 2.85, root density 3.35.

smooth as the observed Seasat geoid suggests and that the long wavelength of the observed Seasat geoid is caused by the subducting slab.

In this analysis the modeling of the anomalous density distribution is done in two steps. First, by taking into account the effect of the crustal density anomaly analyzed in this section and, second, by taking into account the slab density anomaly which will be analyzed in the next section. In determining if a model is considered good the main criteria will be how well the predicted geoid due to the model matches the characteristics of the observed SEASAT geoid. The models will be iterated with changes in the density anomalies and/or changes in the compensating root depth.

Using the constraint mentioned above resulted in the minimum depth of 80 km to produce a relatively constant geoid. The anomalous geoid of this model has undulations within $\pm$ 3 m of the average value. This model can be considered to compensate the bathymetry almost completely, yielding a crustal thickness of about 24 km.

Neither the density of the compensating root nor the crustal density are known. It is generally assumed to be 3.35 gm/cm^3 (Turcotte and Schubert, 1982). Since normal oceanic crust is generally assumed to be 6 to 7 km thick, an additional model was tried to seek a thinner crust. Figure 18 shows a three layer model with the same density for seawater and crust that the model in Figure 17 has. However, the root density is assumed to be .50 gm/cm^3. This gives an oceanic crust of approximately 18 km. In order to decrease the oceanic crust thickness more, it would be necessary to make the compensating root density even larger. An alternative procedure to lower the thickness of the oceanic crust would be to assume a model with a more complicated structure.

C. SLAB-LITHOSPHERE ANALYSIS

In this section the oceanic lithosphere, the subducting slab, and the back-arc lithosphere will be modeled with the objective of yielding a geoid that matches the observed Seasat geoid anomaly. A geophysical model depicting the lithosphere and associated subducting slab is shown in Figure 19. The slab is modeled as subducting at a 45° angle to a depth of 600 km. The thickness of the slab at the top is 100 km, thinning to a thickness of 50 km at the bottom. The oceanic lithosphere is a constant 100 km thick, but increases in density with age. The back-arc lithosphere is 65 km thick, which is the expected thickness for oceanic lithosphere that is 30 M.Y. old (Chapman and Talwani, 1982). The geoid anomalies caused by the model shown in Figure 19 compare very well with the observed Seasat satellite geoid. Figure 20 shows the Seasat satellite geoid anomaly, the geoid anomaly predicted by the model shown in Figure 19, and the difference between the two.

McAdoo and Martin (1984) have shown that oceanic lithosphere generally increases in thickness by a 4 km/my$^{1/2}$ times the square root of the age. For the oceanic lithosphere in this study, with an assumed spreading rate of 100 mm/year and an initial thickness of 100 km, the increase in thickness should be about 14 km. This increase in thickness can be considered equivalent to an increase in density. The Pacific Ocean lithosphere is assumed to have a relative anomalous density of 0.0058 gm/cm^3 with respect to the asthenosphere. The amount of increase in density due to thickening of the lithosphere would be 0.0008 gm/cm^3. The actual modeled increase is 0.0023 gm/cm^3, leaving an increase in density of 0.0015 gm/cm^3 unaccounted for.

The subducted slab is modeled with an initial anomalous density of 0.0086 gm/cm^3 increasing progressively down the slab to 0.0120 gm/cm^3. This increase in anomalous density is approximately twice the increase in density that was unexplained for the ocean lithosphere.

This anomalous increase in density may be due to the lithosphere being compressed as it approaches the trench followed by a further increase as it is forced into the asthenosphere.

Using the model shown in Figure 19 as a standard model, the effect of some of its components on the anomalous geoidal undulation were studied. This is done in order to see

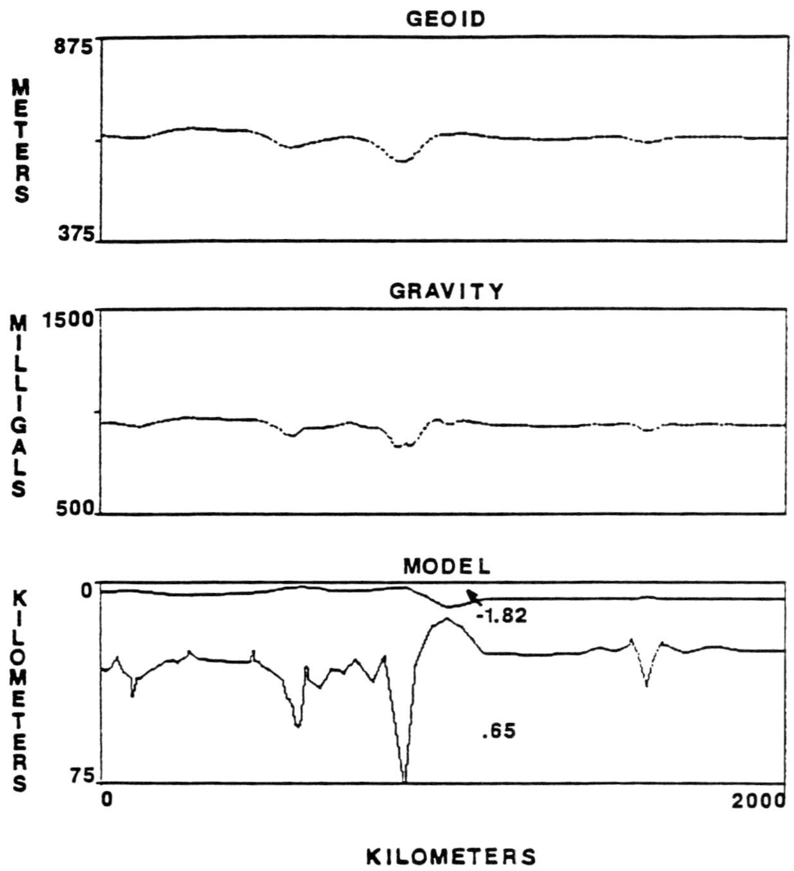

FIGURE 18. Non-isostatic model, seawater density 1.03, crust density 2.85, root density 3.50.

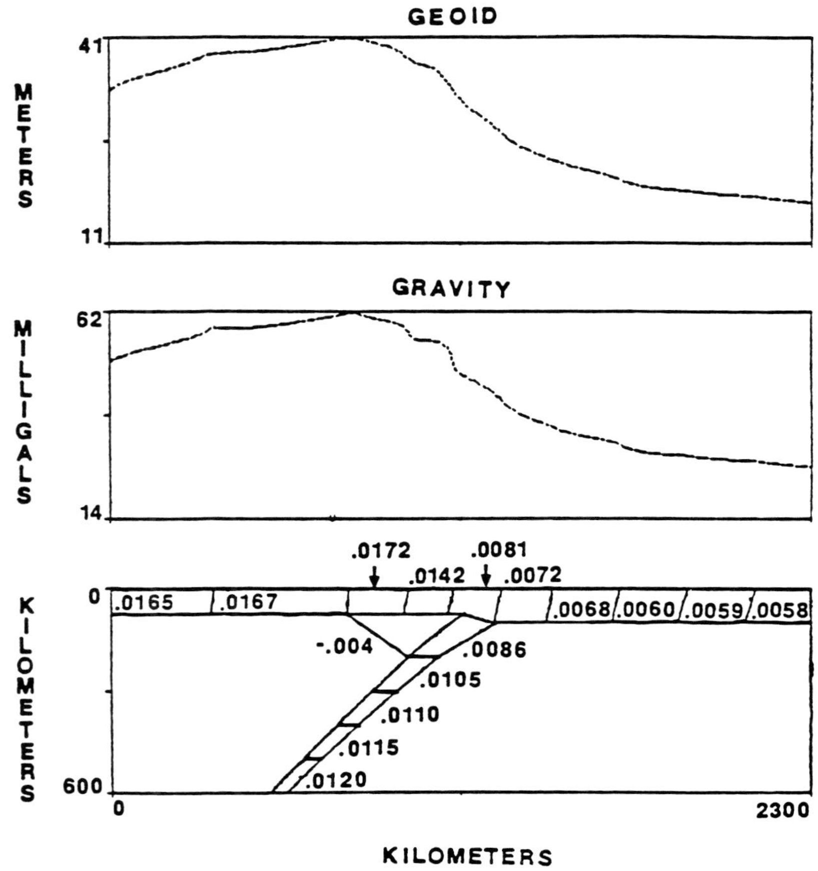

FIGURE 19. Complete slab-lithosphere model, gravity, and geoid profiles.

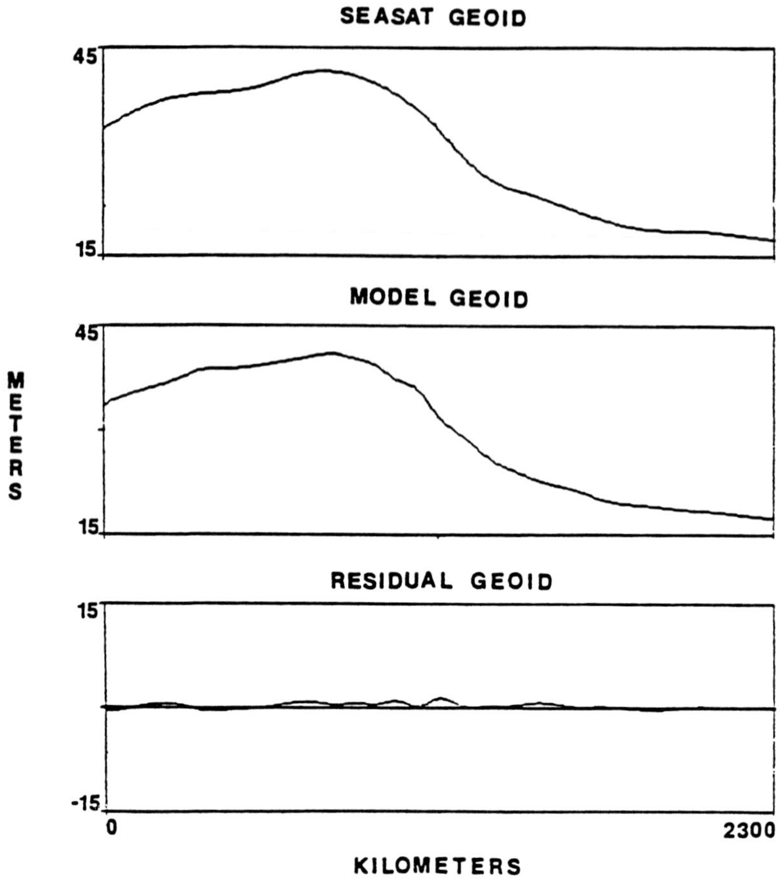

FIGURE 20. Complete slab-lithosphere model, Seasat, geoid, and residual geoid.

their importance in the definition of the geoid. The slab, made up of five parts, with its associated gravity and geoid profiles is shown in Figure 21. The maximum gravity anomaly at the surface is 17 milligals resulting in a maximum geoid anomaly of 12 m. The geoid profile is not quite symmetric, showing a much higher slope opposite to the direction the slab is being subducted.

In Figure 22 the residual anomaly determined by the difference between the Seasat geoid anomaly and the geoid anomaly predicted by the slab is shown. This residual anomain is essentially due to the effect of the non-sinking portion of the oceanic lithosphere and indicates that landward of the trench there is a relative positive anomaly.

Figure 23 shows the oceanic lithosphere (composed of six parts) and the back-arc lithosphere (composed of four parts), along with the gravity and geoid profiles for this model. The geoid profile is composed of two levels, much as the residual geoid profile in Figure 22. The residual geoid profile shown in Figure 24 is more symmetric than the geoid profile due to the slab model as shown in Figure 21.

A model consisting of all the features shown in Figure 19 without the low density zone is displayed in Figure 25. The absence of this low density zone results in the predicted anomalous geoid being too large by approximately four m (Figure 26). This is localized in the area where the low density zone is located.

Figure 27 shows the same model as in Figure 19 with the lowest part of the slab missing. The effect of this part of the model missing is not as great as that of the low density zone,

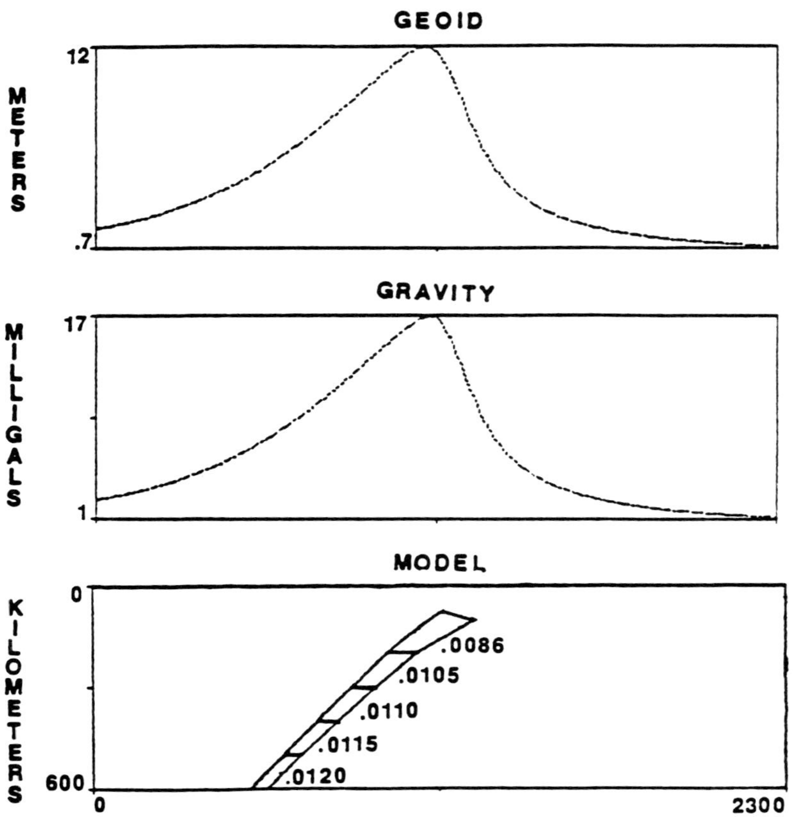

FIGURE 21. Slab model consisting of five individual bodies, gravity, and geoid profiles.

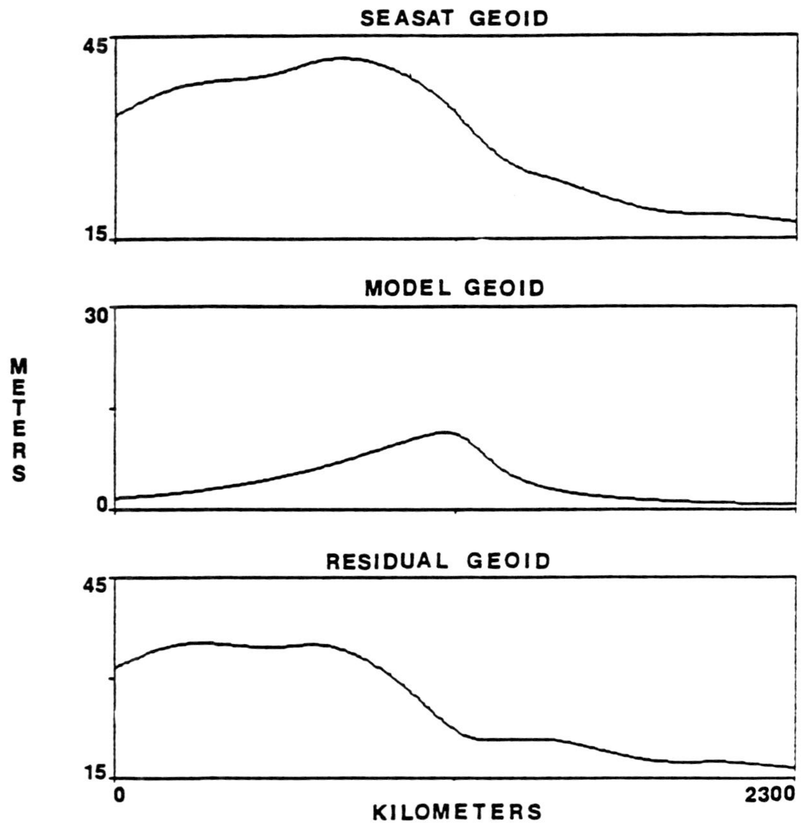

FIGURE 22. Slab model composed of five bodies, Seasat, geoid, and residual geoid profiles.

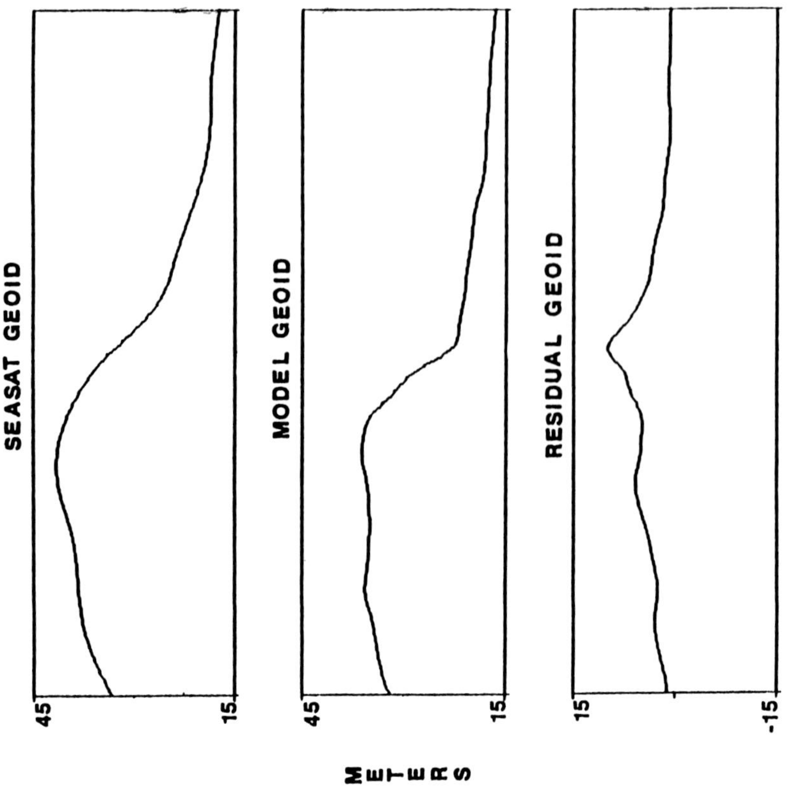

FIGURE 24. Lithosphere model, Seasat, geoid, and residual geoid profiles.

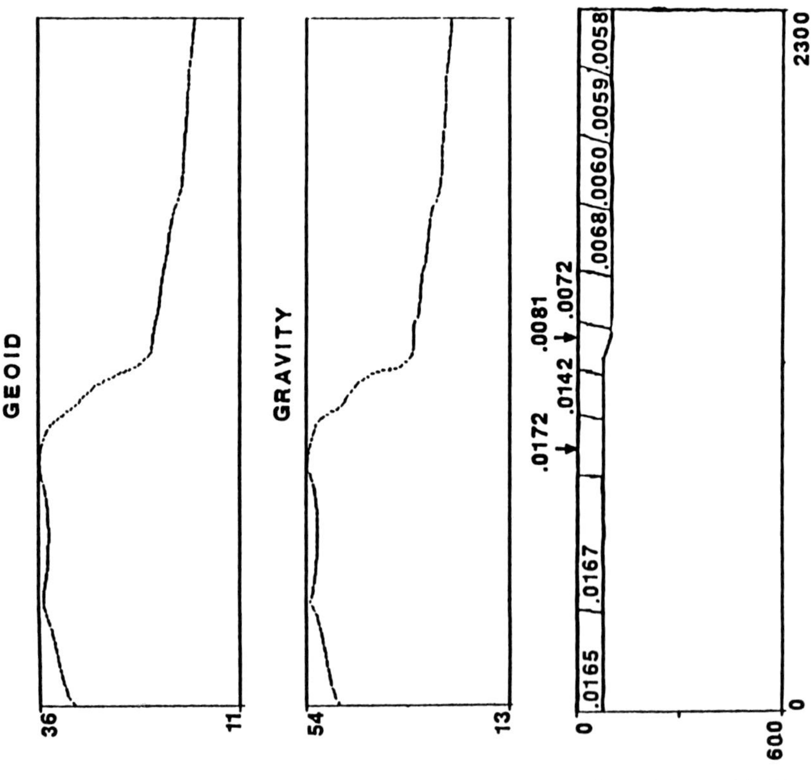

FIGURE 23. Lithosphere model composed of 10 bodies, gravity, and geoid profiles.

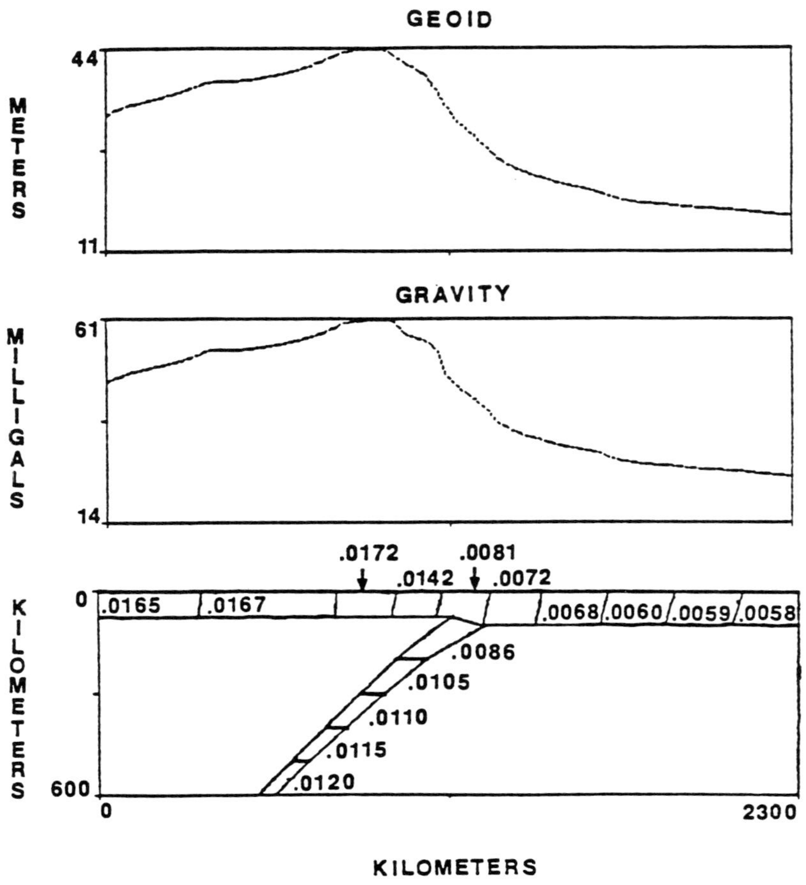

FIGURE 25. Slab-lithosphere model without low density zone, gravity, and geoid profiles.

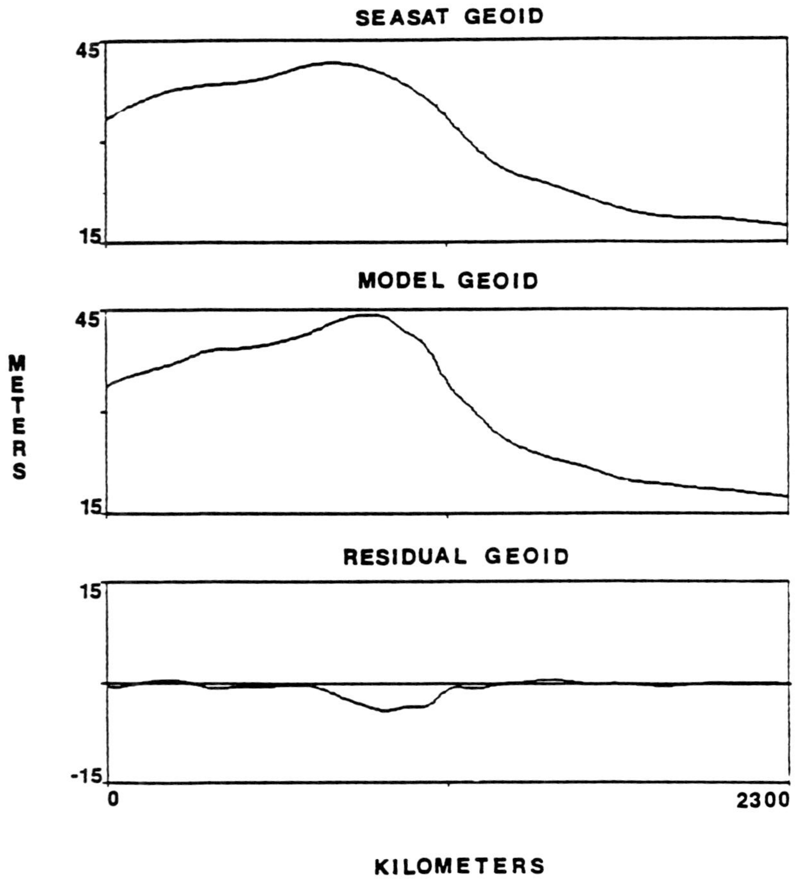

FIGURE 26. Slab-lithosphere model without low density zone, Seasat, geoid, and residual geoid profiles.

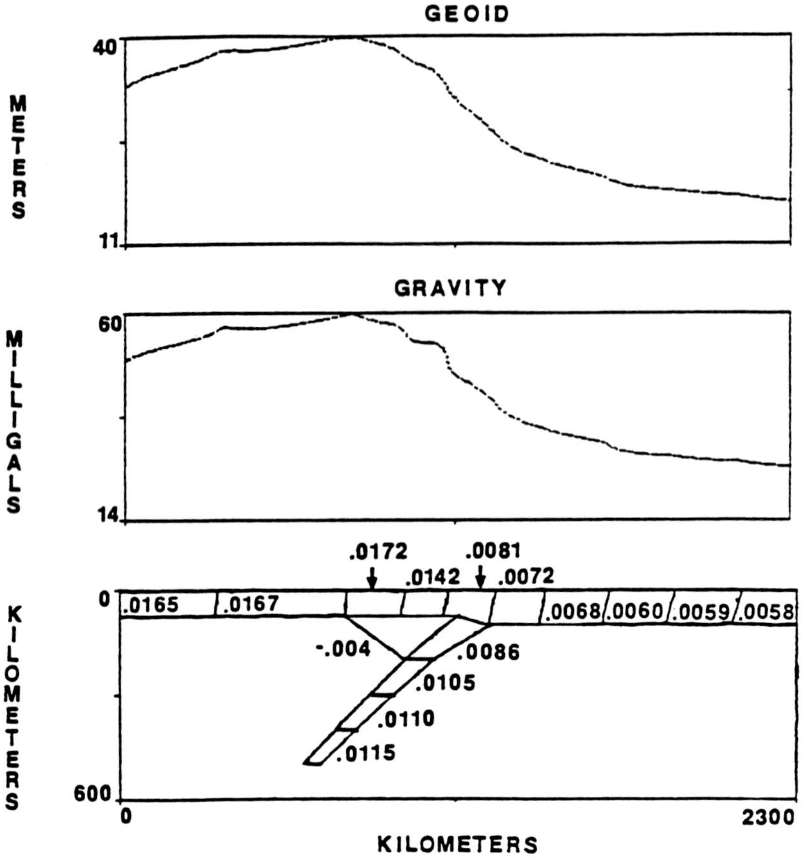

FIGURE 27. Slab-lithosphere model without lowest part of slab, gravity, and geoid profiles.

but does affect to some extent the overall geoid anomaly profile. As shown in Figure 28 there is a general positive increase in the residual geoid anomaly profile. This general increase could be compensated for by increasing the density of the oceanic lithosphere landward of the subducting slab.

Figure 29 shows the gravity and geoid profiles for a model that is missing the lower two parts of the subducting slab. As seen in Figure 30 the difference between the Seasat satellite geoid and the predicted anomalous geoid is larger than the difference shown in Figure 28. This difference again could be compensated for by a larger increase in the lithosphere density.

Just as an increase in the lithosphere density landward of the trench could account for the geoid anomaly that is assumed to be due to the subducting slab, it is possible to decrease the lithosphere density by increasing the dimensions or density of the subducting slab and still maintain an accurate match with the Seasat satellite geoid. It is quite clear that a negative mass is needed to accurately match the Seasat satellite geoid, however, the lithosphere and the slab appear to be linked to each other in that one could be increased (decreased) while the other must then be decreased (increased) in density.

Figure 31 shows an alternate model that is similar to that shown in Figure 19, however the densities assigned to the individual bodies were made more uniform. While this model is simplified, Figure 32 shows it doesn't yield results in as close agreement as the model in Figure 19.

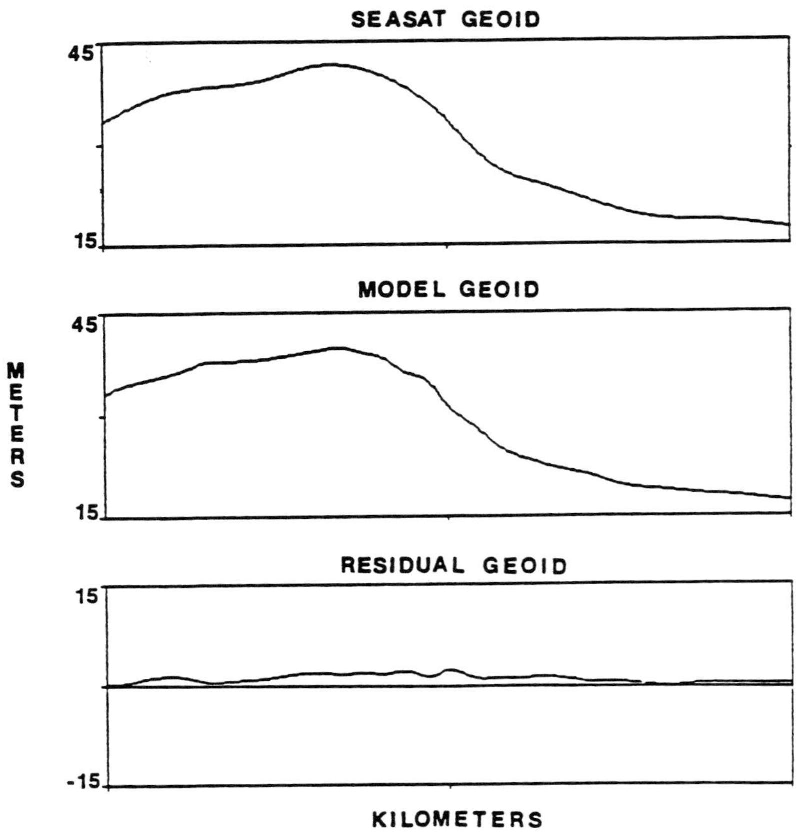

FIGURE 28. Slab-lithosphere model without lowest part of slab, Seasat, geoid, and residual geoid profiles.

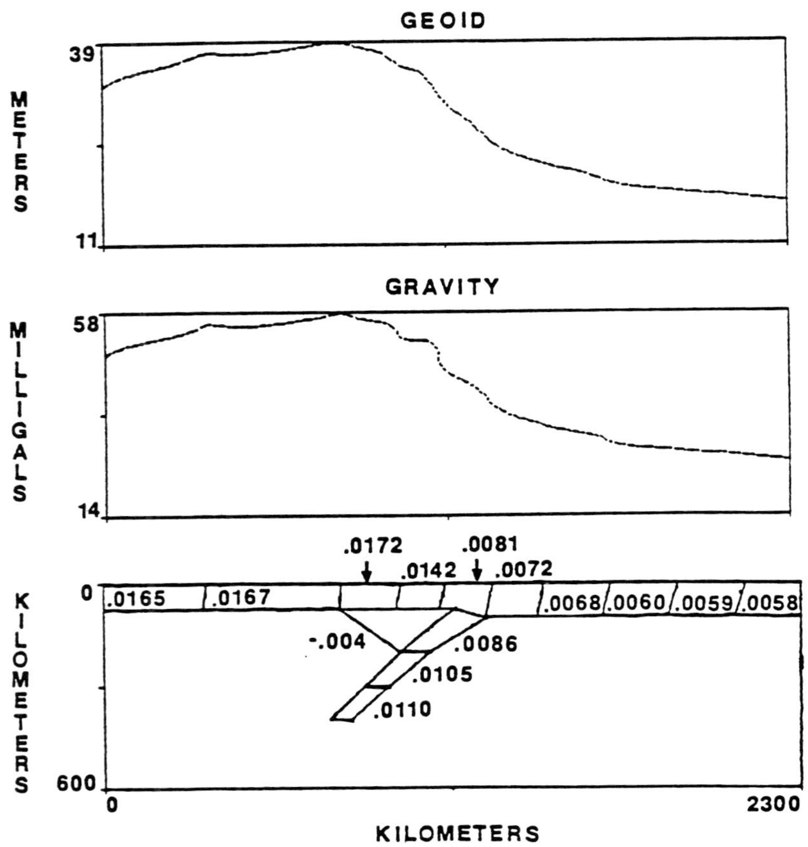

FIGURE 29. Slab-lithosphere model without lowest two-parts, gravity, and geoid profile.

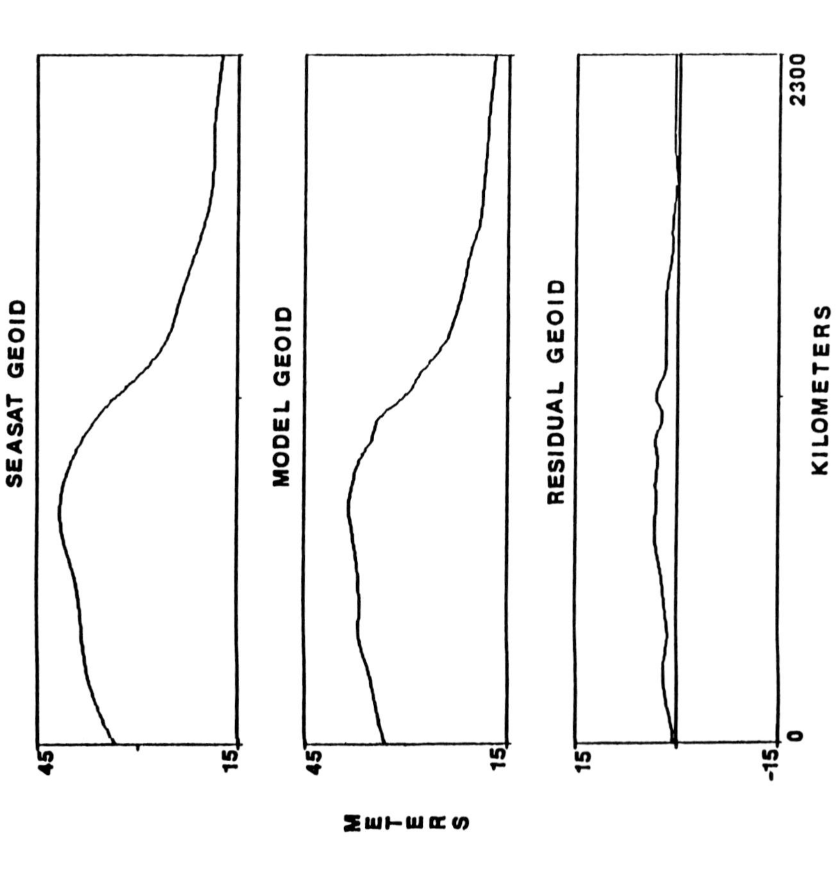

FIGURE 30. Slab-lithosphere model without lowest two parts, Seasat, geoid, and residual geoid profiles.

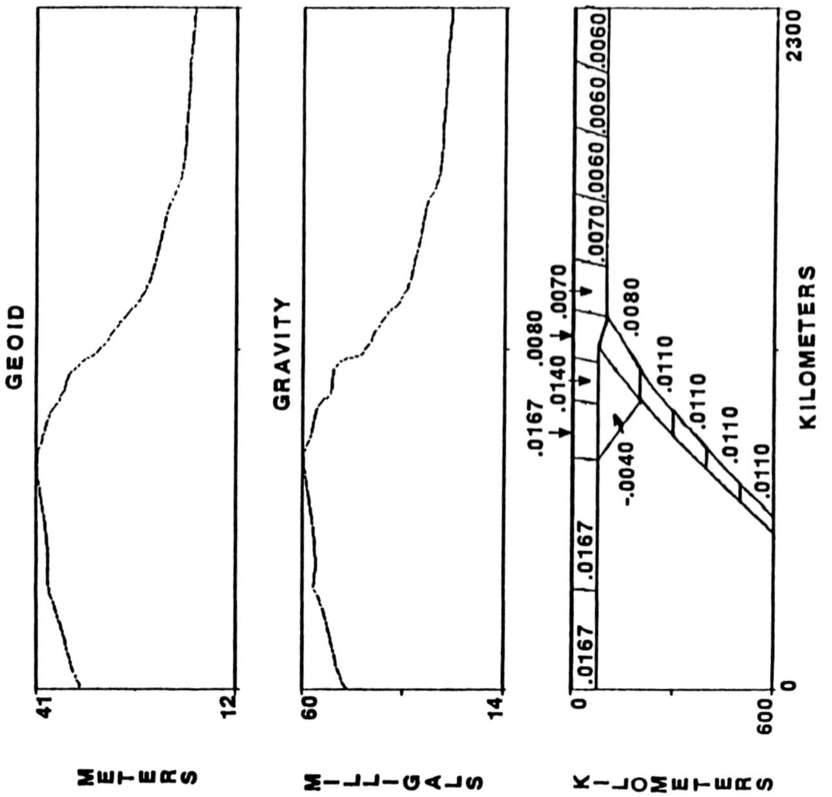

FIGURE 31. Alternative slab-lithosphere model, gravity, and geoid profiles.

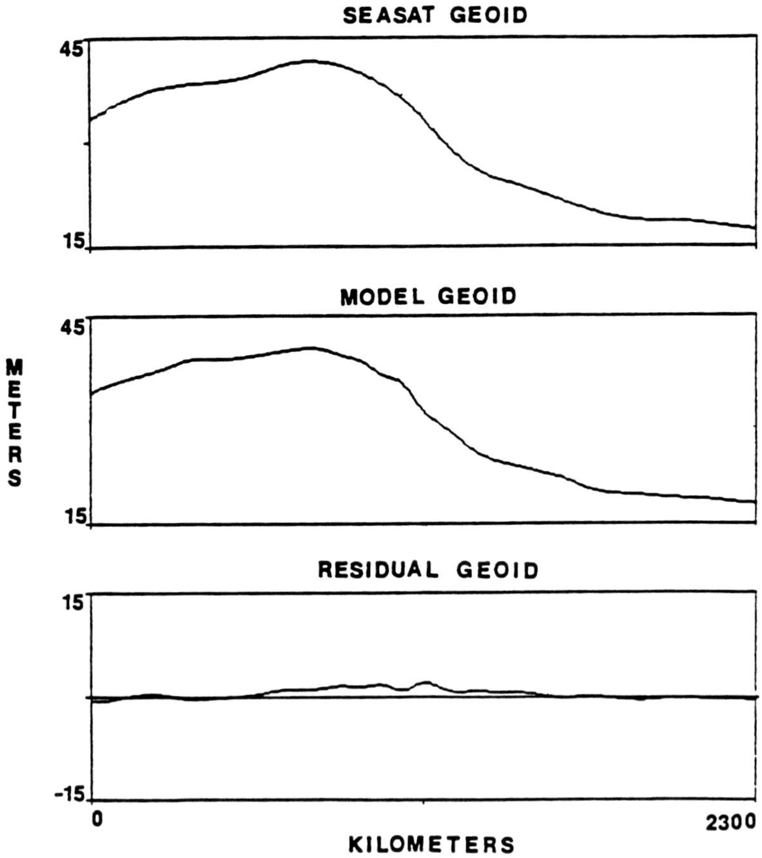

FIGURE 32. Alternative slab-lithosphere model, Seasat, geoid, and residual geoid profiles.

D. STRESS ANALYSIS

Figure 33 depicts the approximate location of the Tonga Trench, the Seasat satellite track, and recorded shallow focus earthquake locations (Sykes et al., 1969). The location of the Tonga Trench was determined from the Synbap Bathymetry data set. The earthquakes are located approximately along an elongate area located northwest of the Tonga Arc. Figure 34 shows the pressure at depths of 9, 31, 59, and 80 km predicted from the non-isostatic model in Figure 17 (Caputo et al., in press) have computed the maximum shear stress (mss) in a region due to the combined load of the topography and a buoyant mass underneath.

Utilizing these results it is possible to estimate the mss at some depths to be expected in the region studied here. Using the results of Caputo et al. (1984) it was found that the mass in the area will be approximately one-third of the load. At 9 km (Figure 34) the maximum stress in the region of the trench may be about 450 bars. In the region around the trench the mss would then increase to approximately 550 bars at 31 km depth, then decreases to about 370 bars at a depth of 80 km. At the approximate location where the shallow focus earthquakes are shown in Figure 33 there is a mss of 180 bars, at the depth between 9 km and 31 km.

V. CONCLUSION AND SUMMARY

After analyzing various geophysical models a model has been found that successfully predicts the anomalous geoid profile that has been observed by the Seasat satellite data

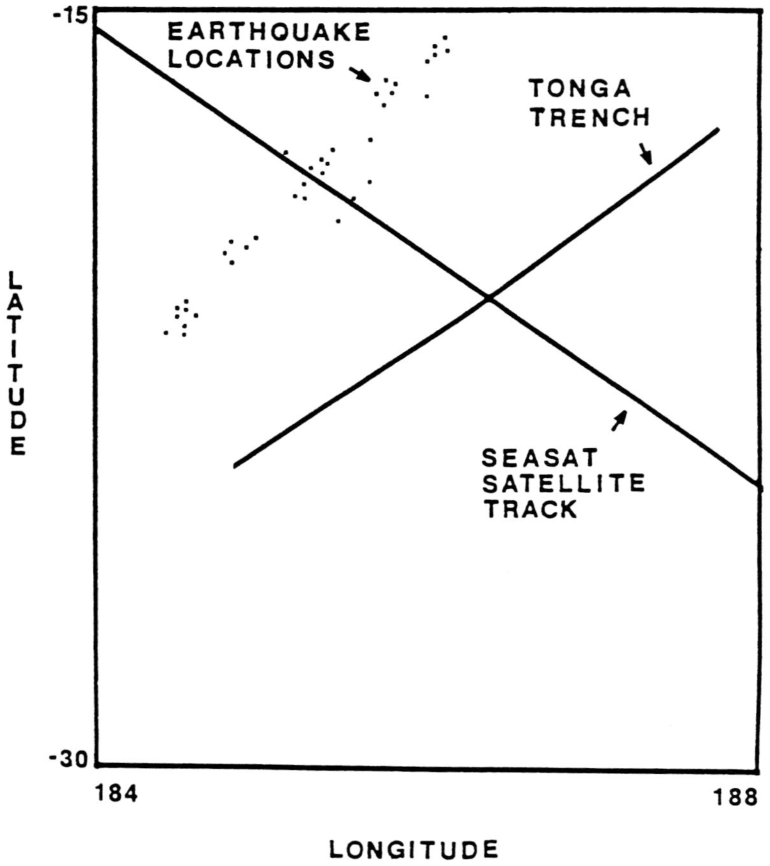

FIGURE 33. Location of Tonga Trench, Seasat data profile, and shallow focus earthquakes to a depth of about 40 km.

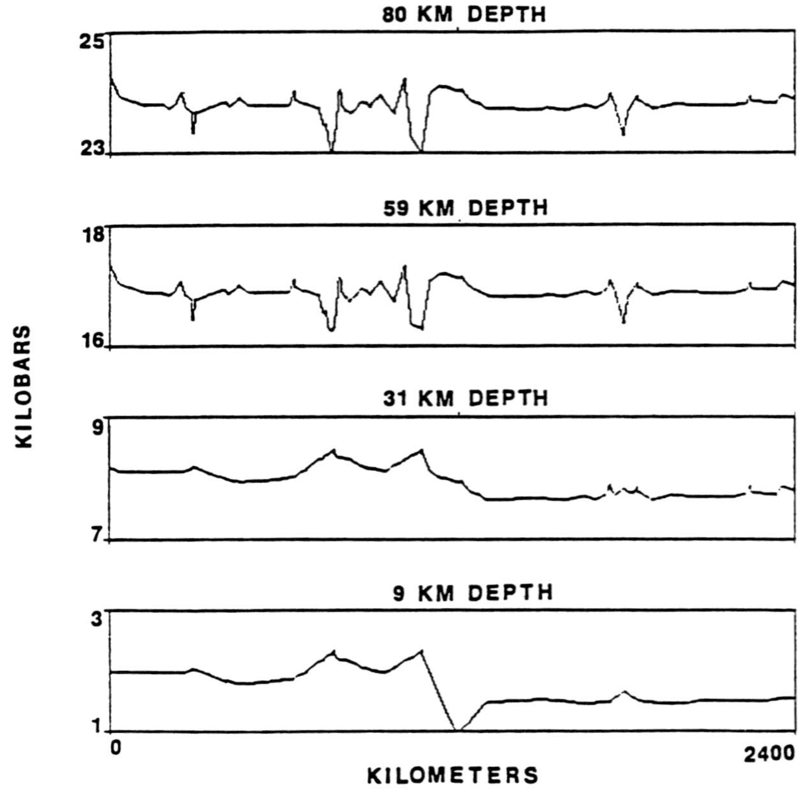

FIGURE 34. Pressure at depths of 9, 31, 59, and 80 km.

acquisition system. Though this model in itself is not unique, by utilizing the available geophysical information presently known, this model should accurately depict the large-scale density features present in this area.

In the analysis of the effects that the bathymetry has on the geoid an isostatic and non-isostaic model were investigated. The assumption of Airy isostasy was made for determining the geophysical model used in the isostatic analysis. It was found that by assuming Airy isostasy the trench was under-compensated, while the Tonga Arc was over-compensated. The compensation of the bathymetry also introduced high-frequency undulations not seen in the Seasat satellite geoid data. Depending on the density parameters used, Airy isostasy predicted a wide range of values for oceanic crustal thickness. The model using a density of 2.85 gm/cm^3 for the crust and 3.50 gm/cm^3 for the compensating root yielded the best agreement between the observed and the calculated geoid. This model predicted an oceanic crust of approximately 11 km, but also predicted that the anomalous geoid in the region should fluctuate by 120 m. Since this large fluctuation is not found in the Seasat satellite geoid data, this would indicate that Airy isostasy does not properly predict the correct compensation of the bathymetry.

During the modeling of the non-isostatic compensation system it was found that to compensate for the Tonga Arc, a mass anomaly distribution to a depth of 80 km was required. This also resulted in an oceanic crust of 24 km. By increasing the compensation root density from 3.35 gm/cm^3 to 3.50 gm/cm^3 it was possible to decrease the oceanic crust to 18 km. These values, an anomaly distribution to a depth of 80 km and an oceanic crust of 18 to 24 km are a direct consequence of the three layer model. The analysis of the subducting slab and lithosphere accurately predicts the observed Seasat geoid data with no high frequency components. In order to achieve a thinner oceanic crust and higher level of the density anomalies a model incorporating more than three layers should be used.

This study has shown that the geoid derived from the Seasat satellite mission accurately reflects the large scale features of the oceanic lithosphere and subducting slab. The actual density variations of these features need to be considered when adjusting the model parameters to fit the observed data. Since the Seasat data over a given trench has minor variations in it, this would allow lateral modeling along the strike of the trench. It appears the Seasat satellite data is sensitive enough to permit this.

As shown in this study all of the individual bodies (16 total) are needed to match the observed Seasat satellite data. Neglecting any of these bodies can introduce an error of at least 2 m if not more.

In modeling the oceanic lithosphere and subducting slab with the existing geophysical evidence available, there appears inherent in the model a feature that suggests the anomalous geoid can be used to indicate the stress in the region. The oceanic lithosphere has an anomalous density increase of 0.0023 gm/cm^3. It seems that 0.0008 gm/cm^3 of this increase can be attributed to the aging of the oceanic lithosphere. The remaining 0.0015 gm/cm^3 may be due to the oceanic lithosphere being compressed as it nears the trench axis. If this is the case then the stress in this region may be affected by this compression. The subducting slab shows an anomalous density increase of 0.0034 gm/cm^3. The subducting slab also seems to indicate a compressional effect and is even more pronounced, as would be expected.

The pressure in the lithosphere was calculated for the depths of 9, 31, 59, and 80 km and assumed to be representative of a load at the corresponding depth. The maximum stress in the area is directly related to the load, resulting in maximum stress in the range of 370 bars to 550 bars for the trench region and 180 bars where shallow earthquakes have been observed. These values are tentative estimates only, but do indicate regions of relatively high stresses where earthquakes have been observed.

VI. APPENDIX

A. DERIVATION OF STOKES' FORMULA

The differential equation of physical geodesy

$$\Delta g = -(\partial T/\partial r) - (2/r)T \tag{34}$$

relates the gravity anomaly on the geoid to the potential and the gradient of the geoid. By using the upward continuation integral

$$\Delta g(r) = (R^2/4\pi r) \int\!\!\int \{(r^2 - R^2)l^{-3} - r^{-1} - 3R\cos\psi r^{-2}\}\Delta g d\sigma \tag{35}$$

gravity anomalies outside the earth may be determined, allowing Equation 35 to be integrated with respect to r. Multiplying by $-r^2$ yields

$$-r^2\Delta g = r^2(\partial T/\partial r) + 2RT = \partial/\partial r\,(r^2T) \tag{36}$$

Integrating

$$\partial/\partial r\,(r^2T) = -r^2\Delta g(r) \tag{37}$$

between the limits ∞ and r yields

$$r^2T = -\int r^2\Delta g(r)dr \tag{38}$$

since the limit of (r^2T) is zero as r goes to infinity. Inserting the upward continuation integral and solving yields

$$T = (R/4\pi) \int\!\!\int \Delta g S(\psi)d\sigma \tag{39}$$

By utilizing Brun's formula, $N = T/\gamma$, finally gives Stokes' formula:

$$N = (R/4\pi\gamma) \int\!\!\int \Delta g S(\psi)d\sigma \tag{40}$$

B. COMPUTER CALCULATION OF GRAVITY ANOMALIES

The gravity effect of a body of cross section S (Figure 35) which is infinite in the Y direction normal to S, can be obtained by integrating the effect of infinite horizontal line elements dA.

The vertical component of gravity of the two-dimensional body at O is,

$$\Delta g = 2k\rho\int\!\!\int[z/(x^2 + z^2)]\,dA \tag{41}$$

where the integration is over the entire surface S.

Using z and θ as the coordinate system Equation 41 becomes,

$$\Delta g = 2k\rho\int\!\!\int d\theta dz \tag{42}$$

again integrating over the surface S. Changing the surface integral to a line integral yields,

$$\xi\Delta g = 2k\rho\Delta\theta\int dz \tag{43}$$

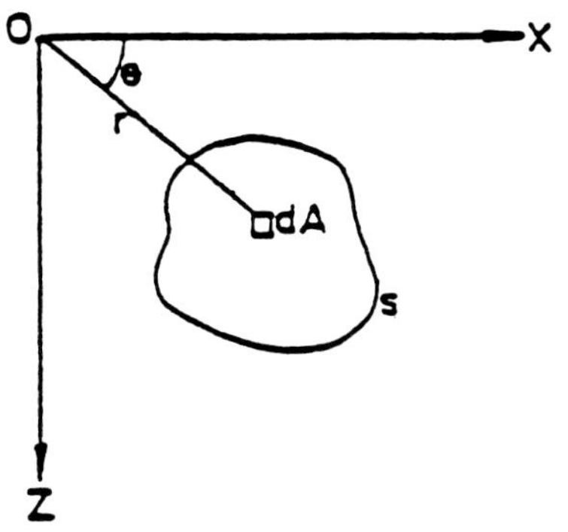

FIGURE 35. Two dimensional body and coordinate system (from Talwani, 1973).

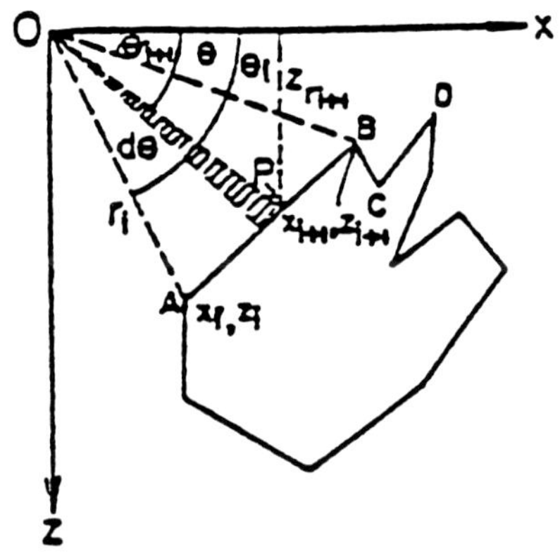

FIGURE 36. Polygonal cross section and coordinate system (from Talwani, 1973).

Since for a point P(z,θ) which is located on the cross section of interest (Figure 36),

$$z = [x_{j+1}(z_{j+1} - z_j) - z_{j+1}(x_{j+1} - x_j)]/[\cot\theta(z_{j+1} - z_j) - (x_{j+1} - x_j)] \quad (44)$$

then,

$$\Delta g = 2k\rho\Sigma[x_j z_{j+1} - z_j x_{j+1}]/[(x_{j+1} - x_j)^2 + (z_{j+1} - z_j)^2]$$
$$x\ [(x_{j+1} - x_j)(\theta_j - \theta_{j+1}) + (z_{j+1} - z_j)\ln(r_{j+1}/r_j)] \quad (45)$$

where,

$$r_j^2 = x_j^2 + z_j^2 \quad (46)$$

and,

$$\theta_j = \text{arc tan } (z_j/x_j) \quad (47)$$

This now gives the vertical component of gravity for an infinite two-dimensional body on a flat Earth by specifying the x and z coordinates of the polygonal coordinates.

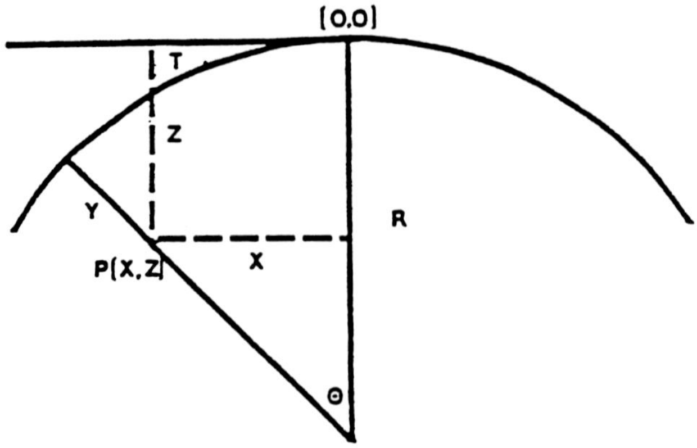

FIGURE 37. Coordinate system for a cylindrical Earth.

To approximate the gravity of a cylindrical Earth the new x and z coordinates need to be found, and then Equation 47 used. By then using the coordinate relocation system shown in Figure 37 we obtain,

$$X = (R - Y)\sin\theta \text{ and } Y = R - [(R - Y)\cos\theta] \tag{48}$$

where $\theta = (T/R)$.

Figure 38 shows three plots of the vertical component of gravity for a infinite horizontal cylinder, computed theoretically (A), for a flat earth (B), and for a cylindrical earth (C). The cylinder has a radius of 25 km, a density of 0.1 gm/cm^3, and is located 75 km below the surface. As can clearly be seen in Figure 38, the three plots are virtually indistinguishable. The theoretical plot has a higher maximum than the line integral methods. This is because the line integral methods are approximating the cross section of a cylinder with 16 points. Shown in Figure 39 are the differences of the three plots shown in Figure 38.

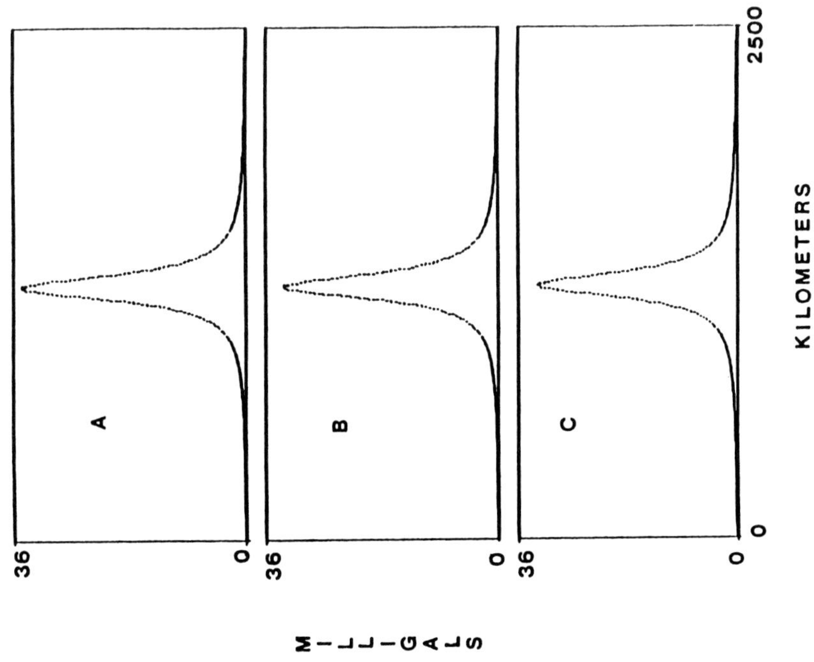

FIGURE 38. Plots of gravity for infinite horizontal cylinder.

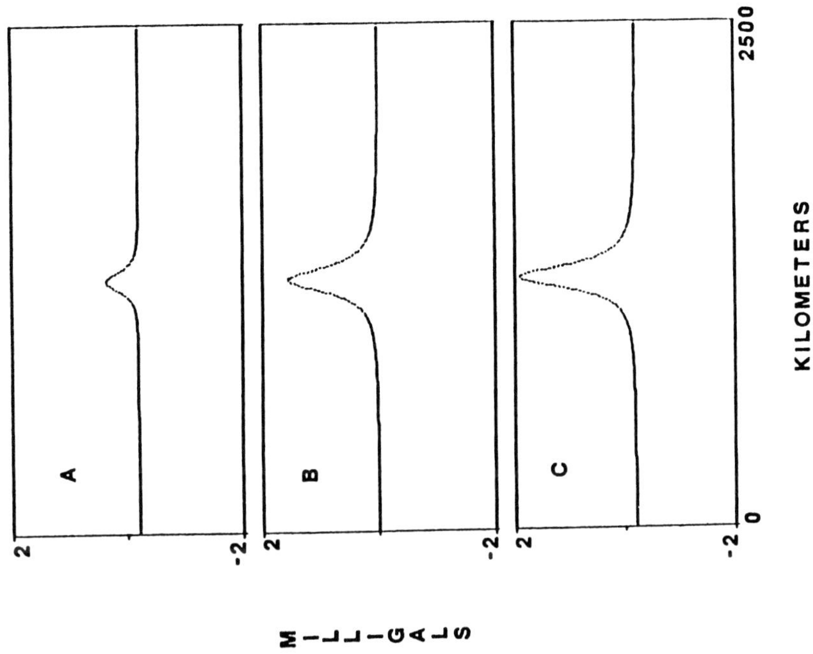

FIGURE 39. Differences of three plots shown in Figure 38.

REFERENCES

Airy, G. B., On the computations of the effect of the attraction of the mountain masses as disturbing the apparent astronomical latitude of stations in geodetic surveys, *Philo. Trans. R. Soc. London,* 145, 101, 1855.

Brace, K. L. and Schultz, M., Ocean-area geoid heights and mean gravity anomalies from GEOS-3 data, *Mar. Geod.,* 3, 141—160, 1980.

Carlson, R. L. and Raskin, G. S., Density of the ocean crust, Nature, in press.

Caputo, M., *The Gravity Field of the Earth,* Academic Press, New York, 1967.

Caputo, M., Manzetti, V., and Nicelli, R., Topography and its isostatic compensation as a cause of seismicity, *Tectonophysics,* in press.

Chapman, M. E., Techniques for interpretation of geoid anomalies, *J. Geophys. Res.,* 84, 3793—3801, 1979.

Chapman, M. E. and Talwani, M., Geoid anomalies over deep sea trenches, *Geophys. J. R. Astron. Soc.,* 68, 349—369, 1982.

Hanus, V. and Vanek, J., Northern part of the Tonga Region: A complicated subduction closure, *J. Geophys. Res.,* 46, 385—395, 1979.

Heiskanen, W. A. and Vening Meinesz, F. A., *The Earth and Its Gravity Field,* McGraw-Hill, New York, 1958.

Heiskanen, W. A. and Moritz, H., *Physical Geodesy,* W. H. Freeman Company, San Francisco, 1967.

Jeffreys, H., *The Earth, Its Origin, History, and Physical Constitution,* Cambridge University Press, New York, 1970.

Jones, G. M., A technique for computing geoid height anomalies over two-dimensional Earth models, *Geophys. J. R. Astron. Soc.,* 69, 329—337, 1982.

Karig, D. E., Origin and development of marginal basins in the Western Pacific, *J. Geophys. Res.,* 76, 2542—2561, 1971.

Lame, D. B. and Born, G. H., SEASAT measurement system evaluation: Achievements and limitations, *J. Geophys. Res.,* 87, 3175—3178, 1982.

Le Pichon, X., Sea-floor spreading and continental drift, *J. Geophys. Res.,* 73, 3661—3697, 1968.

Leitao, C. D. and McGoogan, J. T., Satellite altimetry applied to marine geoid determination, *Acta. Astron.,* 3 (5—6), 459—469, 1976.

McAdoo, D. C. and Martin, C., Seasat observations of lithosphere flexure seaward of trenches, *J. Geophys. Res.,* 89, 3201—3210, 1984.

NASA publication 622—97, *Geophysical Data Record (GDR) Users Handbook, Altimeter,* Jet Propulsion Laboratory, Pasadena, CA, 1980.

Pratt, J. H., On the attraction of the Himalaya Moutains and of the elevated regions beyond the plumb-line in India, *Philo. Trans. R. Soc. London,* 145, 53, 1855.

Sclater, J. G., Hawkins, Jr., Mammerickx, J., and Clement, G. C., Crustal extension between the Tonga and Lau ridges: Petrological and geophysical evidence, *Geol. Soc. Am. Bull.,* 83, 505—517, 1972.

Stanley, H. R., The GEOS-3 project, *J. Geophys. Res.,* 84, 3861—3871, 1979.

Strobach, K., Curvature of island arcs and plate tectonics, 39, Z, *Geophysics,* 819—831, 1973.

Sykes, L. R., bIsacks, B. L., and Oliver, J., Spatial distribution of deep and shallow earthquakes of small magnitudes in the Fiji-Tonga region, *Seismo. Soc. Am. Bull.,* 59, 1093—1113, 1969.

Talwani, M., *Computer Usage in the Computation of Gravity Anomalies in Methods in Computational Physics,* Vol. 13, Adler, B., Academic Press, New York, 1973, 343—389.

Tapley, B. D., Born, G. H., and Parke, M. E., The SEASAT altimeter data and its accuracy assessment, *J. Geophys. Res.,* 87, 3179—3188, 1982.

Turcotte, D. L. and Schubert, G., *Geodynamics,* John Wiley & Sons, New York, 1982.

Vening Meinesz, F. A., Une nouvelle methode pour la reduction isostatique regionale de L'intensite de la pesanteur, *Bull. Geod.,* No 29., 1931.

Wainwright, E., Calculation of isostatic gravity anomalies and isostatic geoid heights using two-dimensional filtering: implications for structure in subduction zones, Ph.D. dissertation, Texas A & M, 1983.

Watts, A. B., On geoid heights derived from GEOS-3 altimeter data along the Hawaiian-Emporer seamount chain, *J. Geophys. Res.,* 84, 3817—3826, 1979.

Chapter 2

CONVERGENCE AND STRIKE-SLIP MOTION AT THE NORTHERN TERMINUS OF THE TONGA TRENCH, SOUTHWEST PACIFIC

Dawn J. Wright

TABLE OF CONTENTS

INTRODUCTION

Oceanic trenches are sites of convergence between two lithospheric plates, where one plate normally moves beneath the other to be eventually destroyed or incorporated into the mantle. These convergent margins are perhaps among the most striking and complex of the Earth's tectonic features. This is due in part to the fact that oceanic lithosphere is consumed at trenches, leaving scant evidence of the material destroyed. Furthermore, there exists a diversity of features associated with trenches, such as the wide variations in seismicity, volcanism and morphology.

The important role that trenches and their accompanying island arcs played in the dynamics of large-scale tectonic activity was not fully understood or appreciated until the pioneering marine gravity work of Vening Meinesz in the early 1930s. His work showed that trench-arc systems were associated with strong perturbations of the Earth's gravitational field (anomalies), centered over the axis of the trench. Meinesz attributed the large negative anomalies to a downbuckling of the crust, which he termed the "tectogene", and considered it to be held down either by compressional forces or by the drag of converging convection currents (Meinesz, 1952). At the same time he also associated the trenches with zones of widespread seismicity throughout the descending lithosphere. Soon afterward, the model was changed to accommodate clear asymmetry of features associated with trenches and island arcs, and mantle convection was proposed as a driving mechanism (Kennett, 1982).

The Tonga Trench-Arc system is one of many similar convergent margins in the Southwest Pacific. This study examines the tectonics associated with the northern terminus of the trench, between 12°S and 18°S (Figure 1). Although other portions of the trench system have been studied, this region has not been investigated in detail. The northern terminus of the Tonga Trench, with its interplay of both subduction and strike-slip motion, is considered a complicated portion of the trench system (Hanus and Vanek, 1979b). This interplay of tectonic motion is not fully understood and has not been addressed frequently in the literature. Several authors have stated that the northern terminus of the trench shoals, bends abruptly to the west, and becomes a transform fault (Johnson and Molnar, 1972; Hawkins, 1974a; Hanus and Vanek, 1979a and b; Natland, 1980; and Kroenke, 1984) but none have addressed in detail the nature or character of such a transition.

The most thorough investigation of the northern terminus of the Tonga region was undertaken by Hanus and Vanek (1979b). They examined only seismicity data and investigated the morphology of the Wadati-Benioff zone. A major finding was the existence of a deep-seated, left-lateral strike-slip feature which forms "the northern tectonic closure of the active Tonga subduction" (Figure 2). The characteristics of this transform fault are described (strike, boundary dimensions, depth of earthquake activity), but no detailed explanation is given of how or exactly where the eastern part of the fault makes a transition to Tonga subduction.

Uyeda and Kanamori (1979) and Uyeda (1982) have classified the Tonga active margin as a "Mariana" type due to a generally weak mechanical coupling between the interacting plates. The stress regime of a "Mariana-type" subduction is tensional, leading to a deep trench, steeply dipping slab and Benioff zone, and a general absence of sediment accretion. These are all features which have been observed in the Tonga region.

It is the objective of this study to provide a detailed characterization of the trench's transition from a subduction zone to a transform fault, based on available bathymetric, seismic reflection, and gravity data. Differences in the morphologic character of the two major parts of the northern Tonga Trench system (subduction on the north-south limb versus strike-slip motion on the east-west limb) are described, and the nature of the feature's transition from subduction to strike-slip motion is evaluated. This involves pinpointing the approximate location of the transition and discerning the tectonic processes in the area.

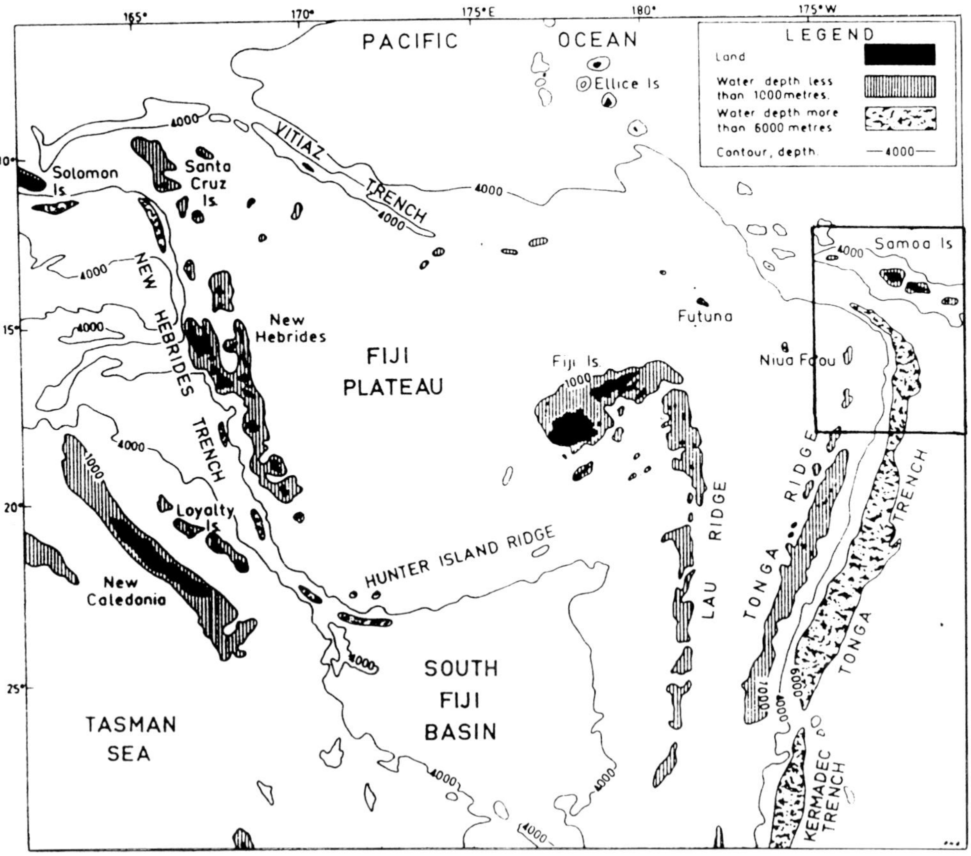

FIGURE 1. Physiographic elements in the northeastern part of outer Melanesia (Gill and Gorton, 1973). Box encloses the area of this study.

A. TECTONIC AND GEOLOGIC SETTING OF THE STUDY AREA

The Tonga Trench (Figure 1) is the northern portion of the linear Tonga-Kermadec Trench-Arc system, which is approximately 1400 km in length, 55 km in width, and has a maximum depth of 10,800 m. The trench marks the subduction of the Pacific Plate beneath the Australian Plate in the south, and beneath the Tonga microplate in the north (Sclater et al., 1972). Convergence between the Tonga microplate and the Pacific Plate along the northernmost 600 km segment of the trench is the major focus of this study. The Tonga microplate is bounded on the west by an active spreading center in the Lau Basin. The spreading center is extruding tholeiitic basalts (Hawkins, 1976) and is spreading at a half-rate of from 3 cm/year (Karig et al., 1978) to 3.8 cm/year (Weissel, 1977; Malahoff et al., 1982). The Tonga Trench marks the eastern and northern boundaries of the Tonga microplate and the western boundary of the Pacific Plate.

The general tectonic features of the Lau Basin-Tonga Trench region are shown in Figure 3. In the study area, these include the Tonga Ridge, which is actually composed of two linear magmatic arcs, the islands of Samoa, and a series of normal faults. The faults have essentially the same strike, occur in an echelon pattern, and may be associated with the eastward migration of the Tonga Ridge and Trench.

It has been established by Sclater et al. (1972) and Carlson and Melia (1984) that in the Lau Basin-Tonga Trench region the overriding plate is advancing, but at a rate that is significantly slower than the retreat of the Tonga subduction zone. This is due primarily to the rapid (10 to 13 cm/year) east-northeastward migration of the Tonga Ridge and Trench

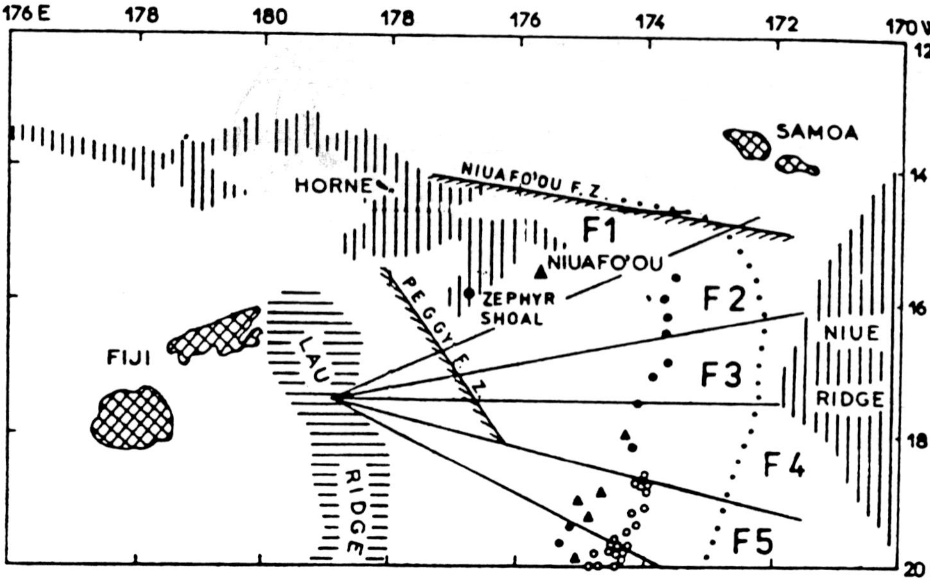

FIGURE 2. Main structural elements and locations of vertical sections used for the study of the northern closure of the Tonga subduction zone by Hanus and Vanek (1979b). The Niua Fo'ou fracture zone is the left-lateral strike-slip referred to in the text. The Tonga Trench is denoted by a dotted line, active volcanoes by solid triangles, active islands and seamounts by solid circles, and coral islands and reefs by open circles.

(Carlson and Melia, 1984). The mechanisms of this migration have recently been debated in the literature. Dewey (1980) and Carlson and Melia (1984) have found that migration is due to a gravitationally induced rollback of the descending slab, whereas Furlong et al. (1982) and Hynes and Mott (1985) maintain that migration is due to the adjustment of the dip of the descending slab, which is caused by an increase in the westward motion of the Pacific Plate. Regardless of the mechanism, the migration of the trench is largely responsible for the formation of a major locus of crustal extension in the central Lau Basin. Additionally, the trench migration is probably the cause of the extensional faulting that was earlier noted to occur east of the Lau Basin.

In the study area, the Tonga Trench separates tectonic features of strikingly different ages. For instance, the newly formed seafloor of the Tonga microplate is of Late Pliocene age (2.5 Ma) according to the dating of fossils from gravity cores (Sclater et al., 1972) and residual magnetic anomalies (Malahoff et al., 1982). The two magmatic arcs of the Tonga Ridge, the Tofua Ridge and Tongatapu Ridge, are of Early Pliocene (3.5 to 5 Ma) and Miocene (17 Ma) ages, respectively, according to K/Ar dating of quartz gabbros and basalts (Ewart et al., 1973 and 1977; Kroenke, 1984). However, the subducting seafloor of the Pacific Plate is of Cretaceous or Jurassic age as inferred from ongoing studies of the age of the oceanic crust and the distribution pattern of magnetic anomalies in the Southwest Pacific (Drummond et al., 1982; Scholl et al., 1985).

A disagreement still exists among workers as to the age progression of the islands of Samoa. The Sr^{87}/Sr^{86} ratios of Hedge et al. (1972) and the petrological studies of Natland (1980) have led those authors to believe that the islands become progressively older toward the east. Hawkins and Natland (1975) and Natland (1980 and 1985) argue against a hot spot model for Samoa and tie the age progression to plate deformation processes rather than to a mantle plume. However, the paleomagnetic data of Keating and Tarling (1985) are broadly supportive of a hotspot origin for Samoa, which strongly suggests that the islands become progressively older toward the west.

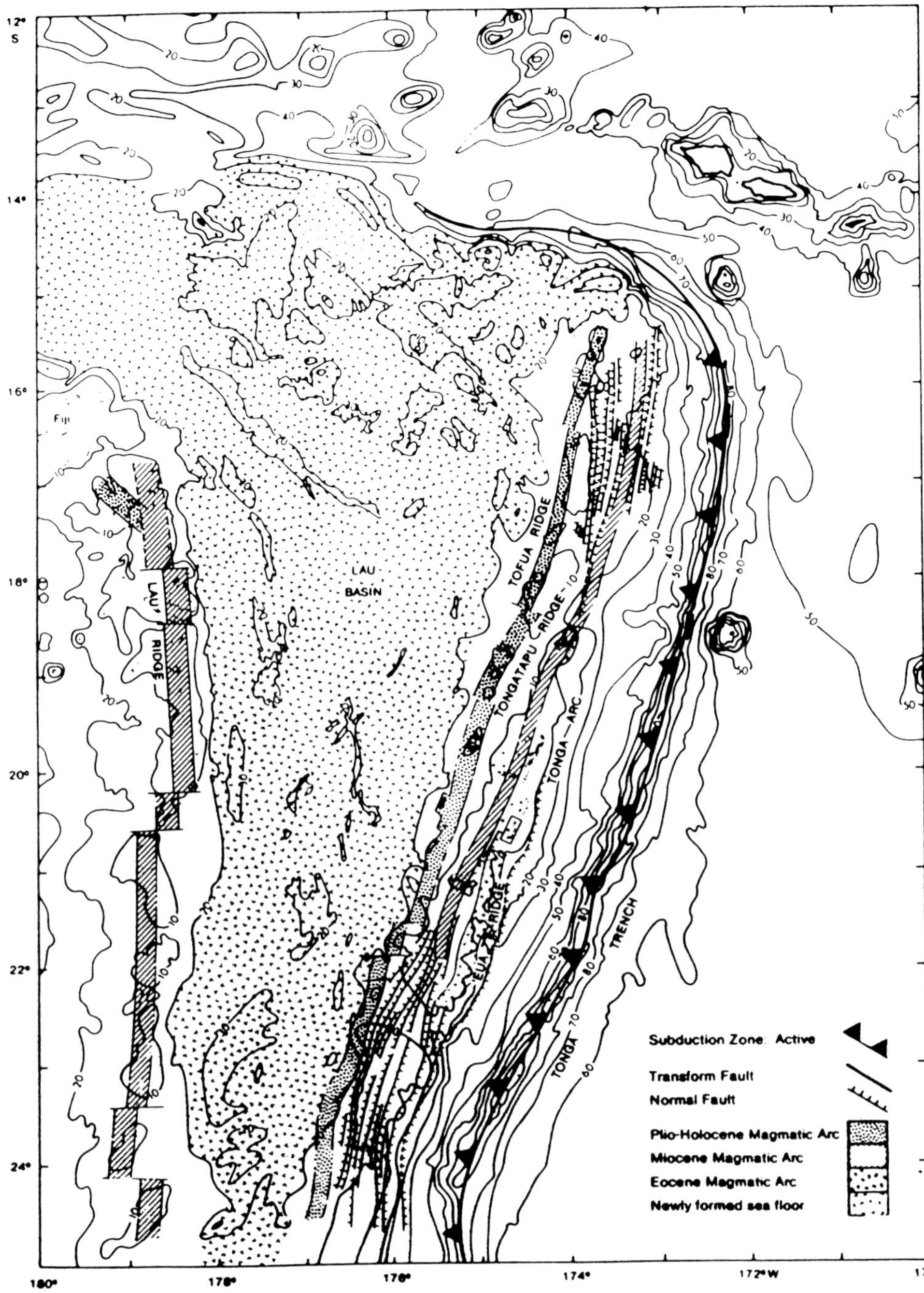

FIGURE 3. Tectonic map of the Tonga Arc, Lau Ridge and Basin, and Tonga Trench area (Kroenke, 1984).

The various tectonic elements of the northern Tonga Trench region (Figure 3), with the exception of the islands of Samoa, are oriented nearly parallel to the trench and appear to have maintained the same general strike as they evolved. The evolutionary history of the Tonga Trench is as follows. Subduction along the trench was initiated during the Early Miocene at approximately 22 Ma as deduced from the documentation of volcanic activity by radiometric dating and drilling results at DSDP Site 205 near the Lau Ridge (Andrews and Packham, 1975; Kroenke, 1984). At this time the Lau and Tonga Ridges were still united, forming a single volcanic axis, and the formation of the Lau Basin had not yet taken place. Igneous activity along this volcanic axis occurred from 17 to 15 Ma and a major influx of volcaniclastic debris took place at 12 Ma (Ewart et al., 1977; Kroenke, 1984). A decrease in accumulation of volcaniclastic sediment at 10 Ma signalled the waning phase of Lau-Tonga Ridge volcanism and the cessation of subduction along the trench (Burns and Andrews, 1973; Packham and Terrill, 1975; and Kroenke, 1984). From magnetic anomalies and seismological evidence, Watts et al. (1977) have suggested that the separation of the Lau and Tonga Ridges and the subsequent eastward advance of the Tonga trench-arc system began at 5 to 6 Ma. Subduction resumed along the present-day Tonga Trench with the re-initiation of volcanism along the Tonga Ridge at about 3.5 Ma in the Early Pliocene (Kroenke, 1984).

B. PREVIOUS GEOPHYSICAL STUDIES

The first geophysical studies of the Tonga Trench were carried out in the early 1950s and 1960s before the acceptance of plate tectonics as a scientific paradigm. Members of the Danish deep sea expedition of 1950–1952, aboard *RV Galathea,* carried out coring and sounding operations in the region (Talwani et al., 1961). The first extensive investigation of the trench was carried out by the Scripps Institution of Oceanography during the Capricorn expedition of 1952–1953. Raitt et al. (1955) published the results of those bathymetric, seismic refraction, and magnetic surveys and were the first to delineate the physical boundaries of the Tonga Trench. Talwani et al. (1961) analyzed gravity and seismic refraction profiles across the trench and observed anomalies of about −200 mgal over the entire length of the trench axis. Their gravity models neglected the existence of a downgoing slab or subducting oceanic crust, and thus required crust of abnormal thickness and density to be present beneath the trench axis and Tonga Ridge.

During the 1967 NOVA expedition of the Scripps Institution, rock samples were obtained from the shoreward flank of the Tonga Trench near the trench floor (20.5°S, 173.5°W; Fisher and Engel, 1969). Dredging at a depth of from 9150 to 9400 m yielded fresh to granulated and serpentinized peridotite and dunite, as well as basalts, tuffs, and tuffaceous agglomerates (Fisher and Engel, 1969). The authors were inclined to believe that the basalts and ultramafics that they recovered were fault-exposed components of conduits associated with the aligned volcanoes of the Tonga Ridge, rather than samples of the underlying mantle. They added, however, that extensive portions of the upper mantle might be identical in texture, structure, and composition to their dredged ultramafics. This is especially true in localities where ultramafics represent the settled residues after removal of basaltic magma (Fisher and Engel, 1969). Dredging also revealed no evidence for sediment accretion. Hilde and Fisher (1979) and Hilde (1983) have suggested instead, that the trench wall may be exposed old ocean crust upon which the Tonga Ridge has developed.

Karig (1970) analyzed seismic reflection and magnetic data, as well as samples from coring and dredging, for the entire Tonga-Kermadec island arc system and was the first to propose the migration of the Tonga Ridge and Trench away from Australia.

The results of bottom photography, seismic and magnetic profiling, and dredging across the northern terminus of the trench (14° to 18°S, 170° to 176°W), were reported by Udintsev et al. (1974). Nearly all of their photographs, at a depth of 5030 to 5070 m, showed the slopes of both the overriding and downgoing plates to be covered with unconsolidated sediments, the thicknesses of which were greatest in the upper parts of the trench slopes. Photographs also revealed evidence of more or less strong bottom currents: clean bedrock outcrops, washouts around rock fragments, ridges, ripple marks, and cumulative trains elongated with the direction of the current. Seismic profiling revealed the thickness of sediments to be 400 to 500 m on the slopes of the overriding plate and 100 to 200 m on the slopes of the downgoing plate. From the brink or slope break of both plates to the trench floor they found these unconsolidated sediments to be "wedged-out", stratified, and deformed. A large chunk of serpentinized peridotite with rounded faces, and showing slickensides, was dredged from the slope of the overriding plate at the trench's northern extremity (14.8°S, 173.5°W), thereby providing further evidence of faulting along that segment.

In 1982, several cruises were conducted on the *RV S. P. Lee* of the U.S. Geological Survey in order to geologically assess the potential of the Tonga Island Arc for petroleum resources. The cruises were also designed to expand on earlier work and included the collection of 24-fold multichannel reflection profiles plus sonobuoy refraction, gravity, magnetics, and samples of submarine and island outcrops. Although the majority of studies were focused on the southern Tonga Ridge, one dredge haul was recovered from the lower reaches of the shoreward slope of the northern Tonga Trench at 22°S (Vallier et al., 1985). Most of the rocks dredged had characteristics of an island arc setting rather than of oceanic crust. Once again, there was no evidence found for a large accretionary wedge of offscraped trench sediment (Vallier et al., 1985). The authors noted the discovery of accreted oceanic crust along the shoreward slope of the trench north of their dredge location (Fisher, 1974) but reported an absence of accreted crustal material from their own haul, due either to their own inability to recover such material or the absence of any accreted crust in that area.

Gnibidenko et al. (1985) carried out seismic reflection and refraction investigations of the Tonga Trench region near Osbourn Seamount (26°S) and reported the existence of horst-graben structures on both the overriding and downgoing slopes of the trench. They found the grabens on the overriding slope to be partially ponded by sediments, unlike those of the downgoing slope which were essentially barren of sediment. The faults on both slopes were determined to be normal in nature, testifying to the extensional conditions within the upper crust of the region (Gnibidenko et al., 1985).

C. SUBDUCTION AT THE TONGA TRENCH

Seismicity studies along the entire length of the Tonga Trench (Isacks et al., 1969; Isacks and Barazangi, 1977; Chen and Forsyth, 1978; Chapple and Forsyth, 1979; Hanus and Vanek, 1978, 1979a and b, and 1981; Giardini and Woodhouse, 1984; Wyss et al., 1984; and Stark and Frohlich, 1985) have shown that the Pacific Plate is underthrusting the Australian Plate at a fairly steep angle (38° to 52°) to a maximum depth of 673 km. At the northern terminus of the trench the Pacific Plate is underthrusting at an angle of 52° to 60°, to a depth of 455 km (Figure 4; Hanus and Vanek, 1979a and b). A major finding of the seismicity studies of Hanus and Vanek (1978, 1979b) was that subduction at the northern terminus of the Tonga Trench can be associated with not just one, but three interacting systems of subducted lithospheric plates: the currently subducting Tonga system T1 in the east, a remnant Lau system L2 in the west, and a remnant Horne system H in the north (Figure 5). The Lau and Horne systems are remnants of the foregoing cycle of subduction which ceased at approximately 10 Ma.

The pronounced bending of the Pacific Plate into the Tonga Trench has led to the development of normal fault structures in the plate, which are considered a standard feature of the subduction process (Chapple and Forsyth, 1979). Another major characteristic is the formation of these fault structures into regular patterns of grabens (Hilde, 1983). Consumption of these grabens, as well as other bathymetric features such as seamounts, appears to have an important influence on the subduction process. The lack of sedimentary accretion complexes in the region strongly suggests that sediments are being tectonically displaced into grabens and subducted almost entirely. Another possibility is the process of tectonic erosion, whereby grabens abrade the leading edge and underside of the overriding plate and subduct the eroded material (Hilde, 1983). These mechanisms imply that erosion and accretion may be linked processes. The dominance of one over the other depends principally on the supply of sediments or crustal fragments to the trench.

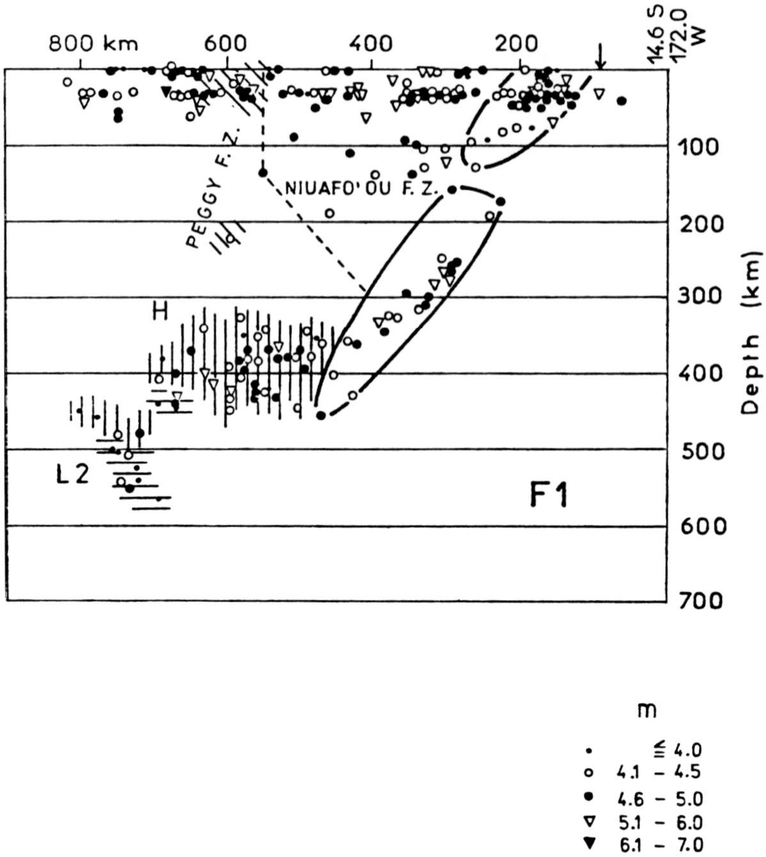

FIGURE 4. Vertical section giving the distribution of earthquake foci, and the configuration and depth of the Benioff zone for the northern Tonga region (Hanus and Vanek, 1979b). Trench axis is denoted by the arrow, m is ISC earthquake magnitude, and the Lau paleoplate (L2), Horne system (H) and fracture zones are denoted by various hatching. Position of section F1 is shown in Figure 2.

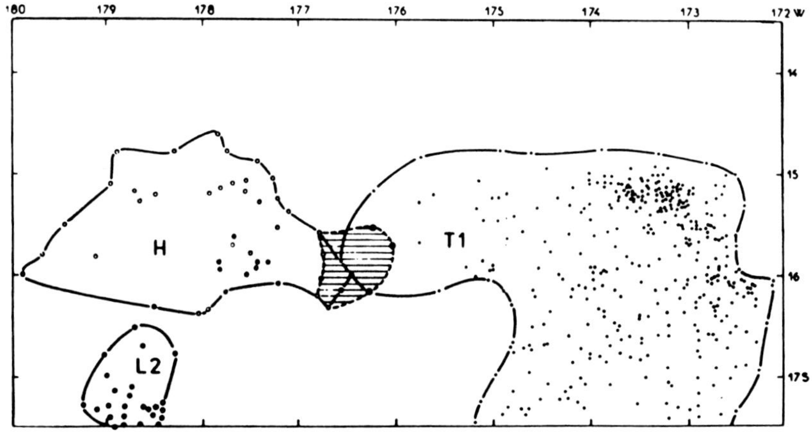

FIGURE 5. Map of earthquake epicenters with delineation of three subduction systems in the northern part of the Tonga island arc (Hanus and Vanek, 1979b). T1: current Tonga subduction; L2: activated Lau paleosubduction; H: activated Horne paleosubduction. Position of the deep collision is denoted by horizontal hatching.

II. GEOPHYSICAL DATA AND METHODS

A. THE GLOBAL MARINE GEOPHYSICAL DATA SET

The Geodynamics Research Program of Texas A&M has obtained the global marine geophysical data set from the National Geophysical Data Center (NGDC) of NOAA. This data set contains bathymetric, magnetic, gravity, and navigational data, times and dates, and identifications of ships' tracks along which seismic reflection, refraction, and swath mapping data were collected. Data are stored, by institution, on 31 6250-BPI tapes with EDCDIC coding. The data for this study consist of all available geophysical data from this collection, for the region 12° to 18°S, 170° to 175°W (Table 1). Individual compilations of ships' tracks, bathymetry, and free-air gravity have been obtained for plotting on 36- and 11-in. Benson electrostatic plotters. Ships' tracks in Figure 6 show the density and distribution of data in the study region. Non-digital seismic reflection profiles from selected cruises crossing the axis of the Tonga Trench have been obtained from the Lamont-Doherty Geological Observatory, the NGDC and the Hawaii Institute of Geophysics. The following sections describe the collection, manipulation, and presentation of the various data types.

B. BATHYMETRY

Bathymetry was obtained by conventional echo sounding techniques with 3.5 kHz and 12 kHz acoustic sources. Values in uncorrected meters had already been computed from two-way-travel-time using a speed of sound in water of 1.5 km/s, and were plotted along ships' tracks for contouring. A new map of the study region was prepared at a scale of 5 in. per degree and with a contour interval of 200 m. Contouring was aided by a comparison to the map of Mammerickx et al. (1974).

In order to illustrate changes in trench morphology, bathymetric profiles were stacked for comparison along the trench. The profiles were plotted with a common vertical exaggeration of approximately 28:1.

C. SEISMIC REFLECTION PROFILES

Individual seismic reflection profiles of trench crossings were obtained with conventional airgun sources at frequencies of 10 to 200 Hz. Seismic reflection profiles are presented as reduced photographic reproductions and as line drawings of the original recordings. As can be seen from Table 1, reflection data were not recorded along all of the ships' tracks, and the quality of existing profiles is variable. For example, most of the data clearly define shallow subsurface features but have poor penetration due to the attenuation of high frequencies. Consequently, only a small percentage of all recorded profiles was used in this study.

The reflection seismograms were examined for the presence or absence of pelagic sediment cover or trench deposits, the locations of oceanic plate basement, the overriding plate, the downgoing plate, possible horst-graben structures, other faults, and possible coherent reflectors within subducting grabens.

D. GRAVITY

Gravity data were digitized along ships' tracks at 2-km intervals. The observations have been reduced to free air anomalies with reference to the International Gravity Formula, 1967 (Hayes et al., 1972). Anomalies were plotted normal to ships' tracks at a scale of 100 mgal/in. The gravity data were too sparse for contouring.

Gravity models of the north-south and east-west segments of the Tonga Trench were constructed with computer programs based on the algorithm of Talwani et al. (1959), assuming a uniform density of 2.90 g/cm^3 for the main crustal layer and a density of 3.35 g/cm^3 for the upper mantle. Density values were constrained by the seismic refraction and

TABLE 1
Summary of Data Used in This Study

Institution	Cruise ID	Bathy. D	Bathy. ND	Grav.	Seism. Reflec.	Year
Scripps Institution of Oceanography	NOVA01AR	X				1967
	NOVA02AR	X				1967
	NOVA07AR	X		X		1967
	NOVA08AR	X		X		1967
	NOVA09AR	X				1967
	NOVA06HO	X		X		1967
	7TOW05WT	X		X		1970
	7TOW06WT	X				1970
	7TOW3BWT	X				1970
	ERDC02WT	X				1974
	ERDC03WT	X	X	X	X	1974
	GECS-FMV	X				1974
	GECS-EMV	X				1973
	AMPH03AR	X				1964
	ANTP14MV	X				1971
	ANTP16MV	X				1971
	ANTP17MV	X				1971
	SOTW09WT	X	X			1972
	SOTW11WT	X				1972
	STYX02AZ	X				1968
	STYX03AZ	X				1968
	STYX04AZ	X				1968
	STYX05AZ	X				1968
	STYX06AZ	X				1968
	STYX08AZ	X				1968
	STYX09AZ	X				1968
	CAPH-BHO	X				1953
	PROA3ABD	X				1962
	PROA3BBD	X				1962
Ocean Res. Inst. Univ. of Tokyo	KH6804-A	X				1968
	KH6804-B	X		X		1968
Hawaii Inst. of Geophysics	72110801	X	X	X	X	1972
	72110802	X	X	X		1972
	70042201	X	X	X		1970
Lamont-Doherty Geological Observatory	ELT31	X		X	X	1967
	C1211	X		X		1968
	C1305	X		X		1970
	V1907	X				1963
NOAA	POL7101	X		X		1971
Miscellaneous	V149	X		X		1970
	KI-80-1	X				1980
	TG-79-1	X				1979
	WSTG80-1	X				1980
	TG771	X				1977
	WS771	X				1977
	WS79-1	X				1979

Abbreviations: D — Digital bathymetric data;
ND — Non-digital bathymetric data (eg., 3.5 kHz echograms)

density determinations of Raitt et al. (1955) and are considered representative of the oceanic crust and upper mantle in the Western Pacific (Carlson, pers. commun., 1986). The computer programs, given any two-dimensional bodies represented by polygons of arbitrary complexity, compute the vertical components of their gravitational attraction at any specified external points (Talwani et al., 1959 and 1961). General models of the north-south subduction

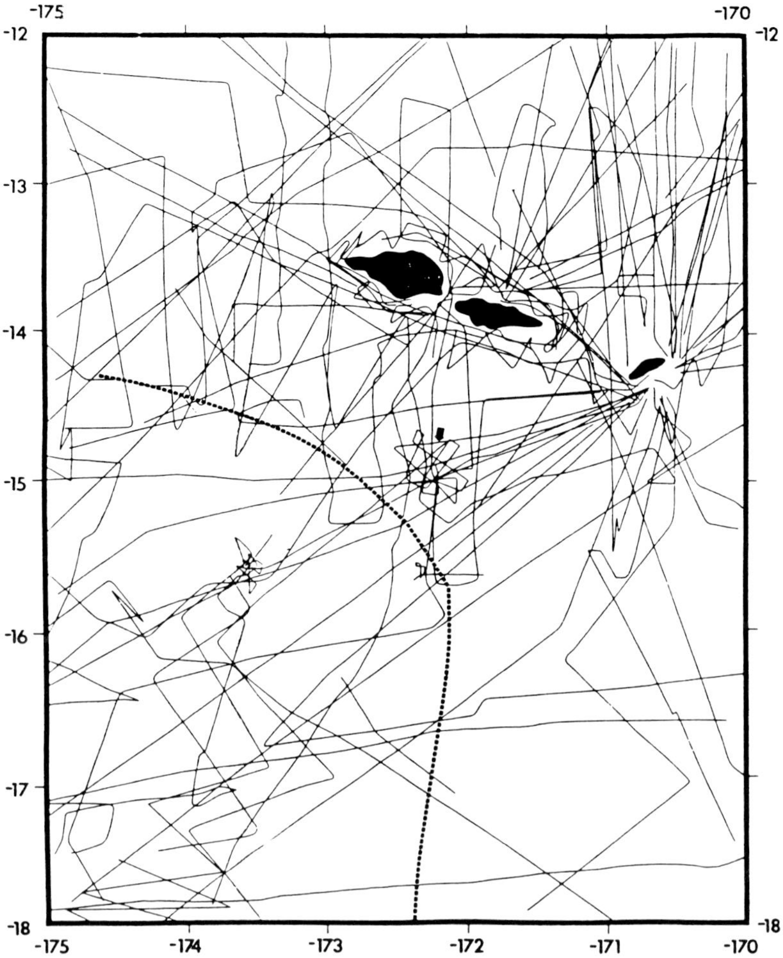

FIGURE 6. Ship tracks showing data coverage in the study area. Trench axis is marked by heavy dotted line and location of an unnamed seamount by heavy arrow.

zone and the east-west transform fault were entered into the computer by digitizing the vertices of polygons representing the mass components of those features. Free-air and isostatic gravity measurements at evenly spaced stations were then computed using the Talwani 2-D programs.

From all of the above interpretations, along with appropriate geologic information, the morphology and nature of convergence at the northern terminus of the Tonga Trench are inferred. Changes in convergent tectonic style (such as individual fault character, dip of downgoing slab, existence and spacing of grabens, changes in pelagic sediment cover, existence of possible tectonic erosion, etc.), are noted and interpreted, beginning from the southernmost crossing of the trench to crossings over the transform fault in the north.

III. RESULTS AND INTERPRETATIONS

This chapter presents the data results and their interpretations with reference to various features. As an aid to evaluating the trench in terms of differing tectonic character (i.e., subduction versus strike-slip motion) the sections on bathymetric and seismic reflection data are further subdivided into two sections: features south of 15.2° latitude and those to the north. This line of demarcation divides the study region into two approximately equal halves.

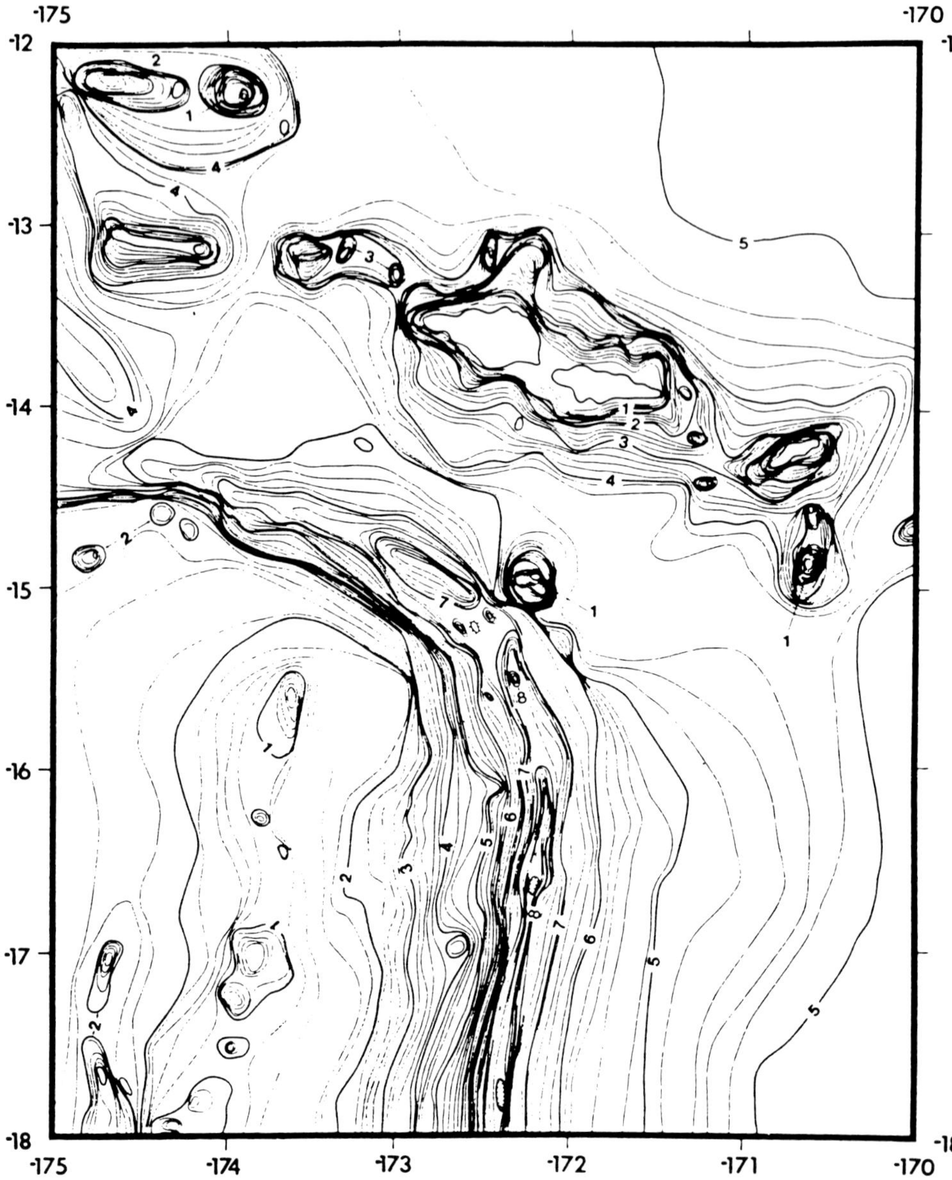

FIGURE 7. Bathymetry of the northern Tonga Trench region. Depths are in uncorrected meters × 1000. Contour interval is 200 m.

A. BATHYMETRY

A new bathymetric map of the study region has been prepared from available data (Figure 7). Bathymetric contours for features on the Pacific Plate and the Tonga microplate (such as the Samoan Islands and the Tonga Ridge) are similar to the earlier mapping by Mammerickx et al. (1974), but more details are shown for the trench axis than in the mapping of Mammerickx et al. (1974), as well as the previous mappings of Chase (1971) and Hawkins and Natland (1975).

B. BATHYMETRIC FEATURES SOUTH OF 15.2°S

1. Tonga Microplate

The Tonga microplate, which in the study region is the northeast corner of the Lau

Basin, is characterized by an average depth of only 1500 m. The major topographic feature of the region is the northern terminus of the Tonga Ridge, which is mapped as a distinct submarine ridge, trending north-south. The ridge is composed of a line of volcanoes (called the Niuatoputapu Group) averaging 350 m in depth and terminating adjacent to the trench's point of curvature. West of the Tonga Ridge the topography is generally very flat with an average depth of 2000 m.

2. Trench Axis Region

East of the Tonga Ridge the contours trend predominantly north-south and become laterally more closely spaced, delineating the shoreward trench slope. At approximately 15.2°S, just to the south of an unnamed seamount, the shoreward trench slope steepens significantly and continues in that manner as the axis of the trench swings to the west. Thus, it is near this latitude that the shoreward trench slope completes a transition to the scarp of a transform fault.

Horst and graben structures are known to exist on both the overriding and downgoing slopes of the trench (Gnibidenko et al., 1985). Possible horsts and grabens on the shoreward trench slope, as well as its greater steepness in comparison to the seaward slope, can clearly be inferred from selected bathymetric profiles shown in Figures 8 and 9. Profiles 1 to 5 are from south of 15.2°S. Horsts and grabens are clearly apparent in Profile 4. A small terrace is evident on the shoreward slope in Profile 2, serving as a trap for sediments.

The trench itself executes a broad arc as it turns from a north to a west trend. It strikes N8°E in the south and gradually swings to N40°W as it approaches 15.2°S. The deepest parts of the trench occur along 172.3°W at 17.8°S, 16.8°S, and 15.5°S (depths of 9036, 9111, and 8620 m respectively). From 15.5° to 15.2°S the trench shoals to a depth of 6000 m. The trench axis throughout is relatively narrow and V-shaped, averaging about 2 km in width, reflecting a lack of sediment or trench turbidite fill (Profiles 1 to 5, Figure 9).

The seaward wall of the trench has a gentler slope than the shoreward wall, as indicated by more widely spaced contours (Figure 7). The subducting Pacific Plate is known to be significantly faulted and to display evenly spaced, deep grabens between the trench axis and the outer high (Hilde, 1983; Gnibidenko et al., 1985; and Raitt et al., 1955). Relief on many of these faults is reflected in the bathymetric contours by isolated groupings of closely spaced contours. The existence of grabens and step faults in the seaward slope can be inferred from Profiles in Figure 9. It is noted, however, that grabens in this region are not as deep and well-developed as those noted on the bathymetric profiles of Fisher and Engel (1969) and Raitt et al. (1955), which are several hundreds of kilometers south of the study region. The grabens appear to be relatively evenly spaced and generally occur 4 to 10 km apart. Fault displacements appear to increase near the trench axis, as the Pacific Plate continues to bend downward into the subduction zone.

Profiles 1 to 3 (Figure 9) exhibit classic subduction zone morphology: pronounced bending of the downgoing plate into the trench and bending-induced normal faulting in the seaward trench slope, as well as normal faulting evident in the shoreward trench slope. Profiles 4 and 5 (Figure 9) mark the beginning of the trench's transition to a transform fault. The lithospheric bulging of the Pacific Plate, as well as deformation of the shoreward trench slope, are less pronounced. A gradual increase in the steepness of the shoreward trench slope is also to be noted.

3. Pacific Plate

There are no major topographic features mapped on the Pacific Plate south of 15.2°S. The sea floor is generally undisturbed in this area, averaging 5200 m in depth. The outer topographic high characteristically observed seaward of trenches (Watts and Talwani, 1974) is rather narrow but fairly well-developed in this region. At the southern boundary of the

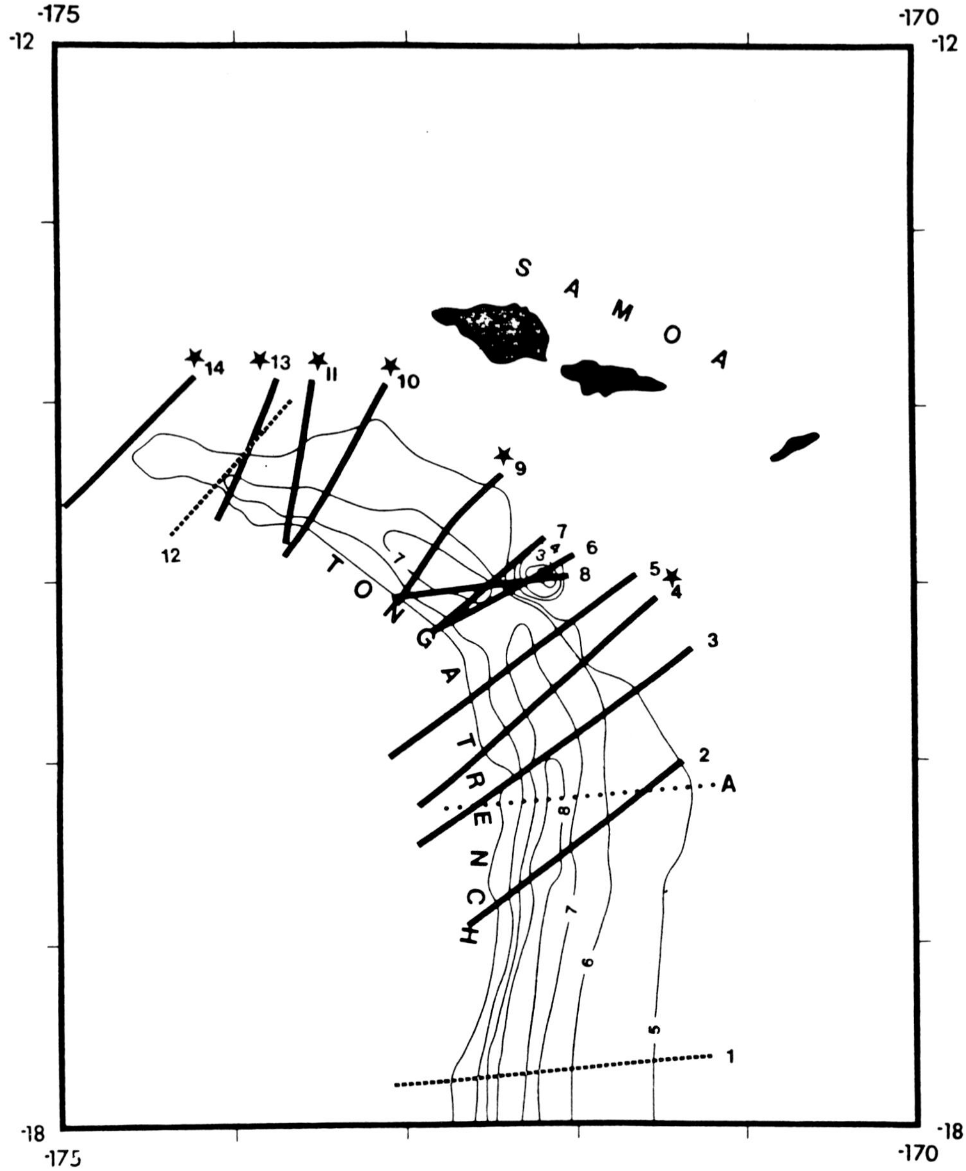

FIGURE 8. Index map for selected bathymetric, seismic reflection and gravity profiles. Solid lines represent profiles of bathymetric data only. Solid lines with a star denote bathymetric and seismic reflection data. Dashed lines represent profiles of bathymetry, gravity, *and* seismic reflection data. Dotted line A represents a profile along which only seismic reflection data was considered.

study region the outer high has an amplitude of 500 m and is located roughly 150 km seaward of the trench axis (Profile 1, Figure 9). At 16.3°S (between Profiles 2 and 3, Figure 9) it reaches a maximum amplitude of about 700 m. The outer high disappears almost completely at 15.2°S as the trench swings to the west. The amplitude and width of the outer topographic high is a consequence of the upward flexure of the lithosphere as it bends into the subduction zone (Watts and Talwani, 1974). The difference in outer high relief north of 16.3°S reflects a decrease in the bending of the plate as it approaches a transition to a transform fault. This is also supported by the seismicity data of Hanus and Vanek (1978 and 1979b). Further details on the outer high are discussed in the gravity section.

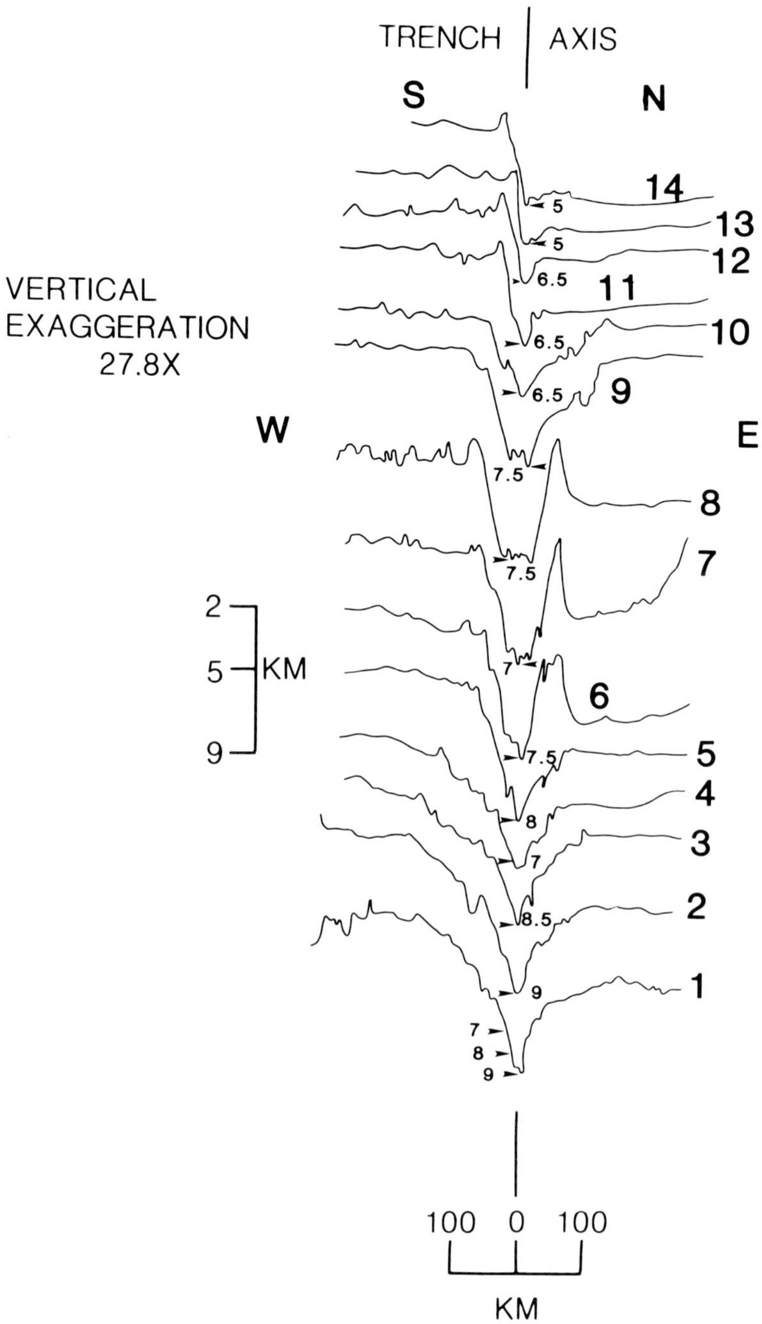

FIGURE 9. Bathymetric profiles across the northern Tonga Trench.

C. BATHYMETRIC FEATURES NORTH OF 15.2°S

1. Trench Axis Region

The scarp that occurs between the Tonga Ridge and the trench axis at approximately 15.2°S continues north of that latitude, paralleling the trench axis and continuing on past the termination of the trench as a bathymetric feature (Figure 7). The contours delineate an abrupt step in the ocean floor or a steep scarp which separates regional depths of 1500 and 5200 m. The trench itself changes trend westward from about N40°W to N80°W, becoming a trough at the base of the scarp.

The slope on the seaward side of the trench is gentler, evidenced once again by more widely spaced contours. Extensive faulting and horst-graben structures are no longer indicated by the contouring.

Morphology of the trench axis region north of 15.2°S exhibits a marked difference from the previous morphology described, as seen in bathymetric profiles 6 to 14 (Figure 9). The shoreward trench slope of the Tonga microplate is considerably steeper than that of the Pacific Plate. The flexure of the Pacific Plate (as indicated by the outer topographic high) continues to lessen, and then disappears completely as the axis of the trench swings westward. The shoreward slopes in general are also considerably less faulted and incised.

As the axis of the trench swings westward, the width of the trench floor increases from 2 to 20 km and flattens greatly at the point of the trench's curvature. The seaward slope in this area of curvature is the westward slope of a small seamount, which is located near the trench axis. The bathymetric contours (Figure 7) and Profiles 7 to 9 (Figure 9) show an irregular sawtooth pattern of small peaks and troughs on the trench floor. As the seamount has not yet reached the axis of the trench, these structures are probably unrelated to the seamount. The peaks may be representative of slivers of crustal material which have been transported obliquely to the trench axis and reworked there without being completely subducted. They may also be remnants of a seamount, which may have been previously subducted at this portion of the trench. It is possible that, as the trench began to make a transition from subduction to strike-slip motion, the remaining portions of this seamount were trapped in the axis of the trench before they could be completely subducted.

The flat trench floor in this area of curvature, due to the impingement of the seamount, suggests the possibility of sediment or trench turbidite fill. This is further suggested by the absence of the 7000 m contour in this area and then the reappearance of closed 7000 m contours northwest of the region (Figure 7).

The width of the trench floor narrows once again northwest of the seamount and returns to a V-shape. It is also in this region that the seaward slopes of the trench change markedly in character. Profiles 9 and 10 (Figure 9) show gentle seaward slopes incised by very small offsets that may represent faults with small displacements. The bulges of the slopes are also considerably less pronounced. Profiles 9 and 10, therefore, are representative of a gradual morphologic transition from subduction to strike-slip motion.

Profile 11 represents the termination of the transition to a transform fault. The most obvious feature in support of this is the level seafloor of the Pacific Plate which trends into a sharp break in the seaward slope and into what appears to be a small escarpment. The axial region in this profile has also narrowed from an average of 20 km in Profiles 9 and 10 to a width of approximately 1.5 km, possibly indicating the region where a principal transform displacement zone is beginning.

Profiles 11 to 14 (Figure 9) display classic transform fault morphology: an asymmetric, V-shaped valley with a high wall or scarp to one side. A marginal valley is noted on the shoreward side of the high scarp in Profile 12. Profiles 13 and 14 are located at the terminus of the trench and west of it respectively, and show a very small, well-defined trough flanked by a shoreward scarp and a seaward series of narrow, shallow ridges running parallel to the trough.

2. Pacific Plate

Several major topographic features are evident on the Pacific Plate north of 15.2°S. The most notable are the Samoan Islands and seamounts which sit on a fairly linear and continuous ridge rising more than 2500 m above the sea floor. The Samoan Islands are primarily thick accumulations of alkalic basalts which formed shield volcanoes on a ridge of unknown composition (Hawkins and Natland, 1975). Both the ridge and the island chain follow the approximately northwest-trending portion of the trench and, according to Hawkins and Natland (1975), are located along the hinge line where the Pacific Plate is downbuckled and probably faulted into the Tonga Trench. The occurrence of hinge faulting south of Samoa, at the westward bend in the trench, has been verified by the focal mechanisms and distribution of earthquakes in the region (Isacks et al., 1969).

Other major bathymetric features of the Pacific Plate also trend east-west. They are Pasco Bank at 13°S, 174.5°W and Taviuni Bank at 12.2°S, 174° to 174°W. Both structures also sit upon ridges roughly 2500 m above the sea floor.

The most pertinent topographic feature to this study is the Machias Seamount at approximately 15°S, 172°W, which is located at the point of curvature in the trench. The seamount rises to a depth of 800 m above a regional depth of 4500 m and is 18 km in diameter. Profiles 6 to 8 (Figure 9) show the morphology of the seamount and the trench axial region adjacent to it. Fryer and Smoot (1985) have found that seamounts 40 to 100 km in diameter may interrupt normal subduction processes. Seamounts less than 30 km in diameter are more than likely carried down with the plate into the trench and probably contribute to only small-scale vertical movement in the outer fore-arc. This may be the case with this seamount, as its diameter is less than this critical value.

However, since the seamount is directly adjacent to the segment of the trench that is undergoing the transition to strike-slip motion, it is more likely that relative plate motion will carry the seamount past the trench rather than into it. The presence of the seamount appears to have little or no effect on the morphology of the trench axis or the overriding plate. It has not significantly disrupted the 6000 m contour which runs adjacent to the seamount-trench intersection on both sides of the trench axis (Figure 7), and very little compressional deformation is evident in the shoreward trench wall opposite the seamount in Profiles 6 to 8 (Figure 9).

Hawkins and Natland (1975) have found that the geology of the Machias Seamount is useful as an indicator of the rate and extent of vertical tectonic motion at the trench. Their petrological studies indicate that rapid subsidence of the seamount is occurring, from which estimates of the extent and rate of downbuckling of the Pacific Plate may be obtained. Samples dredged from near the crest of the seamount include cobbles, gravel, pahoehoe fragments, and coral, indicating that its crest was formerly at wave base or above sea level (Hawkins, 1974b; and Hawkins and Natland, 1975). Hawkins and Natland (1975) have determined a K/Ar date from an anorthoclase-rich phonolite dredged from the seamount to be 940,000 ± 20,000 years, indicating that the Machias Seamount should have formed about 100 km to the east and has been subsiding at a rate of 1.5 mm/year. Hawkins and Natland (1975) and Natland (1980) estimate that downbuckling of the seafloor could account for 80% of this subsidence, with isostatic compensation and a post-glacial rise in sea level accounting for the rest.

It is possible that the seamount formed as a result of the lateral bending and rupturing of the Pacific Plate in this area of transition. Shaw (1973) proposed that linear island chain and seamount volcanism is a consequence of melting in response to a differential shear between moving lithospheric plates and the more rheid asthenosphere. In the study region, in addition to the vertical shear between the lithosphere and asthenosphere, there is also a lateral gradient in shear associated with the onset of strike-slip movement. It follows that shear stresses should be concentrated in the area where the Pacific Plate is bending and tearing to the west. Shear melting of the mantle may therefore be a plausible explanation for the origin of this seamount, as well as other seamounts which may already have been subducted. Natland (1980) maintains that shear melting is also responsible for the post-erosional volcanism of the Samoan Islands.

Bathymetric features thus far have been described mainly in terms of vertical differences in morphology (i.e., changes are noted from the trench slope break down to the trench floor). However, important morphological differences are also evident horizontally. These are to be noted mainly in the area between the trench and the summit of the Tonga Ridge, shoreward of the trench axis. South of 18°S it has been noted that a volcanic arc lies distinctly separate from and west of the summit of the Tonga Ridge (Ewart et al., 1977; Dupont and Herzer, 1985; and Herzer and Exon, 1985). In the study region, however, the Tonga Ridge summit and the volcanic arc are one and the same. A marked difference is observed in the distance between the ridge summit and the trench axis, south and north of 15.2°S. South of that latitude this distance decreases from a maximum of 150 km (Profile 1, Figure 9) to that of approximately 50 km (Profile 5, Figure 9). North of 15.2°S the ridge summit is even closer to the trench axis. The average distance for Profiles 6 to 14 (Figure 9) is only 15 to 25 km. These horizontal differences in trench morphology are also reflective of the trench's transition from subduction to strike-slip motion.

D. SEISMIC REFLECTION DATA

Single-channel seismic reflection profiles over the Tonga Trench are presented as time sections. Locations of these profiles are shown in Figure 8. A total of 9 profiles across the trench were selected. These profiles are fairly well distributed and provide reasonably good coverage of the trench region.

1. Trench Morphology South of 15.2°S

The original records of trench crossings have been reproduced, and are presented with line drawing interpretations directly beneath them (Figures 10 to 12). The seaward slope is covered with a very thin veneer of pelagic sediments (50 to 100 ms two-way-travel-time) to a point approximately 150 km east of the trench axis. From this point westward, however, it is apparent that very little or no sediment is being carried into the trench. What little sediment there is appears to be cut by several small, low angle step faults, with the blocks relatively downdropped toward the trench floor.

Due to the poor resolution of the seismograms, small horsts and grabens can only be inferred from bathymetric offsets in the seaward trench slope. It seems likely that if a subducting plate is dominated by bending-induced normal faults that all have their throw in a common direction — as is the case here — void space may still be created in its surface that is not unlike that of a graben.

The faulted nature of the seaward slope suggests that the lithosphere of the Pacific Plate is in tension near the surface and is being flexed and possibly stretched as it is subducted. This is further evidenced by the lithospheric ''bulges'' or outer topographic highs that are clearly apparent seaward of the trench axis in Profiles 1 and A (Figures 10 and 11). In Profile 4 (Figure 12) the faulting and bulging of the Pacific Plate is less pronounced, thus indicating the onset of the morphologic transition.

The shoreward trench slope is also deformed by relatively steep, normal faulting, with blocks downdropped towards the trench. This is most clearly illustrated by Profiles 1 and 4 (Figures 10 and 12). Oceanic basement is also exposed on the lower part of the shoreward slope. These features, in addition to the morphology of the seaward trench slope described in previous sections, have been cited as evidence for possible tectonic erosion (Hilde, 1983). While these observations do not necessarily establish that the northern Tonga Trench has experienced tectonic erosion, they do suggest that this is an ideal region to test for the effects of that erosion (with multichannel seismic profiling and dredging).

2. Trench Morphology North of 15.2°S

As discussed previously, the morphology of the trench axis region north of 15.2°S exhibits a marked difference from the region to the south. Seismic Profiles 9 and 10 (Figures 13 and 14) continue to show a morphology that is transitional between subduction and transform faulting. The seaward trench slope is offset by normal faults but to a lesser degree than the profiles to the south. Flexing and bulging of the Pacific Plate is also less evident. The sediment supply to the trench is still minimal, though slightly greater than the region to the south. This may be partly due to the proximity of the Samoan Islands and seamounts.

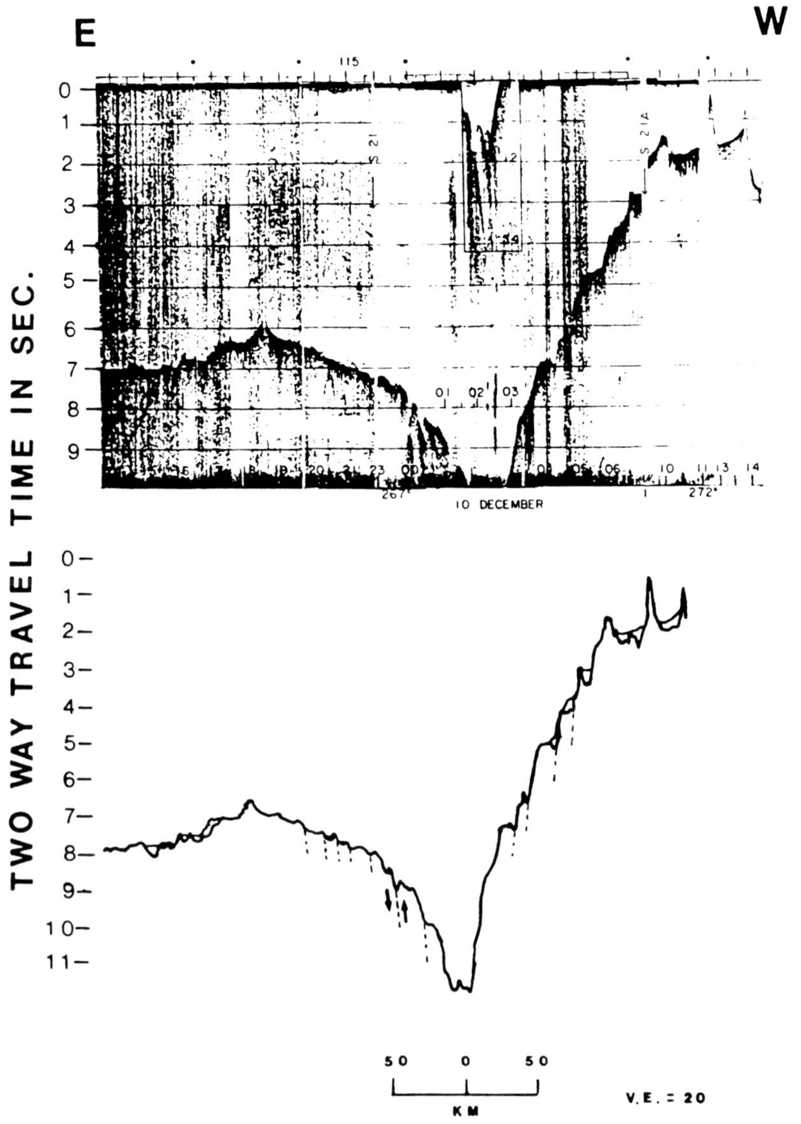

FIGURE 10. Seismic profile from Line 1. Inferred faults shown by dashed lines.

A small amount of sediment is noted in the trench axis on Profile 10 (Figure 14). Small ridges and troughs are noted in the trench axis on Profile 9 (Figure 13). This material may consist of pieces of oceanic crust or of a volcanic seamount which may have been carried obliquely to the trench and have not been completely subducted. It is likely that this material will be carried along the axis with the strike-slip motion that has been initiated. The shoreward trench slope in both of these profiles is virtually undeformed, which could be indicative of a decrease in tectonic interaction between the downgoing and overriding plates.

It appears that more than just "simple" subduction (as illustrated by Profiles 1 and A, south of 15.2°S) is taking place along this portion of the trench. Extensive rotation of the Pacific Plate and a transition to transverse motion may be contributing to rock exposures that reflect mixtures of material from several different structural entities, both shoreward and seaward of the trench axis.

Seismic Profiles 11 to 14 (Figures 15 to 18) illustrate the completion of the trench's

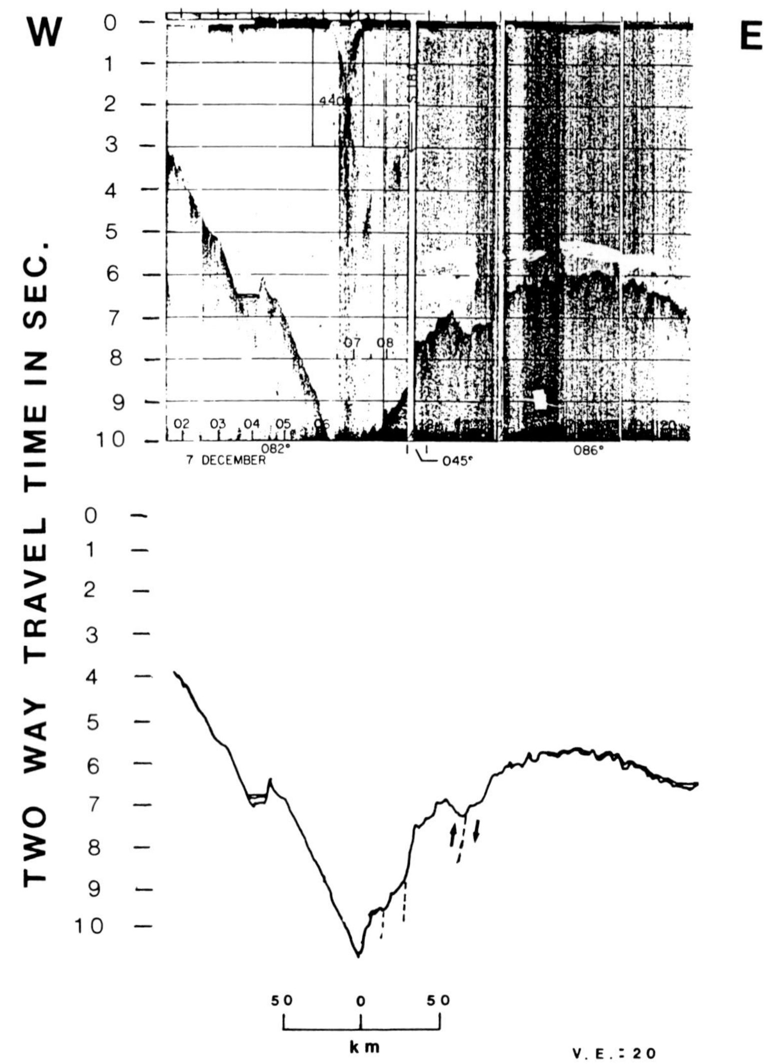

FIGURE 11. Seismic profile from Line A. Inferred faults shown by dashed lines.

gradual transition to transform fault morphology. The general structure of the trench is shown to consist of small ridges and troughs. Bending-induced normal faulting is absent, due to the apparent lack of flexure of the downgoing plate, although small, low angle faults are still evident, particularly in Profile 12 (Figure 16). The small ridges show the reflection character of exposed igneous basement but the troughs, most notably in the transform axis of the last two profiles, are ponded by a thin layer of sediments. These sediments are very likely the pelagics of the Pacific Plate.

It is possible on Profiles 11 and 14 (Figures 15 to 18), to detect flat-lying sedimentary reflectors above acoustic basement. The sediments appear not to be faulted, folded, or significantly deformed. This implies that recent downwarping of the Pacific Plate has not taken place.

E. GRAVITY DATA

Free-air gravity anomalies were plotted in the study region in order to help define the general nature and distribution of densities in the trench region. The large negative anomalies

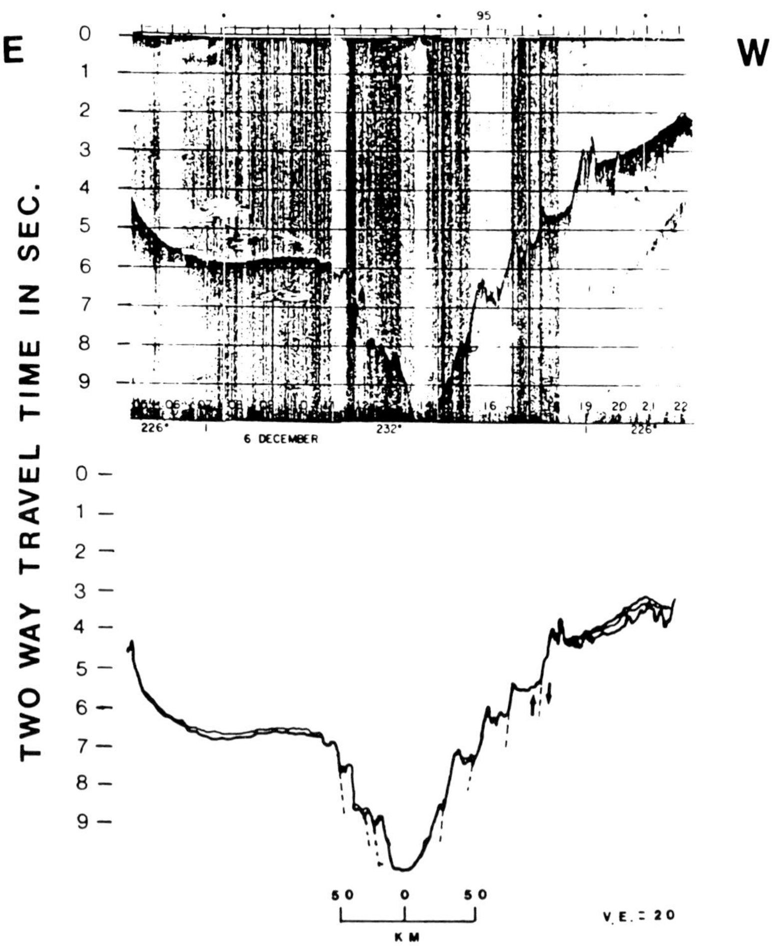

FIGURE 12. Seismic profile from Line 4. Inferred faults shown by dashed lines.

along the Tonga Trench have been studied by Talwani et al. (1961), Watts and Talwani (1974), and Kogan (1975). These studies explain the gravity effect of the trench region in terms of the bending of the Pacific Plate beneath the Indian-Australian Plate, as well as in terms of the gravity effects of the topography, crustal structure, and downgoing slab. The trench axis is marked by large negative anomalies ranging from -240 mgal at the southern boundary of the study region to -100 mgal near the northern terminus of the trench. The shape of the gravity minimum and the location of its axis closely matches the corresponding topographic profiles, as expected for free-air anomaly observations. Although most negative anomalies over Pacific trenches are slightly displaced toward the shoreward wall, suggesting an accumulation of deformed sediments (Radar, 1985), the free-air anomaly in this region is centered over the trench axis.

Gravity values east of the trench are mostly negative except over prominent topographic highs. West of the trench values are primarily positive, particularly over the Tonga Ridge, which shows a large gravity high with a maximum of $+100$ mgal at approximately 18°S.

Two-dimensional gravity models were used to develop crustal structure sections for the north-south and east-west segments of the trench and to make interpretations as to the nature of the gravitational effect of subsurface bodies. These gravity transects are shown by Profiles 1 and 12 in Figure 8. Profile 12 has been projected perpendicular to the axis of the transform for the modelling of that feature. Unfortunately, no gravity data were available for the region of transition from subduction to strike-slip motion.

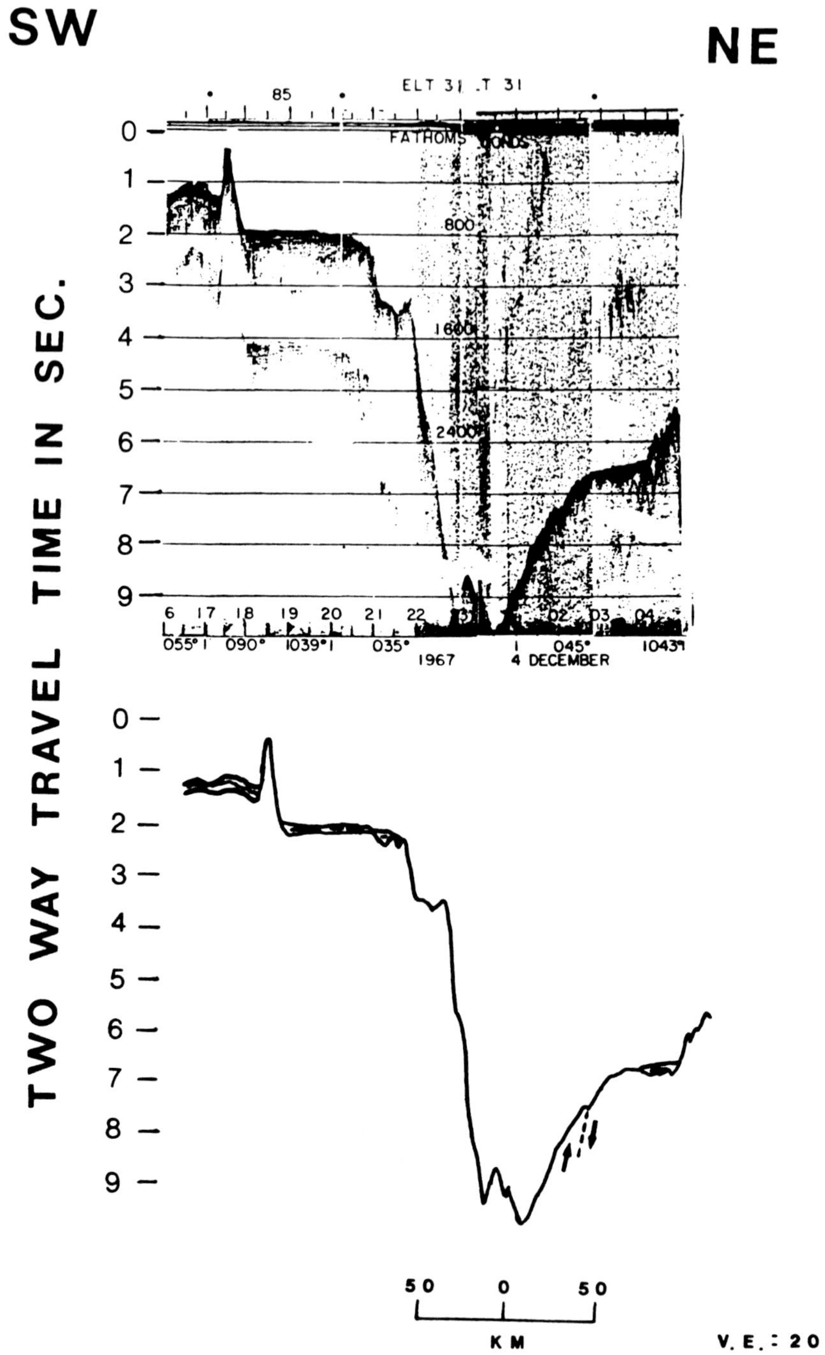

FIGURE 13. Seismic profile from Line 9. Inferred fault shown by dashed lines.

The topography of oceanic basement was constrained by bathymetric and seismic reflection profiles but no seismic refraction data for the study region were available. This creates a problem in that there is no constraint on the Moho, the density discontinuity that, aside from the ocean bottom, has the greatest effect on the gravity anomaly in this study. As there is no unique solution to the problem of finding a mass distribution that will produce the measured gravity field, a number of assumptions must be made before a structure can

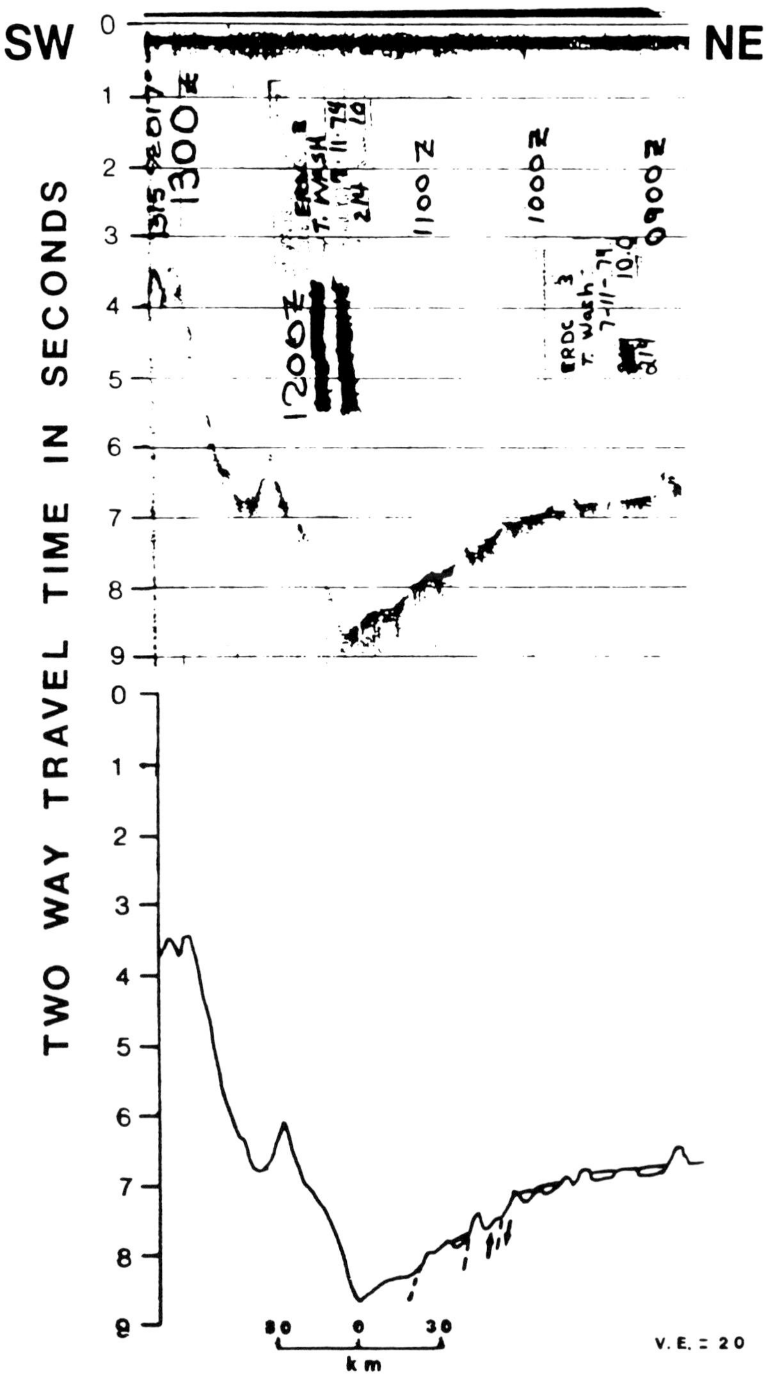

FIGURE 14. Seismic profile from Line 10. Inferred faults shown by dashed lines.

be deduced from gravity anomalies. It is difficult to interpret gravity anomalies in great detail without the help of seismic refraction data, from which information about the density distribution in the crust can be obtained. Therefore, it was necessary to compare the crustal models inferred in this study to those obtained at the closest seismic refraction stations. The only published determinations available to the author were those of Raitt et al. (1955), which

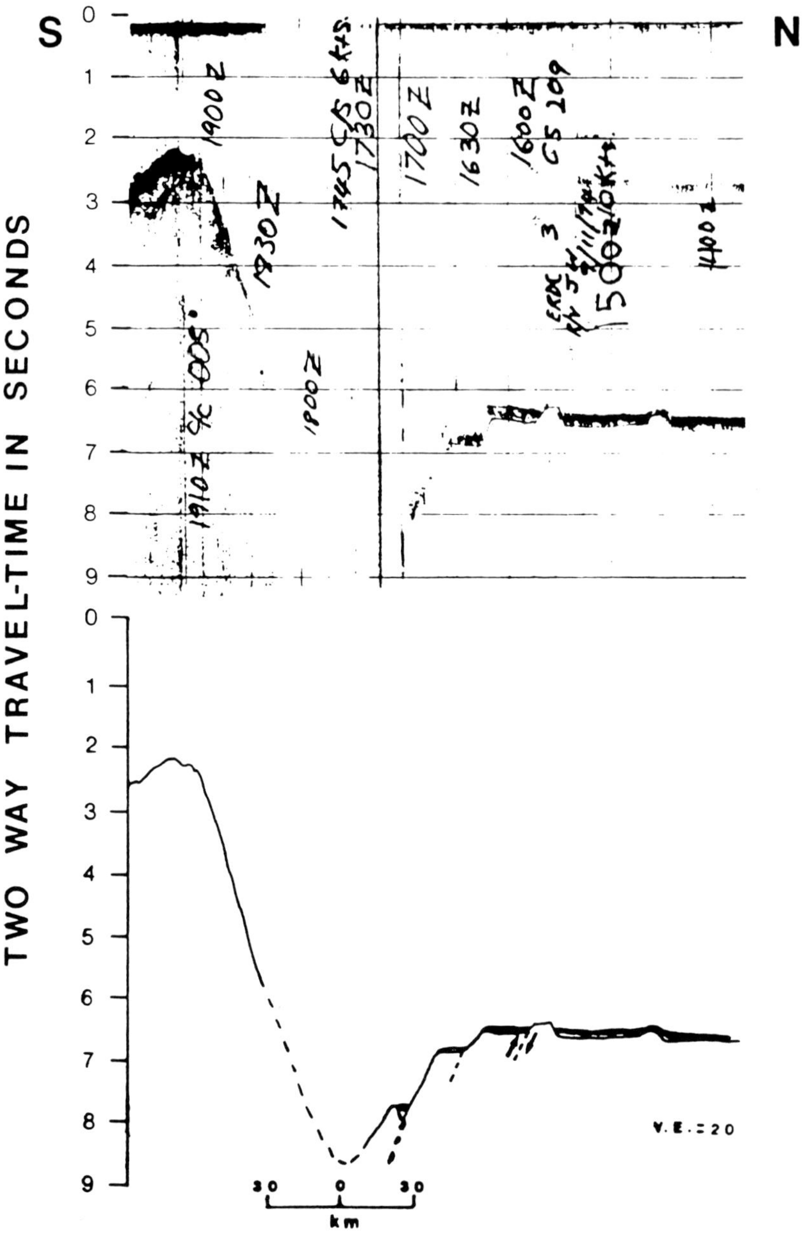

FIGURE 15. Seismic profile from Line 11. Inferred faults shown by dashed lines.

were modified by Talwani et al. (1961), and those of Gnibidenko et al. (1985). Unfortunately, these stations were at a considerable distance from the gravity profiles used in this study. The stations of Raitt et al. (1955) were located at approximately 22°S and those of Gnibidenko et al. (1985) at 26°S.

1. Subduction Zone Model

The two-dimensional model of the north-south subduction zone is divided into four parts: the unsubducted crustal layer, the subducted crust, the subducted crust converted to eclogite, and the downgoing slab. The latter three were treated as density anomalies contrasted to a mantle density of 3.35 g/cm^3. Use of density anomalies for the subcrustal structure makes

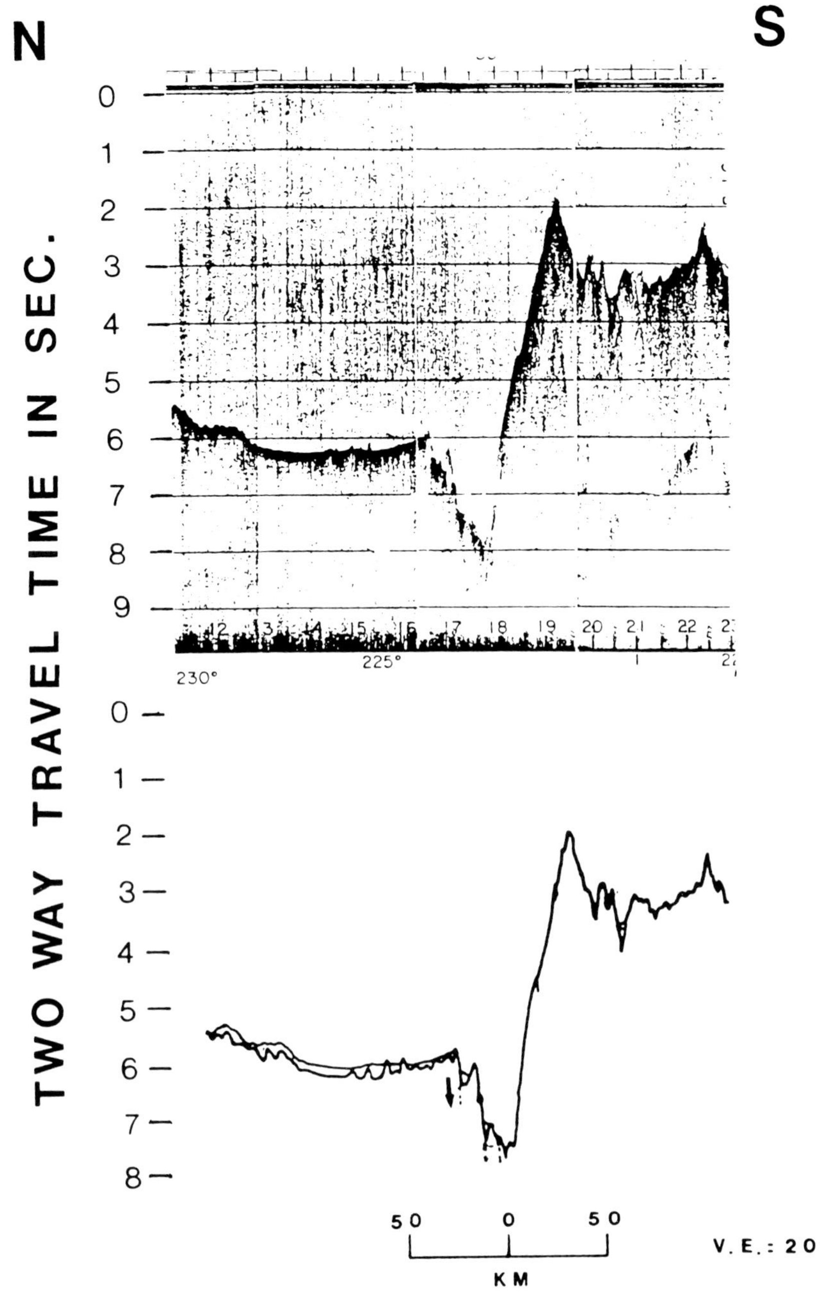

FIGURE 16. Seismic profile from Line 12. Inferred faults shown by dashed lines.

the model easier to modify and allows it to be independent of any assumed density gradient within the mantle (Sager, 1980). Thus it is assumed that density changes in the mantle caused by increasing pressures with depth will occur equally in all parts of the model.

In the absence of any seismic velocity-depth determinations in the study area, the density of oceanic crust was constrained by an average crustal density for the Pacific, which has been used successfully by other authors (Grow, 1973; Grow and Bowin, 1975; Sager, 1980; and Radar, 1985). For simplicity, the crust was assumed to consist of a water layer underlain by a single, homogeneous layer of density 2.90 g/cm^3, but of variable thickness.

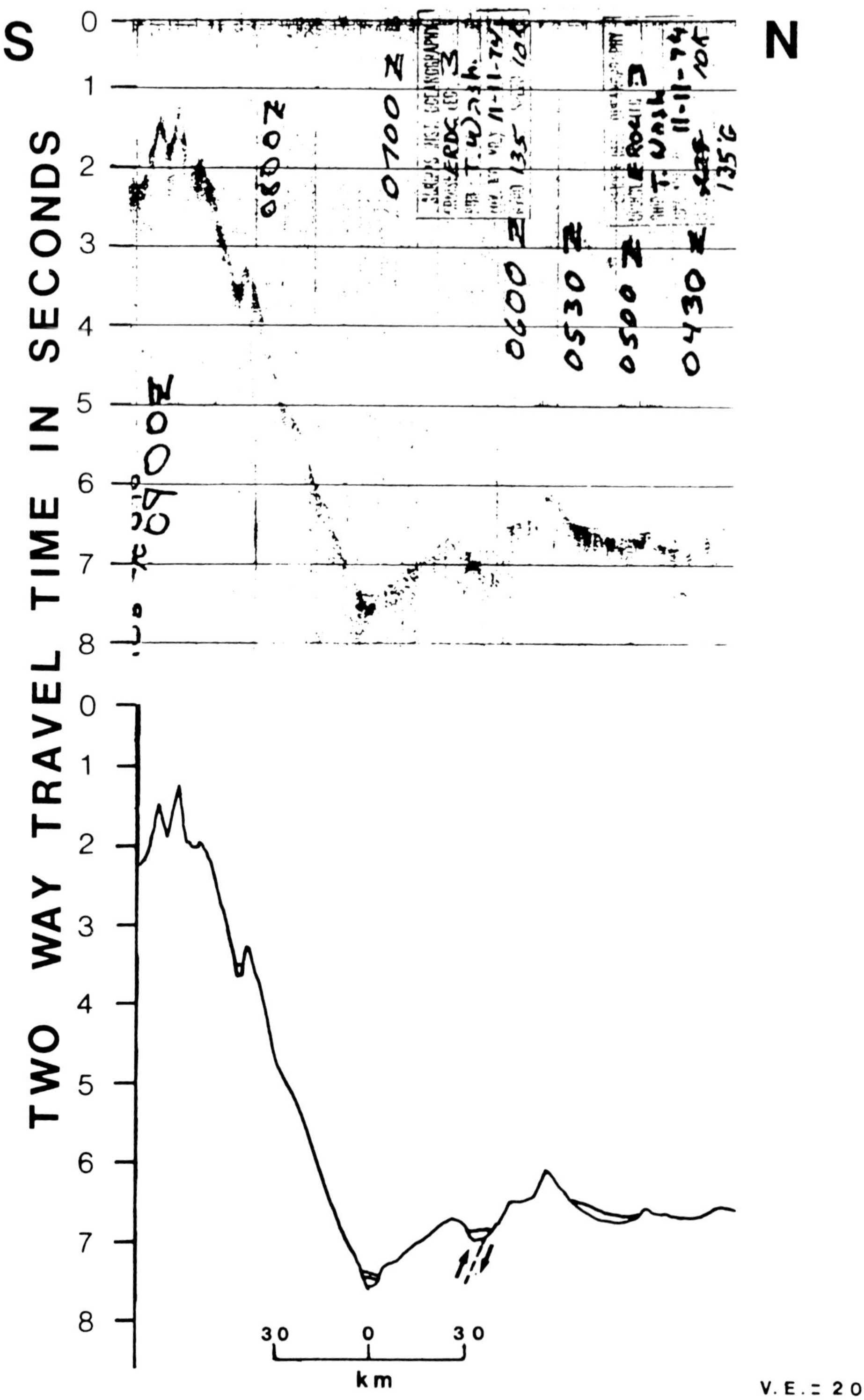

FIGURE 17. Seismic profile from Line 13. Inferred fault shown by dashed line.

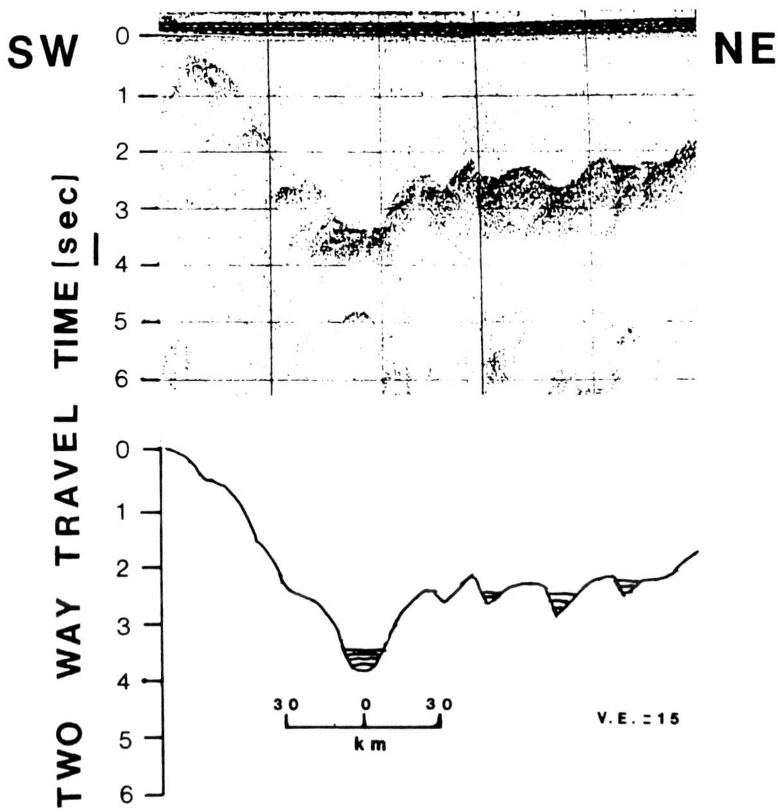

FIGURE 18. Seismic profile from Line 14.

The first density anomalies modelled were the subducted crust, the eclogite layer and the downgoing slab (Figure 19). Because there is no seismic constraint on the subducted crust, it was assumed to occupy the upper 6 km of the downgoing slab. From a depth of 20 to 60 km, this crustal body was assumed to keep its presubduction thickness and density (i.e., a 6 km thickness with a density of 2.90 g/cm^3).

As the subducted crust is carried farther downward with the descending plate, a complex suite of dehydration reactions causes the gabbroic crust, which is rich in plagioclase and hornblende, to turn into pyroxene-rich and garnet-rich eclogite. This occurs at depths of from 60 to 80 km (Anderson et al., 1976; Delaney and Helgeson, 1978). For the purposes of this study, it was assumed that the eclogite phase change occurred at a depth of 60 km as shown in Figure 19. The eclogite in this model has a density of 3.56 g/cm^3, which corresponds to an in situ value for a depth of 60 to 80 km (Grow and Bowin, 1975). The eclogite crust continues as the upper 6 km of the slab to a depth of 130 km. Ringwood (1977) contends that past depths approaching 150 km all or most of the crust has melted away. For simplicity, it has been assumed that the crust has the same density as the rest of the slab past a depth of 130 km. If the eclogite were alternatively continued down the rest of the slab, the increase in the calculated slab anomaly would be small (Sager, 1980).

A constant thickness of 80 km was chosen for the downgoing slab. This value is slightly greater than the width of the Benioff zone (Isacks and Barazangi, 1977; Hanus and Vanek, 1979b) and is in the middle of the range of values chosen for the northern Tonga slab thickness by Hanus and Vanek (1979b) and Chapple and Forsyth (1982).

The slab was positioned to envelop most of the Benioff zone earthquakes as plotted by Hanus and Vanek (1979b) for the northern Tonga region. According to their data, the slab

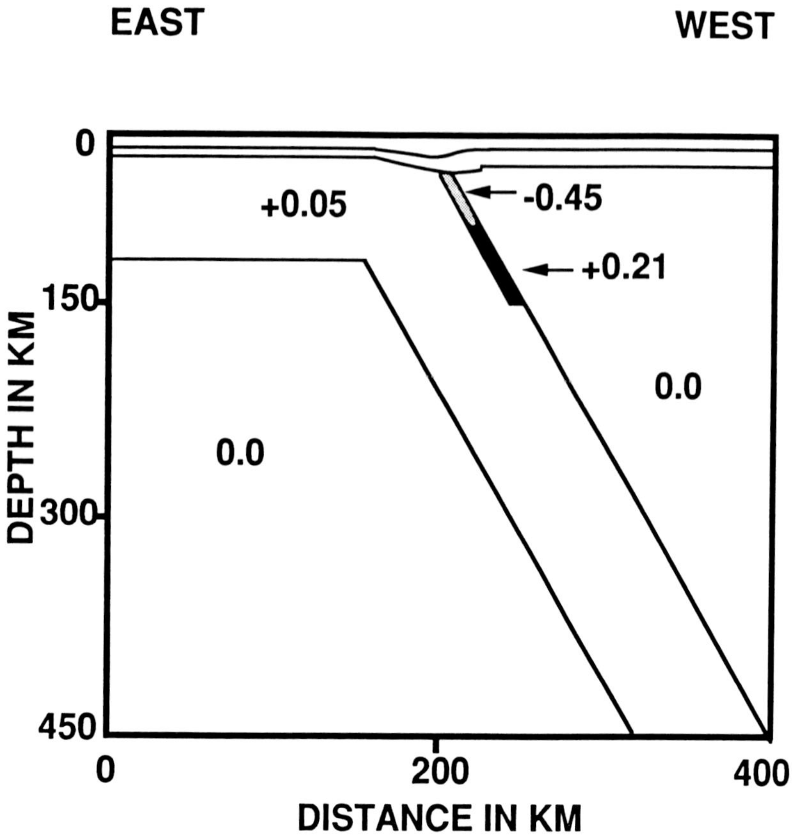

FIGURE 19. Model of the subducted crust (stippled), eclogite (solid), and downgoing slab (shaded). The slab follows the earthquake hypocenters determined by Hanus and Vanek (1979b).

ends at a depth of 455 km, below which no earthquakes are recorded. Continuing the slab deeper would have little effect on the model because of the great depth.

The most commonly used density contrast for the slab is from $+0.04$ to $+0.05$ g/cm^3 (Hatherton, 1970; Grow and Bowin, 1975; and Sager, 1980). A simple, single-density model using a contrast of $+0.05$ g/cm^3 was used in this study.

The gravity contributions of the subducted crust, eclogite and the downgoing slab (below 130 km) are shown in Figure 20. The negative gravity contribution of the subducted crust reaches a minimum of -53 mgal, which is displaced 15 km shoreward of the trench axis. Eclogite contributes a positive contribution which peaks 85 km shoreward of the trench axis. The positive contribution of the downgoing slab reaches a maximum of 50 mgal 200 km shoreward of the trench axis and has a much longer wavelength than the shallower anomalies. Grow and Bowin (1975) have found from thermal and petrologic data that at the trench axis, about 80% of the positive gravity anomaly contribution of the slab is due to density anomalies in the upper 150 km. The combined total of a more than 40 mgal contribution for all of these bodies peaks 100 km shoreward of the trench (Figure 20).

The remaining density anomaly to be modelled was that of the unsubducted crustal layer (Figure 21). Initially, the anomaly due to the downgoing slab was subtracted from the observed free-air anomaly, in order to obtain a residual anomaly. A computed free-air anomaly was then fit to this residual in obtaining a configuration for the unsubducted crustal layer. East of the Tonga Trench the crust in the study area was assumed to have a thickness of 6 km and was continued eastward to eliminate edge effects. West of the trench at 22°S

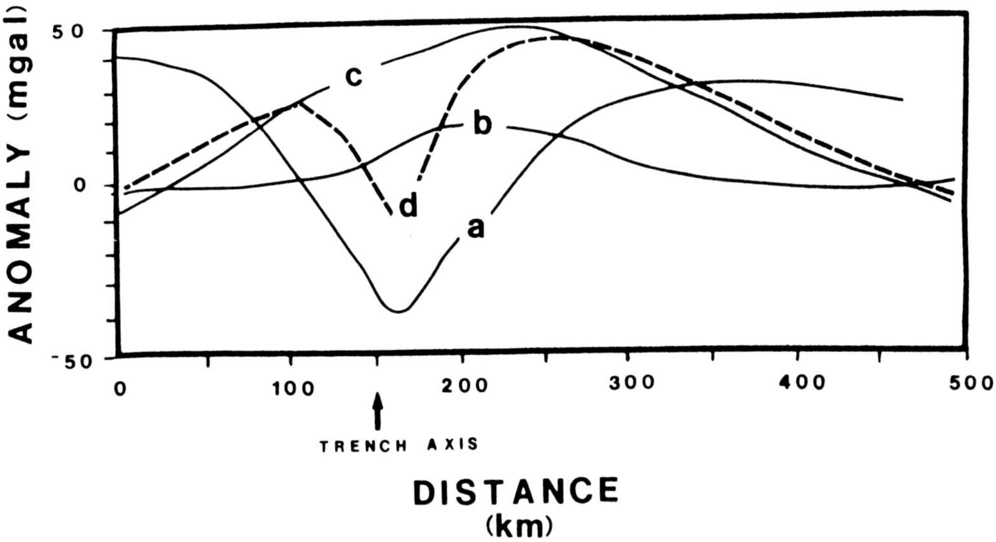

FIGURE 20. Gravity anomalies from the subducted crust (a), eclogite (b), and the downgoing slab (c). Curve d is the sum of curves a through c.

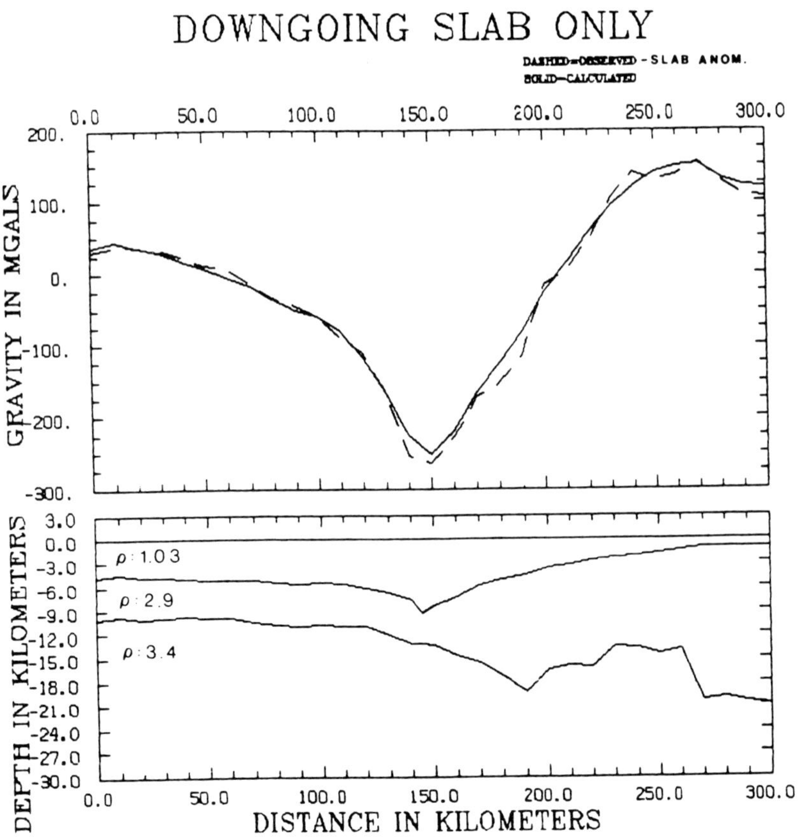

FIGURE 21. Residual and calculated gravity anomalies (top) and a model of the unsubducted crust (bottom), taking into account the downgoing slab only. The apparent thinning of the crust at the trench axis is an artifact of the 3:1 vertical exaggeration.

the refraction data of Raitt et al. (1955) set the Moho at approximately 25 km. In the study area the Moho 150 km west of the trench was also set at this value for a first approximation and continued westward to eliminate edge effects. The Moho was subsequently adjusted to fit the residual anomaly.

Seaward of the trench axis, the residual reaches a maximum of 50 mgal. This maximum is associated with a bulge in the downgoing plate (located approximately 150 km seaward of the trench axis, with a well-developed amplitude of about 500 m), which has been noted in the bathymetric and seismic reflection profiles. The crust seaward of the trench was thinned slightly to obtain a better fit of the gravity data and to account for this flexure. According to Watts and Talwani (1974 and 1975) and Dubois et al. (1975), a bulge occurring in the lithosphere seaward of the trench, in which the low-density crust has a constant thickness, produces a regional positive gravity anomaly whose wavelength corresponds to that of the bulge. Watts and Talwani (1974) have named this the Outer Gravity High, and have found it to be an important factor in explaining the tectonics of convergent margins. The Outer Gravity High has been identified seaward of the Tonga Trench by Watts and Talwani (1974), Dubois (1975), and Kogan (1975), although it has been found to be less developed in terms of wavelength than gravity highs seaward of trenches in the Northwest Pacific.

The larger negative anomaly over the trench axis is due to a large crustal mass deficit. The isostatic compensation model of Hager (1983) estimates that compensation over trenches is only completed when the slab comes to rest, depressing the denser lower mantle.

A significant gravity high is noted 250 km shoreward of the trench axis (Figure 5). The topographic effect of the Tonga Ridge and that of the downgoing slab beneath the ridge contribute to this anomaly. The amplitude and wavelength may also be suggestive of thermally attenuated, ductile lithosphere beneath the ridge. In contrast to "normal" P_n velocities of 8.0 to 8.2 km/s, a mantle velocity of 7.7 km/s was observed from a refraction line atop the Tonga Ridge (Raitt et al., 1955). A similarly low value has been linked to the substantial thinning of a cold, rigid mantle lid, which makes up the bulk of the lithosphere beneath the Fiji Platform (Hamburger et al., 1988). Thermal thinning of the Tonga Ridge lithosphere in the region of the gravity profile is probably associated with back-arc magmatism in the Lau Basin directly to the southwest of the profile, and the transition from subduction to strike-slip motion immediately to the north. This transition may involve high thermomechanical stresses in the lithosphere, resulting in extensional thinning and tearing of the lithosphere as the trench bends to the west.

A notable feature of the model is a mass excess located between 40 and 120 km west of the trench axis. The shallowing of the Moho in this area was necessary in an attempt to match the gravity high. The closeness of the fit indicates only that the selected configuration is a possible solution.

The anomaly due to both the downgoing slab and the subducted crust was subtracted from the observed anomaly to obtain a second residual. The resulting model of the unsubducted crust is shown in Figure 22. A significant mass excess is once again evident east of the trench axis.

The preferred model of the unsubducted crust is shown in Figure 23. For this model, the gravitational effects of all three subcrustal components — the downgoing slab, subducted crust, and eclogite — were removed in order to obtain the residual. West of the trench, the crustal configuration is identical to the previous models, but east of the trench the inclination of the eclogite in the downgoing slab has obviated most of the need for a shallow mass excess. This model is the most realistic in that it is constrained by the presence of all known subcrustal anomalies and is similar to crustal models of other Pacific trenches (Grow, 1973; Sager, 1980; and Radar, 1985).

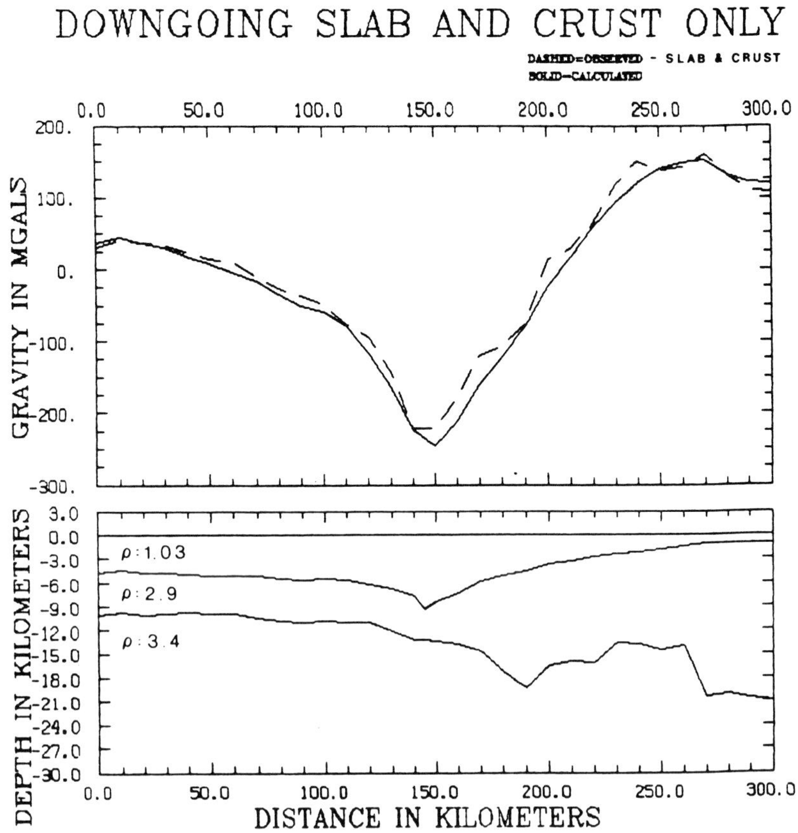

FIGURE 22. Residual and calculated gravity anomalies (top) and a model of the unsubducted crust (bottom), taking into account the downgoing slab and subducted crust. Vertical exaggeration is 3:1.

2. Transform Fault Model

Simplified, layered density configurations of the transform fault were obtained from both isostatic and free-air gravity modelling. As before, there are no seismic refraction determinations across this feature to substantiate crustal thicknesses or densities. The average density for the oceanic crust used in this model was also 2.90 g/cm^3. Sediment thickness across the fault, as derived from the seismic reflection data discussed in the previous section, was determined to be too thin to greatly alter the results, and was neglected from the modelling. A density of 1.03 g/cm^3 was used for the seawater and 3.35 g/cm^3 for the upper mantle material beneath the ocean crust.

The first computations performed were those for an isostatic anomaly, using the principles of Airy isostasy. It was expected that the transform fault would, on the whole, be isostatically compensated, with little or no lateral density inhomogeneities occurring in the mantle (such as those caused by a downgoing slab) at depths greater than 30 km. Buoyancy forces arising from lateral variations in lithosphere temperatures also render such features as transform faults, spreading ridges, and hot-spot swells *thermally* compensated (Haxby and Turcotte, 1978; Sandwell, 1984).

In order to obtain a true isostatic anomaly, calculations were made of the Moho depth along the transform profile, using the densities mentioned previously, and assuming total isostatic compensation at a maximum depth of 33 km, the standard depth of compensation for most isostatic anomaly calculations across passive margins (Holik and Rabinowitz, pers.

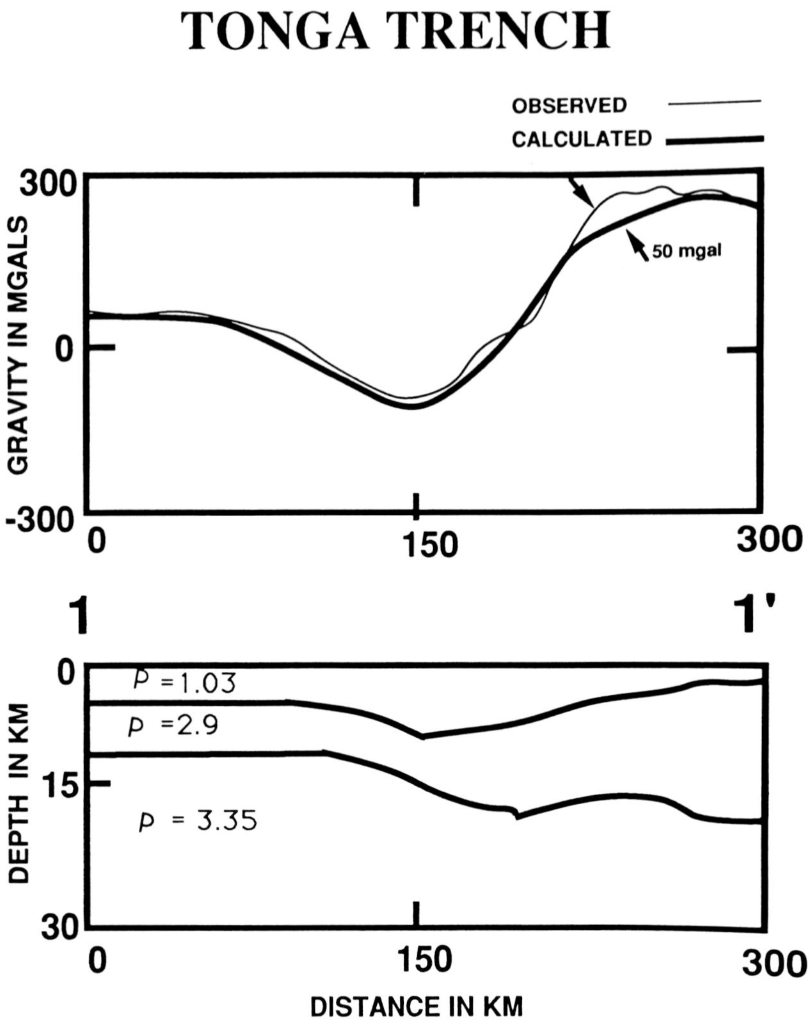

FIGURE 23. Residual and calculated gravity anomalies (top) and the preferred model of the unsubducted crust (bottom), taking into account the downgoing slab, subducted crust, and eclogite. Vertical exaggeration is 3:1.

commun., 1986). In keeping with Airy-type isostasy, the base of the crust as depicted by the Moho is actually an exaggerated mirror image of the surface topography. One notable aspect of the calculations is that the depth to the Moho is predicted to be significantly shallower beneath the transform (Figure 24). The polygon vertices of this compensated density configuration were then entered into a two-dimensional, isostatic gravity modelling program.

Instead of the expected near-zero anomaly across the transform, a significant −30 mgal minimum over the transform axis and a 70 mgal maximum shoreward of the transform were found (Figure 24). This suggests that the transform fault is not in isostatic compensation. Although the crust seaward of the transform is probably compensated, the negative isostatic anomaly over the axis implies crustal overcompensation at that point, which in turn can be attributed to a mass deficiency. Mass deficiencies are normally indicative of some type of strong lateral compression (Talwani et al., 1961). Therefore, although it was expected that the Pacific Plate would be sliding past the Tonga microplate in a pure, left-lateral shear, a

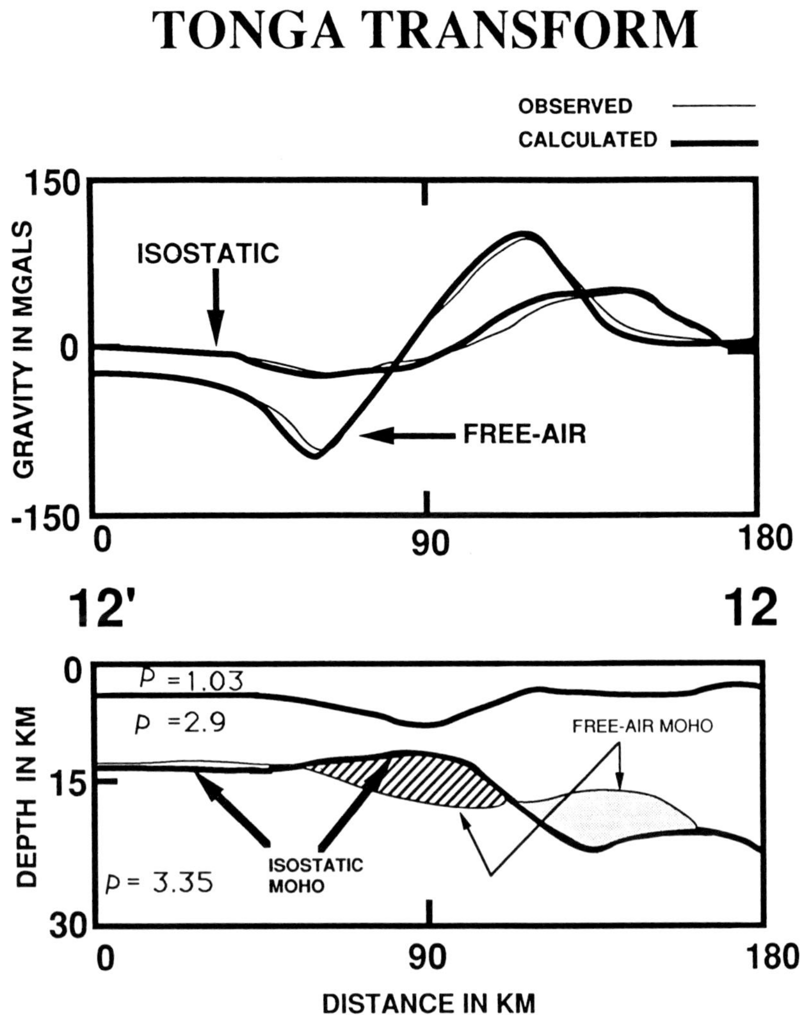

FIGURE 24. Free-air and isostatic gravity anomalies (top) and crustal structure across the Tonga Transform (bottom). Profile 12 (Fig. 8) has been projected perpendicular to the transform axis for this model. Top: The observed isostatic anomaly is represented by a heavy solid line, the calculated isostatic anomaly by a dotted line, the observed free-air anomaly by a dashed line, and the computed free-air anomaly by squares. Bottom: The Moho configuration which fits the observed free-air gravity is represented by a dashed line and is superimposed over the Moho of an isostatically compensated model (solid line). Shaded area shows a mass deficiency and stippled area shows a mass excess. Vertical exaggeration is 3:1.

downbuckling of the Pacific Plate in response to a localized compressional force in the crust may actually be occurring at this location. In addition, the downbuckling may also be a thermomechanical couple in response to changes in the thermal regime beneath the transform, such as horizontal heat conduction from younger, hotter Tonga crust to older, colder Pacific crust. Sandwell and Schubert (1982) have found the effects of thermomechanical coupling to be very important when modelling fracture zone topography. Moreover, the geoid height as revealed by satellite altimeter data across various fracture zones in the South Pacific decreases substantially over colder, older lithosphere (Sandwell, 1984).

Shoreward of the transform axis the crust should return to isostatic equilibrium, due to the absence of ridges, seamounts or other tectonic features. This, however, is not the case, as the isostatic high in this region is indicative of a crustal undercompensation and mass excess. It is difficult to interpret this mass excess uniquely. Part of the downgoing slab at depth may be contributing to the anomaly, but if this were the case the anomaly should continue at a constant, elevated level shoreward of the transform. The crust of the Tonga microplate in this region may be experiencing an upwarp in response to the compression with and subsequent downbucking of the Pacific Plate. The upwarp may also be an isostatic response to the introduction of lower-than-normal density mantle material beneath the shoreward crust of the transform. Possibly a local increase in lithospheric temperature is contributing to asthenospheric upwelling in this region. This especially may be the case if the lithosphere beneath the Tonga Ridge is indeed thinning to the south, as suggested by the subduction zone gravity model (Figure 23).

Regions of mass deficiency and excess are shown in Figure 24. The upper boundary of the mass deficiency and the lower boundary of the mass excess are constrained to be part of the Moho of the isostatically compensated model. The configurations of these bodies were modelled independently in an attempt to match the isostatic anomaly. Having accomplished this, the resulting lower boundary of the mass deficiency and the resulting upper boundary of the mass excess were employed in modelling a crustal structure that needed to match the observed free-air gravity.

In general, the resulting configuration for the transform fault produces a good fit of the observed and calculated free-air gravity anomalies. The crust seaward of the transform was assumed to be 6 to 7 km thick as a first approximation. The transform fault is therefore assumed to be a rift in the crust rather than a zone of extensive crustal thinning as dictated by the isostatically compensated model. There is no evidence of an Outer Gravity High seaward of the transform, and thus little or no evidence of the lithospheric flexure associated with a subducting plate. In the model, the crust appears to bulge slightly at approximately 30 km seaward of the transform axis. While this might be interpreted as evidence for plate flexure, there is no significant rise in the topography and the gravity values are all negative.

A significant gravity high is present shoreward of the transform axis. Its peak is centered over an arc seamount rising 600 m from the surrounding sea floor. The seamount itself is roughly circular in plan view and is best modelled with three-dimensional techniques. Its gravity effect, however, is not expected to be large.

Seismicity data in this region (Hanus and Vanek, 1978 and 1979b) have concluded that a downgoing slab is not evident, therefore, this feature is not a true subduction zone. However, the significant departures from isostasy indicate that this feature may not be a true transform fault either.

The transform gravity profile (Profile 12, Figure 8) is approximately 60 km from the profile at which subduction (Profile 5, Figure 8) was last identified to be occurring by reflection seismics (Wright, 1986) and bathymetry. The downbuckle of the Pacific Plate and possible upwarp of the Tonga microplate are evidence that a pure strike-slip shear may not yet be taking place at this portion of the transform, and that the nature of tectonic motion in this area is partially compressional. Seismicity data have revealed that the transform fault in the study area is actually the eastern end of an extremely complex plate boundary zone, which extends all the way to the central part of the North Fiji Basin (Hamburger, 1986; Hamburger et al., 1988). The zone is characterized by sinistral strike-slip faulting along ESE- to ENE-trending fault planes, as well as active thrusting (Hamburger, 1986).

IV. DISCUSSION

The most outstanding feature of the study area (and perhaps of the entire Tonga convergent margin) in plan view is the marked westward bend in the Tonga Trench at approximately 15.2°S latitude. North and south of this bend there are profound differences in the morphological character of the trench, due to a major change in the interaction of the Pacific Plate with the Tonga microplate.

This study has examined the tectonics of convergence and strike-slip motion along the northern terminus of the Tonga Trench, and has described the comparative morphologies of the trench as it makes a transition from the aforementioned process to that of the latter. It has been found that the trench has undergone important vertical and horizontal surficial changes, as well as subsurface crustal changes, in the process of transitioning from subduction to strike-slip motion. The transition in topography appears relatively gradually, occurring over a distance of approximately 260 km from 16.0°S, 172.3°W to 14.5°S, 173.8°W. Additionally, the trench's subsurface crustal structure suggests that the transition is gradual at depth. A discussion of the major findings of this study and their contribution to the understanding of tectonic processes in the Tonga trench-arc system is presented in the following sections.

A. SUBDUCTION ALONG THE NORTHERN TERMINUS OF THE TONGA TRENCH

1. Structure of the Seaward Trench Slope

The topographic structure of the subducting Pacific Plate is expected to influence the character of the overriding Tonga microplate along the Tonga Trench axis. Analyses of the data have shown that the Pacific Plate along the north-south segment of the trench is gently bulging and cut by normal faults, some of which may define grabens. These features are consequences of the flexure of the Pacific Plate as it bends into the subduction zone.

Extensive development of normal fault structures has also been mapped in other parts of the Tonga Trench region (Hilde and Fisher, 1979; Kroenke, 1984; and Gnibidenko et al., 1985). Normal faults in the Pacific Plate generally develop between the outer topographic high and the trench axis, with displacement on the faults increasing as they move closer to the trench. The average maximum displacement of faults near the trench is 800 to 1500 m. This occurs mainly in the southern half of the study area. The degree of faulting of the plate depends upon the amount of its bending (Jones et al., 1978). Large displacements on the faults in this region are, therefore, consistent with a large amount of bending of the Pacific Plate.

2. Structure of the Shoreward Trench Slope

The structure of the shoreward trench slope provides valuable information on subduction styles along trenches (Warsi, 1983). The bathymetric and seismic reflection data show that the shoreward trench slope is also deformed by relatively steep, normal faults, with blocks downdropped towards the trench. Due to the minimal amount of sediment being carried to the trench, there is no appreciable accreted sediment shoreward of the trench. For the most part, oceanic basement is exposed on the lower reaches of the shoreward slope. This suggests that virtually all sediments arriving at the trench are completely subducted.

Subduction erosion of the leading edge of the Tonga microplate has been suggested as a possible mechanism by Fisher (1974) and Vallier et al. (1985), mainly because little or no sediments are carried to the trench and there is a greater possibility that the subducting plate is abrading and eroding the underside of the overriding plate. The process of subduction erosion involves not only the abrasion or tectonic erosion of the underside and/or leading edge of the overriding plate, but also a subsidence of the shoreward shelf and slope region (Scholl et al., 1980). Although the opening of the Lau Basin and the subduction of the Louisville Ridge have caused an uplift of the southern Tonga forearc in Early Pliocene to Pliestocene time (Packham, 1985), the forearc region north of 18°S has been characterized as "depressed" relative to the southern region (Figure 25; Dupont and Herzer, 1985). In the southern Tonga region a volcanic arc lies to the west of a ridge summit. The ridge

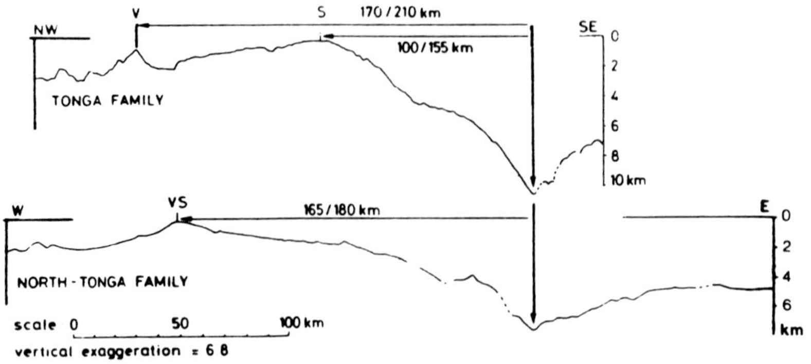

FIGURE 25. Typical bathymetric profiles of the two morphologic families characterizing the Tonga Arc, showing the differences in distance from trench axis to volcanic arc and trench axis to ridge summit (Dupont and Herzer, 1985). The top profile is representative of cross-sectional bathymetry between 21° and 25°S, and the bottom is representative of bathymetry north of 18°S. Note the uplift and doming in the forearc of the top profile in comparison to the bottom profile. V = volcanic arc; S = ridge summit.

summit has been found to be uplifted relative to the volcanic arc to such an extent that the forearc basin dips westward towards the volcanic arc. However, in the study area the volcanic arc and ridge summit are one and the same, and the bathymetric and seismic reflection data do not show a significant "doming" shoreward of the trench, as has been observed to the south. The shoreward trench shelf and slope in this region, therefore, may have undergone recent subsidence rather than uplift, which in turn may be linked to subduction erosion. Additional dredging and drilling on the shoreward trench slopes in the study area are needed in order to confirm the existence of subsidence and/or subduction erosion.

B. STRIKE-SLIP MORPHOLOGY ALONG THE NORTHERN TERMINUS OF THE TONGA TRENCH

Analyses of the bathymetric and seismic reflection data have pinpointed the onset of strike-slip motion at approximately 14.5°S, 173.8°W. The morphology of the seaward and shoreward trench slopes west of this point exhibits a marked difference from the subduction-related morphology to the south. The seaward trench slope break is significantly shallower, as is the trench itself, and the slope is no longer cut by bending-induced normal faulting. The shoreward trench slope is considerably steeper and is also less deformed. The Tonga Ridge is significantly closer to the trench migration relative to Lau Basin spreading. The axial region of the trench is comprised of a seaward series of small, closely spaced ridges and troughs. Closely spaced ridges are sometimes produced along transform faults due to minor adjustments in plate motions (Searle, 1981). The ridges may be representative of a crumpling of the Pacific Plate as a consequence of bending around in the trench and then abruptly grinding past the Tonga microplate. They may also be slivers of a small seamount which has been almost completely consumed at the trench.

C. THE TRANSITION FROM SUBDUCTION TO STRIKE-SLIP MOTION

The previous sections have discussed subduction and strike-slip motion along the Tonga Trench as separate phenomena. However, there is a region along the trench that is undergoing a complex interplay of these two processes. Bathymetric and seismic reflection data have revealed topographic changes in the seaward and shoreward trench slopes that trend gradually from a classic subduction zone morphology to a transform fault morphology. The transition has also been described in terms of the seismicity data of Isacks et al. (1969; Figure 26). The figure illustrates a lithospheric plate model of the termination of a subduction zone by

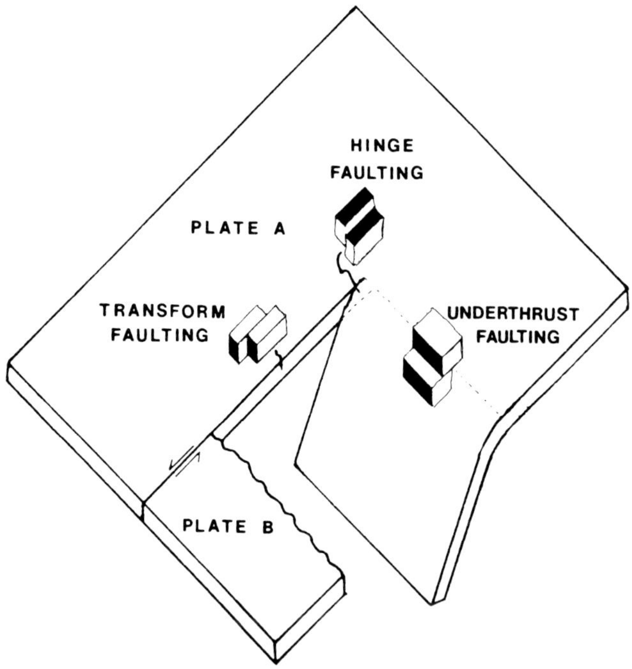

FIGURE 26. Block diagram illustrating a lithospheric plate model of the termination of an island arc by a transform fault (Isacks et al., 1969). The model shows the three types of relative motion between plates A and B. This model and the associated faulting are good approximations to the distribution and focal mechanisms of earthquakes in the Tonga Arc found by Isacks et al. (1969). Plate B is the Pacific Plate that descends into the mantle beneath the Tonga Arc. The hinge faulting occurs south of Samoa where the Tonga Trench bends abruptly westward and then terminates. The transform fault zone extends to slightly north of the Fiji Islands.

a transform fault. In the model, the distribution and focal mechanisms of earthquakes in the region were indicative of underthrust faulting along the subduction zone, left-lateral strike-slip movement along the transform fault, and hinge faulting in the area of transition. Small, low-angle faulting has been noted on Seismic profiles 9, 10, and 12 (Figures 13, 14, and 16). The first two profiles are located directly east of the topographic onset of transform faulting (14.5°S, 173.8°W) and the third profile is located directly west of it. The character of these faults is somewhat step-like and the displacements are small in comparison to displacements further south. Since they have been associated with the cessation of subduction along the trench they may also be linked to a small degree of transitional hinge faulting.

The gravity data suggest that below the surface of the seafloor the transition from subduction to strike-slip motion may also be gradual. The structure of the upper crust along the two profiles which cross the subduction zone and the transform fault exhibit some similarities. For instance, the crust of the Pacific Plate along both profiles exhibits flexure. For the subduction zone, this is due to the pull of a downgoing slab, while at the transform fault a sudden compressional force may be causing a downbuckling of the crust. Regions of mass excess are noted on both profiles shoreward of the axial region. The mass excess shoreward of the subduction zone is minimal, and may in reality be a high density crustal body, while a large mass excess shoreward of the transform fault may be due to unknown tectonic forces or mass anomalies. It may also be due to a flow of mantle material associated with an upwarp of the crust. This is evidenced by a crustal thickness shoreward of the transform that is approximately 5 km thinner than the crust shoreward of the subduction zone.

In contrast to compressional motion at the subduction zone (Profile 1-1′, Figure 8) and the transform fault (Profile 12-12′, Figure 8), the nature of tectonic motion in the transitional area between these regions is extensional. The Machias Seamount located at the bend in the trench is evidence that some kind of crustal extension and magma leakage has occurred. It is possible that the seamount formed as a result of the lateral bending and extensional rupturing of the Pacific Plate in this area of transition. This idea is supported by a K/Ar dating of an anorthoclase-rich phonolite dredged from the seamount that is only 940,000 ± 20,000 years (Hawkins, 1974; Hawkins and Natland, 1975). Shaw (1973) proposed that linear island chain and seamount volcanism is a consequence of melting in response to differential shear between moving lithospheric plates and the morerheid asthenosphere. There is also a lateral gradient in shear associated with the onset of strike-slip movement. It follows that high thermomechanical shear stresses should be concentrated in the area where the Pacific Plate is bending and tearing to the west, resulting in a failure of the lithosphere and magma leakage.

In addition, the models proposed by Kanamori (1976) and Bodine and Watts (1979) predict that trenches exhibiting poorly developed Outer Rises, little or no shallow seismicity, and the presence of an extensional basin are regions of decoupling and tension. Such is the case with this transitional region of the Tonga Trench where: (1) the seaward trench slopes are relatively gentle (Profiles 9 and 10, Figure 9) and an Outer Rise is generally absent (Profiles 6 to 10, Figure 9); (2) deep and intermediate seismicity is observed but shallow seismicity is moderate to absent (Eguchi, 1980; Hamburger, 1986; and Hamburger et al., 1988); and (3) active extension is occurring in the Lau Basin.

D. RELATIVE PLATE MOTIONS IN THE STUDY AREA

Aside from the dissemination of geophysical data, calculations of relative plate motions were considered in order to more clearly delineate the magnitudes and directions of convergence along the trench, and ultimately to help clarify the nature of the transition from convergence to strike-slip motion.

The pole of relative motion between two plates is a mathematical axis through the earth's center about which two plates appear to rotate relative to one another. Each plate has a relative angular velocity, ω, about this pole, which is reckoned as positive when the rotation is clockwise looking from the center of the sphere, and taken to be a positive vector pointing outward along the axis of rotation (McElhinny, 1973). The relative velocity at a chosen point on a plate is, $V_\theta = \omega R \sin\theta$, where ω is the angular velocity of the rotation, R is the radius of the earth, and θ is the angular distance from the chosen point to the rotation pole. The relative velocity is thus greatest at the "equator" of the rotation axis and vanishes at the pole of rotation. The relative velocity vectors must lie along small circles with respect to the pole. If these small circles cross the line of contact between two plates, the line must be either a ridge or a trench, depending upon the sense of rotation (McElhinny, 1973). If, however, the line of contact is itself a small circle, then the contact is a transform fault (McElhinny, 1973).

Rates of relative motion between the Tonga microplate and the Pacific Plate were calculated at several points along the Tonga Trench. The results are shown in Table 2 and Figure 27. It should be noted that the calculations were only first approximations of plate motions in the region, and were made with the assumption that the plates are completely rigid, which may not be true in the Lau Basin. Initially, a rotation pole for the motion of the Tonga microplate with respect to a fixed Indian-Australian Plate was calculated. The calculations involved a consideration of the positions of the Tonga and Lau Ridges. The position of the Tonga Ridge may be restored to its original position relative to the Lau Ridge by a single rotation through θ about a unique axis passing through the center of the Earth. Using two pairs of matching geographic points on the Lau and Tonga Ridges, a relative

TABLE 2
Relative Plate Motion Data for the Study Area

Point	Lat. (°)	Long. (°)	INDI-PAC Vel. (cm/year)	Trend (°)	TONG-PAC Vel. (cm/year)	Trend (°)
1	−18.5	−172.7	9.5	273	22.2	287
2	−17.5	−172.2	9.7	279	23.5	292
3	−16.0	−172.3	9.9	277	25.2	286
4	−15.2	−172.3	10.1	270	26.1	278
5	−14.5	−173.8	10.1	277	26.5	285
6	−14.5	−175.0	10.1	274	26.2	278

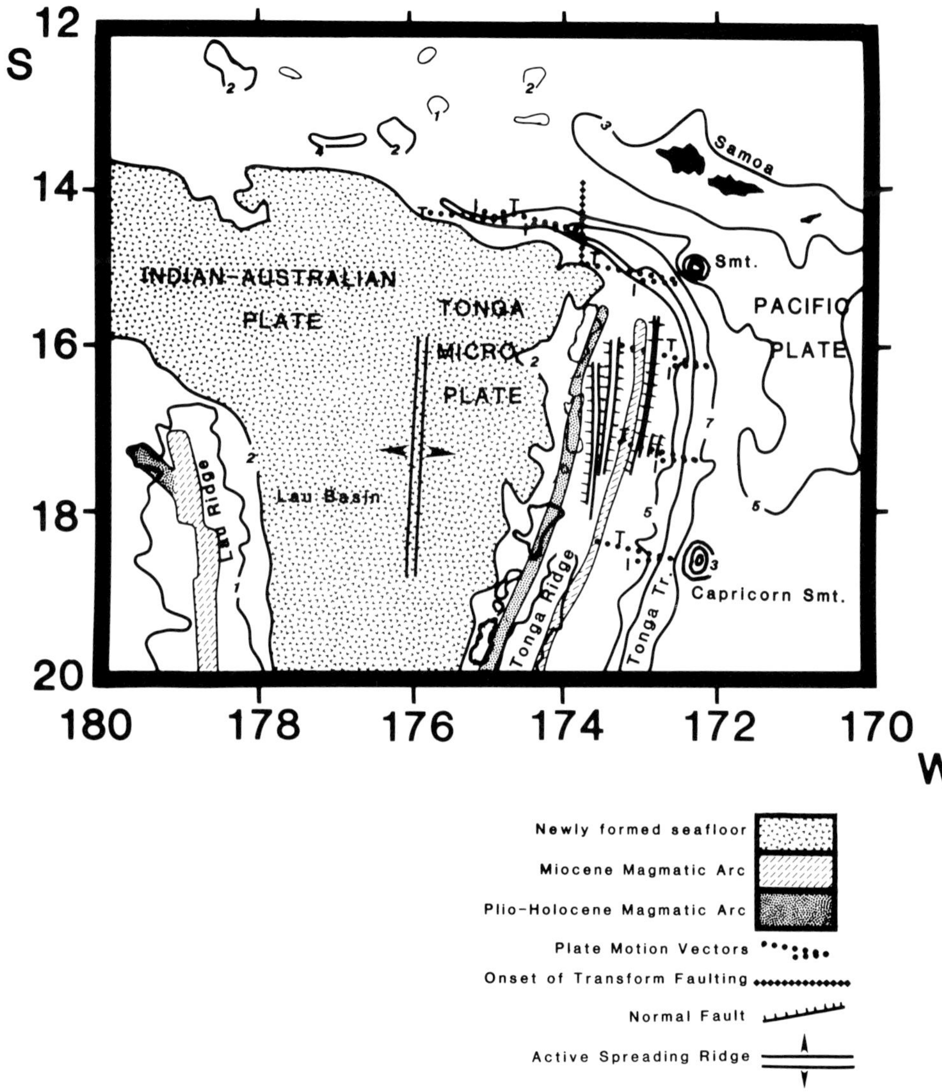

FIGURE 27. Relative plate motions in the northern Tonga region. Bathymetry based on Chase (1985). Contours are in kilometers. Data is from Table 2. Lengths of vectors are proportional to velocity. Numbers beside vectors refer to points in Table 2, T refers to Tonga-Pacific vector, and I refers to India-Pacific vector. Dashed line indicates onset of pure transform faulting as determined by bathymetric and seismic reflection data.

rotation pole was determined by calculating the intersection of two great circles, each of which was equidistant from the respective points on the Lau and Tonga Ridges. This was done by taking the cross product of two vectors located at the chosen points on the Lau Ridge ($P \times P'$), as well as the cross product of the two vectors on the Tonga Ridge ($Q \times Q'$). The cross product of ($P \times P'$) and ($Q \times Q'$) yielded a rotation pole located at 29.3°S, 181.2°E.

The angular velocity of this rotation pole was determined in part from the rate of spreading in the Lau Basin. However, the use of spreading rates for relative rotation pole calculations presents a slight problem. Spreading rates are determined and assumed to be linear between successive pairs of magnetic anomalies that are fitted together by finite relative motion solutions, and are therefore not strictly instantaneous, although the time interval over which the rate is averaged may be small enough to approach an instantaneous solution (Dewey, 1975). Several different spreading rates and ages have been determined for the Lau Basin immediately to the south of the study area. Weissel (1977) determined a full spreading rate of 7.6 cm/year and identified the onset of spreading at 3.5 Ma. The magnetic anomaly data of Cherkis (1980) suggests an age of 4.8 Ma, whereas that of Malahoff et al. (1982) indicates an age of 2.5 Ma. The magnetic data of Malahoff et al. (1982) were used in this study. They are the most supportive of a well-developed unconformity that has recently been identified in the Upper Miocene and Lower Pliocene seismic stratigraphy of the Tonga Ridge (Herzer and Exon, 1985). This unconformity may correspond to a period of thermal uplift that would have immediately preceded spreading of the Lau Basin (Dupont and Herzer, 1985; Herzer and Exon, 1985).

The India-Tonga rotation pole was added vectorially to the India-Pacific rotation pole of Chase (1978), located at 60.7°S, 175.8°E, to obtain a Tonga-Pacific rotation pole (34.7°S, 180.6°E). Using the angular velocities of the India-Pacific and Tonga-Pacific rotation poles, calculations were then made of the magnitude and direction of relative motion at selected points along the axis of the Tonga Trench.

The calculations show the velocity of the Pacific Plate with respect to a fixed Tonga microplate to be strikingly large — over twice that of the velocity with respect to a fixed Indian-Australian Plate. This is a consequence of the relatively rapid opening of the Lau Basin. Furthermore, the relative velocity steadily increases northward, suggesting an increasing rate of convergence between the Tonga microplate and the Pacific Plate. A faster rate of subduction in the northern Tonga Trench area has been cited as a major factor in the formation of the present shape of the Lau Basin, in comparison to a narrower Havre Trough (Hawkins, 1974a; Vogt et al., 1976; and Altman, 1978). It is important at this point to differentiate between the terms "convergence" and "subduction". According to Carlson and Melia (1984), convergence refers to the relative motion between the overriding and downgoing plates, whereas subduction refers to the motion of the downgoing slab in a mantle frame of reference. As subduction ceases in the north an increasing rate of convergence is not to be expected, since the downgoing plate is no longer being subducted, but is merely slipping past the overriding plate.

The change in direction of relative plate motion vectors along the trench is also of note. The India-Pacific plate vectors predict convergence at all six points along the trench (Figure 27). However, the small circles which define the Tonga-Pacific vectors predict convergence only south of the unnamed seamount. They then closely follow the curvature of the trench north of that point and lie directly on the axis of the transform fault, thus predicting strike-slip motion. The divergence of the India-Pacific vector at Point 6 from the trend of the transform fault is 4°, suggesting that strike-slip faulting in its purest form may not be taking place yet. The divergence further indicates that, once again, a transition from convergence to strike-slip motion may still be occurring along this portion of the transform.

E. FUTURE WORK

Although this study has elucidated many aspects of subduction and strike-slip motion along the Tonga Trench, several of the findings and conclusions should be investigated for confirmation. The general paucity of data in the region has led the author to make several inferences and assumptions in the study. Multi-channel seismic reflection and side-scan sonar surveys across the trench will be needed in order to increase bathymetric resolution and to trace and confirm the existence of morphological structures such as grabens. An ocean bottom seismometer survey, as well as multi-channel seismic profiling, will be required to establish the crustal structure and the mode of present-day tectonic activity along the trench. Dredging and possibly drilling of the Machias Seamount and the shoreward slopes of the trench will be helpful in discerning the history of volcanism and tectonic movement in the region and the possible existence of tectonic erosion. Finally, gravity data needs to be collected along the profiles which cross the bend, in order to model the subsurface transition from subduction to strike-slip motion in greater detail.

V. CONCLUSIONS

This study has examined the tectonics of convergence and strike-slip motion along the northern terminus of the Tonga Trench. Analyses of available bathymetric, seismic reflection, and gravity data have aided in describing the topography and crustal structure of the trench as it makes a transition from subduction to strike-slip motion. The major results of these analyses are summarized below.

A primary objective of this study was to describe differences in morphologic character between north-south subduction and east-west strike-slip motion. South of 15.2°S it is noted that a classic subduction zone morphology trends into a morphology transitional between subduction and strike-slip motion. The "transitional" morphology continues north of 15.2°S and progresses into a transform fault morphology. Subduction-related features include lithospheric flexure, bending-induced faulting of the Pacific Plate as it moves into the trench, and the inferred existence of small horsts and grabens. Extensive normal faulting and a general absence of sediments are also noted on the trench slopes of the Tonga microplate. It is speculated that the shoreward trench region may have either undergone a small amount of subsidence or may not have experienced an uplift, as the shoreward trench region to the south of the study area has. Shoreward trench features have also been associated with possible tectonic erosion, which has been suggested by Fisher (1974) and Vallier et al., (1985). This possibility is also supported by a thinning of the crust beneath the transform fault, as deduced from the gravity data.

Strike-slip related features include a sharp trench slope break, a lack of lithospheric bulging and bending-induced faulting in the Pacific Plate, a very steep, undeformed shoreward trench slope, and a ridge and trough province in the axial region.

A second objective of this study was to evaluate the nature and location of the trench's transition from subduction to strike-slip motion. It is concluded that the transition is gradual topographically and at depth. Bathymetric and seismic reflection data show a gradual transition from subduction-related features to those of a strike-slip nature. The topographic transition begins at approximately 16.0°S, 172.3°W and ends at approximately 14.5°S, 173.8°W, over a distance of about 260 km. The gravity data also suggests that the transition may be occurring gradually at depth. A notable finding of the gravity modelling was the departure of the transform fault from isostasy, suggesting that movements along this portion of the fault is more complex than a simple, left-lateral shear. It is concluded that the nature of crustal movement in the study area trends from compressional at the subduction zone to extensional in the area of transition, due to high thermomechanical stresses and possibly localized asthenospheric upwelling, to compressional again at the transform fault, due to localized compressional forces and thermomechanical coupling.

As a first approximation, assuming rigid plate tectonics, calculations of relative plate motion vectors in the study area also support a gradual transition. This is suggested by the Tonga-Pacific vectors that lie directly on the axis of the transform fault, accompanied by India-Pacific vectors that deviate slightly from the trend of the transform, thus indicating a small degree of convergence.

ACKNOWLEDGMENTS

This research was carried out at the Geodynamics Institute, Department of Geophysics, and the Ocean Drilling Program, both of Texas A&M University, under the supervision of Drs. William W. Sager and Thomas W. C. Hilde. Thanks are extended to Chao-Shing Lee, Bruce Bergenbeck, and Jeff Payne, all formerly of the Institute, for valuable discussions and assistance in handling the navigational and bathymetric data.

I am grateful to Paul Matthias, formerly of the Texas A&M Department of Oceanography, for providing much-needed seismic reflection data from Lamont-Doherty Geological Observatory. Drs. Jim Holik, David Divins, and Philip Rabinowitz, of the Department of Oceanography and Ocean Drilling Program, are thanked for their timely input and assistance with the gravity modelling. I thank Drs. Patricia Cooper, Jim Gill, David Divins, Rob Kidd, and Richard Rezak for their thoughtful review of the manuscript.

REFERENCES

Altman, L. W., Distribution of stress in oceanic lithosphere beneath the Lau-Havre Basin. Thesis, Texas A&M University, College Station, 1978.

Anderson, R. N., Uyeda, S., and Ahiko, M., Geochemical constraints at converging plate boundaries, I, Dehydration in the downgoing slab, *Geophys. J. R. Astro. Soc.*, 44, 33—357, 1976.

Andrews, J. E. and Packham, G. H., Initial Reports of the Deep Sea Drilling Project, Vol. 30, U.S. Government Printing Office, Washington, D.C., 1975.

Barazangi, M. and Isacks, B., Lateral variation of seismic-wave attenuation in the upper mantle above the inclined earthquake zone of the Tonga island arc: deep anomaly in the mantle, *J. Geophys. Res.*, 76, 8493—8515, 1971.

Bodine, J. H. and Watts, A. B., On lithospheric flexure seaward of the Bonin and Mariana trenches, *Earth Planet. Sci. Lett.*, 43, 132—148, 1979.

Burns, R. E. and Andrews, J. E., Initial Reports of the Deep Sea Drilling Project, Vol. 21, U.S. Government Printing Office, Washington, D.C., 17—33, 897—907, 1973.

Carlson, R. L. and Melia, P. J., Subduction hinge migration, *Tectonophysics,* 102, 399—411, 1984.

Chapple, W. M. and Forsyth, D. W., Earthquakes and bending of plates at trenches, *J. Geophys. Res.*, 84, 6729—6749, 1979.

Chase, C. G., Tectonic history of the Fiji Plateau, *Geol. Soc. Am. Bull.*, 82, 3087—3109, 1971.

Chase, C. G., Plate kinematics: the Americas, East Africa and the rest of the world, *Earth Planet. Sci. Lett.*, 37, 355—368, 1978.

Chase, T. E., Submarine topography of the Tonga-Fiji region and the southern Tonga platform area, in Geology and Offshore Resources of Pacific Islands Arcs — Tonga Region, Scholl, D. W. and Vallier, T. L., Eds., Circum-Pacific Council for Energy and Mineral Resources Earth Sci. Ser., 2, 21—24, 1985.

Chen, T. and Forsyth, D. W., A detailed study of two earthquakes seaward of the Tonga Trench: implications for mechanical behavior of the oceanic lithosphere, *J. Geophys. Res.*, 83, 4995—5003, 1978.

Cherkis, N. Z., Aeromagnetic investigations and sea floor spreading history in the Lau Basin and northern Fiji Plateau, *U.N. ESCAP CCOP/SOPAC Tech. Bull.*, 3, 37—45, 1980.

Delaney, J. M. and Helgeson, H. C., Calculation of the thermodynamic consequences of dehydration in subducting ocean crust to 100 kb and >800°C, *Amer. J. Sci.*, 278, 638—676, 1978.

Dewey, J. F., Plate tectonics, *Rev. Geophys. Space Phys.*, 13, 326—332, 1975.

Dewey, J. F., Episodicity, sequence and style at convergent plate boundaries, in The Continental Crust and Its Mineral Deposits, Strangeway, D. W., Ed., *Geol. Assoc. Can. Spec. Pap.*, 20, 553—573, 1980.

Drummond, K. J., Nishiwaki, C., Corvalan, J., Doutch, H. F., and Craddock, C., Plate tectonic map of the circum-Pacific region: Pacific Basin sheet, *Am. Assoc. Pet. Geol.*, Tulsa, OK, 1982.

Dubois, J., Launay, J., and Recy, J., Some new evidence on lithospheric bulges close to island arcs, *Tectonophysics,* 26, 189—196, 1975.

Dupont, J. and Herzer, R. H., Effect of subduction of the Louisville Ridge on the structure and morphology of the Tonga Arc, in *Geology and Offshore Resources of Pacific Island Arcs — Tonga Region,* Scholl, D. W. and Vallier, T. L., Eds., Circum-Pacific Council for Energy and Mineral Resources Earth Sci. Ser., 2, 323—332, 1985.

Eguchi, T., Part I, Seismotectonics of back-arc basins; Part II, Some problems at plate boundaries, Thesis, Earthquake Research Institute, University of Tokyo, Tokyo, 1980.

Ewart, A., Bryan, W. B., and Gill, J. B., Mineralogy and geochemistry of the younger volcanic islands of Tonga, S. W. Pacific, *J. Petrol.,* 14, 429—465, 1973.

Ewart, A., Brothers, R. N., and Mateen, A., An outline of the geology and geochemistry, and the possible petrogenetic evolution of the volcanic rocks of the Tonga-Kermadec-New Zealand island arc, *J. Volcanol. Geotherm. Res.,* 2, 205—250, 1977.

Fisher, R. L., Pacific-type continental margins, in *The Geology of Continental Margins,* Burk, C. A. and Drake, C. L., Eds., Springer-Verlag, New York, 25—41, 1974.

Fisher, R. L. and Engel, C. G., Ultramafic and basaltic rocks dredged from the nearshore flank of the Tonga Trench, *Geol. Soc. Am. Bull.,* 80, 1373—1378, 1969.

Fryer, P. and Smoot, N. C., Processes of seamount subduction in the Mariana and Izu-Bonin Trenches, *Mar. Geol.,* 64, 77—90, 1985.

Furlong, K. P., Chapman, D. S., and Alfeld, P. W., Thermal modeling of the geometry of subduction with implications for the tectonics of the overriding plate, *J. Geophys. Res.,* 87, 1786—1802, 1982.

Giardini, D. and Woodhouse, J. H., Deep seismicity and modes of deformation in Tonga subduction zone, *Nature,* 307, 505—509, 1984.

Gill, J. and Gorton, M., A proposed geological and geochemical history of eastern Melanesia, in, The Western Pacific: Island Arcs, Marginal Seas, Geochemistry, Coleman, P. J., Ed., University of Western Australia Press, 543—566, 1973.

Gnibidenko, H. S., Anosov, G. I., Argentov, V. V., and Pushchin, I. K., Tectonics of the Tonga-Kermadec Trench and Ozbourn Seamount, *Tectonophysics,* 112, 357—383, 1985.

Grow, J. A., Crustal and upper mantle structure of the central Aleutian arc, *Geol. So. Am. Bull.,* 84, 2169—2192, 1973.

Grow, J. A. and Bowin, C. O., Evidence for high-density crust and mantle beneath the Chile Trench due to the descending lithosphere, *J. Geophys. Res.,* 80, 1449—1458, 1975.

Hager, B. H., Subducted slabs and the geoid: constraints on mantle rheology and flow, *Calif. Inst. of Technol. Seismol. Lab. Rept.,* 252, 21, 1983.

Hamburger, M. W., Seismicity of the Fiji Islands and tectonics of the Southwest Pacific, Ph.D. thesis, Cornell University, Ithaca, New York, 1986.

Hamburger, M. W., Everingham, I. B., Isacks, B. L., and Barazangi, M., Active tectonism within the Fiji platform, Southwest Pacific, *Geology,* 16, 237—241, 1988.

Hanus, V. and Vanek, J., Tonga-Lau system: deep collision of subducted lithospheric plates, *J. Geophys.,* 44, 473—480, 1978.

Hanus, V. and Vanek, J., Morphology and volcanism of the Wadati-Benioff zone in the Tonga-Kermadec system of recent subduction, *N.Z. J. Geol. Geophys.,* 22, 659—671, 1979a.

Hanus, V. and Vanek, J., Northern part of the Tonga region: a complicated subduction closure, *J. Geophys.,* 46, 385—395, 1979b.

Hanus, V. and Vanek, J., Plate tectonic interpretation of deep earthquakes between the Tonga-Lau and New Hebrides subduction zones, *Tectonophysics,* 75, T19—T28, 1981.

Hatherton, T., Upper mantle inhomogeneity beneath New Zealand: surface manifestations, *J. Geophys. Res.,* 75, 269—284, 1970.

Hawkins, J. W., Geology of the Lau Basin, a marginal sea behind the Tonga Arc. in, *The Geology of Continental Margins,* Burk, C. A. and Drake, C. L., Eds., Springer-Verlag, New York, 505—520, 1974a.

Hawkins, J. W., Petrology of a differentiated alkalic rock series from a seamount near the Tonga Trench, *Geol. Soc. Am. Abstr.,* 6, 153, 1974b.

Hawkins, J. W., Petrology and geochemistry of basaltic rocks of the Lau Basin, *Earth Planet. Sci. Lett.,* 28, 283—297, 1976.

Hawkins, J. W. and Natland, J. H., Nephelinites and basanites of the Samoan linear volcanic chain: their possible tectonic significance, *Earth Planet. Sci. Lett.,* 24, 427—439, 1975.

Haxby, W. F. and Turcotte, D. L., On isostatic geoid anomalies, *J. Geophys. Res.,* 83, 5473—5478, 1978.

Hayes, D. E., Talwani, M., Houtz, R., Pitman, W. C., and Meijer, R., Preliminary Report of Vol. 22, USNS Eltanin Cruises 28—32, March, 1967-March, 1968. Lamont-Doherty Geol. Observ. Technical Report No. CU-1-72, 1972.

Hedge, C. E., Peterman, Z. E., and Dickinson, W. R., Petrogenesis of lavas from western Samoa, *Geol. Soc. Am. Bull.,* 83, 2709—2714, 1972.

Herzer, R. H. and Exon, N. F., Structure and analysis of the southern Tonga forearc, in, Geology and Offshore Resources of Pacific Island Arcs — Tonga, Region, Scholl, D. W. and Vallier, T. L., Eds., *Circum-Pacific Council for Energy and Mineral Resources Earth Sci. Ser. 2,* 55—73, 1985.

Hilde, T. W. C., Sediment subduction versus accretion around the Pacific, *Tectonophysics,* 99, 381—397, 1983.

Hilde, T. W. C. and Fisher, R. L., Graben structure and axial zone tectonics of Tonga Trench, Southwest Pacific, Gen. Assembly 17th Int. Union Geod. Geophys., Canberra, 17, 18, 1979.

Hynes, A. and Mott, J., On the causes of back-arc spreading, *Geology,* 13, 387—389, 1985.

Isacks, B. L. and Barazangi, M., Geometry of benioff zones: lateral segmentation and downwards bending of the subducted lithosphere, in Island Arcs, Deep Sea Trenches, and Back-Arc Basins, Talwani, M. and Pitman III, W. C., Eds., *Maurice Ewing Ser., Am. Geophys. Union,* 1, 99—114, 1977.

Isacks, B. L., Sykes, L. R., and Oliver, J., Focal mechanisms of deep and shallow earthquakes in the Tonga-Kermadec region and the tectonics of island arcs, *Geol. Soc. Am. Bull.,* 80, 1443—1470, 1969.

Kanamori, H., Seismic and aseismic slip along subduction zones and their tectonic implications, in Island Arcs, Deep Sea Trenches, and Back-Arc Basins, Talwani, M. and Pittman III, W. C., Eds., *Maurice Ewing Ser., Am. Geophys. Union,* 1, 163—174, 1976.

Johnson, T. and Molnar, P., Focal mechanisms and plate tectonics of the Southwest Pacific, *J. Geophys. Res.,* 77, 5000—5032, 1972.

Jones, G. M., Hilde, T. W. C., Sharman, G. F., and Agnew, D. C., Fault patterns on outer trench walls and their tectonic significance, *J. Phys. Earth Suppl.,* 26, S85—S101, 1978.

Karig, D. E., Ridges and basins of the Tonga-Kermadec island arc system, *J. Geophys. Res.,* 75, 239—254, 1970.

Karig, D. E., Anderson, R. N., and Bibee, L. P., Characteristics of back-arc spreading in the Mariana Trough. *J. Geophys. Res.,* 83, 1213—1226, 1978.

Keating, B. H. and Tarling, D. H., Paleomagnetic studies of the Samoan Islands, S.W. Pacific Ocean, *EOS, Trans. Am. Geophys. Union,* 66, 1079, 1985.

Kennett, J. P., Marine Geology, Prentice-Hall, Englewood Cliffs, NJ, 1982, 150—151.

Kogan, M. G., Gravity field of the Kuril-Kamchatka arc and its relation to the thermal regime of the lithosphere, *J. Geophys. Res.,* 80, 1381—1390, 1975.

Kroenke, L., Tonga: the Tonga Ridge and Trench; the Lau Ridge and Trough. *U.N. ESCAP CCOP/SOPAC Tech. Bull.,* 6, 77—86, 1984.

Malahoff, A., Feden, R. H., and Fleming, H. S., Magnetic anomalies and tectonic fabric of marginal basins north of New Zealand, *J. Geophys. Res.,* 87, 4109—4125, 1982.

Mammerickx, J., Chase, T. E., Smith, S. M., and Taylor, I. L., *Bathymetry of the South Pacific,* Scripps Institution of Oceanography, La Jolla, CA, Chart #12, 1974.

McElhinny, M. W., *Palaeomagnetism and Plate Tectonics,* Cambridge University Press, New York, 173—175, 1973.

Natland, J. H., The progression of volcanism in the Samoan linear volcanic chain, *Amer. J. Sci.,* 280A, 709—735, 1980.

Natland, J. H., The Samoa-Malaita connection, *EOS, Trans. Am. Geophys. Union,* 66, 1080, 1985.

Packham, G. H., Vertical tectonics on the Tonga Ridge from the Tongatapu oil exploration wells, in *Geology and Offshore Resources of Pacific Island Arcs — Tonga Region,* Scholl, D. W. and Vallier, T. L., Eds., Circum-Pacific Council for Energy and Mineral Resources Earth Sci. Ser., 2, 291—300, 1985.

Packham, G. H. and Terrill, A., Submarine geology of the South Fiji Basin, in *Initial Reports of the Deep Sea Drilling Project, Vol. 30,* Andrews, J. E. and Packham, G. H., et al., U.S. Government Printing Office, Washington, D.C., 617—645, 1975.

Radar, B. L., Sediment trench fill effect on trench age-depth relation, Thesis, Texas A&M University, College Station, 1985.

Raitt, R. W., Fisher, R. L., and Mason, R. G., Tonga Trench, in, *Crust of the Earth. Geol. Soc. Am., Spec. Pap.,* 62, 237—254, 1955.

Ringwood, A. E., Petrogenesis in island arc systems, in *Island Arcs, Deep Sea Trenches and Back-Arc Basins,* Talwani, M. and Pitman III, W. C., Eds., Maurice Ewing Ser., Am. Geophys. Union, 1, 311—324, 1977.

Sager, W. W., Mariana arc structure inferred from gravity and seismic data, *J. Geophys. Res.,* 86, 5382—5388, 1980.

Sandwell, D. T. and Schubert, G., Lithospheric flexure at fracture zones, *J. Geophys. Res.,* 87, 4657—4667, 1982.

Sandwell, D. T., A detailed view of the South Pacific geoid from satellite altimetry, *J. Geophys. Res.,* 89, 1089—1104, 1984.

Scholl, D. W., von Huene, R., Vallier, T. L., and Howell, D. G., Sedimentary masses and concepts about tectonic processes at underthrust ocean margins, *Geology,* 8, 564—568, 1980.

Scholl, D. W., Vallier, T. L., and Maung, T. U., Introduction, in *Geology and Offshore Resources of Pacific Island Arcs — Tonga Region,* Scholl, D. W. and Vallier, T. L., Eds., Circum-Pacific Council for Energy and Mineral Resources Earth Sci. Ser., 2, 3—15, 1985.

Sclater, J. G., Hawkins, J. W., Mammerickx, J., and Chase, C. G., Crustal extension between the Tonga and Lau Ridges: petrologic and geophysical evidence, *Geol. Soc. Am. Bull.*, 83, 505—518, 1972.

Searle, R. C., The active part of Charlie-Gibbs Fracture Zone: a study using sonar and other geophysical techniques, *J. Geophys. Res.*, 86, 243—262, 1981.

Shaw, H. R., Mantle convection and volcanic periodicity in the Pacific: evidence from Hawaii, *Geol. Soc. Am. Bull.*, 84, 1505—1526, 1973.

Stark, P. B. and Frohlich, C., The depths of the deepest deep earthquakes, *J. Geophys. Res.*, 90, 1859—1869, 1985.

Talwani, M., Worzel, J., and Landisman, M., Rapid computations for two-dimensional bodies with an application to the Mendocino submarine fracture zone, J. Geophys. Res., 64, 49—59, 1959.

Talwani, M., Worzel, J., and Ewing, M., Gravity anomalies and crustal section across the Tonga Trench, *J. Geophys. Res.*, 66, 1265—1278, 1961.

Udintsev, G. G., Dmitriyev, A., Sharas'kin, A., Agapova, G., Zenkovich, N. L., Bersenev, N., Kurentsova, N., and Suzyumov, A., New data on trench-faults in the Southwestern Pacific, *Geotectonics*, 65—69, 1974.

Uyeda, S., Subduction zones: an introduction to comparative subductology. *Tectonophysics*, 81, 133—159, 1982.

Uyeda, S. and Kanamori, H., Back-arc opening and the mode of subduction, *J. Geophys. Res.*, 84, 1049—1061, 1979.

Vallier, T. L., O'Connor, R. M., Scholl, D. W., Stevenson, A. J., and Quintero, P. J., Petrology of rocks dredged from the landward slope of the Tonga Trench: implications for Middle Miocene volcanism and subsidence of the Tonga Ridge, in *Geology and Offshore Resources of Pacific Island Arcs — Tonga Region*, Scholl, D. W. and Vallier, T. L., Eds., Circum-Pacific Council for Energy and Mineral Resources Earth Sci. Ser., 2, 109—120, 1985.

Vening Meinesz, F. A., The origin of continents and oceans, *Geol. Mijnbouw*, 31, 373—384, 1952.

Vogt, P. R., Lowrie, A., Bracey, D. R., and Hey, R. N., Subduction of aseismic ridges; effects on shape, seismicity, and other characteristics of consuming plate boundaries, *Geol. Soc. Am., Spec. Pap.*, 172, 1—60, 1976.

Warsi, W. E. K., Convergence tectonics of the Peru Trench: 8°S-15°S latitude, Dissertation, Texas A&M University, College Station, 1983.

Watts, A. B. and Talwani, M., Gravity anomalies seaward of deep-sea trenches and their tectonic implications, *Geophys. J. R. Astro. Soc.*, 36, 57—90, 1974.

Watts, A. B. and Talwani, M., Gravity effect of downgoing lithospheric slabs beneath island arcs, *Geol. Soc. Am. Bull.*, 86, 1—4, 1975.

Watts, A. B., Weissel, J. K., and Larson, R. L., Sea-floor spreading in marginal basins of the Western Pacific, *Tectonophysics*, 37: 167—181, 1977.

Weissel, J. K., Evolution of the Lau Basin by the growth of small plates, in *Island Arcs, Deep Sea Trenches, and Back-Arc Basins*, Talwani, M. and Pitman III, W. C., Eds., Maurice Ewing Ser. Am. Geophys. Union, 1, 429—436, 1977.

Wyss, M., Habermann, R. E., and Griesser, J. C., Seismic quiescence and asperities in the Tonga-Kermadec Arc, *J. Geophys. Res.*, 89, 9293—9304, 1984.

Chapter 3

REGIONAL GRAVITY ANOMALIES AND GEOID UNDULATIONS IN THE NORTH ATLANTIC OCEAN

Woo-Yeol Jung and Philip D. Rabinowitz

TABLE OF CONTENTS

4. Obtain the GEM10C geoid for the area of interest.
5. Calculate the 1° × 1° total geoid by adding the 1° × 1° difference geoid to the GEM10C geoid.

In actual calculations using Stokes' formula, the behavior of the weighting function, S(ψ), poses difficulties near the computation point. Since Stokes' function, S(ψ), changes very rapidly near the computation point, large errors will be present if we simply calculate the effect of 1° × 1° blocks by using the value of S(ψ) corresponding to ψ, the distance from the point of computation to the center of the 1° × 1° block. Hence, the block must be subdivided into smaller blocks and the values of S(ψ) are obtained for each of the subdivided smaller blocks. These values are then averaged for the entire block.

Earlier study (Talwani et al., 1972) pointed out that the same reference ellipsoid must be used to obtain the 1° × 1° difference gravity anomalies. We refer all the gravity anomalies and geoid undulations in this study to the Geodetic Reference System 1967. An error of an amount that is of the order of flattening (1/297) is present in Stokes' formula since the equation is derived based on the assumption that the Earth is a sphere (Heiskanen and Moritz, 1967). If we assume that the maximum 1° × 1° difference geoid obtained following the above procedure is 20 m, the maximum error due to the assumption of sphericity would be less than 0.1 m and can be disregarded.

The software programs for the numerical integration of Stokes' equation are modified versions of those used by Talwani et al. (1972). The geodetic parameters that appeared in Stokes' equation were the same as those of GEM10C, i.e., the radius of equator, R = 6378 km, and normal gravity value, g = 978 cm/sec^2, and ellipticity, e = 1/298.257.

III. DESCRIPTION OF THE REGIONAL GRAVITY ANOMALIES

A. THE 1° × 1° AND 5° × 5° AVERAGED SURFACE SHIP DERIVED GRAVITY ANOMALIES

We first obtained 1° × 1° averaged gravity values from a detailed free-air anomaly map of the North Atlantic Ocean (Jung and Rabinowitz, 1988). Locations of the gravity measurements utilized to construct the detailed gravity map at a 10 mgal contour interval are shown in Figure 1.

The gravity measurements utilized to construct the free-air gravity contour map were referred to the International Gravity Formula 1930 (IGF 1930). The 1° × 1° averaged gravity values after being converted to the Geodetic Reference System 1967 (Bomford, 1980) are presented in Table 1. In order to determine the 1° × 1° averaged values, we subdivided every 1° × 1° block on the gravity map into 9 or 36 smaller size blocks, depending on the steepness or gradients of the gravity contours within each of the 1° × 1° blocks. The free-air anomalies at the center of the subdivided blocks were then determined through interpolation and/or extrapolation, and the averages of the 1° × 1° blocks were computed.

The general trends of the 1° × 1° suface ship gravity map (Figure 2) are very similar to the detailed free-air anomaly map (Jung and Rabinowitz, 1988) with somewhat subdued local peaks and troughs. The most prominent features of Figure 2 are the broad positive anomalies over the Mid-Atlantic Ridge area and negative anomalies over the adjacent oceanic basins. The positive values over the Mid-Atlantic Ridge show a rather distinct change near 32°N. North of 32°N, the 1° × 1° surface ship gravity is more positive than to the south with values in the north, in places, more positive than 40 mgal. Cochran and Talwani (1978) concluded that in order to account for the gravity anomalies in the shallow ridge area to the north of 32°N, a portion of the compensation must be distributed to depths of several hundred kilometers (within the asthenosphere) and that this mass deficiency is relatively uniform throughout the area. However, within this, the more local features are compensated at much shallower depths within the lithosphere.

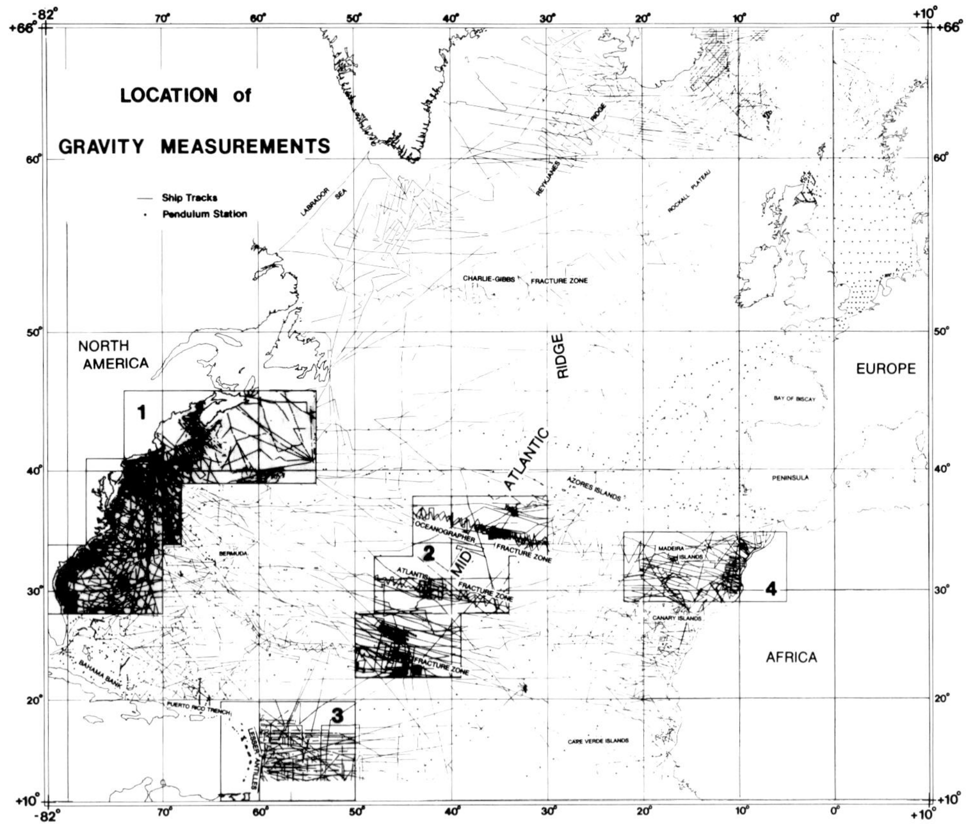

FIGURE 1. Location of gravity measurements used to construct a detailed gravity contour map (Jung and Rabinowitz, 1988). Boxed regions indicate the incorporated area from the Ocean Margin Drilling Program — Regional Atlas Series: 1, Ewing (1984); 2, Rogan and Rabinowitz (1985); 3, Westbrook and Jackson (1984); and 4, Rabinowitz et al. (1984).

Profiles of the 1° × 1° surface ship gravity anomalies and bathymetric depth along the ridge axis are shown in Figure 3. Both on Figure 2 and Figure 3, the Charlie-Gibbs Fracture Zone is clearly manifested by long negative belts. Other major fracture zones such as the Azores and Kane Fracture Zones are not as pronounced gravimetrically as the Charlie-Gibbs Fracture Zone, but they are identifiable. South of 32°N, the gravity field over the Mid-Atlantic Ridge is similar to the gravity field over the other mid-ocean ridges and the compensation is believed to be at depths less than 100 km (Cochran and Talwani, 1978).

The 1° × 1° surface ship gravity values over the western basin are more negative than those over the eastern basin. The largest negative values, more negative than −60 mgal, are observed over the Hatteras Abyssal Plain. From the Hatteras Abyssal Plain toward the Mid-Atlantic Ridge the gravity values gradually increase to near zero at the ridge crest. Over the eastern basin, the gravity values show values near −10 mgal. The locations of extensive volcanic activity away from the Mid-Atlantic Ridge, such as the Bermuda, Cape Verde, Canary, and Madeira Islands, are all associated with relative regional free-air gravity highs with values about 20 mgal. The largest negative surface anomalies in the North Atlantic, more negative than −250 mgal, are observed along the Puerto Rico Trench.

Figure 4 is the 5° × 5° surface ship gravity anomalies. These were obtained by taking averages of the 1° × 1° surface ship gravity anomalies within each 5° × 5° s. Hence, Figure 4 resolves information in the Earth's gravity field with wavelengths longer than about 1100 km. As is expected, the 5° × 5° surface ship gravity map shows that most of short wavelength

TABLE 1
The 1° × 1° Averaged Free-Air Gravity Anomalies (Mgals) of the North Atlantic Referred to Geodetic Reference System 1967 (flattening 1/298.257)

Longitude −90 to −75 (−90, −80, −75); latitude 66 to 10 (66, 60, 50, 40, 30, 20, 10)

						-24
						-32
						7
					11	-16
			-12	-21	9	-64
		-11	-6	-15	-13	-36
	-1	-6	-20	-10	-34	5
	14	-3	-26	-22	-54	-28
	17	-11	-17	-2	-57	-41
		-20	-5	5	-54	-36
		-9	16	-33	-54	-37
		-13	20	53	-71	-30
		-19	32	-42	-24	-61
11	3	-18	39	7	23	-42
-48	-20	-5	33	25	19	13

TABLE 1 (continued)
The 1° × 1° Averaged Free-Air Gravity Anomalies (Mgals) of the North Atlantic Referred to Geodetic Reference System 1967 (flattening 1/298.257)

Lat.	-75					-70									-60
66															
60															
50															
								12	-7			-34	-19	-28	-26
						10	3	6	8	-13	-3	4	11	9	23
					7	9	6	12	4	22	2	-18	-21	-41	-48
					12	4	-1	9	18	-20	-19	-16	-29	-35	-32
		7	-14	-9	7	-4	5	9	-11	-33	-35	-30	-26	-28	-26
40		-10	5	-28	-27	-25	-35	-36	-39	-27	-37	-38	-28	-20	-20
	-4	-14	-61	-31	-14	-15	-30	-35	-34	-24	-8	10	1	-4	7
	10	-45	-34	-27	-37	-41	-48	-37	-36	-27	-25	-20	-18	-24	-21
	3	-33	-27	-31	-27	-33	-34	-36	-31	-26	-25	-27	-17	-14	-21
	-12	-42	-41	-28	-22	-25	-35	-30	-18	-26	-16	-18	-26	-14	-8
	-52	-39	-39	-30	-37	-33	-33	-34	-25	-28	-10	-8	-18	-16	-11
	-57	-49	-50	-44	-37	-31	-26	-27	-37	-35	-15	2	0	-6	3
	-53	-63	-54	-42	-40	-33	-36	-35	-38	-37	64	-1	-15	-14	-14
	-34	-60	-50	-42	-43	-40	-33	-29	-28	-27	-34	-24	-12	-14	-20
	-9	-48	-47	-43	-42	-28	-26	-24	-16	-21	-33	-34	-16	-17	-21
30	-20	-13	-35	-44	-41	-30	-30	-18	-17	-18	-15	-15	-23	-31	-19
	-25	-19	-28	-41	-32	-24	-28	-13	-8	-15	-21	-22	-23	-32	-26
	-19	-26	-36	-44	-35	-35	-27	-17	-11	-9	-22	-29	-36	-35	-39
	-13	-28	-36	-45	-45	-39	-29	-20	-13	-13	-26	-32	-37	-40	-38
	-23	-31	-35	-39	-43	-42	-30	-19	-15	-10	-25	-32	-37	-38	-32
	-31	-32	-33	-40	-40	-40	-37	-27	-17	-16	-22	-35	-36	-40	-35
	-34	-60	-34	-33	-41	-37	-34	-39	-27	-22	-30	-40	-40	-39	-36
	-4	-5	-53	-55	-51	-27	-22	-17	-27	-30	-38	-35	-38	-29	-32
	-68	-25	-66	8	-78	-50	-26	-13	-12	-15	-18	-5	-23	-19	-19
	-15	-104	-149	-104	-98	-18	-27	-11	-8	-33	-4	-6	-12	-1	1
20						-66	-182	-220	-270	-245	-175	-161	-136	-50	-1
												-19	-73	-172	-117
														34	-183
														51	-154
															-56
															42
															26
														82	-75
												53	32	-20	-57
												-24	-67	-74	-66
10	-75					-70									-60

TABLE 1 (continued)
The 1° × 1° Averaged Free-Air Gravity Anomalies (Mgals) of the North Atlantic Referred to Geodetic Reference System 1967 (flattening 1/298.257)

	-60									-50					-45
66															
											18				
60										6	8	11	40	45	60
										8	14	20	15	13	
					12	16	4	-1	9	0	3	8	14	14	
					4	6	3	4	10	11	11	5	11	22	
			-6	-6	-10	-14	-13	1	12	15	12	17	25	24	
					35	42	-14	6	9	11	14	17	26	26	
							26	12	2	1	12	14	14	18	
								30	12	-6	6	18	14	20	
								4	37	11	-23	-21	3	14	
							-1	6	51	24	-19	-16	12	7	
50							0	18	75	18	-25	-24	-20	0	50
							-13	9	71	30	-17	-13	16	37	
							-24	-14	12	41	33	14	41	60	
							-8	5	28	40	39	21	36	51	
					38	33	1	16	38	40	33	8	13	12	
		14	2	-28	-43	24	29	38	41	38	16	1	5	11	
	4	-24	-34	-14	-13	-34	-19	16	37	51	55	47	17	10	1
	-37	-36	-27	-13	-11	-20	-34	-17	20	39	39	33	17	5	-4
	-32	-31	-28	-24	-16	-14	-17	-29	-26	-19	10	7	-5	2	0
	-25	-27	-27	-27	-26	-19	-19	-20	-9	-3	2	6	2	-1	-3
40	-21	-23	-23	-24	-22	-17	-18	-23	-12	-12	-19	-17	-21	-14	-10
	-26	-25	-19	-19	-19	-10	-11	-18	-14	-12	-19	-20	-22	-21	-17
	-19	-23	-18	-20	-23	-23	-15	-10	-14	-16	-15	-12	-8	-9	1
	-19	-27	-24	-24	-23	-14	-17	-4	2	-19	-3	-9	-11	-5	4
	-20	-29	-34	-28	-19	-21	-20	-14	-15	-39	-19	3	-3	-10	4
	-24	-29	-22	-15	-19	-14	-15	-21	-5	-9	-10	5	1	-13	-9
	-26	-15	-9	-23	-26	-20	-17	-23	-22	-23	-23	-13	-3	-9	-14
	-13	-9	-13	-20	-15	-19	-12	-8	-14	-7	-10	-10	-7	-5	0
	-28	-11	-10	-17	-19	-12	-15	-17	-12	-1	2	6	-4	2	4
	-30	-28	-15	-17	-20	-15	-18	-15	-13	-2	0	1	7	3	13
30	-25	-23	-25	-22	-19	-16	-12	-15	-4	-3	-2	0	1	12	15
	-33	-28	-25	-26	-21	-29	-17	-13	-19	-14	-12	-10	-1	3	12
	-41	-37	-34	-35	-25	-27	-20	-14	-16	-15	-11	-7	-1	2	-6
	-38	-32	-38	-43	-40	-32	-25	-23	-20	-21	-15	-5	-4	-7	16
	-38	-39	-31	-31	-25	-28	-25	-26	-21	-16	-9	-8	-9	0	12
	-35	-37	-42	-35	-32	-36	-28	-26	-35	-26	-21	-21	-6	12	7
	-35	-36	-30	-32	-30	-27	-20	-21	-18	-20	-12	-6	-8	-9	6
	-32	-30	-26	-22	-35	-27	-28	-21	-18	-19	-14	-11	-9	-8	3
	-23	-21	-16	-17	-28	-28	-19	-29	-28	-24	-19	-14	-6	4	1
	0	-6	-17	-29	-26	-31	-21	-28	-29	-29	-18	-9	-1	-1	2
20	3	-12	-31	-37	-25	-12	-23	-33	-42	-35	-25	-6	4	9	7
	-31	-29	-23	-28	-23	-4	-9	-26	-28	-29	-19	-15	-1	4	3
	-61	-5	-20	-10	-21	-10	-8	-30	-31	-17	-18	-10	-15	-6	-8
	-121	-28	-22	-22	-38	-41	-29	-24	-25	-17	-22	-19	-3	-7	-12
	-151	-5	-14	-25	-38	-47	-37	-35	-25	-21	-18	-20	-8	6	-15
	-150	-73	-50	-48	-41	-45	-28	-23	-24	-30	-37	-22	-15	-11	6
	-35	-6	-25	-28	-28	-32	-19	-24	-28	-42	-37	-26	-20	-14	12
	-92	-3	-27	-26	-18	-11	-22	-25	-29	-30	-32	-31	-30	-22	-3
	-80	-28	-48	-57	-41	-35	-36	-32	-26	-35					
	-45	-10	-47	-53	-42	-33	-38	-40	-37	-33					
10	-60									-50					-45

TABLE 1 (continued)
The 1° × 1° Averaged Free-Air Gravity Anomalies (Mgals) of the North Atlantic Referred to Geodetic Reference System 1967 (flattening 1/298.257)

Lat.	-45					-40									-30
66								10	32	35	21	10	6	10	9
							31	33	17	5	5	9	13	18	26
						11	6	4	8	13	15	15	21	24	32
					36	14	9	11	15	17	15	14	25	31	37
				47	20	10	13	14	16	18	21	31	38	40	44
60	37	53	45	28	11	10	10	10	8	12	31	38	41	45	44
	37	32	22	18	14	11	10	15	28	35	37	41	44	44	42
	13	18	19	18	17	17	17	32	28	35	37	40	41	40	33
	18	25	22	20	20	10	27	36	39	48	50	42	37	27	24
	25	23	24	26	26	26	22	27	34	50	51	38	27	20	13
	19	25	26	28	32	36	36	35	40	50	39	31	27	21	20
	24	26	25	36	36	30	29	38	35	39	38	34	28	24	25
	16	7	6	12	22	9	6	14	7	6	12	25	27	8	13
	18	10	7	5	17	23	22	24	23	20	23	18	27	38	23
	16	12	14	9	11	12	16	25	19	19	16	12	20	30	25
50	14	12	15	12	15	14	13	21	22	26	20	22	25	37	34
	44	18	12	11	13	15	23	27	31	35	22	18	16	27	30
	61	27	13	10	13	15	19	27	31	35	26	23	21	20	21
	42	16	12	11	13	14	17	20	22	24	27	27	23	22	25
	10	11	9	7	15	18	16	16	16	22	21	25	27	29	33
	15	12	12	8	15	10	8	13	17	19	19	17	25	28	25
	9	13	10	14	10	-1	13	21	18	27	25	19	24	27	26
	-1	4	2	6	8	3	10	18	17	16	17	21	28	29	29
	-6	-7	-6	-1	8	6	10	15	14	15	16	22	26	29	34
	0	0	-4	-10	4	13	14	16	17	14	17	28	31	25	28
40	-12	-10	-14	-17	-10	1	3	13	10	15	25	29	42	36	32
	-17	-13	-11	-7	-7	2	20	17	17	25	33	31	30	45	41
	9	19	18	21	15	10	23	18	17	35	34	52	41	50	37
	6	1	11	23	13	17	13	19	17	26	40	41	39	38	30
	8	11	14	13	14	18	23	22	37	43	40	37	32	29	17
	-3	-4	5	1	25	32	30	20	45	39	38	27	23	13	30
	-4	-5	18	18	19	12	29	36	33	25	21	23	13	13	11
	-2	-2	8	4	15	12	19	21	19	16	16	7	-2	-6	-7
	11	5	6	11	11	24	23	19	16	18	22	4	-9	-10	-10
	5	-3	11	15	22	8	7	6	12	10	-5	-1	-17	-11	-4
30	21	25	12	6	-2	9	4	14	3	8	10	4	-1	-1	-3
	26	32	16	14	6	-6	-6	1	4	-5	2	-1	-2	6	8
	23	18	6	7	-4	-9	-11	-10	-11	-9	-9	-11	-7	0	0
	18	6	6	15	1	-9	-7	-10	-12	-15	-12	-10	-8	-8	-5
	9	0	15	3	4	4	0	-8	-13	-14	-15	-13	-7	-11	-5
	11	10	-2	-7	-10	-10	-5	-11	-3	-9	-8	-10	-10	-6	1
	-2	4	-8	-4	-5	-1	-2	-19	-19	-14	-3	-5	-4	-5	-9
	8	-4	-5	-20	-13	-14	-8	-12	-3	3	12	3	6	4	-1
	-7	-20	-11	-16	-15	-15	-14	-14	-8	-1	4	12	16	19	23
	-8	-2	-4	-16	-20	-21	-18	-15	-9	-8	-2	10	13	22	26
20	-9	-6	-4	-12	-19	-19	-18	-16	-10	-5	-1	1	14	20	21
	-10	-13	-10	-15	-20	-21	-20	-20	-15	-10	-5	-2	3	12	15
	-17	-21	-16	-10	-16	-16	-18	-12	-12	-10	-7	-2	0	4	14
	-12	-15	-9	-8	-22	-16	-15	-5	-2	-2	-2	-3	-2	-2	0
	-31	-18	-12	-7	-13	-22	-17	-12	-4	-5	-7	-6	-6	-9	-10
	2	-23	-27	-17	-16	-18	-19	-16	-17	-17	-15	-13	-11	-12	-11
	4	-13	-12	-20	-15	-16	-19	-17	-15	-15	-14	-14	-14	-13	-13
	-4	-12	-21	-23	-22	-14	-20	-22	-20	-18	-16	-14	-13	-13	-13
10	-45					-40									-30

TABLE 1 (continued)
The 1° × 1° Averaged Free-Air Gravity Anomalies (Mgals) of the North Atlantic Referred to Geodetic Reference System 1967 (flattening 1/298.257)

Lat.	-30										-20				-15
66	29	17	22	40											
	17	26	33	43	41	39									
	31	38	35	42	44	39							35	29	29
	38	39	42	39	43	39	33	30	27	22	14	19	8	7	4
	42	42	42	38	40	38	31	28	28	22	16	13	12	10	4
	44	42	37	34	32	30	26	18	12	13	8	14	11	8	14
60	44	41	38	33	24	19	19	12	9	7	8	10	15	19	25
	37	33	28	21	20	16	10	11	8	7	23	28	10	-3	4
	30	24	23	22	22	13	9	10	18	22	19	-7	-8	0	14
	21	20	23	19	12	10	15	31	37	19	10	7	17	34	45
	18	23	19	10	11	25	26	27	4	6	35	32	36	31	2
	21	26	19	12	32	7	3	12	7	-16	18	24	4	-16	4
	18	17	13	13	19	15	5	-9	-2	-2	8	20	17	6	-1
	13	20	18	20	26	22	9	-16	-13	7	11	-18	-13	-1	10
	38	34	20	5	12	25	22	34	34	9	26	9	-3	-13	-10
	8	26	22	20	12	11	19	29	26	9	15	22	24	0	-11
50	48	39	36	22	22	22	11	19	22	27	11	7	13	6	-6
	33	68	49	42	24	20	17	17	15	15	23	18	19	16	9
	23	29	42	33	32	29	27	17	14	11	18	16	21	19	11
	31	30	30	31	39	43	36	24	12	8	17	34	19	3	17
	35	39	30	30	42	44	40	38	34	16	14	20	5	25	26
	32	32	30	29	33	29	21	19	21	23	7	0	1	39	1
	32	32	25	24	29	34	47	39	20	25	8	28	41	26	5
	50	34	35	33	30	25	26	32	28	10	21	18	12	14	18
	34	42	36	32	29	21	21	16	27	17	-2	-15	-10	1	16
	30	28	28	32	34	22	16	18	17	11	-3	0	15	6	-5
40	29	38	47	24	25	22	17	-3	-4	-6	5	6	2	-17	-11
	41	53	56	33	18	29	27	11	-4	14	5	-14	-14	-16	-6
	38	22	40	25	73	40	30	7	-11	20	-12	-9	-9	-6	3
	12	36	30	23	33	21	6	0	-5	-15	-34	-29	-22	-11	21
	29	29	31	32	15	12	5	5	-13	-17	-17	-14	5	38	47
	5	25	19	-6	-30	-10	-6	-11	-17	-15	-10	-8	5	19	17
	26	21	3	14	2	-2	-19	-22	-23	-17	10	25	9	34	39
	-16	15	29	-17	9	-15	-21	-21	-16	-8	26	22	34	39	-10
	-18	56	-10	-24	-25	-21	-16	0	-5	-3	-2	1	-11	-25	-16
	-15	38	-11	9	6	-5	-7	-2	4	3	-3	-8	-14	-6	19
30	-12	60	-16	4	7	1	-7	-5	1	-1	-12	-16	-9	-20	-22
	-13	-18	-12	-5	-6	-4	-7	-17	-16	-14	-7	1	61	92	-6
	0	-1	-3	3	2	-9	-20	-23	-12	-9	0	9	29	-27	18
	3	3	14	12	7	-4	-11	-8	-4	3	17	21	-4	-49	-52
	-3	6	6	2	-8	-10	-6	-9	-3	3	19	32	-7	-25	-6
	-3	-8	-6	-6	-7	-7	-5	-5	-2	1	12	20	-6	8	
	-13	-9	-17	-15	-5	-4	-2	-1	-4	-5	-6	-8	3	25	
	-12	-17	-19	-20	-17	-7	1	-1	-4	-12	-18	-32	-11		
	14	-1	-5	-10	-8	-9	-4	-2	-9	-20	-28	-36	15		
	18	7	3	2	5	5	0	3	3	-9	-17	-9	30		
20	18	8	7	13	19	15	4	14	19	6	-4	-15	-24		
	12	11	19	23	22	13	12	21	19	11	0	-11	-32	17	
	11	10	14	13	45	25	16	16	9	3	3	-3	-7	27	
	5	7	8	11	39	67	28	72	5	5	6	0	-7		
	2	5	9	13	11	-1	80	40	-7	-7	-5	-8	22		
	-5	-2	3	5	9	11	-3	-20	-19	-15	-10	1	49		
	-11	-15	-7	1	1	-16	-18	-18	-14	-10	-9	-19	11		
	-12	-11	-7	-2	-5	-10	-10	-10	-9	-8	-9	-28	-21		
10	-30										-20				-15

TABLE 1 (continued)
The 1° × 1° Averaged Free-Air Gravity Anomalies (Mgals) of the North Atlantic Referred to Geodetic Reference System 1967 (flattening 1/298.257)

Lat.	-15					-10									0
66		28	46	47	28	25	21	21	23	20	-2	20	10	10	29
		37	46	51	42	29	10	32	3	6	9	16	17	22	34
	17	16	38	42	36	26	46	28	5	-5	-4	5	7	8	4
	4	23	32	33	35	31	45	59	52	31	25	20	-6	-20	11
	8	18	21	22	15	24	35	24	32	11	4	-16	-22	10	26
	21	31	34	16	27	26	26	14	25	-4	-7	17	40	37	2
60	29	10	-16	-19	24	12	-7	24	54	43		38	18	-4	-15
	15	14	-12	-23	-20	-10	7	26					-3	-3	7
	39	29	-28	-17	-4	12									2
	18	-37	-40	-40	-11	-6									
	-15	-11	-32	-47	-31	3									
	-16	-42	-31	-26											
	25	50	38	46											
	55	28	18	44											
	61	27	17	30	30			11	-1						
	7	-4	-3	7	42	30	28	15	16	21					
50	-15	-8	20	40	54	28	26	18	6	11	11				
	-4	-12	-7	10	19	32	26	25	16	13					
	-1	-1	-21	-22	-30	-30	-17	-10	10	28					
	0	4	20	-33	-42	-29	-31	-27	-23	-45	-10				
	7	10	0	-14	43	9	-22	-21	-3	-12	-35				
	-17	10	13	-4	-37	-59	-42	-68	-93	-90	-56	-77			
	5	17	1	-13	-50	24									
	-28	-21	27	46	5	38									
	-4	-7	-9	11	0	-12									
	-12	-34	-33	-30	-39	9									
40	-2	15	10	-28	-2	30									
	-8	59	-1	-14	-34	24									
	28	16	-37	-61	11	26									
	64	15	38	92	-25	15	31	7							
	-3	-25	2	-15	-16	-31	-25	-65							
	12	-6	-9	-13	-16	-49	-58	-41							
	14	-35	-13	-23	-53	-69									
	-20	-12	-1	17	-37	18									
	-1	32	10	23	25										
	32	42	46	-7	-44										
30	-26	40	17	-26	7										
	61	33	15												
	-36	5													
	-11														
20															
10															

TABLE 1 (continued)
The 1° × 1° Averaged Free-Air Gravity Anomalies (Mgals) of the North Atlantic Referred to Geodetic Reference System 1967 (flattening 1/298.257)

Lat	0										10	15	Lat
66	37	36	11	-4	1	7	8	-2	-4	5			66
	30	21	-1	-18	-20	-22	-9	-14	4	16			
	8	6	-1	-19	-25	3	30	14					
	19	14	29	23	12	9							
	27	-7	8	-5									
	-1	-4	9	-13	3								
60	-9	5	3	-12	9								60
	15	5	0	-4	3	-18							
	-4	-5	-6	-8	-7	-5	-11						
	1	8	12	5	-2	3	3						
	-8	-14	-13	-15	-4	4	9						
	-10	-18	-23	-11	-12	-17	-16	-13					
		-11	-18	-16	-11								
			-4	-7									
50													50
40													40
30													30
20													20
10	0										10	15	10

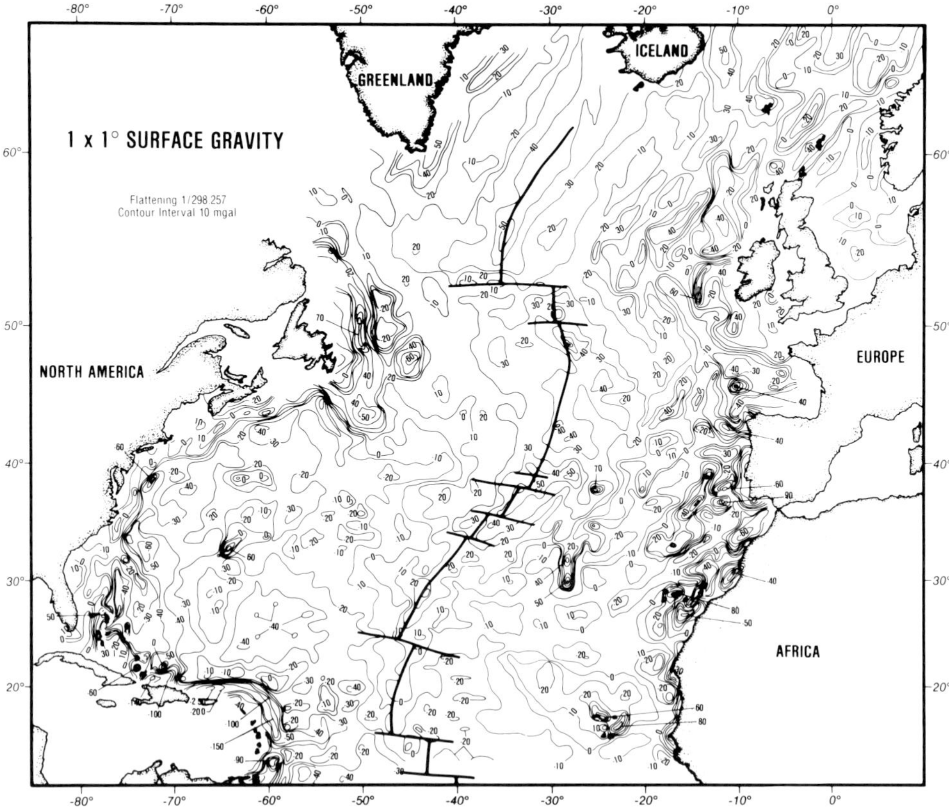

FIGURE 2. Map of the 1° × 1° averaged free-air gravity anomalies determined from the detailed gravity map of Jung and Rabinowitz (1988). This map was constructed based on the 1° × 1° averaged values shown in Table 1 which were referred to Geodetic Reference System 1967. Note the relative positive values along the Mid-Atlantic Ridge and negative values over the basins. Volcanic islands are characterized by high positive anomalies.

features of the 1° × 1° surface ship gravity were filtered out. The outstanding features of the 5° × 5° surface ship gravity map include; (i) positive anomalies with wavelengths of 2000 to 3000 km and amplitudes up to 30 mgal over the Mid-Atlantic Ridge in the region north of about 30°N and (ii) negative anomalies over the western North Atlantic Ocean basin with wavelengths of 3000 km and amplitudes of −40 mgal. The largest negative values (more than −100 mgal) are observed over the Puerto Rico Trench. The intermediate wavelength (200 to 1000 km) gravity highs over the extensive volcanic islands off the Mid-Atlantic Ridge system such as Bermuda, Cape Verde, Canary, and Madeira Islands are characterized with relative positive gravity anomalies in the 5° × 5° regional field. The extinct midocean ridge in the Labrador Sea is expressed as relative negative low in contrast with the present Mid-Atlantic Ridge system. Another important feature of the 5° × 5° surface ship gravity map is a relative negative over the Rockall Plateau, which is a structural high.

Talwani and Le Pichon (1969) noted an apparent strong correspondence between topography and regional gravity in the North Atlantic Ocean. They considered that 5° × 5° surface ship anomalies would be close to zero and would not correlate with topography if isostatic equilibrium were attained locally and at a shallow depth (about 30 km). However, isostatic anomalies in excess of 10 mgal were observed over areas of several million km^2 when an Airy-Heiskanen type of isostatic correction was applied to the 5° × 5° averaged gravity map. Thus Talwani and Le Pichon (1969) concluded that the isostatic mechanism operates to depths much greater than 30 km and there are additional lateral density inhom-

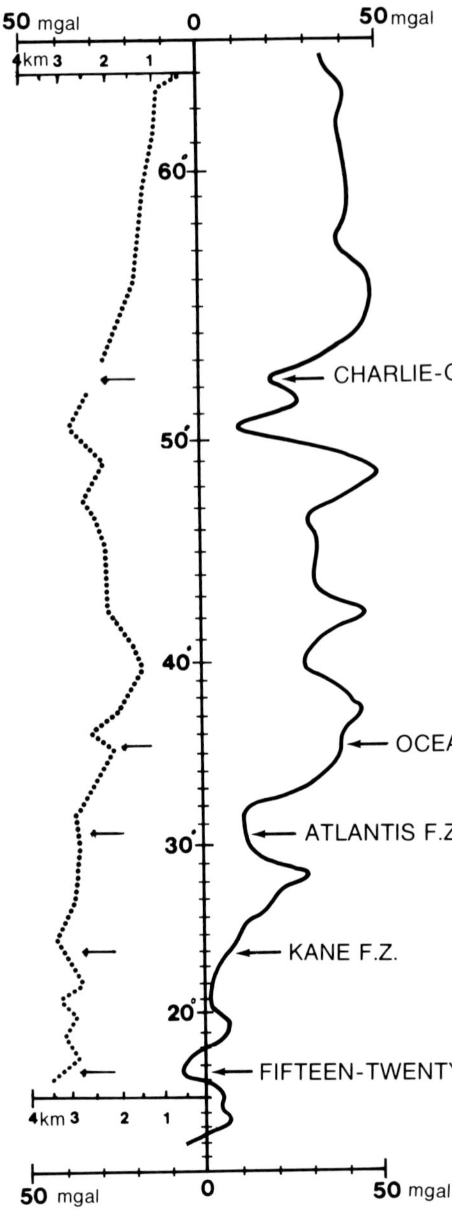

FIGURE 3. The 1° × 1° averaged free-air gravity anomalies along the Mid-Atlantic Ridge axis. North of 30°N shows broad positive values except for the Charlie-Gibbs Fracture Zone. Accompanying 1° × 1° averaged bathymetry (dotted line) was determined from the General Bathymetric Charts (GEBCO) map.

ogeneities in the upper mantle which are nonisostatic in character. They also cited that convective motions might cause the density inhomogeneities in the upper mantle which are nonisostatic in character. They also cited that convective motions might cause the density inhomogeneities in the upper mantle which would explain part of the nonisostatic regional gravity anomalies in the North Atlantic Ocean.

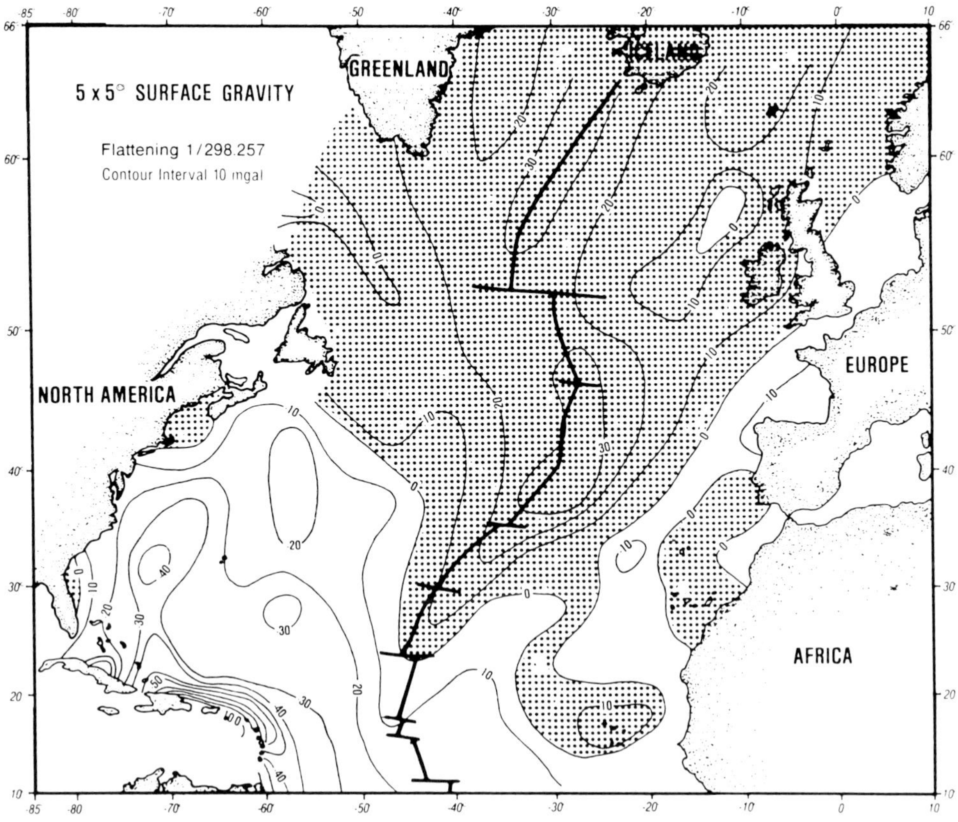

FIGURE 4. Map of the 5° × 5° averaged free-air gravity anomalies determined from the 1° × 1° averages (Figure 2; Table 1). Note smooth regional field with positive values in the north and negative values in the south. Mid-Atlantic Ridge and volcanic islands such as Bermuda, Cape Verde and Madeira are characterized by relative positive values and Puerto Rico Trench by negative values.

B. COMPARISON BETWEEN SURFACE SHIP AND SEASAT DERIVED GRAVITY ANOMALIES

The 1° × 1° SEASAT gravity map (Figure 5) were constructed based on data provided by Dr. Richard Rapp of the Department of Geodetic Sciences, The Ohio State University. It compares favorably with the 1° × 1° surface ship gravity map as might be expected. For example, the major fracture zone such as the Charlie-Gibbs Fracture Zone, the location of extensive volcanism activity such as Bermuda, Cape Verde, Canary and Maderia Islands, the basins on either side of the Mid-Atlantic Ridge and the north-south variations along the Mid-Atlantic Ridge are similar in the 1° × 1° SEASAT altimeter-derived and surface ship gravity maps. However, the gravity anomalies in the 1° × 1° SEASAT gravity map appear smoother than the 1° × 1° surface ship gravity map, although both maps exhibit similar wavelengths and amplitudes. For example, the 1° × 1° SEASAT gravity map does not show "local" features near 50°N, 30°W, while the 1° × 1° surface ship gravity map clearly does. A careful comparison shows that other more "local" features (about 200 to 300 km wavelengths) are observed on the Mid-Atlantic Ridge south of about 40°N in the 1° × 1° surface ship gravity map that are not present in the 1° × 1° SEASAT gravity map. This implies that the surface ship gravity measurements give more information for the short wavelength features than the altimeter derived gravity anomalies. However, it should be noted that relative large discrepancies between the two regional gravity fields are observed over the area where the surface ship gravity data are sparse (e.g., near 48°N, 50°W).

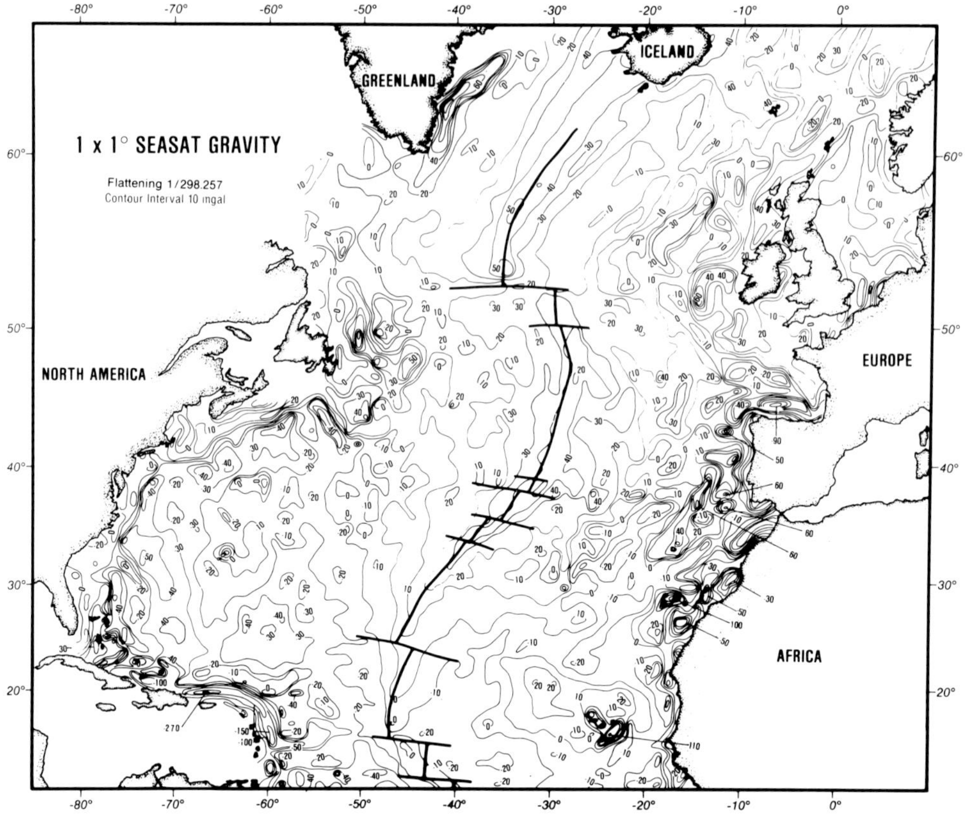

FIGURE 5. Map of the 1° × 1° averaged free-air gravity anomalies derived from the SEASAT altimeter data provided by Dr. Richard Rapp of Ohio State University via magnetic tape. The procedures on how to obtain these values are described by Rapp (1979, 1983). This map shows very similar trends in contours and amplitudes to the surface ship-derived gravity map given in Figure 2. However, the surface ship-derived values show shorter wavelength features in areas where very detailed data coverage is available.

We show in Figure 6 the differences between the 1° × 1° surface ship gravity and SEASAT altimeter-derived gravity anomalies along selected latitude lines. Accompanying bathymetry in Figure 6 was obtained from the General Bathymetric Chart of the Ocean (GEBCO) map by smoothing over 1° interval. One interesting feature of Figure 6 is that the differences are largest over the short wavelength topographic features — mostly smaller than 200 to 300 km wavelength. This is most probably due to the fact that the 1° × 1° SEASAT gravity anomalies are recovered from relatively widely spaced altimeter data (east-west separation ranges from 17 to 130 km; Cheney, 1982). Hence, the process to obtain the 1° × 1° SEASAT gravity anomalies might have inadvertently smoothed out short wavelength features and resulted in lower real values. In addition, the procedure to contour surface ship gravity measurements in poorly surveyed areas surrounding the topographic highs will cause some errors. Nonetheless, the two regional gravity fields obtained independently show close agreement. The areal average of the total 3032 1° × 1° surface ship gravity anomalies is −0.3 mgal while that of the 1° × 1° SEASAT altimeter-derived gravity anomalies is 1.2 mgal. Figure 7 is the histogram of differences between the 1° × 1° surface ship gravity (Figure 2) and SEASAT altimeter-derived gravity anomalies (Figure 5). The root mean square of the differences between two anomalies is 8.7 mgal. As can be seen in Figure 7, about half of the 3032 1° × 1° surface ship gravity anomaly differences are less than ±4 mgal, and three-quarters are less than ±8 mgal. The altimeter measurements can

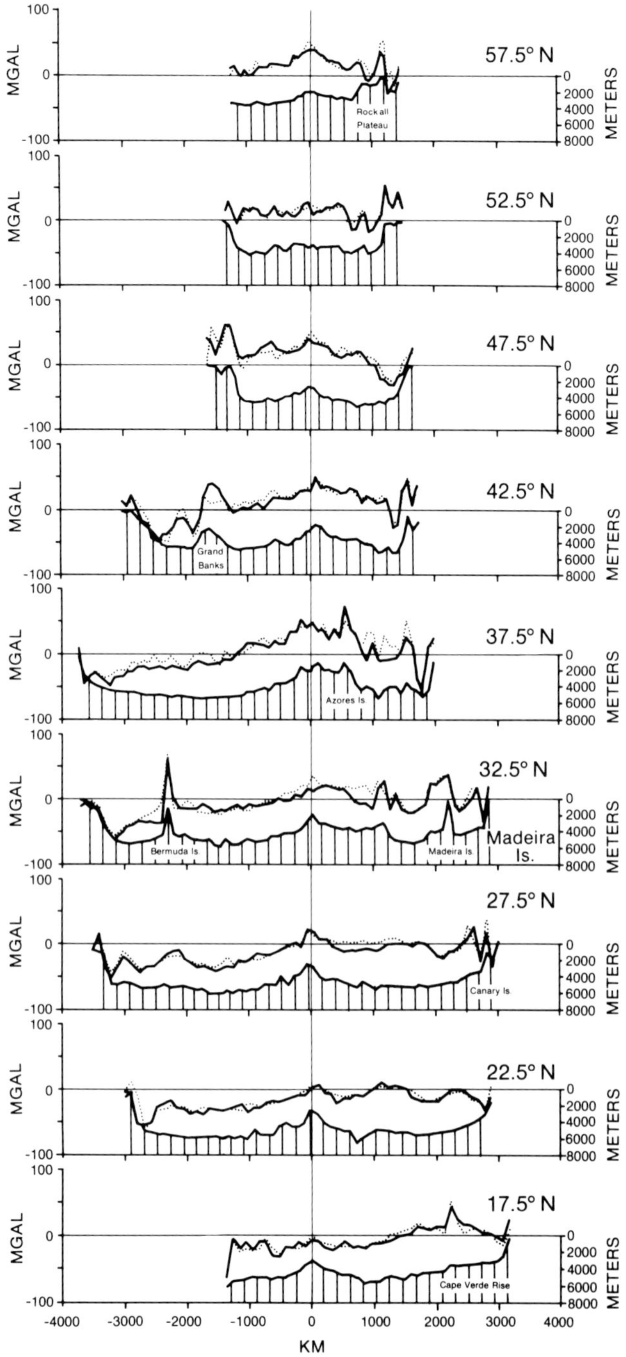

FIGURE 6. Cross sectional profiles of the 1° × 1° averaged surface ship-derived (straight line) and SEASAT altimeter-derived (dotted line) gravity anomalies. Bathymetric profiles were determined from the GEBCO map with 1° interval. All three profiles are aligned with respect to the ridge axis; distances are in kilometers from the ridge axis. Note the major bathymetric features are registered by positive gravity values.

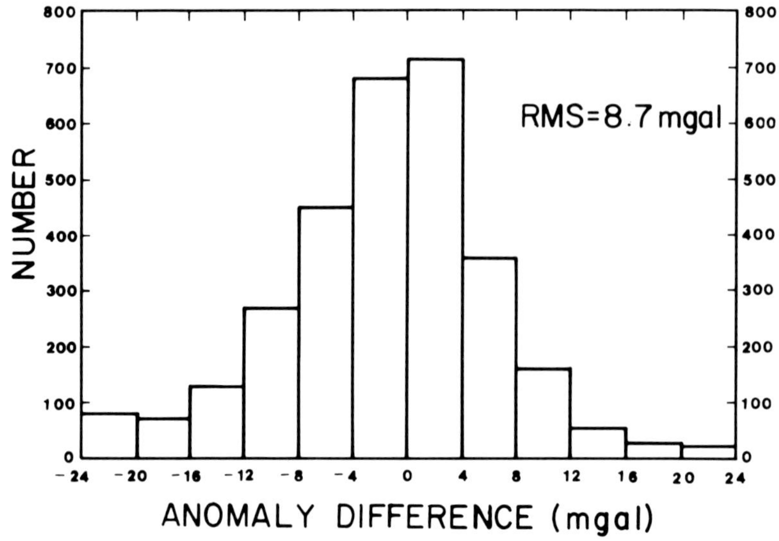

FIGURE 7. Histogram of the differences between the 1° × 1° surface ship and the SEASAT gravity anomalies. About three-quarters of differences are less than 8 mgals.

thus be used with confidence for regional geologic studies in areas where surface ship gravity measurements are not available.

IV. DESCRIPTION OF THE GRAVIMETRIC GEOID

A. THE 1° × 1° DIFFERENCE GEOID

As noted earlier, the 1° × 1° difference geoid undulations were computed based on the 1° × 1° difference gravity anomalies which were determined by subtracting degree 20 gravity field of GEM10C model from the 1° × 1° averaged free-air anomalies. Thus the 1° × 1° difference geoid resolves geoid information with wavelengths of 200 to 2000 km. This range of wavelengths reflect the condition of mass distribution in the upper mantle, possibly the outer 300 km of the Earth (Bowin, 1983). The most positive values (> 14 m) are observed over the Cape Verde Rise and the most negative values (−10 m) over the Puerto Rico Trench.

The 1° × 1° difference geoid along the Mid-Atlantic Ridge generally reflects the long wavelength bathymetric features and show relative positive values ranging between −6 m and 6 m from basin to ridge axis. South of about 30°N this geoid high appears to connect with an arcuate high centered over the eastern portion of the Lesser Antilles at 15°N, 55°W. At 40°N, 33°W, a relative geoid high is observed over the Azores triple junction. This geoid high is associated with mantle upwelling beneath this area. The geoid variations associated with plate cooling will mask the geoid effects associated with deeper mantle processes. Jung and Rabinowitz (1986) thus removed the geoid effects associated with plate cooling and found out the center of geoid high shifts to the Azores Island. They interpreted this result to indicate that the main axis of convective mantle upwellings does not lie beneath the present day spreading center but beneath the Azores Islands.

The ocean basins surrounding the Mid-Atlantic Ridge are characterized by negative geoid values. In the western North Atlantic Ocean, geoid lows more negative than −6 m are observed at 40°N, 45°W. The geoid lows in the western basin is observed trending in the SW-NE and crosses the Mid-Atlantic Ridge near the Charlie-Gibbs Fracture Zone. Unlike

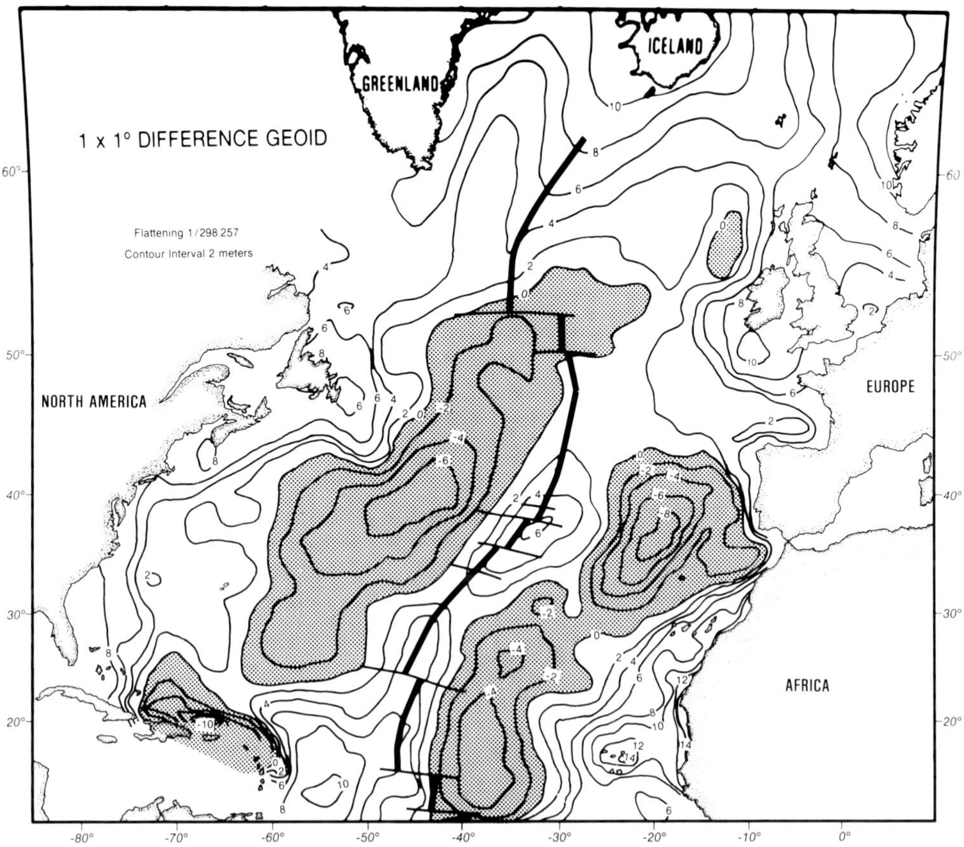

FIGURE 8. Map of the 1° × 1° difference geoid computed from the 1° × 1° difference free-air gravity anomalies by using Stokes' formula. See text for discussion on how geoid were computed. Note the positive geoid high along the ridge axis and over the volcanic islands. Basins and trench are characterized by negative values.

the western basin the geoid lows in the eastern basin are separated into two approximately equal size of depressions centered near 36°N, 20°W and 20°N, 37°W.

The volcanic islands such as Bermuda Rise and the Cape Verde Rise are characterized by relative geoid highs. Haxby and Turcotte (1978) explained the goeid high over the Bermuda Swell in terms of a Pratt compensation scheme having anomalously low densities to depths of approximately 100 km. Crough (1982) concluded that the Cape Verde Rise is in local isostatic equilibrium with an isostatic root of 40 km, supported by a low-density root of altered lithosphere. Although Crough (1978) concluded that hot spot reheating of the lithosphere was the main cause of depth anomalies in the oceans, Watts and Daly (1981) argued that factors other than thermal reheating of the lithosphere by hot spots such as upwelling mantle convection may contribute to the origin of depth and geoid anomalies. Parsons and Daly (1983) and Detrick et al. (1985) also suggested that the observed depth, geoid and heat flow anomalies over the Bermuda Rise are consistent with simple convection models in which the lower part of the thermally defined plate acts as the upper thermal boundary layer of the convection.

It is interesting to note that though the geoid high over the Azores Swell shows a similar wavelength and amplitude ranges from surrounding basins to the geoid peak to those of the Cape Verde Rise, the computed 1° × 1° difference geoid high over the Azores Swell shows less value than that of the Cape Verde Rise. It may imply that the mechanisms responsible for the Azores Swell lie deeper than those of the Cape Verde Rise.

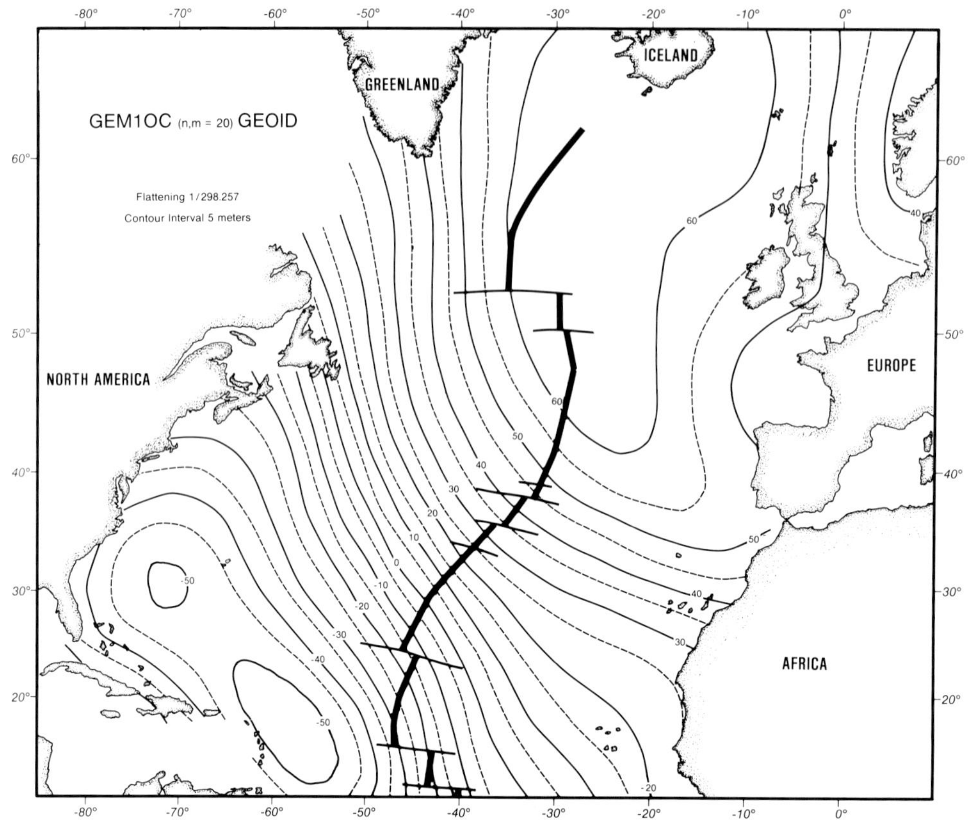

FIGURE 9. Map of the degree 20 geoid of GEM10C model. Two long wavelength geoid extremes are observed: a positive high over the Iceland and a negative low over the north of Puerto Rico.

Relative positive geoid values are observed over the continental margins of the North Atlantic. The largest negative geoid undulation in the 1° × 1° difference geoid field occurs along the Puerto Rico Trench area with value more negative than − 10 m north of Puerto Rico. Similar negative geoid lows are observed along the trenches in the western Pacific Ocean by Watts and Leeds (1977) and in the Indian Ocean by Kahle and Talwani (1973) and Kahle et al (1978).

B. THE 1° × 1° TOTAL GEOID AND ITS COMPARISON WITH THE SEASAT DERIVED GEOID

The 1° × 1° total geoid is obtained by adding the 1° × 1° difference geoid to the GEM10C geoid (Figure 9). Thus the 1° × 1° total geoid can provide additional information on mass distributions and tectonic processes within the deep earth. This total geoid reveals geoidal information with wavelengths greater than 200 km. The 1° × 1° total geoid undulations of the North Atlantic Ocean (Figure 10) are characterized by two geoidal extremes; a NW-SE-oriented low centered over the western North Atlantic Ocean basin and a NE-SW-oriented high centered over Iceland. Between these two extremes, most of the geoid contours trend a NW-SE.

The origin of these two long wavelength features are generally believed due to dynamic effects which are associated with whole mantle convections and/or fluctuations of core mantle boundary. For example, Chase (1979) estimated the maximum possible source depth for the positive anomaly at 2000 or 2900 km (near core-mantle boundary) even though he did not exclude the possibility that a more gradual density variation at a shallower depth over a

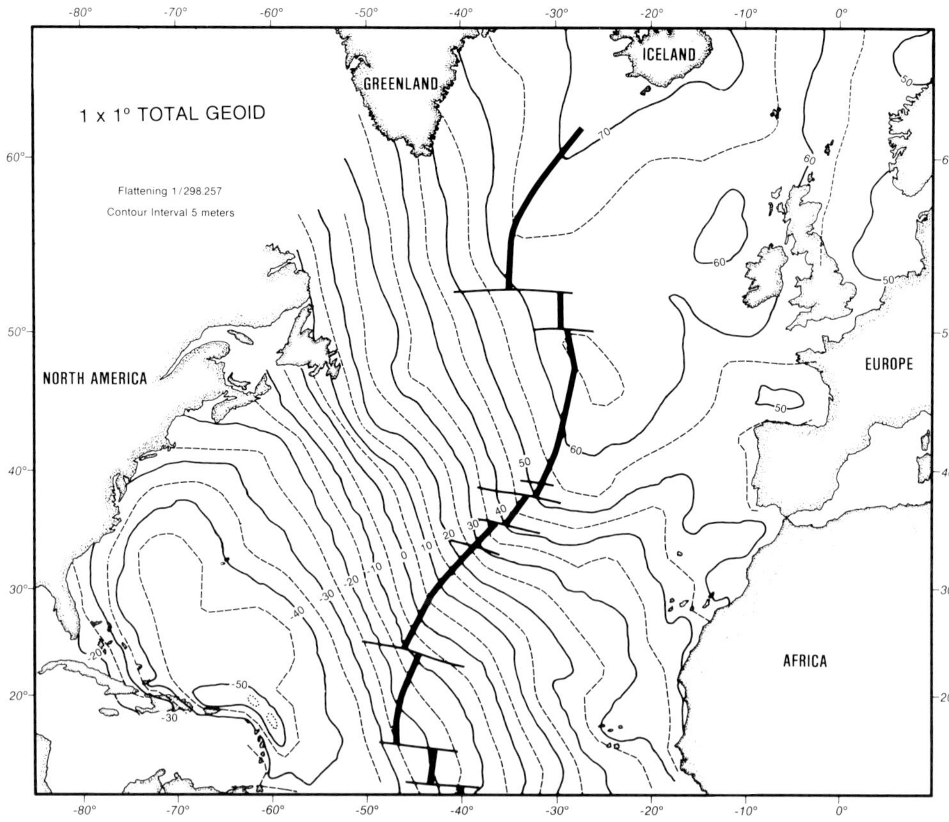

FIGURE 10. Map of the 1° × 1° gravimetric total geoid obtained by adding the 1° × 1° difference geoid (Figure 8) to the degree 20 geoid of GEM10C model (Figure 9). Note many intermediate wavelength (200 km to 1000 km) features are superimposed on the two long wavelength geoid extremes.

broad area can cause the anomaly. He estimated the sources of the negative anomaly could be no deeper than 1200 km. McKenzie (1967) explained that geoid undulations of such long wavelength feature cannot arise from rigidly supported density anomalies from within the lithosphere, since shear stress associated with the density differences which might have caused such long wavelength features is much greater than the lithosphere can support.

Many intermediate wavelength (200 to 1000 km) features are superimposed upon these two long wavelength geoid extremes. Most significant ones are observed over the volcanic islands such as the Bermuda, Cape Verde, and Madeira. The Puerto Rico Trench is characterized by arcuate geoid lows which is the most negative one in the North Atlantic Ocean.

The 1° × 1° SEASAT altimeter-derived geoid shown in Figure 11 were constructed based on the data provided by Dr. Richard Rapp of The Ohio State University via magnetic tape. The procedures on how to recover the geoid undulations from the altimeter data were described by Rapp (1979, 1983). The overall pattern of the 1° × 1° SEASAT geoid is very similar to that of the 1° × 1° total geoid. However, there exists some differences between two geoids. Amplitude differences greater than 10 m are observable over the Iceland and Puerto Rico Trench areas. This is probably due to the fact that the geoid computation near the edges of the study area are influenced by abrupt termination of the 1° × 1° difference gravity. In addition the 1° × 1° difference gravity over steep varying areas such as the Puerto Rico Trench might include errors occurred during the interpolation to obtain averaged values. This error would in turn appear in the numerical integration of Stokes' formula. Over the western North Atlantic Ocean basin, the 1° × 1° SEASAT geoid map shows a

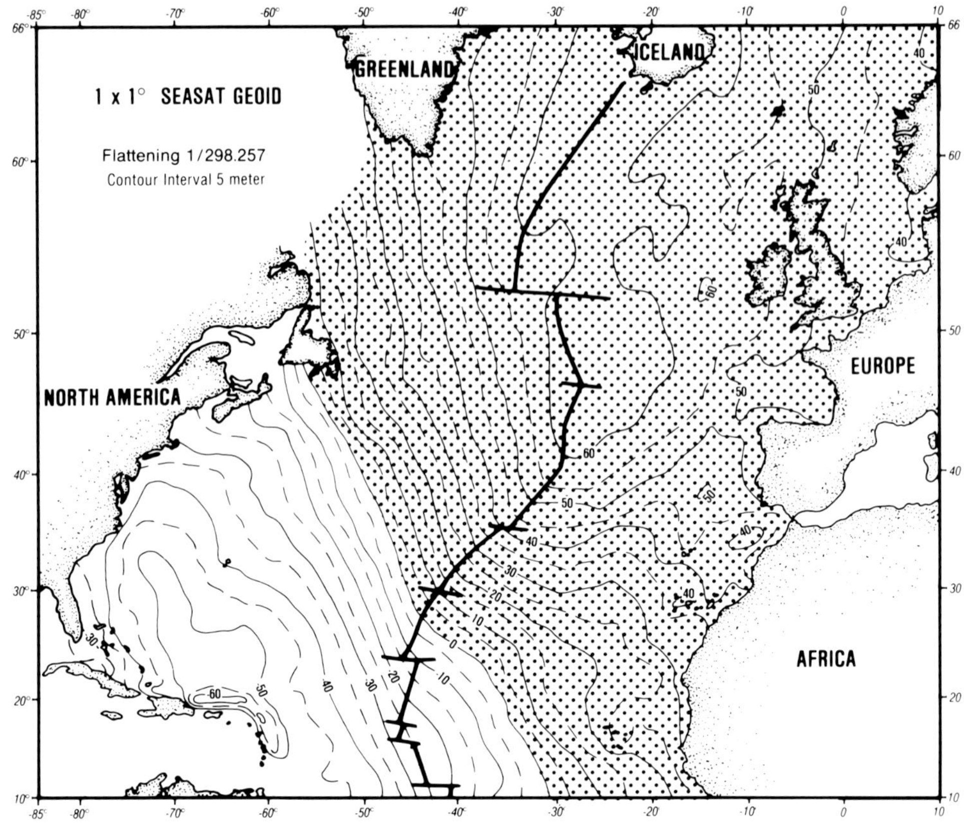

FIGURE 11. Map of the 1° × 1° SEASAT altimeter derived geoid undulations. Overall features are very close to those of the 1° × 1° total geoid (Figure 10), but shows relative large differences over the Iceland and the Puerto Rico Trench areas.

NW-SE elongation with a more uniform pattern than the 1° × 1° total geoid map. The minimum values (−60 m) of the 1° × 1° SEASAT geoid are observed along the Puerto Rico Trench in a similar fashion to the 1° × 1° total geoid.

V. CONCLUDING REMARKS

In this paper the regional gravity anomalies were determined for the North Atlantic Ocean from a detailed surface ship-derived gravity map (Jung and Rabinowitz, 1988) and compared to those obtained from SEASAT altimeter data. Both the 1° × 1° surface ship and the 1° × 1° SEASAT gravity show broad positive values over the Mid-Atlantic Ridge area and negative values over the adjacent basins. There exists a north-south variation in the regional gravity field over the ridge; the positive values to the north of 32°N are still too low to account for the shallow water depth in the north. Cochran and Talwani (1978) suggested a portion of the compensation must be distributed to depths of several hundred kilometers within the asthenosphere and that this mass deficiency is relatively uniform throughout the area. The locations of extensive volcanic activity such as the Bermuda, Cape Verde, Canary, and Madeira Islands are all associated with relative regional (1000 to 2000 km wavelength) gravity highs with values of about 20 mgal.

Although both the 1° × 1° surface ship and the 1° × 1° SEASAT gravity profiles exhibit similar wavelengths and amplitudes, careful comparison shows more local features are present in the 1° × 1° surface ship gravity. The root mean square of the difference between the 1° × 1° surface ship and the 1° × 1° SEASAT gravity anomalies is 8.7 mgal. The differences are the largest over the short wavelength topographic features. This is most probably due to relatively widely spaced SEASAT altimeter tracks.

The practical computation procedures proposed by Talwani et al. (1972) enable us to compute the geoid for the limited area without knowing the gravity anomalies over the whole Earth's surface. We computed the 1° × 1° difference geoid for the North Atlantic Ocean utilizing the 1° × 1° difference free-air anomalies employing Stokes' formula.

Some characteristic observations of the 1° × 1° difference geoid are positive highs along the Mid-Atlantic Ridge and Reykjanes Ridge and negative lows over the adjacent basins. Volcanic islands such as the Bermuda, Cape Verde, and Azores are associated with relative positive highs. Though the wavelength and amplitude range from the surrounding basins to the geoid peak over the Azores Swell and the Cape Verde Rise are similar, the computed 1° × 1° difference geoid shows less value over the Azores Swell. It may imply that the mechanisms responsible for the Azores Swell lie deeper than those of the Cape Verde Rise.

By adding the 1° × 1° difference geoid to the GEM10C geoid, we obtained the 1° × 1° total geoid. The 1° × 1° total geoid undulations are characterized by two long wavelength geoid extremes with many intermediate wavelength features superimposed on them. Comparison between the 1° × 1° total geoid and the 1° × 1° SEASAT geoid shows in general good agreement; near Iceland and the Puerto Rico Trench, relatively large differences reaching up to 10 m are observed.

In conclusion, both the surface ship-derived gravity and geoid maps compare favorably with those obtained from the SEASAT altimeter data, respectively. The SEASAT altimeter-derived maps are especially good for regional study because they are constructed based on altimeter data more evenly spaced and covering remote areas. Thus, they will be particularly useful for studying tectonic problems in areas where surface ship gravity measurements are sparse or not available. However, for areas where abundant surface ship gravity measurements exist, the resolution of the surface ship-derived gravity is better, at present, than that of altimeter-derived maps.

ACKNOWLEDGMENT

We thank Dr. Richard Rapp for kindly providing tapes of the gravity and geoid data used in this study. This work was part of a Ph.D. dissertation (WJ). We thank Drs. W. Bryant, R. Carlson, D. Fahlquist, and T. Hilde, for critical reviews. This research was supported under National Science Foundation grants and Office of Naval Research contracts.

REFERENCES

Bomford, G., *Geodesy,* Clarendon Press, Oxford, 1980.

Bowin, C. O., Depth of Principal Mass Anomalies Contributing to the Earth's Geoidal Undulations and Gravity Anomalies, *Marine Geodesy,* 7, 61—100, 1983.

Chase, C. G., Subduction, the geoid, and lower mantle convection, *Nature,* 282, 464—468, 1979.

Chapman, M. E. and Talwani, M., Comparison of gravimetric geoids with GEOS 3 altimetric geoid, *J. Geophys. Res.,* 84, 3803—3816, 1979.

Cheney, R. E., Comparison data for SEASAT altimetry in the western North Atlantic, *J. Geophys. Res.,* 87, 3247—3253, 1982.

Cochran, J. R. and Talwani, M., Gravity anomalies, regional elevation, and the deep structure of the North Atlantic, *J. Geophys. Res.,* 83, 4907—4924, 1978.

Crough, S. T., Thermal origin of mid-plate hot-spot swells, *Geophys. J. R. Astron. Soc.,* 55, 451—469, 1978.

Crough, S. T., Geoid height anomalies over the Cape Verde Rise, *Mar. Geophys. Res.,* 51, 263—271, 1982.

Detrick, R. S., Von Herzen, R. P., Parsons, B., Sandwell, D., and Dougherty, M., Heat flow observations on the Bermuda Rise and thermal models of midplate swells, *J. Geophys. Res.*, 91, 3701—3723, 1986.

Ewing, V., Free air gravity anomalies in the continental margins of eastern North America, *Ocean Margin Drilling Program — Regional Atlas Ser. 2, 3, 4 and 5,* Marine Science International, Woods Hole, 1984.

Haxby, W. F. and Turcotte, D. L., On isotatic geoid anomalies, *J. Geophys. Res.*, 83, 5473—5478, 1978.

Heiskanen, W. and Moritz, H., *Physical Geodesy,* S. Freeman, San Francisco, 1967.

Jung, W. and Rabinowitz, P. D., A detailed free-air gravity anomaly map of the North Atlantic Ocean and its significant features, *Am. Ass. Petro. Geol., Special Map Series,* 1988.

Jung, W. and Rabinowitz, P. D., Residual geoid anomalies of the North Atlantic Ocean and their tectonic implications, *J. Geophys. Res.*, 91, 10383—10396, 1986.

Kahle, H. G. and Talwani, M., Gravimetric Indian Ocean geoid, *Z. Geophys.*, 39, 167—187, 492—499, 1973.

Kahle, H. G., Chapman, M., and Talwani, M., Detailed 1 × 1° gravimetric Indian Ocean geoid and comparison with GEOS 3 radar altimeter geoid profiles, *Geophys. J. R. Astron. Soc.*, 55, 703—720, 1978.

Parsons, B. and Daly, S., The relationship between surface topography, gravity anomalies, and temperature structure of convection, *J. Geophys. Res.*, 88, 1129—1144, 1983.

McConathy, D. R. and Kilgus, C. C., The Navy Geosat Mission: An overview, *Technical Digest,* 8, Johns Hopkins APL, 170—175, 1987.

Mckenzie, D. P., Some remarks on heat flow and gravity anomalies, *J. Geophys. Res.*, 72, 6261—6273, 1967.

Lerch, F. J., Putney, B. H., Wagner, C. A., and Klosko, S. M., Goddard earth models for oceanographic applications (GEM 10B and 10C), *Marine Geodesy,* 5, 145—187, 1981.

Rabinowitz, P. D. and Jung, W., Gravity anomalies in the western North Atlantic Ocean, in *The Geology of North America, Volume M, The Western North Atlantic Region,* Vogt, P. R. and Tucholke, B. E., Eds., Geol. Soc. Am., Boulder, CO, 1986, 205—214.

Rabinowitz, P. D., Jung, W., and Haxby, W., Comparison of surface ship derived and SEASAT altimeter derived gravity anomaly maps in the North Atlantic Ocean, *EOS Trans. AGU,* 66, 356, 1985.

Rabinowitz, P. D., Hayes, D. E., Fritsch, J., and Dostmann, H., Free-air gravity anomalies off Morocco, Northwest Africa, *Ocean Margin Drilling Program — Regional Atlas Ser.*, 12, Marine Science International, Woods Hole, 1984.

Rapp, R. H., GEOS 3 data processing for the recovery of geoid undulations and gravity anomalies, *J. Geophys. Res.*, 84, 3784—3792, 1979.

Rapp, R. H., The determination of geoid undulations and gravity anomalies from SEASAT altimeter data, *J. Geophys. Res.*, 88, 1552—1562, 1983.

Rogan, A. M. and Rabinowitz, P. D., Gravity anomalies over the Mid-Atlantic Ridge, *Ocean Margin Drilling Program — Regional Atlas Ser. 11,* Marine Science International, Woods Hole, 1985.

Shum, C. K., Altimeter methods in satellite geodesy, Ph.D. dissertation, U. of Texas, Austin, 1983.

Talwani, M., New geophysical techniques for offshore exploration, *Oceanus,* 26(3), 17—22, 1983.

Talwani, M. and Le Pichon, X., Gravity field of the Atlantic Ocean, in *The Earth's Crust and Upper Mantle,* Geophys. Monogr. Ser. 13, P. J. Hart, Ed., Am. Geophys. Union, Washington, D.C., 1969, 341—351.

Talwani, M., Poppe, H., and Rabinowitz, P. D., Gravimetrically determined geoid in the western North Atlantic, Sea Surface Topography from Space, NOAA Tech. Rep. ERL-228-AOML 7-2, part 2, NOAA, Boulder, CO, 1972, 1—34.

Watts, A. B. and Leeds, A. R., Gravimetric geoid in the North Pacific Ocean, *Geophys. J. R. Astron. Soc.*, 50, 249—278, 1977.

Watts, A. B. and Daly, S. F., Long wavelength gravity and topography anomalies, *Annu. Rev. Earth Planet. Sci.*, 415—448, 1981.

Watts, A. B., Cochran, J. R., Patriant, P., and Doucoure, M., A bathymetry and altimetry profile across the Southwest Indian ridge crest at 31°S latitude, *Earth Planet. Sci. Lett.*, 73, 129—130, 1985.

Westbrook, G. and Jackson, R., Free air gravity anomalies of the Lesser Antilles arc and adjacent terrains, Ocean Margin Drilling Program, Regional Atlas Ser. 10, Marine Science International, Woods Hole, 1984.

Chapter 4

THREE-DIMENSIONAL SEA FLOOR AND SUB-BOTTOM VISUALIZATION

Paul K. Matthias

TABLE OF CONTENTS

I. INTRODUCTION

A new technique of applying three-dimensional rendering to an integrated product of side-scan sonar, bathymetry and sub-bottom data in real time has been developed. The computer-generated products are displayed in stereo 3-D. The virtual product is corrected for many geometric and radiometric distortions and supports improved geophysical interpretation.

Several researchers have described their analyses of collocated side scan imagery and swath bathymetry (Mougenot et al., 1984; Mitchell and Somers, 1989; Searle et al., 1990; Jany et al., 1990; Stewart, 1990; Mitchell, 1991; Hussong and Fryer, 1983). The comparison of these two data types is useful for locating features in the side-scan images and comparing the effect of topography on side-scan imagery. This chapter describes the fusion of these data types and the corrections made to the backscatter intensities to create a realistic 3-D sea floor image in real time.

The graphics workstation has become an indispensable problem-solving tool in the geophysical community. By joining analytical, computational, and graphics capabilities, workstations enable the creation of complex models and the interactive visualization of the models' products. Visualization allows modelers to view and manipulate data in spatial rather than numerical format and infer patterns and relationships based on visual interpretation.

Data fusion provides integrated synergistic views of multiple data products. Rather than simply displaying different data sets side-by-side, data fusion products merge these data sets into one interactive display. The resultant combination provides more information than the individual products collectively.

This study culminated in the application of high-speed computers and visualization software to fuse together and display virtual products of side-scan sonar imagery, bathymetry and sub-bottom data.

II. DATA

A. SIDE-SCAN SONAR

Side-scan sonars have been used to provide sonographs of the ocean floor since the early 1960s. This remote sensing device is subject to many radiometric and geometric distortions common to satellite remote sensing. Many of the same pre-processing techniques are used to correct these distortions.

Manufacturers of side-scan sonars have made gains toward correcting the radiometric distortions. The bulk of the geometric corrections and image extraction tools require coupling the sonar with a powerful top-side processor. Post-processing techniques used on the USGS MIPS system and real-time processing on Triton Technology's Q-MIPS™ have been described in the literature (Chavez, 1986; Danforth et al., 1991, and Newton, 1988).

A typical side-scan sonograph (Figure 1) usually employs a grayscale for depicting the relative variations in return signal intensities. The sonograph pixels are not corrected for many radiometric and geometric distortions including speed, crabbing, yaw, slant range, and water column.

The Q-MIPS sonar image processing system digitizes the analog signal at up to 750 kHz sampling rate. Slant range correction and water column removal are performed in real time. The imagery display may also be corrected for ship's speed. The data is stored on high-density 1.25 GByte Write Once Read Many (WORM) optical disks while it is displayed on the screen. Q-MIPS' data format includes the storage of 1024 8- or 16-bit pixels per channel per sonar "ping".

Q-MIPS sonographs yield clues to sea floor geology, feature geometry and material type. The operator may select from a wide variety of spatial and transform domain image

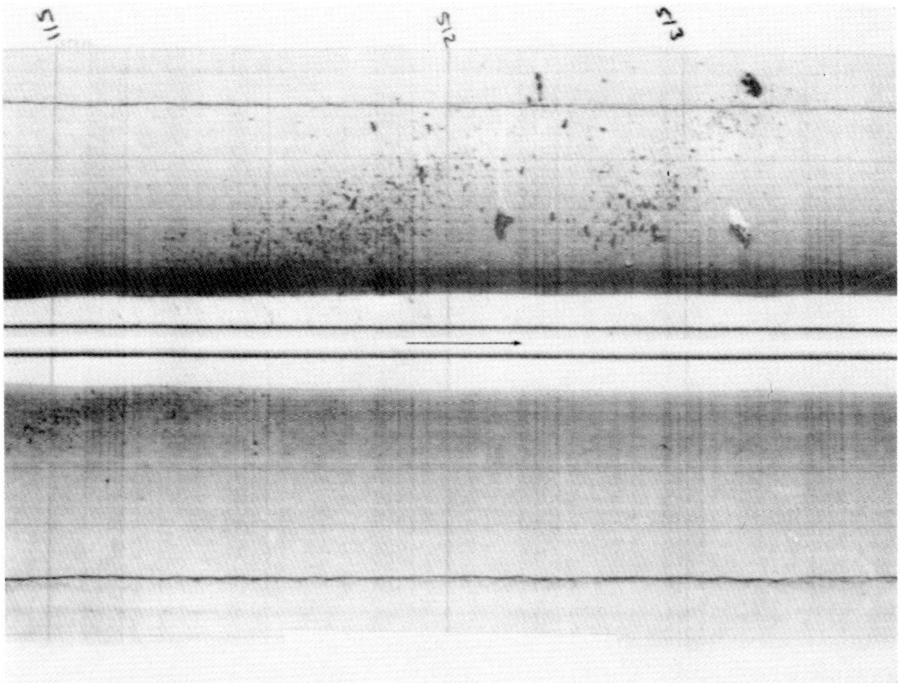

FIGURE 1. Typical side-scan sonograph hardcopy from C-130 aircraft survey.

enhancement routines, color manipulations and equalizations. Accurate estimations of lithology may be made based on return signal strength, represented by color on the Q-MIPS display. Tectonic features such as ridges, hills, and erosional features may also be mapped. Q-MIPS corrects the imagery for slant-range, beam angle compensation, and speed in real time. But 2-D side-scan sonographs are still subject to other radiometric and geometric distortions which have, until recently, made them useful only as a qualitative tool.

Efforts to make sonographs a quantitative geological prospecting tool include efforts to "correct" pixel intensities for variations in slope and grazing angles. The idea is to create a sonograph image whose pixel intensities vary with changes in lithology. This effort is one step toward making side-scan sonar a quantitative geophysical tool where pixel intensities reveal geology.

Historically, two major erroneous assumptions have been made in side-scan imagery systems:

1. The Earth is flat and horizontal.
2. Sonar "pings" are collected exactly perpendicular to the line of travel.

The first assumption leads to errors in the location of sea floor features. The sea floor is rarely flat and horizontal. With no corrections for variations in topography, imaged objects are displayed as if they existed on a flat, horizontal plane. To correct for this assumption, sonographs must be "warped" to a bathymetric model.

The second assumption leads to errors caused by the crabbing, yaw, and heading changes of the sonar "fish". These errors tend to misplace the depiction of sea floor features. It is analogous to cutting a photograph into random slivers and reconstructing it by laying the slivers parallel to each other. To correct this assumption, sonograph imagery pixels must

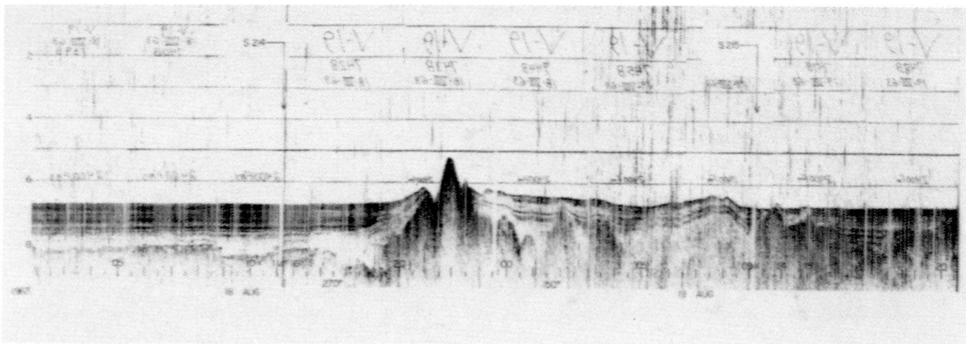

FIGURE 2. Typical sub-bottom profile over Chain Ridge, Indian Ocean.

be geocoded or Earth-referenced and displayed in true coordinates. That is, unlike the typical electrostatic line plotter, the sonograph display must accommodate oblique pings rather than laying them down in a raster line-by-line manner.

B. BATHYMETRY

Bathymetry data has traditionally been presented as depth soundings or contour maps. Modern ocean floor bathymetry data can be acquired through three general methods: depth soundings, multibeam echosounding, or swath acoustic interferometry.

Multibeam echosounding data have recently become available with the release of previously classified coastal Sea Beam data. Multibeam echosounding provides high-resolution data across a swath. The problem of accurately aligning these swaths into a chart, however, is an interactive, iterative process. It can be extremely difficult, costly, and time-consuming (Rosenblum, 1989).

Swath bathymetry has recently gained popularity as a less expensive alternative to multibeam echosounding. The conversion of the constructive interference nodes to bathymetric depths has been described (Lesnikowski, 1989) and is encoded for the Q-MIPS System. Q-MIPS stores bathymetric depths for the nodes on each half of the swath. One major advantage of swath bathymetry is that bathymetry values are gathered concurrent with side-scan imagery along discrete nodes. These nodes represent acoustic travel time to the bottom at increasing angles off vertical. After converting these times to slant ranges, Q-MIPS computes the depth below the towfish and offset distance from the centertrack in real time. Adding these depths to the towfish depth provides bathymetry values at each node across the swath.

C. SUB-BOTTOM

Sub-bottom profiles have been acquired and analyzed since the use of explosives in the 1950s. Seismic reflection profiles gathered with a sparker, boomer or air gun provide detailed information on the shallow sub-bottom. Seismic reflection records clearly show a discernible range of topographic features and sediment deposit character including ridges, troughs, outcrops, mudslides, onlap, offlap, pinchout, ripples, and sand waves. A "typical" seismic reflection profile using an explosive source is shown in Figure 2. This profile depicts Chain Ridge in the western Indian Ocean with Somali Basin to the left (west). Chain Ridge separates generally flat-lying sediments to the west from the more tectonically deformed sediments toward the Central Indian Ocean Basin to the east (Matthias et al., 1988).

Sub-bottom features exist in three dimensions. The geologist usually has to "interpret" the continuity or areal extent of these features profiled in only two dimensions. The geologist can refer to side-scan sonographs and charts of bathymetry, but classical geology requires a merging of all data types into one model.

III. DATA FUSION

Q-MIPS stores 1024 sonograph pixels per channel per ping while interferometric swath bathymetry usually generates a maximum of about 20 values per ping. Additional bathymetry values may be extracted from the sonographs when objects found on the bottom cast shadows. Given the length of the shadow on the ground perpendicular to the beam of the sonar and the range scale to the object, Q-MIPS computes the relative elevation of an object (Newton, 1988). The relative elevation is added to the interpolated depth based on swath bathymetry at the same location. This data set is gridded using an inverse distance squared method such that every sonograph pixel is associated with a bathymetry value. The merging of bathymetry and side-scan imagery is called ''Bathimagery™''.

Fusing the bathymetry data with side-scan imagery from Q-MIPS requires the use of a high-speed graphics workstation, the Silicon Graphics IRIS 210 VGX. The graphic rendering benchmark for this system is 1 million polygons drawn per second.

The most realistic Bathimagery rendering requires a shaded colorized surface in a perspective view. The perspective transformation produces more realistic scenes by making objects farther from the viewpoint appear smaller than closer objects. Clipping planes allow deletion of unnecessary details from the geometry. Double-buffering allows the user to create an alternate image in the background while one is viewing another image on the screen. Double-buffer mode splits the framebuffer into a front- and back-buffer allowing the user to control which buffer is displayed. This is used for generating animation sequences or ''fly-bys'' of the simulated seafloor. While the image is drawn into the back-buffer, the front-buffer is displayed. The buffers are switched in a rhythm of up to 30 frames per second.

The data fusion process requires the display of the Bathimagery pixels to be in real-world coordinates. That is, each sonograph pixel is displayed with a minimum of four attributes: Northing (or latitude), Easting (or longitude), bathymetric depth and sonograph pixel intensity. Other attributes that could also be included on a pixel level are other geophysical measurements including gravity and magnetic values.

The user interface to the Bathimagery product includes x, y, and z, azimuth, inclination, field-of-view, and resolution sliding bar controls. Changes in x, y, and z assume a right-handed coordinate system with z upwards. The azimuth and inclination slidebars change the perspective view window. That is, the operator may look left and right or up and down. Changing the resolution allows moving at greater speed over a decreased resolution Bathimagery product to any point over or under the simulated ocean floor. As soon as the operator stops moving, full resolution is restored automatically.

Vertical sub-bottom profiles may be added to the Bathimagery product. The merging of Bathimagery and sub-bottom data draws heavily on the use of color to aid in interpretation. By adjusting the user interface slidebars, a user may view multiple sub-bottom profiles at the same time. Plate 1* depicts a Bathimagery and sub-bottom profile product from Lake Washington, near Seattle. The profile is a time section gathered with an AMS 3.5 kHz acoustic sub-bottom profiler. The image depicts an area about 200 by 200 m.

IV. APPLICATIONS

The ability to view Bathimagery with sub-bottom profiles speeds geologic interpretation. Pertinent ocean floor data may be incorporated into one synergistic view. Subtle structural and stratigraphic features are enhanced that might otherwise be overlooked.

* Plate 1 appears after page 130.

Bathimagery merged with sub-bottom profiles allows us to map stratigraphic and tectonic features from one survey line to the next while viewing a simulated ocean floor. The areal extent of reflectors or features can be mapped given a simultaneous 3-D view of the ocean floor surface bathymetry, side-scan imagery, and sub-bottom profile.

Interactive Bathimagery provides a view of the sea floor environment for the Remotely Operated Vehicle (ROV) operator. The operator alternately switches from ROV telemetered TV or sonar views to a 3-D Bathimagery view for path planning and obstacle avoidance. For instance, while surveying wreckage or a construction site, the operator can look at the "big picture" while guiding the ROV.

Bathimagery may be used for analyzing and selecting sites for drilling rigs, power plants, and other coastal and offshore structures. The planned construction of channels, docks, breakwaters, bridge piers, and harbor facilities may also be supplemented by a 3-D Bathimagery view of the local terrain. Using a Bathimagery workstation, a variety of potential routes for laying a pipe or cable may be viewed. The coordinates of the route are logged to disk and used in laying the pipe or cable.

V. STEREO 3-D

A more realistic Bathimagery display may be viewed stereoscopically. Stereoscopic Bathimagery requires a stereo pair of images. If one slide is taken of a Bathimagery landscape displayed on the screen and another is taken after the object is rotated some small angle (typically 6°) around the z axis, a stereo pair is produced. This stereo 3-D pair may be displayed on a non-depolarizing screen or viewed with a hand-held viewer. The effect of 3-D is to give depth to the landscape, indiscernible without the stereo pair.

Alternatively, the stereo product can be viewed in real time on the display monitor. Both a left and right eye view are presented on the screen at twice the normal rate of display. An infrared emitter is connected to the frame buffer of the computer such that the alternate views cause the emitter to transmit a signal to a pair of special liquid crystal display (LCD) lens glasses. The LCD glasses have controller cards on both lenses that cause them to turn clear and opaque at the rate of display of the stereo pair of images. The result is that the left eye sees only the left eye image while the right eye sees the right. The resultant stereo 3-D imagery appears to project into and out of the screen. The user is cued to variations in depth using this method that would not otherwise be perceived. This stereo display may be viewed in real time as it is collected.

VI. CONCLUSION

Three-dimensional and stereo 3-D simulated views of the ocean floor Bathimagery and sub-bottom profiles make it possible to view the ocean floor much like a roadside outcrop or canyon wall. The operator can move over the surface, generate animation sequences and interactively make interpretations of the areal extent of geologic features. As collocated data become available, gravity, magnetics and well-log data will be added to the Bathimagery display.

REFERENCES

Chavez, P. S., Jr., Processing Techniques for Digital Sonar Images from GLORIA, American Society for Photogrammetry and Remote Sensing, (August 1986) vol. 52, No. 8, 1133—1145.

Danforth, W. W., O'Brien, T. F., and Schwab, W. C., Near-Real-Time Mosaics from High-Resolution Side-Scan Sonar, *Sea Technol.*, (January 1991) 54—59.

Hussong, D. M. and Fryer, P., Back-arc seamounts and the Sea Marc II Seafloor Mapping System, *EOS Trans. Am. Geophys. Union,* 64 (45), 627-632, 1983.

Jany, I., Scanlon, K. M., and Mauffret, A., Geological interpretation of combined seabeam, GLORIA and seismic data from Anegada Passage (Virgin Islands, North Caribbean), *Mar. Geophys. Res.,* 12, 173, 1990.

Lesnikowski, N. S., Deep Towed Interferometric Swath Bathymetry, Oceans '89 Proceedings (September 1989) 1130—1133.

Matthias, P. K., Rabinowitz, P. D., and Dipiazza, N., Sediment Thickness Map of the Indian Ocean, map and accompanying text, AAPG (1988).

Mitchell, N. C., Improving GLORIA images using Sea Beam data, *J. Geophys. Res.,* 96, 337.

Mitchell, N. C. and Somers, M. L., Quantitative backscatter measurements with a long-range side-scan sonar, *IEEE J. Oceanic Eng.,* 14, 368-374, 1989.

Mougenot, D., Kidd, R. B., Mauffret, A., Regnauld, H., Rothwell, R. G., and Vanney, J. R., Geological interpretation of combined Sea Beam, GLORIA and seismic data from Porto and Vigo Seamounts, Iberian Continental Margin, *Mar. Geophys. Res.,* 6, 329, 1984.

Newton, F. L., Jr., The Consummate Search/Survey System — New Developments in Sonar Image Processing, Under Seas Defense '88 Conference Proceedings (October 1988) 158—167.

Rosenblum, L. J., Visualizing Oceanographic Data, IEEE Computer Graphics and Applications (May 1989) 14—19.

Searle, R. C., Le Bas, T. P., Mitchell, N. C., Somers, M. L., Parson, L. M., and Patriat, P., GLORIA image processing: the state of the art, *Mar. Geophys. Res.,* 12, 21, 1990.

Stewart, W. K., Three-Dimensional modeling of seafloor backscattering from sidescan sonar for autonomous classification and navigation, Proc. Symp. Unmanned Untethered Submersible Technol., 1989, preprint.

Chapter 5

TRACKPOINT II AND ULTRA-SHORT BASELINE NAVIGATION SYSTEM

Byron R. Selman

TABLE OF CONTENTS

I. TYPES OF UNDERWATER ACOUSTIC TRACKING AND NAVIGATION SYSTEMS

There are two basic types of systems used today to track or locate beacons attached to underwater targets, whether the targets are ROVs, sonar towfish, manned submersibles, divers, or any kind of instrumentation packages. These systems are long base line (LBL) and ultra-short base line (USBL) tracking systems. A third type of system — the short base line tracking system (SBL) — has almost completely been replaced by USBL technology. In each type of system, the ''base line'' referred to is the known and fixed distance between the number of acoustic elements of the tracking system. The choice between these two types of systems depends upon the operational scenario and desired results.

An LBL system consists of a number of acoustic transponders arranged in a local grid on or near the sea floor (Figure 1). In this type of system, a surface ship with a single interrogation/receive transducer can navigate itself as well as separate, moving targets in relation to the fixed transponder grid. Essentially, the system continuously calculates the ranges from the object being navigated to the transponders, using the known velocity of acoustic signals through the water. This is also known as a ''range-range'' system, because it measures and uses range information only in solving the triangles required for a position solution. The ''long base line'' in this case is the known distance between the fixed transponders on the sea floor. The major advantage of this type of system is the inherent accuracy advantage of the longer base lines in the triangulation process. The major drawbacks to this type of system are the logistics of installing and maintaining a local sea floor transponder network, and being limited to working only within this grid.

To eliminate the requirement of the network of sea floor transponders, the short base line system came into use (Figure 2). In a typical short base line application, a number of acoustic receivers are suspended from the surface ship or platform, and the system uses the ''range-range'' acoustic triangulation method to determine the relative position of some number of undersea targets. Here, the surface ship or platform becomes the local grid, and the ''short base line'' is the fixed and known distances between the receiving hydrophones. A major advantage of the SBL system over the LBL system was that the system was no longer limited to working within a single fixed geographic transponder grid. The grid is attached to the moving vessel, and all of the undersea tracking is relative to this grid. Problems became evident in the logistics of designing and installing suitable hydrophone mounts on the surface vessel, along with the inherent decrease in the ability to accurately determine the position of targets located at relatively long distances from the acoustic grid.

In the 1980s, the ultra-short base line system came into use, and has today almost completely replaced SBL use. The USBL (Figure 3) system is a direct result of the computer age and utilizes completely different technology than either the LBL or SBL systems. A set of acoustic receiving elements are mounted in a fixed and known pattern within a single hydrophone assembly, physically spaced within a half wavelength of the system's operating frequencies. This means that they are typically a fraction of an inch apart. The close spacing of these elements is the fixed and known ''base line''; hence the definition of this type of system as ''ultra-short base line,'' or ''super-short base line.''

Operationally, the USBL system uses phase comparison techniques to determine both the horizontal and vertical angles at which incoming acoustic pulses impact the receiving elements of the hydrophone. The system is also capable of measuring and using the range to certain acoustic beacons in solving triangles. It is therefore referred to as a ''range-bearing'' system rather than a ''range-range'' system.

The ability of the USBL system to track multiple acoustic targets from a single hydrophone fixture gives it several potential advantages over other types of navigation systems. The basic USBL system consists of a command/display module, a single interrogate/receive

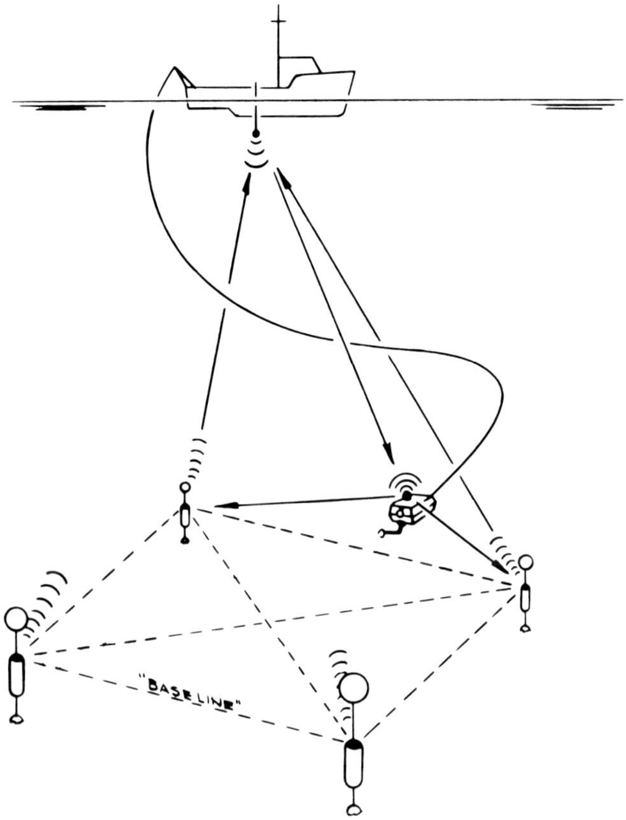

FIGURE 1. Long base line (LBL) system.

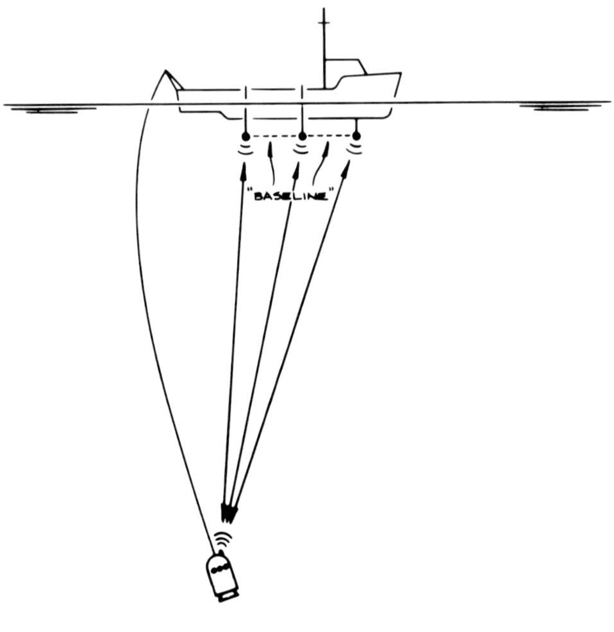

FIGURE 2. Typical short base line (SBL) system.

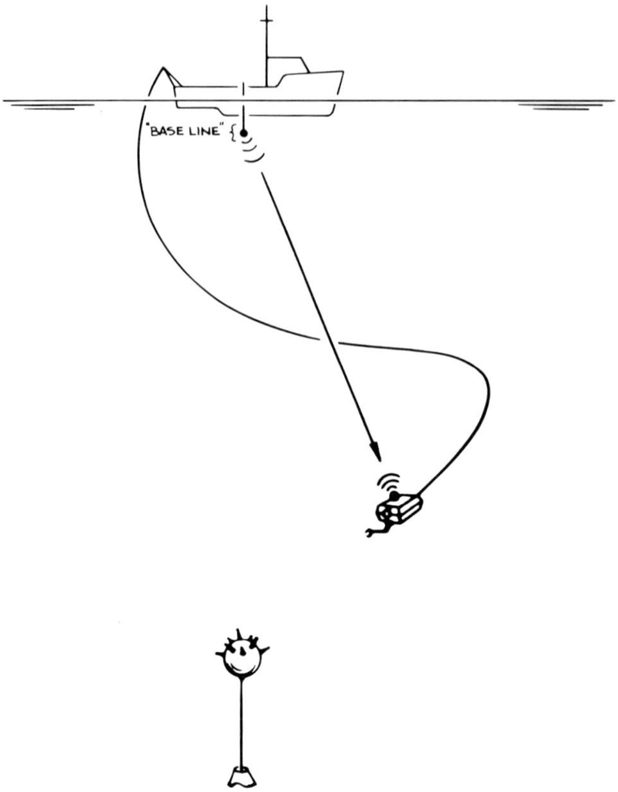

FIGURE 3. Typical ultra short base line (USBL).

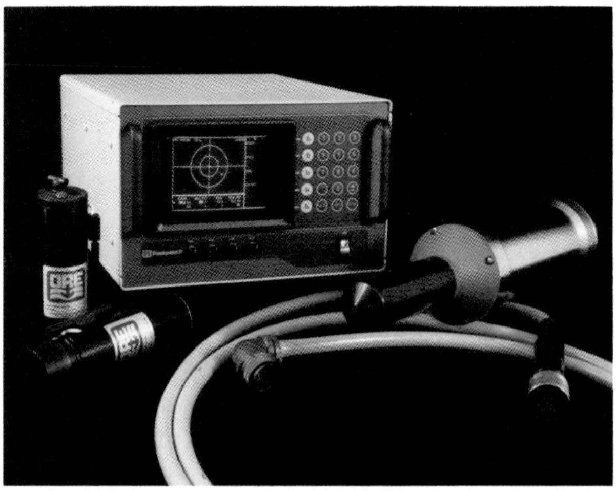

FIGURE 4. Basic USBL system.

hydrophone assembly, an interconnect cable, and the small acoustic transponders, responders, or pingers that are attached to the target (Figure 4). This yields a very compact and mobile system that can quickly and easily be installed on almost any sort of platform or vessel. The Super Phantom O.R.E. Trackpoint II, for instance, is easily transported and installed by one person. This type of system can be operated from small craft as well as

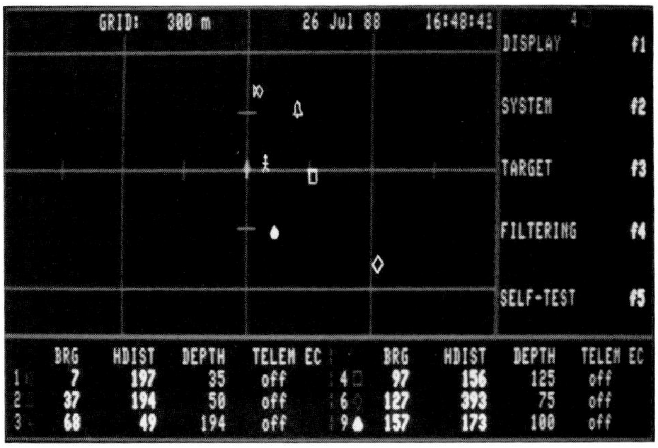

FIGURE 5. Six targets may be tracked and displayed.

from large surface ships. When a compass is installed in the system's hydrophone, the USBL system can even be operated from a hovering helicopter. The installation requirements of the hydrophone are only that it be mounted so that it has sufficient clear water radius (typically on the order of three feet) around the receiving elements, and that the unit be below the keel of the vessel if omnidirectional tracking is desired.

The major advantages of the USBL system are its low cost, portability, flexibility, and ease of operation. Use of an LBL system would be desirable in operations that require the highest achievable accuracy of position of underwater targets located far from the receiving hydrophone, since the USBL accuracy of the horizontal position calculation decreases at longer ranges. However, for the majority of sub-sea tracking, navigation, and relocation tasks, the USBL system is the ideal tool. The Trackpoint II system, for example, will operate with any combination of acoustic transponders, responders and pingers. These acoustic devices are not much larger than a typical beer can and can be attached to almost anything that needs to be tracked. The ability of the system to operate with three different types of acoustic beacons gives the operator a flexibility in choosing the right beacon for a specific task.

II. THE TRACKPOINT II SYSTEM

The Super Phantom O.R.E. Trackpoint II is an integrated, ultra-short base line acoustic tracking and navigation system designed to operate with transponders, responders or free running pingers for a wide range of sub-sea navigation and relocation tasks. The system is microprocessor-based, and consists of a command/display module (C/DM), a combination interrogate/receive hydrophone transducer assembly, and the required length of interconnect cable. The system also decodes and displays simple telemetry information from suitably equipped telemetry transponders. Configurations for ROV and towfish tracking, vessel positioning, and subsea relocation are operator-controlled using a keypad entry system.

Trackpoint II presents the user with a video display of the underwater position of the target or targets relative to a chosen reference point on the surface vessel. In addition to the graphic display of target position, Trackpoint II displays digital values for azimuth and range to each target. Six targets may be tracked and displayed simultaneously (Figure 5).

The C/DM is a one-man portable electronic cabinet and is usually mounted in the electronics operation area, navigation bridge, or Combat Information Center of the ship. It can be mounted to a suitable chart table or bench or in a standard instrumentation rack. The unit contains the system microprocessors, software, and required electronics to perform the acoustic measurements and calculations required for determining the position of tracked targets and generates the system outputs and display. The system operator typically controls the Trackpoint II from this unit, using a combination of system input connectors and the integral operator keypad for data entry and control. The system can also be completely controlled via a standard RS-232C serial data port. Some optional versions of the Trackpoint II are specially packaged in watertight housings for installation in remote vehicles. Those versions require use of the serial data link for control and to receive tracking data.

The interrogate/receive hydrophone assembly is a single, one-man portable cylindrical unit that is mounted in the water below the surface vessel or platform. This unit transmits acoustic pulses to interrogate transponders and receives the acoustic signals from pingers, transponders, or responders. This single assembly actually consists of an array of individual hydrophone elements that are used in the ultra-short base line technology upon which the system operating architecture is based. The assembly is normally mounted on an overboard pole, or through a gate valve in the hull of a surface ship, positioned below the hull so that there is always a direct and clear ''line-of-sight'' between the hydrophone transducer and the acoustically marked target being tracked. The hydrophone is omnidirectional and will receive and process signals arriving from any direction throughout the lower hemisphere of operation.

A vertical reference unit (VRU) is used in conjunction with the hydrophone assembly. The VRU consists of accelerometer-based inclinometers mounted perpendicular to each other. The VRU is used to measure and remove vessel pitch and roll distortion from the tracking data. The hydrophone assembly is available in a number of configurations. One combines the VRU and amplifier with the interrogate/receive functions in one unit with a diameter of 6 in. Another houses the VRU and amplifier in a remote surface unit, which results in a more hydrodynamic and less expensive 3-in. interrogate/receive hydrophone for underwater deployment. The 3-in. assembly can be faired to reduce strumming, allowing the vessel to travel at higher speeds with the hydrophone deployed.

The interconnect cable consists of a cable assembly with sufficient conductors to connect the C/DM to the interrogate/receive hydrophone assembly. This cable can be supplied in various lengths and must be long enough to reach from the C/DM location down to the hydrophone mounted below the ship.

In operation, the C/DM emits an interrogation pulses through the hydrophone assembly which the acoustic transponder receives, replying with its own discrete acoustic signal. When the Trackpoint II receives the transponder's reply, it instantly calculates the slant range to the transponder and uses phase comparison technology to determine both the azimuth and depression angles for the incoming signal. The operator on the surface craft sees a full color video display of the plan view of the area of operation.

All locations or underwater objects marked with transponders are depicted as brightly illuminated symbols, which are continuously updated as they move through the area. The operator can read the range and bearing from the surface vessel to each of the locations and/or objects, and can also choose to have the system display the range and bearing from one object or location to another. The Trackpoint II operates in several frequency bands, including 8 to 14, 22 to 30, and 28 to 40 kHz. This provides a great deal of flexibility in both high acoustic ''noise'' and long range applications. The standard system software is capable of tracking suitable acoustic sources to ranges of 32,000 ft (10 km). If a signal can be ''heard'' by the Trackpoint II, the system can compute a position.

FIGURE 6. USBL integrated with surface system.

III. APPLICATIONS FOR THE ULTRA-SHORT BASE LINE SYSTEM

A. TRACKING ROVS

USBL technology was developed primarily to track and navigate underwater remotely operated vehicles (ROVs). ROVs have been most commonly used as underwater video camera platforms to enable surface operators to visually inspect submerged objects. The ROV has long been used extensively as an inspection tool for pipelines, drilling rigs, and submerged oil well control structures. Lately the ROV has been incorporated into military operations to visually identify and classify suspicious undersea objects located by active sonar, and in some cases to neutralize mines with explosive charges. In all these operations, the USBL system is invaluable in coordinating the ROV operation. Without a system to enable the ROV pilot to know where the submerged and unseen ROV is located relative to the ship, underwater video searches would be largely ineffectual.

B. TRACKING TOWED SONAR ARRAYS

The USBL system is being used extensively now to track various kinds of sonar arrays. To accurately determine the actual position of targets located with side-scan sonar, for instance, it becomes as important to navigate the sonar sensor platform (usually a "towfish") as it is the surface vessel. With the effects of wind and currents on the "crab angle" of a survey ship attempting to navigate a predetermined course through a survey area, the actual position of the side-scan towfish can be significantly outside the desired survey lane. This becomes even more important when sonar operators are tasked with comparing sonar records with records taken on the same routes on previous surveys. Precise navigation of the sonar platform enables the operators to obtain records from the same sensor position and make comparisons of the bottom features much more easily and completely. When integrated with a surface navigation and plotting system (Figure 6), the USBL system also makes certain that the survey has not missed any bottom coverage, which would create unsurveyed blind spots.

C. RELOCATION AND RECOVERY OPERATIONS

In years past, directional hydrophone-type "pinger relocators" were used to obtain bearings to lost objects carying compatible pingers (such objects include ROVs, manned submersibles, instrumentation packages of various types, flight data recorders, etc.). Searching with bearing-only information usually took hours, or even days. Relocation and recovery time is drastically reduced with the USBL system, giving the searcher not only x, y, and z coordinates to the lost object, but also allowing him to visually monitor the respective locations of the object and the recovery device on an easily interpreted video screen throughout the entire operation. This allows the operator to reliably ''home in'' on the object of the search.

D. MARKING SUB-SEA STRUCTURES

When instrumentation is placed on the sea floor for extended periods of time, it is usually desirable to be able to relocate these instruments for purposes of maintenance, replacement, or recovery. By attaching or incorporating a long-life acoustic transponder to the structure or instrumentation package, the installer gives himself the ability to quickly relocate and recover the object using USBL tracking systems. The acoustic transponders remain quiet for periods of months or even years, listening for the correct interrogation pulse from the navigation system before actively replying and allowing the tracking system to receive their reply signal and calculate and display the position of the transponders. Use of the USBL system also allows the subsea instrumentation to be correctly installed in the desired position.

E. COORDINATING DIVING OPERATIONS

A USBL system can prove extremely valuable in coordinating the movement of divers below the surface vessel. Personnel on the surface are able to monitor and direct divers during search and training missions by attaching a small transponder to the diver's harness. This capability greatly enhances the effectiveness of the divers by eliminating wasted bottom time usually spent searching an area in low-visibility conditions.

Ideally, the USBL system is integrated into a surface vessel navigation system, and the ship maintains position over the desired search area and directs the divers with voice communication throughout the operation. If an acoustic beacon is placed somewhere in the search area, redirecting disoriented personnel back to the site becomes extremely simple, and the value of diver tracking and relocation for personnel safety reasons should be immediately apparent.

F. TELEMETERING SUB-SEA SENSOR DATA

The latest generation of small acoustic mini-transponders manufactured by Super Phantom O.R.E., Inc. are called Multibeacons and have an inherent telemetry capability that makes the USBL system even more versatile for a variety of applications. The Multibeacon is capable of taking a signal from a subsea sensor and acoustically sending that information up to the surface ship or platform. In addition to tracking and navigating the transponder, surface personnel can monitor such parameters as pressure depth, pitch and roll, heading, temperature, salinity, tilt, current velocity, or any of a number of desired sensor readings. The software in the Trackpoint II decodes the received telemetry information and supplies it to an external computer or data logging device.

The potential for this capability is enormous. In ROV navigation, the operator can telemeter pressure depth directly to the tracking system and thereby free up valuable conductors in the vehicle's umbilical cable, allowing them to be utilized for other purposes or eliminated entirely. (Decreasing the size of the umbilical is highly desirable when attempting to enhance the performance of the ROV in deep-water or high-current situations.)

In navigating towed sensor platforms or sonar towfish, the operator has the capability of monitoring heading and tilt of the towed array without resorting to hard-wired sensor packages. Knowledge of the orientation of the towed sensor array greatly aids interpretation of the data collected by that platform.

In deploying instrumentation, sensor packages, or undersea structures on the sea floor, personnel can use Trackpoint II's telemetry capability to determine the orientation of the packages on the bottom. Multibeacons can be used to monitor sensor depth, cable strain, platform tilt, and directional hydrophone orientation while the installation vessel is still in a position to modify the deployment. In the case of long-term deployments, the telemetry transponder will allow the USBL system to monitor such concerns as corrosion protection, the amount of sedimentation over a structure, and remaining battery life. With the use of a long-life telemetry package, this information can be monitored on a regular basis by a small vessel passing by, or even during a brief hover by a helicopter equipped with the tracking system.

In diving operations, the physiological condition of divers can be monitored by using telemetered information on heart rate, core temperature, and respiration rate of the divers while underwater, without having to rely upon communication by, or judgement of, the diver himself. This ability of the Trackpoint II with telemetering Multibeacons to both track and monitor the condition of undersea personnel is, of course, most valuable in hazardous conditions.

IV. TRACKPOINT II AT WORK

The potential of the state-of-the-art commercial USBL tracking and navigation system is finally being realized by the offshore commercial, scientific, and military communities. The latest generation of multiple-frequency, microprocessor-controlled, telemetry-decoding systems gives the technology a vastly increased value. It is in the form of a relatively inexpensive yet extremely versatile and portable tool for a vast array of underwater operational applications. Following are some examples in which Trackpoint II has been a successful problem-solving tool.

A. TRACKING SEISMIC HYDROPHONE CABLES

In 1989, John E. Chance and Associates (JECA) of Lafayette, LA, a well known navigation and survey company in the Gulf of Mexico ''oil patch'', devised a unique application. For accurate seismic signal processing, JECA wanted a cost effective way to precisely position the tail end of a seismic hydrophone cable, which is often several kilometers behind the seismic vessel. JECA had developed a platform known as a ''tail buoy'', which is outfitted with a surface positioning system with a telemetry link and an impeller-driven generator (the JECA Buoytrack system). It is towed on the surface some distance behind the end of the hydrophone streamer, allowing the system operator to obtain an accurate geodetic location of the tail buoy on the surface. This was a step forward, but JECA also wanted to determine where the subsurface end of the hydrophone is located. To that end, the Trackpoint II engineering staff completely repackaged a set of system electronics into watertight containers which could be mounted inside the tail buoy, with power supplied by the DC generator on the buoy (Figure 7). The tracking system receives all operator commands and transmits tracking data via the bi-directional radio telemetry link. Trials of the prototype system are underway.

B. TOWFISH TRACKING

Trackpoint II was used for tracking towfish on the Woodside Offshore Petroleum Route Survey for the development of the Goodwyn Field off the northwest coast of Australia in 1988. This was the first time that extensive fish tracking was carried out in this area. The initial results showed that the towfish does not ''fly'' directly behind the vessel in a straight line, as earlier assumed, but rather ''swims'' down the line. Oil companies now prefer that all subtow vehicles have some form of USBL acoustic tracking, so the location of the sub-sea sensor platform can be known at all times.

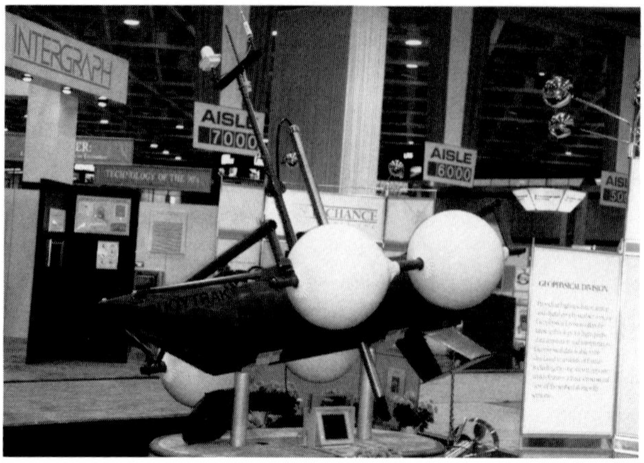

FIGURE 7. USBL mounted inside tail buoy.

C. PIPELINE SURVEYING

In 1986, Marine Technical Services (MTS) of Houston, TX, conducted a pre-route pipeline survey in the Cook Inlet, AK, incorporating the Trackpoint II into their range of instrumentation. In this survey MTS expected both the ship and the sidescan to "crab" along the survey lines to prevent being constantly swept down-current. Accurate towfish tracking was needed to correlate side-scan data to the actual towfish position. A transponder was attached to the tow cable 3 ft above the sidescan towfish, and the Trackpoint II data was transmitted directly to the navigation computer, along with the ship's gyro heading. An x-y position fix of the towfish was calculated and plotted in real time.

D. ARCTIC UNDER-ICE SURVEYING

Arctec, Inc. developed a successful technique to accurately survey under-ice topography using an ROV outfitted with a Trackpoint II system and a responder. The I.S.E. DART ROV incorporates a combination of upward-looking sonar, vehicle depth, and x, y coordinates from the Trackpoint II to generate topographic maps of the underside of Arctic ice floes.

The USBL system with a wired-in responder minimizes acoustic confusion in the extremely high multipath environment between the rough under-surface of the ice and sea floor. After post-processing of the data with Arctec's own integrated software package, RS-232 tracking data supplied by the Trackpoint II results in a navigational accuracy of ± 1.5 ft. The DART has been tracked at ranges out to 850 ft (limited by the ROV tether), even with the vehicle responder sometimes 15 to 20 ft shallower than the Trackpoint II hydrophone. The Trackpoint II has performed without problem for weeks at a time in extremely harsh Arctic conditions.

E. MARINE BIOLOGICAL RESEARCH DEMONSTRATION

In 1987, the Trackpoint II was evaluated by the NOAA National Undersea Research Program at the University of North Carolina at Wilmington, together with representatives from other government agencies. The Trackpoint was part of a super phantom ROV package being demonstrated for applications such as archaeological inspection, marine life surveys, and sea floor inspection. The demonstration site was the wreck of the *Virginias,* a steam powered blockade-runner dating from the Civil War.

As the ROV inspected the wreck, the scientific observers were impressed with the system's unlimited bottom time, cold water capability, and ability to easily generate a permanent video record. The biologists on board also were able to confirm that the ROV did not disturb the native fish population of the wreck. Both the ROV and the Trackpoint II system worked perfectly in the relatively shallow water (45 feet). It would be even more valuable for work in waters too deep for divers. The tracking system repeatedly navigated the ROV back to specific areas of the sea floor after it was deflected from the wreck site by the relatively strong current, and with the pilot hampered by poor visibility.

Chapter 6

AIRBORNE GRAVITY MEASUREMENTS

William R. Gumert

TABLE OF CONTENTS

I. INTRODUCTION

Accurate airborne gravity measurements were an elusive goal for many years. Both government agencies and private industry had requirements that only an airborne system would meet. Numerous tests have been made beginning in the late 1950s.

A system using a helicopter, 30 to 70 knots (50 to 125 km/hr), was perfected in the early 1970s, a slow flying airplane system, 80 to 130 knots (145 to 235 km/hr), became available in the early 1980s, and an over water fast airplane system, 200 to 320 knots (360 to 375 km/hr), was flown in the mid 1980s. Entering the last decade of the 20th century an airborne gravity system is available for every requirement of gravity data from high mountainous regions to deep ocean trenches.

Data from surveys flown using the helicopter and slow fixed-wing system are presented. Accuracies of these systems when compared to ground data are between 1 and 2 mgals.

II. AIRBORNE GRAVITY MEASUREMENT

A. EARLY FIXED WING TEST RESULTS

The first reported tests of airborne gravity measurements in the late 1950s (Thompson et al.) and 1960 (Nettleton et al.) were feasibility trials. Available fixed-wing aircraft were used. Systems were installed consisting of marine type gravity meters, available photogrammetric cameras and/or doppler radar for positioning and some unique hypsometers for vertical variation measurements. All experiments consisted of a few lines flown at relatively high altitudes. In general, the gravity meters performed adequately but the system limitations were in navigation control to measure and calculate the Eotvos correction and in flight stability to limit and/or measure precisely the vertical movement of the aircraft.

Many of the computations were hand calculations of an enormous amount of numbers along the few flight lines. Digital techniques were introduced (Coons et al.) to attempt a faster and more precise computation of the corrections to find free air gravity data.

The results of these tests from 1958 to 1962 indicated that an accuracy of 10 mgals were obtainable by airborne gravimeters.

In January 1965, the APFCRL started an intensive program to evaluate and flight test those existing or easily modified gravimeters that had the capability for accurate gravity measurement. The flight tests were completed in May 1968 (Anthony et al., 1971). Results were reported to be no better than the 10 mgals mentioned in the earlier reports. Again, most error sources were believed to be in the navigation and altimetry systems.

B. HELICOPTER GRAVITY MEASURING SYSTEM

The first attempt at measuring gravity from a helicopter was made in 1963 by the Gravity Meter Exploration Company and Lacoste & Romberg, Inc. No successful results were reported (Nettleton, personal communication). In 1965 the U.S. Naval Oceanographic Office (NAVOCEANO) made helicopter flight tests with a gimbal-suspended LaCoste & Romberg Sea Gravimeter at Fort Hood, TX. Several good gravity values were measured in the hover mode. The values compared to within 5 mgals when reduced from the 15 m hover height to the ground (Gumert, 1970). In 1966, NAVOCEANO equipped a helicopter with a gimbal LaCoste & Romberg Sea Gravimeter and a prototype Aero Service-Spectra-Physics laser altimeter. The helicopter's position was determined by radar tracking from a single NASA radar station located at Wallops Island, VA. Forty hovers were made in Maryland and Virginia with accuracies of 3 mgals. Good results were obtained at speeds from 140 to 280 km/hr which produced line crossing with differences of only a few milligals. The profile tests demonstrated that the helicopter is a stable test bed. This NAVOCEANO test marked the beginning of a tri-service effort to develop a helicopter gravity measuring system (HGMS) (Gumert et al., 1970a).

In November 1968 the U.S. Army Topographic Command (TOPOCOM) made tests using an Air Force CH3E helicopter equipped with a LaCoste & Romberg stable-platform gravity meter, a Spectra-Physics laser altimeter, a Rosemont pressure port calibrator, a 35 mm strip camera, a HIRAN navigation system, a LANCER digital data logger with the necessary interface, and analog recording monitors. The HGMS system was test flown in the Maryland, Virginia, and Delaware areas. About 100 flights were completed in 2 $^1/_2$ months. In 1969 additional test flights were made in the vicinity of Topeka, KS, to further evaluate the HIRAN navigation system and the laser altimeter.

Results of these tests showed cross-tie values of 3 to 5 mgals at line intersections. Adjacent flight lines in the grid pattern showed anomalies through the pattern at the proper spacing and with the correct amplitude. Two lines were flown along the same flight path over a 35-mgal domal anomaly in Virginia. These data showed the anomaly at the amplitude and with the period of the known ground feature.

C. COMMERCIAL TESTING

In 1970, a private company, ENTECH, proposed that it might be possible to modify the HGMS program of the Department of Defense (DOD), to create a system that would meet the more stringent requirements of the oil and mining industries (Iverson, personal communication). In the fall of 1970, an agreement was signed and a program started. A system was constructed and tested during February and March 1971. Flights were flown from 15 to 60 knots in contrast to the DOD program speeds of 70 to 150 knots.

The basic aim of the test program was to create a helicopter-transported gravity and magnetic surveying system to be used for petroleum exploration. A system was required that would measure accurately to two mgals and would respond to anomalies with wavelengths of 3 miles in horizontal dimension. Special evaluation of the altimetry system was required to determine the precision of measurements. Flights were to be made at various speeds to determine where the cut off between accuracy and response, as outlined above, was reached.

The test area was a 20 nautical-mile square, northeast of Rockville, MD, bounded on the east by Baltimore and on the west by Frederick, MD. This area had been surveyed with ground gravity equipment by several U.S. Government agencies; and EnTech established 212 additional land gravity stations (Gumert, 1974).

A viable tool was developed for use in petroleum and mining prospecting in all types of areas. With this airborne system, data could be collected over sea, land, and the sea/land interface. Remote areas and difficult terrain offered no barrier to a quick-look survey using two important geophysical prospecting principles, gravity and magnetics.

Dr. L. L. Nettleton's report on this test concluded:

> Gravity values from the airborne data with this constant shift for each line and at 100-second intervals were placed on a map and contoured. Values from the filtered ground data at the same points were placed on another map and contoured independently. The results were shown on a single map in which both sets of gravity data were shown at each point together with the two sets of contours at 5 milligal intervals. On the whole, the gravity pattern over the area is shown quite well and the same contours are generally parallel and for the most within 1 or 2 milligals of each other. Furthermore, an histogram of the differences in values from air and ground data at the same point shows a generally normal distributon.
>
> On the whole the results of the test have shown that the gravity measuring and accessory instrumentation have been combined into a system capable of mapping moderate to large gravity anomalies possibly indicative of geologic structures.

From August 15 to September 9, 1974, the Geoscience Department of Carson Helicopters, Inc., demonstrated a helicopter gravity system with a S-61 helicopter (Gumert, 1977).

A survey site was selected on the East Coast over a gravity high in southern New Jersey centered around the town of Milville. Carson established 137 land stations at 2-mile intervals in the demonstration area to further define the anomaly for comparison to the airborne information.

The system flown consisted of the LaCoste & Romberg three-axis stable platform gravity meter, laser-pressure altimeter system and the Del Norte Trisponder navigation system all mounted in a Sikorsky S-61 helicopter with flight stabilization system and a special pressure probe. Data were collected on magnetic and printer tape and also on analog charts.

The laser reference altitudes were corrected to mean sea level using readings on the Delaware River at the west side of the survey area and on the Atlantic Ocean on the east side. Other altitude checks were made over the Milville Airport, three rivers and several lakes in the area.

The navigation base line was approximately 70 km long with each transponder on a tower 100 ft above the ground.

A report concluded that their helicopter gravity and magnetic system could be deployed worldwide to measure medium to large anomalies (Gumert et al., 1982). The tests showed that an airborne system now existed that could provide two basic geophysical techniques, gravity and magnetics, with a rapid mapping capability for reconnaissance in remote areas.

D. SUMMARY

From 1963 to 1974, most of the effort for the development of an airborne gravity system came from the mapping and military branches of the U.S. Government. The oil industry, however, did sponsor several small tests, so that it could stay abreast of the development of an airborne gravity system.

Since 1974, only one company has been involved in the development and use of an airborne gravity system for commercial exploration, (Gumert 1985). During the period from mid-1979 to the end of 1990, more than 500,000 line kilometers of airborne gravity and magnetic data have been flown in areas as divergent in terrain and geology as the Southwest Coast of the U.S., the Mexican Gulf Coast region, the Alaskan Arctic Coast, the Atlas Mountains of Morocco, the Amazon Basin of Colombia, the jungles of New Guinea, the Arctic Coastal Plain and the Bahamas Island Group (Gumert et al., 1982).

To complete this brief history of airborne gravity development the tests and program of the Naval Research Lab (NRL) (Brozena 1984; 1988; 1990) are referenced. Using a larger P-3A aircraft, NRL has collected large amounts of data over ocean areas. Universities and other government agencies are using this information to study regional geology and large-scale tectonic structures.

III. SURVEY RESULTS

Numerous papers have been given (see references) to describe systems used to collect airborne gravity and magnetic data, and many studies have been made to determine the system accuracies by both private oil companies and government agencies. The final outcome is that airborne gravity data has been collected in 80 prospective areas around the world and that these data have been used to evaluate geological structures with petroleum potential and to guide the collection of more expensive exploration information, including the collection of seismic data.

A. OFFSHORE CALIFORNIA

In early 1983 a helicopter gravity and magnetic survey was flown over the Santa Maria Basin located along the southern Pacific Coast of California (Figure 1). The survey was flown as a 1.5 mile (2.78 km) grid at 2500 ft (762 m) above sea level with a ground speed of 50 knots (90 km/hr). The observed data had an intersection mistie array with a standard deviation of 2.2 mgals. After systematic first order correction, the approximately 1400 intersections have a standard deviation of 1.4 mgals.

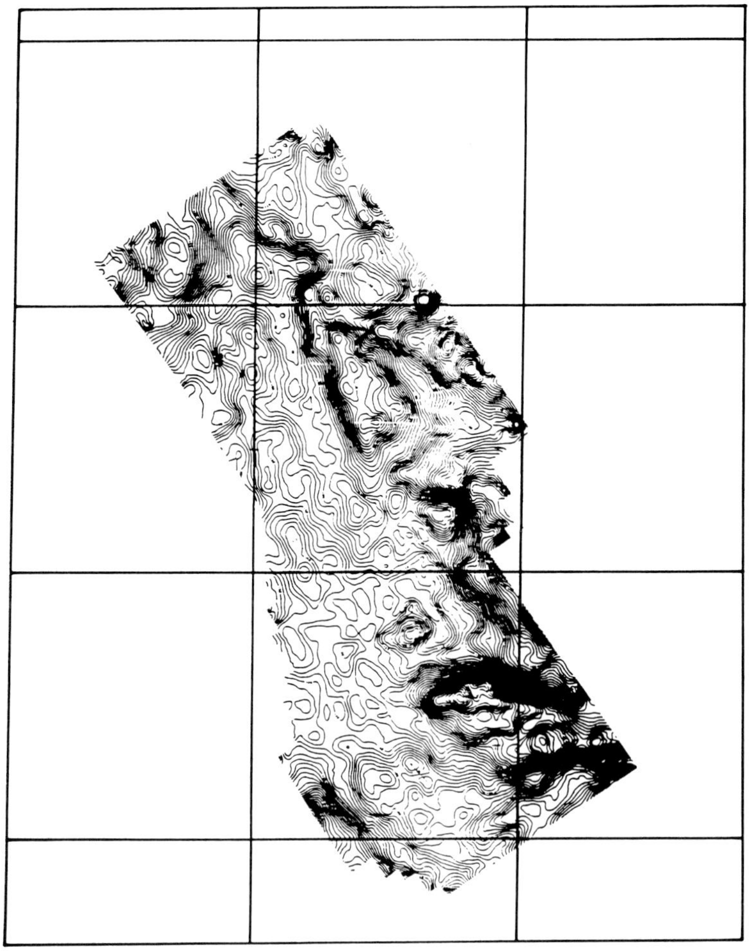

FIGURE 1. Bouguer gravity density 2.0 G/CC.

This area is a mature petroleum exploration region. Several existing oil fields are producing (Figure 2), and there have been many geophysical exploration efforts using all disciplines. Questions remained about the geological setting, primarily in connecting the known onshore structures with the known offshore trends. The land-water interface needed to be surveyed and the airborne system was the only acceptable technique.

Reports from oil companies who used these airborne data concluded that the airborne gravity data were better than the existing marine gravity data and a report (Harrison, verbal communication) from a government consultant comparing the airborne data to existing land and marine data gave a precision of 1 to 2 milligal.

A third-degree residual Bouguer gravity map was produced to best define the anomalous structures (Figure 2). This map delineates a sharp gravity low that is straddled by the existing oil production and certainly suggests the areas for further exploration. Clearly defined is the change of structural trend from major onshore faults previously mapped to the offshore pattern that cuts obliquely across these known trends.

B. MOROCCO

In a three-month period in 1983, the Moroccan Western Atlas Basin was surveyed with more than 10,000 miles of helicopter gravity and magnetic profiles. These data were used

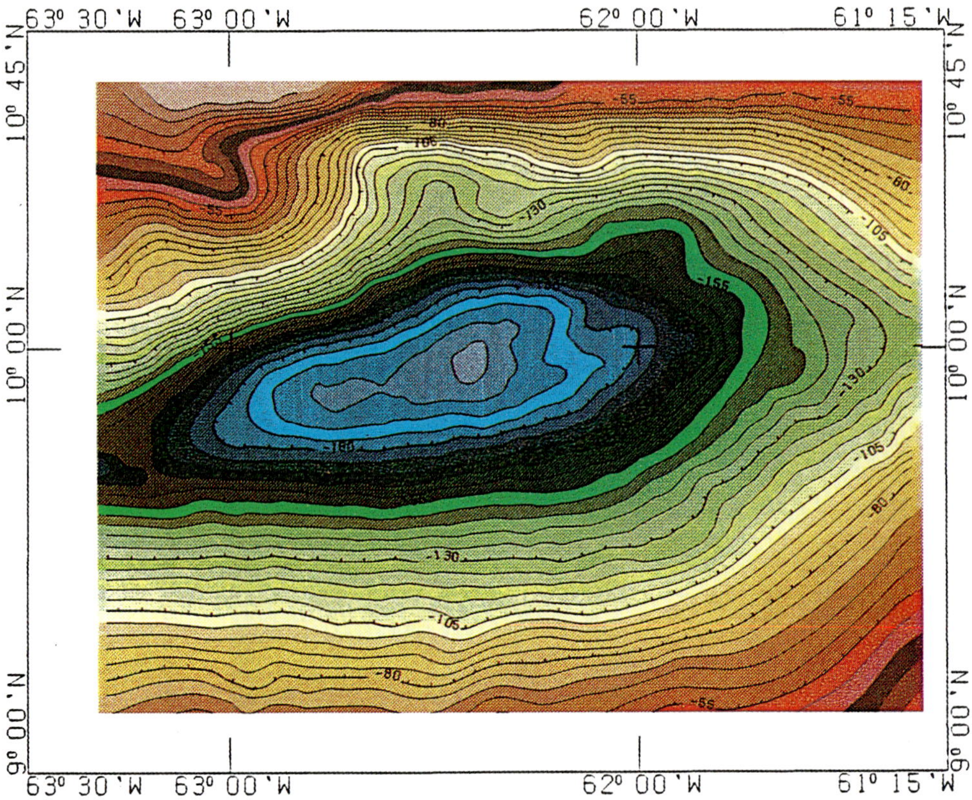

SCALE 1:1,300,000

BOUGUER GRAVITY

10 0 10 20 30 40 50 60 70 80 90 100 120 140 160 180
KILOMETERS

PLATE 2.

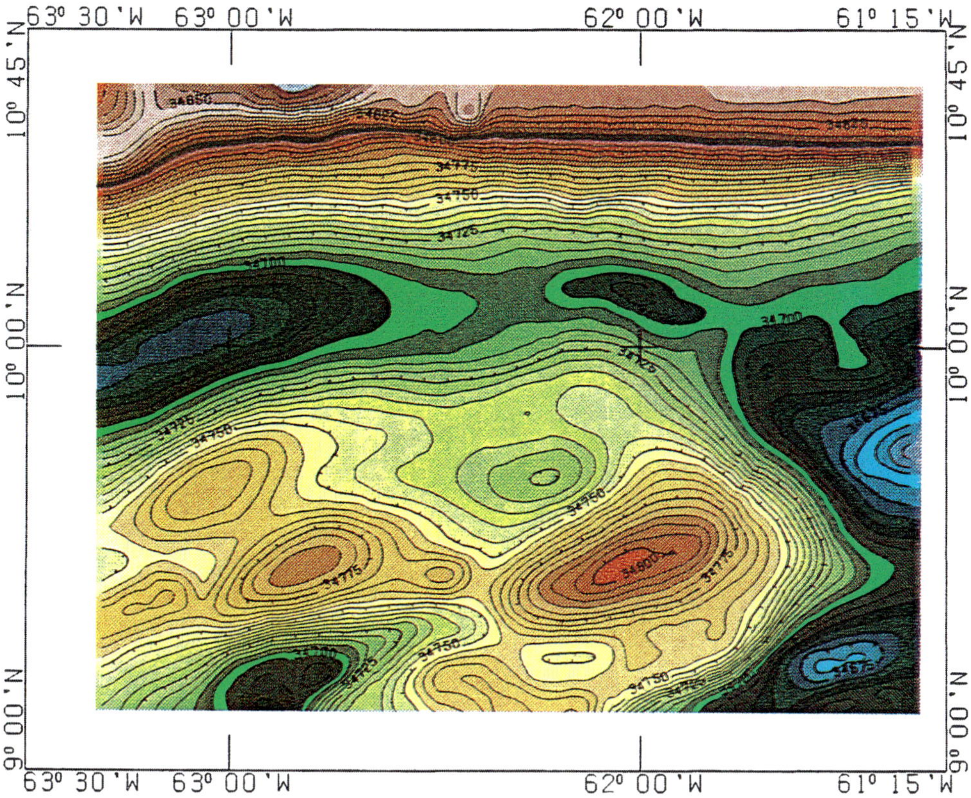

SCALE 1:1,300,000

TOTAL MAGNETIC INTENSITY

10 0 10 20 30 40 50 60 70 80 90 100 120 140 160 180
KILOMETERS

PLATE 3.

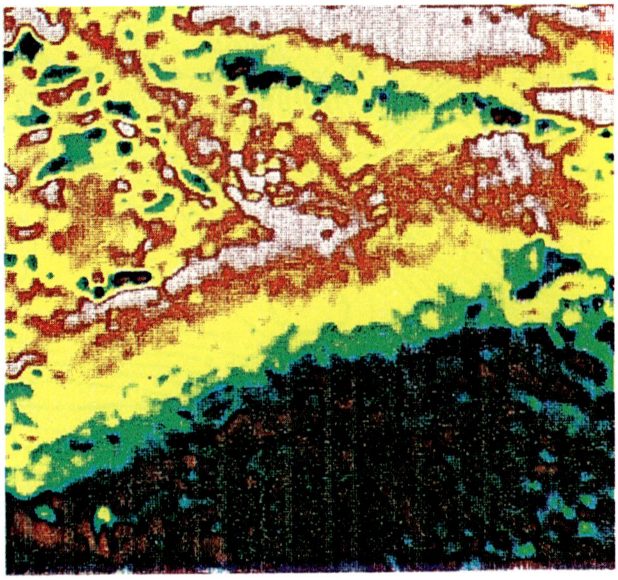

PLATE 4. Shadow graph created from gravity data.

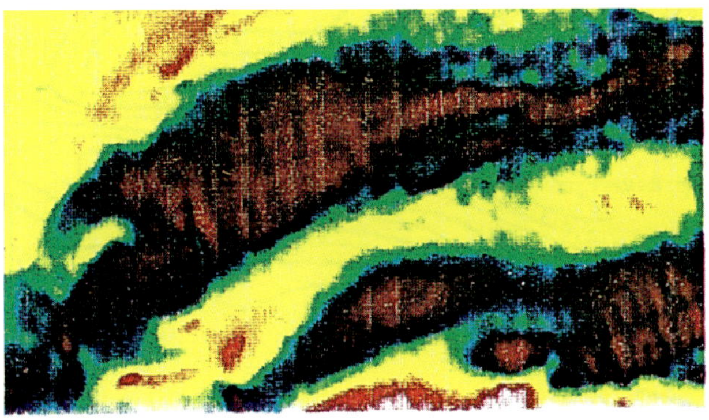

PLATE 5. Shadow graph created from magnetic data.

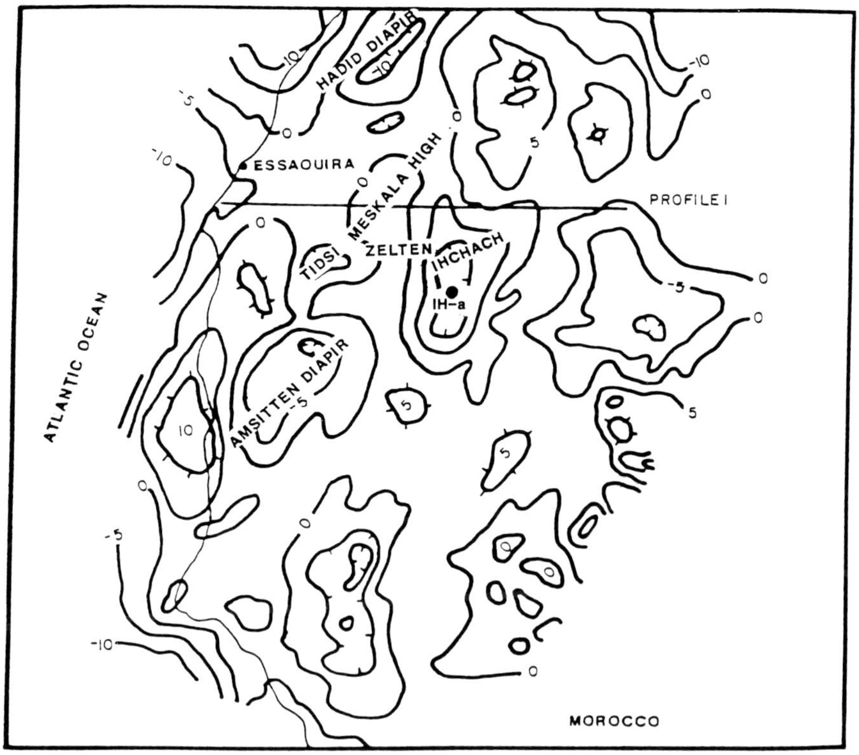

FIGURE 3. Large anomalies seen on the residual anomaly map.

Figure 4, Line 1 shows a model across the western part of the Essaouira Basin. The Meskala block and the Tidsi diapir are well defined. The general model is based on seismic data down to the top of the Trias or base of Lias. The gravity profile was then used for modeling a surface overlying the Paleozoic unconformity.

In Model "A", Figure 4, an average density of 2.4 was assumed for the Triassic layers. The same section was modeled using two layers with 2 densities: 2.3 for the upper Trias and 2.6 for the lower Trias. The comparison of these two models shows basically the same shape of the Paleozoic structures.

A report stated (Demnati, 1987) that a major contribution of the airborne gravity and magnetic survey has been to outline regional basement geology in a relatively short time and especially:

- To delineate and delimit the extent of the Meskala structure which was already under seismic and drilling appraisal.
- To show the Meskala structure is a horst block situated between two northeast trending faults and bounded on both sides by salt filled grabens. This was not clear from the seismic data because of the weak penetration below the salt.
- To locate new structures
- To show Zelten structure farther southeast separated from the Meskala structure by a fault with an east/west offset.

C. VENEZUELA

From late 1987 to the end of 1988 three survey areas were flown in Venezuela for the three operating companies located there. One area was in eastern Venezuela in the state of Monagas.

techniques were modified to address the elevation and filter effects on the computed Bouguer gravity values that were mapped.

In the Maturin area a series of regional and residual Bouguer gravity maps were produced with detailed input from company geologists, and the correct version was selected to best match the intended targets. The Bouguer data (Plate 2*) represents one of the largest continental negative gravity anomalies in the world. This feature was created by a series of overthrusts from the collision of the Caribbean plate with the Venezuelan precambrian shield. Magnetic data (Plate 3) were also collected and mapped. Shadow graphs were created from both data sets (Plate 4) and (Plate 5), and a qualitative structural pattern interpretation was made. These shadow graphs are produced by regridding the data set at a fine grid interval usually 500 m, applying the proper smoothing function, and manipulating the grid in the software. Various vertical angles above the horizontal and a complete 360 degree rotation can quickly find the proper combination in the interactive software to define the faults and structural trends. The general and detailed structural trends are easily defined in a quick time frame (Figure 5) (Graterol 1988, internal report).

With these data and the geologists input, models of the geological structures can be made and modified to match the observed information.

From this data set and two other surveys carried out in other geological provinces of Venezuela, the president of PDVSA, Juan Chacin Guzman, pointed out at the inauguration of the Fourth Geophysical Congress in Caracas (1988) that gravimetric survey methods have simplified and perfected the task in such a way that a "many-year survey" now takes only a few months to complete. To quote only one case: the information recently obtained in a two-month aerogravimetric survey of the areas of eastern Venezuela, Guarumen, and the North Andes Front would not have been possible with the old techniques in less than 2 decades and would not have had the obtained accuracy.

D. KALIMANTAN INDONESIA

A fixed-wing survey was made in the Pembuang Basin of Kalimantan, Indonesia. Approximately 4600 km of data were collected at 90 knots (170 km/hr) on a 15-by-15 km grid at an elevation of 2500 ft (760 m) above sea level. A histogram showing intersections of the computed free air gravity appears in Figure 6.

The survey delineated narrow basins created by a tensional faulting system. Following is a brief description of the system used.

IV. POSITIONING

A Motorola's Falcon 484 System is an electronic positioning system which provides accurate line of sight distance information from a master solution to any one of several reference stations. The master receiver/transmitter is mounted in the aircraft and is used to interrogate the reference stations which are set up at known positions. The elapsed time between the transmitted interrogation signal and each of the reply signals is used to determine the range to each reference station. Each range measurement requires only 2 to 10 msec and the indicated average is updated once per second. A unique coding system is used in the system to minimize any false range readings that may be caused by radar interference.

Range/range data is observed from two to four remote stations on the range console and may be reduced to x-y position by trilateration using one unique solution or a combination of solutions. As an example, for the Pembuang Basin survey each data point was fixed by at least three individual ranges for 80% or greater of the number of data points along any one line.

The range values from the Falcon system are converted to Universal Transverse Mercator

* Plates 2 to 5 appear after page 130.

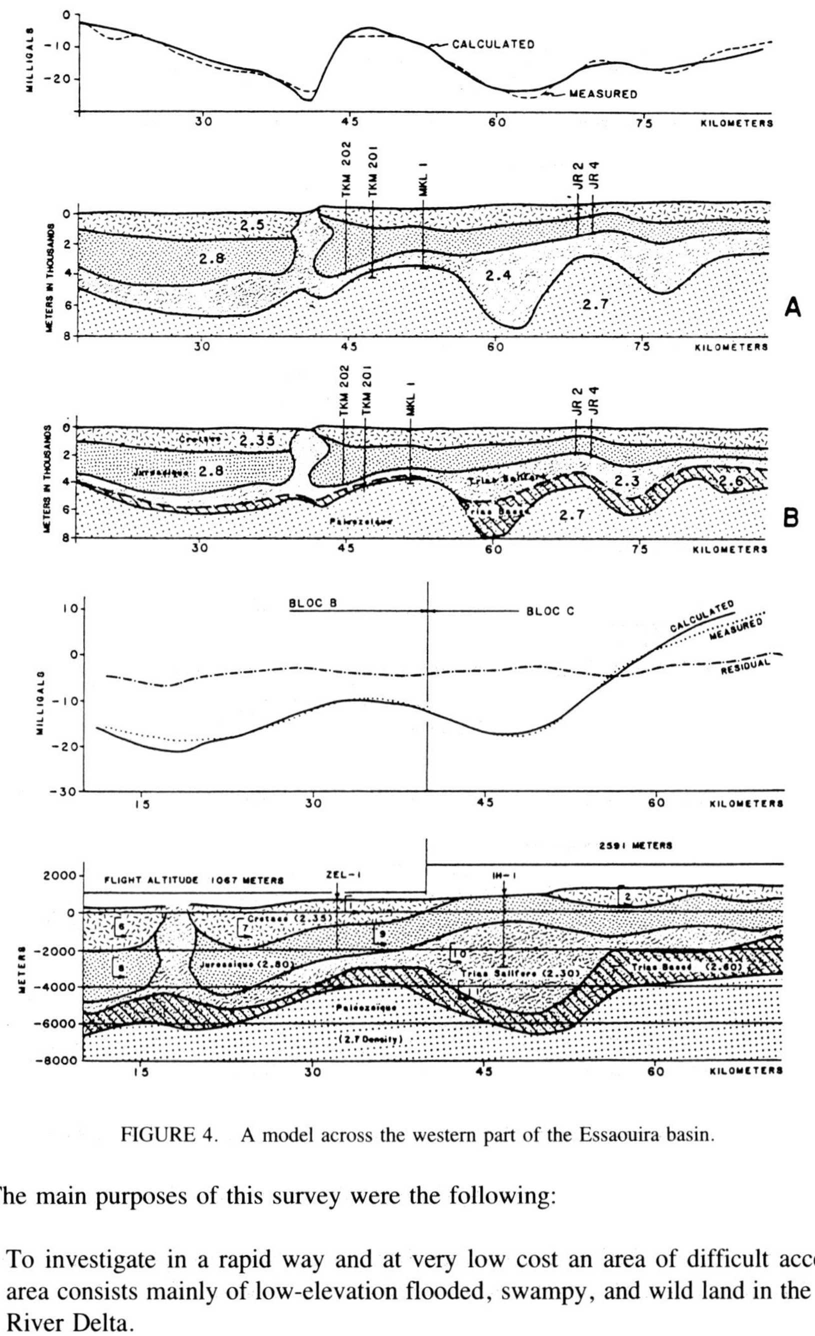

FIGURE 4. A model across the western part of the Essaouira basin.

The main purposes of this survey were the following:

- To investigate in a rapid way and at very low cost an area of difficult access. The area consists mainly of low-elevation flooded, swampy, and wild land in the Orinoco River Delta.
- To correlate the results of this survey with the Venezuelan structural patterns. These structures are presently under very intense exploration, especially west of the city of Maturin, Monagas State.
- To select, by means of the potential field methods of gravity and magnetics, areas of good possibilities for the accumulation of oil and gas deposits to be surveyed systematically in detail to define possible drilling sites.

This survey in Monagas was flown by helicopter at 50 knots (90 km/hr) with a 2-by-4 km grid. 10,587 km of data were collected in January and February. Standard interpretation

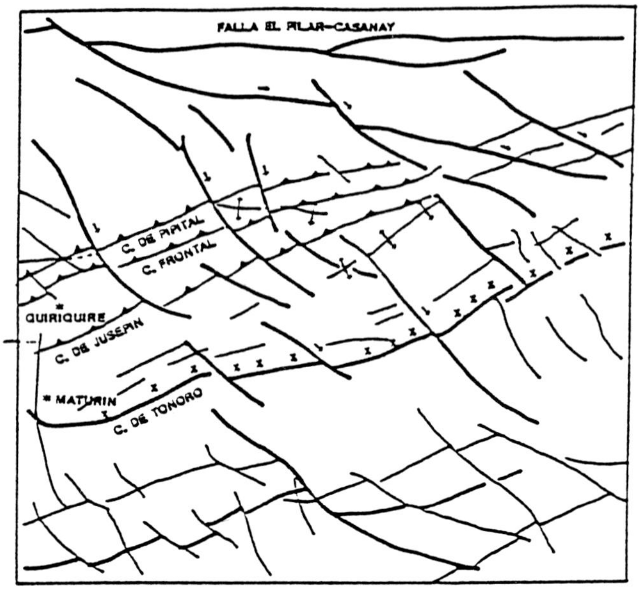

FIGURE 5. Structural interpretation.

(UTM) northings and eastings using simple trigonometry and these in turn are converted in latitudes and longitudes using some standard spheroid as the basis for all computations.

The data for each of the geodetic points was established by Doppler survey unit. The information consists of a number of recorded signals received by each station during a satellite orbital pass.

A. GRAVITY METER

The LaCoste & Romberg (L&R) Air/Sea Gravity Meter (Model S) is a highly over-damped, spring-type gravity sensor mounted on a gyro-stabilized platform with associated electronics for gravity measurement and for recording the values on both strip chart and magnetic tape.

The Model S gravity meter has a measurement range, without resetting, of 12,000 mgals and a static accuracy of better than 0.01 mgal. However, for practical purposes, readings are recorded only to the nearest 0.10 mg. Before delivery to the purchaser, each gravity meter is laboratory tested. During these tests, each meter is subjected, by machine, to various combinations of horizontal and vertical accelerations from 0 to $\pm$100,000 mgals. These tests show that the meters register errors no greater than $\pm$0.25 mgals even at maximum acceleration.

The basic gravity meter unit as designed by L&R consists of two components: a gravity sensor and an inertial system used to control the heading of the meter. The gravity sensor is mounted on a stabilized platform which is controlled by two servomotors to maintain stable position by producing rotation about two horizontal axes that are perpendicular to each other. This constant rotation in response to movement of the meter insures proper sensor orientation at all times. The inertial system utilizes three Honeywell GG 49 gyros, two Donner accelerometers and three corresponding servos.

The analog recording system used on conventional marine gravity systems was changed to incorporate recorders with adjustable gains and scale factors. This modification allows the beam signal, average beam, and both cross and longitudinal horizontal accelerations to be monitored more precisely, thus allowing faster stabilization and leveling of the meter at the beginning of data collection.

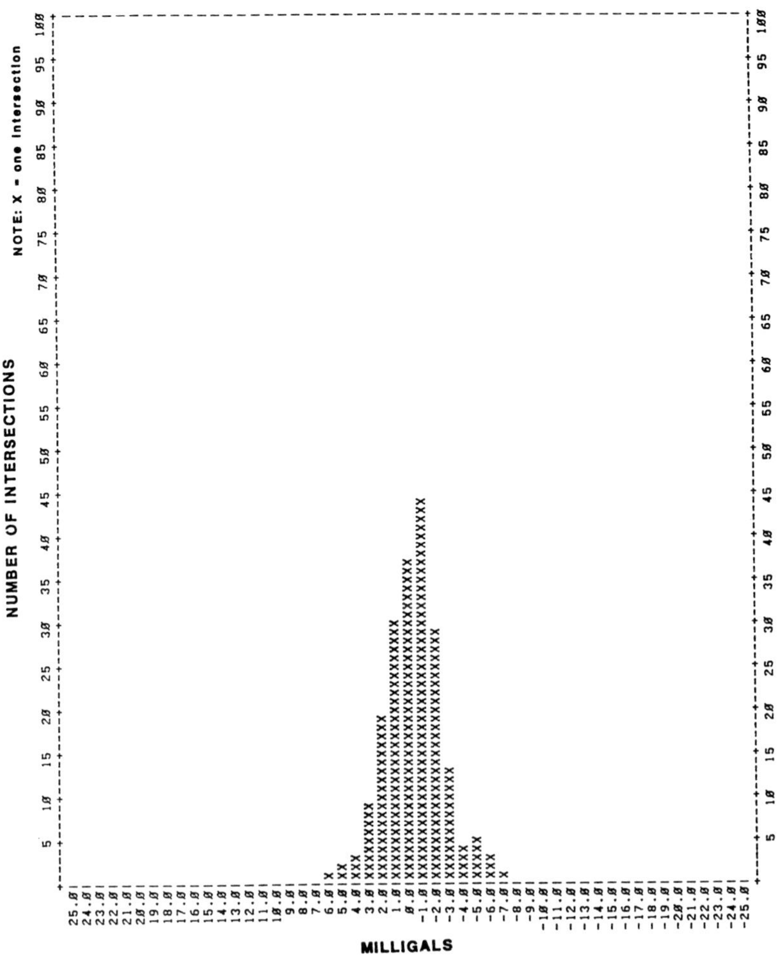

FIGURE 6. Intersection differences of the free-air gravity after first-order systematic shifts.

Real-time filtering in the gravity meter circuits was altered to provide recording of data with only a 6.6 sec delay. Additional switching circuits were added to the meter to assist the operator in platform stabilization during flight. Provisions were also made to record on strip charts all the analog signals from the gravity meter that were previously recorded only on tape. These channels are selectable and can be monitored periodically during flight.

B. MAGNETIC INTENSITY

In addition to gravity data, magnetic data for each flight is collected. The magnetic data is output to the same field tape containing the gravity data. The processing of the magnetics follows the same principles as the other parameters. It is checked first for spikes and discontinuities. Once these spikes are removed, the raw data are then corrected to conform to the base magnetic datum for the survey area.

The magnetometer detector is mounted in a pod on the wing-tip of the Twin Otter aircraft. Three compensation coils are located in the wing-tip and are adjusted to compensate the magnetometer for aircraft roll, pitch and yaw variations that can affect the magnetic field around the sensors. Compensation flights were made to verify the adjustments.

The variations of the Earth's magnetic field are recorded during flight in units of 0.10 of a gamma at a 1-sec sample rate on magnetic tape. For monitoring during flight, the magnetic data are also recorded on an analog strip chart.

C. VERTICAL CONTROL

Vertical positioning of the aircraft is obtained from the altimetry system consisting of the radar and the three independent barometric altimetry systems. All are carefully calibrated before each flight to provide the processor with vertical control to within 1 foot. Drifts in the barometric systems due to linear slope in the barometric plane are detected and corrected by correlating horizontally the aircraft's position over ground structures of known elevation. Non-linear slopes in the barometric plane (normally found in point-to-point distances of many miles or large differences in time) are detected and corrected by using ground-based barometric systems that keep track of changes in pressure with respect to time.

V. DATA REDUCTION AND PROCESSING

A. FIELD TAPE

Reduction of the airborne gravity data begins with field tapes produced aboard the aircraft. Each tape contains a series of records written at 1-sec intervals.

B. DATA EVALUATION

Using various computer programs, detailed analyses of data and additional consistency checks are performed. These consistency checks indicate areas of missing or degraded data which have been caused by adverse flight conditions or equipment malfunctions. Critical parameters can be recovered by digitizing the analog recording. Noncritical parameters can be recovered through short interpolations. Extreme cases of unusable data will result in a line being broken into two or more line segments of acceptable data. Should a line still prove unacceptable, it is reflown to ensure good quality of the data set.

From the processed data, a profile is made for each line of the topography, total magnetic intensity with International Geomagnetic Reference Field (IGRF) removed, horizontal magnetic derivative, Eotvos spring tension in milligals, vertical accelerations from the pressure altimeter, average beam times "K" (a constant), and three versions of the free-air gravity with various levels of filtering. This plot reveals the acceptability of line data.

Periodically throughout the survey a track plot is made which reveals navigation errors and lines that do not fall within the navigational or contractual specifications.

When the survey is completed, line intersection analysis is performed on free-air gravity, magnetics, topography, terrain correction and Bouguer gravity. Intersection analysis identified accuracy of individual lines and confirms regional agreement of data. Any mis-ties are corrected through level shifting, residual error analysis (REA), and a 3-D inspection of trends by a trained geophysicist. The gravity and magnetic data can now be drafted into map form. A flowchart of the entire data reduction process can be found in Figure 7.

C. FILTERING AIRBORNE DATA

The basic digital version of the LaCoste & Romberg filter developed by Dr. Lucien LaCoste for application in marine and airborne gravity computations is the differential equation:

X = Unfiltered Data $\qquad \Delta t$ = Sample Time Interval

$\overline{X}$ = Filtered Data $\qquad W_0$ = Constant (1/Time Constant)

$$A = \frac{W_0 \Delta t}{2}$$

$$\overline{X}_{N+1} = \frac{1 - A}{1 + A}\overline{X} + \frac{A}{1 + A}(X_{N+1} + X_N)$$

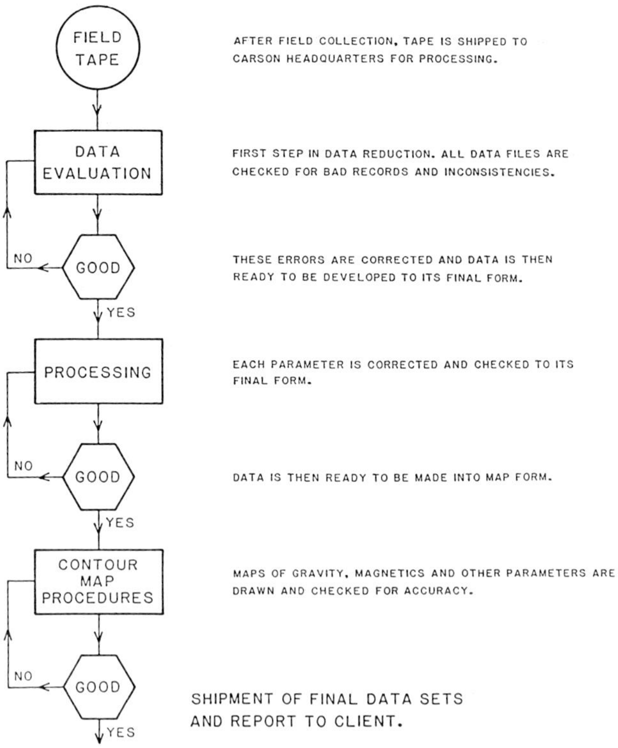

FIGURE 7. Data reduction flow chart.

The following initial conditions are required for the differential equation:

$$\bar{X}_0 = X_0 \qquad \bar{\bar{X}}_0 = X_0 \qquad \bar{\bar{\bar{X}}}_0 = X_0 \text{ etc.}$$

Passing the filter over the data with W = .05 will cause a time lag of 20 sec to occur. This is known as one stage of one way 20-sec filtering (1,1,20). Passing the same filter over the data again, but in the opposite direction, will reverse the time lag, resulting in a total time lag of 0 sec. Using the same terminology as above, the total effect is known as one stage of two-way 20-sec filtering (1,2,20).

The curves of Figure 8 show the amplitude response curves in the frequency domain for three commonly used filters in the processing of airborne gravity data.

The terminology in the preceding paragraphs has been discussed in several meetings during the past several years and the following excerpt from comments by Dr. L. L. Nettleton in October 1971 best define these parameters:

> " . . . a great deal of attention was given to the filtering in terms of filter design, periods of noise and of 'signal'. The 'signal' was considered as a 'lump' which is represented as a half cycle of a sine wave; the 'wave length' is then twice the width of this 'lump'. The 'period' is the time required at the aircraft speed to traverse the wave length distance."

It has been proven that those wavelengths of interest to industry are retained in the proper use of these LaCoste & Romberg filters. An aircraft speed of 50 knots will allow anomalies of 1 $^1/_2$ to 3 miles in horizontal dimensions to be detected and observed in final contour maps and profiles.

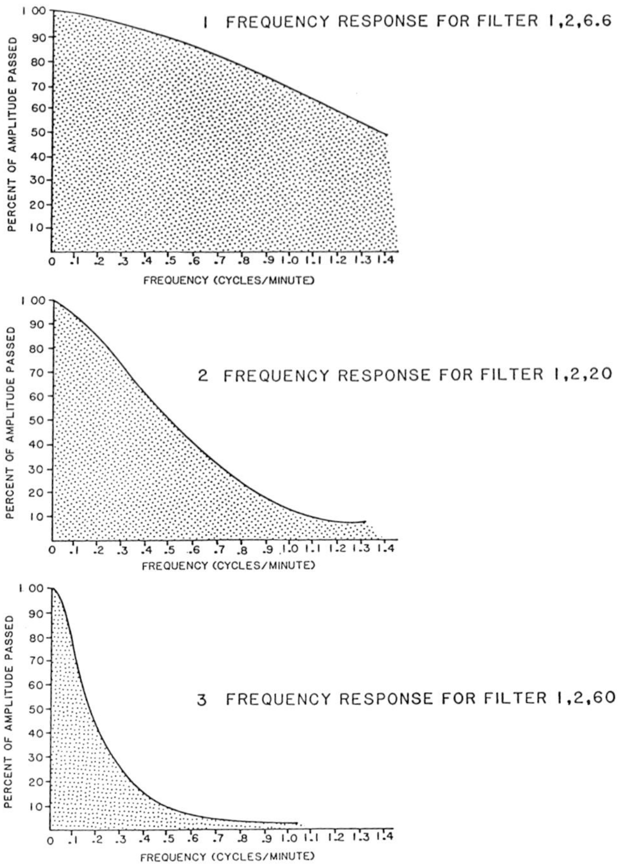

FIGURE 8. Amplitude response curves.

VI. CONCLUSION

In conclusion, the technological advances in navigation have approached the accuracy of the gravity meters so that airborne gravity measurement can be made that are more than accurate enough to be useful to exploration.

The two principal groups performing surveys (Brozena, Gumert) conclude that it still is not a simple task. Computing dynamic corrections to a precision of 1 mgal is an endeavor that requires rigorous and constant attention from the pre-planning survey stage through the data collection, and to the final computations.

REFERENCES

Anthony, D., *Navigation requirements for dynamic gravimetry by means of gradient measurements,* Proc. Symp. Dynamic Gravimetry, Fort Worth, TX, 1970.

Anthony, D., *Gradiometer Applications and Status of Sensor Development,* AFCRL 71-0314, 1971.

Anthony, D. and Perry, R., *AFCRL's Experimental Aerial Gravimetry Program,* AFCRL 71-0411, 1971.

Behannon, K. W. and Ness, N. F., *The Design of Numerical Filters for Geomagnetic Data Analysis,* NASA Technical Note TN-D-3341, Washington, D.C., 1966

Bell, C. B., et al., *Development of a Gradiometer for Measuring Gravity Gradients Under Dynamic Conditions,* AFCRL-71-003, 1970.

Bomford, G., *Geodesy,* 2nd ed., Oxford University Press, Oxford, 1962.

Browne, B. C., Second order corrections to pendulum observations at sea, *Mon. Notices R. Astron. Soc. Geophys. Suppl.*, 4, 271—279, 1937.

Brozena, J. M., A preliminary analysis of the NRL airborne gravimetry system, *Geophysics,* Vol. 49, No. 7, 1984.

Brozena, J. M., The NRL Airborne Gravimetry System, *Geophysics,* Vol. 55, No. 3, 1990.

Brozena, J. M. and Peters, M. F., An airborne gravity study of eastern North Carolina, *Geophysics,* Vol. 53, No. 2, 1988.

Carson, F. and Gumert, W., Airborne Gravity Surveying, French Patent 83 16899, 1983.

Carson, F. and Gumert, W., Airborne Gravity Surveying, United States Patent 4,435,981, 1984.

Carson, F. and Gumert, W., Airborne Gravity Surveying, United Kingdom Patent 2148016B, 1985.

Carson, F. and Gumert, W., Airborne Gravity Surveying, Canadian Patent 1 204 158, 1986.

Carson, F. and Gumert, W., Airborne Gravity Surveying Method, French Patent 86 01237, 1986.

Carson F. and Gumert, W., Airborne Gravity Surveying Method, Canadian Patent 1 256 537, 1989.

Coons, R. W., Strange, and Woollard, G. P., *Evaluation Study of Airborne Gravimeter Operational Test,* Ref. No. 62-2, Geophysics and Polar Research Center, Department of Geology, University of Wisconsin, Madison, 1962.

Demnati, Ahmed Exploring Moroccan western Atlas basin with airborne gravity, magnetic survey data, *Oil Gas J.*, 1987.

Durboraw, I. N., III Test Results of a Differential GPS Receiver in a Dynamic Environment, Int. Congr. Fed. Int. Geodesy, Toronto, 1986.

Eotvos, L., Experimenteller Nachweis der Schwereanderung, die ein auf normal geformter Erdoberflache in Ostlicher oder westlicher Richtung bewegter Korper durch diese Bewegung erleidet, *Gesammelte Arbeiten,* pp. 293—305, Budapest, 1953.

Frey, E. J. and Harlan, R., *Airborne Gravimetry Program,* AFCRL-60-0316, I, 1969.

Frey, E. J. and Harlan, R., *Design Configuration and Error Analysis Study of an Airborne Gravimetric System,* AFSWC-TR-71-6, 1971.

Glicken, M., Eotvos corrections for a moving gravity meter, *Geophysics,* Vol. 27, No. 4, 1960.

Graterol, V., Interpretation of Eastern Venezuela Basin, Carson Services, Inc., 1988.

Gumert, W., Helicopter Gravity System for Geophysical Exploration, paper presented at the Society of Exploration Geophysicists 44th International Meeting, Dallas, 1974.

Gumert, W., Helicopter Gravity System Over the Outer Continental Shelf Off the Virginia Coast, paper presented at the Society of Exploration Geophysicists 47th Annual International Meeting, Calgary, 1977.

Gumert, W., Advantages of the Continuous Profiling Airborne Gravity Surveys, paper presented at the International Meeting on Potential Fields in Rugged Topography, Switzerland, 1985.

Gumert, W. and Cobb, G., Helicopter Gravity Measuring System, paper presented at the Symposium on Dynamic Gravimetry, Texas, 1970.

Gumert, W. and Cobb, G., Helicopter Gravity Measuring System, paper presented at the 6th Meeting of the International Gravimetric Commission, Paris, 1970.

Gumert, W. and Navazio, F., Airborne Gravity of the 1980s, *Oil Gas J.*, 1982.

Gumert, W. and Stancato, T., Current Status of Airborne Gravity Surveys, SEG, paper presented at the 55th Annual International Meeting, Washington, D.C., 1985.

Hammer, S., Airborne Gravity Is Here! Geophysics: The Leading Edge of Exploration, 1982.

Hammer, S., Airborne Gravity Is Here! *Geophysics,* Vol. 48, No. 2, 1983.

Hammer, S., A note on airborne gravity terrain corrections, *Geophysics,* Vol. 48, No. 3, 1983.

Harlan, R. B., Eotvos corrections for airborne gravimetry, *J. Geophys. Res.*, 14, 73, 1968.

Henry, T. J. G., Determination of Topographic Profiles by Use of Radar and Pressure Altimeters, Meteorological Division, Dept. of Transport of Canada, 1949.

Hicks, F. L., Data Reduction of Airborne Gravity Data, Massachusetts, 1967.

Krabill, W. and Martin, C., Aircraft Positioning Using Global Positioning System Carrier Phase Data, Vol. 34, No. 1, 1987.

LaCoste, L. J. B., Measurement of Gravity at Sea and in the Air, *Rev. Geophys.*, Vol. 5, No. 4, 1967.

LaCoste, L. J. B. and Harrison, J. C., Some theoretical considerations in the measurement of gravity at sea, *Geophys. J. R. Astron. Soc.*, 5, 89—103, 1961.

Nettleton, L. L., Lacoste, L. J. B., and Harrison, J. C., Tests of an airborne gravity meter, *Geophysics,* Vol. 25, No. 1, 1960.

Nettleton, L. L., Lacoste, L. J. B., and Harrison, J. C., Quantitative evaluation of precision of airborne gravity meter, *J. Geophys. Res.*, 67, 4395—4410, 1962.

Porter, W., Kruczynski, L., Abby, D., and Weston, E., Global positioning system differential navigation tests at the Yuma Proving Ground, J. Inst. Navigation, Vol. 32, No. 2, 1985.

Rowe, R., Salvermoser, F., and Beruff, R., Developmental Efforts to Improve the Accuracies of Geodetic and Geophysical Surveys, Venezuela, 1983.

Thompson, L. D. G. and LaCoste, L. J. B., Airborne gravity measurements, J. Geophys. Res., Vol. 65, No. 1, 1960.

Chapter 7

THE LACOSTE & ROMBERG AIR/SEA GRAVITY METER: AN OVERVIEW

Herbert D. Valliant

TABLE OF CONTENTS

I. INTRODUCTION

The objective of this chapter is to provide an overview of the LaCoste & Romberg air/sea gravity meter system which is used for gravity measurements from a moving platform, including but not limited to ships, aircraft, and helicopters. The intention is to summarize the theory behind the gravity meter's operation as it has been described over many years in the open literature. Consequently, I shall attempt to avoid mathematical developments as much as possible. In the few cases where the mathematics have not been previously published, they will be included in appendices. The objective is to enable end users and others to gain a basic understanding of the system from a single source of information. A description of the operation of early gimbal-mounted systems (before 1965) is not included as none of these systems are currently in operation. The chapter concerns itself only with the instrumentation and does not attempt to include a discussion of survey methodology.

II. HISTORICAL SKETCH

The LaCoste & Romberg (LAR) air/sea gravity meter was hatched at a scientific meeting in St. Louis (circa 1950) when Drs. J. L. Worzell and A. L. McCahan independently asked Drs. Lucien LaCoste and Arnold Romberg if they could produce an instrument to measure gravity in a submarine. Dr. LaCoste returned to his laboratory where, over the next several years, he developed a highly successful gravity meter that not only functioned in a submarine but also on a ship. Other marine meters at the time, principally the Askania meter, utilized heavy damping to avoid excessive motion of the sensitive element. Dr. LaCoste developed a unique method for reading heavily damped meters that avoided the long time constant normally expected with heavy damping.

The first meter was hung from gimbals like a pendulum. Although the gimbal suspension eliminated the effects of horizontal accelerations on the gravity meter itself, a correction due to the motion of the pendulum, called the Browne Correction, was required. This correction could only be made with reference to an accurate horizontal reference. Therefore a concomitant development was a horizontal reference in the form of an inertia bar, which was based on an earlier design of Vening-Meinesz. The inertia bar was refined to the point where resonant periods of 2 min could be routinely achieved. Even so, such a short period was only adequate for marine measurements and inadequate for measurements in an airplane. A gyro-stabilized reference was not developed until later because of the high cost and low reliability of gyroscopes at the time.

Early tests with this new meter (e.g., Harrison, 1959) showed great promise and its use became more widespread as its development continued. The next major development came in 1965 with the advent of relatively inexpensive and reliable liquid-floated gyroscopes. This allowed a vertical reference to be constructed that could, at least in theory, have as long a resonant period as desired. It was a natural step to make the gravity meter a part of the inertial platform thereby keeping it vertical at all times. This eliminated the Browne Correction, but introduced other errors known as cross-coupling errors that result from the effect of horizontal accelerations on the gravity-meter mechanism. A sensor that had been redesigned to reduce cross-coupling effects as much as possible and mounted on a gyro-stabilized platform was introduced in 1965. This meter has had widespread acceptance and to this date 95 systems have been constructed.

The principal term in the cross-coupling correction arises because of the arcuate motion of the proof mass in a beam-type sensor. The next significant development (LaCoste, 1983) was a sensor in which the proof mass was constrained to move in a vertical line, rather than in an arc of a circle. This eliminates the principal term of the cross-coupling error and the symmetry of the new instrument also helps reduce other cross-coupling effects to negligible levels. An additional error source arises from non-linearity of the sensor response to large accelerations. The main cause of this non-linearity is the air dampers in beam-type gravity meters. Another feature of the straight-line gravity meter was the introduction of liquid damping, which reduced damper non-linearity to the point where this source of error is also negligible. To date, three of these new systems have been constructed and successfully deployed.

Along with improvements in the sensor and vertical reference, the electronics that supported the system underwent slow evolution. Initially, digital electronics were not available so that the entire electronic system for the gyro-stabilized meter was analog. This included the controller that nulled the gravity meter beam, the platform erection loops, and the computation of cross-coupling corrections. Much of this development predated the introduction of operational amplifiers, so that simple arithmetic circuits, such as multipliers were not available. Therefore, rather complicated electro-mechanical methods of analog computation were developed. Eventually analog multipliers and operational amplifiers became available to simplify the cross-coupling computation. The automatic gravity meter nulling system (auto-reader), a rather elegant electro-mechanical device, remains in use on many systems to this day.

Analog electronics later gave way to digital electronics and computer control. Early systems used digital electronics only for logging data to magnetic tape. One early version incorporated a mini-computer to filter the data before logging it to magnetic tape. This gave way to a hybrid system in which a computer was used to replace the electro-mechanical auto-reader as well as filter and log the data. The platform erection loops and cross-coupling calculations were still performed with analog electronics. Five of the hybrid systems were produced. Concurrent with the development of the hybrid system at LaCoste & Romberg, others (Valliant et al. 1985) were developing a full computer controlled digital system. This system eliminated all of the analog arithmetic. A personal computer controls the gravity meter and erection loops, calculates cross-coupling, filters the data, and logs the data on a suitable storage medium. This system has proven to be more reliable and accurate than the hybrid system and Lacoste & Romberg adopted it in 1989. Four digital systems have been constructed and are currently in regular deployment.

III. MARINE GRAVITY SENSORS

A. THE BEAM SENSOR

The air/sea gravity sensor is based on the zero-length spring principle discovered by Dr. L. J. B. LaCoste, circa 1930. Use of a particular geometry with a special spring permitted the development of a vertical suspension having very long periods. It can be shown (e.g., LaCoste, 1934, 1935, and 1988) that such a system can have infinite period. When the period is infinite, and the torque exerted by the spring exactly balances the torque exerted by gravity, the gravity meter beam will remain stationary, at least in theory, at any position it is placed. When this condition is achieved, the smallest change in gravity causes the beam to rotate to one stop or the other. Thus infinite period corresponds to infinite sensitivity. If the period is less than infinite and the beam is displaced from its equilibrium position, a restoring torque will return it to that equilibrium position, which is the condition for land meters. Attempting to increase the period beyond infinity results in an unstable condition corresponding to a negative period, and the gravity meter beam will tend to seek one stop or the other. LAR air/sea gravity meters are operated at infinite period with heavy damping.

Figure 1 illustrates a simple mechanical analog to help explain these conditions. Consider a ball rolling on a smooth surface. In Figure 1a, the surface is concave, which corresponds to the case of finite period. If the ball is displaced from its equilibrium position and released, it will return to that position after oscillating (simple harmonic motion) back and forth a

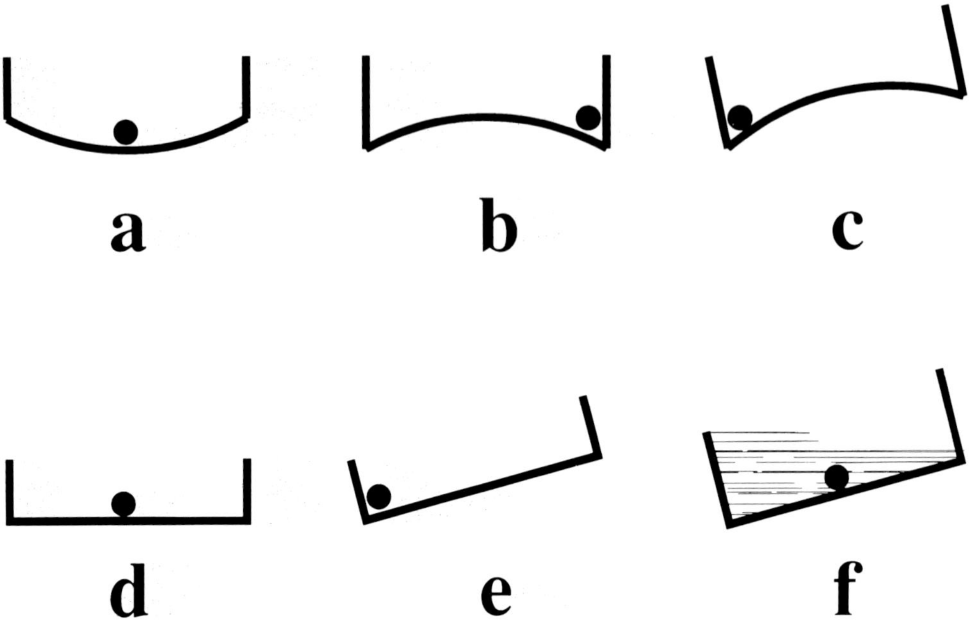

FIGURE 1. A ball rolling on various shaped surfaces illustrates the effect of tilting the vertical axis of a gravity meter to adjust its natural period. (a) corresponds to a finite positive period, typical of land gravity meters. (b) and (c) correspond to negative periods and represent unstable conditions. (d), (e) and (f) correspond to infinite period which is employed for air/sea gravity meters. In (f), the ball is surrounded by a viscous liquid.

few times. The period of oscillation is the same as for a pendulum whose length equals the radius of curvature of the surface. In Figure 1b, the surface is convex corresponding to a negative period. The ball will come to rest at one side or the other unless it happens to be balanced exactly at the top of the curve. If the surface is tilted as in Figure 1c, the ball remains at one end until the angle of tilt is sufficient for it to suddenly roll to the other end. In Figure 1d, the surface is plane and horizontal corresponding to infinite period. The ball will remain wherever it is placed. If the surface is tilted, as in Figure 1e, corresponding to a small change in gravity, the ball will roll to the low side. If, however, the ball is surrounded by a viscous liquid, as in Figure 1f, which is equivalent to damping the gravity meter, the ball will roll to the bottom at a constant velocity.

It can be seen that the position of the sphere on the flat plate cannot be used to measure the tilt of the plate. However, the velocity of the sphere through the liquid is directly proportional to the sine of the tilt angle. Thus, by analogy, small changes in gravity can be observed by measuring the velocity of the beam motion in a heavily damped gravity meter. The time constant for such a system is the time required for the sphere to achieve a velocity equal to 0.6 of its terminal velocity. The time constant becomes shorter as the damping is increased because higher damping makes the velocities smaller and therefore steady state velocities are reached sooner. The time constant of the LAR air/sea gravity meter is typically less than 1 msec. This means that the sensor's response to gravity changes is nearly instantaneous, and any time delay in the data is due to external filtering. This is an important point as it is a common misconception that the LAR sensor has a long time constant due to its heavy damping.

The proof of infinite period is developed in detail in the previous references by LaCoste. Perhaps the most succinct explanation was given in a Sprengnether Inc. manual (anonymous, undated) as follows. Consider the diagram in Figure 2 which describes the zero-length geometry. The diagram shows a system such that the lengths AO and BO are equal and the

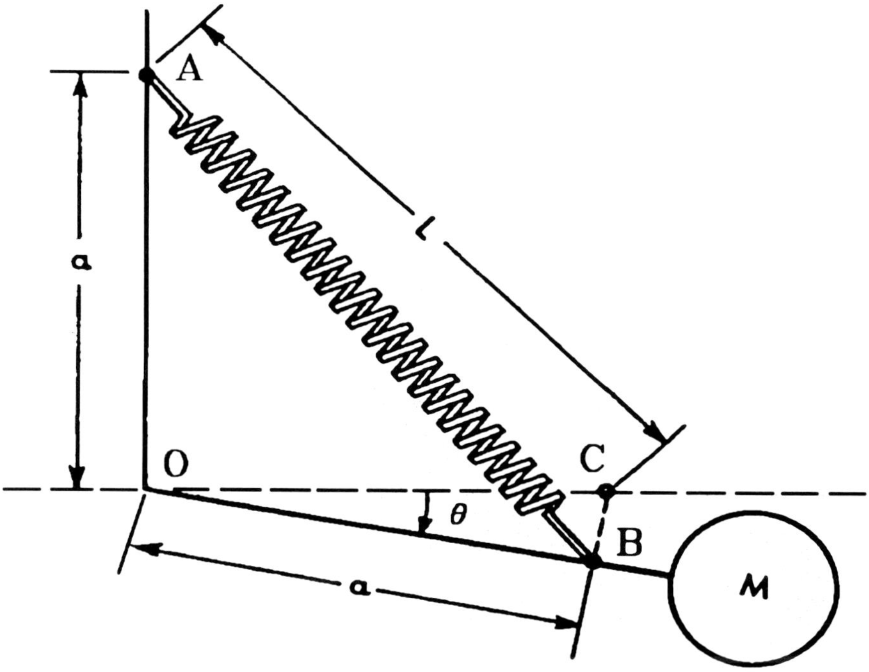

FIGURE 2. The principle of the zero-length suspension system. A mass M is attached to the movable beam OB, which is free to rotate about O. The beam is supported by a zero-length spring attached at points A and B. (From Sprengnether, Inc. manual. With permission.)

line segment AO is exactly vertical. That is, the spring is at an angle of 45° when the beam is horizontal. This, however, is not a general restriction and the more general formulation of LaCoste shows that the system works equally well with unequal lengths for AO and BO. Verticality of AO is required, in theory, to attain infinite period. (In practice, the restoring force of the hinges does require some labilizing force). Let the rest-length of the spring be L_o and its stretched length be L. Its elongation is then $(L - L_o)$. The spring tension is then given by $T = k(L - L_o)$, k being the spring constant. The magnitude of the restoring moment due to deflection Θ from horizontal is:

$$M_r = \frac{kaL_0}{\sqrt{2}}\left(\frac{\theta}{2} - \frac{3}{8}\theta^2 + \cdots\right) \tag{1}$$

It is immediately evident that if $L_o = 0$ the system exerts zero restoring moment and thus is in a state of neutral equilibrium (infinite period). The condition that $L_o = 0$ implies that the spring's elongation equals its length. That is to say, the spring behaves as if it had zero unstretched length, and hence is dubbed a zero-length spring. Equation 1 ignores damping, but there is no loss of generality in doing so.

The natural period of the zero-length suspension is adjusted by tilting the meter in the direction of the beam (longitudinal direction). Tilting the top of the meter from the vertical toward the mass M decreases the natural period. Tilting it in the opposite direction yields an unstable system. The period is exactly infinite when AO is vertical. This is the condition for air/sea gravity meters. For land gravity meters the period is normally adjusted from 15 to 18 sec by tilting the axis forward slightly.

The beam is nulled (restored to its horizontal position) in LAR gravity meters by moving the point of attachment of the spring A up or down as required. If the dimension AO is approximately 2.5 cm as is the case for LAR meters, the motion of point A to compensate for a 0.1 mgal (0.1 ppm) change in gravity is 2.5×10^{-7} cm; 0.1 mgal represents the standard resolution of the LAR air/sea meter recorded data. This point emphasizes that the design (LaCoste, 1945) and manufacture of the lever system that controls the displacement of point A is critical.

The nature of the motion of the gravity meter beam can be described by the following differential equation (LaCoste et al., 1967) which is general for all systems experiencing simple harmonic oscillation:

$$g + z'' + bB'' + fB' + kB - cS = 0 \quad (2)$$

The single and double primes refer to single and double differentiation with respect to time, respectively. The term B is a measure of the beam displacement from its horizontal position, usually in volts. Using B or the angle Θ in Equation 2 is equivalent, but we use B here to be consistent with the terminology in the LAR manual. In the case of the LAR air/sea meter, the first three terms result from gravitational and acceleration forces on the mass; "z" is the vertical acceleration of the gravity meter case. The coefficients b,f,k,c must be constants if the gravity meter is to be linear. The coefficient b may be assumed constant for small motions. The coefficient f results from damping; k is the restoring force of the spring and cS represents the vertical force per unit mass exerted by the spring S, often referred to as the spring tension. S is adjusted by moving the point of attachment of the spring on the gravity meter case (A in Figure 2) with a micrometer screw acting through a lever system (LaCoste 1945). For air/sea meters k = 0 (infinite sensitivity) and f is chosen to be very large. If f is very large then b can also be ignored and the equation is simplified to:

$$g + z'' + fB' + kB - cS = 0 \quad (3)$$

This is a first order differential equation whose steady state solution is given by:

$$g = S + KB' \quad (4)$$

The coefficient K in Equation 4 converts from units of beam velocity to units of gravity (mgal/mV/sec). Thus, the gravity meter is read by observing only the spring tension setting, S, which is proportional to the position of the micrometer screw, and the velocity of the beam motion. It is very important to realize that the position of the beam is not needed. However, the beam velocity is calculated by taking successive differences (numerical differentiation) of the beam position. The process of differentiation has the effect of enhancing higher frequency noise. In effect, we are trying to obtain a very accurate small number from the difference of two relatively large numbers. This constraint implies that although beam position is not directly required, it must nevertheless be determined with great precision.

Although the beam position is unimportant for measuring gravity, it must be prevented from running to the limit of its travel. Heavy damping prevents it from moving too quickly and assures that the beam stays within its reading range under large higher frequency accelerations that result from wave motion. The LAR meter has a dynamic range greater than 200,000 mgal at these frequencies. Nevertheless, the beam will still drift to a stop for long period accelerations or large gravity changes such as would be encountered by sailing north or south. Therefore the beam is continuously restored approximately to its reading-line (null, or horizontal position) by an automatic feedback system that slowly adjusts the spring tension S. Exact nulling is not required because the beam velocity is also measured as described above. The automatic nulling systems will be described in detail later. The dynamic range for slow changes is determined by the length of the micrometer screw and the geometry of the lever system. It is either 20,000 or 12,000 mgal, depending on the meter serial number, and is ample to accommodate world-wide gravity measurements.

If $B' = 0$ in Equation 4, $g = cS + 0$. This condition is achieved by adjusting S until the beam remains stationary and is referred to as a "still reading". A still reading is useful to determine the factor K. If a still reading is obtained, and then S is altered a fixed amount ΔS, the change in reading $\Delta g = c\Delta S - KB'$. As gravity was not changed $\Delta g = 0$ and, therefore, $KB' = c\Delta S$ which determines the factor $K = c\Delta S/B'$. The usual procedure, referred to in the LAR manual as a K-check, is to obtain a still reading, then slew the spring tension by 30 mgal and finally adjust the factor K so that the gravity reading remains unchanged.

The coefficient c must also be determined empirically. On most air/sea meters the lever system is adjusted so that c is approximately constant for all values of S. For these meters c is determined by making a measurement over a known gravity interval. The nearest locale to the LAR factory is Cloudcroft, NM, which is 1000 km from Austin, TX. The accepted value for the test range (241.9 mgal) is now known to be too small by 100 ppm. Also the magnitude of the test difference requires a very large extrapolation for meters having a 12,000 mgal range. Consequently, most serious gravity-meter users, especially those contemplating measuring large gravity differences, recalibrate their meters over a larger gravity range which encompasses the same latitude (or gravity) difference as expected for the survey. A few older Model S (beam-meters) and all Model SL (straight-line) meters have non-linear calibration factors, which are functions of S. These are calibrated in the same manner as land-meters (Valliant 1990) and the calibration function is expressed in the form of a calibration table. A constant factor was important at one time because data were hand calculated, but the contemporary proliferation of personal computers obviates that need.

B. THE STRAIGHT-LINE SENSOR

The straight-line sensor differs from the beam sensor insofar as the proof mass is constrained to move in a vertical line. The principle of the straight-line suspension is described in detail by LaCoste (1983). Figure 3 illustrates the principle. The proof mass is a cylindrical shell closed at the center. A vertical cross-section would be in the form of an H. The mass is supported by two anti-parallel zero-length suspension systems as shown. The attachment of five linkages constrains the motion to a translation and a slight rotation with respect to the vertical axis of symmetry. The rotation has no effect except to prevent the linkages from binding. Stationary pistons inserted into the hollow ends of the mass provide damping.

The straight-line sensor is also filled with a viscous (silicon) fluid. This fluid is used as the damping medium instead of air. The higher viscosity of the liquid relieves the close tolerances required for the dampers. This, along with the linear motion, substantially improves the linearity of the beam response to vertical accelerations. It will be shown in the section on error sources that such non-linearities are extremely important as they can cause large errors. The error due to non-linearity is negligible in the case of the straight-line meter. Other sources of error, called cross-coupling errors (which will be described later) are also negligible. Using a liquid also eliminates the effects of the compressibility of air which is present in air dampers. Finally, the fluid damps unwanted modes of vibration in the various mechanical components, thereby greatly improving the instrument's susceptibility to shock and vibration.

Reduced tolerances on the damping mechanism and its adjustment is a major convenience in the manufacture of straight-line meters. This advantage is offset by two added complications. In order to null the gravity meter, the attachment point of both springs are moved identically, increasing the complexity of the lever system. Also, the meter cannot be tuned to infinite period by tilting as described for the beam meter. Tilting the meter serves only

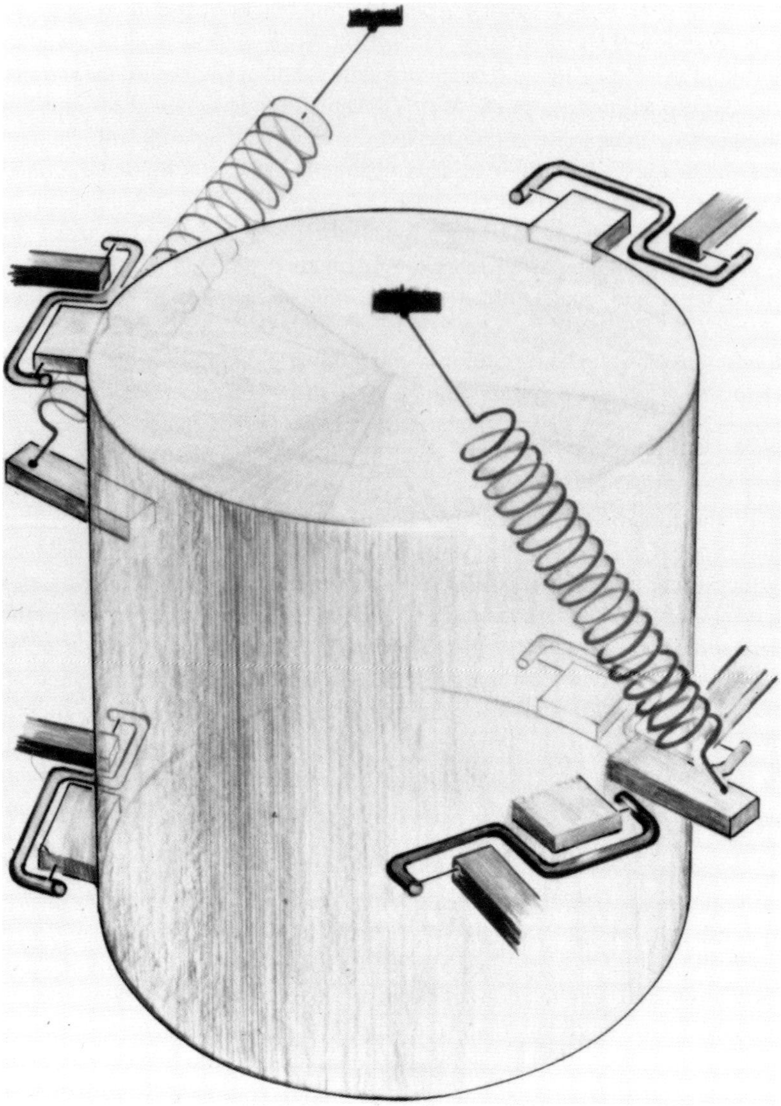

FIGURE 3. The principle of the straight-line gravity meter. A cylindrical proof-mass is supported by two anti-parallel zero-length suspension systems and constrained by five linkages. It therefore moves in a vertical straight line with a slight rotation. (Illustration by J. Geuer, Almonte, Ontario, Canada.)

to increase the period of one suspension in a positive sense while increasing it for the other in a negative sense. Of course, if the springs are constructed to have exactly zero unstretched length, or if their positive and negative lengths exactly balance each other, the period will be infinite. The meter can be tuned by adjusting the unstretched length of one or both springs by lengthening or shorting the connecting linkage. This is difficult to achieve in practice as the adjustment must be made after the meter is flooded with liquid. The insensitivity of the period adjustment to tilt is a major functional advantage that more than offsets the manufacturing inconvenience.

In Figure 4, a single beam suspension is used to illustrate the principle for adjusting the straight-line gravity meter to infinite period. A weak labilizing spring (S) is attached to the beam at its center of gravity (B) and anchored slightly behind the pivot point (0) of the

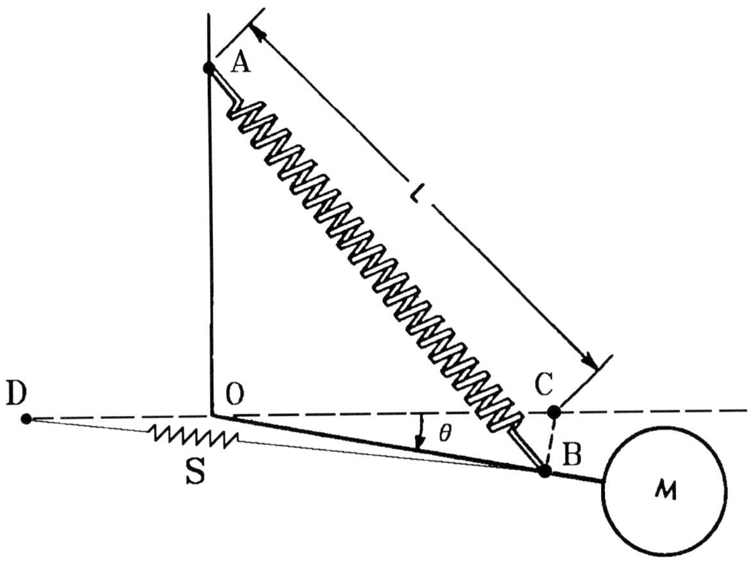

FIGURE 4. Illustration of the mechanism to tune the straight-line meter to infinite sensitivity. A weak spring S is attached to the beam at B and to the frame at D. Point D is horizontally adjustable. The labilizing force of spring S is used to tune the gravity meter to the desired period. (Adapted from illustration in Sprengnether, Inc. manual.)

beam. We assume that the weak spring is mounted to one side so that it will clear the beam as it moves. We will also make the anchor point adjustable so that it can be moved from slightly behind (labilizing condition) to slightly ahead (stable condition) of the pivot point. Suppose, for example, the position of the anchor is adjusted so that it is collinear with the beam pivot, the weak spring will exert no torque on the beam for any beam angle.

If, on the other hand, it is adjusted ahead (mass side) of the pivot, the weak spring will exert a restoring torque assisting the main spring and thereby decreasing the period. When the anchor is behind the pivot, the weak spring will oppose the main spring and increase the period. The same principle is used in the straight-line meter. The weak spring is not shown in Figure 3, for the sake of clarity, but it is attached to the proof mass in the same manner as the five fixed links.

At the present time, three straight-line meters have been produced. The first unit, SL-1, has been in continuous operation by the Geological Survey of Canada since 1983. Valliant (1983) has shown that cross-coupling errors are indeed negligible for SL-1. He has also shown that the accuracy of the two meter types are equivalent when the beam meter is perfectly adjusted so that the cross-coupling errors are exactly compensated. Cross-coupling adjustments are apt to change unexpectedly in beam meters, usually as a result of rough handling during transportation. A method known as cross-correlation, discussed below, can be used to detect and correct for these changes in the cross-coupling calibration. For accurate results with model S meters, it is essential to perform cross-correlation analysis on a regular basis. Although the straight-line meter has negligible cross-coupling error, it is still good practice to do regular cross-correlation in order to monitor for malfunctions.

IV. THE PLATFORM

A. REVIEW OF INERTIAL THEORY

The LAR inertial platform utilizes gyros and accelerometers to provide a horizontal reference accurate to 8 sec of arc (LaCoste et al., 1967) in the presence of 50,000 mgal horizontal accelerations. Its mechanization is nearly identical to precise inertial survey systems. Mechanically, it differs from such systems insofar as it utilizes less accurate horizontal accelerometers than inertial systems, but it uses one of the world's most accurate vertical accelerometers (the gravity meter). Less accurate horizontal accelerometers are acceptable because the platform is used primarily as a vertical reference rather than for position measurements.

An ideal gyro maintains a constant angular orientation in space. A practical gyro will exhibit a certain amount of drift. The ideal gyro's output signal is proportional to its angular displacement from some reference position. This signal can be amplified, filtered and used to operate a servo motor to restore the gyro (and platform) to its original position. A gyro and servo is required for each axis to be stabilized. Only two axes are required in a vertical reference. Once the platform is able to maintain a constant orientation in space we can add three orthogonal accelerometers and measure accelerations in three axes, which can be integrated to obtain velocities and integrated again to provide positions. Such a system is said to be space stable and is suitable for extra-terrestrial vehicles.

As the Earth rotates about its axis, a space-stabilized reference will appear to tumble once per day when referred to the Earth. This is somewhat of a nuisance in terrestrial vehicles and a method to align the platform with the Earth's vertical is desirable. Consequently, gyros are equipped with an electrical input which can modify their output signal. That is we can make the gyro "think" it has rotated when it has not. Assume that a system with two horizontal axes is attached to the Earth and oriented in the cardinal directions. The platform will tumble in the east-west direction and remain stationary in the north-south direction. Therefore we must supply a torquing signal to the east-sensing gyro proportional to the Earth's angular velocity to counteract Earth rotation. (Integrating gyros are used on the LAR platform and, therefore, the input signal is proportional to velocity.)

This platform will now be Earth-stabilized and the accelerometer signals can be used to obtain positions in the cardinal directions as before. Conventional inertial survey systems compute the torquing signal according to an Earth model. This model is usually computed with respect to a smooth mathematical surface called the reference ellipsoid. Thus, all position measurements are made with respect to the ellipsoid. We previously mentioned that real gyros tend to drift. This drift can cause severe errors if allowed to accumulate. For that reason these systems must be frequently calibrated against a known standard. Zero velocity is an easily achieved accurate standard. Consequently, this type of mechanization must stop every few minutes for a zero-velocity update (ZUPT). This condition is satisfactory on land but awkward at sea.

The LAR platform derives its torquing signals from the accelerometers. In the absence of horizontal accelerations the platform is driven so that the accelerometer outputs are zero. That is the condition for the platform to be exactly orthogonal to the gravity vector and it is therefore oriented to the local vertical and gyro drift is automatically compensated. Measurements made with this mechanization will be referred to the geoid instead of the ellipsoid. In the presence of horizontal accelerations the platform will tend to be oriented perpendicular to the resultant of the horizontal and vertical acceleration vectors. Thus, horizontal accelerations generally cause leveling errors.

It will be shown later that this mechanization behaves like a pendulum. Increasing the period, or the length of the pendulum, reduces the error due to horizontal accelerations. The LAR platform is normally tuned to a 4 min period, although 18 min and even longer have been used. Such errors are negligible when sailing survey lines, but become appreciable when sustained horizontal accelerations occur as the ship alters course to acquire a new line. Normally, one plans the turn to provide a straight approach to the survey line of 6 to 10 min to allow the platform to settle. This characteristic tends to be a drawback when surveying in heavy traffic, or where other obstacles demand frequent course changes. Methods for correcting these platform errors will be discussed in a subsequent section.

FIGURE 5. Photograph of the LAR air/sea gravity meter, shown with the full digital control system.

If the length of an undamped equivalent pendulum is made equal to the radius of the Earth, it can be shown that horizontal accelerations cause no errors. A pendulum of this length would have a period of approximately 84 min. This condition is known as Schuler tuning. Scarborough (1958) details the theory of Schuler tuning in an appendix. Schuler tuning is ideal but difficult to achieve in practice. Any electronic noise or improper initial conditions will cause the platform to oscillate with an 84 min period. As the platform is not damped the amplitude of these oscillations tend to grow with time. Commercial inertial navigation systems utilize ancillary velocity information with a Kalman filter to control Schuler errors. A three-axis LAR platform has been operated in Schuler-tuned mode for short intervals of time. Dr. N. Clarkson (personal communication, 1989) reports having successfully operated a Schuler-tuned LAR platform in an airplane, though he also reported that it required considerable care.

It will be shown in the section on errors that large transient errors can be induced by an azimuthal rotation of the platform. Adding a third, vertical axis eliminates this source of error.

B. MECHANIZATION OF THE TWO-AXIS LAR PLATFORM

The standard LAR platform comprises two orthogonal gyros and two orthogonal horizontal accelerometers mounted on a rigid base. This assembly, referred to as the platform, is in turn bolted to the top of the gravity sensor. The sensor is supported by a two-axis gimbal driven by two torque motors. Figure 5 is a photograph of the platform and control electronics. We will now analyze the system for one axis; the second axis is identical.

Figure 6 shows the control system. One servo loop feeds the gyro output signal back

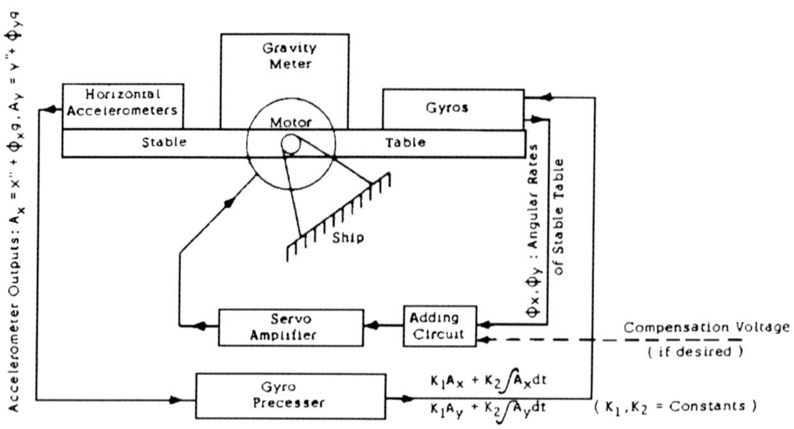

FIGURE 6. Block diagram of the controller for one axis of the platform. (LaCoste, 1967, Figure 7.)

to the platform via the torque motors. This loop locks the platform to the gyro and is sufficient to create a space-stable platform as noted above. The gyro output is also slightly filtered en route to the torque motor. Too much filtering will create a phase delay causing instability. The stiffness of the platform can be controlled by the loop gain. Too little gain yields a mushy platform; too much gain will cause vibrations. A summing circuit is incorporated so that an additional torquing voltage can be applied to the platform. This optional torquing voltage, known as the compensation voltage, can be utilized to correct for certain small platform errors. For LAR meters a fraction of the accelerometer signal is used for compensation. It is said that the compensation voltage corrects for small lags in the accelerometer signal.

A second loop processes the accelerometer signal which is fed back to the gyro input. This is the loop that aligns the platform to the gravity vector. LaCoste has shown that a good signal to feedback is one comprising a term proportional to the accelerometer signal plus a term proportional to its integral. The equation of motion for this type of feedback is given by LaCoste (1967, Equation 32) as:

$$\phi' + 2fw_o\left(\phi + \frac{x''}{g}\right) + w_0^2\int\left(\phi + \frac{x''}{g}\right)dt = 0 \quad (5)$$

where ϕ = platform error angle, and χ' = horizontal acceleration.

This equation is the same equation that describes the motion of a damped pendulum. The coefficient of the integral term (w_0^2) determines the period of oscillation and the coefficient of the proportional term ($2fw_0$) determines the damping. We can set these coefficients to any value we desire. LaCoste also shows that errors due to horizontal accelerations will be minimized when the damping is set at 0.707 times critical. LAR platforms are therefore adjusted for this amount of damping.

For the sake of clarity in the discussions below, I will comment on terminology for identifying gyro orientation. Gyros are normally labeled with a sensitive axis. This is the axis about which the gyro measures angular velocities or displacements. Thus if the sensitive axis of a gyro is oriented east-west, it measures angular rotation in a north-south plane. I find it less confusing to think of the gyros as being oriented in the direction they actually sense rotation. I therefore refer to them as east-sensing or north-sensing gyros. Outputs are positive for rotations east or north and negative for rotations west or south. Their sensitive axes are, of course, orthogonal to these directions.

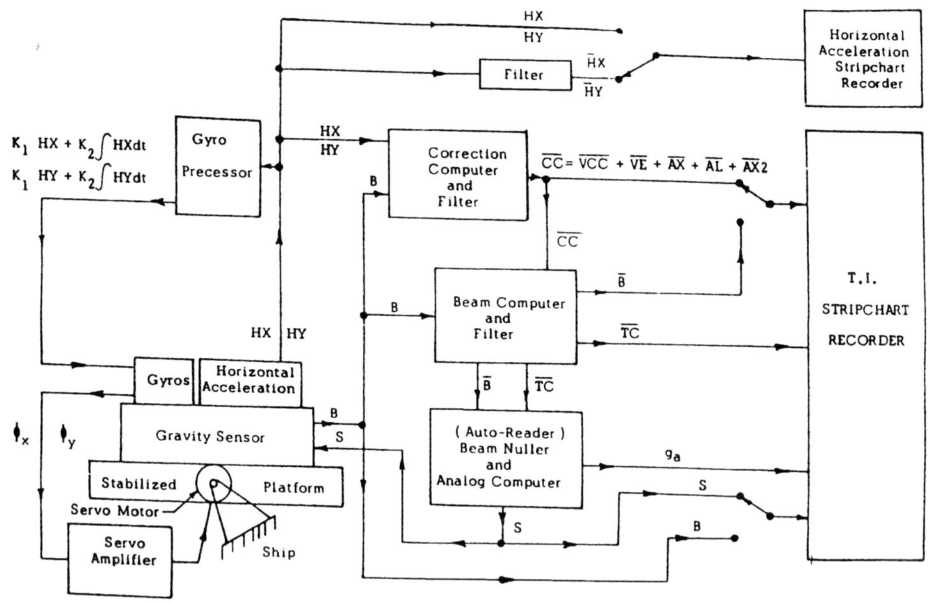

FIGURE 7. Block diagram of the basic analog air/sea control system. HX and HY are horizontal accelerations; CC is cross-coupling; VCC, VE, AX, AL, and AX2 are cross-coupling monitors; B is beam position; S is Spring tension; g_a is analog gravity; ϕ_x and ϕ_y are platform error angles; K_1 and K_2 are constants. A bar over the symbol indicates that the value is filtered. (LAR manual. With permission.)

C. MECHANIZATION OF THE THREE-AXIS LAR PLATFORM

A third vertical axis and torque motor can be added to the system described above. A third gyro is also required to sense azimuth rotations. It is desired to torque the platform about its vertical axis so as to maintain a constant (or slowly changing) azimuth. The azimuth torque motor is locked to the azimuth-sensing gyro exactly as described above except that no provision is made for a compensation voltage. Again, this produces a space-stable system. It will appear to rotate as the Earth turns unless it is located exactly on the equator. We note that if the two-axis platform were oriented in the cardinal directions, the torquing signal applied to the east-sensing gyro is maximum while the signal applied to the north-sensing gyro is zero. If the signal from the north-sensing gyro is not zero we desire to rotate the platform until it is, in order to stabilize the third axis relative to north. We can accomplish this by feeding some of the north-torquing signal to the azimuth gyro input, thereby rotating the platform azimuth to keep the north-torquing signal zero. This type of control is called gyro-compassing. An error in azimuth proportional to the north velocity of the vehicle is inherent with this type of mechanization and is known as the gyro-compassing error. It is small at ship speeds. The period and damping of the azimuth motion is controlled by including integral and proportional terms as described above. LAR platforms normally have an azimuth period of about 90 min with damping set at 0.707 critical.

D. CONTROL SYSTEMS

1. Analog Electronics

Early LAR control systems were entirely analog. As many of these are still in operation a brief description is in order. Figure 7 is a block diagram of this system. It is self explanatory. Note that all computation is performed using analog or mechanical arithmetic. Signals from the accelerometers and gyros are processed as described above and fed back as required. Signals are filtered and recorded as required. The cross-coupling correction (described in

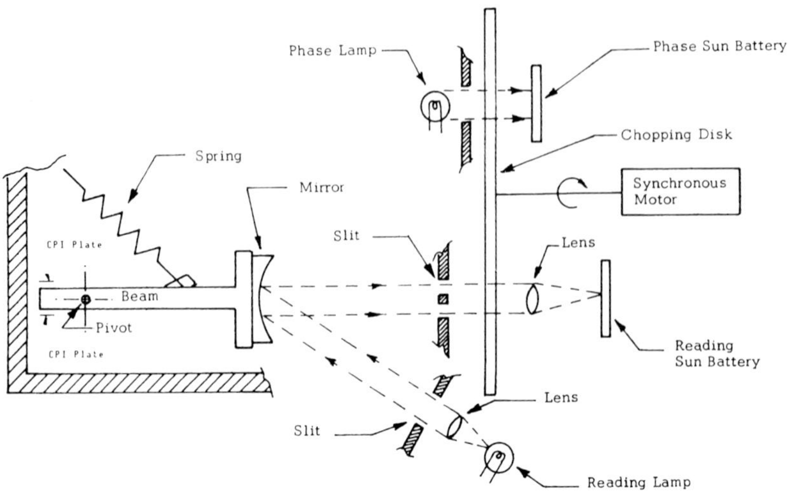

FIGURE 8. Diagram of the photoelectric system for reading beam position. The location of fixed plates for CPI is also shown. (LAR manual. With permission.)

detail later) is computed and recorded. The beam derivative is computed and added to the cross-coupling to form the total correction (TC). An electro-mechanical device controls the spring tension and computes gravity. Note that spring tension S in the diagram is a shaft rotation and not an electrical signal. A digital data acquisition system was added circa 1970 to digitize the signals and store the data on magnetic tape. A few systems at that time also included a mini-computer for additional filtering of the gravity data.

Originally, the beam position was detected optically. Figure 8 illustrates this mechanism. A beam of light, reflected from a mirror attached to the end of the beam illuminates a double slit. The intensity of the illumination from each slit varies with the position of the beam. When the beam is at the reading line (horizontal) both slits are illuminated equally. A chopper disk rotating in front of the two slits alternately images one slit and then the other onto a photocell (sun battery). The phase and amplitude of the photocell output indicates the direction and amplitude of the beam motion respectively. A second lamp, slit and photocell assembly provides a phase reference. The signals from the photocells are processed according to the block diagram shown in Figure 9.

In 1984, the photoelectric detector was replaced by a capacitance position indicator (CPI). Two fixed plates are added on either side of the beam near its tail as shown in Figure 8. The capacitance between the beam and each fixed plate varies with the beam position. These two variable capacitors are then connected to an AC bridge circuit to generate a voltage proportional to beam position. The CPI circuit has proven to be more reliable and convenient to use in the field as it eliminates the critical adjustments required after changing a lamp.

Block diagrams for the accelerometer, gyro, and cross-coupling processing circuits are shown in Figures 10 to 12 respectively. These are self explanatory. Figure 13 is a diagram of the autoreader and gravity computer. This diagram is complicated and requires some explanation. It is a combination of mechanical, electro-mechanical, and electronic components.

Let us start the explanation with the spring tension servo motor. The servo motor is driven by the beam signal B and the average spring tension velocity $\langle S' \rangle$ (where the angle bracket denotes average value). The beam signal B is obtained from the beam signal circuits shown in Figure 9. S' is obtained from a tachometer mechanically coupled to the servo motor. It provides velocity damping for the spring tension servo loop. The output of the

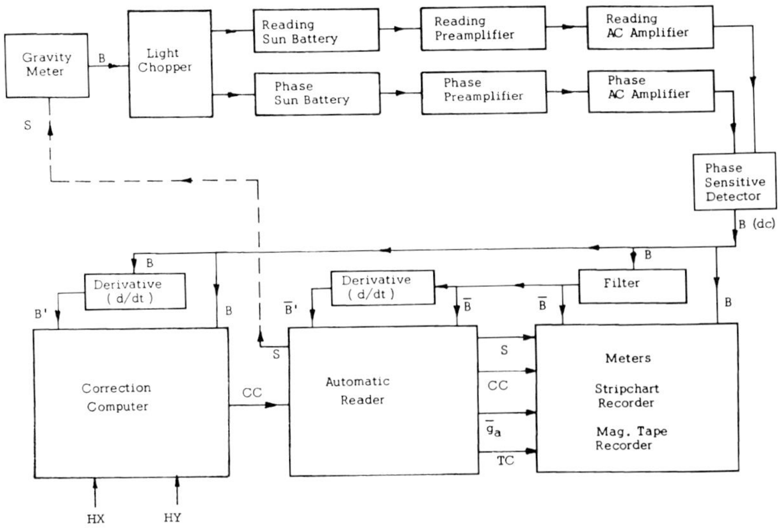

FIGURE 9. Block diagram of the beam position circuits. The dotted line indicates shaft rotation. Symbols are defined in Figure 7. (LAR manual. With permission.)

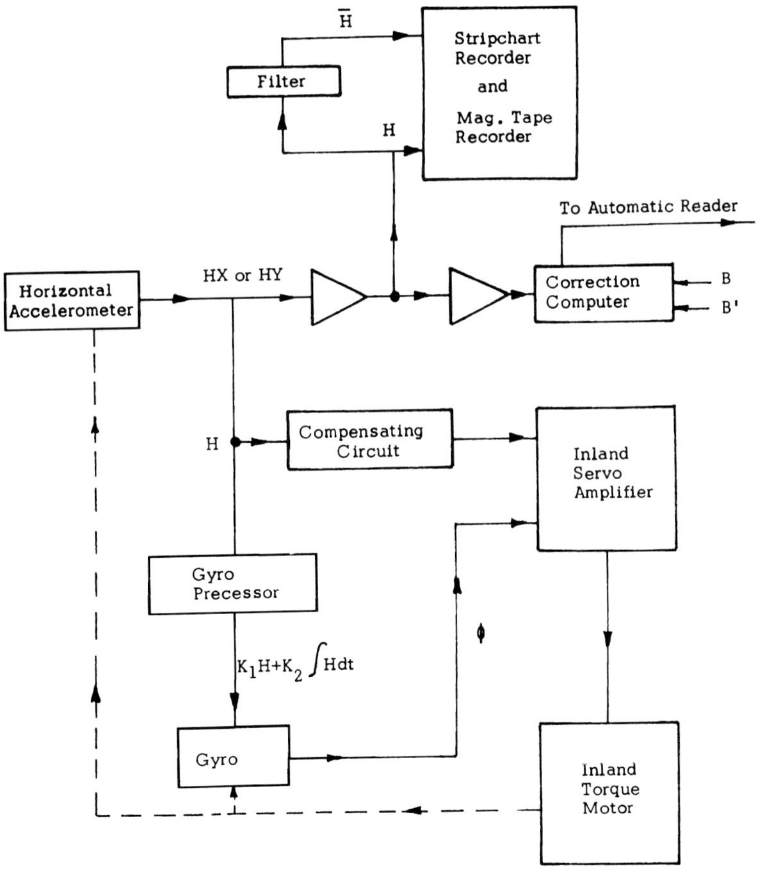

FIGURE 10. Block diagram of the accelerometer processing circuits. The dotted line implies coupled rotary motion. Symbols are defined in Figure 7. (LAR manual. Permission requested.)

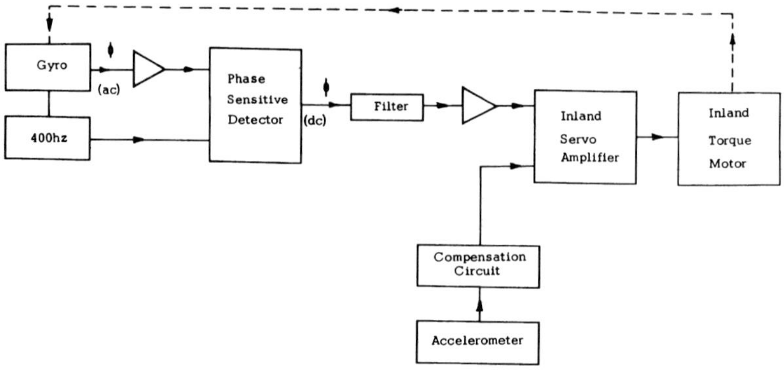

FIGURE 11. Block diagram of the gyro processing circuits. The dotted line implies coupled rotary motion. (LAR manual. With permission.)

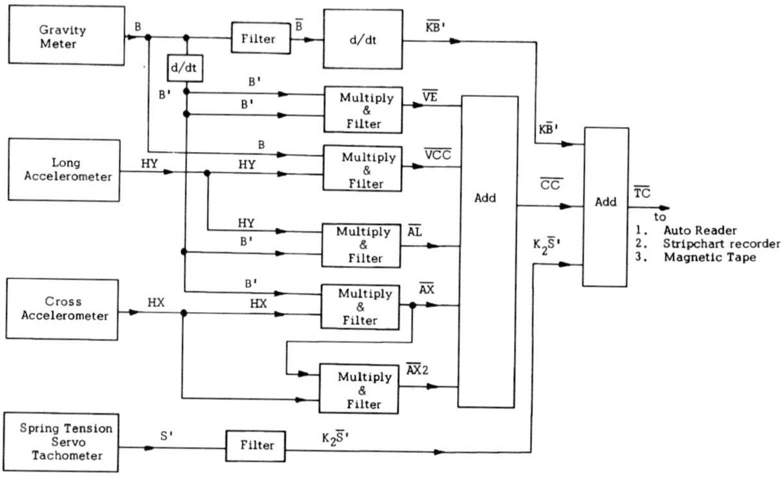

FIGURE 12. Block diagram of the cross-coupling circuits. Symbols are defined in Figure 7. Vertical bars indicate the absolute value of the enclosed parameter. (LAR manual. With permission.)

servo motor is a shaft rotation proportional to the spring tension value. This shaft connects to one input of a differential. A differential is a three shaft gear system arranged so that the rotation of the output shaft equals the sum of the rotations of the two input shafts. Its other input shaft is connected to the slew motor. The slew motor permits manual adjustment of the spring tension value.

The output of this differential, which is the sum of the rotations of the servo and slew motor rotations, is connected to a second differential, a shaft-encoder and to the gravity meter micrometer screw. The shaft-encoder digitizes the shaft angle for recording on a digital data acquisition system. The micrometer screw moves the upper end of the zero-length spring in the gravity meter to null the beam. The purpose of the second differential will become clear shortly.

The required algorithm to compute gravity is the sum of spring tension S, and total correction TC; both must be identically filtered. As the total correction was filtered during its computation as diagrammed in Figure 12, S must be filtered identically. The usual filter is a three section RC filter, each section having a time constant of 20 sec, cascaded with two more sections having a 60 sec time constant each. The overall time delay is therefore

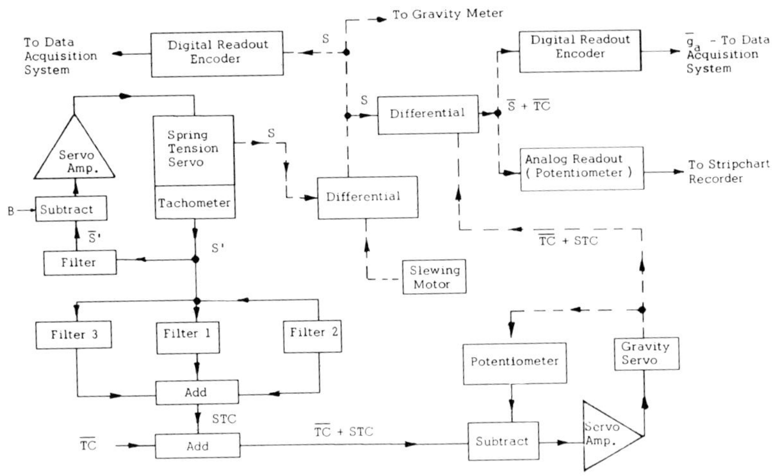

FIGURE 13. Block diagram of the autoreader and gravity computer. The dotted lines imply coupled rotary shaft motions. Symbols are defined in Figure 7. STC is the spring tension correction. (LAR manual. With permission.)

approximately 3 min. (On some systems the final stage had an optional 120 sec time constant, instead of 60 sec, for an overall time delay of approximately 4 min.) The spring tension S is in the form of a shaft rotation and cannot be filtered directly.

The circuits at the lower left-hand corner of the diagram comprising filters 1, 2 and 3 and the summing circuit generate a signal from the tachometer output (S′) called the spring tension correction STC. STC has the property that the sum S + STC = ⟨S⟩. The theory explaining how STC is generated from S′ is described in Appendix A. ⟨TC⟩ is added to STC in the next adder whose output when added to S will be gravity. Unfortunately TC + STC happens to be an electrical signal and S is a shaft rotation. The gravity servo motor shown in the lower right-hand side of the diagram rotates a shaft by an amount proportional to ⟨TC⟩ + STC. This shaft connects to the unused (so far) input of the second differential. The rotation of the second differential output shaft is then proportional to S + STC + ⟨TC⟩ which equals ⟨S⟩ + ⟨TC⟩: the desired result. This shaft is then connected to a digital shaft encoder and to a potentiometer to obtain digital and analog signals for data storage. In analog-only systems the shaft encoders are omitted and the data is recorded only on a strip chart. The autoreader, co-invented by Drs. Lucien LaCoste and Neal Clarkson, is indeed an elegant and beautiful device.

2. The Hybrid Analog-Digital System

Figure 14 is a block diagram of the hybrid system. It replaces the electro-mechanical autoreader with a personal computer. The computer also performs the final gravity filtering and logs the data to magnetic tape. A few systems also perform real-time cross-correlation. Most of the original analog circuits for processing data remain. A few systems called ''bird-dog'' systems omitted the auto-reader function, retaining the mechanical autoreader.

3. The Full Digital System

The prototype system (Valliant et al., 1985) was developed at the Earth Physics Branch of the Canadian Department of Energy, Mines and Resources. This system was built around a DEC LSI-11 microcomputer. The prototype has been in continuous service since about 1983. LAR recently commercialized the design, paying particular attention to making the system easy to use. The only changes that are made to the sensor and platform are to replace

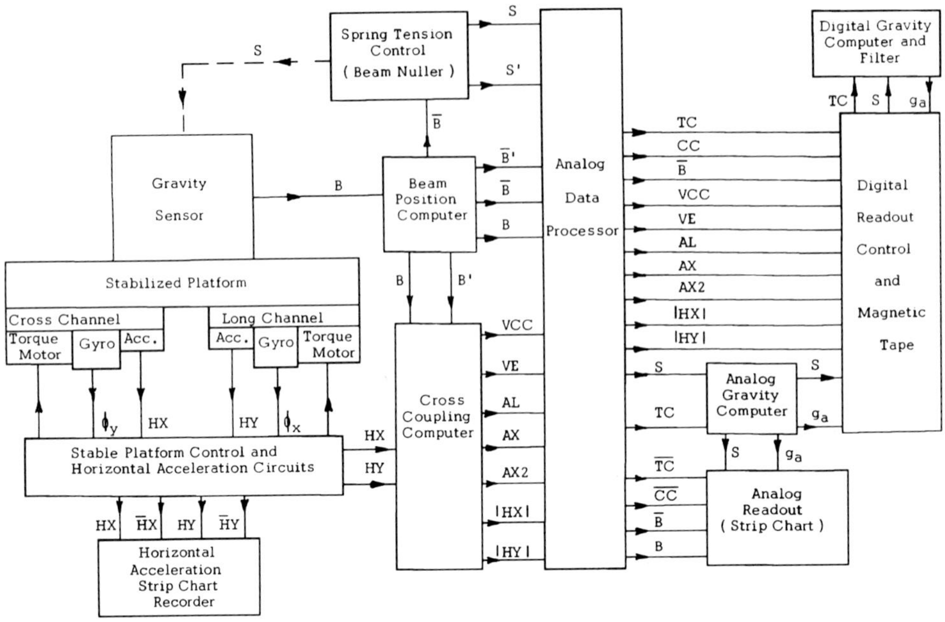

FIGURE 14. Block diagram of the hybrid system using partial analog and digital circuits. Symbols are defined in Figure 7. (LAR manual. With permission.)

the spring tension synchro motor with a stepping motor and some minor wiring changes. Figure 15 is a block diagram of the new system. It shows that all the analog electronics have been replaced by a computer and two small printed circuit boards. The platform interface board contains the heater circuits and the final amplifier for the beam signal along with its zero and gain controls. The computer interface board contains the stepping motor driver, interrupt signal generator, and the routing of wires from the main cable to the various computer cables. New, smaller (4 $^1/_2$ lb) servo amplifiers have been designed to operate the platform torque motors. A precise-frequency 200 Hz power source supplies the gyro spin motors and provides the system timing. The heart of the system is an IBM/AT clone with three expansion boards: an analog to digital converter (A/D); a digital to analog converter (D/A); and a stepping motor controller (SMC). The computer is equipped with a 40 Mbyte hard disk, a floppy-disk drive, a dot-matrix color printer, and a serial port.

The gyro, accelerometer, and beam signals are sampled 200 times a second by the D/A. The sample timing is precisely controlled by interrupting the computer with a 200 Hz signal derived from the precision oscillator. The interrupt service routine (ISR) also computes the voltages fed back to the torque motors and to the gyro inputs as explained in the theory section above. The ISR uptdates the D/A outputs 200 times a second, and the system clock once a second. The D/A outputs connect to the torque motors, via the servo (power) amplifiers, and directly to the gyro inputs. There are no analog circuits in the platform control loops except for the A/D, D/A and servo amplifiers. The CPI and heater circuits are the only other analog circuits. Finally, the ISR stores the sampled data in a data register for access by the background task. The ISR operates as a foreground task, to control the time-critical functions, executing every time an interrupt is received by the computer.

The background task runs continually in an endless polling-loop whenever the foreground task is inactive. It computes cross-coupling, gravity, filters the data, maintains a real-time clock, services keyboard activity, formats and routes data to the hard disk, printer, and/or the serial port. It also operates the spring tension servo loop. The desired spring tension rate is calculated from the input data and the SMC is updated at 10 sec intervals to turn the

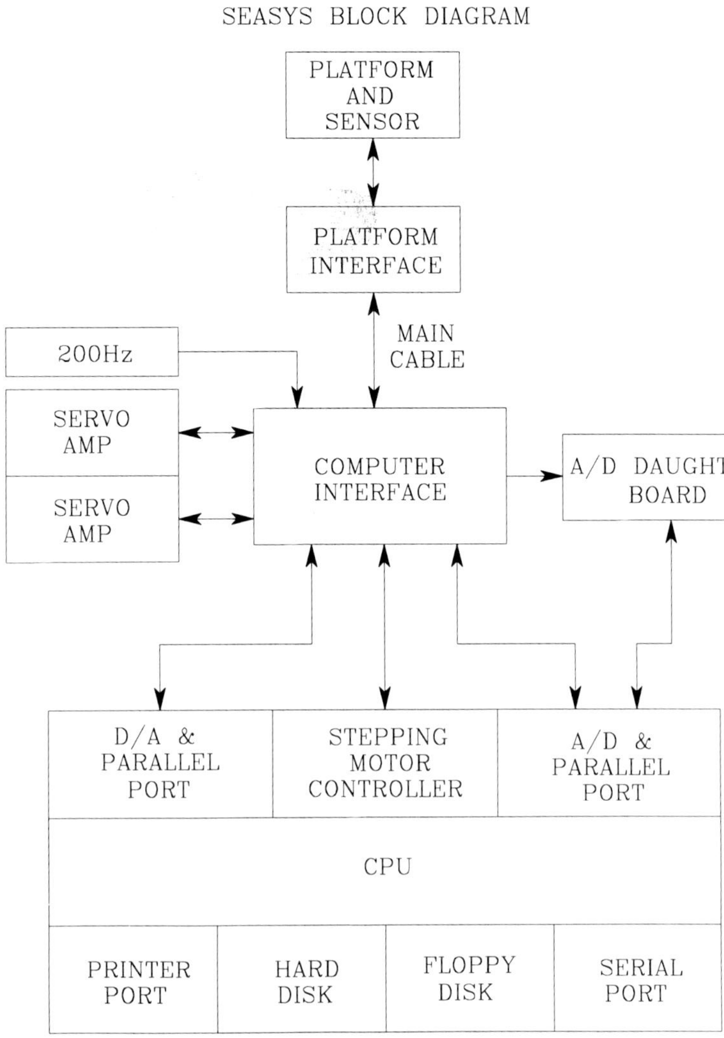

FIGURE 15. Block Diagram of the full digital system. (LAR SEASYS manual. With permission.)

stepping motor. The maximum data logging rate is 0.1 Hz. Communication with the system is by menu selection.

The system is fully automatic. It self-starts and operates without human intervention whenever power is applied. The automatic feature can, however, be aborted at any time and the system fully controlled manually. Data files are automatically created once a day at midnight and are named according to year and day number. Manual file naming is also supported. Five additional analog inputs and 24 bits of digital input are available to append user-supplied data to the data stream. An alarm system monitors critical functions and shuts the system down in the event of a major malfunction. If the alarm is triggered, one of the digital outputs is turned on which can be used to activate a remote alarm.

V. ERROR SOURCES AND THEIR CORRECTION

A. NON-LINEAR ERRORS

LaCoste et al. (1967) examined the effects of non-linear response in detail. He considered non-linearities in all of the coefficients of Equation 2 and concluded that the only significant source of error was due to non-linear damper response. LaCoste (1967, Equation 10) showed that the average error $\langle e_g \rangle$ can be expressed as follows:

$$\langle e_g \rangle = (f_1K + f_3K^2)\langle (z'')^2 \rangle \tag{6}$$

where the angle bracket denotes an average value, K is the previously defined K-factor, and the damping coefficient is defined as:

$$f = f_0 + f_1z'' + f_2B + f_3B' \tag{7}$$

The last three terms of f represent non-linear terms proportional to z'', B, and B', respectively. Equation 6, may be further simplified as:

$$\langle e_g \rangle = f_4\langle (B')^2 \rangle \tag{8}$$

where:

$$f_4 = Kf_1 + f_3 \tag{9}$$

Equation 8 is important because it shows that even if this error cannot be completely eliminated by careful adjustment, it can be easily modeled as a function of the beam velocity provided the coefficient f_4 can be determined. The average value of the square of the beam velocity is usually designated as VE. The coefficient is determined by giving the gravity meter a sinusoidal vertical displacement and observing the resulting error. The displacements and periods of the applied motion correspond to vertical accelerations of 50,000, 100,000, and 150,000 mgal, respectively. Although, strictly speaking, the VE term is not due to cross coupling, it is included as a term in the cross-coupling correction to be described next.

It is fortunate that a correction can be made because LaCoste further shows that a linearity of 1/99,999 is required to keep this error below 1 mgal for acceleration amplitudes of 100,000 mgal. The major source of non-linearity is the dampers which have two effects: First, the depth of mesh of the two damper halves varies as the beam displaces. This effect is reduced by keeping the beam motion relatively small (typically less than 1.3 mm) and by using two sets of dampers located above and below the beam. Thus, as the mesh increases on one damper, it decreases on the other. Second, the movable halves of the dampers have an arcuate motion, so that the damper clearance is constantly changing. This effect is reduced by tilting the damper so that the transverse motion is symmetrical. The error can be minimized by carefully adjusting the damper clearance. This is a tedious, time-consuming, trial-and-error procedure whereby the meter is tested, the damper adjusted minutely and the meter re-tested. This sequence is repeated as often as necessary to obtain a satisfactory error level, typically less than 4 mgal at accelerations of 100,000 mgal.

As straight-line meters utilize viscous fluid damping, damper clearances are large and therefore vary by smaller percentages. The effects of compressibility of air are also nearly eliminated. The error due to damper non-linearity with these meters is typically less that 1 mgal at accelerations of 100,000 mgal.

B. CROSS-COUPLING ERRORS

The classical references describing cross-coupling errors and their correction are LaCoste (1967), LaCoste and Harrison (1961), and Bower (1966). Cross-coupling errors are often classified in two categories: inherent and imperfection. Inherent cross-coupling is a fundamental property of the beam type gravity meter and cannot be eliminated, except by reducing the beam motion to zero. It can, however, be reduced somewhat by restricting the beam motion through increased damping. There is a practical limit to this approach as the beam velocity is also reduced and thereby the resolution of the gravity measurement. Inherent cross-coupling is not present in straight-line meters due to the different geometry, however.

Imperfection cross-coupling arises because the suspension is not stiff enough. An ideal system is one that has very low compliance in the vertical direction and is perfectly rigid in all other directions. This ideal can never be realized in practice since all mechanical systems must exhibit some elasticity. This is especially true for instruments as delicate as gravity meters which are expected to have a resolution of 0.1 mgal in the presence of 100,000 mgal accelerations in both the vertical and horizontal directions. This is equivalent to a signal-to-noise ratio of 10^{-6}.

We will first look at systematic errors in a general sense and then consider specific cases. LaCoste (1967, Equation 64) showed an interesting way to look at these errors. He considered that they could be described as a power series of velocities and accelerations. Furthermore he was able to disregard all first order terms because they either averaged to zero or were elements of the Eotvos correction. The power series is then:

$$e_g = a_1\langle(x'')^2\rangle + a_2\langle(y'')^2\rangle + a_3\langle(z'')^2\rangle + a_4\langle(x')^2\rangle + a_5\langle(y')^2\rangle + a_6\langle(z')^2\rangle + a_7\langle x''z'\rangle + a_8\langle y''z'\rangle + a_9\langle x''z''\rangle + a_{10}\langle y''z''\rangle \ldots + a_n\langle(x'')^2z''\rangle + \ldots \quad (10)$$

where the x, y, and z axes are transverse to the beam, along the beam, and vertical, respectively.

The first two terms arise from platform leveling errors to be described later. The third and sixth terms arise from damping non-linearity as described previously. The fourth and fifth terms have negligible effect on the gravity meter and can be disregarded.

The sixth term is negligible for LAR meters. The seventh and eighth terms are due to inherent cross-coupling. The seventh term, due to horizontal accelerations transverse to the beam, is negligible and can be disregarded. The ninth and tenth terms are due to imperfection cross-coupling. The nth term arises from second order imperfection cross-coupling. Eliminating the negligible terms we are left with:

$$e_g = a_1\langle(x'')^2\rangle + a_2\langle(y'')^2\rangle + a_3\langle(z'')^2\rangle + a_8\langle y''z'\rangle + a_9\langle x''z''\rangle + a_{10}\langle y''z''\rangle \ldots + a_n\langle(x'')^2z''\rangle + \ldots \quad (11)$$

Although not all of these terms arise from cross-coupling effects, they are generally lumped together in one correction called the cross-coupling correction. Equation 11 is important because it indicates that a correction can be calculated as a linear combination of observable quantities. In order to calculate a correction, the terms in angle brackets need to be computed and averaged in real time. These values form the seven cross-coupling monitors that are recorded as part of the gravity meter's data stream. The coefficients are determined empirically or statistically through cross-correlation analysis. Each of the monitors have been assigned names:

$\langle x''^2\rangle = \langle XACC\rangle^2 \propto$ ⟨square of cross accelerometer⟩
$\langle y''^2\rangle = \langle LACC\rangle^2 \propto$ ⟨square of long accelerometer⟩
$\langle z''^2\rangle = VE \propto$ ⟨square of beam velocity⟩
$\langle y''z'\rangle = VCC \propto$ ⟨long accelerometer X beam position⟩
$\langle x''z''\rangle = AX \propto$ ⟨cross accelerometer X beam velocity⟩
$\langle y''z''\rangle = AL \propto$ ⟨long accelerometer X beam velocity⟩
$\langle x''^2z''\rangle = AX2 \propto$ ⟨(cross accelerometer)^2 X beam velocity⟩

Although the first two of these monitors are computed and recorded, their coefficient is normally set to zero. They are recorded so that they can be included in cross-correlation analyses to determine if platform errors are present. The last five terms comprise the normally calculated cross-coupling correction. VE has been discussed in the section on non-linear effects. We will now attempt to relate the last four in terms of the physics of the beam-type gravity meter. The following discussion is not pertinent to the straight-line meter.

1. Inherent Cross-Coupling

LaCoste (1967) shows that we must add a term $y''\tan\Theta$ to the left-hand side of Equation 2 to account for horizontal accelerations when they are present. As this is the only term added, it also represents the error in gravity resulting from the horizontal accelerations. Thus:

$$e_g = y''\tan\theta \simeq y''\theta \tag{12}$$

He further shows that the average error becomes:

$$\langle e_g \rangle = \frac{1}{2}(y_1''\theta_1\cos\phi) \tag{13}$$

where y_1'' and Θ_1 are the amplitudes of the sinusoidal longitudinal acceleration and beam motion respectively and ϕ is the phase difference between them. We have shown previously that $\Theta' \propto z''$ and hence $\Theta \propto dz''/dt$. Therefore, for sinusoidal accelerations Θ and z' are proportional to each other with a $\pi/2$ phase difference. We can then express Equation 13 as:

$$\langle e_g \rangle \propto y_1''z_1'\cos\left(\phi + \frac{\pi}{2}\right) \tag{14}$$

This equation shows that when $\phi = \pi/2$ (i.e. y'' and z' are in quadrature) the error is maximum. This phase relation occurs if the motion is circular. This condition generally exists to some extent at sea because the motion of water particles in waves is generally elliptical. If the meter is mounted off the ship's roll axis, it will be subjected to diagonal motion when the ship rolls. The phase angle ϕ is 0 for this type of motion and therefore $\langle e_g \rangle$ is also zero. Diagonal motion does not contribute to inherent cross-coupling. It should also be noted that if the meter is located off the roll-pitch axes, additional vertical accelerations will be generated that can exacerbate the non-linear error described above. Finally, Equation 14 shows that the error is proportional to the product of the longitudinal acceleration and beam position which is the cross-coupling monitor previously identified as VCC. There is no inherent cross-coupling in the straight-line meter as horizontal accelerations are always orthogonal to the axis of the mass displacement.

2. First-Order Imperfection Cross Coupling

This form of cross-coupling error arises almost entirely from small horizontal displacements of the gravity meter beam induced by horizontal accelerations due to the elasticity of its suspension. LaCoste (1967) describes two possible sources: both deal with longitudinal horizontal accelerations. In one instance, the stretching of the beam suspension causes a shift in the relative location of the center-of-gravity of the beam. In the second instance he describes the effect of wires attached to the beam to provide stiffness in the transverse direction.

These causes have turned out to be minor and LaCoste (personal communication, 1990) has subsequently found that damping is the dominant source of imperfection cross-coupling. As the theory of LAR air dampers has not been previously published, it is included in Appendix B. The dampers are formed as interleaving concentric cylinders whose cross section is shown in Figure 1, Appendix B. The clearance is typically 0.1 mm. Appendix B states that a horizontal acceleration of 30,000 mgal will deflect the damper by approximately 0.0026 mm. If the dampers are exactly centered the resulting error will alternate positive and negative and the average error will be zero. Exact centering is utopia. Appendix B shows that if the dampers are eccentric by 0.0025 mm, an error of 57 mgal is generated in the presence of 30,000 mgal vertical and horizontal accelerations. This discussion shows that this form of cross-coupling error can be minimized, even possibly eliminated, by very careful adjustment of the dampers. As this error can arise from both longitudinal and transverse horizontal accelerations, the dampers must be carefully adjusted in two directions. Appendix B also shows that this error can also be reduced by making the ratio of the dimensions (Figure 1, Appendix B) A/B = 1.8. The damper deflection is proportional to x'' in one direction and y'' in the other so that AX and AL are suitable monitors to correct for this error.

Because the straight-line meter uses liquid damping, the damper clearances are much larger and these errors are negligible.

3. Second-Order Imperfection Cross Coupling

Appendix B also describes a different form of damping error. It arises because the damper response is not linear, but has a second-order term. Accordingly an error is generated when the vertical acceleration has twice the frequency of the horizontal acceleration. The appendix shows that for a vertical acceleration of 15,000 mgal and a horizontal acceleration of 50,000 mgal the error will amount to 20 mgal. This error is shown in Appendix B to be proportional to the product of x''^2 or y''^2 and z''. Therefore, AX2 is the appropriate monitor to correct for this form of cross coupling. Appendix B further shows that the error can also be minimized by making the ratio of the clearances A/B (Figure 1, Appendix B) equal to 1.8.

Again, due to liquid damping, this error is negligible in the straight-line gravity meter.

4. Calibrating the Cross-Coupling Correction

Calibrating the cross-coupling correction requires the determination of the coefficients a_1 to a_n in Equation 11 above. The method is to generate appropriate gravity meter motions that can induce each term of the cross-coupling correction independently. The coefficient is then determined directly as the magnitude of the error divided by the magnitude of the monitor. The previous discussions suggest the type of motions required. To calibrate VE a vertical motion is appropriate. The calibration of VCC is done by moving the meter in a circular motion in a vertical plane. AX and AL requires a diagonal motion generated by moving the meter up and down a ramp. AX2 is calibrated by moving the gravity meter over an arc of a circle. $XACC^2$ and $LACC^2$ are not calibrated as they are only recorded and not normally used in the cross-coupling correction.

The compensation voltage should be adjusted before calibrating the cross-coupling coefficients. This is done by subjecting the gravity meter and platform to sinusoidal horizontal accelerations and adjusting the fraction of the accelerometer output used for compensation to minimize the error in observed gravity.

It is appropriate to perform these calibration tests at acceleration amplitudes consistent with those found at sea. Consequently the tests are performed with acceleration amplitudes up to 100,000 mgal. However, generating large accelerations with periods that are experienced at sea would require huge tests machines. To be practical, smaller machines operating at higher frequencies are used. Periods from 3 to 5 sec are usually employed. The results at these shorter periods must be extrapolated to those expected at sea (dc to say around 10 sec). The frequency response of the system is designed to extend beyond the test frequencies. The extrapolation is therefore not expected to present any difficulties and, indeed, many years of experience clearly indicates that the method is satisfactory.

5. Calibration of Cross-Coupling by Cross-Correlation

A statistical method for calibrating the cross-coupling correction using real data observed at sea was developed by LaCoste (1973). The method is summarized in Appendix C. The advantage of this technique is that it can monitor any temporal changes in the calibrations. Laboratory calibrations are performed under ideal conditions and the meter shipped to its destination. The critical nature of the various mechanical adjustments have been previously described and these are subject to change as a result of rough handling during shipment. AX is particularly prone to this problem. Also, the calibration tests described above utilize well-defined motions, whereas motions at sea are random combinations of all of these motions and others. Therefore, testing with real data can lend a degree of confidence to the cross-coupling corrections.

The technique uses real data to determine optimal values of coefficients a_1 to a_n by multiple linear regression. These are the coefficients that minimize the square of the curvature (roughness) of observed gravity. The coefficients determined by this method normally agree well with those determined empirically. When they do not agree, a change in the calibration or some other problem is indicated. When there is disagreement, the gravity data can be recomputed with the new coefficients. All the data required to perform cross-correlation and to compute corrected gravity is included in the digital data recorded by the gravity system. Many operators avoid recording the digital data. Some are using obsolete analog-only systems that pre-date digital recording; others utilize their own data-logging systems that only record gravity; still others log the complete data record but at a slower rate. These operators forfeit the opportunity for quality control that cross-correlation affords. It is extremely desirable to cross-correlate some or all of the data from each and every cruise to keep an on-going record of a meter's performance. Cross-correlation software is readily available (e.g. from LAR, L. LaCoste, Geological Survey of Canada, among others) and its use relatively inexpensive.

Choosing the correct data to correlate is important. First of all, correlation is a statistical process that requires a large data set to give correct results. Also, the results need to be determined for average conditions. Therefore, data from as many sea states and conditions as possible should be included. Data acquired while the ship is turning or during other abnormal events should not be used as it is not representative. Data acquired under extremely smooth conditions (small curvature) neither contribute to nor harm the solution. The practice of performing cross-correlation one line at a time is incorrect and should be discouraged. This practice can lead to excessive smoothing of gravity as well as unwanted noise.

Another approach suggested by LaCoste (personal communication, 1989) is to determine the coefficients required to minimize the errors at line crossings. A preliminary test indicates that the idea works, although the test data had too few cross-over points. The method does require that the survey be well structured, with a large and well-distributed number of line crossings.

6. Real-Time Cross-Correlation

It is possible to perform regression analysis iteratively and thereby do cross-correlation in real time. Real-time cross-correlation is available on computer equipped air/sea systems. The main criticism appears to be that the computed coefficients tend to vary with time. If the time-constant of the solution is too short, one is in danger of performing cross-correlation on a line-by-line basis which was criticized in the previous section.

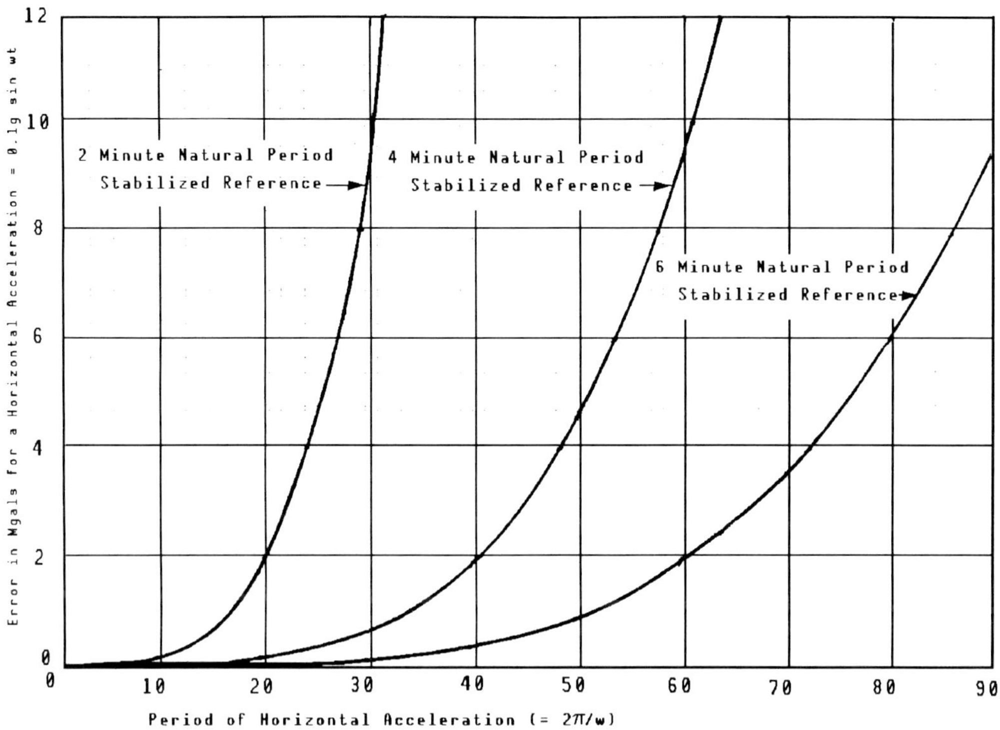

FIGURE 16. Variation of gravity meter error with period of horizontal accelerations and period of the stable platform for damping equal to 0.707 critical. (LaCoste 1967, Figure 8.)

C. PLATFORM ERRORS

We previously noted that the platform is accurate to 8 sec of arc with 50,000 mgal horizontal accelerations (LaCoste et al. 1967). The reference did not indicate the period of the horizontal acceleration, though it can be inferred from the LAR horizontal test machine whose period is approximately 3.5 sec for 50,000 mgal. The period of the acceleration is important. LaCoste (1967, Equation 38) gives the following equation for the average error due to sinusoidal horizontal accelerations for a flat Earth approximation as follows:

$$\langle e_g \rangle = \frac{(w_0/w)^4}{(w_0/w)^4 + 1} \left\langle \frac{(x'')^2}{2g} \right\rangle \tag{15}$$

where w_0 = natural period of the platform (see Equation 5), and w = period of horizontal acceleration.

The equation shows that the error is greater when the period of the horizontal accelerations is long or the natural period of the platform is short. Figure 16 (LaCoste 1967, Figure 8) shows this relation. Valliant (1979) extended the analysis to a spherical Earth. The results are not significantly different for the marine environment and relatively short platform periods but may be of interest for airplane measurements. Equation 15 shows that the average error is proportional to x''^2 and for the case of horizontal accelerations in both directions to $(x''^2 + y''^2)$. This equation indicates that $XACC^2$ and $LACC^2$ are good monitors to check for platform errors.

One possible approach to correct for errors due to horizontal accelerations is suggested by Equation 5. We can rewrite the equation in Laplace notation as follows:

$$\phi = -\left[\frac{2fw_0s + w_0^2}{s^2 + 2fw_0s + w_0^2}\right]\left(\frac{x''}{g}\right) \tag{16}$$

where s is the Laplace operator and the portion of the equation in square brackets is called the system transfer function. It is a simple procedure to develop a digital filter (using the Z transform, for example) having the same transfer function. We can then filter the horizontal accelerations to estimate the platform level error. The error in gravity can then be calculated from LaCoste's (1967) Equation 30:

$$e_g \simeq -x''\phi - g\frac{\phi^2}{2} \tag{17}$$

Equations 16 and 17 offer the possibility of computing platform level errors is real-time. The horizontal accelerations need to be derived from navigation information, although the accelerometer signals might suffice for a first approximation. Navigation data have generally lacked sufficient accuracy to make the idea attractive. The advent of GPS makes the idea more promising, however.

Another approach to calculate a correction for horizontal accelerations is developed in Appendix D. A correction can be determined as follows:

$$C = \frac{(A^2 - a^2)}{2g} \tag{18}$$

where $A = \sqrt{A_x^2 + A_L^2}$ = resultant accelerometer output, $a = \sqrt{a_E^2 + a_N^2}$ = resultant true horizontal accelerations, and C = the correction to gravity.

This equation shows that a correction can be deduced from a knowledge of real horizontal accelerations and accelerometer outputs. Precise navigation is required to determine true horizontal accelerations. Furthermore, the two observables in this equation must be filtered identically. The frequency and phase response of the navigation system must be known so that it can be compensated. If the navigation filtering is less than is applied to the accelerometer filter, an additional compensating filter can be designed. If the navigation filter is longer than the accelerometer filter, the navigation must be deconvolved and then filtered with the same filter (3 × 20 sec) as applied to the accelerometer. As squaring is a nonlinear operation, the amount of filtering applied to the observables can have a large effect. Also, the order of filtering, before or after squaring, will effect the results.

At the present time LaCoste (personal communication, 1990) is attempting to devise an appropriate filtering scheme. When used in conjunction with a three-axis platform, this method has the potential to compute gravity corrections in real time and permit gravity observations during ship maneuvers. LaCoste (personal communication, 1990) has found that these two methods appear to give equivalent results. Indeed, their equivalence is demanded by theory, and can be used to test filtering schemes.

The foregoing discussions do not include errors due to azimuth rotations. These can be large and are reduced by conducting the survey in relatively straight lines. Suppose for example, the platform is sitting still on the laboratory floor and oriented according to the cardinal directions. The east-sensing gyro is being torqued with a voltage proportional to the Earth's angular rate of rotation. As explained above, the torquing voltage includes the integral of the accelerometer output. Furthermore, suppose that the platform is instantaneously rotated 180° in azimuth without translation. No horizontal accelerations are generated. The platform is now being torqued in the wrong direction about its north-south axis to counteract the Earth's rotation. The only method to restore equilibrium is for the integral of the accelerometer signal to reverse sign, which requires a large accelerometer error signal. Therefore, a large platform transient will be observed. The time to restore equilibrium is about 1.5 platform periods (normally 6 min) unless the accelerometer signal saturates (which is likely), substantially increasing the duration of the transient. The problem of azimuth rotation is eliminated with the three-axis platform which maintains a constant azimuth. However, with appropriate modeling of the transient from observed azimuths, it is possible to use these equations for two-axis platforms as well. Although the three-axis platform eliminates transients due to azimuth rotations it does not correct for sustained horizontal accelerations usually associated with heading changes. These errors must be dealt with as described above.

VI. NAVIGATION WITH THE THREE-AXIS PLATFORM

The mechanization of the three-axis platform is very similar to commercial inertial survey systems. Accordingly, it is possible to derive navigation information from the platform. The theory was developed by Valliant and LaCoste (1976) and involves the simultaneous solution of two differential equations describing the east and north velocities using the accelerometer and gyro signals as forcing functions. Unless the equations are solved with correct initial conditions, the solutions will oscillate with an 84 min period. Therefore, the equations are solved iteratively, starting with approximate initial conditions, until the correct initial conditions are determined. The equations are finally solved using the predetermined initial conditions to calculate east and north velocities. Infrequent and relatively inaccurate ancillary navigation is required to control gyro drift and offset. Valliant (1981) indicates that navigation accurate to between 200 and 400 m is feasible depending on the quality and frequency of the ancillary navigation updates.

Of more interest to marine gravity surveys, Valliant and LaCoste (1976) showed that the Eotvos correction (Glicken, 1962) can be obtained directly from the platform data. They further showed that the accuracy of the inertially determined Eotvos correction is intermediate between those calculated using LORAN "C" and HI-FIX navigation data.

VII. CONCLUDING REMARK

I have had the good fortune to be closely associated with Dr. L. LaCoste since about 1972. I dedicate this inadequate summary of his work to this true gentleman, great physicist, ingenious engineer, and good friend.

I would be remiss to omit the contribution of Dr. Neal Clarkson, whose wizardry with electronics and mechanical devices translated Dr. LaCoste's concepts into functioning hardware. Also, Mr. Larry Burris contributed much to the mechanical design of the straight-line sensor.

APPENDIX A

EXPLANATION OF SPRING TENSION CORRECTION

For simplicity, we will examine the problem for a single-stage filter only. The extension to multiple-stage filters would follow the same development. We will use x to represent the spring tension instead of S to avoid confusion with the Laplace operator s. The single stage filter equation is given in Laplace form as:

$$\langle x \rangle = \frac{w_0}{s + w_0} x \tag{1}$$

where x = spring tension, w_0 = 2π/filter time constant, s = Laplace operator, and <> denotes the filtered value of the enclosed parameter.

We can take the derivative of both sides of Equation 1 in Laplace notation as follows:

$$s\langle x\rangle = \frac{sw_0}{s + w_0}\, x \tag{2}$$

We can also rewrite Equation 1 as follows:

$$s\langle x\rangle + w_0\langle x\rangle = w_0 x \tag{1'}$$

or

$$\langle x\rangle = x - \frac{s}{w_0}\langle x\rangle \tag{1''}$$

Substituting from Equation 2 we get:

$$\langle x\rangle = x - \left(\frac{s}{s + w_0}\right)x = x - \left(\frac{1}{s + w_0}\right)x' \tag{3}$$

where the prime denotes differentiation with time. Equation 3 shows that

$$\langle x\rangle = x + STC \tag{3'}$$

where

$$STC = -\left(\frac{1}{s + w_0}\right)x' \tag{4}$$

The term enclosed in brackets represents a filter transfer function and Equation 4 indicates that STC can be obtained by appropriately filtering x′ which is the unfiltered spring tension velocity or tachometer output. Three such filters are used in the autoreader circuit to accommodate the multiple stages of filtering that are employed.

APPENDIX B

DAMPER THEORY

The following is reproduced verbatim from previously unpublished notes of Dr. L. LaCoste:

STUDY 5-65

CALCULATION OF DAMPING FOR AN INTERLEAVING CONCENTRIC CYLINDER TYPE OF FLUID DAMPER

The type of damper that has been used in LR S-gravity meters consists of interleaving concentric cylinders such as is shown in cross section in Figure 1. In previous dampers, the clearances A and B have been made equal as nearly as practical. The present study will attempt to calculate the damping of this type of damper both for A and B equal and unequal and to calculate the effect on the damping of shifting one damper slightly laterally (in the x direction) relative to the other damper.

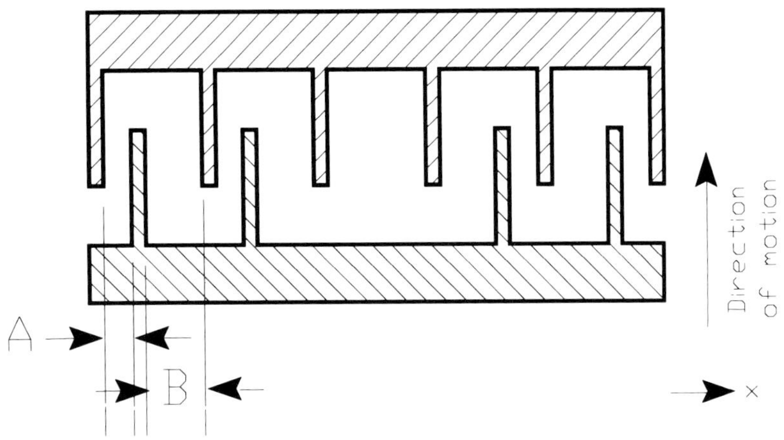

FIGURE 1. The cross section of the air damper used on Model S meters.

In order to simplify the analysis, the following approximations will be made:

(a) The flow of air (or other fluid) thru the damper will be assumed to be normal to the rings. Air flow parallel to the rings will be neglected.
(b) Laminar flow will be assumed.
(c) Rather than performing the calculations for circular grooves in the dampers, they will be performed for rectangular grooves. This approximation will later be eliminated for the case in which the dampers are approximately centered.

It is believed that the above simplifications will not affect the significant features of the results obtained.

In order to be able to calculate the damping, it is necessary to use an equation giving the volume of air that will flow per second thru a rectangular aperture for a given pressure. Such an equation was derived in Study 2-65; it is eq (15). It can be written as:

$$V = KpA^3 \tag{1}$$

where:

V = the volume per unit time
K = a constant
A = the width of the rectangular aperture, and
p = the pressure difference

(Author's note: The form of this equation can be verified in the Handbook of Chemistry and Physics, CRC Press)

Since there are two different size apertures in Fig 1, it is necessary to derive a suitable equation from Equation 1. Referring to Fig 2 this can be done as follows: From Equation 1 we have:

$$V = Kp_AA^3 = Kp_BB^3 \tag{2}$$

therefore

$$p_B = \frac{p_AA^3}{B^3} \tag{3}$$

and

$$p = p_A + p_B = p_A + p_A\frac{A^3}{B^3} \tag{4}$$

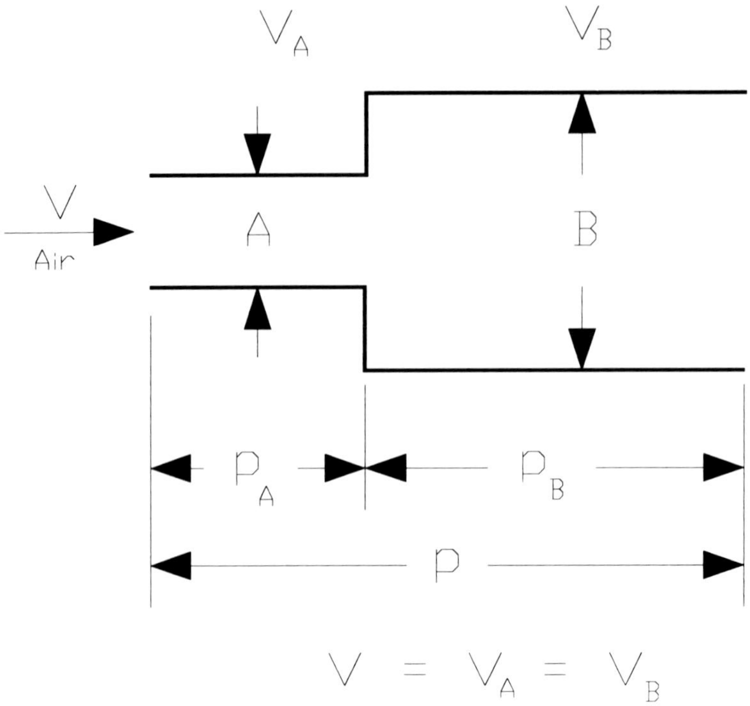

FIGURE 2. Cross-section of fluid path.

or

$$p = p_A\left(\frac{B^3 + A^3}{B^3}\right) \tag{4'}$$

Eliminating p_A from Equations 2 and 4′ gives:

$$V = Kp\left(\frac{A^3B^3}{A^3 + B^3}\right) \tag{5}$$

We will now apply Equation 5 to the damper of Fig 1. First, however, we will assume that the lower part of the damper is displaced a distance x to the right of the symmetrical position shown. The clearances on the right will then become

$$A - x \text{ and } B + x$$

and those on the left will become

$$A + x \text{ and } B - x$$

Using Equation 5 the total volume of air escaping from the damper per second can be calculated, Including all four directions it is:

$$V = Kp\left[\frac{(A + x)^3(B - x)^3}{(A + x)^3 + (b - x)^3} + \frac{(A - x)^3(B + x)^3}{(A - x)^3 + (B + x)^3} + \frac{2a^3B^3}{A^3 + B^3}\right] \tag{6}$$

or

$$V = Kp\left[\frac{[AB + (B - A)\,x - x^2]^3}{(A^3 + B^3) + 3\,(A^2 - B^2)\,x + 3\,(A + B)x^2} + \frac{[AB + (A - B)\,x - x^2]^3}{(A^3 + B^3) + 3\,(B^2 - A^2)\,x + 3\,(A + B)\,x^2} + \frac{2A^3B^3}{A^3 + B^3}\right] \quad (7)$$

For convenience we will take

$$B = kA \quad (8)$$

and

$$s = \frac{x}{A} \quad (9)$$

then Equation 7 becomes

$$V = KA^3p\left[\frac{[k + (k - 1)\,s - s^2]^3}{(k^3 + 1) - 3\,(k^2 - 1)\,s + 3\,(k + 1)\,s^2} + \frac{[k - (k - 1)\,s - s^2]^3}{(k^3 + 1) + 3\,(k^2 - 1)\,s + 3\,(k + 1)\,s^2} + \frac{2k^3}{k^3 + 1}\right] \quad (10)$$

In subsequent work we will be using as a reference for damping, the damping when s = 0, or when the dampers are centered. Therefore it will be convenient to use an expression for damping which is normalized to equal 1 when s = 0. We can obtain such an expression from Equation 10 as follows. We will define D from the relation

$$V \equiv kA^3p\left(\frac{4k^3}{k^3 + 1}\right)\left(\frac{1}{D}\right) \quad (11)$$

Then from Equation 10

$$\frac{1}{D} = \left(\frac{k^3 + 1}{4k^3}\right)\left(\frac{[k + (k - 1)\,s - s^2]^3}{(k^3 + 1) + 3\,(k^2 - 1)\,s + 3\,(k + 1)\,s^2} + \frac{[k - (k - 1)\,s - s^2]^3}{(k^3 + 1) + 3\,(k^2 - 1)\,s + 3\,(k + 1)\,s^2}\right) + \frac{1}{2} \quad (12)$$

Equation 11 shows that D is proportional to p/V, therefore D represents damping. Also Equation 12 shows that D = 1 when s = 0. Accordingly we will use it to represent normalized damping.

Since the dampers used in previous overdamped LR gravity meters had A = B or k = 1, we will evaluate Equation 12 for the case. We obtain:

$$\frac{1}{D_1} = \left(\frac{1}{2}\right)\left(\frac{[1 - s^2]^3}{1 + 3s^2}\right) + \frac{1}{2} \quad (13)$$

For small values of s, Equation 13 can be approximated by:

$$\frac{1}{D_1} \doteq 1 - 3s^2 \quad (14)$$

Equation 14 shows that for s = 0 the first derivative of $1/D_1$ is zero but the second derivative is not. This curvature gives rise to cross coupling errors when the vertical accelerations has twice the frequency of the horizontal accelerations. An estimate of the magnitude of this effect can be made for one of the present S gravity meters. For meter S19, which is 10 times stiffer than earlier meters, the beam was found to deflect 0.00075″ for 222 gals of horizontal acceleration. Therefore for 50 gals the deflection will be 0.00017″. Since the damper clearance is ±0.004″, the preceding deflection corresponds to:

$$s = 0.00017'' \div 0.004'' = 0.042''^{*} \tag{15}$$

Substituting this value of s into Equation 14 gives:

$$\frac{1}{D_1} = 1 - 0.0053 \tag{16}$$

This corresponds to a damping change of 0.53% or ± 0.27%. If the vertical acceleration has a peak amplitude of 15 gals, then the average error will be:

$$\bar{e} = \frac{1}{2}(15{,}000 \text{ mgal})\ 0.0027 \tag{17}$$

where the factor 1/2 is introduced because it is the average value of $\sin^2\Theta$. The error corresponding to a 50 gal horizontal acceleration and a 15 gal vertical acceleration of twice the frequency of the horizontal acceleration therefore calculates to be:

$$\bar{e} = 20 \text{ mgal} \tag{18}$$

which is of the same order of magnitude as that found experimentally for S19.

The preceding example shows that curvature of the damping characteristic is very undesirable because of the double frequency cross coupling error. It is also very undesirable because of the need for very accurate damper centering in order to make the ordinary cross coupling effect negligible. A second calculation for S19 will illustrate this. We will assume both horizontal and vertical maximum accelerations of 30 gals and that the damper is off center by 0.0001″, which corresponds to a value of s = 0.025 also. Therefore the two values of s to be substituted into Equation 14 are:

$$s_1 = 0.025 + 0.025 = 0.05, \text{ and}$$

$$s_2 = 0.025 - 0.025 = 0$$

substituting into Equation 14 gives

$$\left(\frac{1}{D_1}\right)_2 - \left(\frac{1}{D_1}\right)_1 = 3(0.05)^2 = 0.0075 = 0.75\% \tag{19}$$

or ± 0.0038. This corresponds to a cross coupling error of:

$$e = \frac{1}{2}(30{,}000 \text{ mgal})\ 0.0038$$

$$= 57 \text{ mgal} \tag{20}$$

It is important to determine how the two preceding cross coupling errors are affected by the stiffness of the beam suspension. The ordinary cross coupling error described in the last paragraph is proportional to the lateral yielding of the beam times the slope of the damper characteristic curve. Therefore, stiffening the suspension by a factor n will reduce the ordinary cross coupling errors by a factor 1/n. This reduction in error will make accurate damper centering less critical by a factor n also, because the slope of the parabolic damper characteristic is proportional to the amount of uncentering of the dampers.

The initially described double frequency cross coupling error, however, is affected even more by stiffening the suspension. Since this error is proportional to the ordinate of the damper characteristic, it is decreased by a factor of $1/n^2$ if the beam suspension is stiffened by a factor n.

The preceding examples and considerations show that it is very desirable to reduce the curvature of the damper characteristic at least in the vicinity of s = 0. In order to determine whether this is feasible, we can proceed as

* *Note:* Here the double prime is used to indicate inches.

follows. One way to determine the curvature is from the second derivative, however, it is also possible to determine the second derivative by expressing Equation 12 as a ratio of two polynomials in s. Since such a polynomial expression might be very instructive, this method will be used, Adding the two terms in the parenthesis of Equation 12 gives:

$$\frac{1}{D} = \left(\frac{1 + k^3}{2k^3}\right)\left(\frac{k^3(1 + k^3) + 3k(1 - k^2 - k^3 + k^5)s^2 + 3k(-3 + k + k^2 - 3k^3)s^4 + (-1 + 9k + 9k^2 - k^3)s^6 - 3(1 + k)s^8}{(1 + k^3)^2 + 3(-1 + 2k + 6k^2 + 2k^3 - k^4) + 9(1 + k)^2 s^4}\right) + \frac{1}{2} \tag{21}$$

Neglecting powers of s higher than the second, Equation 21 can be approximated by:

$$\frac{1}{D} = \left(\frac{1}{2}\right)\left(\frac{1 + \dfrac{3k(1 - k^2 - k^3 + k^5)}{k^3(1 + k^3)}s^2}{1 + \dfrac{3(-1 + 2k + 6k^2 + 2k^3 - k^4)}{(1 + k^3)^2}s^2}\right) + \frac{1}{2} \tag{22}$$

The condition for Equation 22 to be independent of 5^2 is:

$$\frac{1 - k^2 - k^3 + k^5}{k^2} = \frac{-1 + 2k + 6k^2 + 2k^3 - k^4}{1 + k^3} \tag{23}$$

or

$$1 - 2k^3 - 6k^4 - 2k^5 + k^8 = 0 \tag{24}$$

Solving Equation 24 numerically gives:

$$k = 1.805 \tag{25}$$

APPENDIX C

CROSS-CORRELATION MATHEMATICS

The mathematics behind the process are most succinctly described in matrix notation. Matrix notation will be used here. Refer to LaCoste (1973) for a more detailed development.

We wish to find, in a least-squares sense, the optimum amounts of certain recorded monitors to be added to gravity to eliminate any correlation between the curvatures (first derivative) of the monitors and gravity.

For any given observation the observables are the monitor values and gravity. These are first filtered appropriately and then differentiated to obtain the curvatures for each monitor and for gravity. LaCoste (1973) gives details of the desired filtering.

If we represent the observed curvature of the monitor values as the matrix **M** and the vector of unknown fractions as **H** and the vector of gravity curvatures as **G**, we can write a set of observation equations as:

$$\mathbf{MH} = -\mathbf{G}$$

The negative sign occurs because we desire a correction to be added rather than an error to be subtracted. As there are many more observations than unknowns we need to solve this set of equations using the least-squares principle. To do this, normal equations are formed as follows:

$$\mathbf{M^TMH} = -\mathbf{M^TG}$$

The solution to the normal equation is then found as:

$$\mathbf{H} = -(\mathbf{M^TM})^{-1}\mathbf{M^TG}$$

The vector **H** contains the desired monitor fractions to be used to correct gravity. This process is multiple linear regression. However the elements of matrix $\mathbf{M^TM}$ are the sums of all possible cross-products of the monitors and the vector $\mathbf{M^TG}$ turns out to be all possible cross products between the monitors and gravity. Hence, for this case, the term cross-correlation is applicable.

APPENDIX D

CORRECTION FOR HORIZONTAL ACCELERATIONS

The following is reproduced verbatim from previously unpublished notes of Dr. L. LaCoste:

June 1981

HORIZONTAL ACCELERATION CORRECTION FOR AN AIRPLANE GRAVITY METER

We will consider the case in which we know:

g_m = gravity readings,
A_c = cross accelerometer readings,
A_L = long accelerometer readings,
a_E = east horizontal accelerations (from navigation system),
a_N = north horizontal accelerations (from navigation system),

and we know that the sensitive axes of g_m, A_c, and A_L are orthogonal.

We will let:

$$A = \sqrt{A_c^2 + A_L^2} = \text{resultant accelerometer output}$$

$$a = \sqrt{a_E^2 + a_N^2} = \text{resultant horizontal accelaration}$$

The resultant acceleration R can be expressed in the following two ways:

$$R^2 = g_m^2 + A^2 = (g + z'')^2 + a^2 \tag{1}$$

where g = the acceleration corresponding to gravity, and z″ = the vertical acceleration.

From the above equation we obtain:

$$(g + z'')^2 = g_m^2 + A^2 - a^2 \tag{2}$$

or

$$g + z'' = \sqrt{g_m^2 + A^2 - a^2} \tag{3}$$

So far we have made no approximations. Since A^2 and a^2 are small compared to g_m^2 in the case in which we are interested, we can aprpoximate (3) as follows:

$$g + z'' = g_m\sqrt{1 + \frac{A^2 - a^2}{g_m^2}}$$

$$\doteq g_m\left(1 + \frac{A^2 - a^2}{2g_m^2}\right)$$

$$\doteq g_m + \frac{A^2 - a^2}{2g_m} \tag{4}$$

The horizontal acceleration correction is then

$$C \doteq \frac{A^2 - a^2}{2g_m} \tag{5}$$

or very nearly

$$C = \frac{A^2 - a^2}{2g} \tag{5'}$$

We should note that C depends only on A^2 and a^2, not on their components.

If we let g_m = the gravity meter reading, then the corrected gravity is

$$g \doteq g_m + \frac{A^2 - a^2}{2g} \tag{6}$$

assuming that other necessary corrections have been made. As previously stated, Equation 6 is based on the assumption that the sensitive axes of the accelerometers are orthogonal to the sensitive axis of the gravity meter. We will now let ΔA_c and ΔA_L represent any bias errors in A_c and A_L respectively. Equation 6 then gives an incorrect value for gravity, namely:

$$g' = g_m + \frac{(A_c + \Delta A_c)^2 + (A_L + \Delta A_L)^2 - a^2}{2g}$$

$$= g_m + \frac{A^2 - a^2}{2g} + \frac{2A_c\Delta A_c + 2A_L\Delta A_L + (\Delta A_c)^2 + (\Delta A_L)^2}{2g} \tag{7}$$

or

$$g' = g + \frac{(\Delta A_c)^2 + (\Delta A_L)^2}{2g} + \frac{A_c\Delta A_c + A_L\Delta A_L}{2g} \tag{8}$$

We will now show that ΔA_c and ΔA_L can be determined by crosscorrelating the time variations of g′ with the corresponding variations of A_c and A_L. To show this we will consider each of the three terms on the right of Equation 8, the first term g is correct gravity, whose variations should not correlate appreciably with A_c and A_L, since gravity is nearly perpendicular to the sensitive axes of the accelerometers. The second term on the right of Equation 8 is a constant and therefore does not correlate with any variations in A_c and A_L. Therefore the third term in Equation 8 gives very nearly the entire correlation between the variations in g′ and those in A_c and A_L.

A crosscorrelation analysis will therefore determine what fractions of A_c and A_L must be subtracted from g′ to remove the crosscorrelations. These fractions are equal respectively to A_c/g and A_L/g. It is therefore possible to make accurate corrections for level biasing errors.

REFERENCES

Anon. Theory of operation long period vertical seismometer: W. F. Sprengnether Instrument Co., Form No. DW-600A, undated.

Bower, D. R., The determination of cross-coupling errors in the measurement of gravity and sea, *J. Geophys. Res.*, 71, 487—493, 1966.

Glicken, M., Eotvos corrections for a moving gravity meter, *Geophysics,* 27, 531—533, 1962.

Harrison, J. C., Tests of the LaCoste & Romberg surface ship gravity meter, *J. Geophys. Res.*, 64, 1875—1881, 1959.

LaCoste, L. J. B., A new type long period vertical seismograph, *Physics,* 5, 178—180, 1934.

LaCoste, L. J. B., A simplification on the conditions for the zero-length-spring seismograph, Bull. Seismo. Soc. Am. 25, 1935.

LaCoste, L. J. B. and Romberg, A., Force measuring instrument, US Patent 2,377,889, 1945.

LaCoste, L. J. B., Crosscorrelation method for evaluating and correcting shipboard gravity data, *Geophysics,* 38, 701—709, 1973.

LaCoste, L. J. B., LaCoste & Romberg straight line gravity meter, *Geophysics,* 48, 606—610, 1983.

LaCoste, L. J. B., The zero-length spring gravity meter, *Geophysics,* 7, 20—21, 1988.

LaCoste, L. J. B., Clarkson, N., and Hamilton, G., LaCoste & Romberg stabilized platform shipboard gravity meter, *Geophysics,* 32, 99—109, 1967.

LaCoste, L. B. J. and Harrison, J. C., Some theoretical considerations in the measurement of gravity at sea, *Geophys. J. R. Astron. Soc.*, 5, 89—103, 1961.

Scarborough, J. B., *The Gyroscope Theory and Applications,* Interscience Publishers Inc., New York, 1958.

Valliant, H. D., Reevaluation of the effects of horizontal accelerations on the LaCoste & Romberg gravimeter platform, *Geophysics,* 44, 826—829, 1979.

Valliant, H. D., Marine navigation with the LaCoste & Romberg inertial platform: *Geophysics,* 46, 1469—1470, 1981.

Valliant, H. D., Field trials with the LaCoste & Romberg straight-line gravity meter, *Geophysics,* 48, 611—617, 1983.

Valliant, H. D., Gravity meter calibration at LaCoste & Romberg, *Geophysics,* 56, 705—711, 1990.

Valliant, H. D. and Cooper, R. V., Position measurements with the LaCoste & Romberg air/sea gravimeter, *Geophysics,* 46, 40—44, 1981.

Valliant, H. D., Halpenny, J., and Cooper, R. V., A microprocessor based controller and data acquisition system for LaCoste & Romberg air/sea gravity meters, *Geophysics,* 50, 840—845, 1985.

Valliant, H. D. and LaCoste, L. J. B., Theory and evaluation of the LaCoste & Romberg three-axis inertial platform for marine gravimetry, *Geophysics,* 41, 459—467, 1976.

Chapter 8

HIGH DEFINITION SEISMIC STRATIGRAPHY: CONTINENTAL SLOPE, NORTHERN GULF OF MEXICO

Peter K. Trabant and Clyde Lee

TABLE OF CONTENTS

I. INTRODUCTION

High definition multifold seismic reflection data provide an effective tool for detailed analysis of the internal facies or system tracts within stratigraphic sequences deposited over the past half-million years. The term "high definition geophysics", or HDG, is hereby applied to seismic reflection recordings in the frequency range of 60 to 1000Hz, which display the strata geometries and genetically related sedimentary packages associated with depositional sequences. The term does not apply to those high-resolution geophysical systems such as sparkers, which by virtue of their pulse signatures do not reveal fine stratigraphic details.

The principles of seismic stratigraphy developed during the 1970s by Vail and others (Mitchum et al., 1977) have proven an invaluable tool toward the interpretation of marine stratigraphic sequences from seismic reflection data. The method is based on the acoustic response of seismic reflections from sedimentary strata deposited as sequence system tracts, which are the product of interactions between subsidence, compaction, eustatic fluctuations in sea level, and spatial variations in sediment supply. In petroleum exploration the technique deals with time scales of a 100,000 years and greater associated with 4th and 3rd order glacio-eustatic cycles.

Shallow penetrating (1 to 2 s) high-resolution (HRG) seismic reflection survey methods are used primarily to investigate the sea bed for hazards to drilling and production operations (Trabant, 1984). HRG can provide a valuable tool for the detailed analysis of shorter term depositional sequences and events; including those associated with 4th and 5th order para-cyclic eustatic fluctuations in sea level on time scales of less than a hundred thousand years, and even very short term depositional events such as slumping. The ability to identify such smaller "sequences" has been discussed by Lopez (1990) in terms of resolution with respect to time. His time versus depositional features chart emphasized the existence of an observational gap covering events on time scales of 100 to 40,000 years. It is precisely this gap which is covered by HDG seismic profiles, albeit not for sequences as deep as the Gulf Coast Tertiary but very adequate for the upper Pleistocene.

The following discussion covers the interpretation of several HDG/HRG profiles from the continental slope, northern Gulf of Mexico. The application of the seismic stratigraphic method in the interpretation of such profiles greatly enhances our present understanding of system tract geometries within sedimentary sequences, providing practical models for the interpretation of deeper seismic reflection data; particularly in the evaluation of the seismic stratigraphic method and its application to sedimentological processes and accumulations, the modeling of petroleum reservoirs, and the deciphering of global climatic fluctuations during the past half-million years.

A detailed biostratigraphic study covering the entire Pleistocene was performed on a 297 m (975 ft) core from the lower slope of the northwestern Gulf of Mexico by Beard et al. (1982) in which they recognized the correlation between glacial cycles and seismic sequences on HRG profiles. They concluded that most major Pleistocene depositional sequences are related to eustatic cycles of sea-level change, and that their age and depositional character may be inferred from seismic reflection couplets corresponding to low stand/high stand systems tracts.

On the basis of exploration seismic reflection data from the Mississippi Fan, Feely et al. (1990) mapped 13 sequences encompassing the past 3.5 million years on the basis of correlations between seismic stratigraphy and fluctuations in sea level based upon oxygen isotope data.

Coleman and Roberts (1990), working with shallow (less than 600 ft) engineering borehole data, which included oxygen isotope chronostratigraphy and HRG survey profiles, have correlated seismic condensed sections associated with sea-level high stands across the continental shelf and uppermost slope offshore Louisiana.

Piper and Perissoratis (1991), using shallow cores in conjunction with a 3.5 kHz sub-bottom profiler and Boomer records, described the shallow seismic stratigraphy of the past 150ka for the northern Aegean continental shelf and upper slope to depths of 33 m (100 ft).

II. HIGH DEFINITION GEOPHYSICS

The interpretation of multifold exploration seismic reflections by the seismic stratigraphic method yields information primarily on 3rd order depositional sequences for ages greater than 3 million years and 4th order sequences for younger deposits associated with global cyclic fluctuations in sea level. HDG profiles, on the other hand, depict those stratigraphic sequences associated with 4th (100,000 years) and 5th order (paracycles) fluctuations in sea level. These include such short term events as slumps and deposits associated with rapid spatial variations in sediment supply.

The difference between these two types of data is strictly a function of the resolution of the techniques which are a function of source frequency and reflection data sampling rate. While standard exploration geophysical methods employ frequencies centered in the range of 10 to 40 Hz, HDG involves frequencies greater than 100Hz. The resultant trade off is the limited penetration achieved by HDG data, which is generally less than 2 s. Thus, exploration seismic reflection data with sampling rates of 2 to 4 msec is geared toward objectives at time-depths of 2 to 6 s and a theoretical resolution of approximately 3 to 8 m (10 to 25 ft) based upon velocities of 1750 to 3650 m/s (5740 to 12,000 ft/s). However, HDG multifold sections sampled at rates of 0.25 to 1.0 msec provide an observed resolution on the order of 1 to 2 m (3 to 6 ft) within the uppermost second.

Aside from their widespread application for engineering "geohazards" investigations and as a single-channel seismic system with academia, high resolution geophysical tools have been of little value to date in offshore petroleum exploration, due to their limited penetration to depths required for exploration objectives. However, an improved understanding of the seismic response associated with recent sedimentary deposits should prove a valuable asset to the seismic stratigraphic method when applied to deeper objectives. The following discussion centers on the application of the seismic stratigraphic method toward the interpretation of HRG and HDG profiles in terms of identification of stratigraphic sequences, systems tracts and the correlation of reflectors to local and global fluctuations in sea level inferred from oxygen isotope paleotemperature curves.

III. CONTINENTAL SLOPE: NORTHERN GULF OF MEXICO

The upper Cenozoic geology of the northern Gulf of Mexico is relatively complex as a result of interaction between salt tectonism (Martin, 1978; and West, 1989) and cyclical sedimentary depositional phases associated with changes of sea level induced by global climatic fluctuations (Woodbury et al., 1978; and Feely et al., 1990). The effect of these interactions has been to shift repeatedly sedimentary depo-centers between the inner continental shelf and the upper continental slope areas. The resulting Cenozoic stratigraphy of the slope thus consists of cyclical sedimentary sequences associated with alternating rapid and slow rates of sediment accumulation (Coleman and Roberts, 1990), causing a seaward progradation of the continental shelf edge over time. In addition, the sea floor topography of the upper slope area is very hummocky (Gealy, 1956) and includes water depths ranging from 200 to over 2000 m (Bouma et al., 1989).

During sea level lowstands, the supply of sediment from the Mississippi River Valley drainage system and coastal plain is discharged closer to the shelf edge and upper slope, developing numerous relatively thick shelf-margin deltas and submarine fans (Suter, 1986).

These depositional systems produce the low stand systems tracts (LSTs of Posamentier and Vail, 1988) and expanded units (Coleman and Roberts, 1990) identifiable on HRG seismic sections. Sediment loading from the shelf edge depo-centers stimulates isostatic adjustment of the underlying salt, producing diapirism which form and deepen inter-diapiric basins on the continental slope (West, 1989). During high stands in sea level, deposition retreats shoreward of the shelf edge, producing transition systems tracts (TSTs) while reducing sediment input to the outer shelf and slope, where condensed sections are formed in association with the high stand systems tracts (HST) on seismic reflection records.

In their discussion of processes on the Mississippi Delta, Coleman and Prior (1981) have suggested that instabilities resulting from the rapid deposition of deltaic sediments on the outer shelf and upper continental slope during low stands of sea level also lead to sediment mass failures. Chaotic slump and debris flow deposits (Mass Transport Complexes of Weimer, 1990) are thus common on the upper continental slope and can be quite extensive when associated with delta-canyon-fan systems as well as salt tectonics (Sangree et al., 1988). The presence of salt diapirs, shelf-edge deltaic depo-centers and associated meandering rivers, and cyclical fluctuations in sea level provide mechanisms by which sediments accumulate on the upper continental slope. A number of anomalous features unique to upper slope areas have been discovered in the review of HRG profiles from the upper slope. These include brine pools (Trabant, and Presley, 1978; MacDonald et al., 1990); bottom simulating reflectors (BSRs) and gas hydrates (Shipley et al., 1979; Brooks et al., 1984); mud volcanoes (Neurauter and Bryant, 1989); and natural gas eruptions or blow-outs (Prior et al., 1989).

The lack of available paleostratigraphic and sedimentologic data from depths to 1000 ft below the sea floor on the upper slope, with the exception of that of Beard et al. (1982), precludes precise timing of shallow seismic stratigraphic sequences at this time. Published data from engineering boreholes and Leg 96 of the Ocean Drilling Project (ODP) reach a depth of less than 200 m and cover only a few locations along the shelf break and uppermost slope (Coleman and Roberts, 1990, Bouma et al., 1986). These published chronostratigraphic data may, however, be judiciously extrapolated when approximating the ages of seismic stratigraphic units on the slope.

IV. SEQUENCE STRATIGRAPHY: UPPER CONTINENTAL SLOPE

Sequence stratigraphic models provide a rational basis for the analysis of shallow seismic stratigraphic sequences. The nomenclature and resulting depositional system tracts associated with a single cycle in sea level are depicted in Figure 1, from Vail and Wornardt (1990). Seismic reflectors generated by specific depositional environments and events may be matched to phase variations of global or local sea level curves based on the oxygen isotope record and tied to biostratigraphic time markers, such as those of Trainor and Williams (1987).

The validity of this method has been shown for the Gulf of Mexico's lower slope (Beard et al., 1982), continental shelf (Coleman and Roberts, 1990) and Mississippi Fan (Feely et al.; 1990), and for the northern Aegean Sea (Piper and Perissoratis, 1991). The method is illustrated by the profile in Figure 2, in which the global oxygen isotope curve of Trainor and Williams (1987) is matched to a HDG seismic section.

The methodology for performing seismic stratigraphic analyses on high resolution profiles consists in picking the condensed sections associated with the base of reflector couplets. These reflectors are identified by their relatively strong acoustic character and continuity, and frequently occur as one of a number of closely spaced parallel reflectors. Where erosion has taken place the sequence follows the top of the unconformity. These horizons represent maximum flooding stages (mfs) and denote sequence boundaries (SBs) associated with sea-level high stands (HSTs). Once picked, the sequence boundaries mark the base of each individual eustatic cycle, with variations in thickness attributable to either variations in the

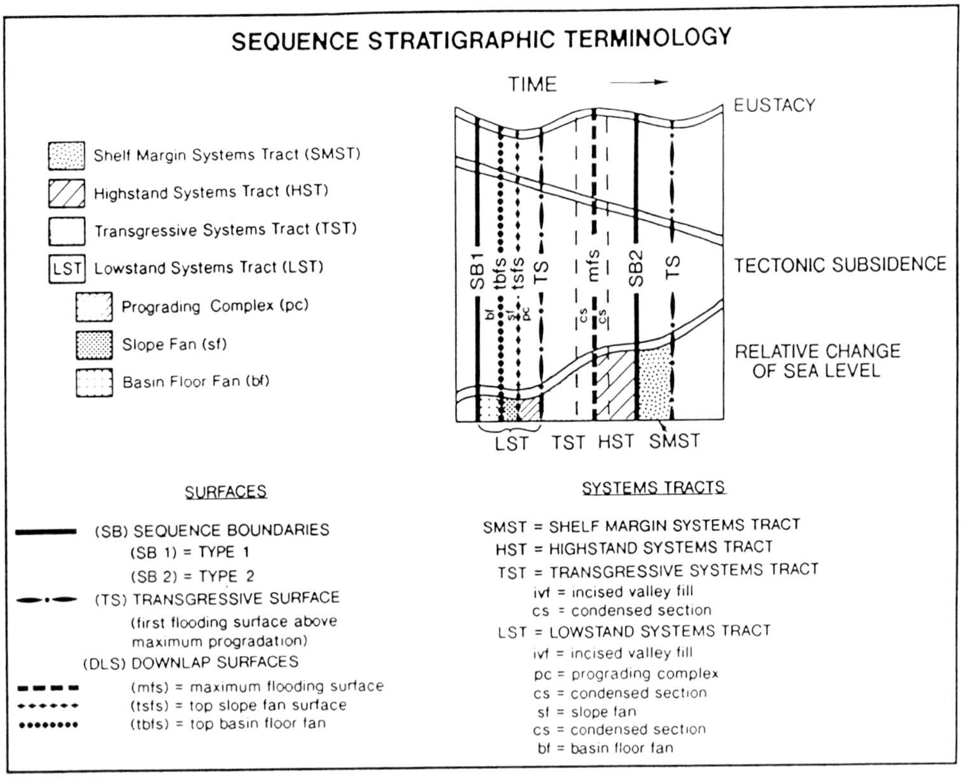

FIGURE 1. Sequence stratigraphic terminology for a single eustatic cycle. (From Vail and Wornardt, GCSSEPM Foundation 11th Annu. Res. Conf. Abstracts, 1990. With permission.)

duration of eustatic cycles or sediment deposition. Overlying the sequence boundaries, one frequently finds acoustically transparent zones or chaotic reflections, indicative of more rapid and sometimes localized sedimentary deposits. These interboundary features represent increased deposition rates associated with low stands in sea level (LSTs), and may be subdivided into systems tracts. However, when working with small line segments on the order of a few miles, the precise nature of a systems tract may not be readily apparent, and one must be content in recognizing the presence of LSTs by their seismic signature alone.

The following interpretations are thus approximations where sequence stratigraphic ties and date assignments serve more to illustrate the possibilities rather than act as a perfectly tuned model to the eustatic curve and associated events, until borehole information and a composite oxygen isotope curve for the northern Gulf of Mexico become available.

V. CASE STUDIES

The following high-resolution seismic sections were obtained during the course of geoengineering investigations for hazards to the emplacement of drilling and production structures. The sections provide an exercise in the application of the seismic stratigraphic method to the interpretation of the uppermost sedimentary sequences while revealing some of the complexities associated with individual system tracts.

A. LINE A: SHELF EDGE GROWTH FAULT

The high resolution record illustrated in Figures 3 and 4 has been obtained just south of the continental shelf break. The line is oriented from west to east in water depths of 200

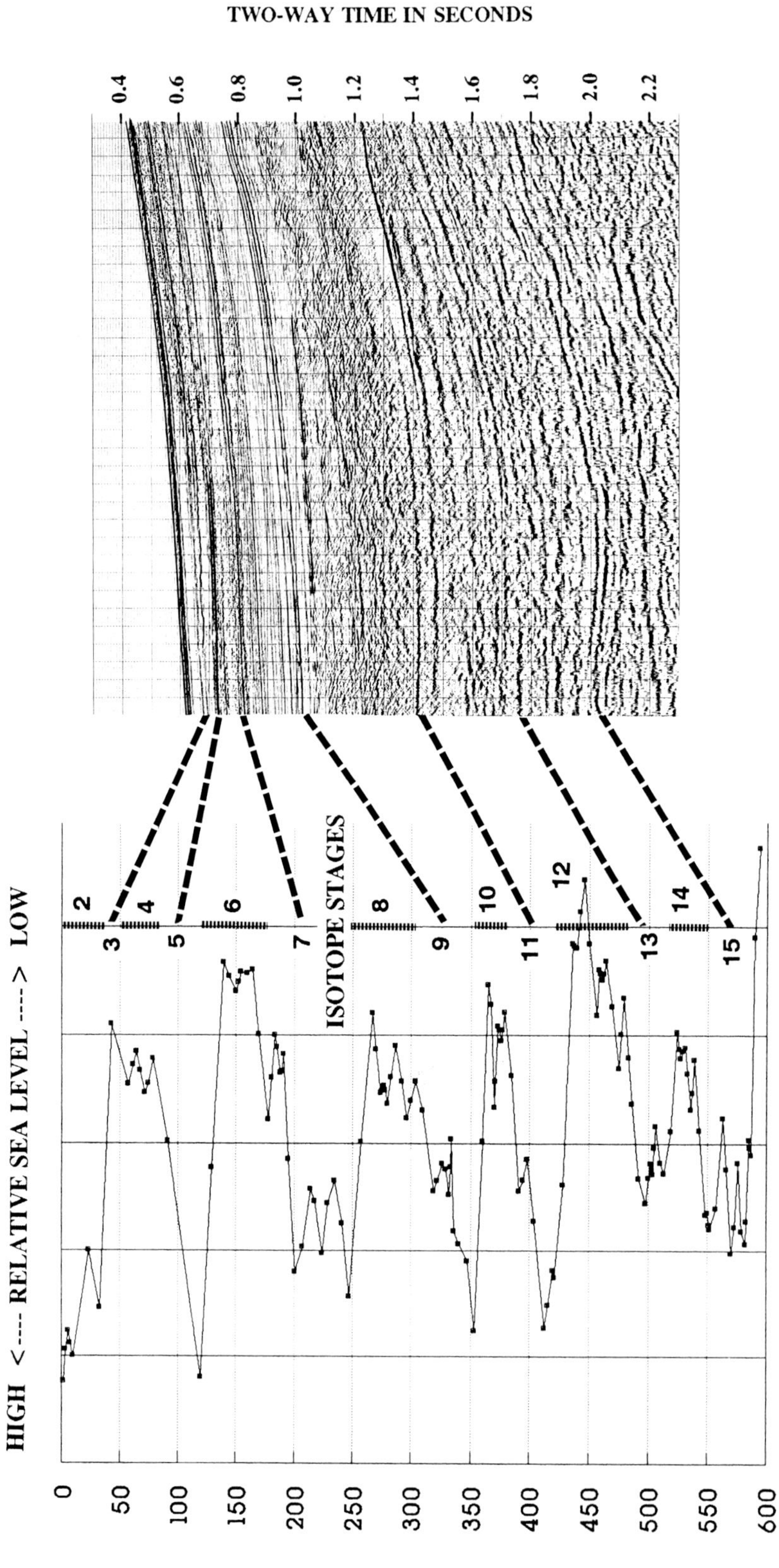

FIGURE 2. Correlation of oxygen isotope paleotemperature curve and HDG seismic section. (From Trainor and Williams, *Trans Gulf Coast Assoc. Geol. Soc.*, 37, 247, 1987. With permission.)

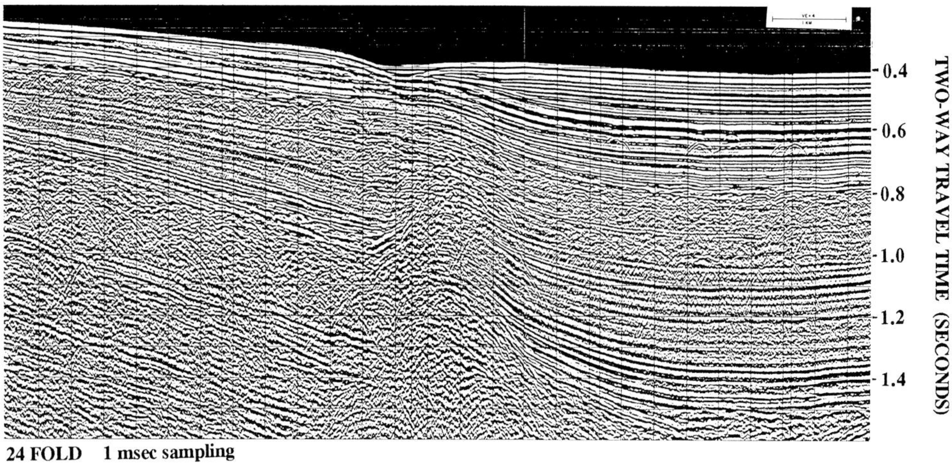

FIGURE 3. Line A: High resolution multifold migrated seismic profile, upper continental slope, Northern Gulf of Mexico. (Courtesy of BP Exploration Inc. With permission.)

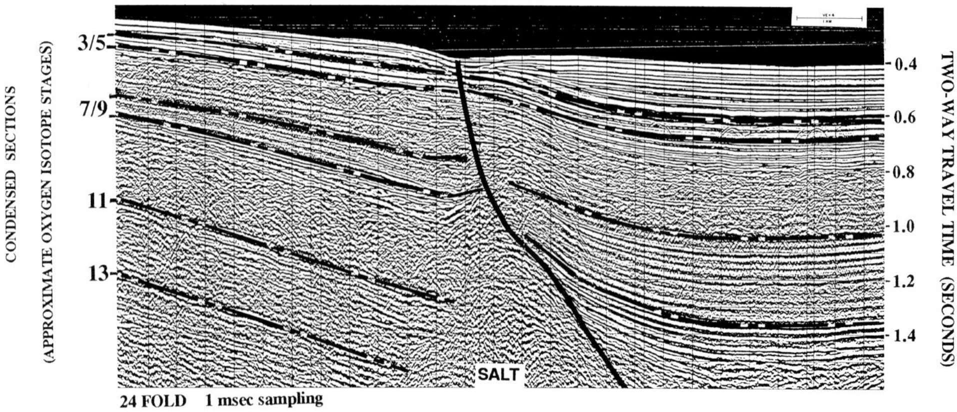

FIGURE 4. Interpretation of Line A (Figure 3), with nomenclature as per Figure 1. (Courtesy of BP Exploration Inc. With permission.)

to 320 m (656 to 1050 ft), and is a high resolution (HRG) section by virtue of the seismic source and sampling rate.

The profile was obtained using a single 80 in^3 Watergun® source and a 48-channel hydrophone streamer. The data was digitally recorded at sample rate of 1 msec to a total time-depth of 2 s. Processing was performed for 24-fold coverage to produce a migrated section for optimum definition of the shallow stratigraphy.

Line A represents a dip section basinward of the shelf edge. A salt diapir is offset from the center of the record with an associated radial growth fault. The fault's offset increases away from the salt structure producing a sea floor scarp greater than 10 m (30 ft) and greatly expanded sections on the down-thrown block. The major sequences are apparent and consist of repeated cycles of stronger continuous reflectors, characteristic of condensed sections (odd-numbered sequence boundaries in Figure 4), representing deposition during maximum flooding stages (mfs) associated with sea-level high stands (HSTs); which overlie thick chaotic sequences, representing expanded sections associated with more rapid deposition during low stands in sea level (LSTs). The section has been correlated with a shallow boring

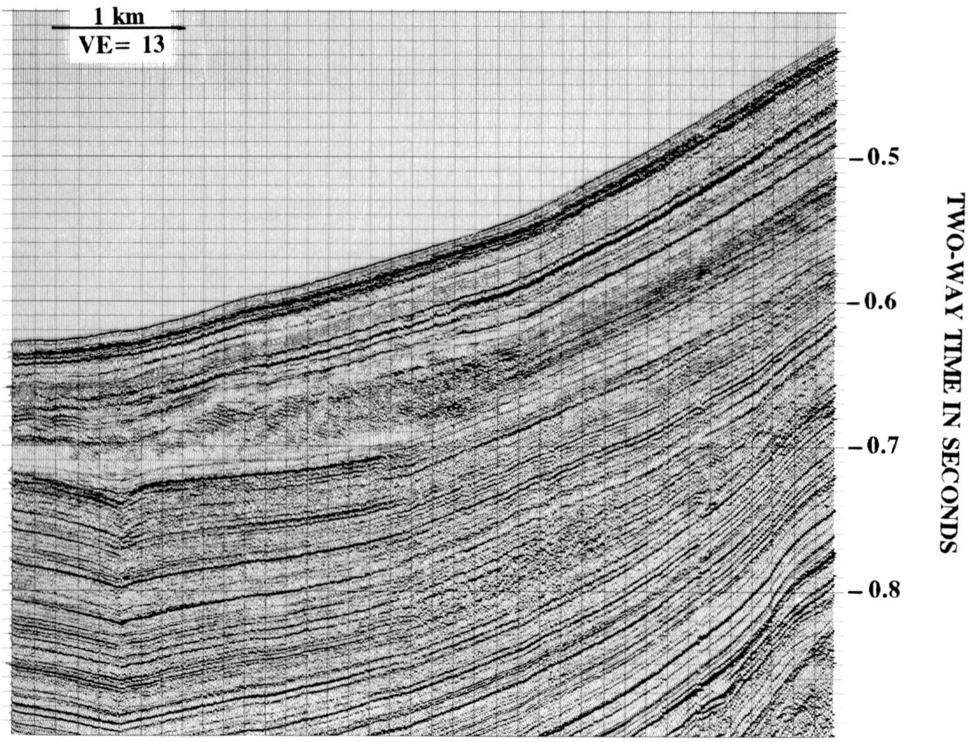

FIGURE 5. Line B: High definition multifold seismic profile, continental slope, Northern Gulf of Mexico. (Courtesy of SeaScan Inc. With permission.)

located 7 miles to the west (Coleman and Roberts, 1990) to produce the tentative time-sequence correlation shown in Figure 4, where the odd-numbered sequences correspond to the warmer oxygen isotope stages of Figure 2.

The increased sequence thickness on the down-thrown block of the active growth fault are the result of continuous subsidence through the present, such as described for deeper sections by Mitchum et al. (1990). One notes the well defined chaotic systems tracts associated with rapid deposition during sea-level low stands and the continuous reflectors representing reduced deposition associated with higher sea levels.

B. LINES B AND C: UPPER SLOPE

The HDG profile illustrated in Figures 5 through 8 is located to the south of the continental shelf break in water depths from 305 to 480 m (1000 to 1575 feet).

The data were obtained using an array of four 8 in^3 air guns (Tri-cluster® seismic source), and a 48-channel hydrophone streamer. The data were digitally recorded at sample intervals of $^1/_4$ msec for the six near-channels and at 1 msec rates for the farthest 24 channels to a total time depth of 3 s. Processing involved the production of two sections; one designated Line B (Figures 5 and 6) consisted of the $^1/_4$ msec sampled data processed in a two-fold format, while the second (Line C, Figures 7 and 8) comprised the 1 msec sampled data processed for a 48 fold display. It is the combination of such a source and processing procedures which produce optimum high definition or HDG seismic profiles. The uppermost 100 msec of the $^1/_4$ msec recordings closely match those produced by a deep-towed 3.5 kHz subbottom profiler system over the same area.

Line B (Figures 5 and 6) reveals the fine details which may be obtained to depths of 1

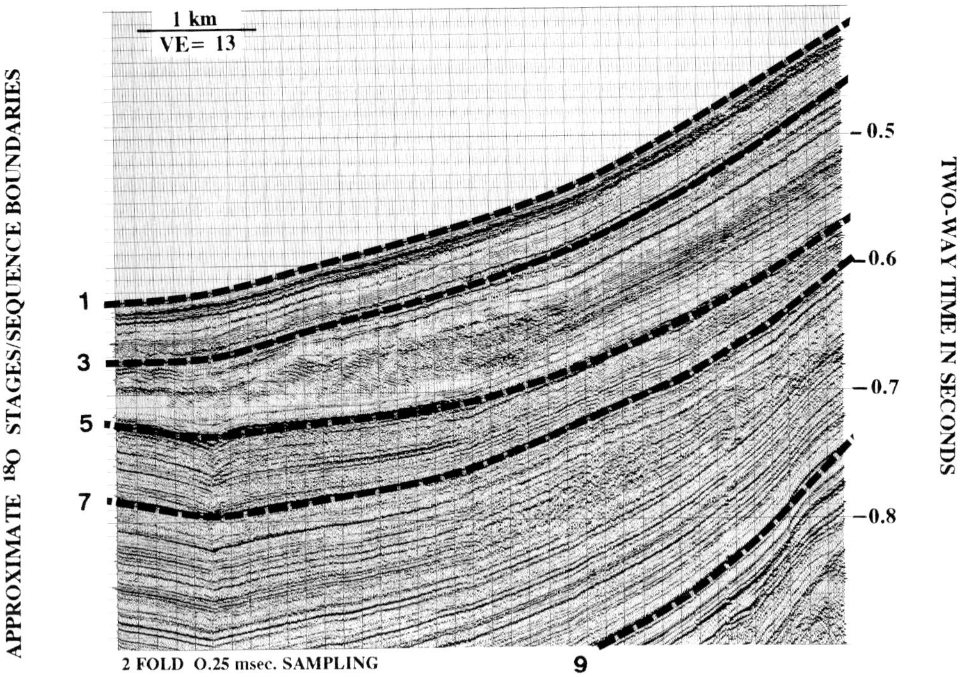

FIGURE 6. Interpretation of Line B (Figure 5), with nomenclature as per Figure 1. (Courtesy of SeaScan Inc. With permission.)

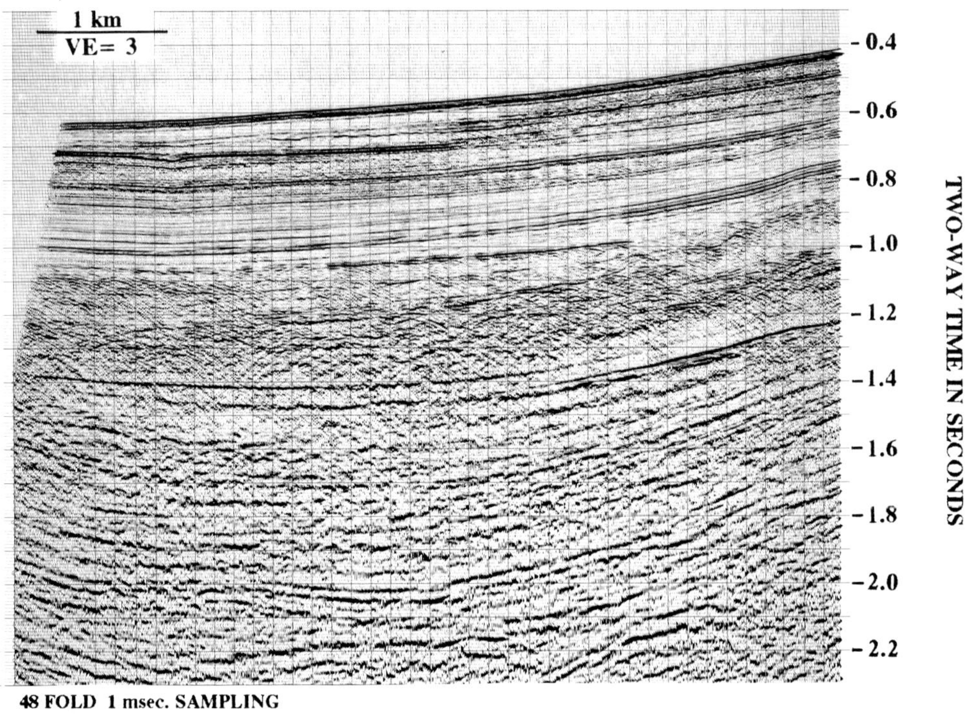

FIGURE 7. Line C: High definition multifold seismic profile, continental slope, Northern Gulf of Mexico. (Courtesy of SeaScan Inc. With permission.)

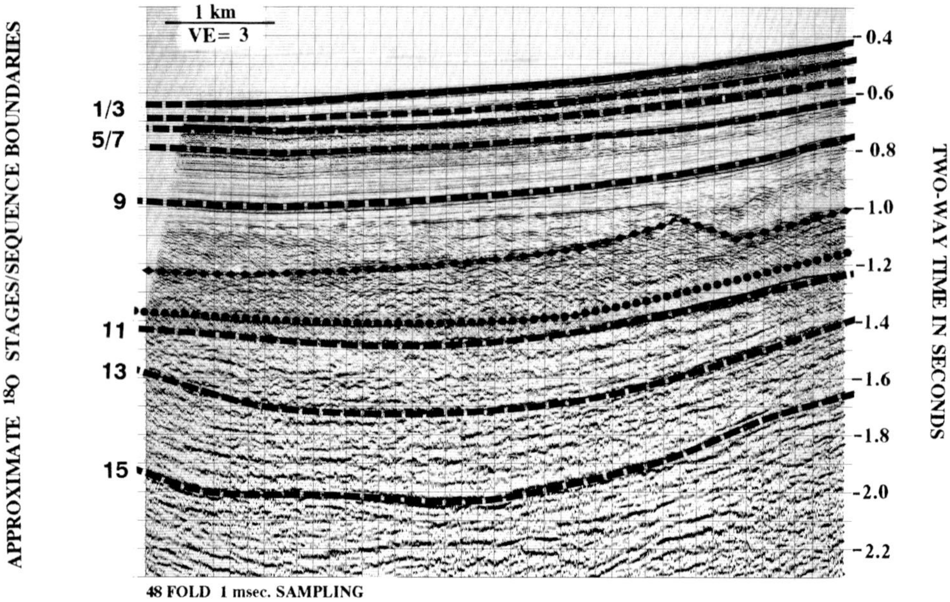

FIGURE 8. Interpretation of Line C (Figure 7), with nomenclature as per Figure 1. (Courtesy of SeaScan Inc. With permission.)

second with HDG techniques. This becomes readily apparent when comparing Figures 5 and 7, where the different processing and sampling rates bring out such details as the slump feature (Line B) and high frequency reflectors while increasing the vertical exaggeration.

The interpretation presented in Line B (Figure 6) includes the odd numbered sequence boundaries separating expanded sections and containing the following features: Sequences 1 through 9 consist of well defined sub-parallel reflectors representing sedimentary deposits associated with the three major eustatic cycles of the past 350,000 years (Figure 2). The uppermost sequence boundary (present sea floor) is in the process of forming a condensed section by virtue of its long exposure and paucity of sediment deposition, which consists mainly of pelagic ooze. An area of slumped sediments within the expanded section between Stages 3 and 5 indicates that slumping occurred during a period of lowered sea level. An Early to Late Wisconsinan delta has been reported by Berryhill et al. (1986) a few miles up-slope of this location and may be synchronous with the slumping.

The reflectors of Line B accumulated over the past 350ka, indicating that on average the reflectors represent periods of approximately 4000 years. Aside variations in the character of the reflectors, which should be correlative with both fluctuations in sea level and changes in sediment source distribution over the short term represented, the Line has few features.

Line C (Figure 7) is a 1 msec, 24-fold processed version of Line B. This HDG profile displays continuous reflectors to depths of 2 s below the sea floor, with good delineation (Figure 8) of the condensed sections or sequence boundaries down to stage 15 (550ka). While the uppermost sequences depicted consist of subparallel packets of reflectors, the Sequence 10 system tract consists of a thick jumbled assemblage interpreted as a possible slope fan complex. Downlapping is apparent to the north within this unit as well as the characteristic "concave upward" signatures frequently associated with such deposits (Figure 8). At greater depths, within sequences 11 through 15, one observes evidence of erosional surfaces intermixed with chaotic sequences attributed to mass sediment movement and deposition.

The detailed mapping in three dimensions of the individual system tracts from such

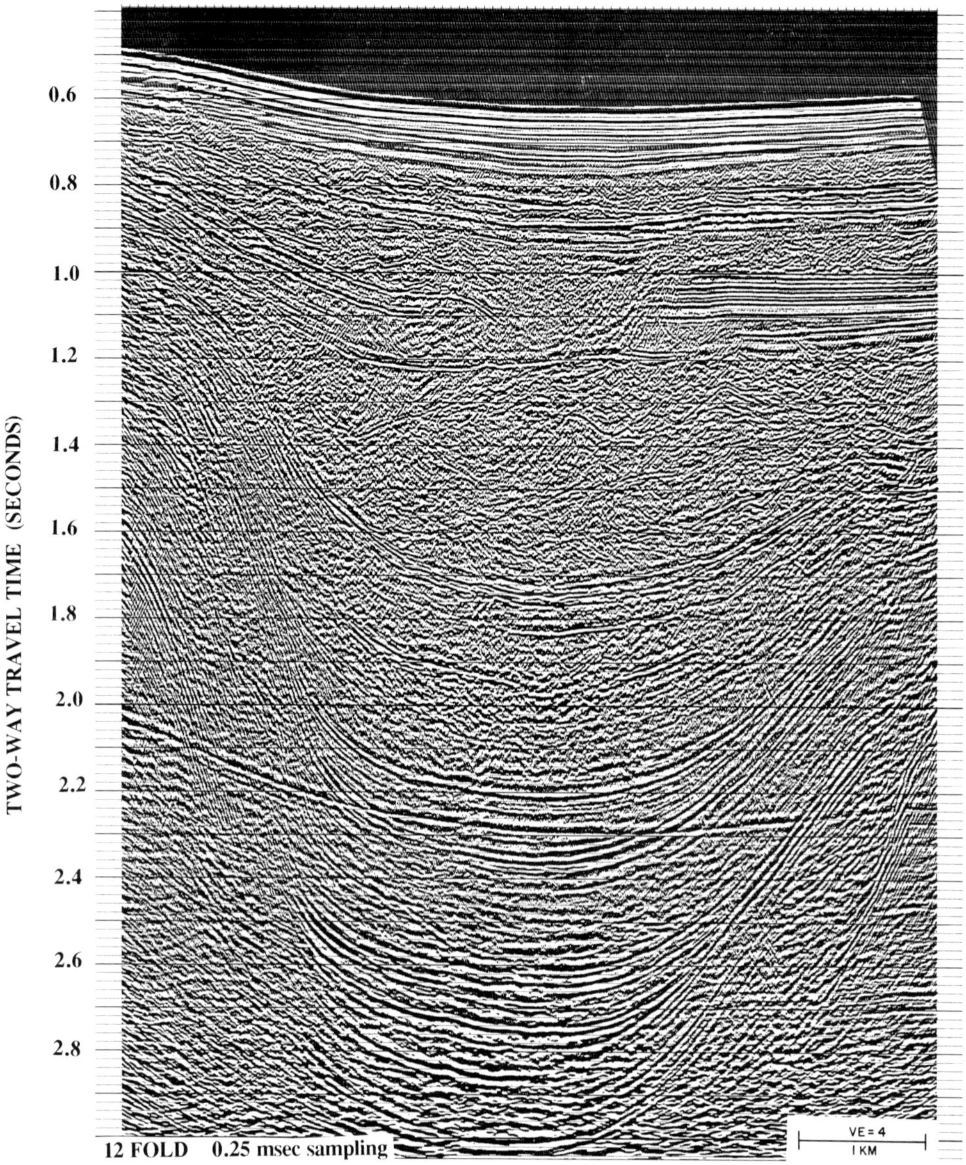

FIGURE 9. Line D: High resolution multifold seismic profile across infilled valley, continental slope, Northern Gulf of Mexico. (Courtesy BP Exploration Inc. With permission.)

HDG data should provide a good model for the specific systems tracts sought at depth as petroleum reservoirs.

C. LINE D: CANYON/VALLEY COMPLEX

The high resolution profile illustrated by Figures 9 and 10 was obtained in the vicinity of the western rim of the Mississippi Canyon. The line is oriented from northwest to southeast in water depths of 375 to 465 m (1230 to 1526 ft). The section is normal to a tributary channel to the Mississippi Canyon and illustrates cyclic incised valley fill (ivf) systems tracts.

The profile was obtained using a single 80 in.^{3} Watergun® seismic source and a 48-channel hydrophone streamer. The data was digitally recorded at a sample interval of 1

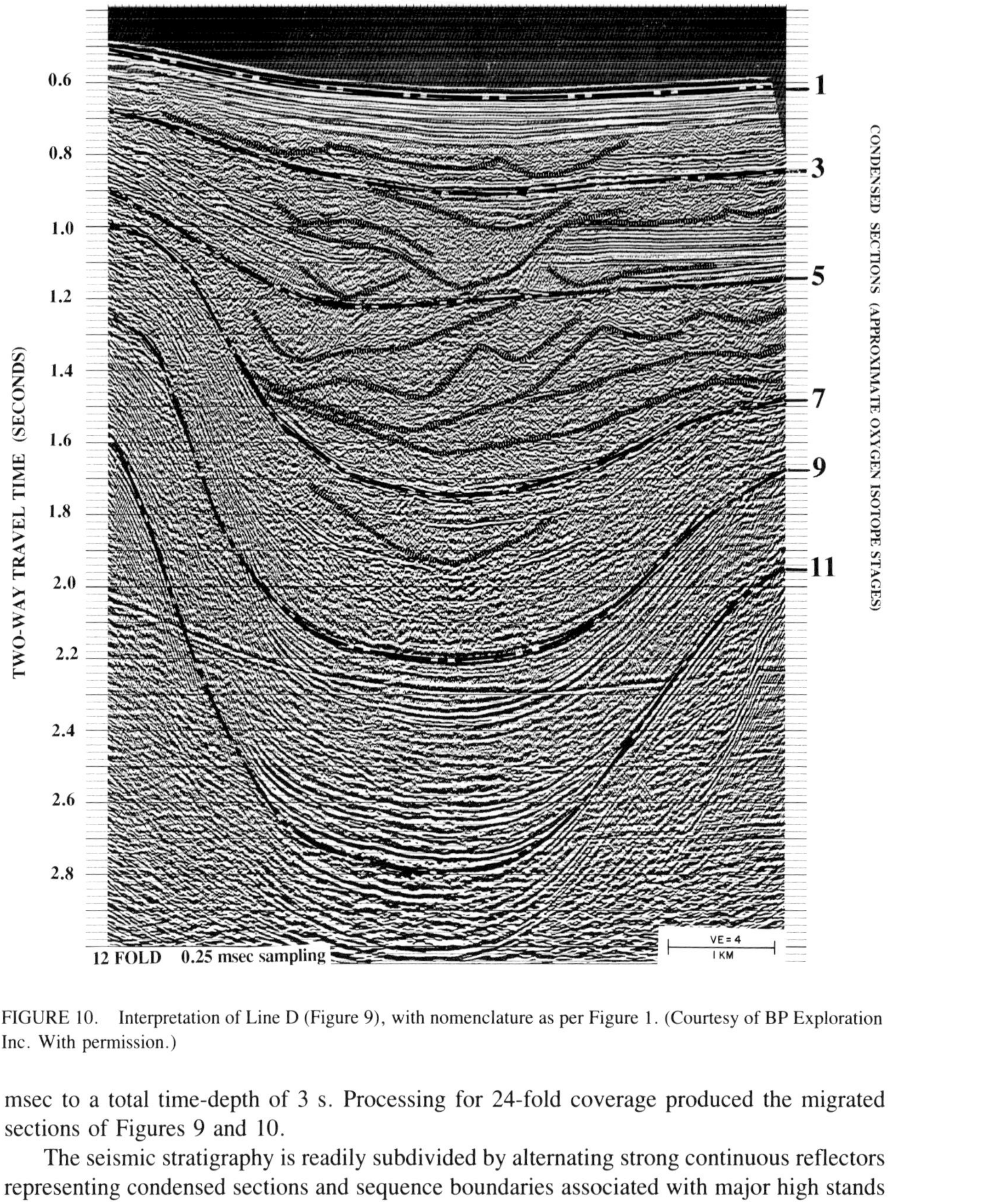

FIGURE 10. Interpretation of Line D (Figure 9), with nomenclature as per Figure 1. (Courtesy of BP Exploration Inc. With permission.)

msec to a total time-depth of 3 s. Processing for 24-fold coverage produced the migrated sections of Figures 9 and 10.

The seismic stratigraphy is readily subdivided by alternating strong continuous reflectors representing condensed sections and sequence boundaries associated with major high stands in sea level, and alternating periods of channel erosion, sediment transport and infilling.

Within the expanded units there are a large number of paleochannels associated with previous meanders, buildups, and erosion as sediments were carried down slope to be added to the major Mississippi Canyon system deposits or deposited as channel fill. Several of the larger channels exhibit evidence of lateral migration, buildup and erosion, and frequent occurrences of slumped material. Subsidence of the valley may have occurred in response to fill and withdrawal of underlying salt.

The numbered sequences are based exclusively on the interpreted seismic sequences and should approximate the equivalent oxygen isotope stages.

VI. CONCLUSIONS

Current technology can produce high definition seismic reflection data from which one can obtain the detailed 3-D geometry of sedimentary depositional sequences and system tracts associated with short period fluctuations in sea level, and variations in the spatial distribution of fluvial and submarine sedimentary depositional systems. Detailed surveys employing HDG techniques should permit mapping the distribution of upper Pleistocene sedimentary sequences associated with short-term 5th and 4th order eustatic fluctuations in sea level on the continental slope of the northern Gulf of Mexico. The mapping of the distribution of stratigraphic sequences and associated systems tracts should provide petroleum explorationists an extensive insight into the causes, effects and seismic signatures at depth, which produce reservoir quality deposits.

ACKNOWLEDGMENTS

The author wishes to acknowledge BP Exploration Inc. and SeaScan, Inc. for providing the illustrative HRG and HDG seismic sections, and the Sytech Corporation for the processing of the displayed seismic sections.

REFERENCES

Beard, J. H., Sangree, J. B., and Smith, L. A., Quaternary chronology, paleoclimate, depositional sequences, and eustatic cycles, *AAPG Bul.*, 66(2), 158, 1982.

Bouma, A. H., Coleman, J. M., Stelting, C. E., and Kohl, B., Influence of relative sea level changes on the construction of the Mississippi Fan, *Geo-Mar. Lett.*, 9, 161, 1989.

Bouma, A. H., Coleman, J. M., and Meyer, A. W., Eds., Initial reports, deep sea drilling project, leg 96, U.S. Govt. Printing Office, Washington, D.C., 1986.

Brooks, J. M., Kennicutt, II, M. C., Fay, R. R., McDonald, T. J., and Sassen, R., Thermogenic gas hydrates in the Gulf of Mexico, *Science*, 225, 409, 1984.

Brooks, J. M. et al., Gulf of Mexico hydrocarbon seep communities, IV. Descriptions of known chemosynthetic communities, OTC preprint 5954, 663, 1989.

Coleman, J. M. and Prior, D. B., Subaqueous sediment instabilities, the offshore Mississippi River Delta, in *Offshore Geologic Hazards;* AAPG education course note series 18, Sec. 5, 1981, p.112.

Coleman, J. M. and Roberts, H. H., Cyclic sedimentation of the northern Gulf of Mexico shelf; in *Sequence Stratigraphy as an Exploration Tool; Concepts and Practices in the Gulf Coast,* GCSSEPM Foundation 11th Annu. Res. Conf. (Abstr.), 1990, p.113.

Feely, M. H., Moore, T. C., Loutit, T. S., and Bryant, W. R., Sequence stratigraphy of Mississippi Fan related to oxygen isotope sea level index, *AAPG Bul.* 74(4), 407, 1990.

Gealy, B. L., Topography of the continental slope in northwest Gulf of Mexico, *Bull. Geol. Soc. Am.*, 66, 203-227, 1956.

Lopez, J. A., Temporal resolution of depositional processes in the Gulf Coast Tertiary, in *Sequence Stratigraphy as an Exploration Tool; Concepts and Practices in the Gulf Coast,* GCSSEPM Foundation 11th Annu. Res. Conf. (Abstr.), 1990.

MacDonald, R. I., Reilly, J. F., II, Guinasso, N. L., Jr., Brooks, J. M., Carney, R. S., Bryant, W. R., and Bright, T. J., Chemosynthetic mussels at a brine-filled pockmark in the northern Gulf of Mexico, *Science*, 248, 1096, 1990.

Martin, R. G., Northern and eastern Gulf of Mexico continental margin: stratigraphic and structural framework, in *Framework, Facies, and Oil-Trapping Characteristics of the Upper Continental Margin,* AAPG Studies in Geology No. 7, 1978, 21.

Mitchum, R. M., Vail, P. R., and Thompson, S., III, Seismic stratigraphy and global changes in sea level, II. The depositional sequence as a basic unit for stratigraphic analysis, in *Seismic Stratigraphy—Applications to Hydrocarbon Exploration,* Payton, C., Ed., AAPG Memoir 26, 1977, 53-62.

Mitchum, R. M., Sangree, J. B., Vail, P. R., and Wornardt, W. W., Sequence stratigraphy in late Cenozoic expanded sections, Gulf of Mexico, in *Sequence Stratigraphy as an Exploration Tool; Concepts and Practices in the Gulf Coast,* GCSSEPM Foundation 11th Annu. Res. Conf. (Abstr.) 1990, 237.

Neurauter, T. W. and Bryant, W. R., Gas Hydrates and their association with mud diapir/mudvolcanoes on the Louisiana continental slope, *Offshore Technol. Conf.*, preprint 5944, 599, 1989.

Piper, D. J. W. and Perissoratis, C., Late Quaternary sedimentation on the north Aegean continental margin, Greece, *AAPG Bul.*, 75(1), 46, 1991.

Posamentier, H. W. and Vail, P. R., Eustatic controls on clastic deposition II—Sequence and systems tract models, in *Sea Level Changes: An Integrated Approach,* Wilgus, C. K., Hastings, B. S., Kendall, C. G. St. C., and others, Eds., SEPM Spec. Publ., 42, 1988, 125.

Prior, D. B., Doyle, E. H., and Kaluza, M. J., Evidence for sediment eruption on deep sea floor, Gulf of Mexico, *Science,* 243, 517, 1989.

Sangree, J. B., Vail, P. R., and Sneider, R. M., Evolution of facies interpretation on the shelfslope: applications of the new eustatic framework to the Gulf of Mexico, *Offshore Technol. Conf.*, preprint 5695, 133-144, 1988.

Shipley, T. H., Houston, M. H., Buffler, R. T., Schaub, F. J., Mcmillen, K. J., Ladd, J. W., and Worzel, J. L., Seismic evidence for widespread possible gas hydrate horizons on continental slopes and rises, *AAPG Bul.*, 63(12), 2204-2213, 1979.

Suter, J. R., Ancient fluvial systems and Holocene deposits, southwestern Louisiana continental shelf, in *Late Quaternary Facies and Structure, Northern Gulf of Mexico: Interpretations from Seismic Data,* AAPG Studies in Geology No. 23, 1986, 81.

Trabant, P. K., *Applied High Resolution Geophysical Methods; Offshore Geoengineering Hazards,* Prentice-Hall, Englewood Cliffs, NJ, 1984, 230.

Trabant, P. K. and Presley, B. J., Orca basin, anoxic depression on the continental slope, northwest Gulf of Mexico, in *Framework, Facies, and Oil-trapping Characteristics of the Continental Margin,* AAPG Studies in Geology No. 7, 1978, 303.

Trainor, D. W. and Williams, D. F., Isotope chronostratigraphy: high resolution stratigraphic correlations in deep-water exploration tracts of the northern Gulf of Mexico, *Trans. Gulf Coast Assoc. Geol. Soc.*, 37, 247, 1987.

Vail, P. R., Seismic Stratigraphy Interpretation Procedure, I. In *Atlas of Seismic Stratigraphy,* (1) AAPG Studies in Geology 27, 1987, 1.

Vail, P. R. and Wornardt, W. W., Well log seismic sequence stratigraphy: a new tool for exploration in the '90s, in *Sequence Stratigraphy as an Exploration Tool; Concepts and Practices in the Gulf Coast,* GCSSEPM Foundation 11th Annu. Res. Conf. (Abstr.,) 1990, 379.

Vail, P. R., Mitchum, R. M., Jr., and Thompson, S., III, Seismic Stratigraphy and Global Changes of Sea Level, IV. Global cycles of relative changes of sea level, in *Seismic Stratigraphy—Applications to Hydrocarbon Exploration,* Payton, C., Ed., AAPG Memoir 26, 1977.

Weimer, P., Sequence stratigraphy, facies geometry, and depositional history of the Mississippi Fan, Gulf of Mexico, *AAPG Bull.*, 74(4), 425, 1990.

West, D. B., Model for salt deformation on deep margin of central Gulf of Mexico basin, *AAPG Bull.*, 73(12), 1472, 1989.

Woodbury, H. O., Spots, J. H., and Akers, W. H., Gulf of Mexico continental-slope sediments and sedimentation, in *Framework, Facies, and Oil-Trapping Characteristics of the Continental Margin,* AAPG Studies in Geology No. 7, 1978, 117.

Chapter 9

SEISMIC REFLECTION DATA ANALYSIS OF THE ORIENT AND SWAN FRACTURE ZONES BOUNDING THE CAYMAN TROUGH

Mary Norris Tinker

TABLE OF CONTENTS

I. INTRODUCTION

This study focuses on the Oriente and Swan Fracture Zones which form the active plate boundary of the eastern and western Cayman Trough, respectively. The objectives of this study include: (1) detailed study and interpretation of published and unpublished seismic reflection profiles to compare the eastern and western Cayman Trough, (2) delineation of compressional and extensional sedimentary structures along the Swan and Oriente Fracture Zones, (3) correlation of these structures and their relative sequence in time with changes in the spreading direction recorded in the spreading fabric of the Cayman Trough floor and (4) comparison of the results of this study with published tectonic spreading models of the Cayman Trough.

The Cayman Trough (Figure 1) is a seismically active transform plate boundary that extends 1600 km eastward from the Gulf of Honduras to the Windward Passage and likely to the Puerto Rico Trench. Westward from the Gulf of Honduras, it extends through Guatemala as the Polochic-Motagua fault zone and likely to the Middle America Trench. Topographically, the Cayman Trough is a depression that lies between the Cayman Ridge to the north and the Nicaraguan Rise to the south, which together separate the Yucatan Basin on the north from the Colombian Basin on the south. The Cayman Trough marks the boundary between the North American and Caribbean plates (Molnar and Sykes, 1969). The narrow Swan and Oriente Fracture Zones bound the trough to the south and west, and to the north and east of the Mid-Cayman Spreading Center (MCSC), respectively. These fracture zones are left-lateral, strike-slip faults which trend generally east-west.

Molnar and Sykes (1969) determined from focal mechanisms of earthquakes that motion along the transform faults of the Cayman Trough is left-lateral. Perfit (1977) suggested that the MCSC developed in response to the divergence of the Cayman Ridge and Nicaraguan Rise. There is wide disagreement in estimates of the total amount of spreading and rates on the MCSC, but the present slip rates along the Cayman Trough fall between 0.4 cm/year and 6.0 cm/year (Sykes et al., 1982; Holcombe et al., 1973; Malfait and Dinkelman, 1972; Schwartz et al., 1979; Mann and Burke, 1984). Based on magnetic anomalies, Macdonald and Holcombe (1978) determined historical spreading rates of 2 cm/year from the present to 2 Ma and 4 cm/year from 2 to 6 Ma for the Cayman Trough. The age of initial spreading in the Cayman Trough has been variously calculated as sometime between Late Eocene and Late Oligocene in age (Perfit, 1977; Minster and Jordan, 1978; Macdonald and Holcombe, 1978).

The outline for the procedures to be followed throughout this study are to: (1) briefly describe pertinent published geological and geophysical data, (2) describe the seismic reflection data sets available for this study and the methods of analysis of these data sets, (3) interpret the data for the definition of sedimentological and structural features within the Cayman Trough, the Cayman Ridge and Nicaraguan Rise, (4) map distribution of extensional and compressional sedimentary structures within the Swan and Oriente Fracture Zones, (5) integrate the interpreted data for the fracture zones into a tectonic relative motion model for the North American and Caribbean plates along the Cayman Trough, and (6) compare the results of this study with published tectonic spreading models of the Cayman Trough.

II. PREVIOUS WORK

A. GEOLOGY

The Cayman Trough, a 1600-km plate boundary between the North American and Caribbean plates (Figure 1), has a maximum width (140 km) at the MCSC. Sediment cover is slight or absent (<200 m) (Ewing; J., et al., 1960; Edgar et al., 1971) in the central region of the trough and thickens progressively toward the western end (>1.5 sec thick,

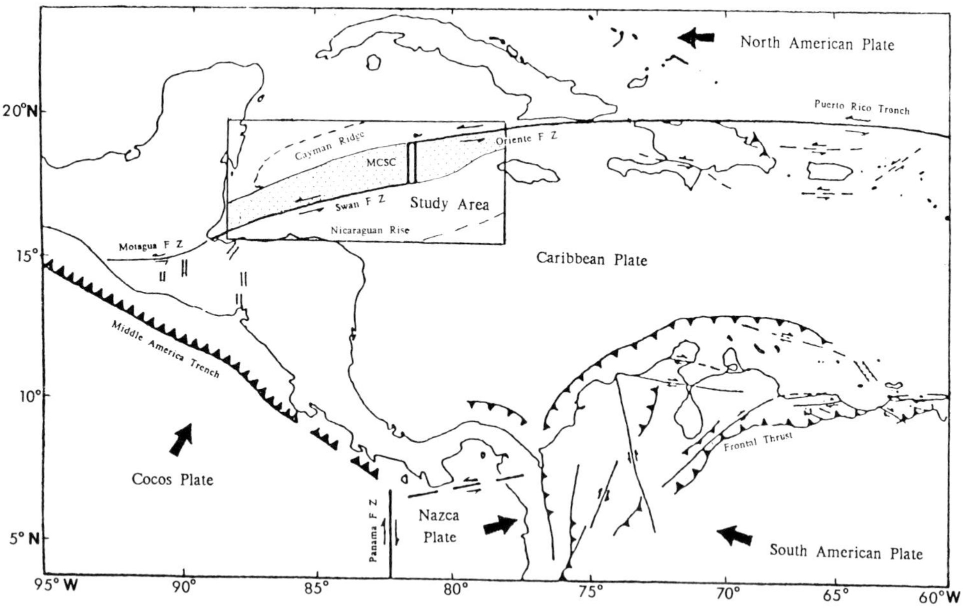

FIGURE 1. Study area, on Caribbean plate boundary chart. The stipled area within the study area is the Cayman Trough. (After Jordan, 1975.)

using two-way travel time; Erickson et al., 1972), and less so toward the eastern end. Erickson shows as great as 2.25 sec of sediment in the western Cayman Trough. The sediment cover has been disturbed in the Swan Fracture Zone (Holcombe et al., 1973) indicating recent tectonic movement. While no specific geologic studies have been done on the Oriente Fracture Zone, seismic reflection data show that the sediment has been recently disturbed. The eastern Cayman Trough contains isolated sediment ponds in the topographic lows.

The sea floor, which separates the northern and southern fracture zones of the Cayman Trough, is rough and irregular. East of 83°W the southern margin of the Cayman Ridge is steep, with the highest elevations on the ridge immediately adjacent to the deepest parts of the Cayman Trough (Erickson et al., 1972). The northern margin of the Cayman Trough east of 81°W is fairly straight and uniform. The southern margin west of 82°W is not as steep as the northern margin, and bathymetric irregularities are more common (Bowin, 1968). The floor of the trough between 80°W and 83°W is characterized by an area of north-south trending ridges and valleys. Several of these ridges stand greater than 2000 m above the trough floor. The detailed bathymetric map of the MCSC by CAYTROUGH (1979) shows two major trends in the spreading fabric in the floor of the Cayman Trough. One trend is N 37°W and the other trend is north-south.

Analysis of dredge hauls from the MCSC indicate a complex suite of rocks (Case, 1975), which include basalts, gabbros and ultramafic rocks. Perfit (1977) dredged pelagic limestone, Miocene to Pliocene in age, from the crest of a ridge just east of the MCSC. Many of the metamorphic features shown in rocks from the MCSC can be attributed to the combined processes of contact, dynamic and hydrothermal metamorphism that occurs at active spreading centers (Bonatti et al., 1975; Stern et al., 1976).

Eggler et al. (1973) took dredgings within the trough and recovered serpentinite, orthopyroxene, clinopyroxene, spinel and probably pseudomorphs after olivine. This indicated to them that the serpentinite samples were serpentinized spinel iherzolite. This original mineral assemblage is stable in the mantle at depths of 25 to 70 km (Green and Ringwood, 1967). The samples indicated that they were crystallized from a melt in the crust or very shallow mantle.

B. GEOPHYSICS

Many different spreading rates have been obtained for the Cayman Trough: Holcombe et al. (1973) obtained a spreading rate of 2.2 cm/year ± 50% by fitting to the bathymetric data a smooth regional curve of depth as a function of distance from the spreading center, and comparing the resultant curve to the age-depth relationship deduced by Sclater et al. (1971). On the other hand, field work by Schwartz et al. (1979) along fault zones in Guatemala lead to Holocene slip rates of 0.45 cm/year to 1.8 cm/year, with the higher slip rates favored. These slip rates were based on offsets of stream terraces and other geomorphic features. Macdonald and Holcombe (1978) document with magnetic anomalies a more or less continuous spreading on the MCSC back to 8.3 Ma. Within the estimated error, the spreading rate is 2.0 cm/year (total opening rate) from 0 to 2.4 Ma and 4.0 cm/year from 2.4 to 6.0 Ma. With these spreading rates, Macdonald and Holcombe (1978) extrapolated the initial opening to be 30 Ma or Late Oligocene. Minster and Jordan (1978) obtained a spreading rate of 2 cm/year for the last 2.4 Ma and calculated a Late Eocene age for the initiation of the Cayman Trough from slip rates.

Ewing, J., et al. (1960) conducted a seismic refraction survey of the Cayman Trough region which included seven seismic refraction profiles across the Cayman Trough. Their results showed an extremely thin crust under the floor of the Cayman Trough, with the Mohorovicic discontinuity only 11 to 12 km below sea level. The minimum crustal thickness measured at the MCSC was 4.1 km. Mantle depth increases progressively from the MCSC to the east (14 km) and west (20 km). Bowin (1968) developed a detailed crustal model based on seismic refraction, bathymetry and gravity. This model confirmed the presence of a thin crust beneath the Cayman Trough. The crustal thickness according to Bowin agrees with the measurements of Ewing. The Cayman Trough overlies mantle of normal seismic velocity (8.0 to 8.3 km/sec). Ewing suggested that the thin crust indicated that the Cayman Trough is a tensional feature and that the crust beneath the Cayman Trough is oceanic in nature.

Ewing also determined that the Cayman Ridge is underlain by continental basement with velocities ranging from 6.3 to 6.5 km/sec and increasing to 7.2 km/sec at a depth of 17 km. These authors obtained a crustal thickness at the Cayman Ridge of 20 to 25 km. The absence of intense magnetic anomalies led Ewing to suggest that little or no basic or volcanic rocks were present near the surface of the Cayman Ridge. Ewing found that the Nicaraguan Rise is underlain by continental basement with velocities ranging from 6.3 to 6.7 km/sec. Seismic refraction arrivals from the Mohorovicic discontinuity were not obtained, but minimum depth calculations placed the discontinuity at least 20 to 25 km below sea level.

Seismic events with well-constrained depths within the Cayman Trough are shallow (depths less than 70 km) (Sykes et al., 1962). Maps of well-located epicenters (accurate to better than 25 km) show concentrations of earthquakes along the Swan and Oriente Fracture Zones with east-west trending clusters near 75°W to 77°W and 85°W to 88°W at the distal ends of the Cayman Trough (Sykes et al., 1982; Holcombe et al., 1973; Mann and Burke, 1984). Earthquakes are limited to within approximately 50 km of the MCSC and the Oriente and Swan Fracture Zones. Molnar and Sykes (1969) have determined focal mechanisms for three earthquakes: one south of Grand Cayman Island adjacent to the Oriente Fracture Zone (81°19′W; 18°90′N) and two north of the Bay Islands (85°90′W; 16°82′N and 85°60′W; 16°96′N) in the Swan Fracture Zone. All yielded fault plane solutions which are consistent with left-lateral, strike-slip movement parallel to the walls of the Cayman Trough. The slip vector azimuths were 261° for the earthquake south of Grand Cayman Island and 250° and 75° respectively for the earthquakes north of the Bay Islands.

Holcombe and Sharman (1983) proposed a post-Miocene evolution for the Cayman Trough based on the variation in orientation of spreading fabric and magnetic lineations.

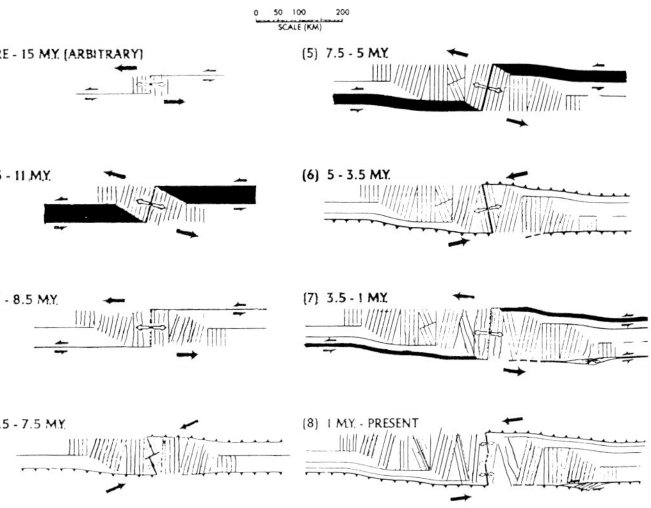

FIGURE 2. Post-Miocene tectonic development of the Cayman Trough. (After Holcombe and Sharman, 1983.)

Their 8-stage model assumes: (1) the tectonic spreading fabric is perpendicular to the spreading direction, (2) cross-trough extension both widens the trough and lengthens the spreading center and (3) cross-trough compression narrows the trough but does not shorten the spreading center. Combining these assumptions with the spreading fabric lineation data, Holcombe and Sharman were able to see a cross-trough width variation. Their tectonic development model for the Cayman Trough (back to 15 Ma) is shown in Figure 2. Holcombe and Sharman thought their model was self-consistent and that it had implications for variations in relative motion between the North American and Caribbean plates. The Holcombe and Sharman model will be used later in this study to compare and contrast the results obtained from this study of the Cayman Trough.

III. DATA AND METHODS

A. DATA

The study area includes all of the Cayman Trough, the Cayman Ridge and the Nicaraguan Rise. Within this study area seismic reflection and bathymetric data were compiled from published and unpublished surveys. A summary of the data and their sources are given in Table 1. Included are seismic reflection data acquired in 1968 and 1971 by Texas A&M University (*R/V Alaminos*) and in other years (see Table 1) by the University of Texas Marine Science Institute (*R/V Ida Green, R/V Fred Moore,* and the Humble Oil Co.) and Lamont-Doherty (*R/V Conrad*). The tracks for the seismic reflection profiles used in this analysis are shown in (Figure 3).

Because of the rugged topography and the general lack of magnetic data, seismic reflection data were considered to be a primary source of information with which to examine the structure of the Cayman Trough. The seismic reflection data consist of single-channel and multichannel time sections. The sediment that is present is clearly seen in the seismic refection profiles; however, the basement is difficult to define (within the resolution of the seismic reflection data) in the western third of the trough. The assumed velocity for calculations of the water depth is 1500 m/sec. The assumed average velocity for calculations of the sediment thickness from sea floor to basement is 1800 m/sec. This number was obtained by taking the average of the stacking velocity of the multichannel seismic data of

TABLE 1
Data Sources used for this Study

Cruise ID	Institution	Ship	Sound Source	Year	Data Type
68-A6-III	Texas A&M Univ.	*Alaminos*	40 KJ Sparker	1968	Ss M B
71-A4-Leg I	Texas A&M Univ.	*Alaminos*	120 in^3 Airgun	1971	Ss M B
71-A4-Leg V	Texas A&M Univ.	*Alaminos*		1971	Ss M B
C1203	Lamont-Doherty Geol. Obs.	*Conrad*	unknown	1966	Ss B G
IG-29	Univ. of Texas Marine	*Ida Green*	4 Bolt	1978	Sm M B
CT1-1	Science Inst.		Airgun	1978	Sm M B
CT1-4			Array	1978	Sm M B
FM-05	Univ. of Texas Marine	*Fred Moore*	4 Bolt	1980	Sm M B
CT2-1	Science Inst.		Airgun	1980	Sm M B
CT2-2			Array	1980	Sm M B
Pecos C	Univ. of Texas Marine Science Inst.	Humble Oil Co.	Sleeve Exploder	1969	Sm M B

Ss = single-channel seismic Sm = multichannel seismic
M = magnetics B = bathymetry G = gravity

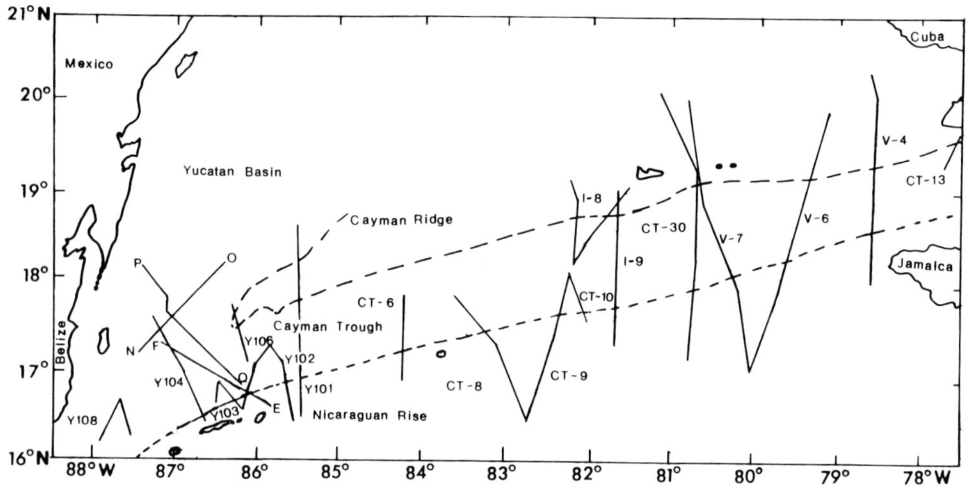

FIGURE 3. Seismic reflection profiles discussed in this study. The symbols for the seismic reflection lines are: Y = *R/V Alaminos* 1968 cruise, I = *R/V Alaminos* 1971 cruise, Leg I, V = *R/V Alaminos* 1971 cruise, Leg V, CT = *R/V Fred Moore* and *R/V Ida Green* cruises, and other letters represent the *R/V Conrad* cruise. See Table 1 for the specifics of these cruises.

the thick sediment accumulation of Profile CT1-8A. This value was compared with other average velocity values for other pockets of sediment, and was found to compare favorably.

The acoustic sources and the recording scale employed in the seismic reflection data sets are diverse (Table 1). The source used for recording the seismic reflection records of the *R/V Conrad* is unknown. Seismic reflection records obtained from the *R/V Alaminos* in the 1968 cruise used a 40 K joule sparker with a firing interval of 13.14 sec. The 1971 *R/V Alaminos* cruise used a 120 in^3 airgun as a sound source. Acoustic signals were detected with a multi-element hydrophone array towed approximately 150 m behind the vessel. The resulting data clearly defines the shallow subsurface features, but most of these data have limited penetration due to the relatively low-energy sources. The UT-MSI *R/V Ida Green* and *R/V Fred Moore* both used a 4 Bolt airgun array as a sound source towed at a depth of 10 m. The UT-MSI and Humble Oil Co. cruise used a sleeve exploder as a sound source towed at a depth of 7.62 m. The resulting data clearly defines the shallow and deep subsurface features, and basement is defineable by the hyperbolic diffractions arising from the basement surface.

B. METHODS

The seismic reflection profiles were examined for sediment distribution and features of the sediment deformation. The procedure for interpretation of the seismic reflection data was to first identify reflector terminations and patterns to define fault distributions and sedimentary structures within the trough and to define sediment sequences when possible. Finding the reflector terminations limits the relative age of depositional and tectonic events. These steps were applied to the data on a line-by-line basis. Minimum depth to basement was determined, and a sediment isopach map (minimum thickness of sediment between sea floor and acoustic basement) was constructed using a 25 msec contour interval. Minimum sediment to appear on seismic reflection data was determined by using one-quarter of the seismic source wavelength, which is usually 100 m. Therefore, 25 m is the minimum sediment accumulation to be seen on the seismic reflection data.

The reflection profiles are presented in stacked form to illustrate the structure of various provinces and individual structural and sedimentary features, and detailed line-drawing interpretations are presented for selected key profiles across important features. The *R/V Ida Green* and the *R/V Fred Moore* cruises provided the best information on the internal structure at depth because the data was obtained with a multichannel seismic system.

The seismic reflection data were examined in sections, province by province, for the Cayman Trough and associated fracture zones, the Cayman Ridge, and the Nicaraguan Rise. The general aspects of each individual province were first described using a series of seismic reflection profiles. Key features were then examined in detail using line-drawing interpretations to illustrate the features. Figure 3 shows the seismic reflection profiles discussed in this study. The seismic reflection data consist of single-channel and multichannel time sections with the vertical scale given in two-way travel time in seconds. Data from the fracture zones were examined in detail to determine extensional and compressional features throughout. A tectonic spreading model of the Cayman Trough through time is developed based on patterns in the distribution of extensional and compressional features.

IV. SEISMIC REFLECTION DESCRIPTION

A sediment isopach map (minimum two-way travel time to basement) was prepared to delineate the sediment distribution and trough boundaries. Ship tracks are also superposed on this map. Individual letters and numbers represent positions along the seismic reflection profiles which will be discussed within this chapter. An assumed average velocity of 1800 m/sec is used throughout this study for calculations of sediment thicknesses. Sediment accumulations reported in this discussion are minimum depths to basement.

In the eastern Cayman Trough, isolated pockets of sediment are found in topographic lows between basement fault blocks. No sediment is detectable on the topographic highs. The resolution of the seismic reflection data suggests that there is less than 25 m of sediment on the topographic highs. One-quarter of the wavelength for the seismic source is generally the lowest quantity of sediment accumulation that can appear on the seismic reflection data (Telford et al., 1976). Most seismic source wavelengths are on the order of 100 m.

The thickest sediment accumulation (1350 m) in the eastern Cayman Trough occurs in an area between 79°W and 80°W and 18°N and 19°N. Small isolated pockets of sediment are visible along the Oriente Fracture Zone. Sediment accumulations within the Oriente

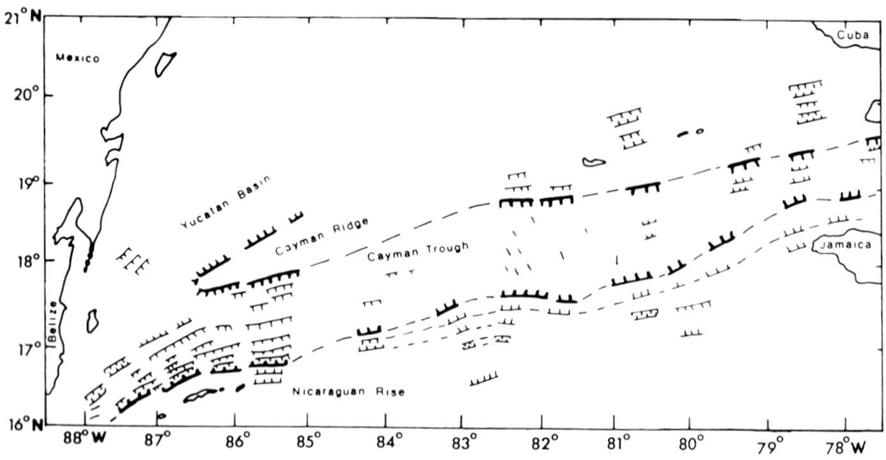

FIGURE 4. The fault distribution pattern of the Cayman Trough region.

Fracture Zone are thickest (900 m) closest to the MCSC at 80°40′W and 19°N. Sediment accumulations thin eastward along the fracture zone. The central Cayman Trough contains isolated ponds of sediment in some topographic lows where sediment accumulations are usually less than 225 m. Sediment accumulations thicken progressively westward from the MCSC. West of 85°W the trough floor is entirely covered with sediment. The thickest sediment accumulation (2025 m) occurs north of Roatan Island. Another large area of thick sediment accumulation occurs south of the Cayman Ridge (1575 m) at 86°W and 17°25′N.

Turning to the boundaries of the Cayman Trough, the thickest sediment accumulation (900 m) along the Cayman Ridge occurs between 85°20′W and 86°10′W and 17°50′N. Sediment is also seen on the margins of the Cayman Ridge. In the central and eastern Cayman Ridge, sediment occurs in isolated ponds (less than 225 m thick) on the slopes north of the Cayman Trough. North of the Cayman Ridge, sediment thickens toward the Yucatan Basin. On the Nicaraguan Rise small ponds of sediment occur between 83°W and 86°W. From Swan Island to Jamaica, sediment accumulation on the Nicaraguan Rise everywhere exceeds 1800 m, and reaches as great as 2025 m at 81°W and 17°10′N.

Fault distribution patterns (Figure 4) were mapped to aid in the discussion of the seismic description and sediment distribution of each province. The orientation of the Cayman Trough is N 60°E from 88°W to 86°W and N 75°E east of 86°W. The axis of the Cayman Ridge and the northern margin of the Nicaraguan Rise parallel the axis of the Cayman Trough. The heavy lines on Figure 4 represent basement fault scarps that form the boundaries of the Cayman Trough.

The finer lines within Figure 4 represent faults within the sediment that are seen within the Cayman Trough, the Cayman Ridge and the Nicaraguan Rise. The faults within the eastern and western Cayman Trough are seen down to basement, indicating that the faults within the sediment are due to the movement of the individual basement blocks. As the faults become more numerous toward the Swan Fracture Zone in the western Cayman Trough, the faults cease to reach basement and die out at a shallower depth within the sediment accumulation. This indicates possible rotation and extension along the Swan Fracture Zone.

Within the Cayman Trough, the faults trend in two distinct directions (Figure 4). At the distal ends of the trough, the faults trend parallel to the axis of the trough. Between 80°W and 83°W, these faults trend at an angle from the perpendicular to the trough boundaries. In the eastern Cayman Trough, the direction of displacement on most of the faults is down toward the Oriente Fracture Zone, while in the western Cayman Trough, the direction of displacement on most of the faults is down toward the Swan Fracture Zone. This is probably due to extension of the trough between the Cayman Ridge and Nicaraguan Rise. While the faulting is predominantly down towards the fracture zones, in the western Cayman Trough, there appears to be a small horst-and-graben system from 86°W to 88°W and 17°N. This horst-and-graben system probably resulted from antithetic faulting.

There are faults near the MCSC from 80°W to 83°W (Figure 4) that have undetermined motion from the widely spaced seismic reflection data alone. When the fault distribution map was compared with the Holcombe and Sharman (1983) map of the spreading fabric trends of the Cayman Trough, these fauts near the MCSC define ridges and valleys which trends north 20°W. These faults, which define the ridges and valleys of the MCSC, were also compared with the detailed bathymetric map of the MCSC by CAYTROUGH (1979). These faults lie along basement trends (which also define the ridges and valleys of the MCSC) which trend N 37°W. The bathymetric map indicates a possible north-south trend, and both the north-south trend and the N 37°W trend probably resulted from a recent spreading direction change. The faults near the MCSC on the fault distribution map (Figure 4) were drawn from an average of these two trends at N 28°30′W.

To facilitate the discussion of the seismic reflection data, the study area will be divided into five provinces: the eastern, central, and western portions of the Cayman Trough, the Cayman Ridge, and the Nicaraguan Rise.

A. CAYMAN TROUGH

As shown by the fault distribution map (Figure 4), the width of the Cayman Trough varies along its 1600-km length, resulting in an asymmetrical appearance. The central portion of the trough is widest at the MCSC (130 km), and the width decreases progressively to the east and west. The western portion has a width of 115 km, while the eastern portion has a width of 80 km. This variation in width may indicate that the spreading rate has varied and/or that spreading was initiated at different times (Holcombe et al., 1973).

1. Eastern Cayman Trough

The eastern portion of the Cayman Trough, for purposes of this discussion, lies east of 81°W and includes the Oriente Fracture Zone. The basement in the eastern Cayman Trough is block faulted, resulting in high, irregular topography. Profiles V-4, V-6 and V-7 (Figures 5, 6, 7) illustrate the horst-and-graben fault-block topography of this area. The basement fault scarps are characterized by 10° to 15° slopes and display no internal reflectors within the basement (within the resolution of the seismic reflection data). This area is characterized by deep water depths, irregular topography, and small ponds of sediment. Water depths vary from 2200 to 5100 m within the trough. Seismic reflection profiles discussed below include the Texas A&M University *R/V Alaminos* Profiles V-4, V-6, and V-7, the UT-MSI *R/V Ida Green* Profile CT1-30B and the UT-MSI *R/V Fred Moore* Profile CT2-13B.

In one of the valleys on Profile V-4 at region A (Figure 5), the sediment has been deformed. There is an angular unconformity within the sediment section at shallow depth. This is indicated by the mildly deformed, nearly horizontal reflectors which onlap onto a dipping reflector in the subsurface. The sediment underlying the unconformity was folded prior to the deposition of the overlying sediment. The overlying sediments have been subsequently mildly deformed and folded. On Profile V-4 (region B), a thick (1125 m) wedge of sediment has been deposited against the northern margin of the Nicaraguan Rise. This wedge of sediment can be traced continuously to the east and west for a considerable distance. The reflectors within this region dip at a 3° to 4° angle. The slope of the sediments is due to deposition, rather than tectonic movement after deposition. Some deformation of the sediment is apparent at shallow depth. The sediment within this wedge probably accumulated from the Nicaraguan Rise.

Profie V-6 (Figure 6) at region C displays a thick (675 m) sediment accumulation resting

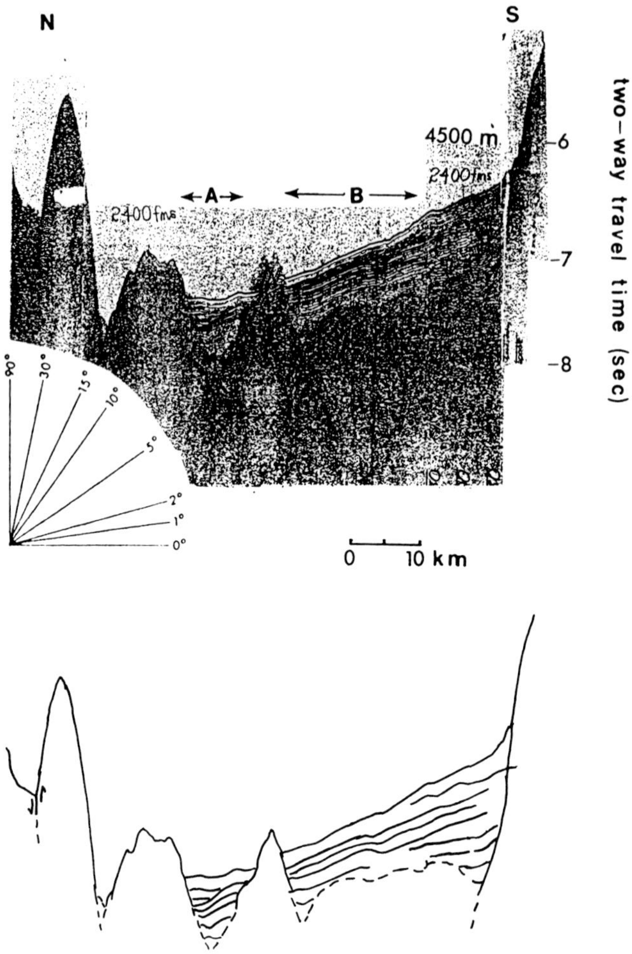

FIGURE 5. Profile V-4. The Cayman Trough.

upon a basement fault block. This fault block has been uplifted and possibly tilted down toward the Oriente Fracture Zone. The sediment was deposited prior to the movement of the fault block, otherwise there would not be a substantial accumulation of sediment on the peak. In cross section, the sediment layers increase in thickness from north to south.

South of this fault block at region D (Profile V-6), a valley containing 450 m of folded sediment is present. Note that the present sea floor is flat. Beneath these recent sediments, the reflectors are concave up in the center and terminate against the adjacent basement blocks. Extensional movements of the fault blocks probably resulted in collapse and down folding of the sediments.

Two of the seismic reflection profiles from the eastern Cayman Trough show a central region bounded by fault blocks to the north and south (Profiles V-7 at region F and V-6 at region G; Figure 3). These blocks give a symmetrical appearance to these portions of the profile. The central region on both profiles (F and G on V-7 and V-6) is interrupted by additional faulting creating an irregular valley floor. These symmetrical regions could be locus for extension, which is apparently unrelated to the MCSC and the Swan and Oriente Fracture Zones. An attempt to trace this feature farther to the east within the Cayman Trough failed. This could indicate a more complex model of the Cayman Trough in this area than was previously considered.

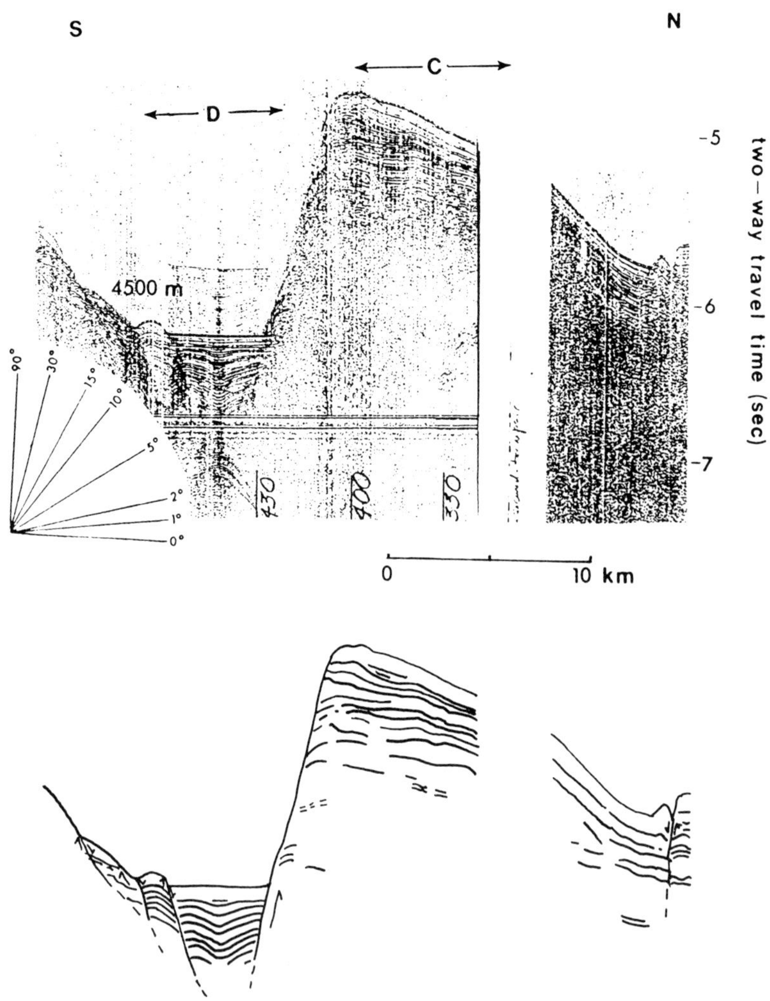

FIGURE 6. Profile V-6. The Cayman Trough.

2. Oriente Fracture Zone

The Oriente Fracture Zone is a narrow (3.0 to 10.0 km) deep (Profiles V-7, V-6 and V-4; (Figures 5, 6, 7)) bounded on the north and south by fault scarps that are sediment free (within the resolution of the seismic reflection data). Water depths reach as great as 6750 m in this fracture zone. Small pockets of sediment occur within the fracture zone and on the slopes (Profile V-6 at region E).

Within the Oriente Fracture Zone, sediment has been deformed both by extension and compression. Profile V-7 at region H1 (Figure 7), near the MCSC, shows at least 400 m of subparallel reflectors in the sursurface that dip toward the south along the profile. This feature resulted from a component of opening (or extension) across the fracture zone which also resulted in the basement block to the north moving up relative to the basement block to the south. Flat-lying reflectors cover the dipping reflectors, indicating that the movement of the adjacent basement blocks has ceased. Along the southern margin of the fracture zone on Profile V-7, there is an area of chaotic reflectors 360 m thick, which possibly resulted from recent compression. This period of compression could have destroyed the evidence of extension on the southern side of the fracture zone.

Profile CT1-30B at region H2 (Figure 8) shows reflectors dipping down toward the

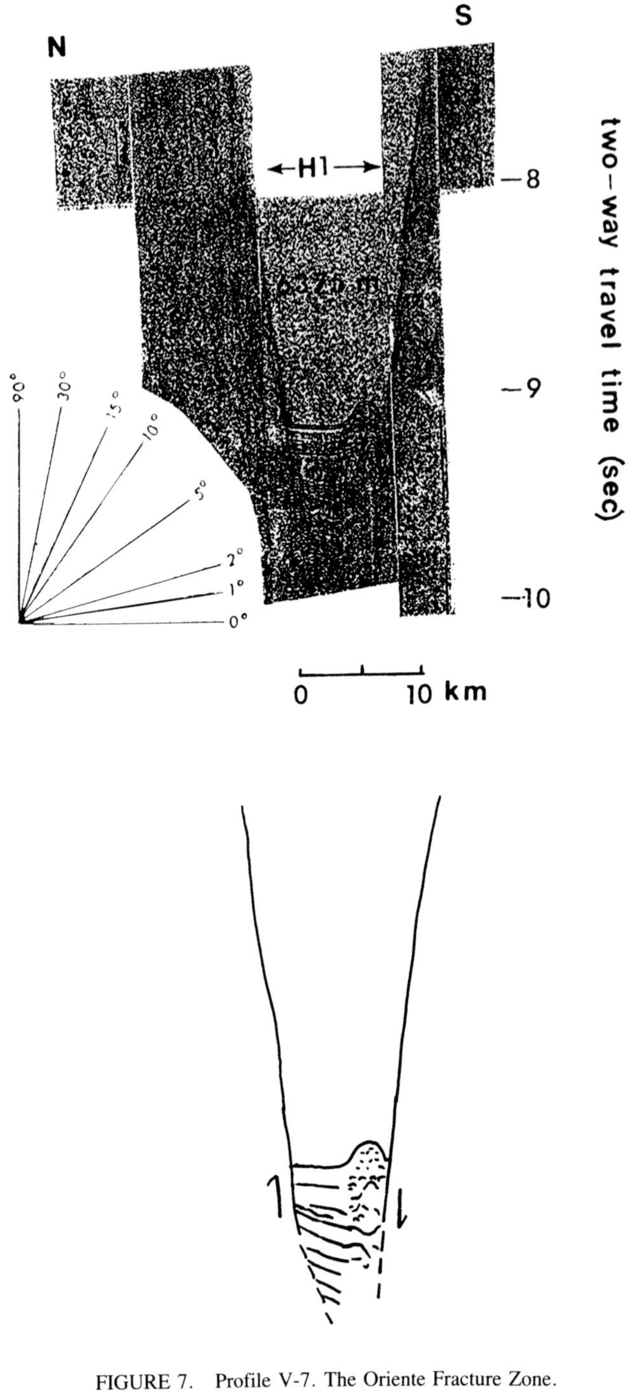

FIGURE 7. Profile V-7. The Oriente Fracture Zone.

south (along the profile) in the subsurface 90 m below the sea floor. This result of extension was probably due to the basement block to the south moving down relative to the basement block to the north. More recent sediment has been deposited subsequent to the tilting of the earlier sediments. Note that the present sea floor appears to be folded, indicating that the fracture zone has undergone compression. This compression, acting upon this fracture zone, has resulted in an area of chaotic reflectors on the southern margin to a depth of 450 m.

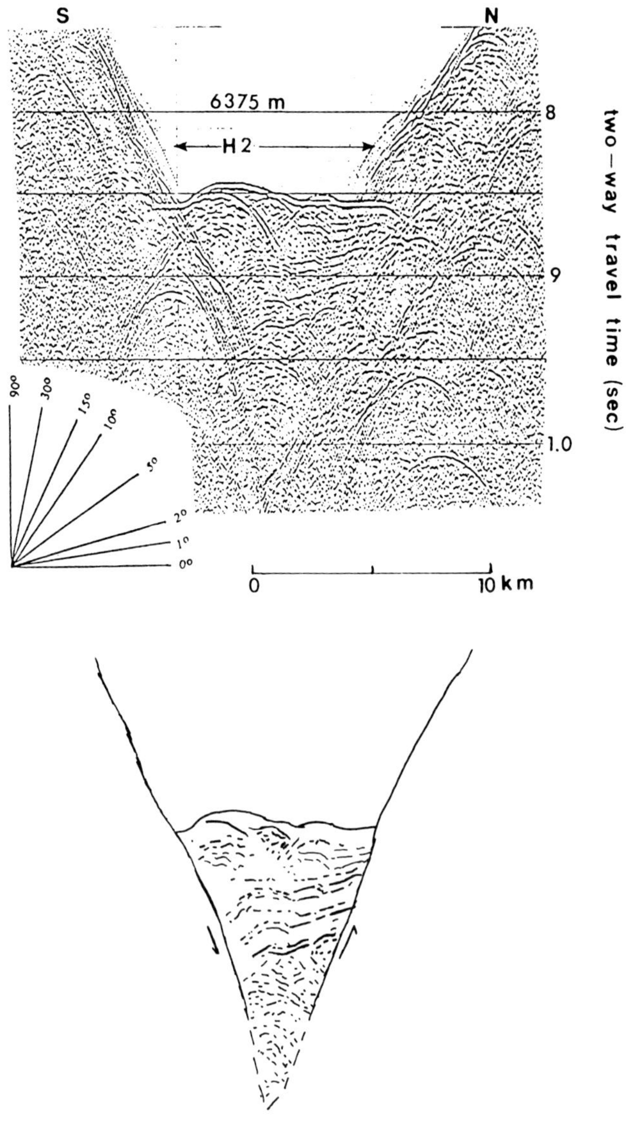

FIGURE 8. Profile CT1-30B. The Oriente Fracture Zone.

The result of compression appears to have destroyed earlier evidence of extension within the southern portion of the fracture zone.

Between longitudes 79°30′W and 78°20′W, sediment observed on Profile V-6 (at region H3) (Figure 9) appears to be highly deformed with only a few subsurface reflectors that can be traced along the profile. Subsurface reflectors on the northern margin, along this profile, appear to dip toward the south. This is an indication of extension which probably pulled apart the fracture zone in this area resulting in the movement of the basement block to the south moving down relative to the basement block to the north. Note that the present sea floor has been folded, which indicates compression. Recent compression seems to have destroyed the evidence of earlier extensional features along the southern margin of the fracture zone along this profile. Sediment to either side of the central upbowed region dips toward the fault scarps to either side. Only profile V-6 clearly shows a recent compressional feature.

Near Cuba at 77°30′W and 19°49′N, sediment dips toward the northeast (Profile CT2-

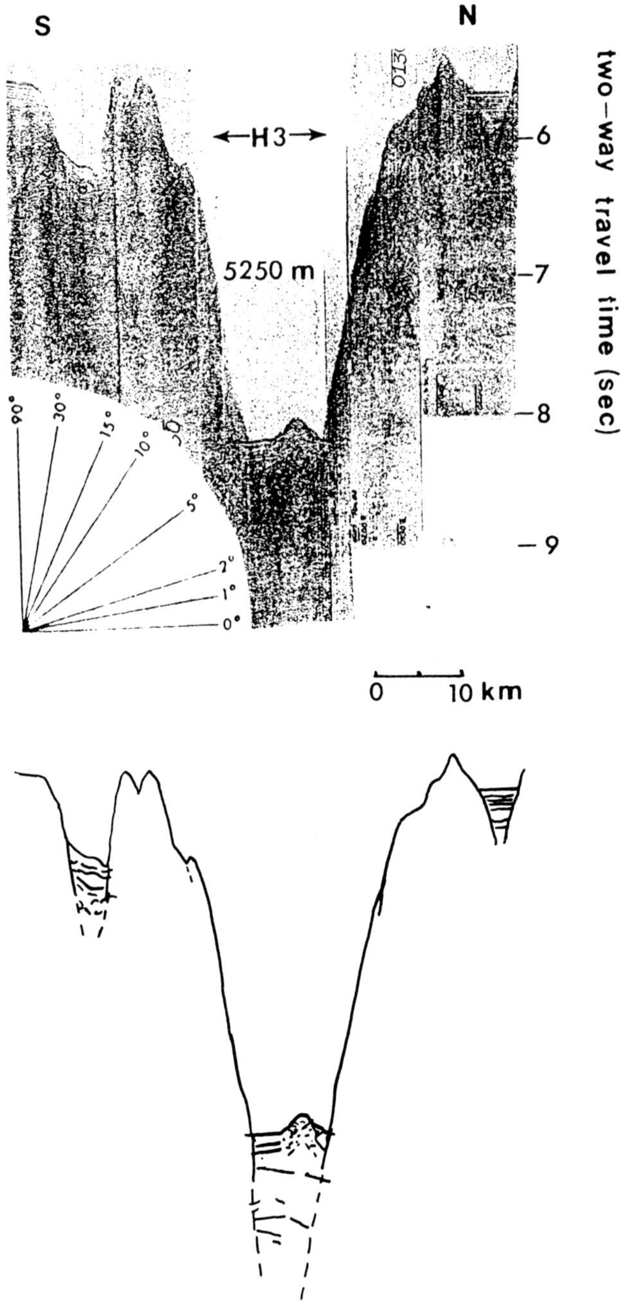

FIGURE 9. Profile V-6. The Oriente Fracture Zone.

13B at region H5; Figure 10). The subparallel reflectors were deposited prior to tilting due to the relative downward movement of the fault block to the northeast. This movement was probably due to extension of the fracture zone. Because the reflectors all dip at the same angle, this indicates that all of the sediment was deposited prior to the movement of the basement within the fracture zone. Only profile CT2-13B shows evidence for extension.

3. Central Cayman Trough

The central portion of the Cayman Trough lies between longitudes 81°W and 83°W, is

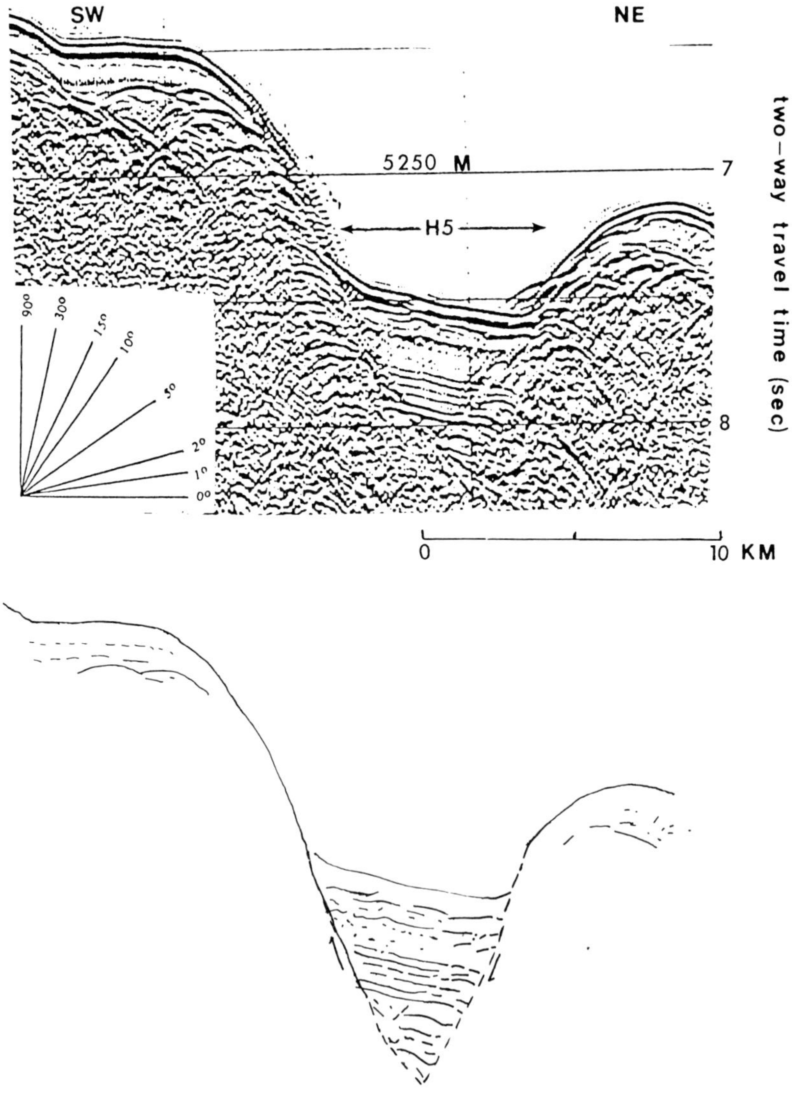

FIGURE 10. Profile CT2-13B. The Oriente Fracture Zone.

topographically higher than the eastern or western Cayman Trough, and, therefore, has shallower water depths (3200 m near the MCSC). The basement in the central Cayman Trough has been broken into individual basement fault blocks (Profile I-9, Figure 11). Some fault blocks, as in Profile I-9 at region J, have been down dropped to the north and tilted to the south as viewed along the profile. Profile I-9 contains no recognizable internal basement reflectors, given the resolution of the seismic reflection data. The rough and irregular basement surface gives rise to numerous hyperbolic diffractions at the basement/water interface.

The central Cayman Trough is virtually sediment free, with some isolated pockets of sediment in a few topographic lows (for example, Profile I-8 at region K; Figure 12) as noted by Eggler et al. (1973). Profile I-8 begins south and west of the Cayman Islands and extends southward from the Cayman Ridge to the central portion of the Cayman Trough. The water depth of most of these topographic lows in this area (Profile I-8) is 3.5 to 4.0 km. Along Profile I-8 sediment is present in some topographic lows but not in others. If

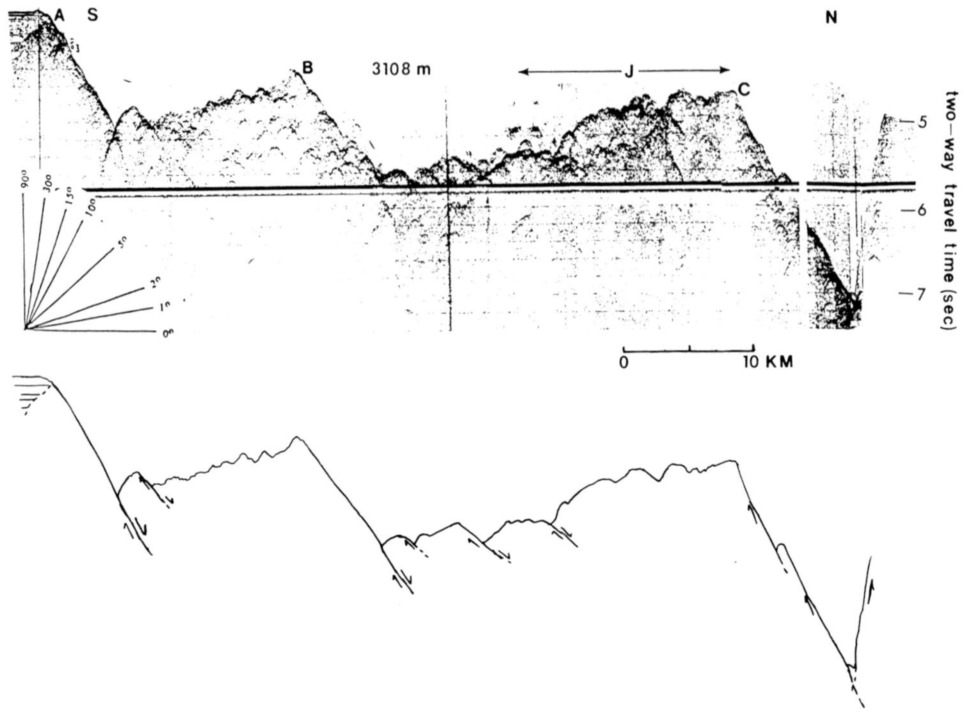

FIGURE 11. Profile I-9. The central Cayman Trough. Letters A, B, and C represent fault scarp locations for the bathymetric map of the MCSC (Figure 13).

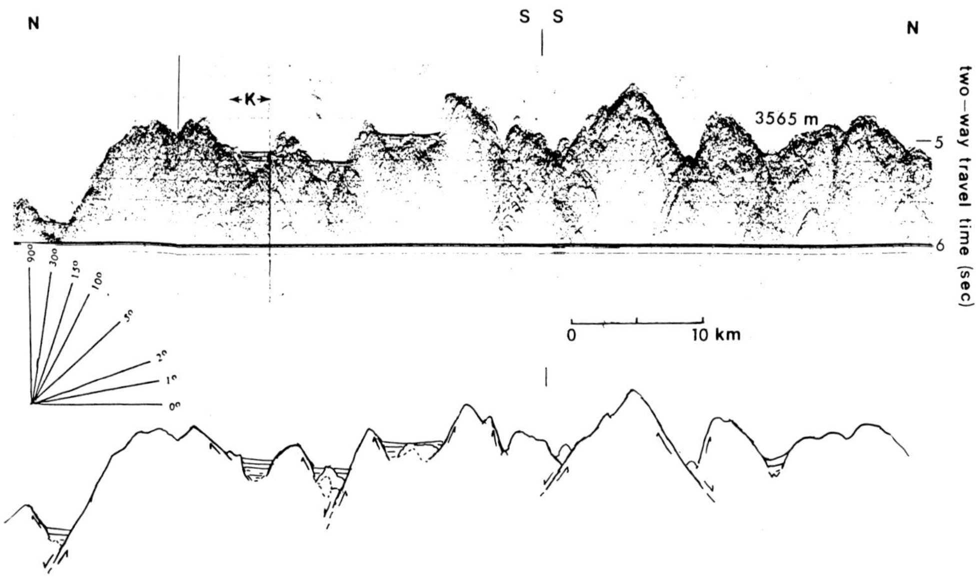

FIGURE 12. Profile I-8. The central Cayman Trough. (After Eggler et al., 1973.)

the only sediment source was pelagic, then all of the valleys of the central Cayman Trough would have sediment accumulation. El'tsina and Lozovaya (1979) stated that the CCD level (Carbonate Compensation Depth) for the Atlantic Ocean is 4 to 6 km. Therefore, all of the topograpic lows are within the range to accumulate carbonate pelagic material, so there must be another sediment source. A possible sediment source is from turbidites off the slopes of the Cayman Ridge. The sediment cannot come from the Nicaraguan Rise because it would be trapped within the Swan Fracture Zone. Sediment accumulations within the central trough show no sign of deformation. This implies that the sediment deposition occurred after the latest movement of the basement faults within the MCSC.

Profile I-9 crosses the Cayman Trough in a north-south direction immediately to the east of the MCSC. This profile was compared with a detailed bathymetric map of the MCSC (Figure 13) to determine the recent spreading direction. From the detailed bathymetric map, the fabric trends in two major directions north-south and N 37°W). The faults shown in Profile I-9 correlated with fault scarps oriented in a N 37°W trend. Profile I-9 cuts the major trends of the MCSC at low angles.

4. Western Cayman Trough

The western portion of the Cayman Trough lies west of 83°W and includes the Swan Fracture Zone. This portion of the Cayman Trough is characterized by shallower water depths than the eastern Cayman Trough and a thick, really large sediment accumulation which completely buries the basement west of 86°W (Profiles Y101, Y102, Y104 and Y106, Figure 3). Water depths vary from 4600 m in the western trough to 5400 m in the Swan Fracture Zone. The western trough consists of basement blocks which have apparently moved, resulting in the faulting of the sediment within this portion of the trough. These basement blocks are down dropped toward the Swan Fracture Zone (Profiles Y102 and Y106; Figures 19, 25). Note that the present sea floor along all profiles has a hummocky surface expression and appears to be highly faulted. This is probably the result of extension on the trough.

The southern margin of the Cayman Ridge (Profile Y106; Figure 25), however, appears to be more deformed than the northern margin of the Cayman Ridge. The present sea floor of the Yucatan Basin (north of the Cayman Ridge) is flat, indicating that there has been no disturbance of the sediment after deposition. The reflectors of the Yucatan Basin end abruptly against the Cayman Ridge. The reflectors of the Cayman Trough, on the other hand, on the southern margin of the Cayman Ridge, appear highly disrupted and folded, indicating possible compression of the sediment against the Cayman Ridge.

Seismic reflection profiles discussed include the Texas A&M University *R/V Alaminos* Profiles Y101, Y102, Y104, Y106 and Y108, the UT-MSI *R/V Ida Green* Profiles CT1-6B, CT1-8A and CT1-10A and the Lamont-Doherty *R/V Conrad Profiles* E-F, N-O and P-Q.

Diapiric structures were mapped throughout the Cayman Trough region (Figure 14). Several diapirlike structures appear on the seismic reflection profiles of the western Cayman Trough (for example, Conrad Profile P-Q at regions O and P and Profile E-F at region Q; Figure 15). Many of the diapirs seen within this study area have the same characteristics: (1) the sides are steep (20° to 30°), (2) many flanking reflectors are upturned and disrupted, (3) the overburden is commonly domed or faulted and (4) some rim synclines are apparent against the diapiric structures on the Nicaraguan Rise. Pinet (1972) observed the presence of diapirlike features on the Nicaraguan Rise south of the Bay Islands. Pinet interpreted these features as diapirs for the reasons as stated above. On the other hand, the lack of rim synclines against most of these features, the small body diameters, and the lack of upturning of many of the flanking reflectors appear to argue against the diapiric interpretation. Schneider and Johnson (1970) suggested that with less vertical exaggeration, these structures may be irregular basement.

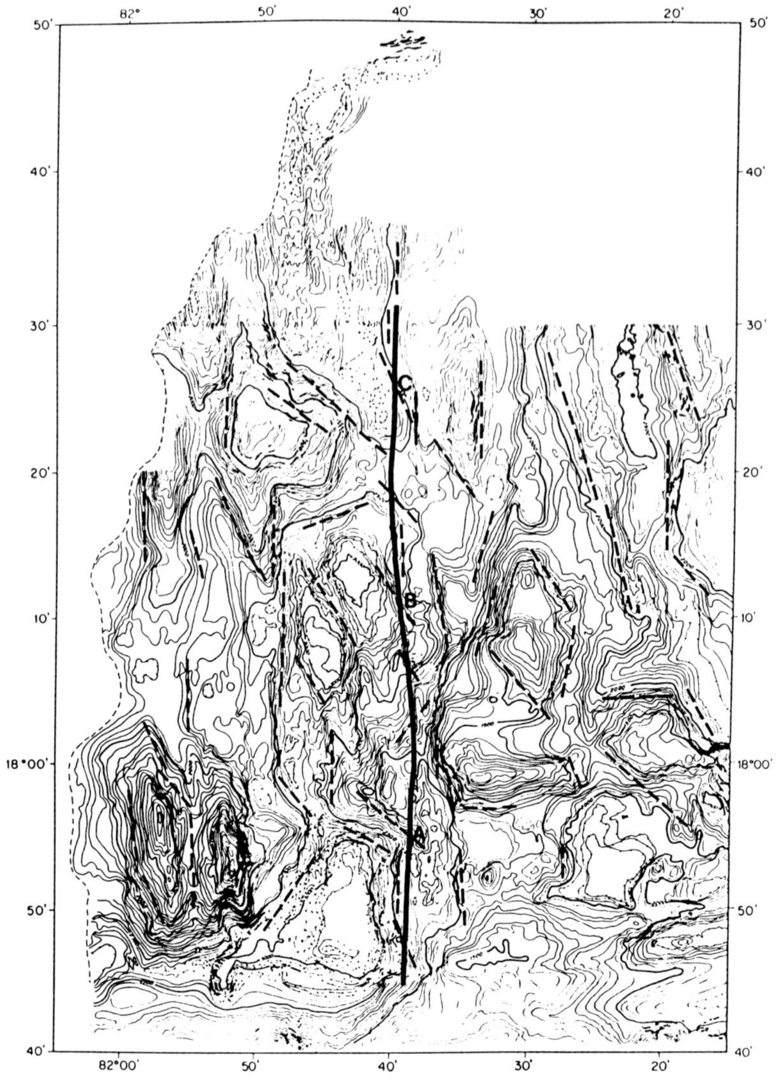

FIGURE 13. The detailed bathymetric map of the MCSC. The solid line running north to south at 81°39′W is Profile I-9. Letters A, B, and C represent corresponding fault scarps on Profile I-9 (Figure 11). (After CAYTROUGH, 1979.)

No magnetic anomalies, however, were observed over the diapirlike features along Profiles Y101, Y102, Y104 and Y110. Therefore, the features are probably irregular basement or diapirs which have been buried under a large sediment accumulation. More closely spaced seismic reflection data is necessary to delineate the geometry of these diapirlike features.

One major submarine canyon has been traced in the western Cayman Trough. Figure 16 illustrates the submarine canyon path trending southwest-northeast, starting directly east of Guatemala at 16°N and 88°W and ending northwest of the Cayman Ridge at 17°60′N and 87°W. The submarine canyon then divides into two canyons; one branch continuing to the northeast, the other curving to the north. This submarine canyon branches downstream, which is unlike most submarine canyons that have minor branches which feed into the main channel farther upstream. This is probably due to some structure that blocked part of the main canyon, creating a branching of the canyon. Another possibility for the canyon to

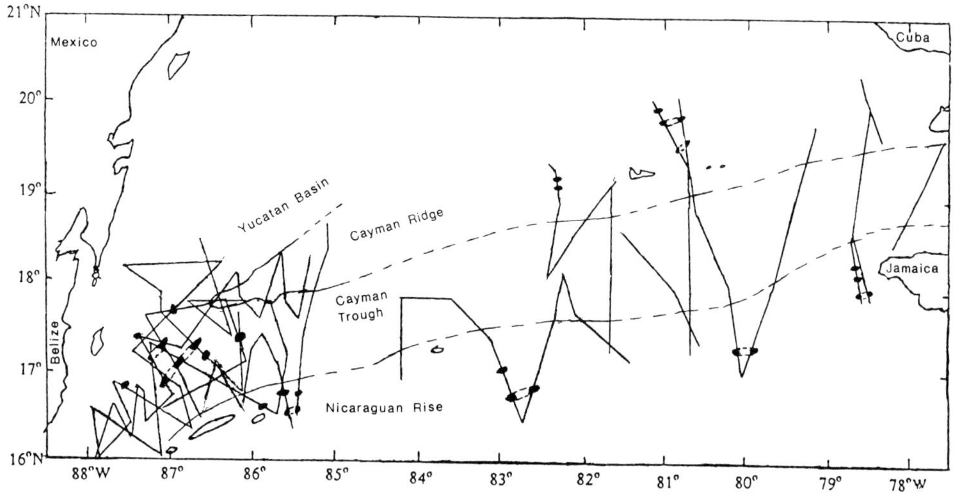

FIGURE 14. The locations of the diapirlike features of the Cayman Trough region.

branch downstream would be for the right-most branch to be an older outlet that is no longer used by the submarine canyon. Seismic reflection data (for example, Profile Y108 at region R; Figure 17) display the canyon characteristics of steep (greater than 30°) sides with occasional sediment deposition at the canyon bottom. Chaotic reflectors indicate possible slump material or debris flow deposits beneath the canyon floor.

Fan structures related to submarine canyons appear on seismic reflection data as lenslike bodies (Profile N-O at region T, Figure 18). The location of the fan (Figure 16) is at the northern end of the south-north trending canyon at 17°60′N and 87°W. Only one lense associated with the submarine canyon outlet was discovered on the seismic reflection profiles of this study. The general pod shape of the fan can be seen on Profile N-O, and some channel and overbank facies may be identified as well by the internal reflections within this lense.

5. Swan Fracture Zone

From 85°10′W to 86°40′W, the Swan Fracture Zone is well defined. The width of the fracture zone in this area is 7.0 to 13.5 km. East of 84°20′W the present location of the Swan Fracture Zone becomes difficult to define and may have a width of 33.0 km. It is bounded on the south by the sediment-free fault scarp of the Nicaraguan Rise (Profile Y102A at region V; Figure 19). Water depths reach as great as 5625 m in this fracture zone. As a result of extension, the sediment within the western Cayman Trough and Swan Fracture Zone has become faulted and subsequently folded. The faults within the sediment of the western Cayman Trough (Profile Y102A at region U, Figure 19), on the northern margin of the Swan Fracture Zone, have been down faulted in a stair-step fashion and these faults become more numerous to the south. Acoustically, well-defined reflectors can be seen on Profile Y102A at region U (Figure 19) to depths of 520 m in the Swan Fracture Zone.

The sediment within the Swan Fracture Zone has been deformed in two different ways throughout its length. Closest to the MCSC (for example, Profile CT1-10A at region S1; Figure 20), the Swan Fracture Zone has been subjected to extension as evidenced by the down-dropped fault blocks. Along this profile sediment within the fracture zone appears to have undergone compression. The reflectors overlying the basement within the fracture zone are chaotic and the overall sediment dip along the profile and within the fracture zone is to the southeast. This dip could have been the result of a sediment slump that occurred before the compression. The sediment has a hummocky surface expression, which was probably the result of the recent compression.

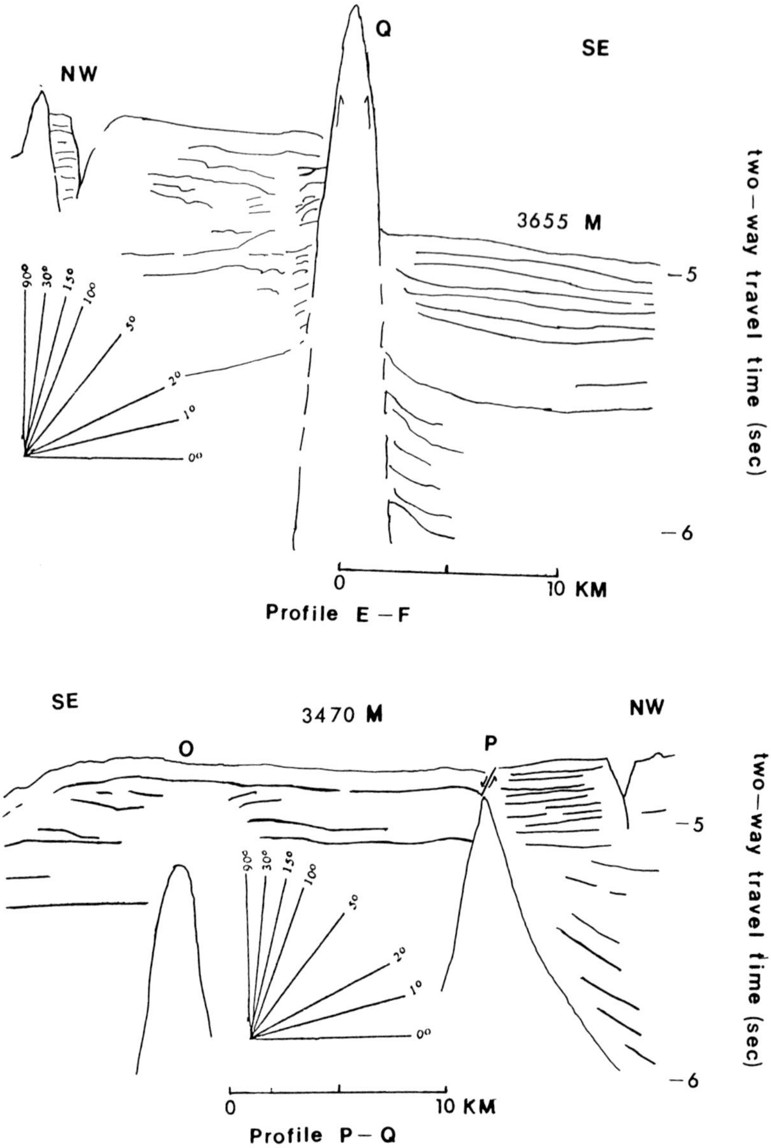

FIGURE 15. Profiles E-F and P-Q. Diapirs in the western Cayman Trough. (After Ewing et al., 1974.)

Profile CT1-8A at region W (Figure 21), on the other hand, shows both compressional and extensional features. Subsurface reflectors (at region S2) are tilted down toward the southeast along this profile. This is probably the result of extension, when the basement block to the northwest moved up relative to the basement block to the southeast. Note that the angle of dip of the reflectors increases with increasing depth, indicating that the relative downward motion of the basement block to the southeast has been contemporaneous with the deposition of sediment. Flat-lying reflectors overlay the tilted reflectors, indicating that the relative movement of the adjacent basement blocks has stopped. The fracture zone in this area seems to have been subjected to recent compression. On the southeastern margin of the fracture zone (at region W), there is an area that has a hummocky surface expression and contains chaotic reflectors. Note that to the northwest of this compressed area, 450 to 1250 m within the subsurface, the tilted reflectors have become folded.

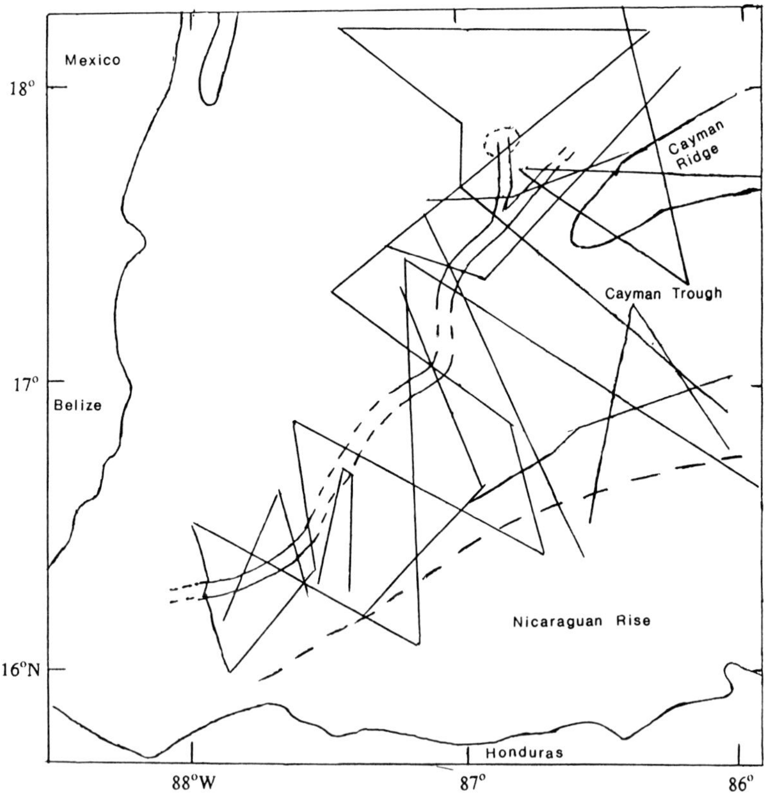

FIGURE 16. The submarine canyon and fan lense in the western Cayman Trough. The dashed area at 86°50′W and 17°50′N is the fan lense.

Sediment within the Swan Fracture Zone on Profile CT1-6B at region S3 (Figure 22) has also been subjected to recent compression. The reflectors appear folded and chaotic. Note that the present sea floor is folded, which indicates compression. No sediment deposition has occurred since the last compressive movement of the basement within and surrounding the fracture zone.

Between 85°50′W and 86°40′W, the sediment within the Swan Fracture Zone has been subjected to extension (for example, Profile Y102A at region S4, Figure 19). The sediment within the Cayman Trough forms a faulted wedge at the northern margin of the Swan Fracture Zone. The uppermost units within this wedge thicken southward along this profile toward the fracture zone. Within the fracture zone, the present sea floor is tilted down slightly to the north. This indicates that the Nicaraguan Rise has moved upward relative to the Swan Fracture Zone. The angle of dip of the reflectors within the sediments of the fracture zone increases with increasing depth beneath the sea floor. This indicates that the extensional movement in the area has been contemporaneous with deposition. No horizontal, undeformed sediment overlays the dipping sediment. The faults north of the Swan Fracture Zone die out with depth within the top 540 to 720 m of sediment. Only evidence of extension is seen along Profile Y102A. Present day tectonic movements within this segment of the fracture zone are evidenced by recent earthquake activity (Molnar and Sykes, 1969).

Within the sedimentary structures of the Swan Fracture Zone, some evidence for slumping of sediment (Profile Y102B at region S5, Figure 23) is evidenced by the hummocky surface expression and chaotic reflectors within the sediment. Subparallel reflectors are seen in the subsurface, further substantiating that the hummocky topography is the result of recent slumping.

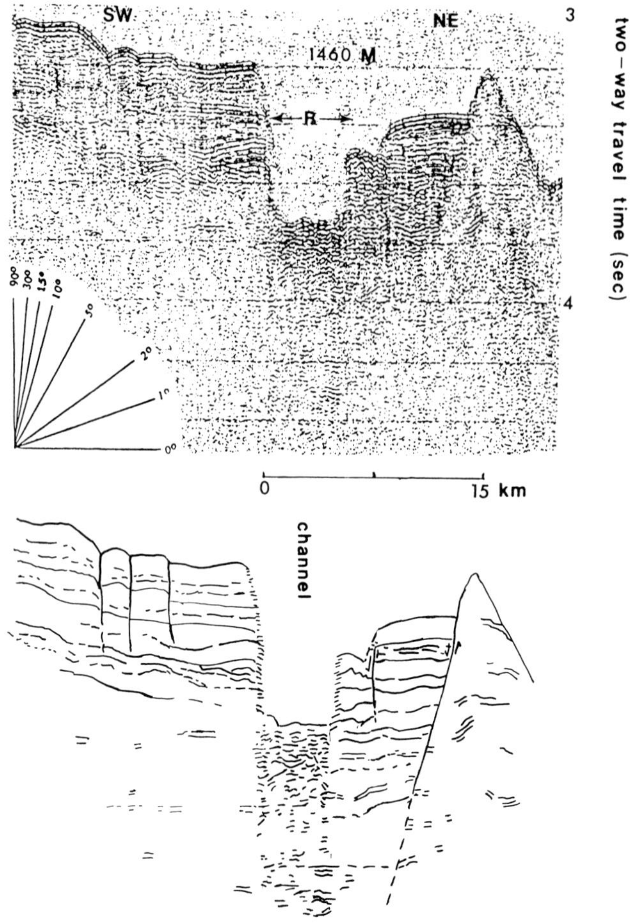

FIGURE 17. Profile Y108. The submarine canyon in the western Cayman Trough.

B. CAYMAN RIDGE

The individual basement blocks of the eastern Cayman Ridge are down dropped toward the Oriente Fracture Zone (Profile V-6 at region E; Figure 9). The fault distribution map (Figure 4) shows that the faults, which separate these basement blocks, parallel the Oriente Fracture Zone. On Profile I-8 at region Y (Figure 24) there is a diapiric structure present on the Cayman Ridge. There are no magnetic anomalies associated with this diapiric feature. More seismic reflection data, though, is necessary to define the geometry of these structures.

The Cayman Ridge is continental basement (Profile Y106 at region X; Figure 25) and has been moved up relative to the Cayman Trough. Sediment is found on the northern and southern margins along the western third of the Cayman Ridge (Profile Y106 at region Z; Figure 25). Near the Cayman Islands sediment accumulation is seen only on seismic within the valleys and not on slopes or ridges (Profile I-8 at region AA; Figure 24). Some sediment within the central portion of the Cayman Ridge (Profile I-8 at region AA) appears to dip to the south, which would indicate relative downward movement of the fault block to the south.

C. NICARAGUAN RISE

The Nicaraguan Rise is a continuous feature from Honduras and Nicaragua to Jamaica. The Nicaraguan Rise is bounded on the north by the Cayman Trough and on the south by

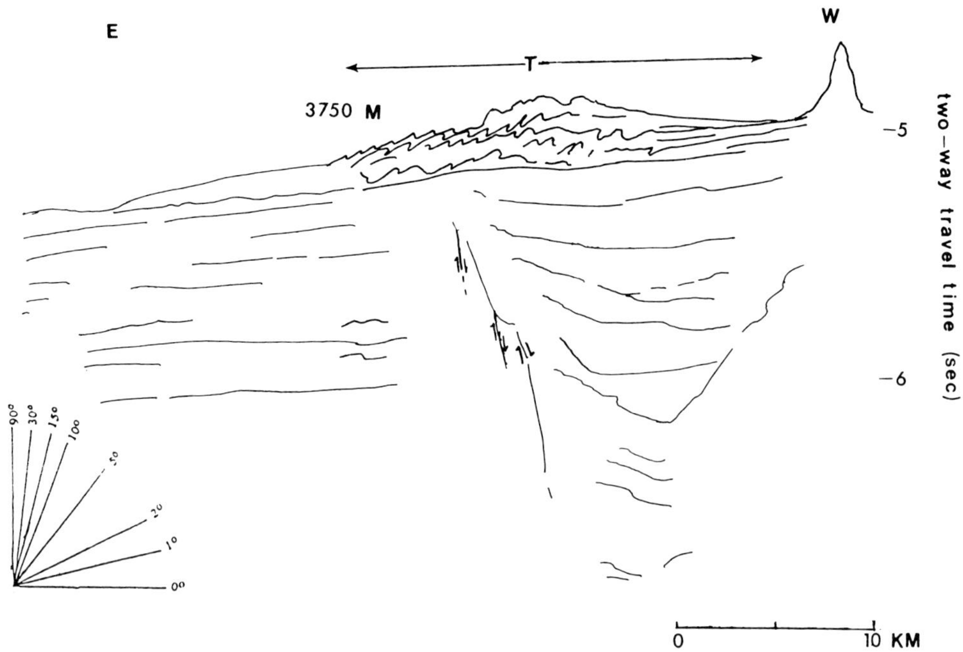

FIGURE 18. Profile N-O. The fan lense in the western Cayman Trough. (After Ewing et al., 1974.)

the Hess Escarpment. Water depths along the crest of the Nicaraguan Rise attain depths of up to 2625 m.

The Nicaraguan Rise is underlain by basement fault blocks that are down dropped in a stair-step fashion toward the Cayman Trough (Profile CT1-9A; Figure 26). The northern fault scarp of the Nicaraguan Rise is sediment-free along Profile Y102 at region V (Figure 19). When the basement fault blocks were mapped (fault distribution map; Figure 4), the faults trend N 65°E. The acoustic basement is defineable on the multichannel seismic reflection data by the diffractions resulting from the faulted basement (for example, Profile CT1-8B at region CC; Figure 27).

Some basement fault blocks cannot be clearly defined on the seismic reflection data. For example, Profile CT1-8B at region CC1 (Figures 14 and 27) shows a diapirlike feature which appears to intrude the sedimentary strata. The reflectors at the southeast side of this structure are upturned against the structure, while the reflectors at the northwest side of this structure are downturned against it. No magnetic data is available to aid in the interpretation of these diapirlike structures. Because there is not enough seismic reflection data available, no direct interpretation of these features can be made.

Throughout the Nicaraguan Rise, the sediment overlying the basement blocks have been severely deformed. For example, Profile CT1-9A (Figure 26) contains sediment accumulations that have been deformed as follows: (1) at region DD, sediment appears to dip toward the north apparently as a result of the down-faulting of the underlying basement blocks, (2) at region EE, sediment becomes faulted due to extension of the underlying basement, (3) at region FF, sediment has been folded due to the individual movements of the fault blocks, and (4) at region GG, the sediment in grabens appears to be highly deformed. There are numerous surface diffractions resulting in a hummocky surface expression within the graben, and, as a result, the top 165 m of sediment are seen as chaotic reflectors that are not highly reflective. The sediment underlying this deformed layer has been gently folded, but not faulted.

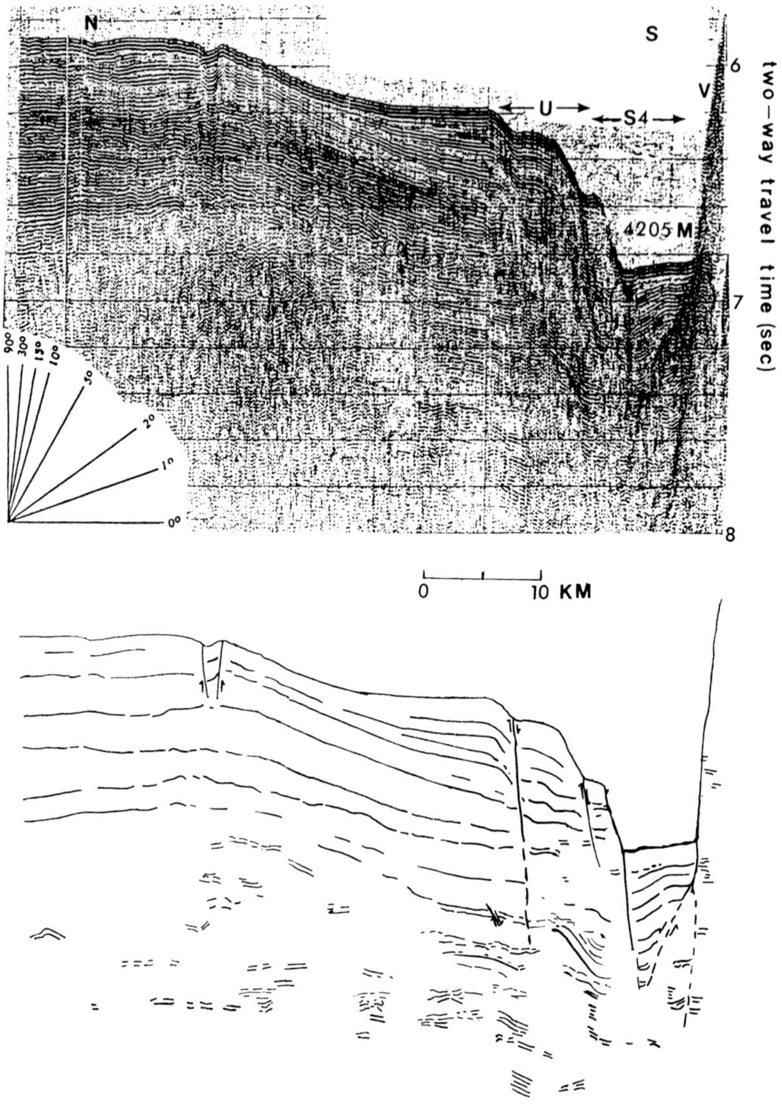

FIGURE 19. Profile Y102A. The Swan Fracture Zone.

V. RESULTS OF THE INTERPRETATION OF THE SEISMIC REFLECTION DATA

The objectives of this chapter are to: (1) map distribution of extensional and compressional sedimentary structures within the Swan and Oriente Fracture Zones, (2) integrate the interpreted data for the fracture zones into a tectonic relative motion model for the North American and Caribbean plates along the Cayman Trough and (3) compare the results of this study with published tectonic spreading models of the Cayman Trough.

As a result of mapping the extensional and compressional features throughout the sediment of the Swan and Oriente Fracture Zones (Figure 28), it is appropriate to relate these features to the recent spreading history and possible variations in the spreading history of the Cayman Trough. In this analysis a spreading rate of 2 cm/year between zero to 2 Ma and 4 cm/year between 2 to 28 Ma (Macdonald and Holcombe, 1978) was used to calculate

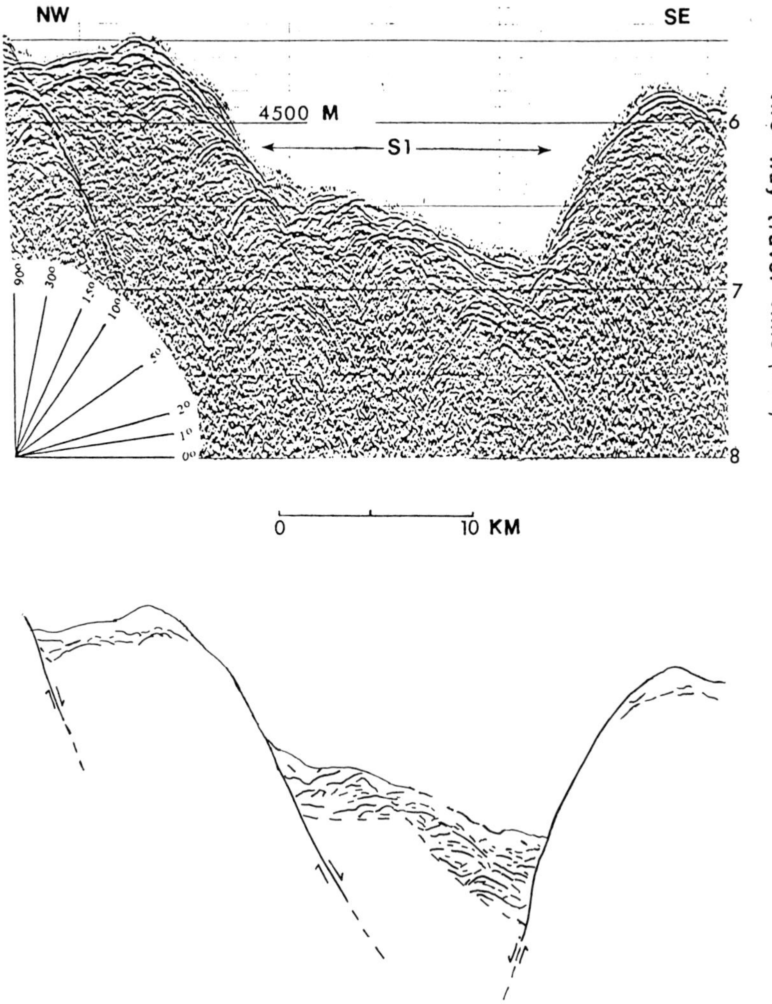

FIGURE 20. Profile CT1-10A. The Swan Fracture Zone.

the age of changes in the spreading fabric. Spreading fabric changes possibly occur due to changes in the pole of rotation between the Caribbean and North American plates.

Most of the published geological data have been collected from the MCSC of the Cayman Trough, the Cayman Ridge, and the Nicaraguan Rise. No geological data have been published for the Swan Fracture Zone, and no geological ages were given for the rock units dredged from the Oriente Fracture Zone by Perfit and Heezen (1978). Because no cores have been obtained from the fracture zones, no geological ages are known for the sediment within the fracture zones.

The assumptions used in interpreting the distribution of compressional and extensional structures along the fracture zones in terms of a relative motion history between the Caribbean and North American plates are: (1) once the Cayman Trough began opening, the overall geometry of the North American-Caribbean plate boundary in the Cayman Trough did not change despite changes in the relative motion, (2) spreading has been symmetrical about the MCSC through time, (3) relative changes in the poles of rotation for the Caribbean and North American plates are reflected in changes in orientation of the spreading fabric, (4) the spreading fabric within the Cayman Trough has been produced along the MCSC normal

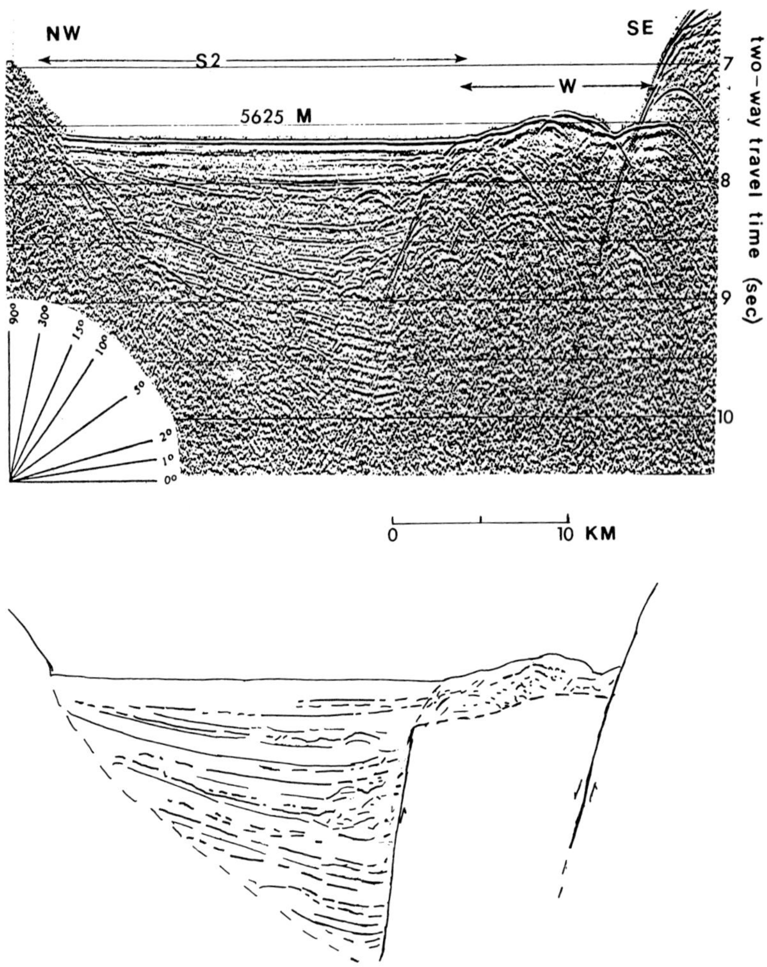

FIGURE 21. Profile CT1-8A. The Swan Fracture Zone.

to the direction of relative plate motion, (5) as spreading direction changes have occurred, extensional and compressional components of the relative plate motion would be primarily taken up along the transform boundaries, (6) these changes in stress conditions at the trough boundaries should be reflected in sediment deformation style along the fracture zones and (7) these extensional or compressional features within the sedimentary structures in the fracture zones, although superimposed except for the latest period of relative motion, have not been destroyed, but moved farther away from the MCSC.

Zones of extension and compression within the Swan and Oriente Fracture Zones are shown on the map (Figure 29) of the tectonic spreading fabric trends of Holcombe and Sharman (1983). Because spreading is assumed to be symmetrical, any extensional and compressional features within the sediment of the fracture zones produced on the western side of the MCSC should have counterparts that appear on the eastern side of the MCSC. Because only compressional features within the sedimentary structures of the fracture zones are seen closest to the MCSC, this indicates that the most recent phase of spreading has been oblique to the overall strike of the Cayman Trough.

Using the distribution of deformational features, a model can be constructed for the motion between the Caribbean and North American plates along the Cayman Trough which shows alternating periods of oblique divergent and convergent components along what has

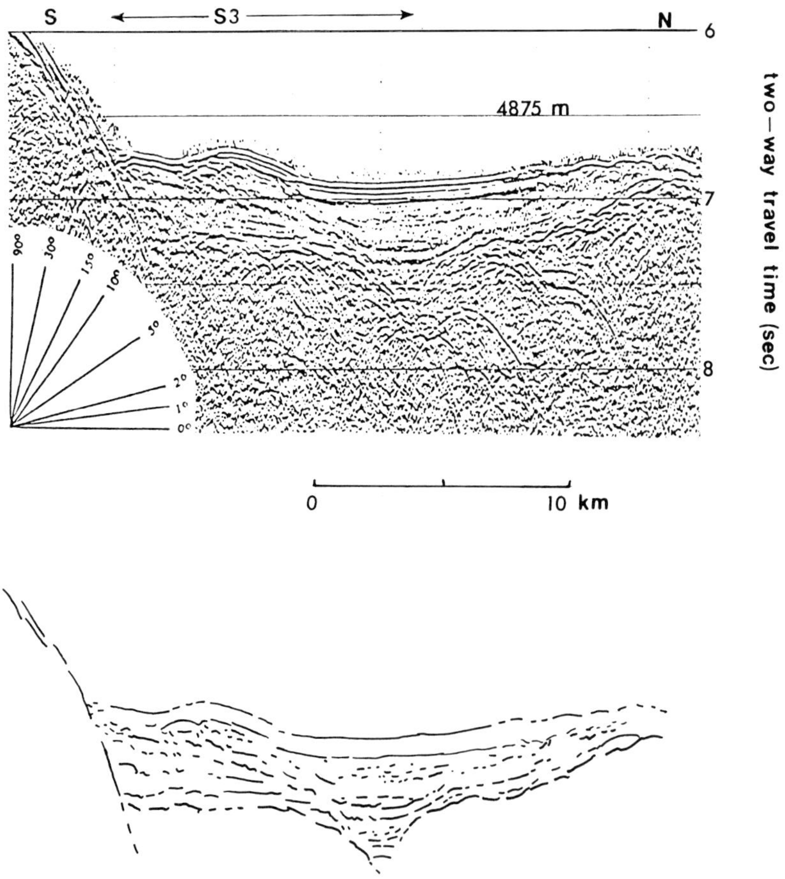

FIGURE 22. Profile CT1-6B. The Swan Fracture Zone.

been primarily a transform boundary. Approximate boundaries between the zones of extensional and compressional features within the sedimentary structure of the Swan and Oriente Fracture Zones were determined. Using the ages for the floor of the Cayman Trough from the magnetic modeling of Macdonald and Holcombe (1978) to assign age limits to the distribution of the variation in the fracture zone structures, the lateral time limits of these boundaries were calculated (Figure 29). Using the distribution of extensional and compressional features and the age limits, the suggested period of oblique compression and extension can be tested.

A model was constructed using the distribution of extensional and compressional features and the age limits of these features to compare with the model. The model from this study (Figure 30) begins with pure transcurrent movement along the transforms (prior to 28 Ma). Then a shift in relative motion between the Caribbean and North American plates occurs which results in continued transcurrent movement plus a small component of opening, or widening, across the trough at 28 Ma. (Stage A reflects the time of the change in the pole of rotation). Extensional features are the dominant sedimentary structure of the Swan and Oriente Fracture Zones (Figure 29) from 28 to 4 Ma. Some compressional features are seen, involving the most recent sediments, which are probably the result of the latest spreading phase. However, the abundant extensional features from 28 to 4 Ma (Figure 30) indicate that the relative motion between the Caribbean and North American plates shifted at about 28 Ma and that up until 4 Ma, motion between the two plates involved a component of oblique extension.

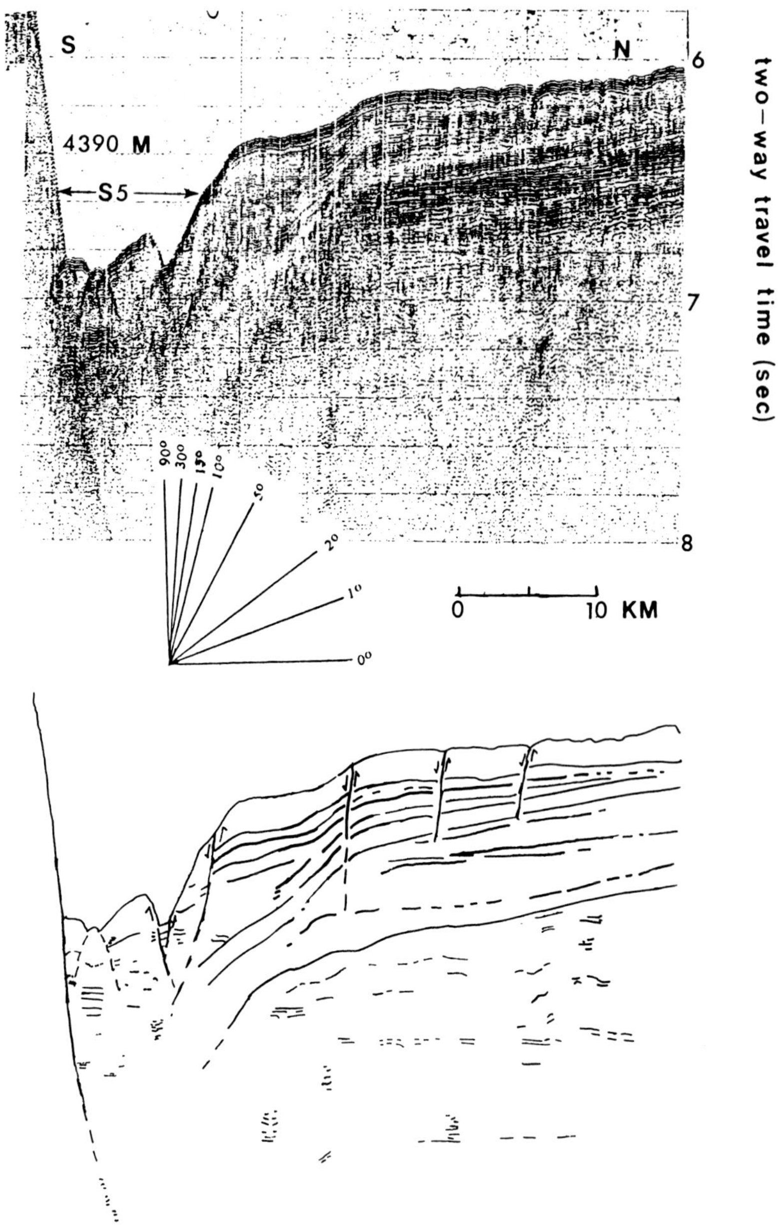

FIGURE 23. Profile Y102B. The Swan Fracture Zone.

From the period 4 Ma to present, only compressional features are present in the sedimentary structures along the Swan Fracture Zone (Figure 29). This indicates that at stage B (4 Ma; Figure 30), another pole shift occurs, resulting in a shift in the spreading direction that is observed in the spreading fabric. This transform motion along the fracture zones included a component of oblique convergence. A small differential decrease in width should have occurred across the trough.

This most recent period of compression along the fracture zones agrees nicely with the relative plate motion suggested from the spreading fabric orientation seen in the detailed bathymetry map of the MCSC (CAYTROUGH, 1979; and Figure 13), restricted to a small area between 81°17′W and 82°W and 17°48′N and 18°28′N, which shows a spreading fabric

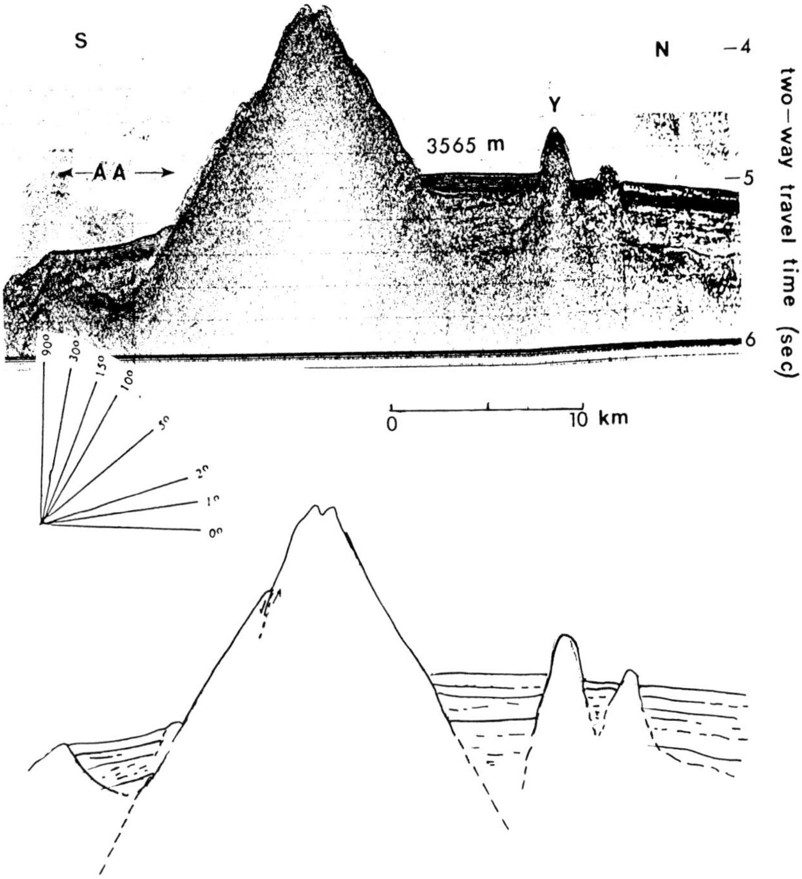

FIGURE 24. Profile I-8. The diapiric structures of the Cayman Ridge.

strike of N 37°W. Compression along the fracture zones in this region agrees with the model constructed by Holcombe and Sharman (Figure 2) which suggests oblique transform motion with a compressive component for the most recent spreading in the Cayman Trough.

In a study of earthquake focal mechanisms of the Cayman Trough, Hilde (personal communication) completed fault plane solutions for an earthquake located on a central ridge of the MCSC near the intersection of the Swan Fracture Zone with the MCSC (81°65′W and 17°85′N at a depth of 33 km) and one located along the Oriente Fracture Zone (80°73′W and 19°05′N at a depth of 33 km). The fault plane solutions suggested right-lateral movement rather than the expected left-lateral movement. However, Hilde felt that the fault plane solution at the intersection of the MCSC and the Swan Fracture Zone may represent left-lateral movement along a nearly-vertical fault plane with a strike of 336°. This indicates that the ridge is not perpendicular to the long axis of the trough and suggests a small component of oblique convergence along the fracture zones. This fault plane solution for the earthquake at the intersection of the MCSC and the Swan Fracture Zone is in agreement with the findings of the seismic reflection data close to the MCSC.

The Holcombe and Sharman model shows that the initiation of spreading within the Cayman Trough began shortly before 15 Ma. Their tectonic spreading model is based on: (1) the lineation trends of the physiography defining specific periods of symmetrical spreading (Case and Holcombe, 1980), (2) magnetic lineation trends and (3) the spreading rates of Macdonald and Holcombe (1978). Their model defines seven changes in the relative plate motion between 15 Ma and the present. The model obtained from this study defines only

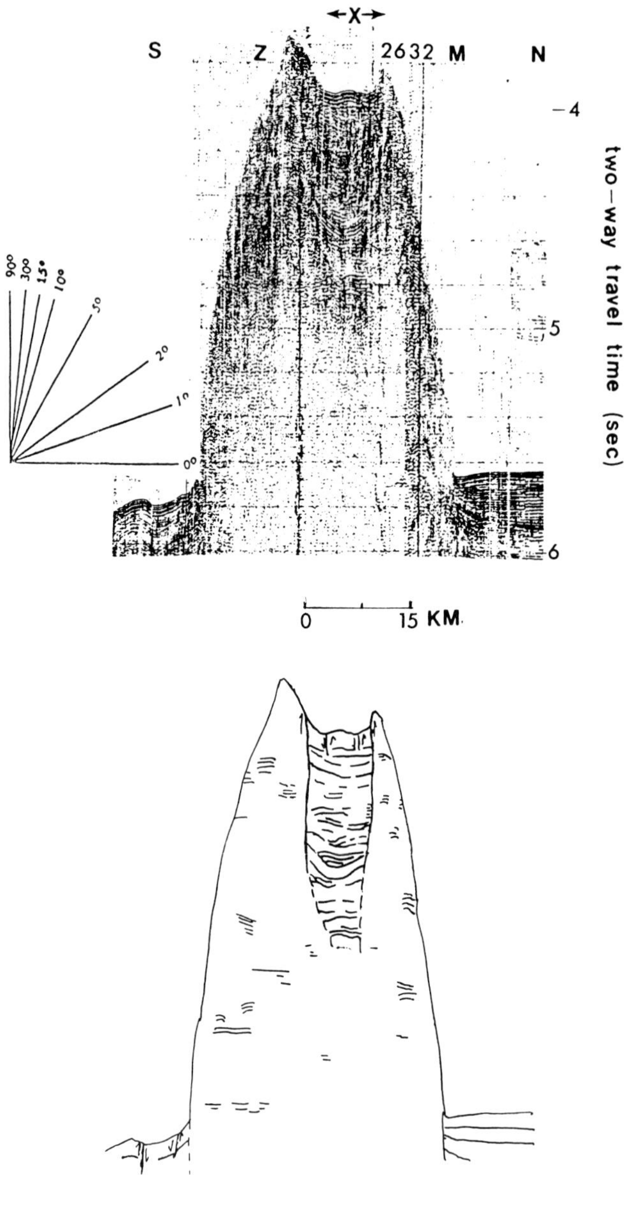

FIGURE 25. Profile Y106. The Cayman Ridge. (After Fahlquist and Davies, 1971.)

two changes in relative plate motion between 28 Ma and the present. While more closely spaced seismic reflection data might possibly allow delineation of additional changes, it is expected that sorting out the detailed variations in relative plate motion from the sedimentary structures for the periods older than the latest phase of relative plate motion would still be very difficult. Spreading fabric orientation is preserved, while the compressional and extensional structures along the fracture zones are superimposed. Therefore, dated spreading fabric orientation changes, assuming they do reflect plate motion changes, may be a much more sensitive means of obtaining the relative plate-motion history. Nevertheless, the structures along the fracture zones do show a variable relative plate-motion history involving alternating periods of slightly oblique convergent, then slightly oblique divergent transform

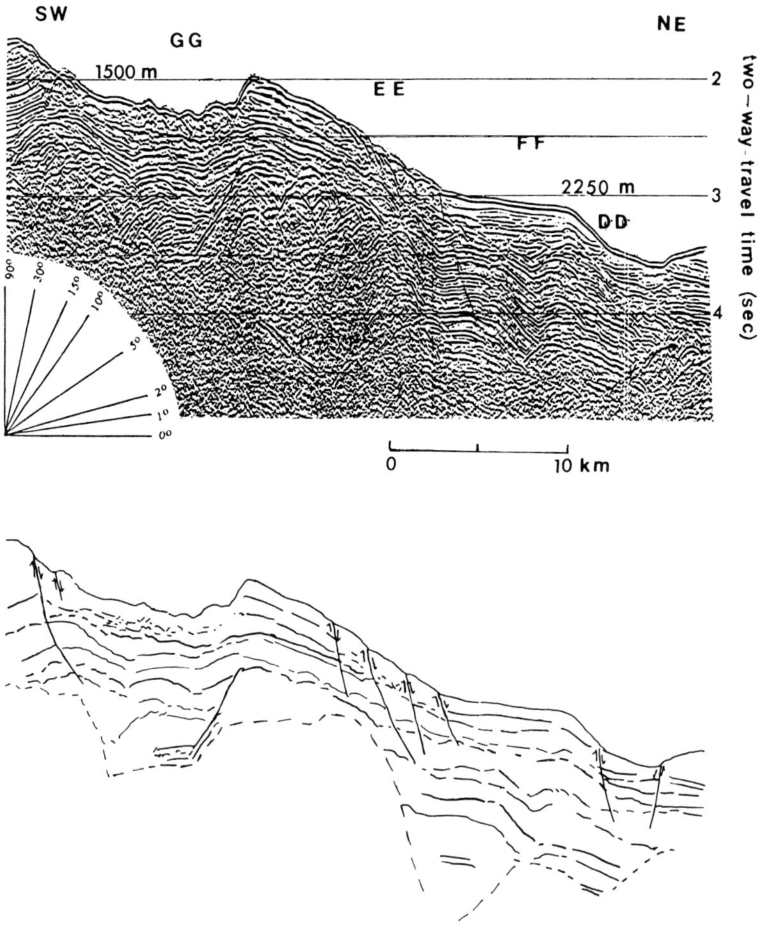

FIGURE 26. Profile CT1-9A. The Nicaraguan Rise.

motion along the Cayman Trough. Within the limits of the seismic reflection data, this study supports the findings of Holcombe and Sharman.

As the Cayman Trough forms a significant proportion of the boundary between the North American and Caribbean plates, details of the sense and history of the plate movement along this boundary also should be recognized in other areas such as Guatemala, Dominican Republic, and the Puerto Rico Trench. Extensional features such as faulting and compressional features such as folding should be seen along these areas to directly correlate with the history recorded in the Cayman Trough.

Burkart (1983), by studying the geomorphology, structure, and radiometric ages along the Polochic-Motagua Fault Zone in Guatemala, concluded that about 130 km of left-lateral offset occurred along this fault in Guatemala between 6 to 12 Ma. Because the Polochic Fault trends about N 70°E, a significant shift in the pole of rotation is suggested at about 5 to 6 Ma (Hilde, personal communication). The trend of the Polochic-Motagua Fault Zone can be connected with the Cayman Trough east of 86°W in geology and structural trends (Burkart, 1983). This pole shift is compatible with the model of this study for the opening of the Cayman Trough, as based primarily on spreading fabric trends.

Evidence for north-south compression has been cited in the Bay Islands by McBirney and Bass (1969), and evidence for north-south compression in the Oriente Province has led Goreau (1981) to propose late Miocene to Holocene north-south compression along the

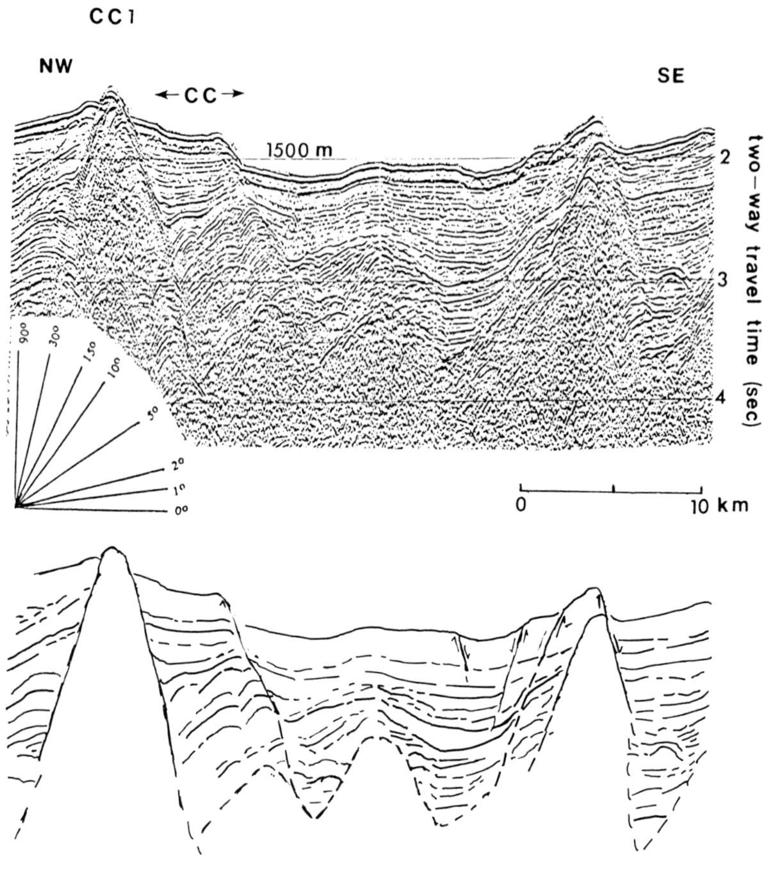

FIGURE 27. Profile CT1-8B. The Nicaraguan Rise.

Oriente Fracture Zone, together with a large component of shear. Recent field work has yielded evidence that pure transcurrent movement may have occurred along the Enriquillo-Cul de Sac trend in Hispaniola in Oligocene time, whereas compression is indicated for Miocene and younger (Mann and Burke, 1984). All of these examples seem to agree with the results obtained from studying the Cayman Trough through seismic reflection data.

The results of this study are significant. One significance of the results is that it is possible to construct a tectonic spreading model of the Cayman Trough by using seismic reflection data and age control (from a reliable source) for the timing of changes in the pole of rotation between the Caribbean and North American plates. Another significance of these results is that the same general steps applied in studying the Cayman Trough (i.e., studying the structural and stratigraphic features with seismic reflection profiles, delineating zones of extension and compression with the sedimentary structures of the associated fracture zones, and using this information to develop a tectonic spreading model of the trough) could be applied to other similar plate boundaries to aid in the structural and stratigraphic understanding and possible modeling of these plate boundaries.

VI. CONCLUSIONS

Based on seismic stratigraphic analysis and mapping of the seismic reflection profiles across the Cayman Trough, numerous sediment accumulations and structural features were identified. The Swan and Oriente Fracture Zones were carefully examined in an effort to

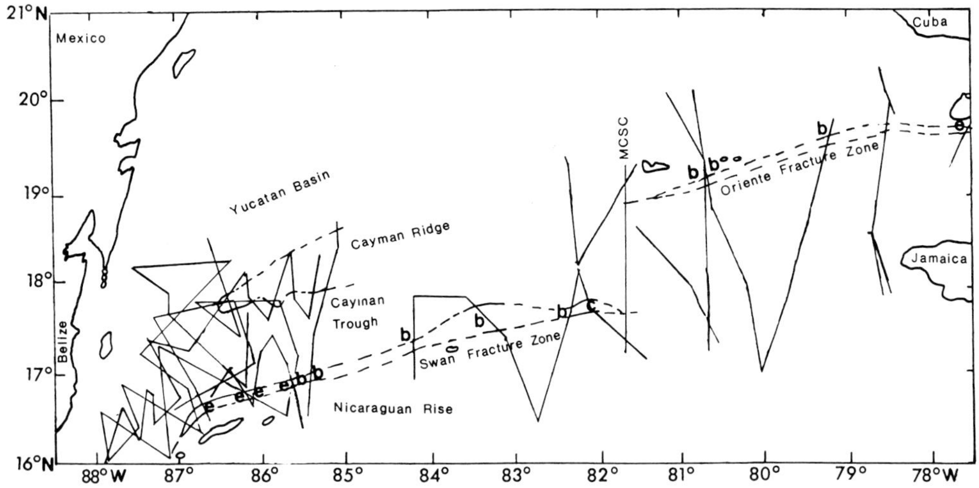

FIGURE 28. Identified compressional and extensional features within the Swan and Oriente Fracture Zones. c = compressional features, e = extensional features, and b = both compressional and extensional features. The dashed lines indicate the width of the fracture zones.

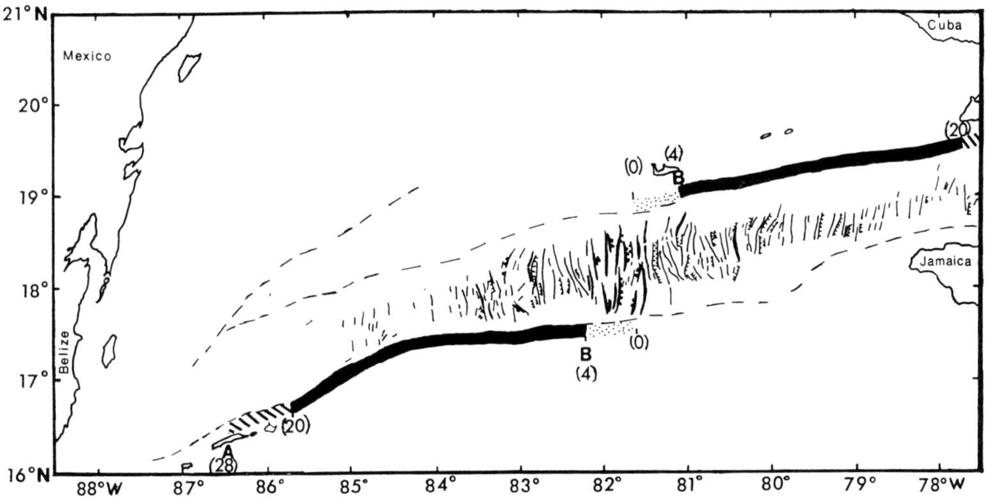

FIGURE 29. The extensional and compressional features as mapped against the tectonic spreading fabric trends. Dashed area indicate extensional features. Stipled areas indicate areas of compressional features. Shaded areas indicate both extensional and compressional features. Capital letters indicate approximate boundaries between the extensional and compressional features. Numbers indicate ages (in Ma) of the spreading stages of the Cayman Trough.

delineate extensional and compressional features and to relate these features to the tectonic spreading history. A relative plate motion model of the North American and Caribbean plates about the Cayman Trough was then generated based on the zones of compression and extension within the sediment of the fracture zones to compare with other tectonic spreading models of the Cayman Trough.

The Oriente and Swan Fracture Zones have moved down relative to the Cayman Ridge and Nicaraguan Rise, respectively. Changes in the relative plate motion between the North American and Caribbean plates were evidenced by the faulting and folding of the sedimentary structures within the fracture zones. Two deformational sequences are seen along the Oriente

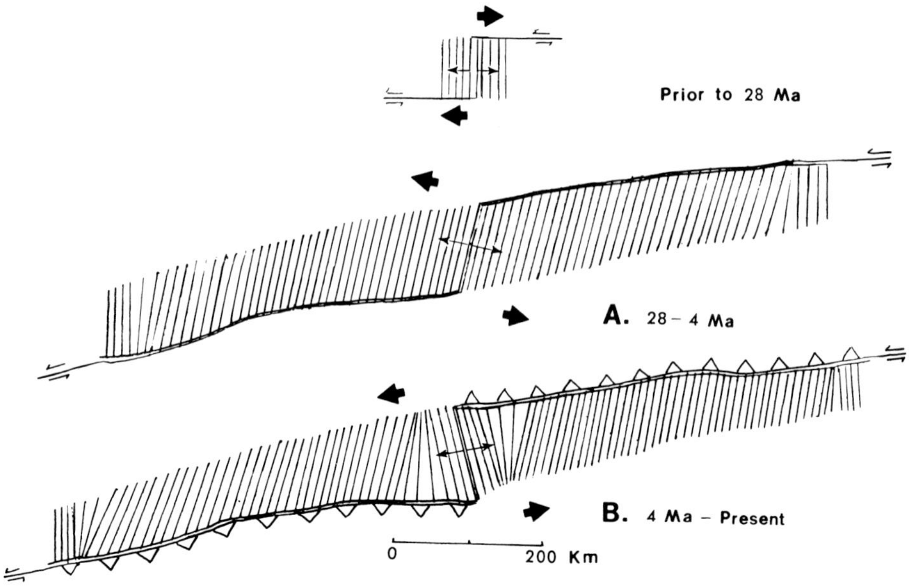

FIGURE 30. The model of the Cayman Trough.

Fracture Zone. Between 78°30′W and 81°W, extensional and compressional features are seen within the sedimentary structures of this fracture zone. Between 77°30′W and 78°W only extensional features are recognized. Three regimes of deformation are present along the Swan Fracture Zone. Between 81°40′W and 82°10′W, only compressional features are present within the sedimentary structures of this fracture zone. Between 82°10′W and 85°30′W, compressional and extensional features are seen, while between 85°30′W and 86°40′W, only extensional features appear.

As a result of mapping extensional and compressional features within the sedimentary sequences of the Swan and Oriente Fracture Zones, it is possible to relate these features to recent spreading history and possible variations in spreading history of the Cayman Trough. In this analysis, a spreading rate of 2 cm/year between zero Ma to 2 Ma and 4 cm/year between 2 Ma to 28 Ma (Macdonald and Holcombe, 1978) was used to calculate the age of changes in the spreading fabric. Spreading fabric changes probably occur due to changes in the pole of rotation between the North American and Caribbean plates.

A model was constructed to test the suggested periods of oblique compression and oblique extension. Prior to 28 Ma, the oldest period of the model of this study begins with pure transcurrent movement along the transforms. This is followed by a shift in relative motion between the Caribbean and North American plates which results in continued transcurrent movement plus a small component of opening, or widening, across the trough from 28 to 4 Ma. At stage B, another pole shift between the North American and Caribbean plates occurred resulting in a shift in the spreading direction that is observed in the spreading fabric. This transform motion along the fracture zones included a component of oblique convergence. This oblique convergence produced compressional features within the sedimentary structures of the fracture zones back to 20 Ma.

This latest period of compression along the fracture zones agrees with the relative plate motion suggested from the N 37°W spreading fabric orientation seen in a detailed bathymetry map of part of the MCSC (CAYTROUGH, 1979). Evidence of compression along the fracture zones from 4 Ma to the present is in agreement with the model constructed by Holombe and Sharman which suggests oblique transform motion with a compressive component for the latest spreading in the Cayman Trough. This also agrees with the fault plane solution for an earthquake located at the intersection between the MCSC and the Swan Fracture Zone (Hilde, personal communication).

The Holcombe and Sharman spreading model of the Cayman Trough shows an initiation of spreading shortly before 15 Ma, and defines seven changes in spreading direction since that time. On the other hand, this study defines only two changes in relative plate motion that began as early as 28 Ma. While more closely spaced seismic reflection data might possibly allow delineation of additional changes in the relative plate motion, it is expected that sorting out the detailed variations in relative plate motion within sedimentary structures from periods older than the latest spreading direction would still be very difficult. Spreading fabric orientation is preserved, while compressional and extensional structures along the fracture zones are superimposed. Therefore, dated spreading fabric orientation changes, assuming they do reflect plate motion changes, may be a much more sensitive means of obtaining the relative plate-motion history. Nevertheless, the structures along the fracture zones do show a variable relative plate-motion history, involving alternating periods of slightly oblique convergent, then slightly oblique divergent transform motion along the Cayman Trough. Within the limits of the seismic reflection data, this study supports and extends the findings of Holcombe and Sharman.

The results of this model indicate that it is possible to produce a tectonic spreading history of the Cayman Trough from seismic reflection data when spreading rates are known. It would be useful to apply the steps taken in studying the Cayman Trough (i.e. studying the structural and stratigraphic features with seismic reflection profiles, delineating zones of extension and compression with the sedimentary structures of the associated fracture zones, and using this information to develop a tectonic spreading model of the trough) to similar plate boundaries to aid in the structural and stratigraphic understanding and possible modeling of these plate boundaries.

ACKNOWLEDGMENTS

I wish to express my gratitude to those institutions which provided me the seismic reflection data to complete this study: Texas A&M University, the University of Texas Marine Science Institute, and the Lamont-Doherty Geological Observatory, as well as those who provided valuable assistance including Drs. Davis A. Fahlquist, Thomas W. C. Hilde, and Mr. Robin G. Lighty.

REFERENCES

Anderson, T.H. and Schmidt, V.A., The evolution of Middle America and the Gulf of Mexico-Caribbean Sea region during Mesozoic time, *Geol. Soc. Am. Bull.*, 94, 941-966, 1983.

Bonatti, E., Honnerez, J., Kirst, P., and Radicati, R., Metagabbro from the Mid-Atlantic at 06°N: Contact-hydrothermal-dynamic metamorphism beneath the axial valley, *J. Geol.*, 83, 61-70, 1975.

Bowin, C. O., Geophysical study of the Cayman Trough, *J. Geophys. Res.*, 73, 5159-5173, 1968.

Burkart, B., Neogene North American-Caribbean plate boundary across northern Central America: offset along the Polochic fault, *Tectonophysics*, 99, 251-270, 1983.

Case, J. E., Geologic framework of the Caribbean Region, in Geology, Geophysics and Resources of the Caribbean, a report the IDOE workshop on the geology and marine geophysics of the Caribbean Region and its resources, Kingston, Jamaica, February 17-22, 1975, 27-62.

Case, J. E. and Holcombe, T. L., Geologic-tectonic map of the Caribbean region, U.S. Geological Survey, Miscellaneous Investigation Map 1-1100, 1980.

CAYTROUGH (Ballard, R., Bryan, W., Dick, H., Emery, K. O. et al.), Geological and geophysical investigation of the Mid-Cayman Rise spreading center: Initial results and observations, in Deep Drilling Results in the Atlantic Ocean: Ocean crust, Talwani, M., Harrison, C. G., and Hays, D. E., Eds., copyright by the *Am. Geophys. U., Maurice Ewing Ser. 2,* 66-93, 1979.

Edgar, N. T., Ewing, J. J., and Hennoin, J., Seismic refraction and reflection in the Caribbean Sea, *Am. Assoc. Pet. Geol. Bull.*, 55, 833-870, 1971.

Eggler, D. H., Fahlquist, D. A., Pequegnat, W. E., and Herndon, J. M., Ultrabrasic rocks from the Cayman Trough, Caribbean Sea, copyright by the *Geol. Soc. Am. Bull.*, 84, 2133-2138, 1973.

El'tsina, G. N. and Lozovaya, N. G., Recent sediment of Puerto Rico and Cayman Trenches, *Lithol. Miner. Resour. (USSR)*, 14, 6, 667-675, 1979.

Erickson, A. J., Helsley, C. E., and Simmons, G., Heat flow and continuous seismic profiles in the Cayman Trough and Yucatan Basin, *Geol. Soc. Am. Bull.*, 83, 1241-1259, 1972.

Ewing, J. I., Antoine, J., and Ewing, M., Geophysical measurements in the western Caribbean Sea and in the Gulf of Mexico, *J. Geophys. Res.*, 65, 4087-4126, 1960.

Ewing, J., Ewing, M., Windisch, C. and Aitken, J., Lamont-Doherty survey of the World Ocean, in *Seismic Reflection Profiles*, Manik Talwani Ed., Lamont-Doherty Geological Observatory of Columbia University, New York, 224-225, 1974.

Fahlquist, D. A. and Davies, D. K., Fault block origin of the western Cayman Ridge, Caribbean Sea, *Deep Sea Res.*, 18, 243-253, 1971.

Goreau, P., Unpublished Ph.D. dissertation, 1981.

Holcombe, T. L., Caribbean bathymetry and sediments, in *Geology, Geophysics and Resources of the Caribbean*, a report of the IDOE workshop on the geology and marine geophysics of the Caribbean and its resources, Kingston, Jamaica, February 17-22, 1975, 27-62.

Holcombe, T. L. and Sharman, G. F., Post-Miocene Cayman Trough evolution: A speculative model, *Geology*, 11, 714-718, 1983.

Holcombe, T. L., Vogt, P. R., Matthews, J. E., and Murchison, R. R., Evidence for sea-floor spreading in the Cayman Trough, *Earth Planet. Sci. Lett.*, 20, 357-371, 1973.

Jordan, T. H., The present-day motions of the Caribbean plate, copyright by the *J. Geophys. Res.*, 80, 4433-4439, 1975.

Ladd, J. W., Relative motion of South America with respect to North America and Caribbean tectonics, *Geol. Soc. Am. Bull.*, 87, 969-976, 1976.

Macdonald, K. C. and Holcombe, T. L., Inversion of magnetic anomalies and sea-floor spreading in the Cayman Trough, *Earth Planet. Sci. Lett.*, 40, 407-414, 1978.

Malfait, B. T. and Dinkelman, M. G., Circum-Caribbean tectonic and igneous activity and the evolution of the Caribbean plate, *Geol. Soc. Am. Bull.*, 83, 251-271, 1972.

Mann, P. and Burke, K., Neotectonics of the Caribbean, *Rev. Geophys. Space Phys.* 22, 4, 309-362, 1984.

McBirney, A. R. and Bass, M., Geology of Bay Islands, Gulf of Honduras, in Tectonic relations of northern Central America and the western Caribbean — the Bonacca expedition, McBirney, A. R. Ed., *Am. Assoc. Pet. Geol., Memoir 11*, 229-243, 1969.

Minster, J. B. and Jordan, T. H., Present-day plate motions, *J. Geophys. Res.*, 83, 5331-5354, 1979.

Molnar, P. and Sykes, L. R., Tectonics of the Caribbean and middle America regions from focal mechanisms and seismicity, *Geol. Soc. Am. Bull.*, 80, 1639-1684, 1969.

Perfit, M. R., Petrology and geochemistry of mafic rocks from the Cayman Trench: Evidence for spreading, *Geology*, 5, 105-110, 1977.

Perfit, M. R. and Heezen, B. C., The geology and evolution of the Cayman Trench, *Geol. Soc. Am. Bull.*, 89, 1155-1174, 1978.

Perry, R. K., Bathymetry map of the Gulf of Mexico and the Caribbean Sea, *Am. Assoc. Pet. Geol.*, 1:3,289,263 at 20°N, 1984.

Pinet, P. R., Diapirlike features offshore Honduras: Implications regarding tectonic evolution of Cayman Trough and Central America, *Geol. Soc. Am. Bull.*, 83, 1911-1921, 1972.

Schneider, E. D. and Johnson, G. L., Deep ocean diapiric structures, ICSU/SCOR Working Party 31 Symposium, Cambridge: The geology of the east Atlantic continental margin, Delaney, F. M. Ed., *Geol. Sci. Rep.*, 13, 153-175, 1970.

Schwartz, D. P., Cluff, L. S., and Donnelly, T. W., Quaternary faulting along the Caribbean-North American plate boundary in Central America, *Tectonophysics*, 52, 431-445, 1979.

Sclater, J. G., Anderson, R. N., and Bell, M. L., Elevation of ridges and evolution of the central eastern Pacific, *J. Geophys. Res.*, 76, 7888-7915, 1971.

Stern, C., deWit, M. J., and Lawrence, J. R., Igneous processes associated with the formation of Chilean ophiolites and their implication for ocean floor metamorphism, seismic layering and magnetism, *J. Geophys. Res.*, 81, 4370-4380, 1976.

Sykes, L. R., McCann, W. R., and Kafka, A. L., Motion of Caribbean Plate during last 7 million years and implications for earlier Cenozoic movements, *J. Geophys. Res.*, 87, 10656-10676, 1982.

Telford, W. M., Geldart, L. P., Sheriff, R. E., and Keys, D. A., *Applied Geophysics*, Cambridge University Press, New York 1976, 220-434.

Chapter 10

SEISMIC STRATIGRAPHY AND SALT TECTONICS ALONG THE SIGSBEE ESCARPMENT, SOUTHEASTERN GREEN CANYON REGION

Alan Mark Swiercz

TABLE OF CONTENTS

I. INTRODUCTION

The lower continental slope in the northern Gulf of Mexico represents a unique geomorphic and geologic province. It is generally subdivided into two areas, the Sigsbee Bulge area and the Central Slope/Mississippi Fan Area (Figure 1). The Sigsbee Bulge forms a broad, arcuate, plateau-like feature approximately 50 to 75 km wide which bulges to the south. Its southern boundary is characterized by a large physiographic expression (up to 900 m of relief) on the sea floor which is known as the Sigsbee Escarpment.

The Sigsbee Escarpment represents the leading edge of a mobil layer of salt which has been emplaced at least 10 to 15 km seaward as a result of sediment loading up dip by the Mississippi River. The tabular or lobate nature of salt in this region is markedly different from the typical domes and ridges associated with the Central Slope and upper Mississippi Fan areas where salt movement is presumed to be predominantly vertical.

A. LOCATION

The study area encompasses a 2340 km^2 area between 27° to 27°20′N latitude and 89°55′ to 90°20′ in the southeastern corner of the Green Canyon region (Figure 2). The northeast to southwest-trending Sigsbee Escarpment runs through the lower portion of the study area dividing it into two distinct regions — the lower Continental Slope, with a semi-continuous salt mass and relatively thin sediment cover to the north, and the Continental Rise, with gently seaward-dipping sediments found to the south. The study area is in water depths which range from 1061 m north of the scarp, to 2275 m seaward of the scarp into the deep basin.

B. OBJECTIVES

The main objective of this study was to develop a regional geologic depositional model by studying the interrelationship between salt migration and clastic sedimentation in the area. This includes: (1) a seismic-stratigraphic analysis; (2) a structural and stratigraphic description of fault patterns and diapiric trends; (3) speculation on the ages of depositional sequences and the timing of major events; and (4) a complete seismic facies analysis. Velocity analysis, reprocessing/salt imaging techniques, and a geologic model for the evolution of Green Knoll were also developed.

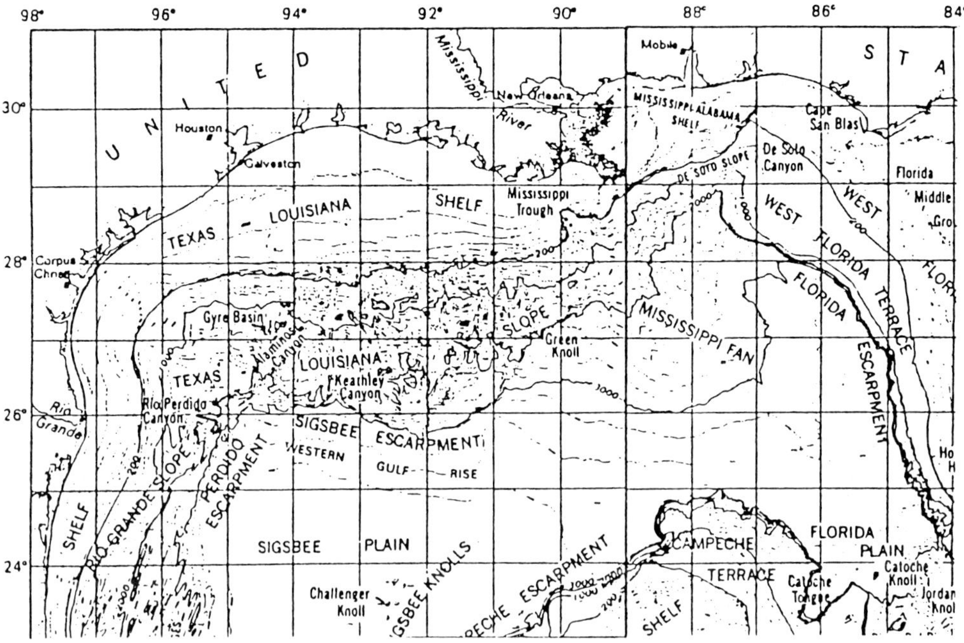

FIGURE 1. The northern Gulf of Mexico province.

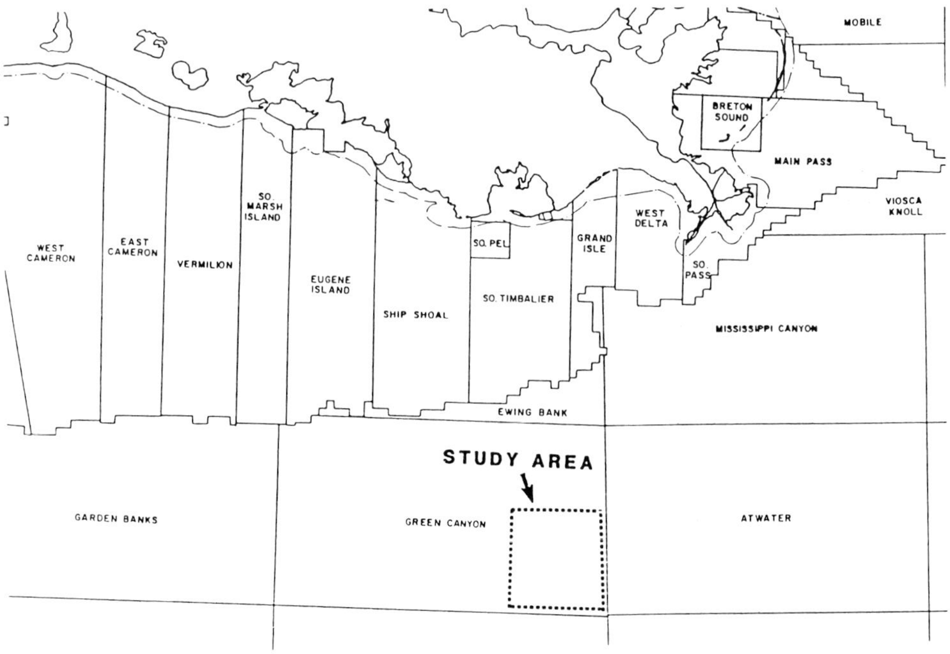

FIGURE 2. Location of study area.

C. PREVIOUS STUDIES

Lehner (1969) was first to suggest that the structural grain and topography of the upper slope in the northern Gulf of Mexico was controlled primarily by salt tectonics. He also concluded that: (1) the top of the salt is easily identifiable as a high-amplitude reflector or an envelope of diffraction patterns; (2) the flat tops of the salt pillows on the lower slope reach a common elevation and create a flat, terrace-like topography; (3) on the lower slope, broad sedimentary troughs appear to subside into a large salt mass; (4) the salt is displaced laterally under the differential load of overburden: and (5) the Sigsbee Scarp is the surface expression of a salt wall (Lehner, 1969).

Amery (1969) recognized a tongue of high velocity material, probably salt, underlying the scarp and extending up to 10 km seaward over continental rise sediments. He concluded that the topography of the scarp was a result of downslope movement of salt and sediments.

Garrison and Martin (1973) added to the previous findings of Lehner (1969) and Amery (1969) and concluded that: (1) the salt mass beneath the lower slope is overlain by a relatively thin sediment cover; (2) large reentrants in the Sigsbee Scarp suggest a lobate nature for the salt front; and (3) uplifted continental rise sediments indicate an upward component of growth of the salt front.

Shih et al. (1977) traced the extension of the Sigsbee Scarp northeastward beneath the Mississippi Fan. They also mapped two sets of subparallel salt ridges and concluded that the ridges were formed due to sediment loading during an eastward shift of the Mississippi River depocenter during the Pleistocene, and that the ridges become younger in a basinward direction.

Humphris (1978) noted a relationship between the major Pleistocene depocenter and the position of the Sigsbee Bulge, and suggested that lateral salt flowage from sediment loading is a primary mechanism for initiation of structures seen on the continental slope (Figure 3).

Watkins et al. (1978) showed that the base of the salt tongue truncates the underlying reflections, suggesting a thrusting relationship as opposed to a decolement along a bedding plane or a lateral flow atop a sedimentary sequence. They also concluded that the underlying Pleistocene and older continental rise sediments extend for at least 10 km beneath the tongue.

Buffler et al. (1978) felt that the salt masses along the scarp probably represent large allochthonous lobes of salt thrust into and over younger sediments distances of 55 to 75 km as a result of a lateral pressure gradient caused by sediment loading updip.

Martin (1980) found that he could group salt structures in the Gulf of Mexico into four categories: (1) near-surface and deeply buried salt structures distinctly outlined on seismic data or confirmed by drilling; (2) anticlines believed to be cored by salt; (3) deeply buried, non-diapiric salt swells; and (4) laterally extruded salt tongues. Martin (1984) modified his previous work by classifying diapiric trends in the Gulf of Mexico basin into seven salt provinces. He suggests that the Lower Texas-Louisiana Slope province is characterized by very broad extrusive salt lobes covered by moderately thick to very thin overburden. The extrusive masses have irregular to lobate planar shapes and are generally wedgelike in cross section and thin toward the continental rise (Martin, 1984).

D. GEOLOGIC HISTORY

The present position of salt in the Gulf of Mexico reflects the complex history and evolution of the basin. Buffler (1984), in consensus with numerous Gulf workers, summarizes the early evolution of the deep Gulf in several stages (Figure 4).

The initial stages of rifting during Late Triassic-Early Jurassic time resulted in the formation of wide-spread rift basins which were filled with nonmarine sediments and volcanics, such as in the northern Gulf Coast (Eagle Mills) and in northern Mexico (Buffler, 1984).

Formation of broad basins and arches continued throughout the waning phase of rifting (Middle Jurassic). In many of the basins, extensive evaporites were deposited, including thick salt in the deeper, more rifted parts of the basins (Buffler, 1984).

A short period (5 to 10 my) of oceanic crust formation during the Late Jurassic separated the salt basins. Oceanic crust is defined by both seismic refraction data (5 to 6 km thick; 6.8 to 7.2 km/sec velocities; Watkins et al., 1978) and seismic reflection data (onlap of

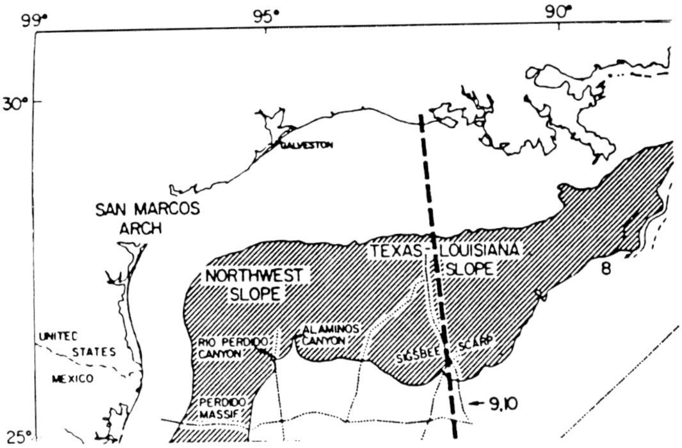

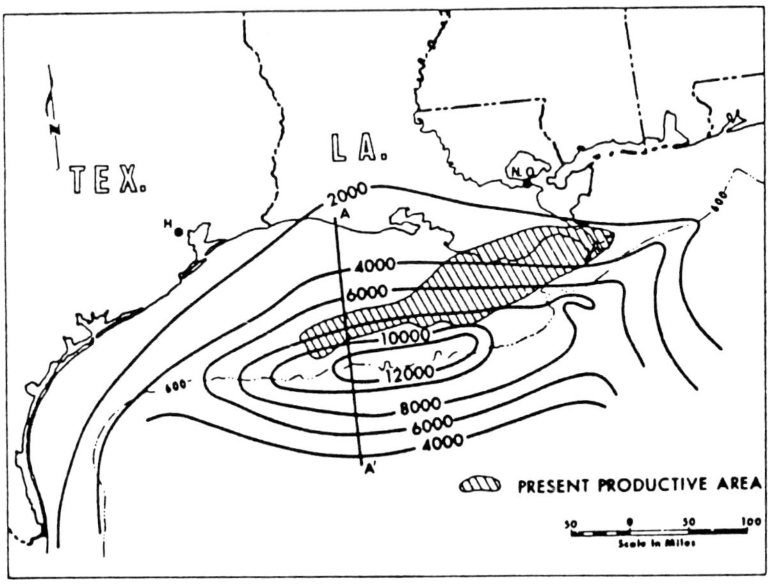

FIGURE 3. Relationship between the major Pleistocene depocenter (as suggested by Woodbury, 1973) and the position of the Sigsbee Bulge. (From Humphris, *Am. Assoc. Pet. Geol. Studies in Geology,* 7, 69, 1978. With permission.)

older Jurassic salt and sediments). It is over 300 km wide in the central and western Gulf but narrows to the east and southeast (Buffler, 1984). As the crust began to cool and contract, the basin subsided and continued to subside throughout the Early Cretaceous. Deeper water sediments were deposited on the oceanic crust while carbonate reefs fluorished around the rim of the basin.

During the Middle Cretaceous there was a rapid drop in sea level (resulting in the formation of a prominent Mid-Cretaceous unconformity), followed by a subsequent rise which drowned the peripheral reefs; thus marking a major turning point in Gulf history (Buffler, 1984). Since the end of Cretaceous time, clastic sedimentation has dominated the Gulf of Mexico basin and has played a major role in the formation of the Sigsbee Escarpment and its salt related features.

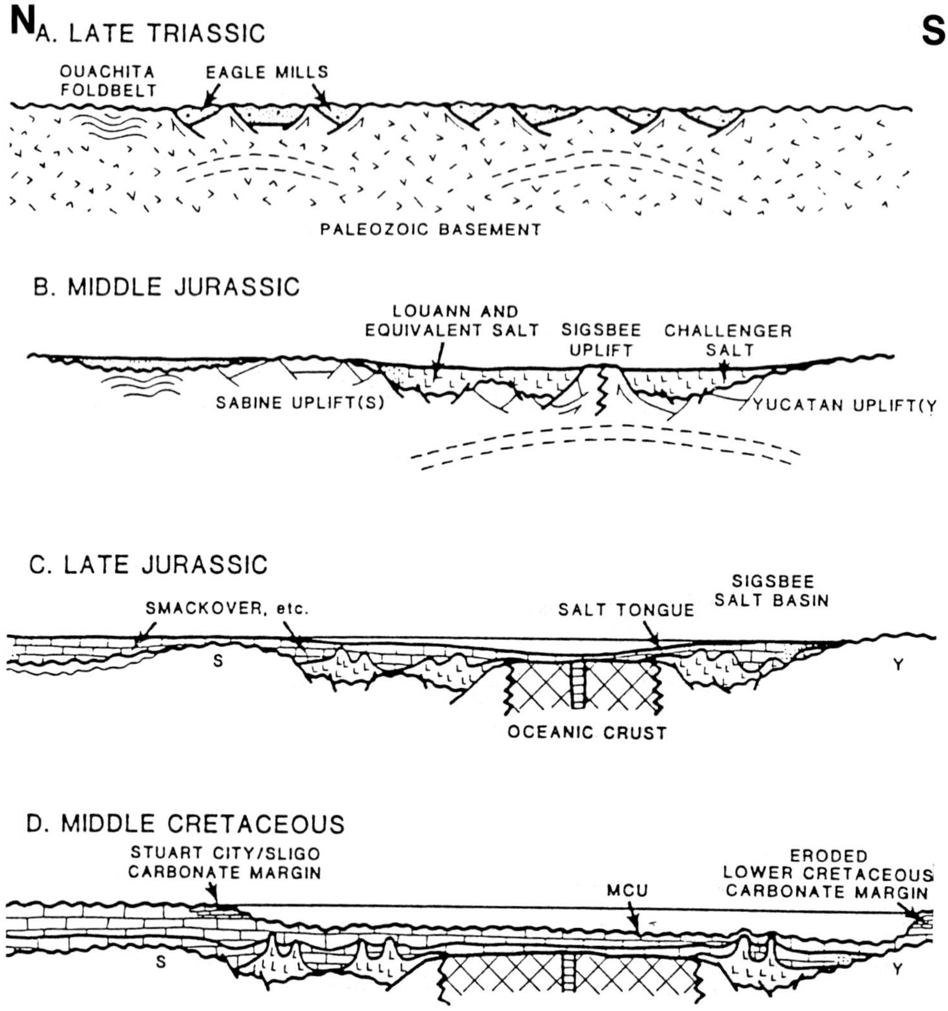

FIGURE 4. Summary of the early evolution of the deep Gulf in several stages. (From Buffler, 5th Annu. Gulf Coast SEPM Res. Conf., Proc. 31-34, 1984. With permission.)

The maximum sediment influx during the Late Paleocene occurred in east Texas, probably associated with the Laramide uplift in the southern Rock Mountains (Dupre, 1985). The Eocene was marked by a major shift of sediment influx to the south Texas region during early Eocene time, followed by a period of little or no shelf progradation during Middle to Late Eocene. During Late Eocene and Oligocene time, deposition increased to the northeast, suggesting the ancestral Colorado, Brazos, Sabine, and Mississippi Rivers were increasing in importance (Dupre, 1985).

The end of Miocene time is characterized by a sudden decrease in the amount of sediment entering the Rio Grande embayment, along with an increase in sediment supply in southwest Texas and Louisiana. The increased amounts of sediment to the northeast reflects the increased sediment load of the Mississippi River, perhaps associated with the rejuvenation of the Appalachian and Rocky Mountains at this time (Dupre, 1985).

The rates of sedimentation continued to increase in the Pliocene (approaching 14 cm/100 years), and appeared to peak (20 cm/100 years) during the Pleistocene (Dupre, 1985). The major Plio-Pleistocene depocenters are located primarily in offshore Louisiana and were controlled mainly by delta switching of the Mississippi River.

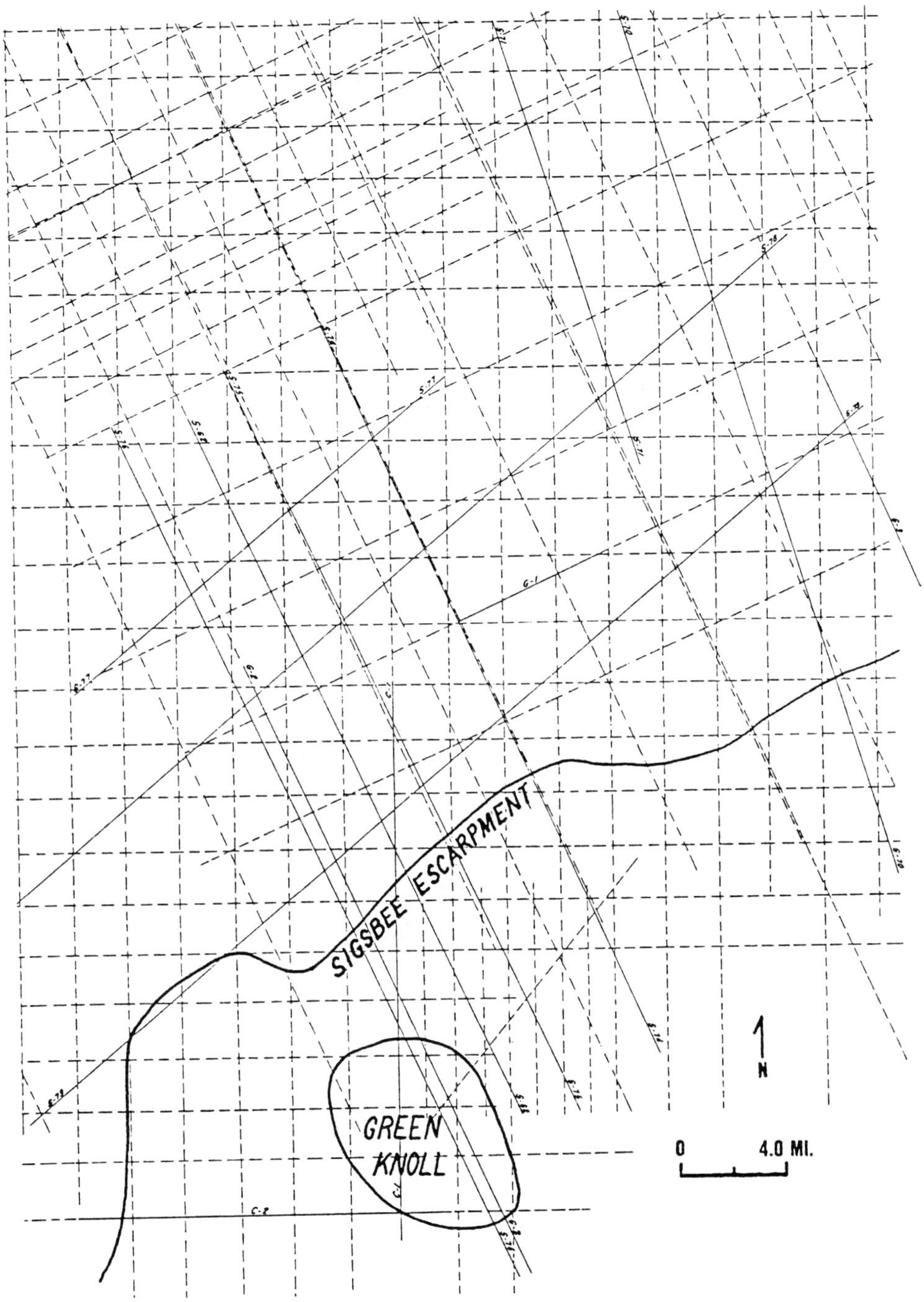

FIGURE 5. General location of seismic data set. (Reference grid in solid lines, additional data dashed.)

E. DATA SET

The data set for this study consists of 53 multichannel, stacked, and migrated seismic lines (Figure 5). Approximately 2400 km of seismic data within the study area provided a tight grid which facilitated seismic mapping.

The original data set consisted of 10 lines provided by Shell Offshore Inc. of New Orleans. This included four northeast to southwest-trending strike lines and six northwest to southeast-trending dip lines. The spacing between these lines varies from 3.6 to 14.4 km. The data set was expanded to 53 lines with Shell in-house contract data and additional lines obtained from Geco Geophysical Company.

The Geco data set consists of seven northwest to southeast-trending dip lines and five northeast to southwest-trending strike lines, spaced evenly by 6.4 km. The proprietary Shell contract data forms a tight 3.2-km grid composed of 31 north to south- and east to west-trending lines.

The proprietary nature of this data necessitates that neither exact line locations (i.e. reference to specific block boundaries) nor shot point information be given. However, all maps and seismic sections will be shown in relation to the seismic reference grid (solid lines) shown in Figure 5.

F. INTERPRETATION PROCEDURE

The first step in the interpretation procedure was to identify and correlate salt structures within the seismic grid. Salt bodies were recognized on seismic sections by a high-amplitude, hard-kicking, top-salt reflector in addition to a chaotic, incoherent, internal reflection pattern. The lower slope is characterized by salt which is often tongue-like in form, and which does not extend vertically downward to great depth. Thus, both top and base salt reflectors were identified in order to generate a salt isopach map. Base salt was not apparent on all of the seismic data and was therefore interpolated between areas of high confidence.

Seismic sequences were then defined according to principles outlined by Vail et al. (1977). Reflection terminations (i.e., downlap, onlap, toplap, etc.), in addition to regional continuity and interval reflector configuration were used to delineate sequence boundaries.

Four major seismic sequences were identified and sequence boundaries were mapped with high confidence throughout the study area. Two-way travel times were converted to depth using a (water-depth corrected) linearly increasing velocity function of $v_z = 5500$ ft/sec $+ 0.5z$ in order to produce structure maps in depth below sea level for each of the four sequence boundary horizons. In addition to the four structure maps, four isopach interval maps and a bathymetry map were constructed.

A seismic sequence analysis was completed using procedures outlined by Kirk (1985), Mitchum et al. (1977), Sangree and Widmier (1977), and Stuart and Canghey (1977).

II. VELOCITY ANALYSIS AND DATA REPROCESSING

A. VELOCITY ANALYSIS

A velocity analysis was completed in order to (1) determine an accurate velocity function to be used in converting isochron and time-structure maps to depth, (2) determine typical depths and velocities for over-pressured sections, and (3) establish velocity trends within the proposed study area.

There are no wells within the study area, so only stacking velocity data (time-depth pairs) were used. Linear increase velocity functions were calculated at 18 locations which included deep sedimentary basins between the salt masses, shallow basins above tabular salt bodies, and basinward of the Sigsbee Escarpment.

Velocity functions in the Gulf of Mexico can be closely approximated by a linear increase velocity function,

$$V(z) = V_o + kz$$

where V_o is the velocity at the surface (in this case, the sea floor), k is the velocity gradient or acceleration factor, and z is the depth (Dobrin, 1976). An in-house Shell program was used to compute best-fit linear increase velocity functions to the time-depth pairs at each of

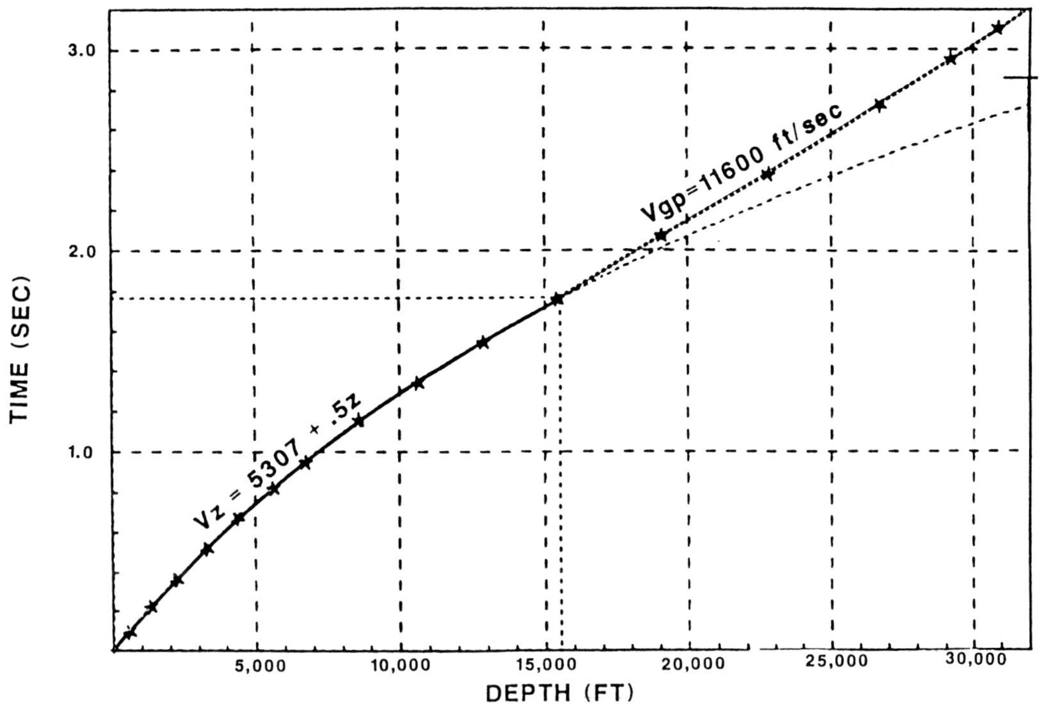

FIGURE 6. Typical velocity function plot for study area. Vz represents the linear increase function for the normally pressured section while Vgp represents velocity for the geopressured section.

the 18 locations. The program allows the user to correct all data to sea bottom, choose a velocity acceleration factor (k), and then plot a velocity function to any portion of the data points. A typical plot is shown in Figure 6.

Figure 7 shows the locations of the velocity function calculations in relation to the major salt formations. The linear increase function (Vz) as well as the velocity (Vgp) and depth to geopressures at each location are listed in Table 1.

The main purpose of the velocity analysis was to determine an accurate velocity function in order to convert isochron and time-structure maps to depth. Ten locations (1—10 on Figure 7) were chosen in the deep sedimentary basins where the reflectors are generally flat-lying. An acceleration constant of $k = 0.5$ was found to yield very low RMS errors in depth (less than 50 ft), and was therefore kept constant. The values of V_o at each of the ten locations were averaged (excluding the high and low values), yielding an average V_o of 5391 ft/sec. Hence, the velocity function used for time-depth conversion in this study area was determined to be Vz = 5400 ft/sec + 0.5 z.

The onset of over-pressured sections can be recognized by a sharp deflection (increase in slope) of the data points from the normal velocity function (Figure 6). Velocities in over-pressured or geopressured sections are generally constant or nearly constant, and can greatly effect time-depth map conversion if not accounted for. Therefore, an average depth (z) and velocity (Vgp) of the over-pressured zone was calculated and incorporated into the averaged linear increase function used in mapping. The average velocity (Vgp) of the geopressured section was calculated to be 11,415 ft/sec, whereas the average depth to the onset of geopressures was found to be 11,025 ft. Therefore, time-depth conversion was accomplished using Vz = 5400 ft/sec + 0.5 z from zero (sea bottom) to 11,000 ft, and Vgp = 11,500 ft/sec below 11,000 ft.

Velocity functions were calculated at eight additional locations (11—18 on Figure 7) in

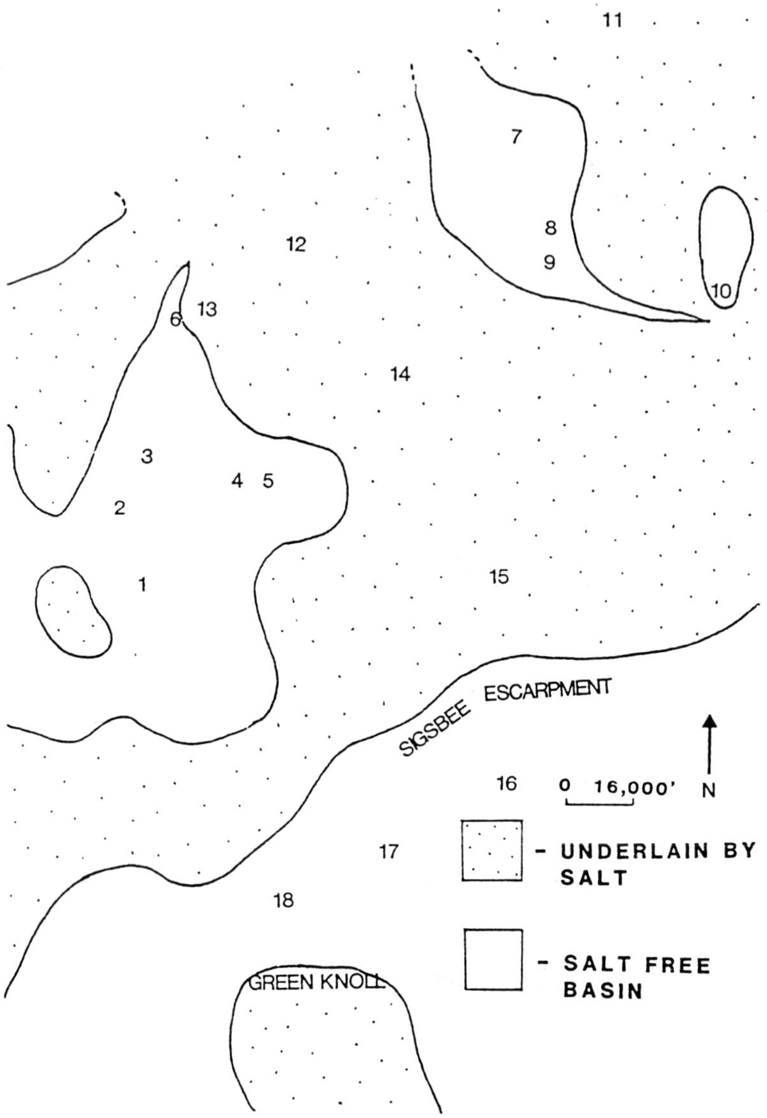

FIGURE 7. Locations of velocity function calculations.

order to discern velocity trends throughout the study area. Three of these locations (16—18) were basinward of the Sigsbee Escarpment. They yielded an average velocity function of $Vz = 5671 + 0.5\,z$, which is slightly higher than that found within the lower slope basins. The average depth to geopressures was also higher, (12,000 ft as opposed to 11,025 ft), while the average velocity in geopressures was slightly lower, 11,387 ft/sec compared to 11,415 ft/sec in the lower slope basins.

Five velocity functions were calculated within shallow basins above tabular salt bodies (11—15 in Figure 7). Values for both V_0 and k were found to be consistently higher than those encountered within the deep basins and also basinward of the scarp. Sediments recovered from a thin sediment package above salt along the outer Sigsee Bulge (Leg 10, Hole 92 of the Deep Sea Drilling Project) show anomalously high bulk densities at shallow burial depths along with an "over-consolidated" appearance (Worzel et al., 1973). This agreement between the DSDP and velocity analysis data suggests that the vertical component

TABLE 1
Velocity Functions, Locations

Location #	Best fit velocity function	Velocity/depth to geopressures
1	Vz = 5307 + 0.5z	Vgp = 11365 l/sec @ 12500 ft
2	Vz = 5420 + 0.5z	Vgp = 12820 l/sec @ 14000 ft
3	Vz = 5435 + 0.5z	Vgp = 13200 l/sec @ 18800 ft
4	Vz = 5422 + 0.5z	Vgp = 11700 l/sec @ 10000 ft
5	Vz = 5427 + 0.5z	Vgp = 8000 l/sec @ 7000 ft
6	Vz = 5427 + 0.5z	Vgp = 11765 l/sec @ 11000 ft
7	Vz = 5365 + 0.5z	Vgp = 11760 l/sec @ 10800 ft
8	Vz = 5326 + 0.5z	Vgp = 9700 l/sec @ 8500 ft
9	Vz = 5300 + 0.5z	Vgp = 9750 l/sec @ 8500 ft
10	Vz = 5470 + 0.5z	Vgp = 12740 l/sec @ 13000 ft
11	Vz = 5073 + 0.55z	Vgp = 13513 l/sec @ 15000 ft
12	Vz = 5849 + 0.5z	Vgp = 11625 l/sec @ 12000 ft
13	Vz = 5620 + 0.5z	Vgp = 9000 l/sec @ 8000 ft
14	Vz = 5494 + 0.7z	Vgp = 14700 l/sec @ 19400 ft
15	Vz = 6153 + 0.55z	Vgp = 13600 l/sec @ 12000 ft
16	Vz = 5711 + 0.5z	Vgp = 11200 l/sec @ 12500 ft
17	Vz = 5540 + 0.5z	Vgp = 11600 l/sec @ 12500 ft
18	Vz = 5762 + 0.5z	Vgp = 11360 l/sec @ 11000 ft

of movement of the allochthonous salt lobes tends to over-compact the thin, overlying sediment package, thus yielding higher velocities than those found at corresponding depths within the deeper, salt-free basins.

B. DATA REPROCESSING

Salt bodies are easily recognized on seismic sections by a high-amplitude, hard-kicking top-salt reflector and a chaotic internal reflection pattern. However, using conventional data processing techniques, it is very difficult to resolve the salt-sediment interface along both the sides of salt domes as well as along the base of salt layers or tongues.

The salt within the study area is generally tabular or tongue-like in nature (as opposed to the predominantly vertical salt spines and domes found on the upper slope and shelf). Since salt thickness is very important in understanding the emplacement process and the general geologic evolution of the area, a 6-mile portion of seismic line S-79 (Figure 8) was reprocessed as an example for improving base salt resolution. (Note: the location of the seismic lines shown in Figures 8, and 11 to 45 are given in Figure 5.)

A first step was to accurately identify the top-salt reflector on the seismic section. Six intersecting tie-lines aided in the correlation process and yielded a high confidence top-salt pick (Figure 8).

The original velocity spectra displays were then analyzed at 1000-ft intervals. A typical velocity spectra display is shown in Figure 9. The upper portion of Figure 9 shows the raw geophysical data. At very small time intervals (less than 5 msec), the data is summed within numerous velocity "windows" from top to bottom. The spikes on the lower graph indicate the RMS velocity which had the best fit or highest coefficient at each corresponding time.

These spikes (coefficients) were then re-picked and digitized at 2000-ft intervals in order to build a velocity function to be used in stacking the upper sediment package data. A salt velocity of 14,750 ft/sec was implemented at top salt and was kept constant to the bottom of the section. Although this does not represent true geologic conditions, it is hoped that the base-salt reflector would stack, while the slower sediments beneath the salt tongue would not.

Prior to the initial stack, the ΔT (NMO) corrected points were checked to ensure that

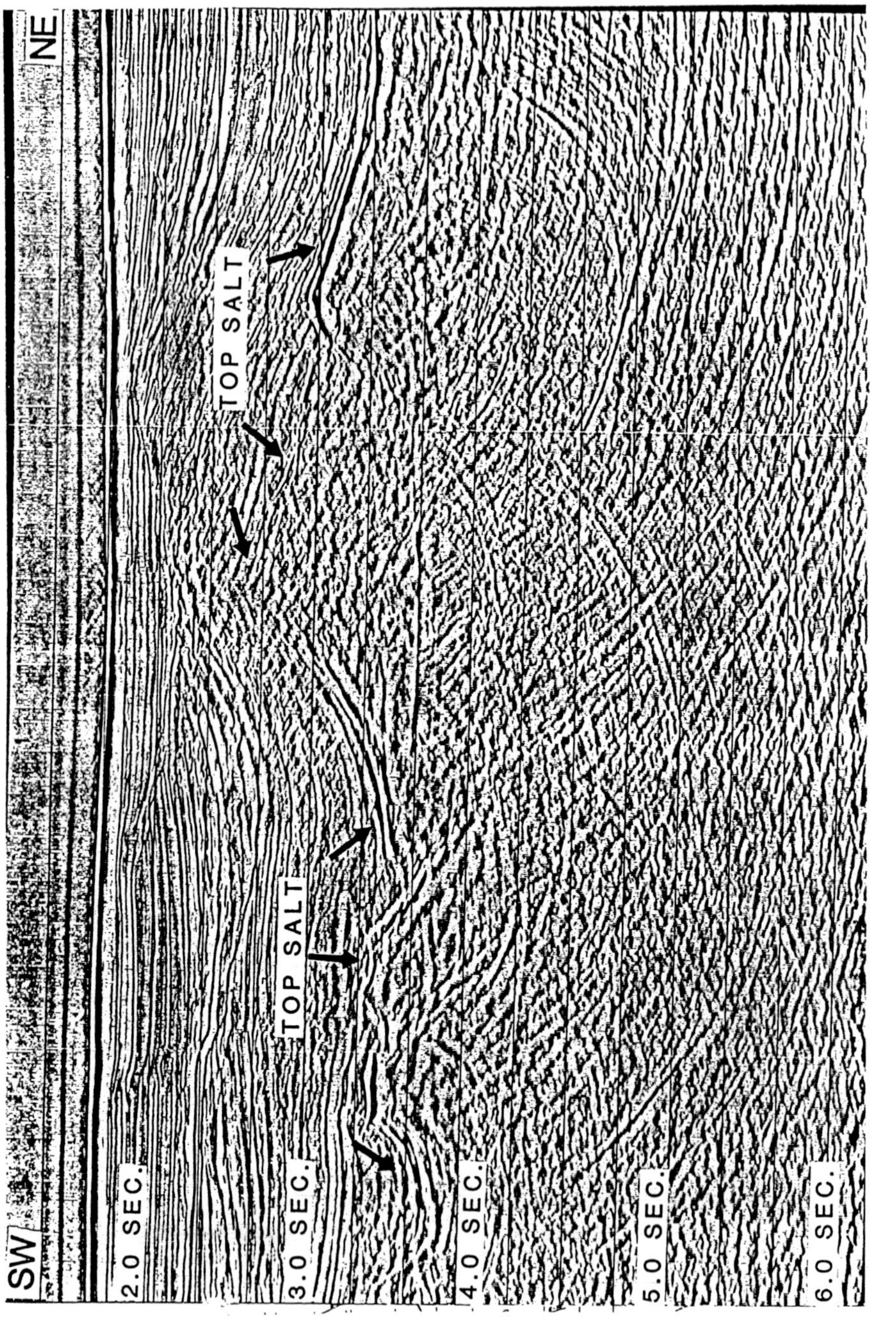

FIGURE 8. Portion of seismic line S-79 to be reprocessed in order to improve poor base salt resolution. (Courtesy of Shell Offshore Inc.)

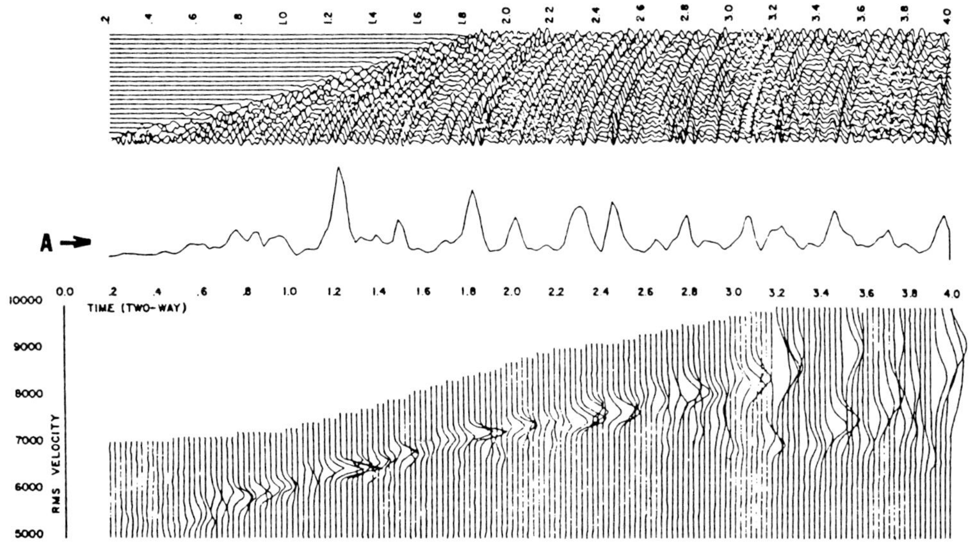

FIGURE 9. Typical velocity spectra display. The peaks on the vertical lines represent the best estimate of stacking velocity. (From Waters, Reflection Seismology: A Tool for Energy Resource Exploration, John Wiley & Sons, 1981. With permission.)

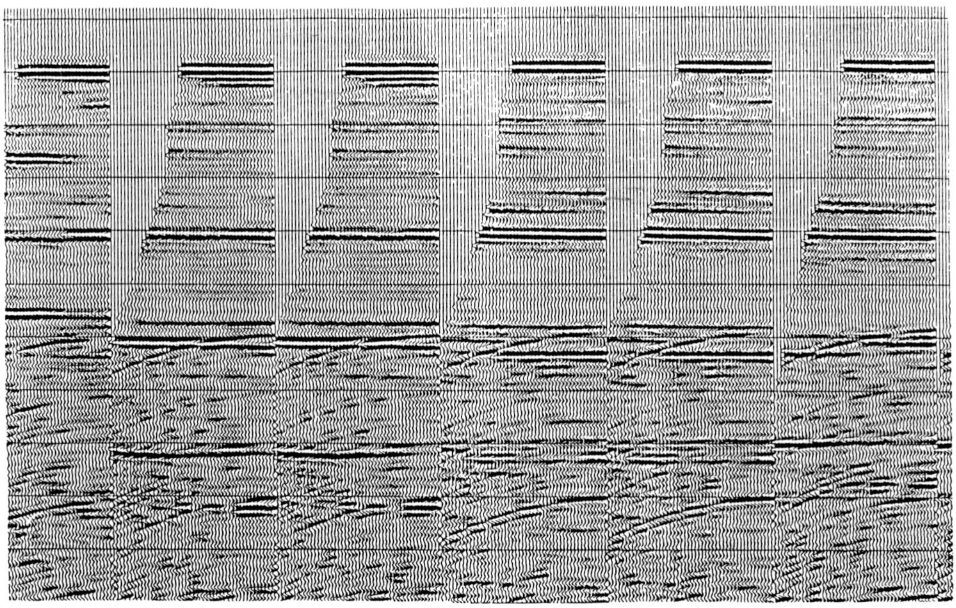

FIGURE 10. Sample of CDP gathers (Δ-t corrected points). Each gather consists of 40 traces with the left side being the zero-offset case.

the stacking velocities were accurately chosen and digitied. Figure 10 illustrates a small portion of the corrected points. Each gather consists of 40 traces, with the left-hand side being the zero-offset case. If the correct velocities have been applied the reflectors will be horizontal, and a horizontal stack will yield a high-amplitude reflector. However, if the reflectors are sloping upward (in which case the velocities chosen are too slow), or downward (velocities too high), then it is obvious that a horizontal stack would yield a much lower amplitude reflector. The reflectors in Figure 10 are predominantly flat-lying, suggesting that the velocities were accurately chosen. Note that the downward-dipping hyperbolic curves found deeper in the section are multiples, and therefore do not represent poorly chosen velocities.

The initial stack, using a sediment velocity function above salt, and a salt velocity (14,750 ft/sec) everywhere below top salt, yielded some interesting results. In Figure 11, it is apparent that the base-salt reflector is beginning to stack; especially at the southwestern end of the section where the upper salt surface is somewhat smoother. Using the stacked section in addition to the intersecting tie lines, the base-salt reflector was identified with some degree of confidence.

The data was then stacked a second time — this time implementing sediment velocities below the salt layer. The same function for the sediments above salt was used as in the initial stack. Likewise, a constant velocity of 14,750 ft/sec was used within the salt layer. However, a typical over-pressured shale velocity of 11,500 ft/sec (constant) was inserted for the section below salt. This sediment-salt-sediment velocity structure more accurately represents the true geologic conditions than did the sediment-salt velocity structure used in the initial stack.

To implement this type of velocity scheme, the effective velocity at the base of salt is calculated using the Dix equation:

$$Vi_{12} = \sqrt{\frac{Ve_2^2 to_2 - Ve_1^2 to_1}{to_2 - to_1}}$$

where Vi is the interval velocity (14,750 ft/sec for salt), and Ve_2, Ve_1 and to_2, to_1 are the effective velocities and two-way travel times at the top and bottom of the (salt) layer, respectively. The only unknown quantity is Ve_2 and it is solved for by:

$$Ve_2 = \sqrt{\frac{Vi^2(to_2 - to_1) + Ve_1^2 to_1}{to_2}}$$

The results of the second stack (Figure 12) were somewhat disappointing in that there was very little improvement of base salt resolution in comparison to the first stack. However, both the first- and second-run stacked sections showed better base-salt resolution than did the original section. This demonstrates that in areas of tabular or tongue-like salt bodies, base-salt resolution can be modestly improved through simple reprocessing techniques.

III. SEISMIC STRATIGRAPHY AND FACIES ANALYSIS

A. SEISMIC DEPOSITIONAL SEQUENCES

Four major seismic-stratigraphic sequences were identified and the sequence boundaries were mapped throughout the study area following principles outlined by Vail et al. (1977). From oldest to youngest, the depositional sequences are labled as Intervals I—IV. The sequence boundaries are labeled a—e, with ''a'' representing the lower boundary of Interval I, and ''e'' representing the sea floor. The relative ages and typical reflector configuration of depositional Intervals I—IV and sequence boundaries a—e are illustrated in Figure 13. This is considered the ''type'' section for Intervals I—IV within the study area.

B. SEISMIC FACIES TYPES

Five different facies types were recognized from analysis of Shell seismic lines. These include: parallel continuous, parallel discontinuous, shingled, reflection free, and chaortic/

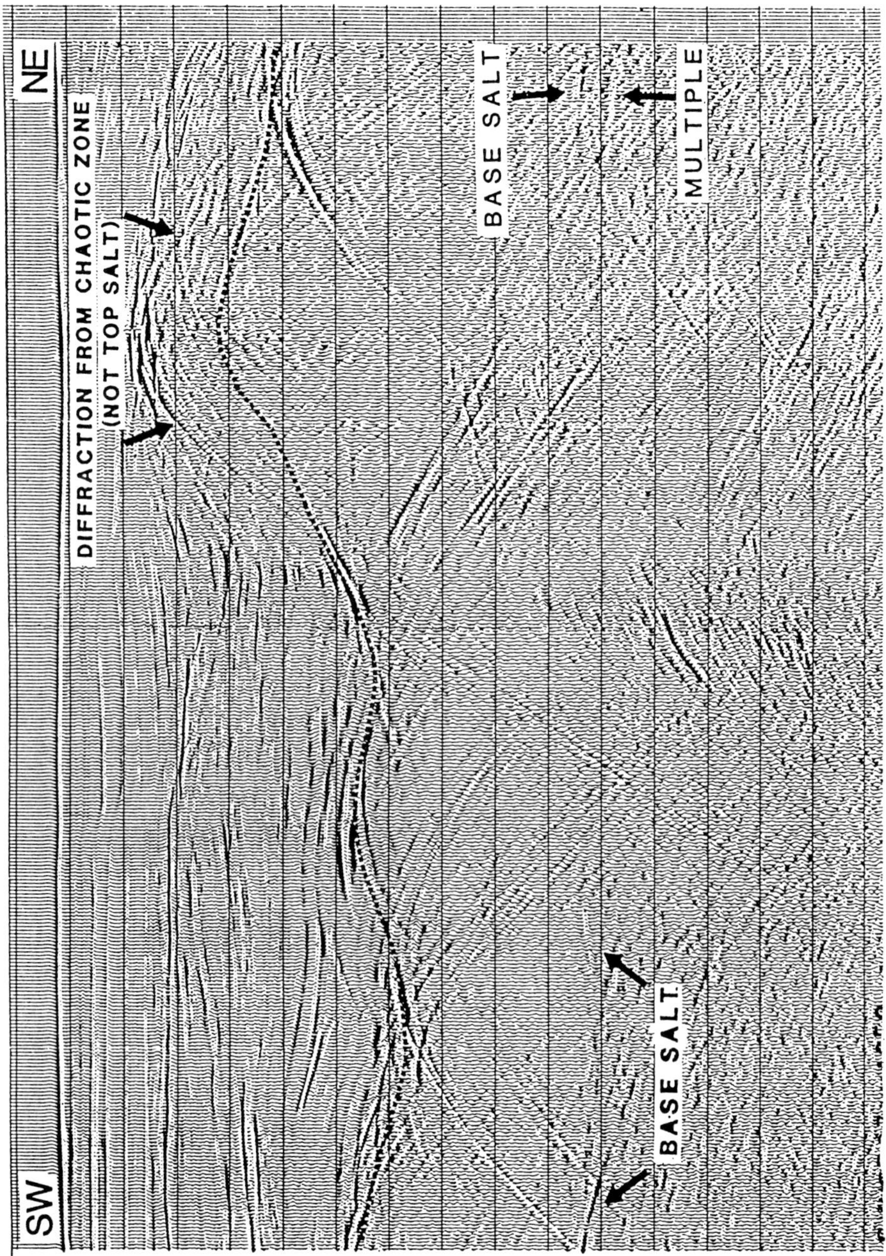

FIGURE 11. Unmigrated portion of seismic line S-79 illustrating faint base-salt reflector after re-stacking data.

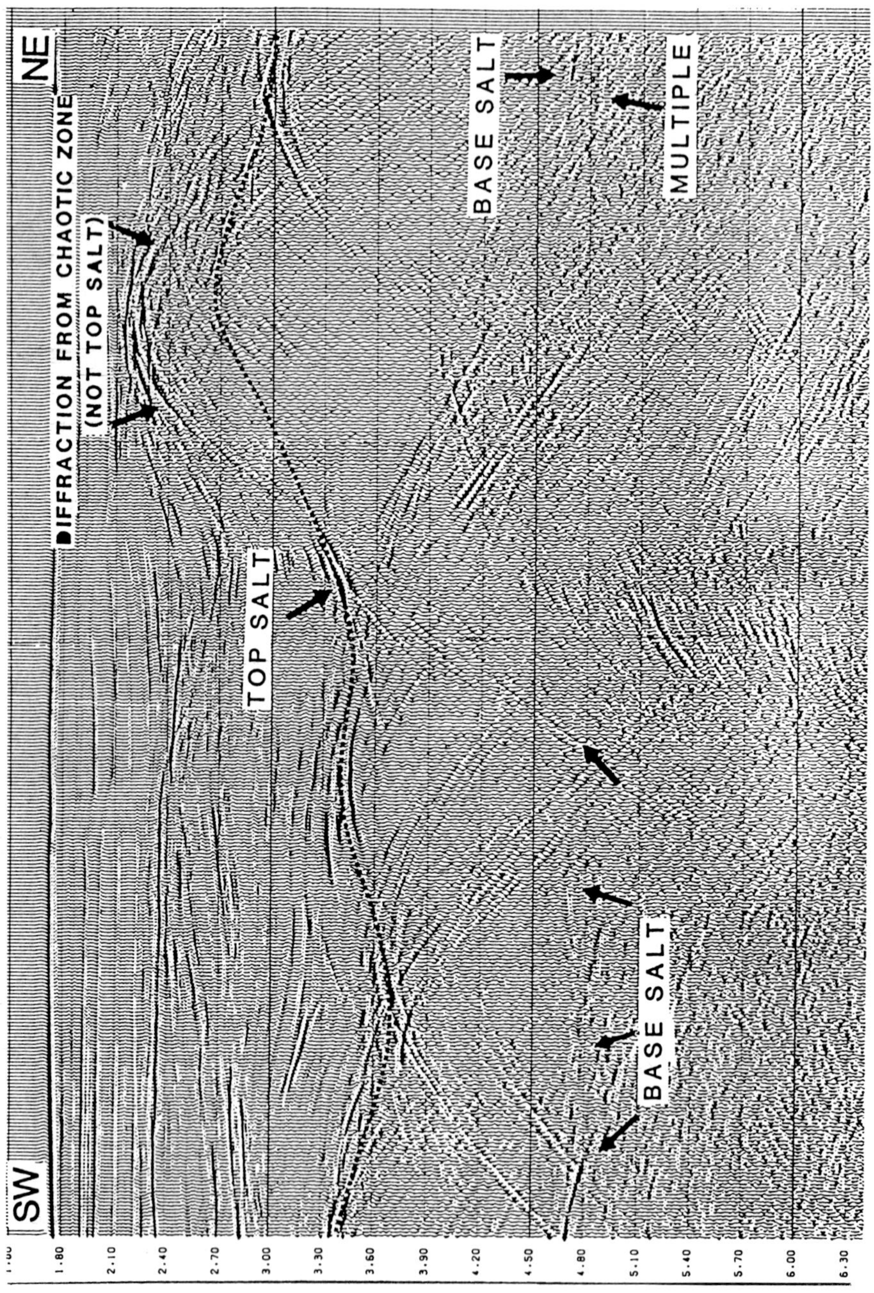

FIGURE 12. Unmigrated portion of line S-79 illustrating faint base-salt reflector after second re-stack.

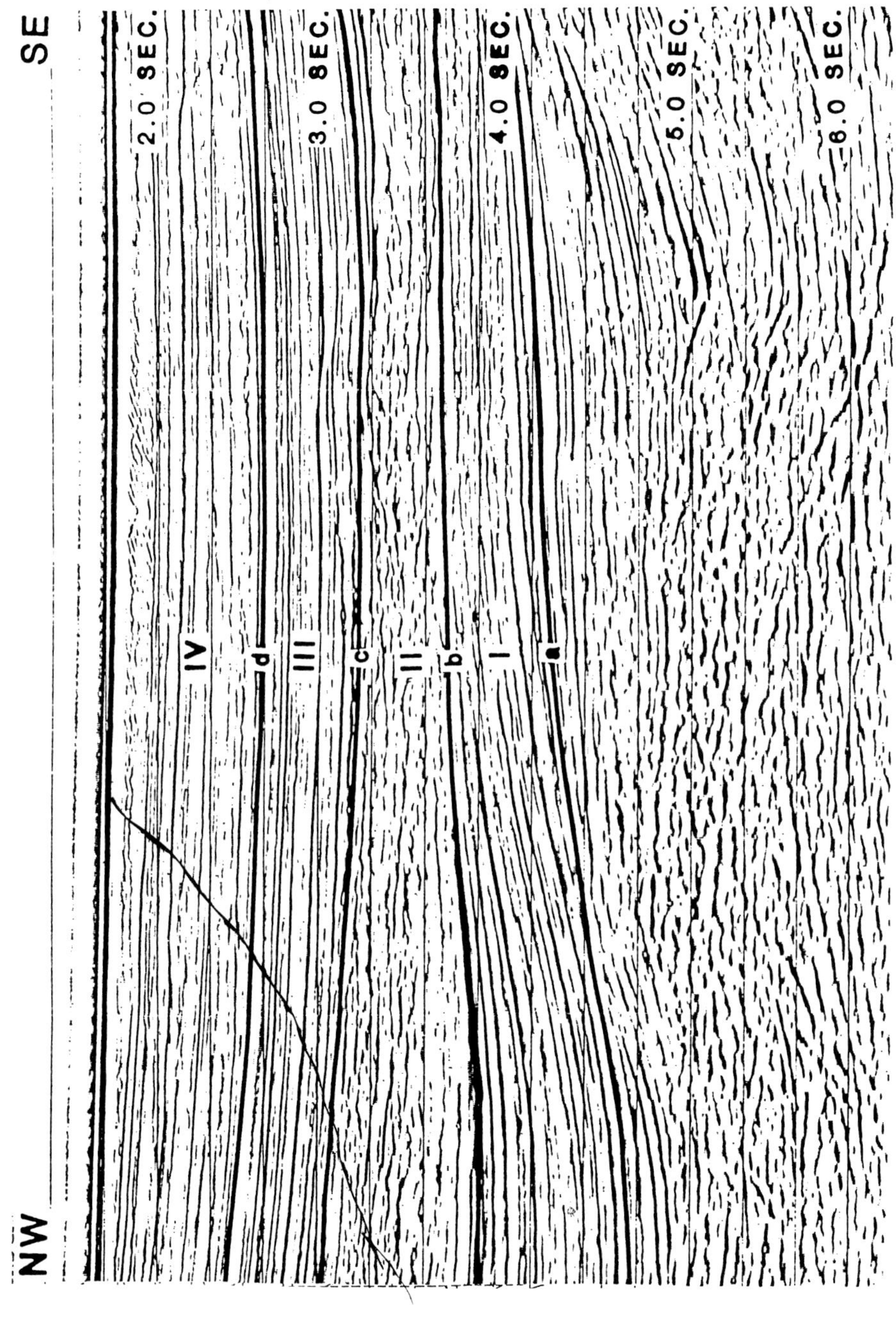

FIGURE 13. Portion of seismic line S-75 illustrating ''type'' section for study area illustrating major depositional sequences (Intervals I—IV) and sequence boundaries (horizons a—e) in ascending chronostratigraphic order. (Courtesy of Shell Offshore Inc.)

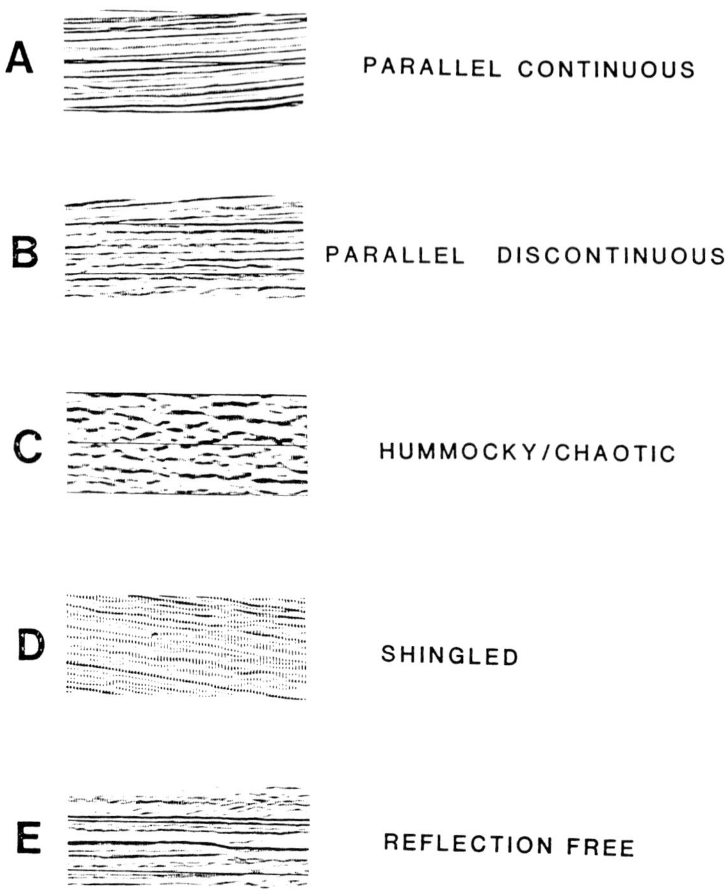

FIGURE 14. Five seismic facies types (A—E) recognized within study area.

hummocky (Figure 14). Of course, variations within each facies type (and within each depositional interval) occurred with respect to reflector configuration, amplitude, frequency content, and lateral and vertical gradation into other facies types.

Parallel Continuous — The parallel continuous facies is characterized by parallel to subparallel reflectors which have moderate to high amplitudes (Figure 14a). The facies pattern is interpreted to be a result of low density turbidites perhaps interbedded with hemipelagic sediments, which are deposited over relatively large areas during periods of sea level high stands (Watkins, 1986).

Parallel Discontinuous — The parallel discontinuous facies is characterized by subparallel to slightly hummocky reflectors of variable amplitude (Figure 14b). This facies seems to fall somewhere between a typical parallel continuous facies and a truly chaotic facies. Although this facies is not common in the study area, it may suggest a highly variable depositional environment.

Chaotic/Hummocky — The chaotic/hummocky facies is characterized by moderate to high amplitude highly discontinuous reflectors (Figure 14c). The nature of this facies suggests that deposition occurs through a mass transport process such as slumping. Small to large scale slumping is common in areas of salt movement. The chaotic/hummocky facies type is predominant during sea level low stands, and especially so during active sea level lowering (Watkins, 1986).

Shingled — The shingled facies is characterized by moderate to high amplitude reflectors of low to moderate continuity (Figure 14d). It is generally a thin prograding seismic pattern, commonly with parallel upper and lower boundaries, and with gently dipping parallel oblique internal reflectors that terminate by apparent toplap and downlap (Mitchum et al., 1977).

Reflection Free — Reflection-free zones on seismic sections are generally associated with homogeneous, nonstratified geologic units. This type of facies (Figure 14e) generally consists of homogeneous hemipelagic sediments which are deposited during sea level high stands (Sangree and Widmier, 1977; Watkins, 1986).

C. FACIES ANALYSIS OF DEPOSITIONAL INTERVALS I-IV

Seismic facies maps were constructed for Intervals I-III using the basic principles outlined by Kirk (1985). Some of the intervals showed more than one facies type at a particular location, but only the dominant facies type is indicated on the map. The areal extent of individual facies types are shown on the maps along with onlap/downlap directions. A facies map was not made for Interval IV (upper interval) due to the highly variable nature of this depositional sequence.

Facies Analysis, Interval I — Facies analysis of Interval I reveals that moderate to high amplitude, parallel, continuous reflectors are the predominate facies type, as shown in Figure 13. This is considered the "type facies" of Interval I. Figure 15 shows the facies distribution for Interval I. It should be noted that only the upper portion of the interval is present in most of the eastern area of the map, whereas the full interval is present in the deep basin associated with the western portion of the map. It is in this western basin that the most striking reflector terminations are found. The direction of apparent downlap is toward the east, ranging from northeast to southeast, while there is simple onlap in all other directions around the periphery of the basin. This suggests a major source direction from the west-northwest.

As is the case with the western portion of the facies map, the zone of parallel, continuous facies in the eastern area is primarily associated with the deeper sedimentary basins. The blank area between these zones of parallel facies is associated with a structural high and represents an area of non-deposition or erosion of Interval I.

A large area of chaotic facies type is found in the southern region of the facies map (Figure 15). This area is just north of the Sigsbee Scarp and is characterized by shallow salt and a relatively thin total sediment package, with only the upper portion of Interval I being represented. The chaotic nature of Interval I may be directly attributed to the disruptive forces of vertical and lateral salt movement common to this region.

Facies Analysis, Interval II — Interval II is characterized by low to high amplitude, very discontinuous to chaotic reflectors. This reflection character of Interval II is markedly different from that of the surrounding intervals, I and III (Figure 13).

The facies map of Interval II (Figure 16) shows extensive distribution of chaotic facies. Downlap is generally to the southeast in the western portion of the map. There is also a perferred southeastern orientation of onlap in the western map area, suggesting a northern or northwestern sediment source. Some of the small basins found in the central, southern, and northeastern regions show a more random onlap pattern, with the onlap indicators pointing to the periphery of the individual basins. Several of these small basins in the central and eastern area are characterized by parallel, continuous facies patterns.

A northwest to southeast trending zone of non-deposition or erosion of Interval II separates the northwestern and northeastern areas of the map. This zone is similar to that found on the Interval I facies map (Figure 15) in that it is associated with an underlying salt ridge which creates a structural high. As one might expect, the deep structural high has less of an effect on the intervals higher in the section, thus the zone of nondeposition for Interval II is much smaller than that of Interval I.

Facies Analysis, Interval III — The facies map of interval III (Figure 17) indicates a

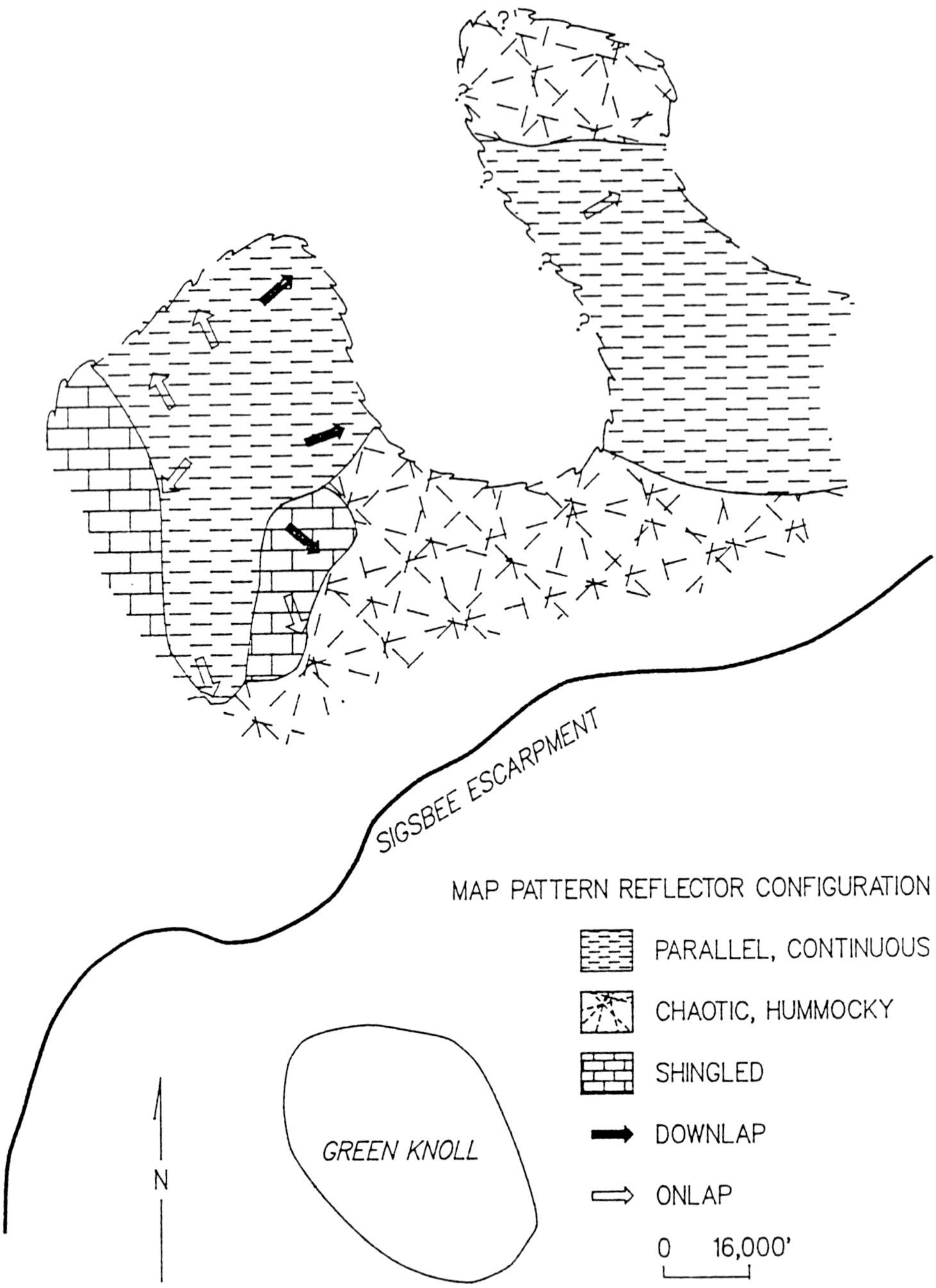

FIGURE 15. Seismic facies map for Interval I.

wide variety and distribution of facies types. The most common facies type is parallel, continuus. Figure 13 shows the moderate to high amplitude, very continuous, parallel reflector configuration which is considered to be the "type facies" for Interval III.

The effects of the deep structural ridge, which resulted in zones of erosion or nondeposition in the lower Intervals I and II, are not as apparent in Interval III. Figure 17 shows the orientation of the ridge axis which extends beneath the center of the map in a northwest to southeast direction. Although Interval III shows no erosional or nondepositional effects of the deep-seated ridge, the differential facies distribution on either side of it suggests that the ridge played a large role in controlling deposition of Interval III sediments.

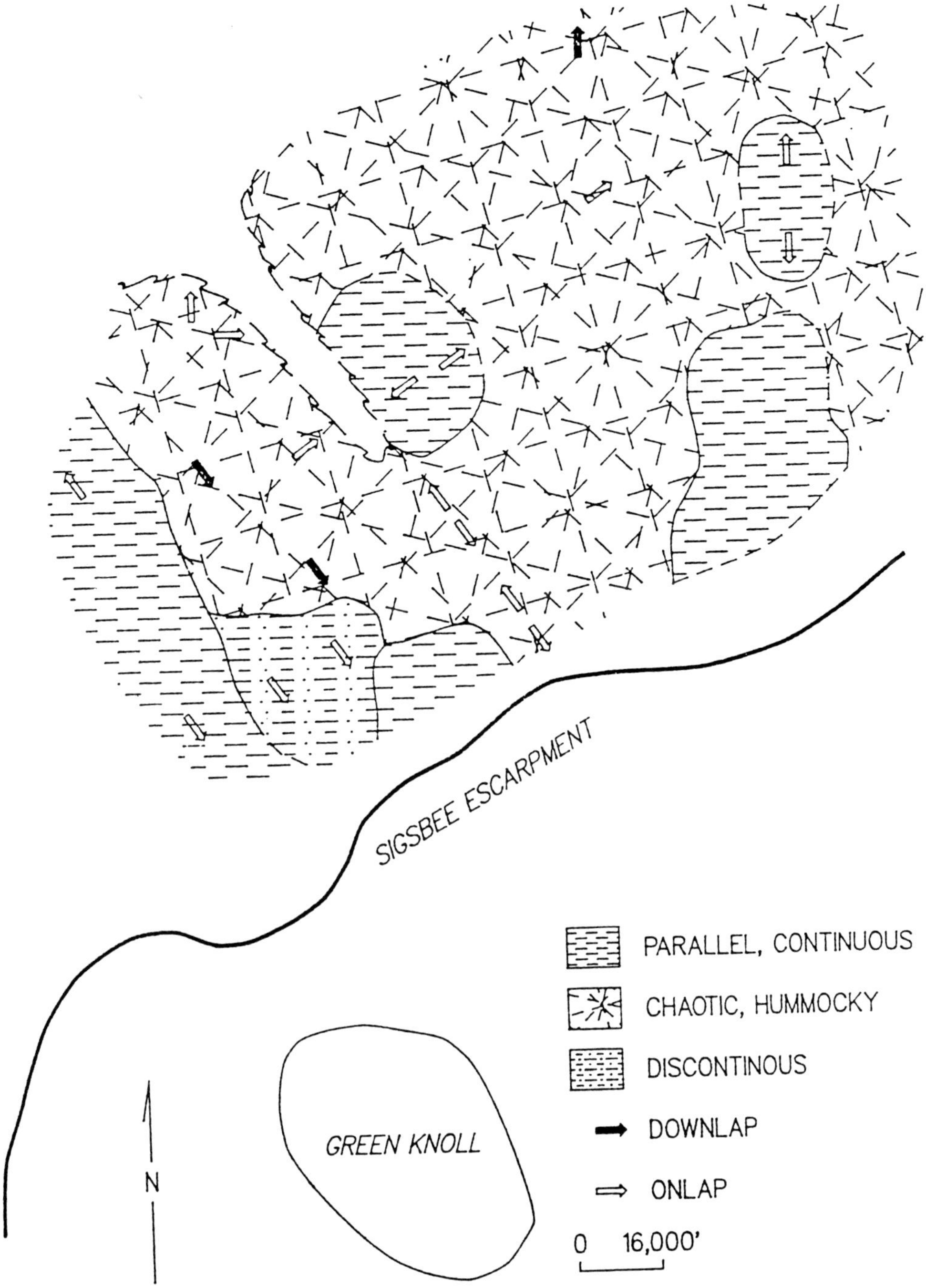

FIGURE 16. Seismic facies map for Interval II.

To the southwest of the ridge axis one finds parallel, continuous reflectors almost exclusively. This area is associated with the deep western basin which is devoid of salt. Onlap directions are again predominently to the east and southeast.

Northeast of the ridge axis there is a variety of facies types including chaotic, parallel, reflection free, and shingled. This indicates a much different depositional (and probably post-depositional) history. The semi-continuous to continuous salt mass which underlies this

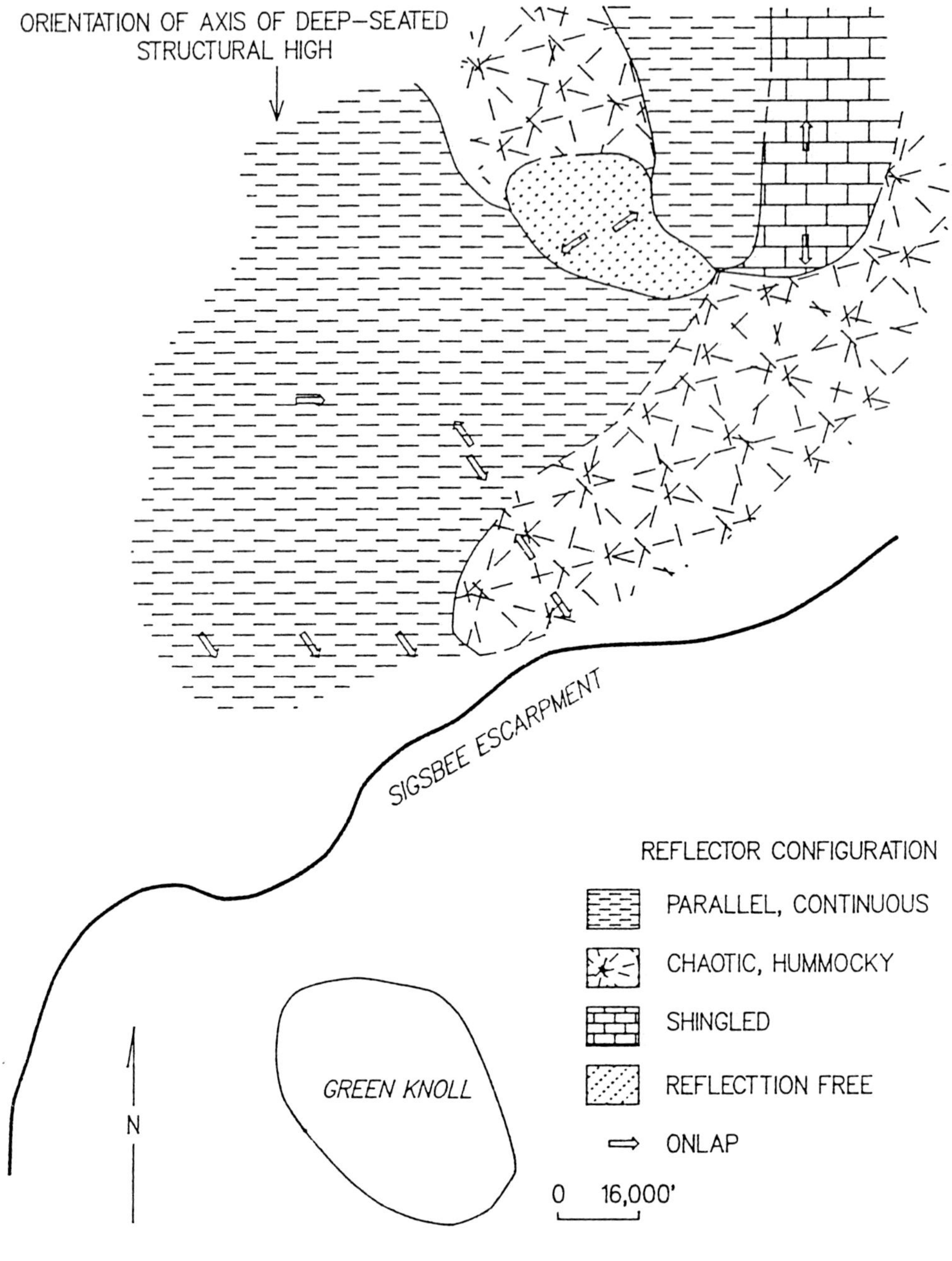

FIGURE 17. Seismic facies map for Interval III.

area may have affected sediment deposition locally and, therefore, seismic facies character varies on a local level.

The most uncharacteristic facies type in terms of Interval III standard is the reflection-free zone found in the eastern portion of the facies map (Figure 17). This localized seismic configuration pattern of Interval III is illustrated in Figure 18. It is apparent that the interval is not completely reflection free in that there are some very low amplitude, discontinuous, chaotic reflectors found within the sequence. The presence of these faint interval reflections, plus the fact that the dominant facies type for Interval III is parallel-continuous, would suggest that this localized reflection-free zone is a product of a mass wasting process

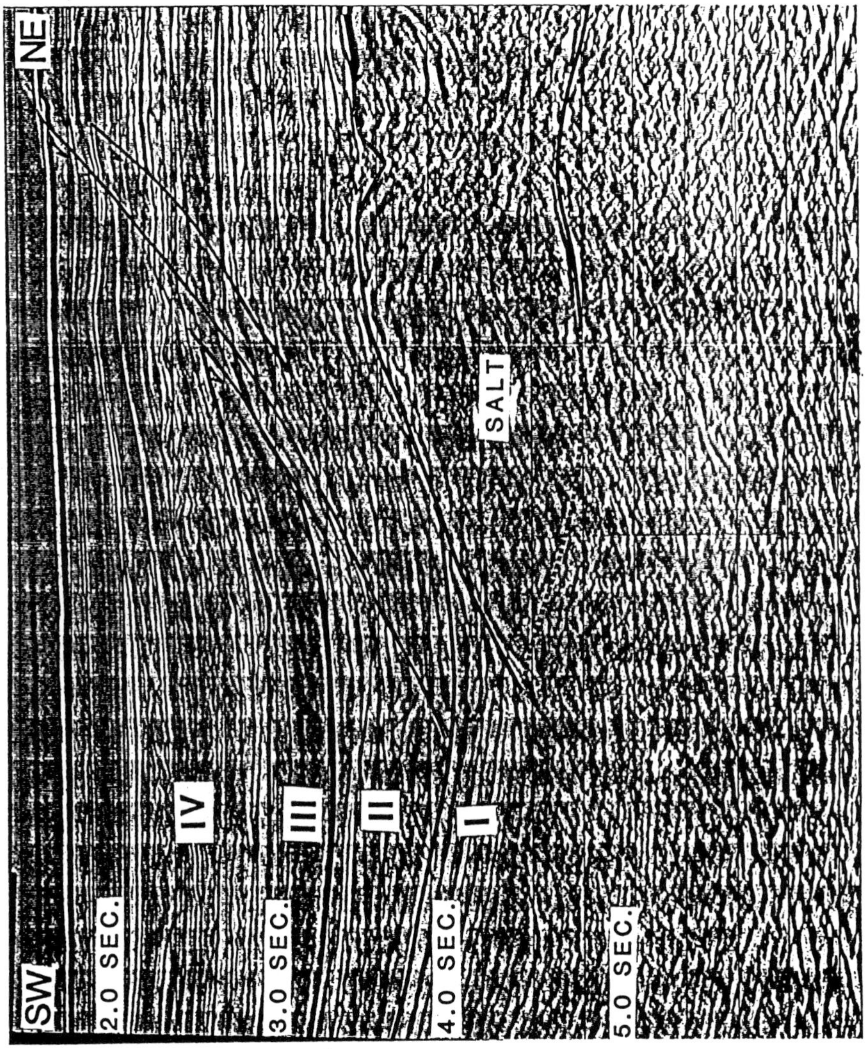

FIGURE 18. Portion of seismic line S-78 illustrating the uncharacteristic reflection-free nature of Interval III. (Courtesy of Shell Offshore Inc.)

(slumping) rather than a result of the pelagic deposition so often associated with reflection-free facies.

The south-central to southeastern area of the facies map (Figure 17) is characterized by chaotic facies. This is generally true with all the facies maps (Intervals I—III). Since there are obviously variations in sea level and therefore depositional styles throughout Interval I-Interval III time, it suggests that perhaps the chaotic facies pattern along the southern boundaries of the facies maps can be attributed to salt tectonics. Salt is very shallow in this area and the overlying intervals (I—IV) make up a relatively thin sediment package. Syn-depositional and/or postdepositional salt deformation along the scarp could result in slumping or perhaps disruption of the overlying strata.

Facies Analysis, Interval IV — Interval IV represents a relatively thick section of Upper Pleistocene to recent sediments. It shows numerous variations in seismic character, both vertically and laterally. The sequence could be divided into several subsequences, but due to the limited nature and scope of this study, that was not possible. Since a general facies map of the total interval would not prove useful, only a complete facies description was made for Interval IV.

Although Interval IV is quite variable from basin to basin, it generally has three distinct reflector configurations from top to bottom (Figure 13). The lower portion consists of low to medium amplitude, shingled to chaotic facies, with some evidence for slumping. Above that, there is a zone of generally continuous, low to high amplitude reflectors. The uppermost section consists of predominantly reflection-free to very low amplitude chaotic facies with intermittant high amplitude, very continuous reflectors.

The variability of Interval IV is illustrated by comparing Figures 13 and 19. In Figure 13, the lower two-thirds of the interval shows lateral variation from moderate amplitude, chaotic to discontinuous zones into low to high amplitude, very continuous facies types. The upper third of the sequence shows a very low ampliitude continuous to reflection-free zone, which is characteristic of pelagic deposition. Both the reflection-free and chaotic zones are absent from the interval in Figure 19. In this case, essentially the whole interval is composed of moderate amplitude, continuous reflectors.

Two channels were recognized in the upper sequence, Interval IV, although neither could be mapped with confidence for long distances. The first channel (Figure 20), shows typical onlapping-fill facies. These tend to be uniform, parallel, continuous reflections which onlap the channel sides. The base of the channel is marked by high amplitude, chaotic reflections which are characterstic of coarse-grained, channel bottom deposits. On the exterior of the channel, there are reflector terminations on both sides. The presence of a diapiric high (salt) directly beneath the channel initially suggested the apparent channel could be a low angle to view of a fault. However, intersecting tie-lines do not support this idea.

The second channel is shown in Figure 21. Although there are similar reflector terminations on the exterior of this channel, the channel fill is much different than that previously described. Low amplitude, mounded chaotic facies at the base of the channel grade upward into low amplitude, wavy, continuous reflections at the top. (Another possibility is that the feature represents a slump deposit.)

D. SUMMARY

1. Five basic facies types were recognized within the study area, including parallel continuous, parallel discontinuous, chaotic/hummocky, shingled, and reflection free.
2. Amplitudes ranged from low to high in all of the facies types, except reflection free.
3. Onlap and downlap directions were oriented randomly toward the periphery of basins found in the eastern portions of the study area, while the deep western basin showed a preferred southeastern orientation. This suggests that the major source was from the northwest.

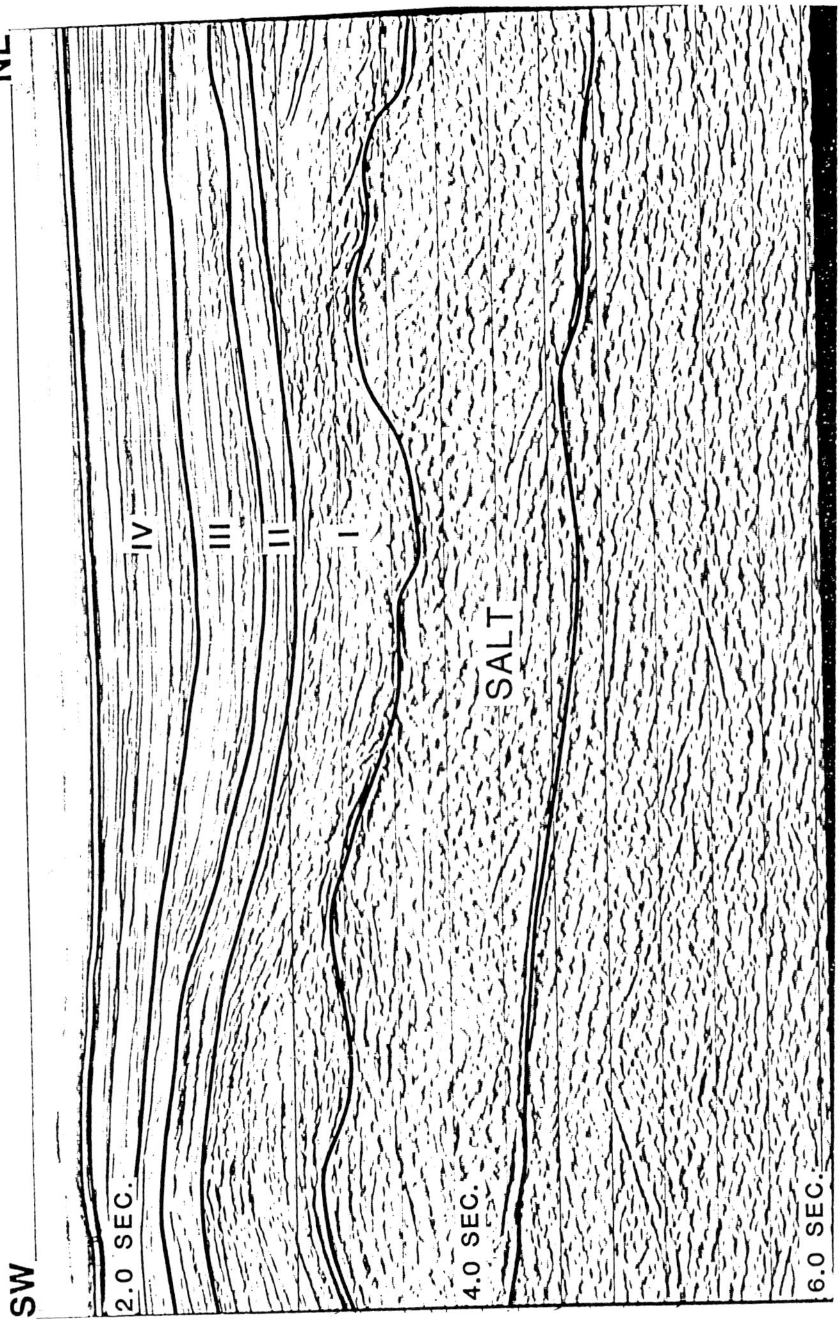

FIGURE 19. Portion of seismic line S-79 illustrating variable nature of Interval IV in comparison to Figure 13. (Courtesy of Shell Offshore Inc.)

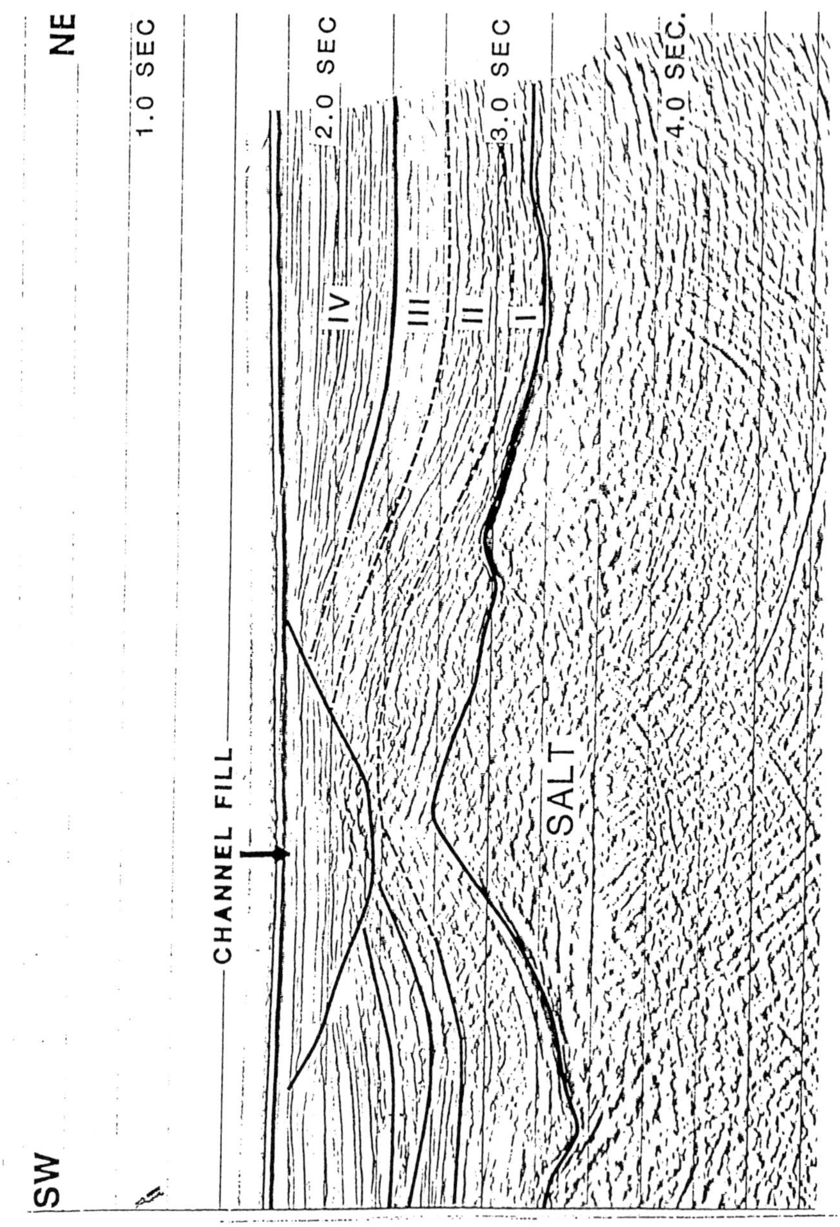

FIGURE 20. Portion of seismic line S-79 illustrating possible channel-fill facies. (Courtesy of Shell Offshore Inc.)

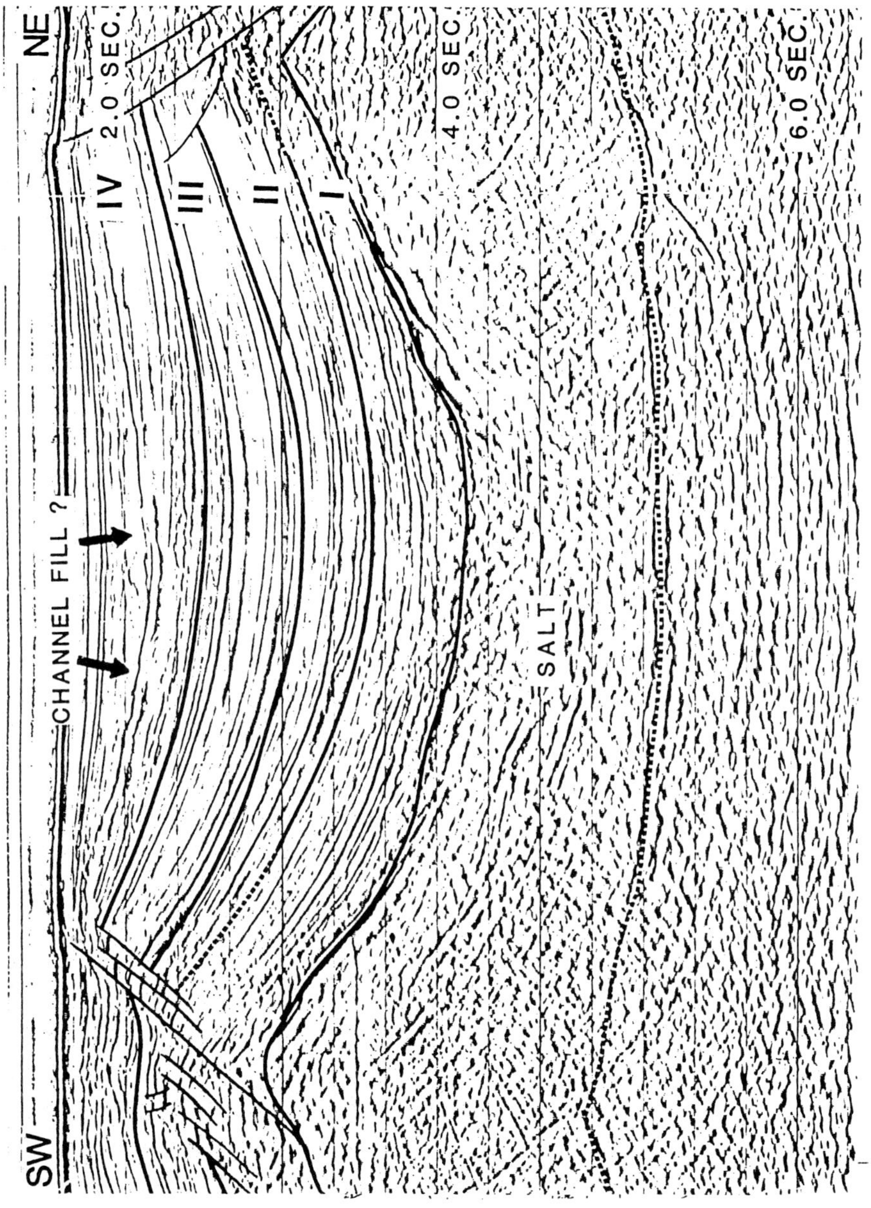

FIGURE 21. Portion of seismic line S-78 illustrating possible channel-fill facies. (Courtesy of Shell Offshore Inc.)

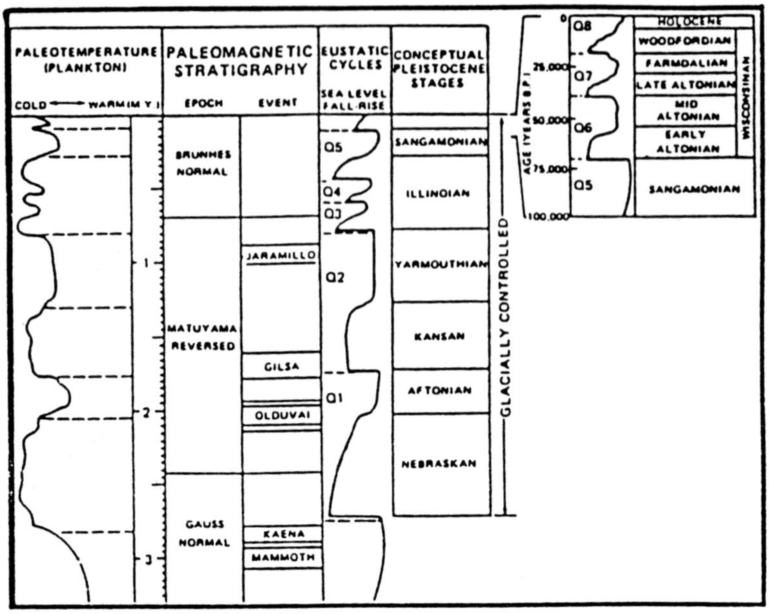

FIGURE 22. Pleistocene eustatic cycles model. (From Beard et al., *Am. Assoc. Pet. Geol. Bull.*, 66, 158, 1982. With permission.)

4. A northwest-southeast trending diapiric ridge appears to have at least partially controlled deposition, and thus had some effect on facies character in addition to the obvious effects of sea level fluctuations.
5. The general presence of chaotic facies in the southern margins of the study area near the Sigsbee Scarp suggests that the relatively shallow mobile salt mass was responsible for slumping or disruption of the overlying strata.
6. There is evidence for both slump and channel features within the study area.

E. QUATERNARY CHRONOLOGY VS. AGE OF MAPPING HORIZONS

The absence of wells within the study area precludes accurate age dating of mapping horizons. Therefore, the only plausible method of age dating mapped horizons is by correlating seismic depositional sequences to eustatic cycles. The problem of assigning reasonable ages to sequences using this method is compounded as there is still some controversy among workers as to the precise age of Pleistocene stage boundaries, as well as the Tertiary-Quarternary boundary. Despite these inherent problems, the ages assigned to the mapped horizons are reasonable, and serve to generally date timing of geologic events within the study area.

Beard et al. (1982) used glass fission track dates, combined with nannofossil dating at a proposed Pliocene-Pleistocene boundary stratotype at Urica, Italy, to confirm an age of 2.8 m.y. before present at this boundary. The chronostratigraphic subdivision of this interval (2.8 m.y. ago to recent) was achieved using biostratigraphy, climatic inferences (population changes in plankton, etc.), eustatism, magnetic history, and isotopic indices (Beard et al., 1982). They suggested that a combination of these methods allows a basis for correlating marine and continental depositional units and the application of conceptual Pleistocene stage terminology to North American and Gulf Coast stratigraphy (Figure 22).

Seismic Intervals I—IV were correlated to the eight major Gulf Coast eustatic cycles defined by Beard et al. (1982) solely on the basis of seismic facies character (Figure 23). Watkins (1986) proposed that seismic facies character is directly related to these sea level

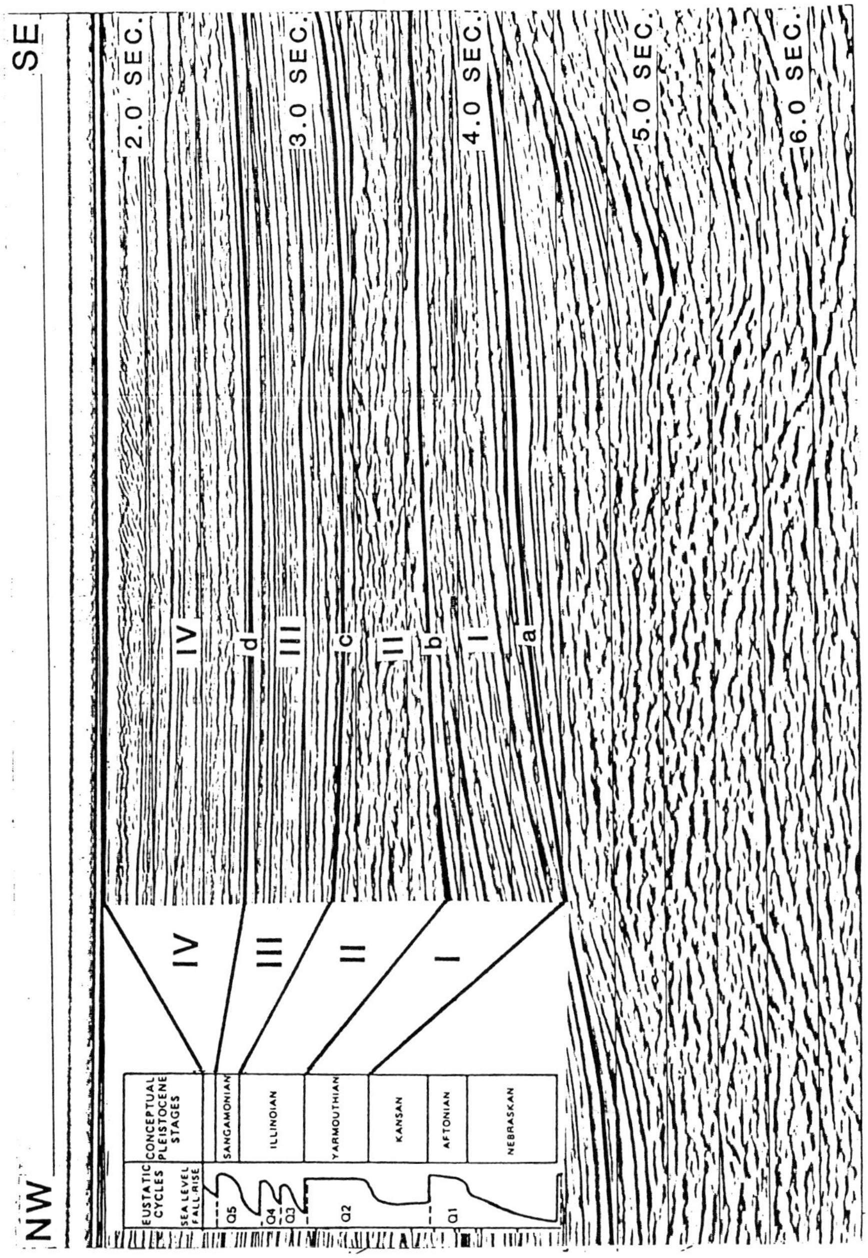

FIGURE 23. Correlation of Intervals I — IV to Pleistocene stages on "type" section shown in Figure 13.

TABLE 2
General Correlation of Depositional Intervals (I — IV) to Pleistocene Stages

	Mapping horizon	Depositional interval	Pleistocene stage	Approximate duration
Sea floor				
	D	IV	Wisconsian	0—70,000 year BP
	C	III	Sangamonian	70,000—300,000 year BP
	B	II	Illinoian	300,000—750,000 year BP
	A	I	Yarmouthian	750,000—1,280,000 year BP

fluctuatins. He suggested that: (1) reflection-free zones are associated with homogeneous pelagic deposition which occurs during sea level high stands, (2) hummocky/chaotic facies are generally associated with mass transport (debris flows and slumping) during falling sea level and sea level low stand, and (3) strong continuous reflectors are generally associated with turbites being deposited during rising sea level and sea level high stands (Watkins, 1986).

Interval IV is somewhat cyclic, having three characteristic facies types — reflection free, parallel continuous, and small zones of somewhat chaotic facies. This appears to correlate with the eustatic cycles during the Wisconsinian (Figure 22).

Sangamonian time was marked by a sea level high stand. This period correlates well with Interval III in that the zone of low to high amplitude continuous reflectors is probably representative of intermittant turidites mixed with pelagic sediments associated with a sea level rise.

There was a rapid drop in sea level at the end of the Yamouthian Period. The Illinoian Period was thus marked by a general sea level low, with frequent fluctuations that were characterized by gradual rises, and abrupt sea level falls. The chaotic nature of Interval II corresponds very well to this eustatic cycle pattern during Illinoian time.

Interval I is similar to Interval III in that is is characterized by low to high amplitude, very continuous reflections. Thus, Interval I can also probably be associated with a sea level high stand, most likely during the Yarmouthian. The uniform frequencies of Interval I and the virtually non-changing level of the oceans during the Yarmouthian, may serve as further evidence for this correlation.

Previously defined depositional sequences (Intervals I—IV) were correlated to eustatic cycles on the basis of seismic facies character. Intervals I—IV were found to correlate with Yarmouthian, Illinoian, Sangamonian, and Wisconsinian Periods respectively, as illustrated in Table 2.

F. APPARENT THICKNESS VS. TRUE THICKNESS OF DEPOSITIONAL UNITS

In measuring thicknesses of seismic intervals for isopach purposes, it became obvious that steeply dipping units would yield apparent thicknesses which were larger than the true thicknesses of the units (Figure 24). If the velocity of the sediments was constant, the true thickness could be easily attained by measuring the interval perpendicular to the dip angle. However, the velocity function of $V_z = 5400$ 1/sec + $0.5z$, which was used in converting time-structure maps to depth, assumes a vertical velocity gradient.

A plot of the ratio of apparent to true thickness vs. angle of dip is illustrated in Figure 25. Dips as high as 10° yield a relative thickness error of only 1.5%, which is negligible in terms of seismic time-depth conversion. The steepest dips within the study area are on the order of 20° and have a relative thickness error of 6.4%. This amount of error is not insignificant, therefore, a 5% thickness correction was applied to areas of steeply dipping (15 to 20°) reflectors.

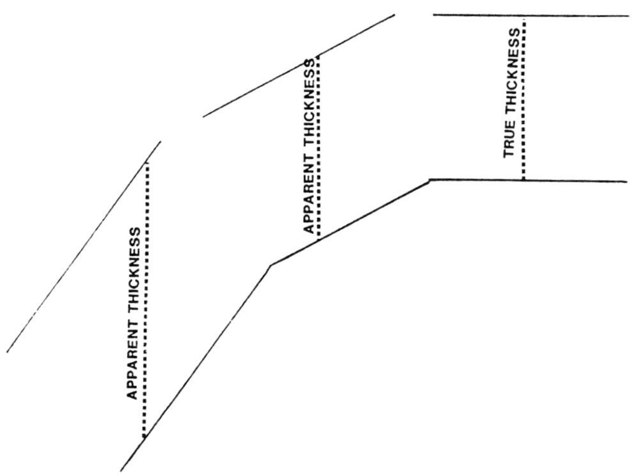

FIGURE 24. Plot illustrating increase in apparent thickness with increasing geologic dip when using vertical velocity gradient.

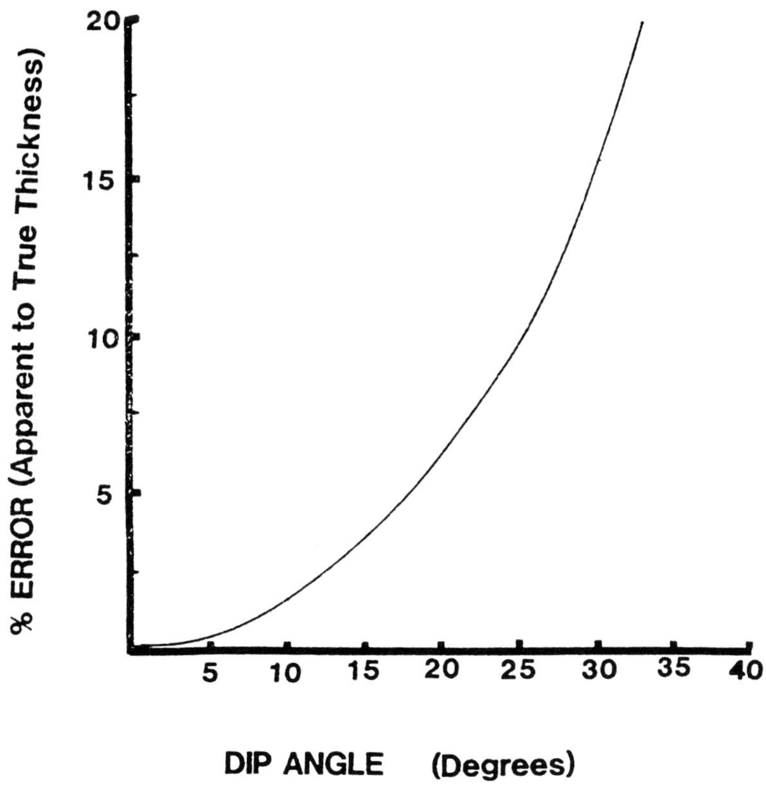

FIGURE 25. Plot of the ratio of apparent to true thickness, vs. dip angle illustrating measured thickness error.

IV. SALT DISTRIBUTION, THICKNESS, MORPHOLOGY, AND EFFECTS ON DEPOSITION

A. SALT DISTRIBUTION, THICKNESS

The offshore region in the northern Gulf of Mexico can be generally divided into four salt provinces (Figure 26). The upper shelf is characterized by small, oslated salt structures

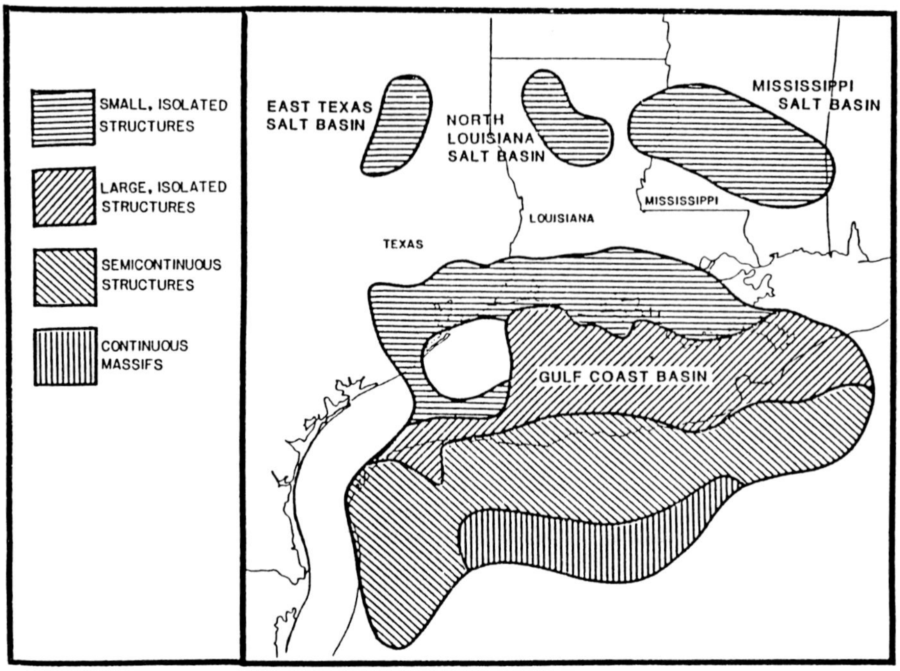

FIGURE 26. Major salt provinces in the northern Gulf of Mexico. (From Yorston and Fox, Houston Geological Soc., Cont. Ed. Series, Oct. 23, 1985, Houston, TX. With permission.)

whereas the lower shelf/upper slope area has somewhat larger, isolated structures. Then there is a drastic change in salt morphology between the Upper and Lower Texas-Louisiana Slope. While the upper slope and shelf are characterized by isolated, vertically emplaced salt structures, the lower slope is composed of very broad, semicontinuous extrusive salt lobes covered by a moderately thick to very thin overburden (Martin, 1984). The southernmost portion of the Sigsbee Bulge is composed of continuous massifs or salt tongues. Figure 27 illustrates the distribution and relative thicknesses of salt within the study area.

The salt within the study area is a semicontinuous to continuous mass, with very few isolated salt wings or salt domes. The salt is relatively thin, and does not extend downward to great depths. There are several deep basins, the largest being in the western portion of the study area, where the salt has apparently been mobilized due to sediment loading.

The salt attains a maximum thickness of approximately 22,000 ft (2.0 sec) in the northeastern portion of the map (Figure 27). There is another salt thick in the central region which varies in thickness from 11,000 to 13,000 ft and has an inverted "u" shape.

The edges of the lobate salt bodies are generally wedge-like or tongue-like in nature. Figure 28 and 29 illustrate the relation of the allochtonous salt tongue with respect to the Lower Slope and Continental Rise sediments. The laterally extruded salt creates a bathymetric scarp which has over 2000 ft of relief. The salt is covered by a relatively thin sediment layer which gradually thickens landward. As a result of this thin overburden, the upper surface of the salt is characterized by poorly developed salt structures (incipient diapirs). Figure 29 illustrates the nature of the salt along a strike section.

B. EMPLACEMENT MECHANISM

The reflectors along both the top and bottom of the salt tongue (Figure 26) show some terminations against salt, but for the most part they are concordant with the salt layer. Watkins et al. (1978) suggested that the truncation of underlying sediments indicates a

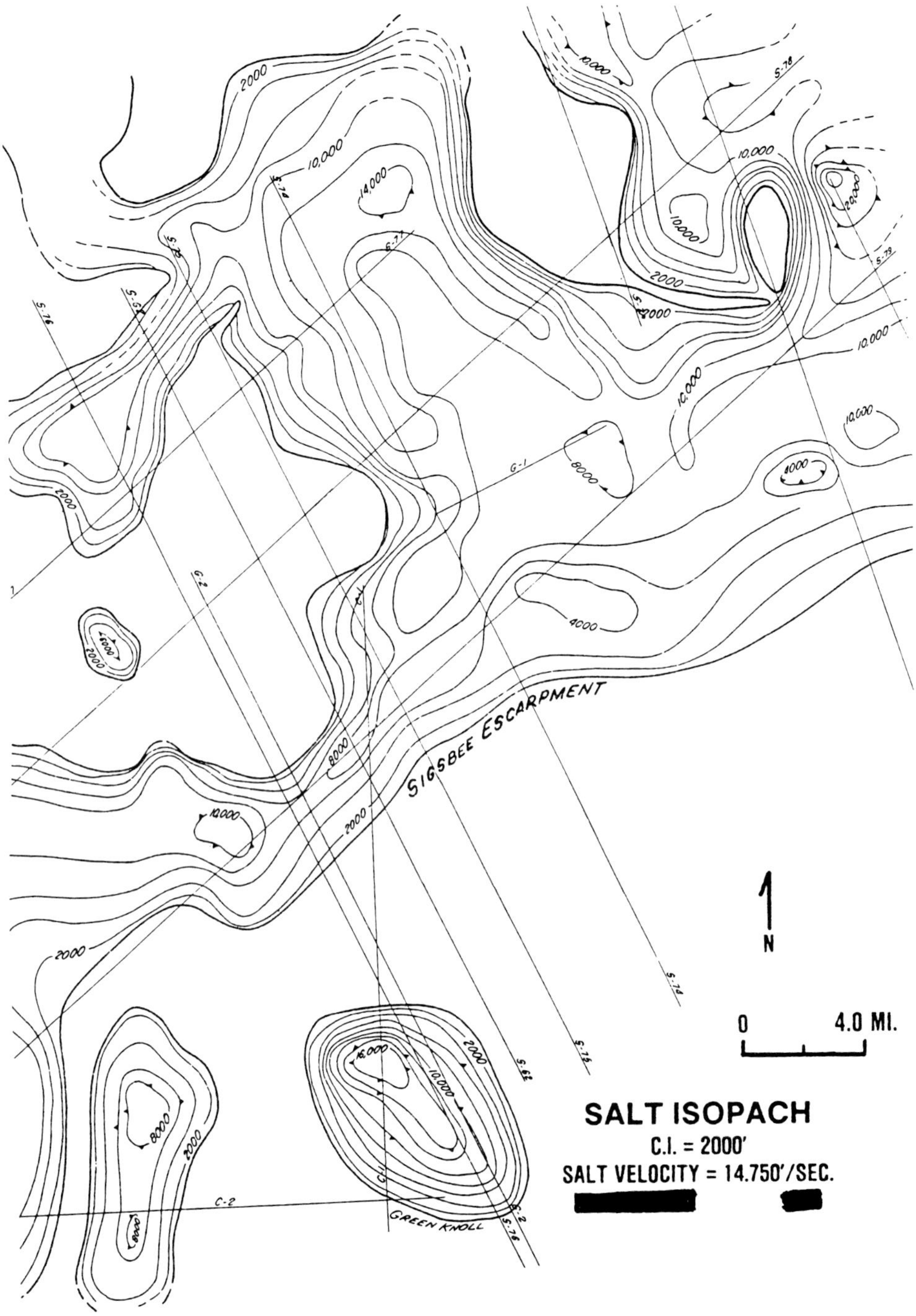

FIGURE 27. Salt isopach map.

"thrust" relationship as opposed to a decollement along a bedding plane or a lateral flow of salt across the top of a sedimentary sequence. However, as in the case of Figure 26, it appears as if the salt is mobilized parallel to bedding. Conversely, the simple observation of Jurassic salt being juxtaposed against Pleistocene sediments would support the thrusting model. There is apparently a combination of the two processes. Perhaps, the allochthonous

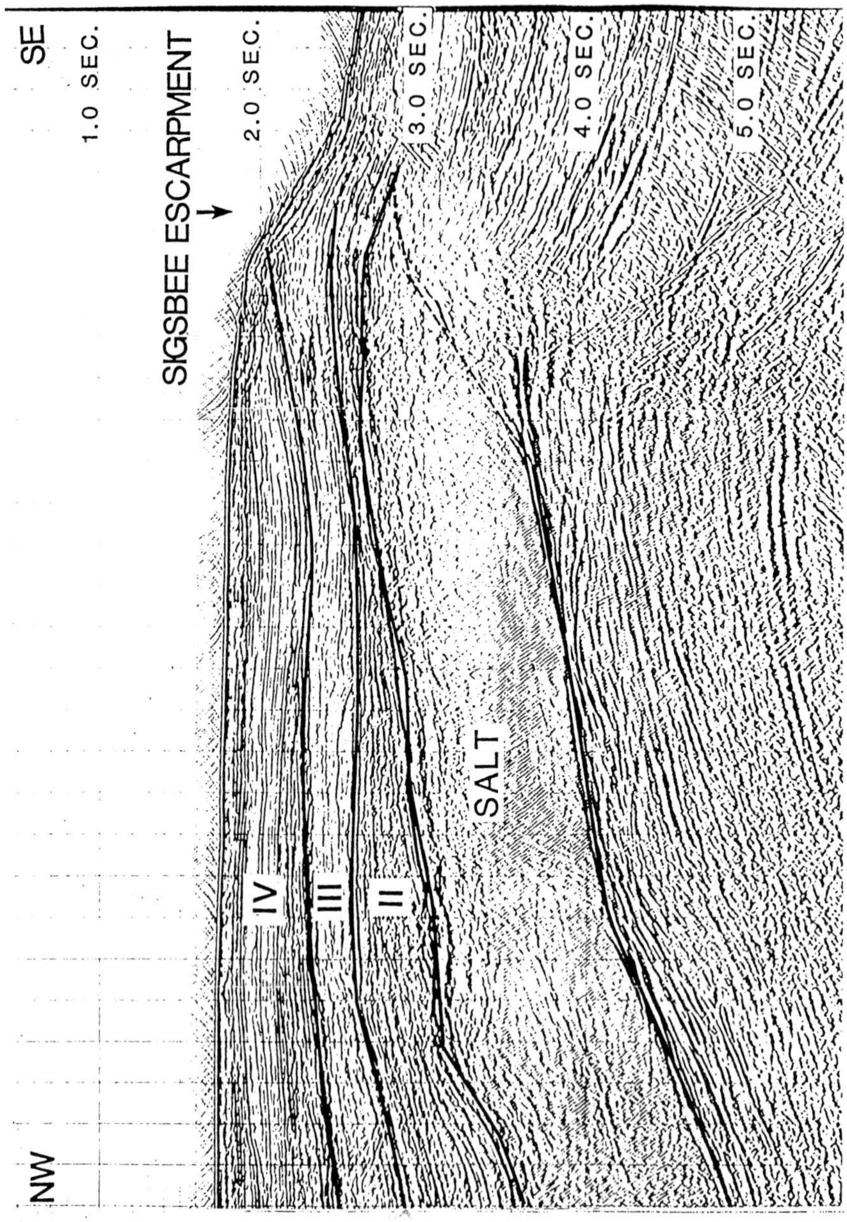

FIGURE 28. Portion of seismic line G-3 illustrating the nature of allochthonous salt tongue. (Courtesy of GECO.)

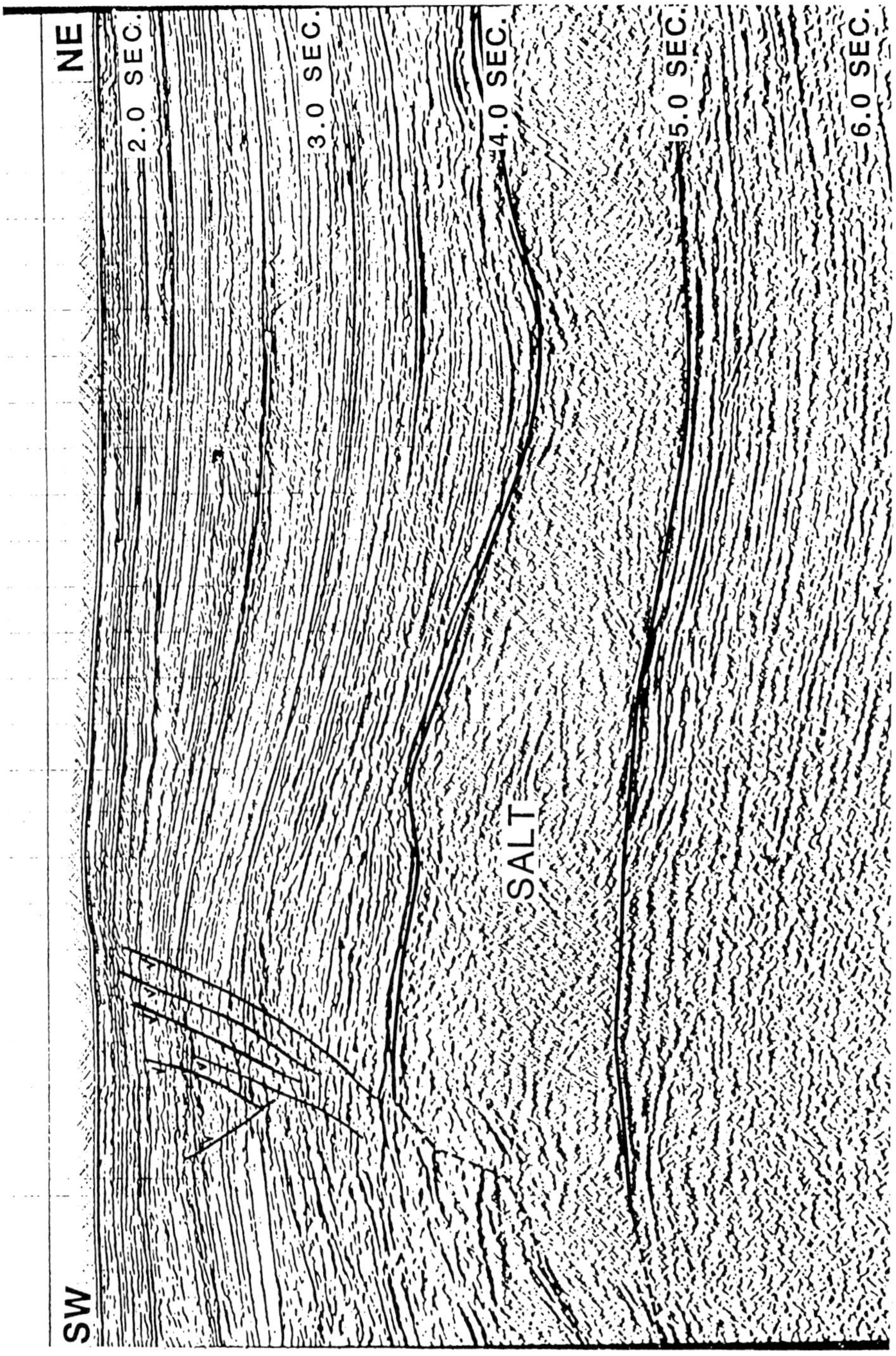

FIGURE 29. Portion of seismic line G-1 illustrating the nature of allochthonous salt tongue along a strike line. (Courtesy of GECO.)

salt moves with a larger component of vertical motion when there is ample overburden to create a density inversion, and thus drive the system with the buoyancy model; otherwise, when the system depends predominantly on the lateral pressure gradient caused by sediment loading updip, the salt tongue may exploit bedding plane weaknesses and move with a larger component of lateral motion (Watkins, 1986).

C. SALT MOBILIZATION SURFACES

Deep, strong reflectors found within the "salt free" basins may represent salt withdrawal or mobilization surfaces (Figure 30). These reflectors are unusually continuous and strong for being so deep in the section (generally between 5.0 to 6.0 sec or 23,000 to 30,000 ft). They are often characterized by a chaotic zone above them, and at the margins of the basin they appear to curve upward toward known salt structures (Figure 31). It is not unreasonable to believe that salt which is withdrawn from a basin may leave a salt migration surface. The high amplitude reflections may result from residual salt which is left behind as the salt is mobilized, or from differential sediment velocities and densities above and below the surface.

D. EFFECTS OF SEDIMENT LOADING

There are three isolated salt wedges or domes which are not attached to the main salt mass (Figure 27). One is a very small wedge located within the deep western basin while the two larger isolated salt structures (one being Green Knoll) are found seaward of the Sigsbee Scarp. All three of the isolated salt structures appear to have once been attached to the main salt mass but have subsequently been segmented as a result of localized sedimentation. (A detailed geologic model for the origin and development of Green Knoll and the detached wedge found to the west of Green Knoll is presented in Section V.) As is the case for the main salt mass, the three isolated salt structures do not show evidence for purely vertical emplacement (i.e., the presence of rim synclines), but rather a combination of vertical and lateral movement.

There appears to be a delicate balance between the effects of initial salt structural highs and lows (which locally tend to funnel or block sediment transport), and the direction and rate of sedimentation, and its ability to load and mobilize salt. Figure 32 is a total sediment isopach for the sediments above top salt, or essentially, it represents a top-salt structure map with the sea floor being the datum. The deep salt-free basins are obviously representative of the thickest isopach units, being greater than 3.5 sec or 15,000 ft in depth. Other than these deep basins, and the areas directly adjacent to them which are characterized by the wedge-shaped terminations of the main salt mass, the deepest "above salt" basin is found in the very northeastern corner of the map. This basin, which attains a depth of 10,500 ft, is flanked on either side by somewhat discontinuous salt ridges which may have served to funnel sediments into the basin as well as southward into the even deeper, oval-shaped, salt-free basin (Figure 32). The much larger snakehead-shaped basin in this region is also flanked by discontinuous ridges or highs. Both the ridge axes, and the basin axis trend in a northwest-southeast direction. The ridge found to the west of "snakehead" basin, and a parallel ridge found even farther to the west, make up the legs of the inverted-"u" salt thick which is found on the salt isopach map (Figure 27). The salt thick in the northeast corner of the isopach map is also associated with a ridge, suggesting that, in general, salt isopach thicks are associated with structural highs. It is apparent that the structural ridges associated with salt highs played a major role in controlling deposition. The northwest-southeast trending ridge which parallels line S-74 appears to have had the most effect, as evidenced by the seismic facies variation across the ridge (see Section III).

E. EVIDENCE FOR RECENT SALT MOVEMENT

The salt in the study has by no means reached static equilibrium as evidenced by the

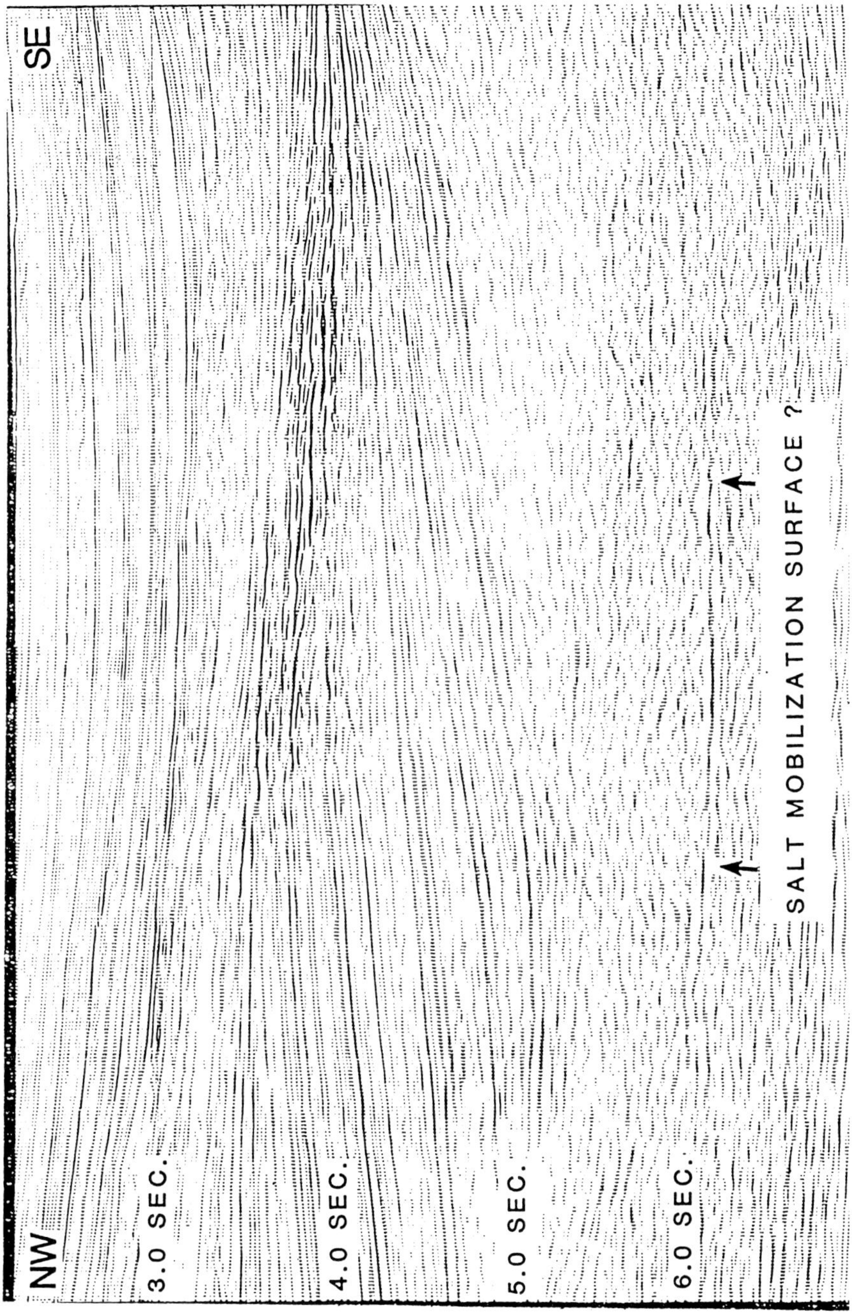

FIGURE 30. Portion of seismic line S-62 illustrating possible salt migration surface. (Courtesy of Shell Offshore Inc.)

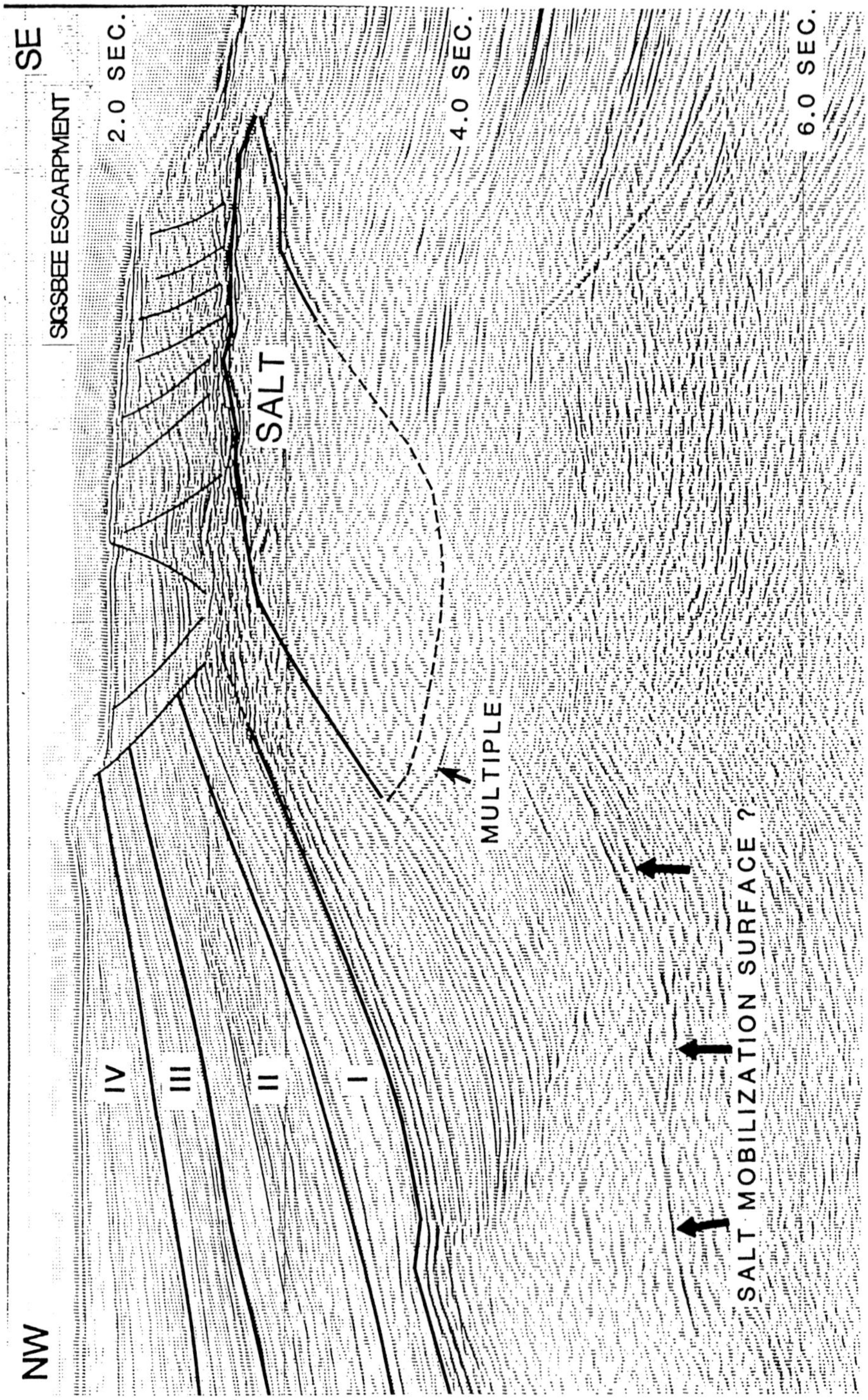

FIGURE 31. Portion of seismic line S-62 illustrating possible salt migration surface near Sigsbee Escarpment. (Courtesy of Shell Offshore Inc.)

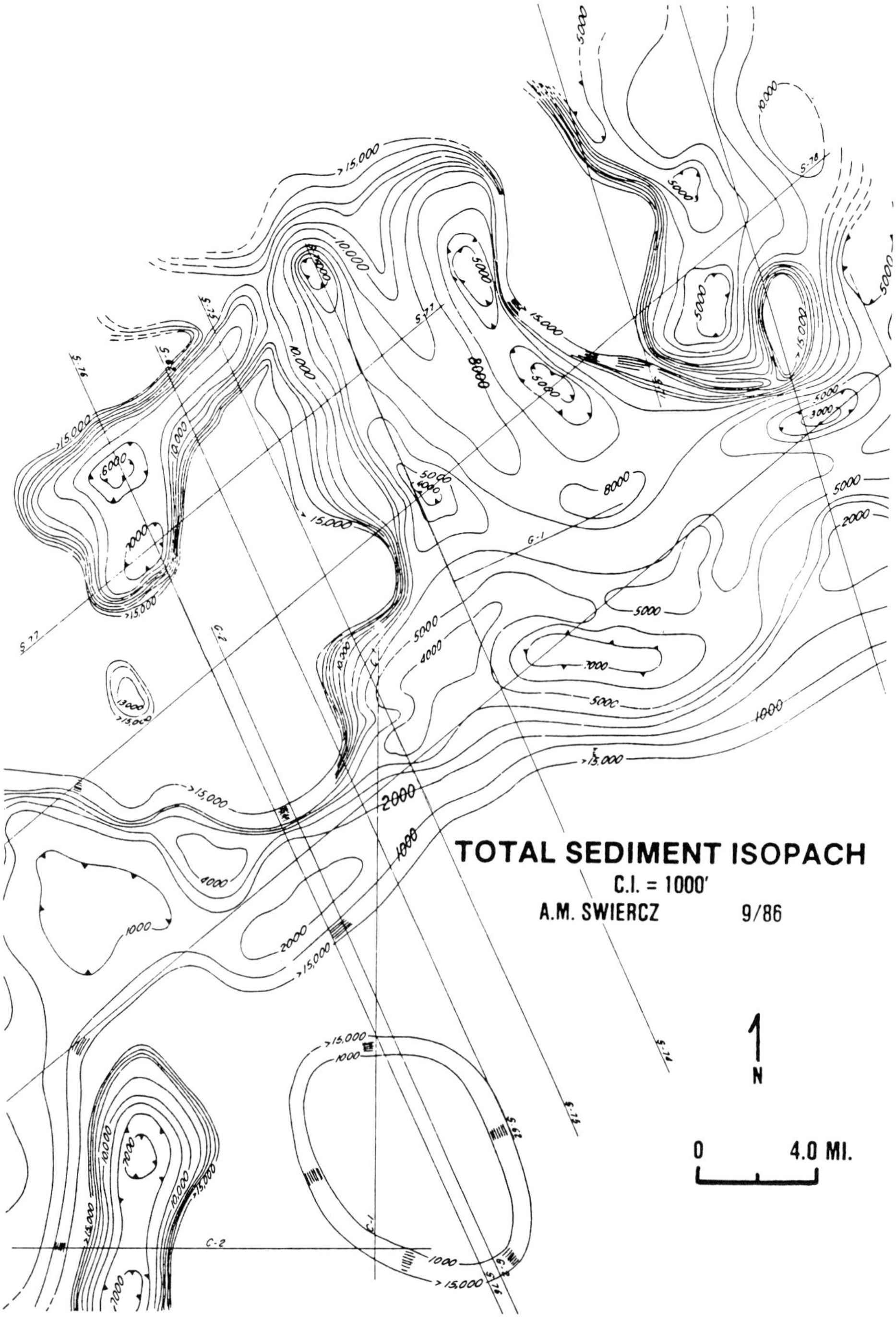

FIGURE 32. Total sediment isopach map.

physiographic expression on the sea floor in Figure 33. Furthermore, there are several faults that reach the sea floor which also indicate recent salt movement. Despite the evidence for recent salt movement, seismic interval thickness changes on the flanks of salt structures suggest an episodic history of salt growth. In Figure 33, Interval I thins to the southwest across the salt structure. This indicates that upward salt movement was contemporaneous

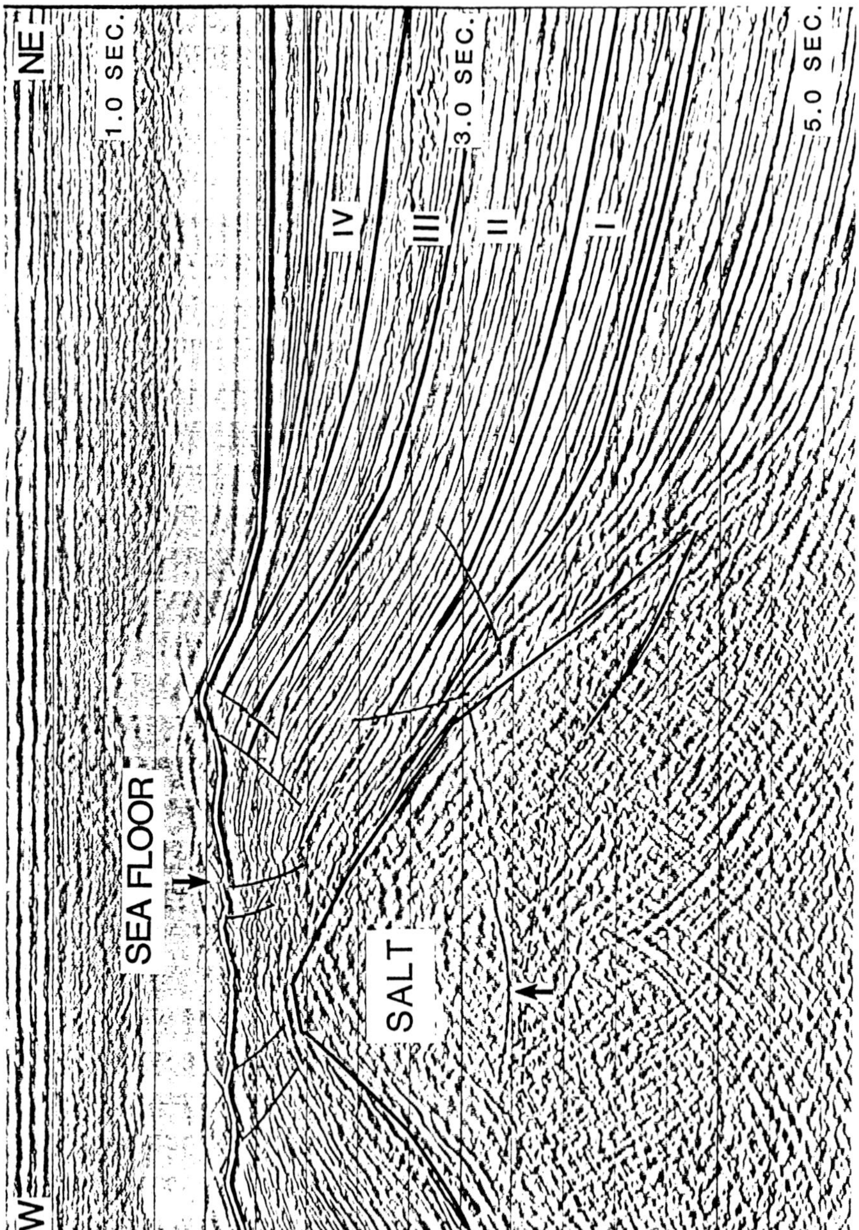

FIGURE 33. Portion of seismic line S-78 illustrating evidence for recent salt movement. (Courtesy of Shell Offshore Inc.)

with Interval I deposition. Interval II shows very little thinning, suggesting a cessation of diapiric growth during Interval II (Illinoian?) time. The small amount of thinning that does occur is probably a result of Interval II being deposited across a pre-existing structural high. Intervals III and IV show thinning and even erosional truncation which would correlate with renewed diapiric uplift during this time. Such sporadic salt movement can be attributed to variable sedimentation rates resulting from glacial-interglacial cycles and major changes in source direction associated with the Mississippi River.

F. FAULT PATTERNS ASSOCIATED WITH SALT MOBILIZATION

There are several types of fault patterns which are found within the study area. The relatively thin sediment package found above the mobile salt tongue along the edge of the Sigsbee Scarp is often characterized by a series of closely spaced, steeply dipping, down-to-the-basin, normal faults (Figure 34). Although DSDP results from Leg 10, Hole 92 reported overcompacted sediments above the mobile salt tongue, this fault pattern suggests that there exists an extensional regime which is attempting to accommodate the large component of horizontal salt movement. The vertical component of salt movement may be responsible for the overcompacted sediments.

Just south of the Escarpment the sediments appear to be under compression. Figure 35 illustrates a series of landward-dipping thrust faults which are found just basinward of the salt tongue. The large allochthonous lobes of salt which are loaded updip and thrust seaward are undoubtedly responsible for the compressional nature of faulting seaward of the scarp.

Another type of compressional feature is the large anticlinal structures found seaward of the scarp and very deep in the section (Figure 36). These structures tend to form ridges or folds which parallel the salt front suggesting that they are also associated with the advancing salt wedge.

Landward of the escarpment, within the interior of the study area, the faulting is extensional. Figure 37 illustrates a "traditional" graben which has formed above a salt structure. There are relatively few grabens found in the study area, probably because much of the salt movement is lateral. Traditional grabens with typical radial fault patterns are most often found above vertically emplaced diapirs (Atwater and Forman, 1959).

The most common and characteristic faults found within the study area are located along the periphery of the interior salt-free basins. They are generally found above the edges of the salt lobes (which flank the basins), and are downthrown into these basins. The fault planes are often listric and parallel to the duckbill-shaped salt lobes which tend to "pinch-out" toward the isolated basins. Figures 38 and 39 illustrate these relationships as well as the expansion of the section (especially Interval IV) across the faults. This type of faulting is obviously associated with basinal development. (Note that the salt structures in Figures 38 and 39 make up the inverted-"u" salt ridge discussed earlier.)

A somewhat unique case exists in Figure 40 in that a series of five growth faults are seen to dip away from the large salt-free basin. Furthermore, only one fault is truely listric and tends to parallel the salt structures. Interval III (Sangamonian?) is the first interval that expands downthrown across the main (northeastern) growth fault, and thus probably marks the initial stage of lateral salt withdrawal (Larberg, 1983).

G. CONCLUSIONS

1. Salt structures within the study area are in the form of a semicontinuous to continuous mass.
2. There are several deep, salt-free basins from which the salt has apparently been mobilized due to sediment loading.
3. The salt mass attains a maximum thickness of 22,000 ft although, in general, it is relatively thin (5000 to 10,000 ft) and does not extend vertically to great depths.

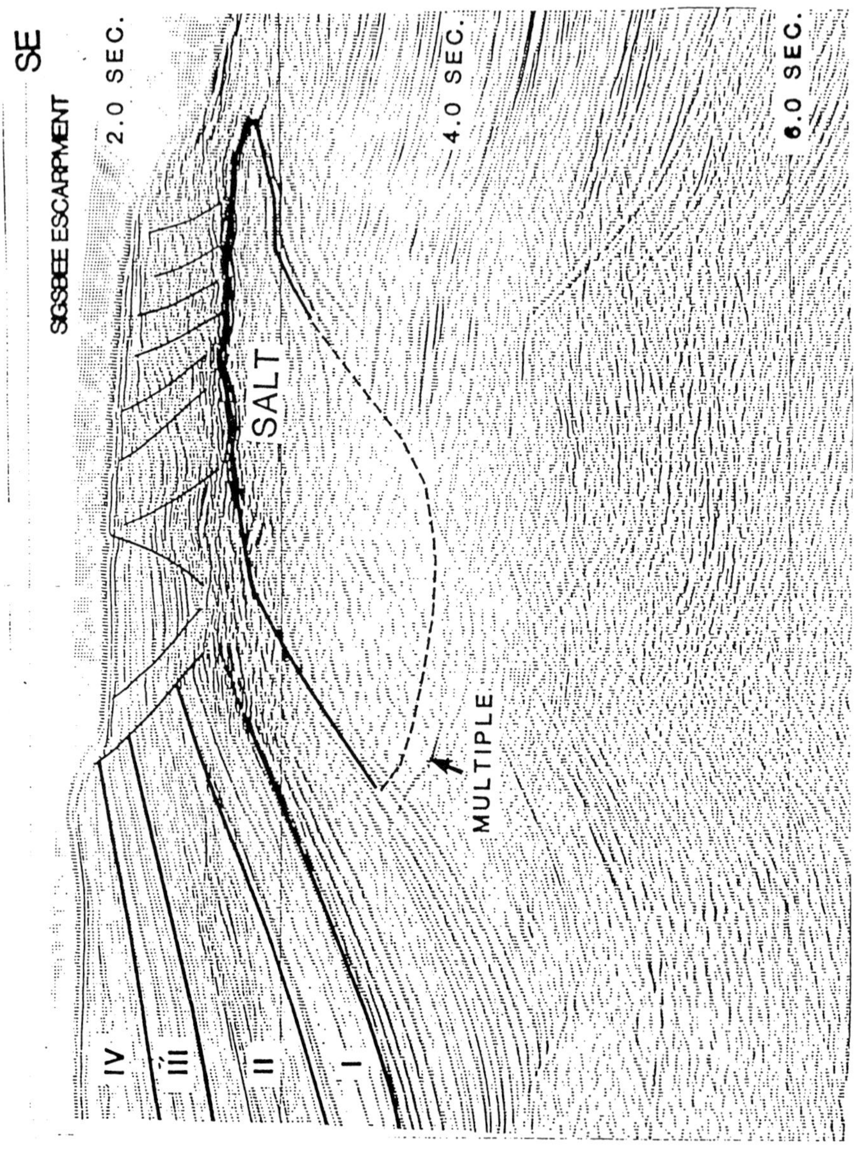

FIGURE 34. Portion of seismic line S-62 illustrating series of normal faults found above mobile salt tongue. (Courtesy of Shell Offshore Inc.)

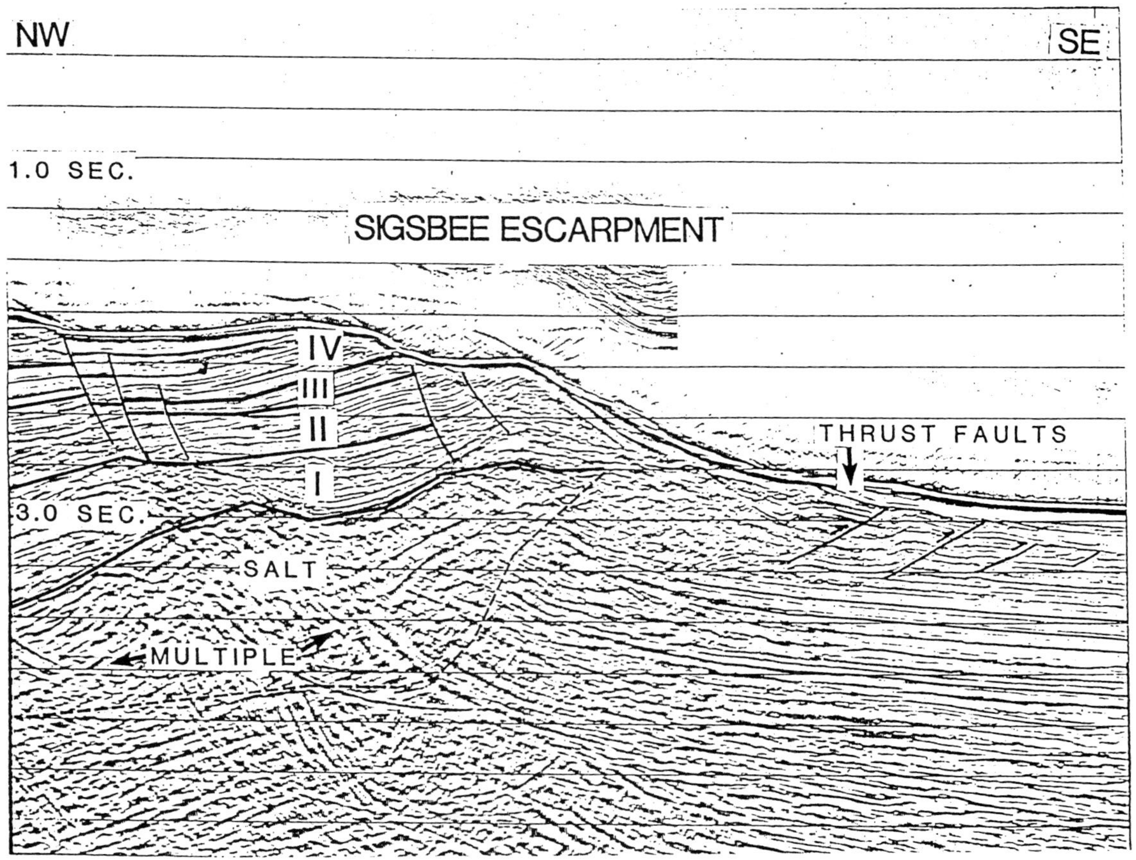

FIGURE 35. Portion of seismic line S-75 illustrating series of small-scale thrust faults found basinward of the Sigsbee Scarp. (Courtesy of Shell Offshore Inc.)

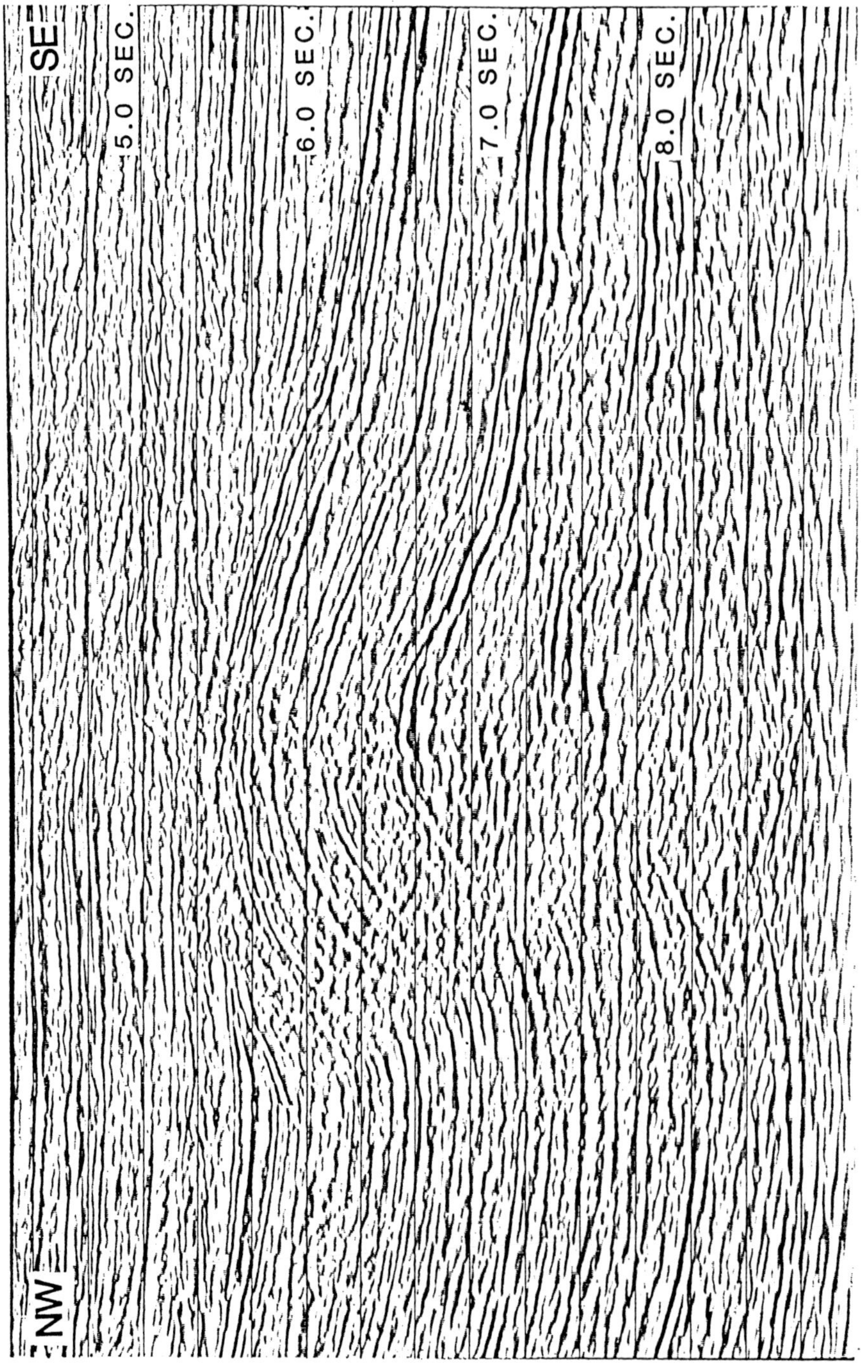

FIGURE 36. Portion of seismic line S-75 illustrating the presence of large compressional features basinward of the Sigsbee Scarp, and very deep in the section. (Courtesy of Shell Offshore Inc.)

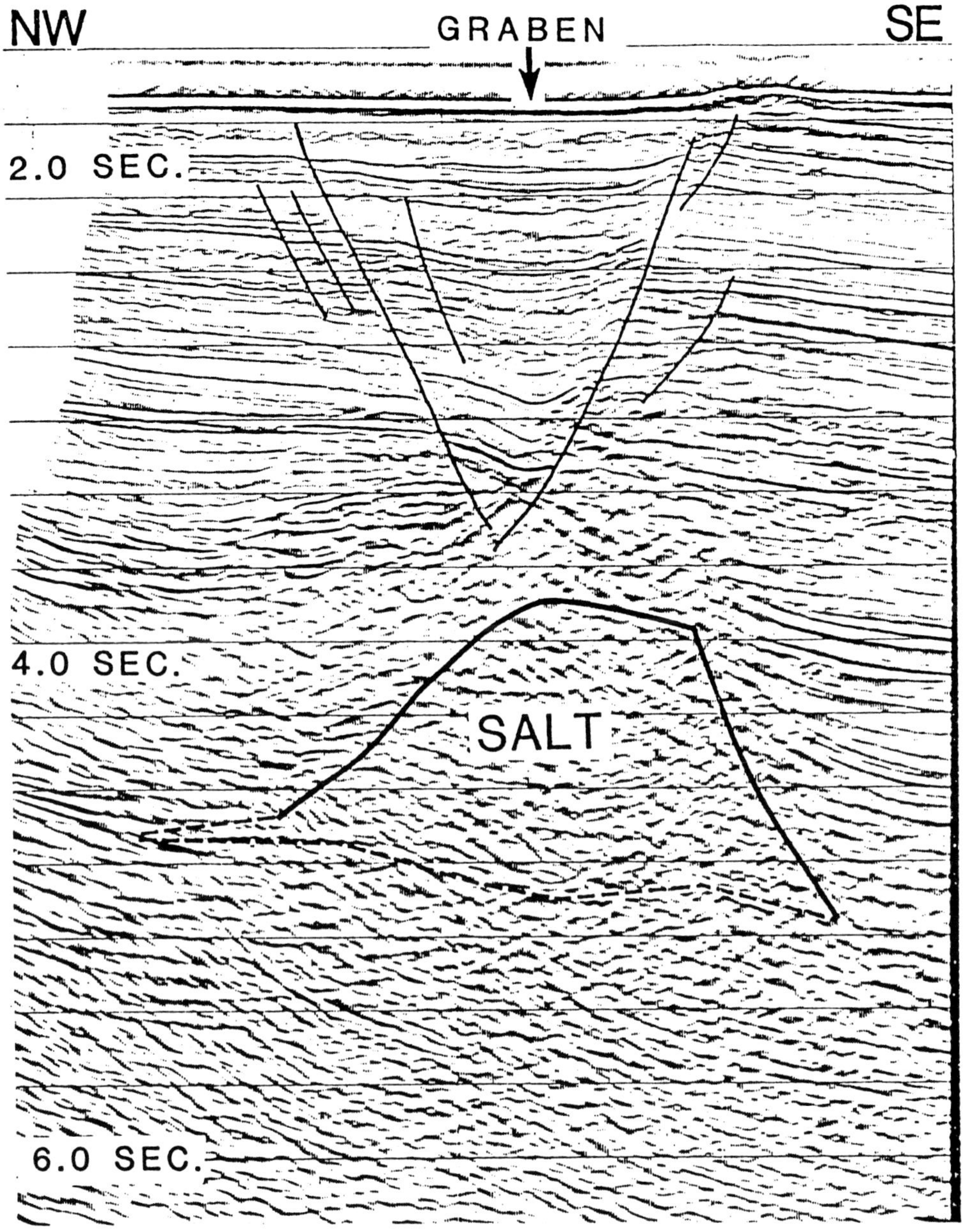

FIGURE 37. Portion of seismic line S-75 illustrating a typical graben found above a salt structure. (Courtesy of Shell Offshore Inc.)

4. Movement of the allochthonous salt tongue appears to vary between "thrusting" (buoyancy) and lateral exploitation of bedding weaknesses, depending on the overburden thickness and also probably on the force of the lateral pressure exerted by updip loading.
5. Deep, strong reflectors found within the "salt-free" basins may represent residual salt left behind by mobilized salt along salt migration surfaces.
6. The three detached salt structures (including Green Knoll) appear to have been segmented from the main salt mass by localized sedimentation. They were not vertically emplaced although they have experienced significant vertical uplift since their separation from the main salt mass.

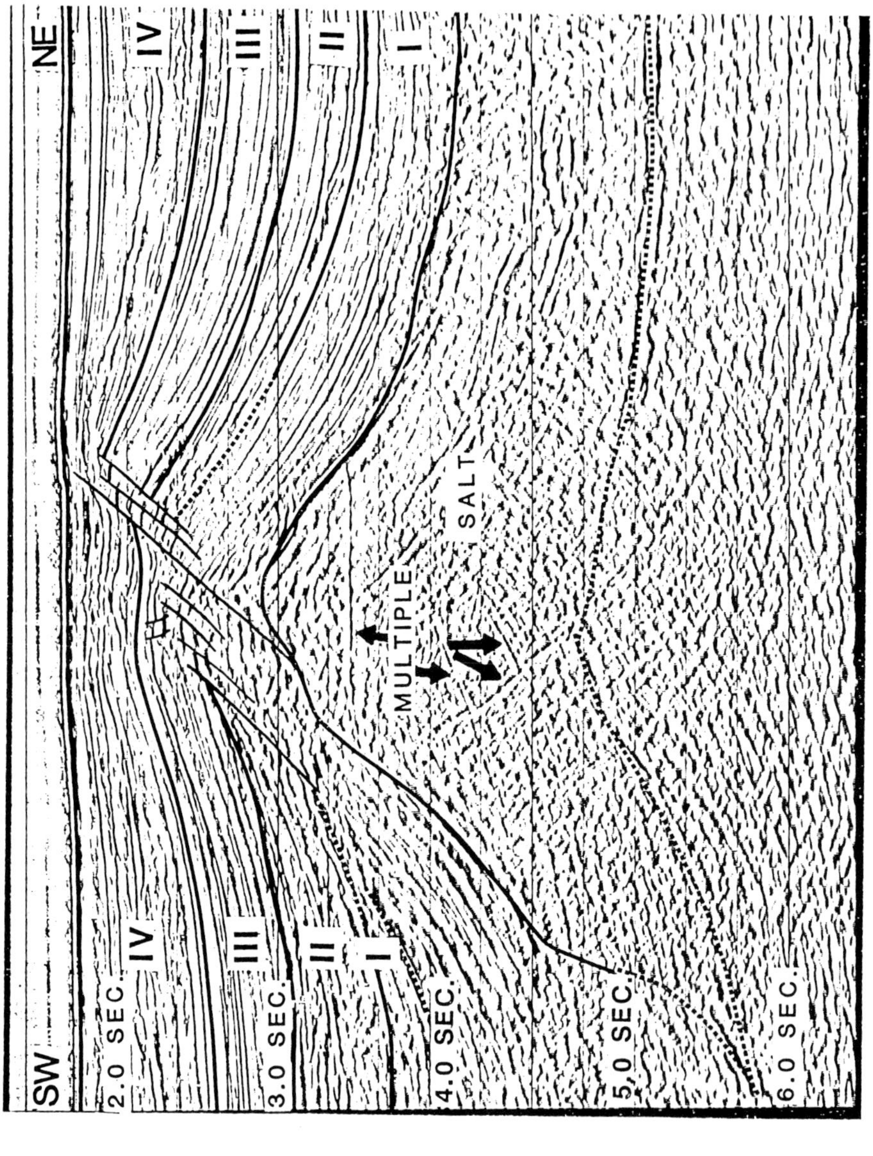

FIGURE 38. Portion of seismic line S-78 illustrating listric normal and growth faults dipping to the southwest above edge of salt body. (Courtesy of Shell Offshore Inc.)

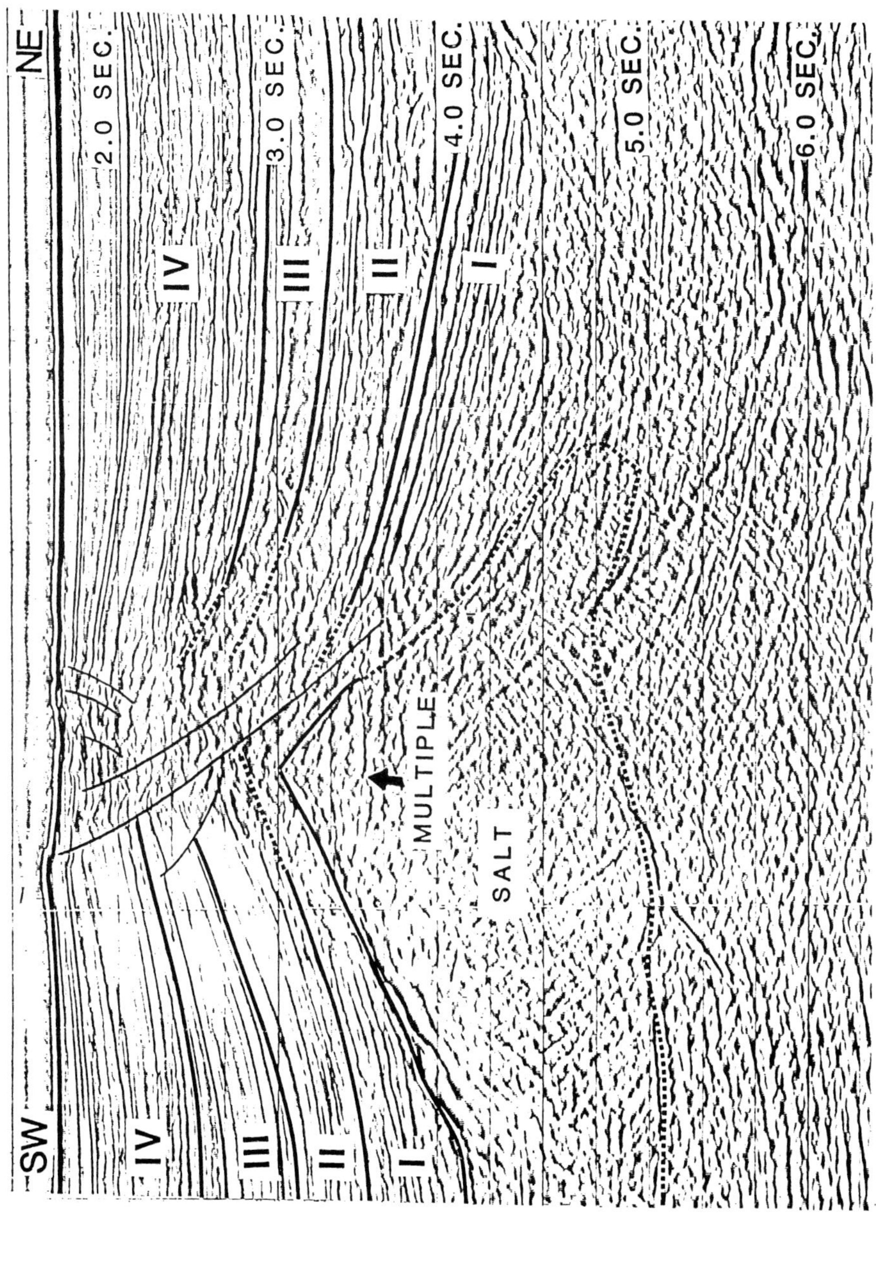

FIGURE 39. Portion of seismic line S-78 illustrating listric normal and growth faults dipping to the northeast above edges of salt body. (Courtesy of Shell Offshore Inc.)

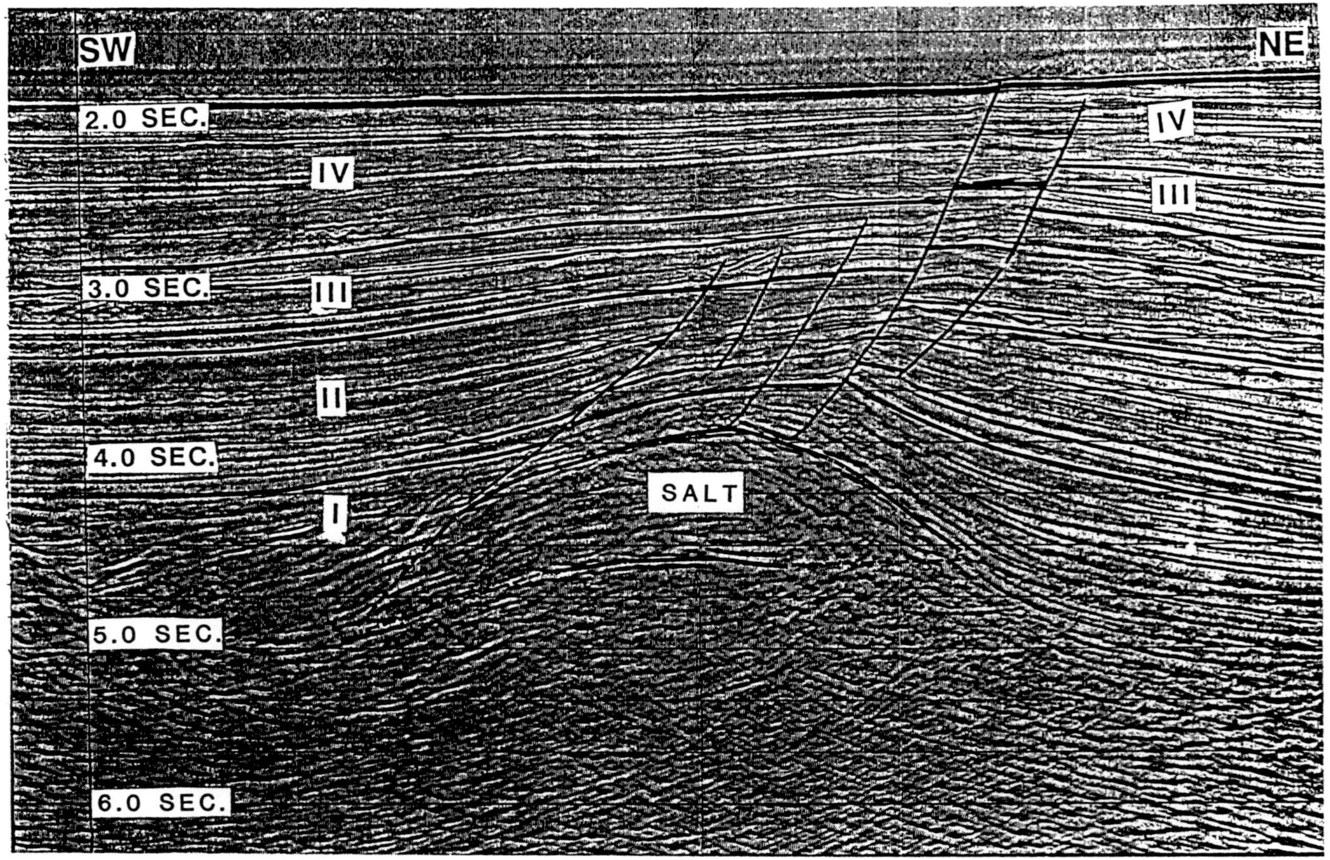

FIGURE 40. Portion of seismic line S-77 illustrating growth faults above salt. Interval III is the first interval that expands downthrown across the main (northeastern) growth fault and thus probably marks the initial stage of lateral salt withdrawal. (Courtesy of Shell Offshore Inc.)

7. The presence of discontinuous salt ridges appears to have funnelled sediments into the deep basins which in turn squeezed more salt into the adjacent ridges. Thus, the structural highs or ridges are associated with salt isopach thicks.
8. Bathymetric expression on the sea floor above salt structures suggests recent salt movement.
9. Variable interval thickness changes on the flanks of salt structures and across growth faults suggest an episodic history of salt growth.
10. The thin sediment package above salt along the toe of the scarp is often characterized by a series of down-to-the basin, normal faults. This suggests an extensional regime which is attempting to accommodate the section overlying the horizontally mobilized salt.
11. Seaward of the scarp there are two types of compressional events; a series of small-scale thrust faults just basinward of the salt tongue, and a deep-seated anticlinical ridge which parallels the salt front.
12. Normal and growth faults are often found above the edges of salt lobes. These faults tend to be listric and downthrown into the deep, salt-free basins.

V. ORIGIN AND DEVELOPMENT OF GREEN KNOLL

A. LOCATION

Green Knoll is located in the northern Gulf of Mexico at approximately 90°20′W, 20°N (see Figure 1). It is asymetrical in shape, being 13.8 km in length and 9 km in width, and its physiographic expression on the sea floor constitutes 136.5 m of relief. The location of Green Knoll (basinward of the Sigsbee Escarpment) suggests that the Sigsbee Escarpment does not represent the southern extent of salt in the northern Gulf of Mexico as is believed.

B. COMPARISON TO OTHER KNOLLS AND DOMES

Seismic data can help suggest the composition of Green Knoll. Early workers often alluded to its presence (and that of other knolls found in the Gulf), but they failed to make conclusive statements as to their physiography. Buffler, Worzel, and Watkins (1978) suggested that Green Knoll was indeed cored with salt on the basis of its proximity to the Sigsbee Escarpment and also that the symmetry of the deformed sediments around the knoll attested to the vertical nature of salt movement in this area. Their conclusion that Green Knoll is composed of salt appears to be correct. The seismic profile C-1 (Figure 41) shows evidence of dissolution occurring on the surface of the knoll. This phenomenon is characteristic of salt diapirs which come in contact with sea water. Shale diapirs do not exhibit dissolution surfaces.

C. EVIDENCE FOR NON-VERTICAL EMPLACEMENT OF GREEN KNOLLS

While the assumption that Green Knoll is composed of salt appears valid, the vertical nature of salt movement in this area as proposed by earlier workers can be contested. A comparison of Green Knoll to other knolls in the Gulf serves in drawing several conclusions which will be supported and supplemented by additional data presented later in this chapter.

Figures 42 and 43 represent seismic lines over Challenger Knoll, and an isolated diapir found to the southwest of the Florida Escarpment, respectively. Challenger Knoll is a confirmed (DSDP-2) salt diapir, and the isolated diapir is also believed to be cored by salt. Both the Challenger Knoll and the isolated diapir show well-developed rim synclines which have formed in response to the vertical nature of their growth. The reflectors adjacent to Green Knoll show no such development of a rim syncline. In fact, the deep reflectors show virtually no deformation at all other than velocity pull-up, which is merely an artifact of

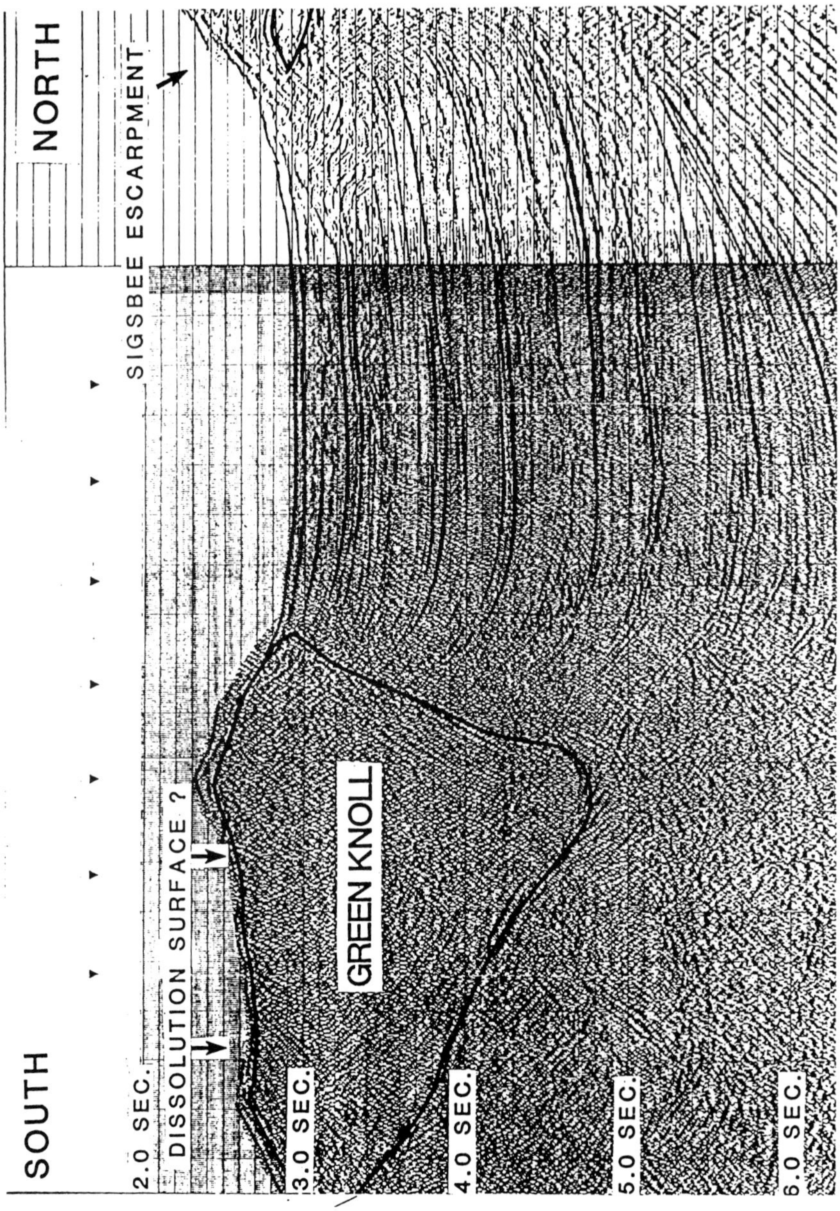

FIGURE 41. Portion of seismic line C-1 illustrating possible dissolution surface on Green Knoll. (Courtesy of Shell Offshore Inc.)

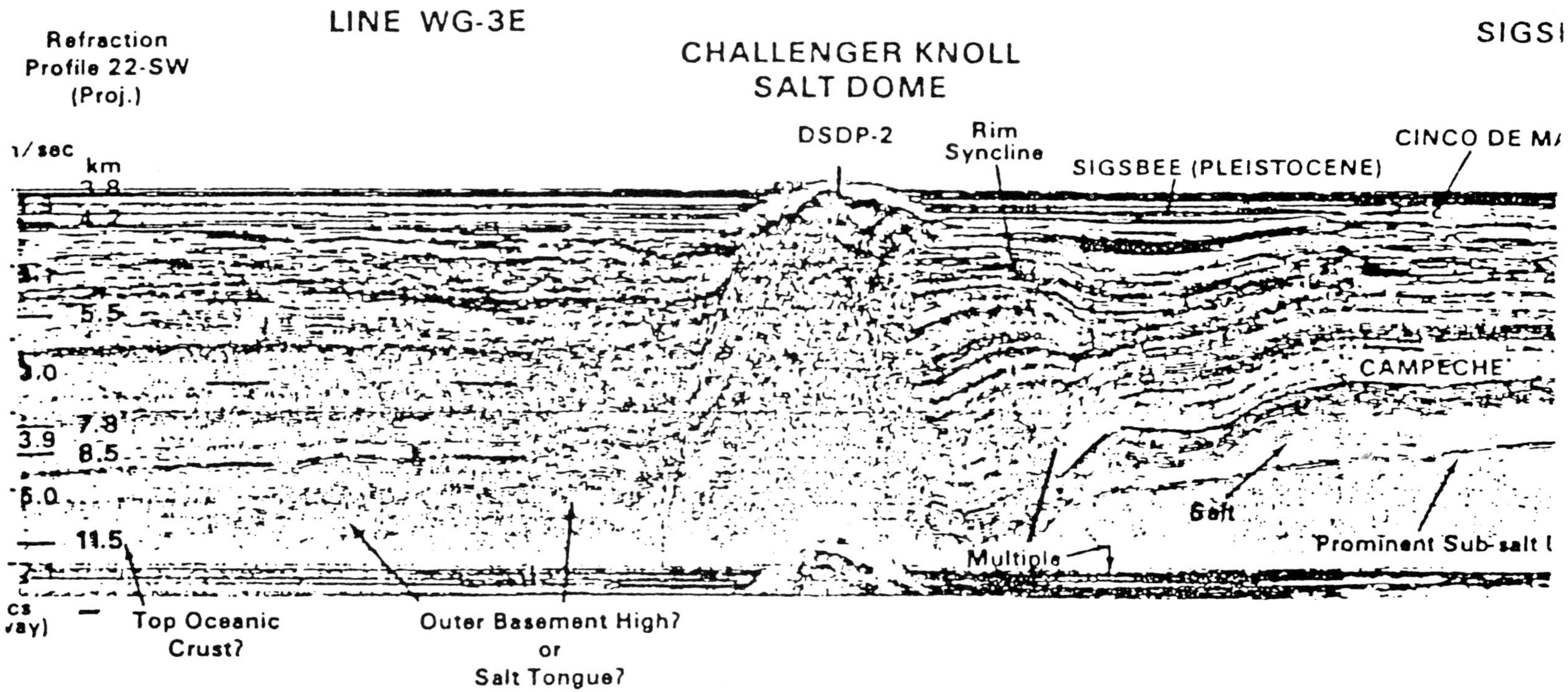

FIGURE 42. Seismic line over Challenger Knoll illustrating development of rim syncline. (From Buffler, 5th Annu. Gulf Coast SEPM Res. Conf., Proc., 31-34, 1984. With permission.)

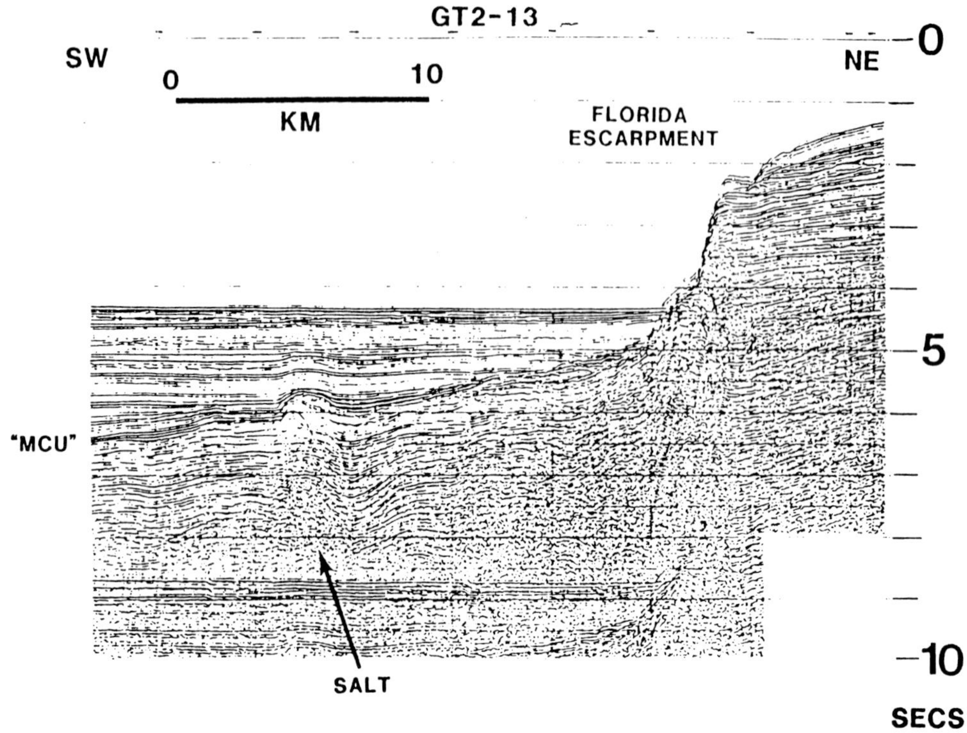

FIGURE 43. Seismic line over an isolated diapir found to the southwest of the Florida Escarpment illustrating development of rim syncline. (From Buffler, 5th Annu. Gulf Coast SEPM Res. Conf., Proc., 31-34, 1984. With permission.)

seismic data processing and not a true representation of geologic structure. If not for this pull-up effect, the deep reflectors could possibly be observed beneath Green Knoll as the strong base-salt reflector (Figure 44) indicates that the knoll is a shallow, detached salt structure, and not a vertical spine stemming upward from a mother salt bed.

Since a large component of vertical movement can be ruled out in the Green Knoll case, one must ask how this sizeable salt structure formed, and how, and from what direction was it fed? The salt Isopach map (Figure 27) shows the proximity of Green Knoll to the Sigsbee Escarpment as well as relative salt thicknesses.

The knoll is asymmetrical, attaining its maximum thickness (2.25 sec or 16,875 ft) toward the northwest. This suggests that the knoll may have been fed from this direction. However, a northwest-to-southeast-trending seismic line (Figure 44) demonstrates that the reflectors to the northwest of Green Knoll are continuous and undisturbed, showing no evidence of salt migration from this direction. A similar case is found for the reflectors adjacent to the eastern and southern margins of Green Knoll.

A more plausible conclusion is that the salt was mobilized (locally) from a westerly direction. This hypothesis is supported by (1) the presence of a deeper salt wedge found between the escarpment proper and Green Knoll, and (2) the presence of highly deformed reflectors found at depth westward of Green Knoll (Figure 45).

The detached wedge is probably a remnant of a salt ''feeder'' or tongue which supplied salt to the Green Knoll area in response to localized sediment loading to the west during Middle Tertiary(?) time. The disturbed reflectors (below 6.0 sec) found to the west of Green Knoll (Figure 45) may represent a migration surface along which salt moved as localized sedimentation segmented the feeder.

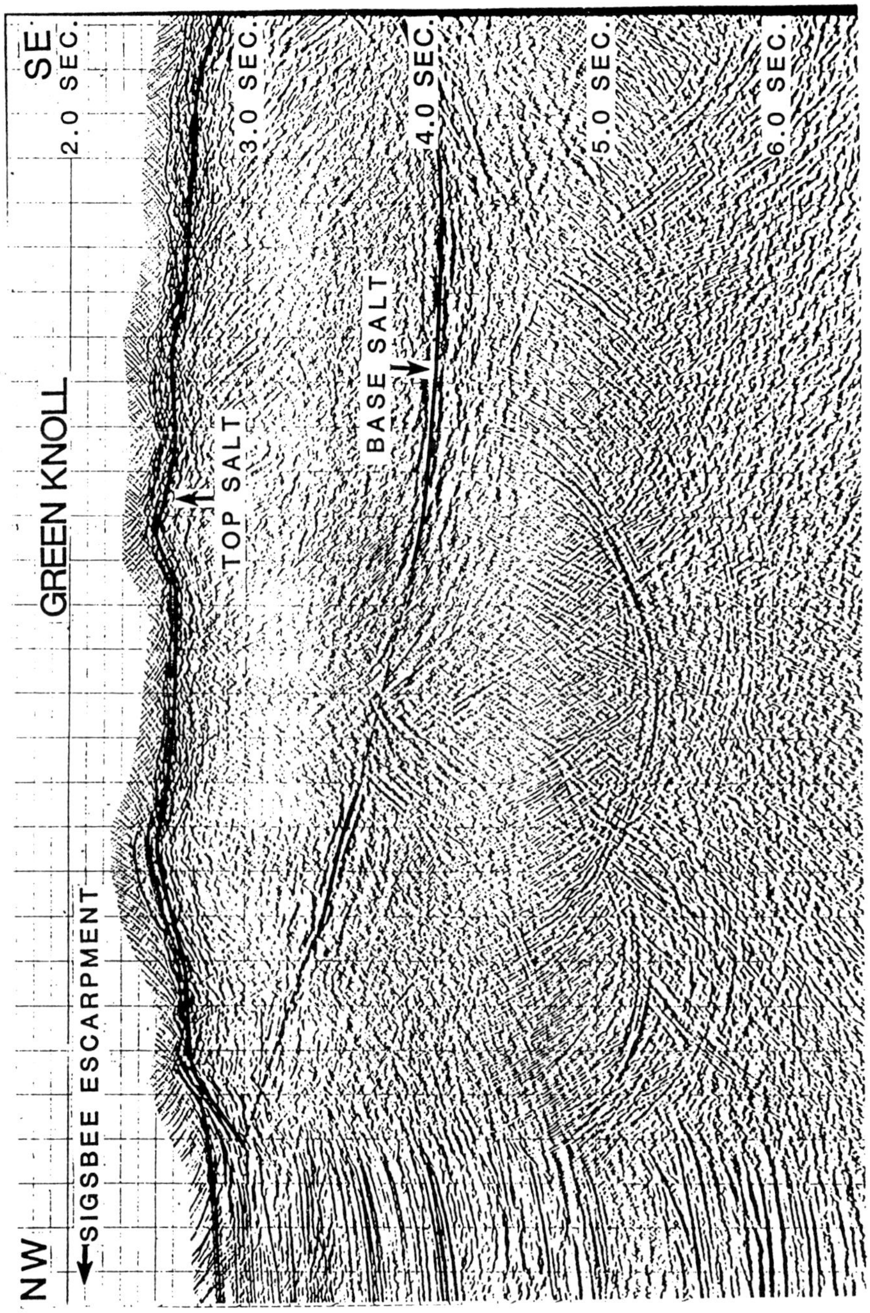

FIGURE 44. Portion of seismic line G-2 illustrating the nature of Green Knoll. (Courtesy of GECO.)

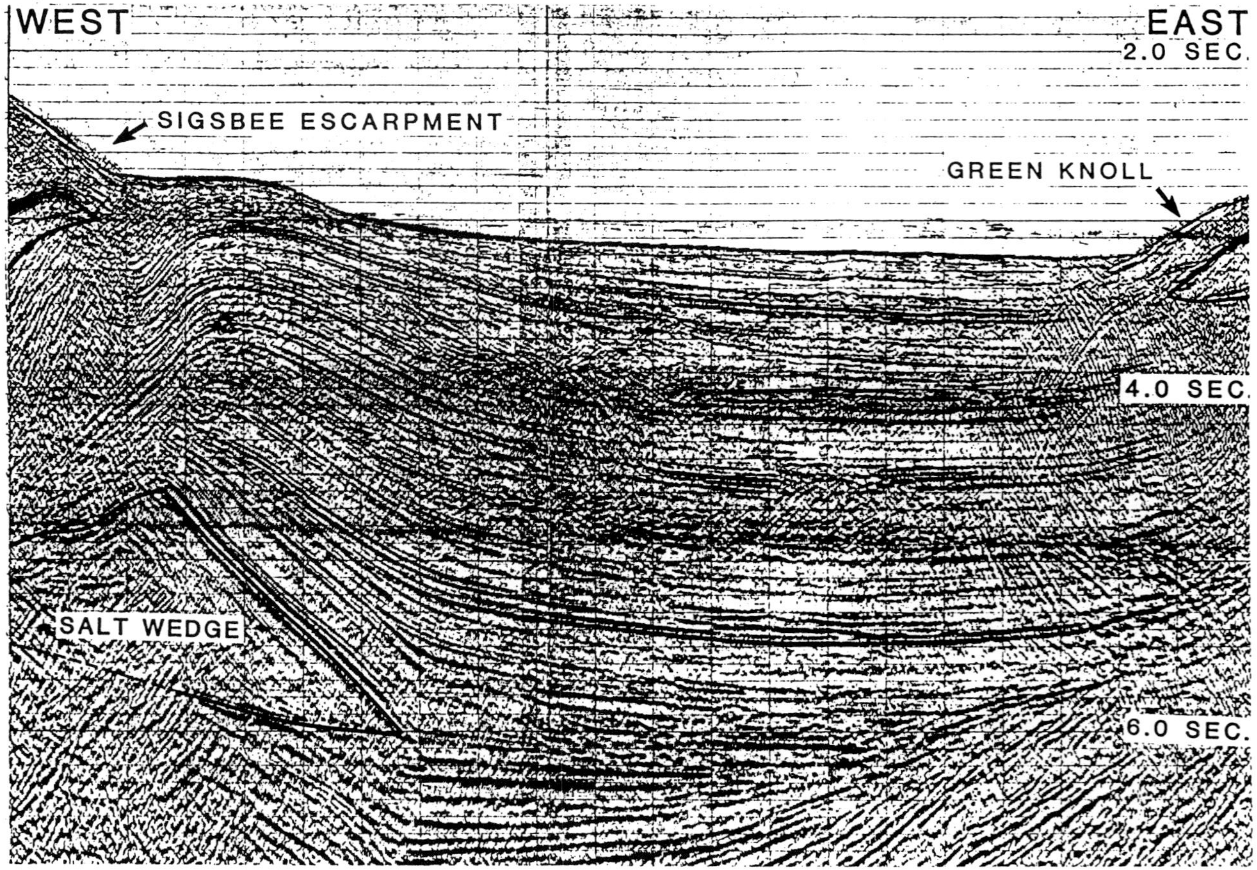

FIGURE 45. Portion of seismic line C-2 illustrating the presence of a deep, detached salt wedge found between the Sigsbee Escarpment and Green Knoll, and also the presence of highly deformed reflectors found at depth between the detached wedge and Green Knoll. (Courtesy of Shell Offshore Inc.)

Just to the northwest of Green Knoll the orientation of the Sigsbee Escarpment changes (Figure 27). North of Green Knoll the scarp trends in an east-northeast direction. However, west of the knoll the scarp trends in a south-southwest direction. Therefore, it seems plausible that salt could have been fed in an easterly direction (at least locally) to feed Green Knoll.

Reflectors above and directly adjacent to both Green Knoll and the detached wedge suggest that they have experienced significant vertical movement since their original emplacement.

D. CONCLUSIONS

Main conclusions concerning the evolutionary history of Green Knoll are:

1. The southern extent of Louanne salt deposition is believed to be tens of kilometers landward from the present-day Sigsbee Escarpment; therefore, it is unlikely that Green Knoll represents a vertically extruded salt mass. Furthermore, Green Knoll does not have a rim syncline associated with it as do most vertically emplaced Gulf salt domes.
2. The asymetrical nature of Green Knoll suggests that it may have been fed from the west or northwest.
3. The refectors found along the northwestern northern, eastern, and southern margins of Green Knoll ar relatively undisturbed.
4. The presence of a detached salt wedge and disturbed reflectors west of Green Knoll (between the knoll and the Sigsbee Scarp) suggests that salt was fed from west to east.
5. The south-southwest orientation of the escarpment west of Green Knoll supports the eastern migrating salt model proposed in forming Green Knoll.
6. Green Knoll and the detached wedge were probably connected as part of an allochthonous salt lobe or tongue which was subsequently segmented by localized sedimentation.
7. Both Green Knoll and the detached salt wedge have experienced significant vertical movement since their emplacement during mid-Tertiary(?) time.

VI. STRUCTURAL EVOLUTION

A. ANALYSIS OF STRUCTURE AND ISOPACH MAPS

Analysis of structure and isopach maps indicates that the region has undergone a complex geologic history as a result of differential loading, variations in direction and rates of sedimentation, complex faulting and basinal development, and a combination of lateral and vertical salt movement.

Four isopach maps, three structure maps, and a bathymetry map are analyzed in ascending chronostratigraphic order. The isopach maps represent seismic Intervals I—IV, with I being the oldest (see Section III, Table 2 for ages). There appears to be one or perhaps even two unconformities below the base of Interval I, but due to correlation difficulties and limited regional extent, they could not be mapped in detail. The structure maps represent horizons b through e, with horizon e being the sea floor. A structure map for horizon a was not useful due to its limited regional extent.

B. INTERVAL I ISOPACH

The Interval I isopach (Figure 46) is restricted to the large salt-free basin found in the western portion of the study area. This interval is most likely present in the smaller deep basins to the east, but jump correlations over such a distance would be very speculative. The upper portion of Interval I is present above salt in some of the shallower basins.

Interval I attains a maximum thickness of 4500 ft in the northern tip of the map, and thins gradually to the south. There is also a small sediment pod (2500 ft thick) to the west.

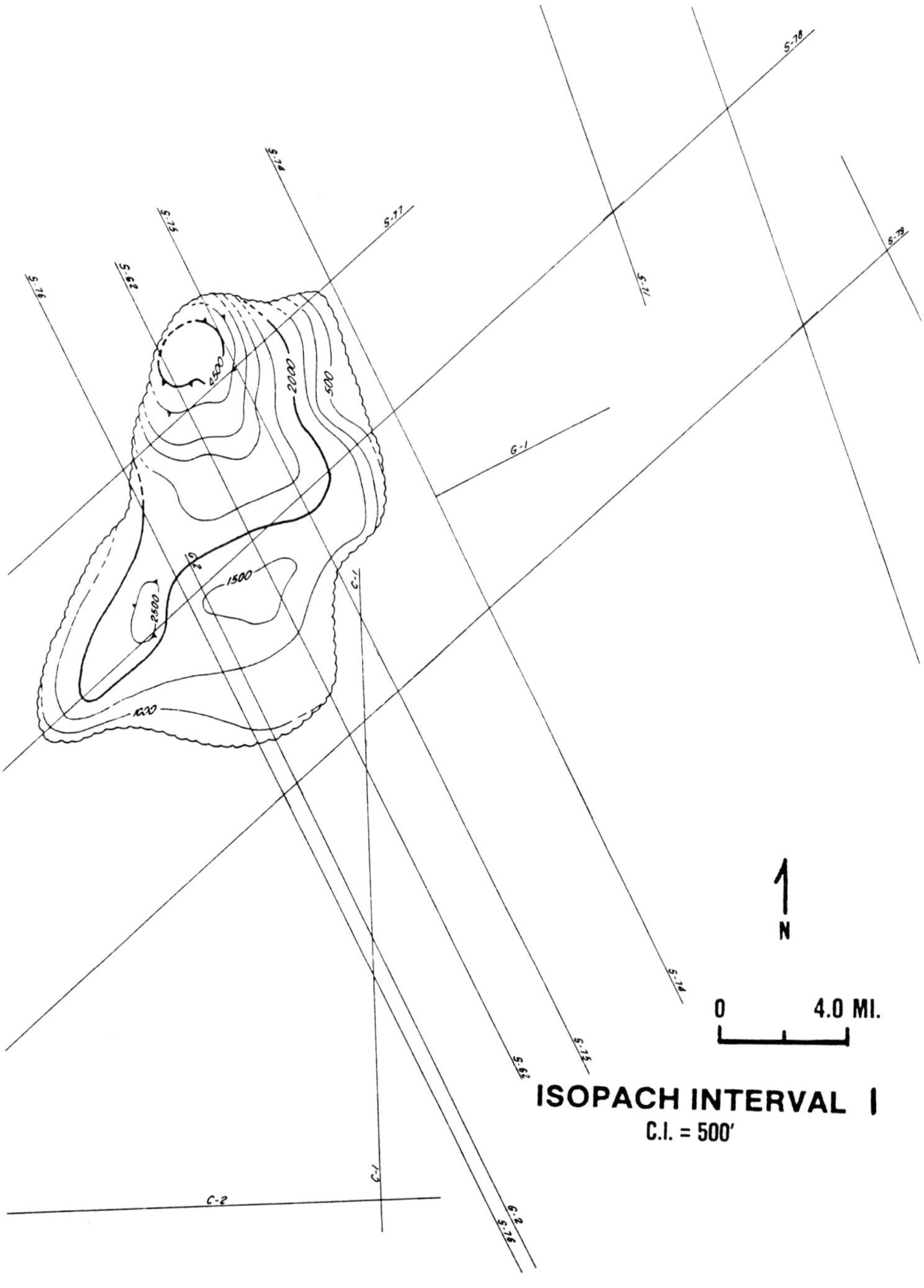

FIGURE 46. Interval I isopach map.

This thickness distribution, along with the downlap/onlap pattern (see Section III), suggests that sediments entered the basin predominantly from the north-northwest during Interval I (Yarmouthian) time.

C. HORIZON 'b' STRUCTURE MAP

The structure map for horizon 'b' (Figure 47) represents the top of Interval I (Yarmouthian-Illinoian boundary?). The map is divided into two regions, east and west, by a

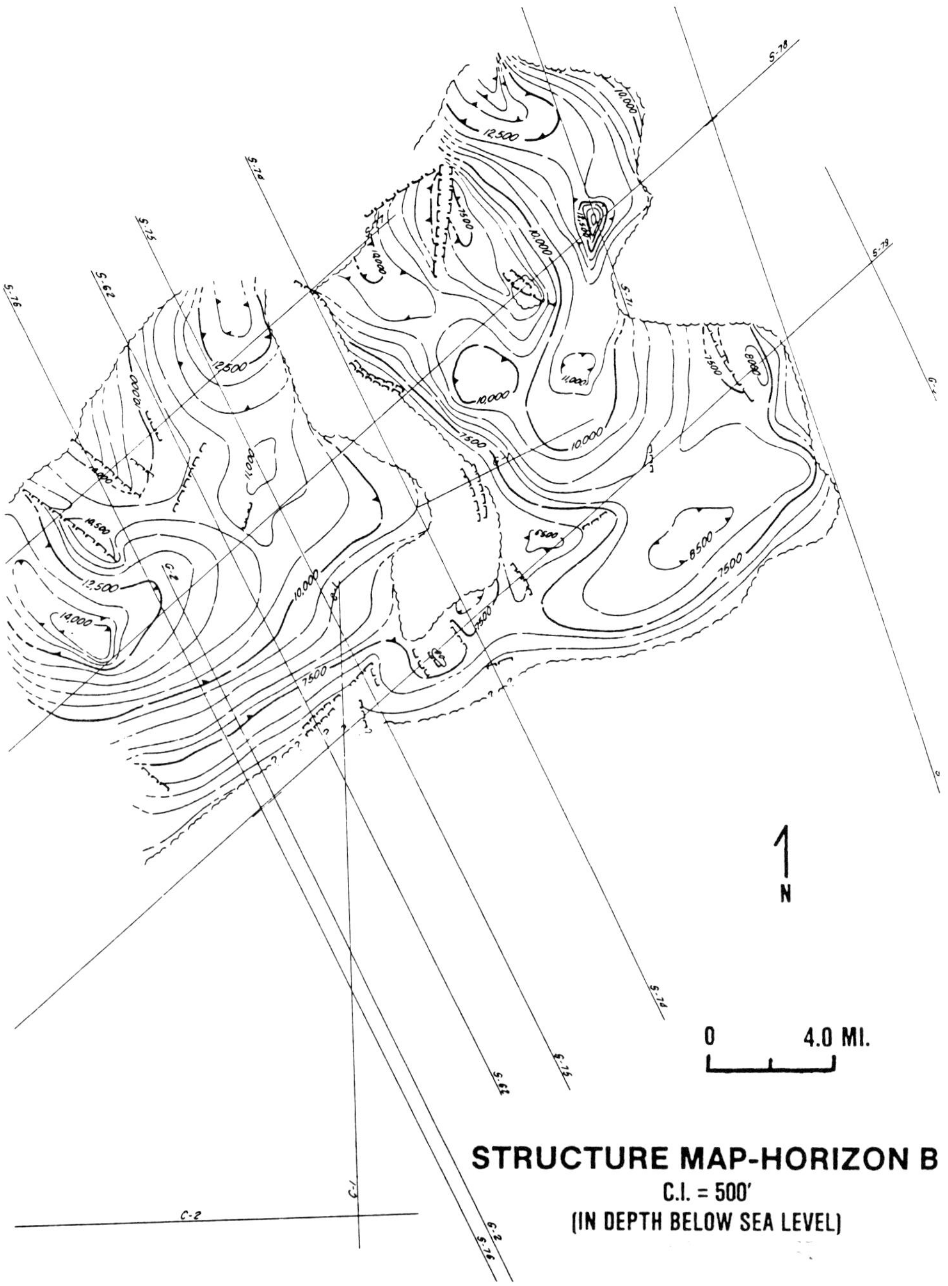

FIGURE 47. Horizon 'b' structure map.

northwest-southeast trending zone of non-deposition or erosion which is associated with an underlying salt ridge.

The western portion of the map is characterized by two deep basins, one to the north and one to the west. The northern basin attains a depth of 13,000 ft below sea level while the western basin reaches 14,000 ft in depth. There is erosion of horizon b at the southern margin along the toe of the Scarp as the allochthonous salt acts to uplift the section. This is noted in both the western and eastern portions of the map.

The eastern portion of the map is characterized by two deep (14,000 ft), isolated, salt-

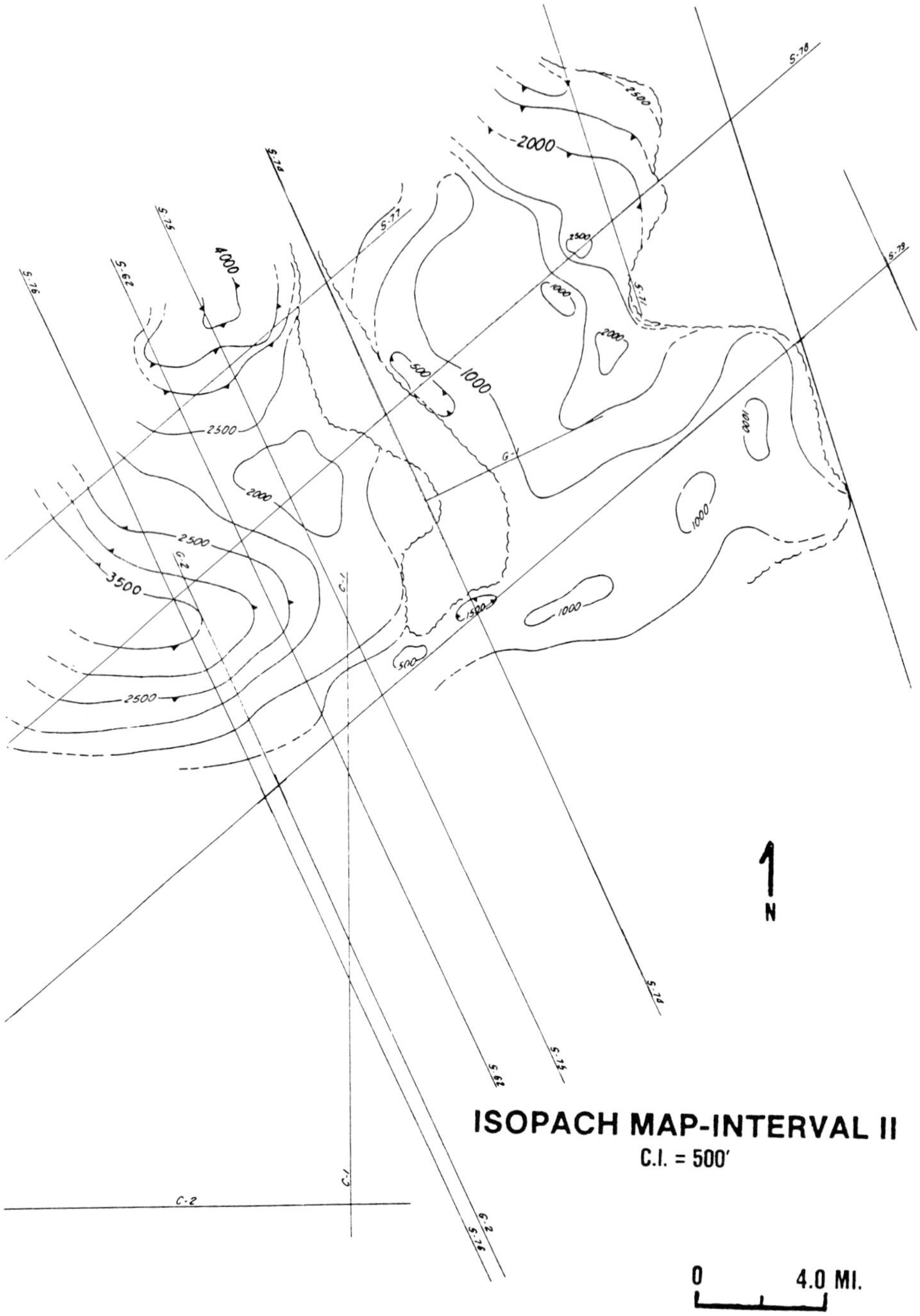

FIGURE 48. Interval II isopach map.

free basins in the northeast corner of the map and a structural high in the north-central region. The structural high is associated with the eastern limb of the inverted ''u'' salt ridge discussed in Section IV. There are two grabens above the ridge as well as a small circular zone of non-deposition or erosion. There are several relatively shallow basins found above the continuous salt mass.

D. INTERVAL II ISOPACH

Similar to the horizon b structure map, the Interval II isopach map (Figure 48) is also divided in half by the northwest-southeast trending zone of erosion or non-deposition. The western portion of the map shows an isopach thick to the north as did the Interval I isopach. However, the zone of maximum thickness had migrated slightly to the west during Interval I–Interval II time. The most prominant difference between the Interval I and II maps is the development of a large isopach thick (3500 ft) in the west-southwest corner of the Interval Map II. This suggests that the maximum sediment influx had shifted from north-northwest, to west-northwest.

In general, the Interval II unit is thinner in the eastern portion of the map. This serves as further evidence that the major sediment source during Interval II (Illinoian?) time was from the northwest. However, in the northeast corner of the map there remains an isopach thick which is associated with the deep basins discussed previously under "horizon b".

E. HORIZON 'c' STRUCTURE MAP

The structure map for horizon 'c' (Figure 49) represents the top of Interval II and marks the boundary between Illinoian-Sangamonian (?) time. The northwest to southeast trending zone of non-deposition or erosion which is prominent on the previous structure and isopach maps is no longer present. This suggests a possible cessation in the diapiric growth of the salt ridge, or more likely, an increased sediment supply (associated with sea level lowering during Illinoian time) which overwhelmed the ridge. The presence of the ridge is still evident as a structural high and with the presence of a series of parallel-trending, down-to-the-southwest, normal faults. To the northeast a second parallel ridge represents the other leg of the inverted-"u" salt thick. Another structural high exists in the northwest portion of the map. This high also is salt controlled.

The deep basin (9500 ft) in the far western part of the map had migrated eastward in comparison to the horizon 'b' map. Likewise, the deep basins in the northeastern corner of the map show a similar eastward migration. This suggests that perhaps the source had begun to switch from northwest to northeast.

F. INTERVAL III ISOPACH

The Interval III isopach (Figure 50) indicates that the sediment source had indeed shifted during Interval III (Sangamonian?) time. The prominant isopach thicks which were found in the western portions of the Interval I and II isopach maps are no longer present on the Interval III map. The maximum thickness found in the western region of the Interval III map is only 2000 ft, and it's location is eastward of the related isopach thicks found in Intervals I and II.

The central portion of the map is characterized by northwest to southeast trending contours which show an alternating pattern of thicks and thins. This suggests that the underlying salt ridges still played a major role in controlling deposition during Interval III time. A good example of the effect of the salt ridges is found parallel to the main salt ridge along seismic line S-74 as the northwest to southeast trending isopach thin (1000 ft contour) is flanked on either side by 2000-ft intervals.

The maximum thickness for Interval III (3500 ft) is found in the upper northeast corner of the map. Again, this suggests that the main sediment source during Interval III (Sangamonian?) time was from the northeast.

G. HORIZON 'd' STRUCTURE MAP

The structure map for horizon 'd' (Figure 51) represents the boundary between Interval III and Inverval IV (Sangamonnian-Wisconsinian?) time. The 'd' horizon map is remarkably similar to the 'c' horizon structure map in that all of the major basins and ridges are present, and in approximately the same location. This suggests that the northwest to northeast switcn in sediment input had occurred prior to horizon 'c' time, probably during the latter stage of Interval II (Illinoian?) time.

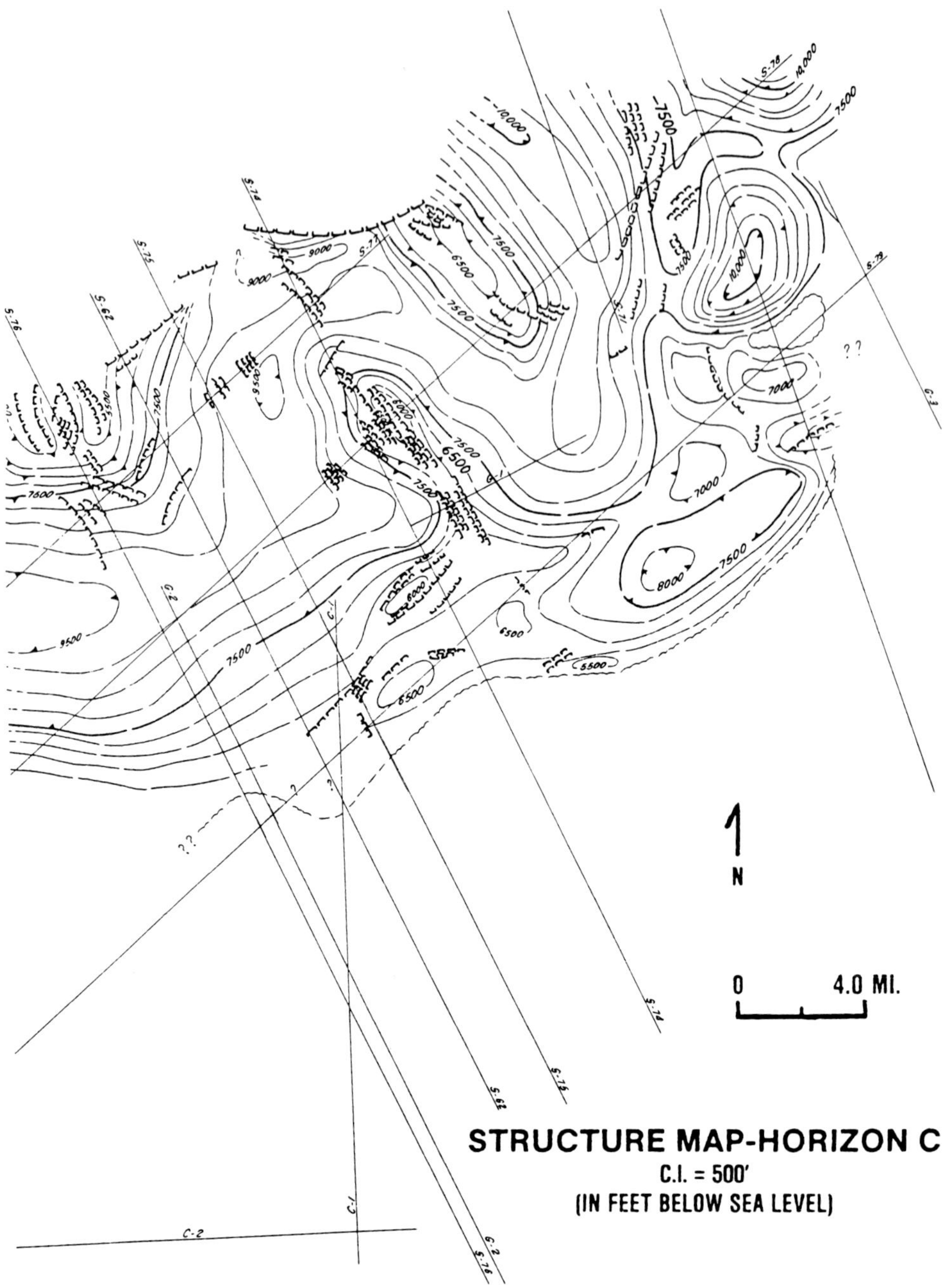

FIGURE 49. Horizon 'c' structure map.

Faulting at the 'd' horizon is more prominent than at any of the previous horizons, 'a—c'. The upper northwest corner of the map shows the development of a graben. The structural high which parallels line S-74 shows a different fault pattern — a series of northwest to southeast trending, down-to-the southwest, normal faults along the crest of the ridge. In the northeast region of the map, the faults generally trend in a northeast to southwest direction.

The central and northeast central region of the map is characterized by elongate basins and ridges which trend in a northwest to southeast direction. The deepest basin is in the

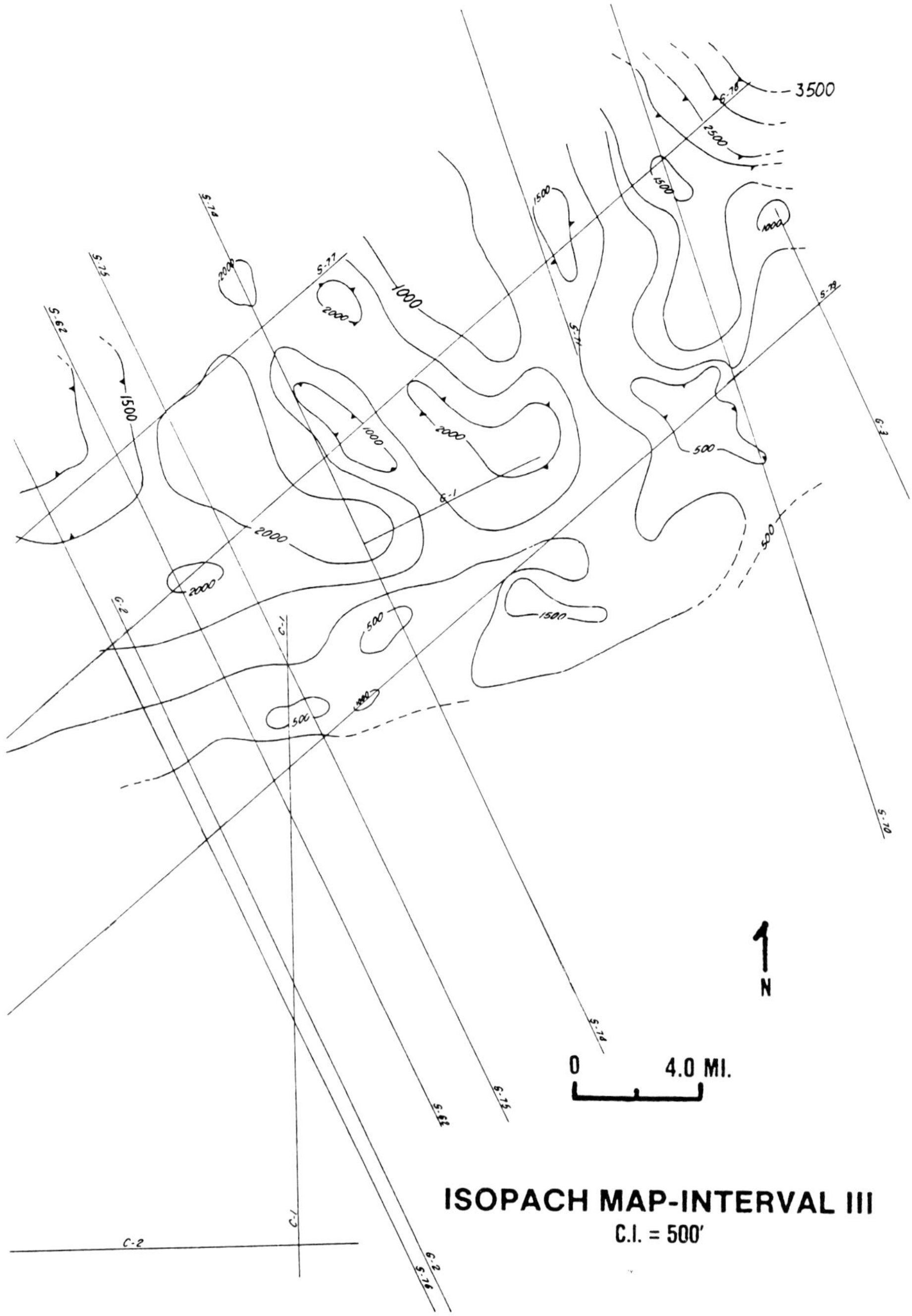

FIGURE 50. Interval III isopach map.

north central portion of the map, attaining a maximum depth of 8500 ft. This basin is flanked on either side by salt-related structural highs, the most prominent being to the northeast.

H. INTERVAL IV ISOPACH

Interval IV (Figure 52) represents the thickest overall unit. The zone of maximum thickness (4500 ft) is in the north central region of the map. The axis of maximum sediment

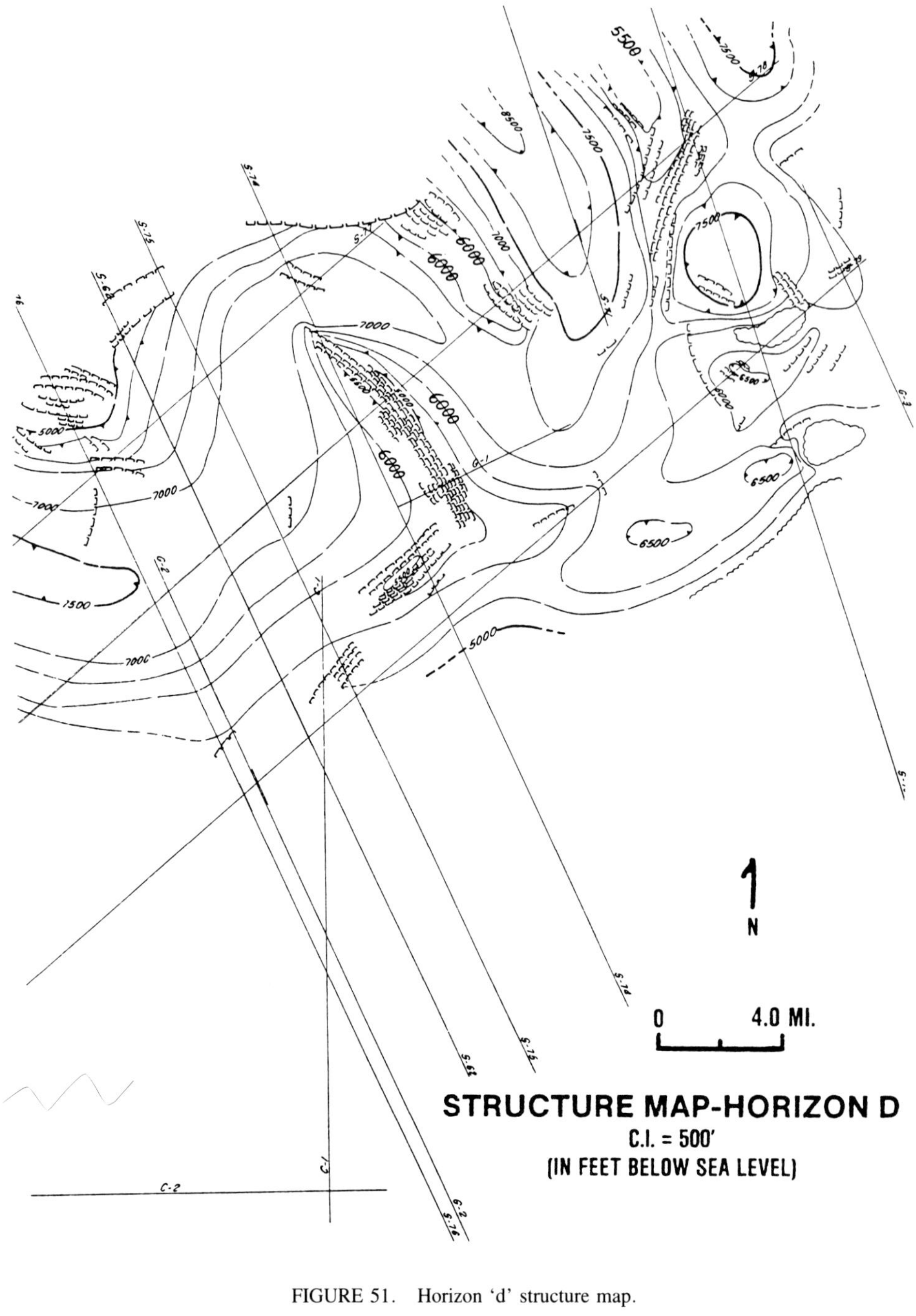

FIGURE 51. Horizon 'd' structure map.

input appears to have shifted slightly from a northeast direction during Interval III (Sangamonian?) time to a more northerly direction during Interval IV (Wisconsinian?) time, although there is still considerable sediment input in the northeast (3500 ft isopach thickness) at this time.

As was the case for Intervals II and III, the central portion of the Interval IV map is characterized by generally northwest to southeast trending isopach thicks and thins. There is drastic thinning across the main salt ridge (which parallels line S-74) as the 1000 ft contour would indicate. This suggests that the salt was active during deposition of Interval IV.

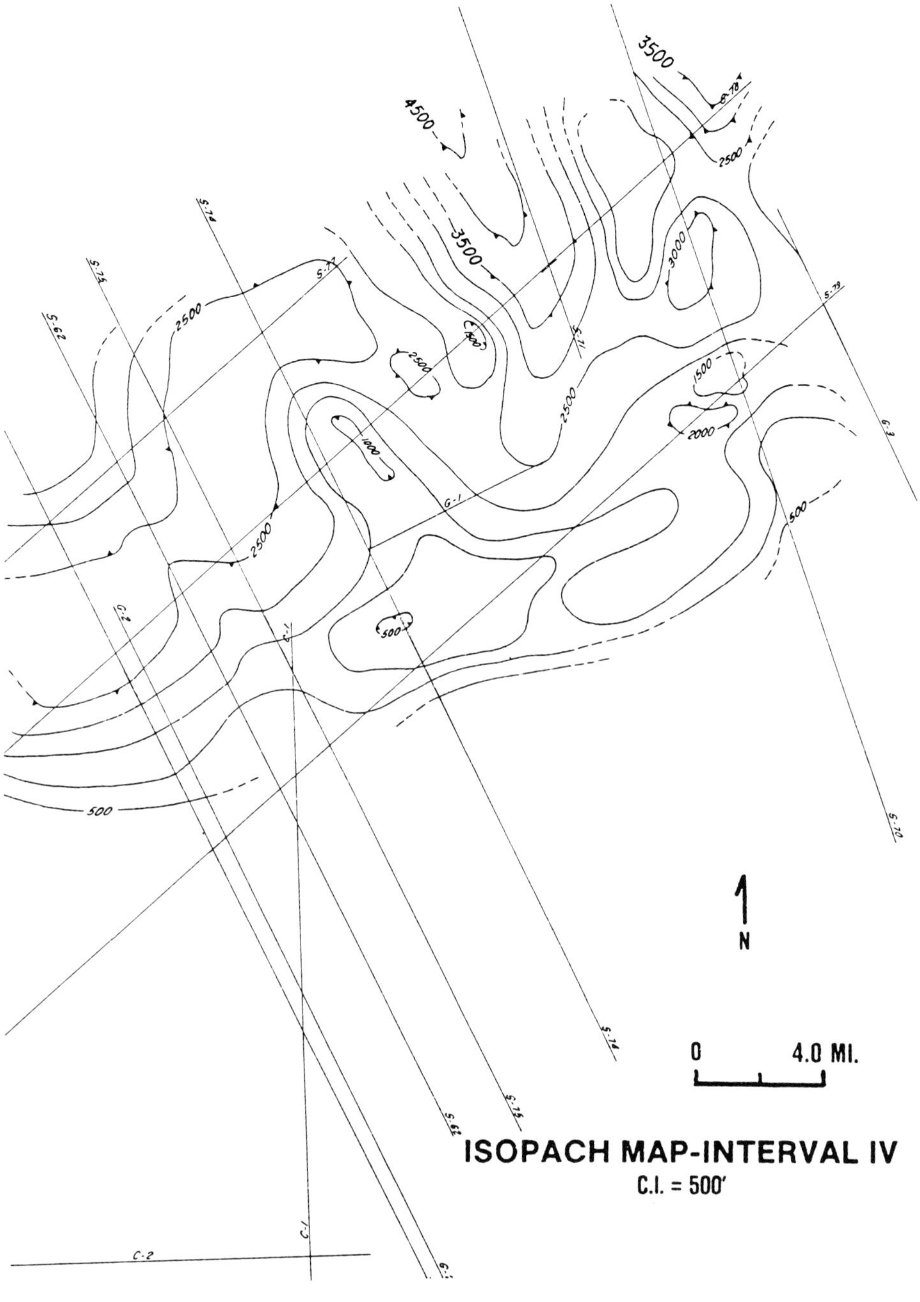

FIGURE 52. Interval IV isopach map.

The western portion of the Interval IV isopach map attains a maximum interval thickness of only 2500 ft. However, the 2500 ft contour represents a rather broad basin whose axis trends in a northeast to southwest direction.

I. HORIZON 'e' (BATHYMETRY) MAP

The bathymetry map (Figure 53) illustrates the water depths landward of the Sigsbee Escarpment. The greatest water depths are found in the south-southwest portion of the map and are approximately 5100 ft. (Of course just seaward of the scarp, water depths are on

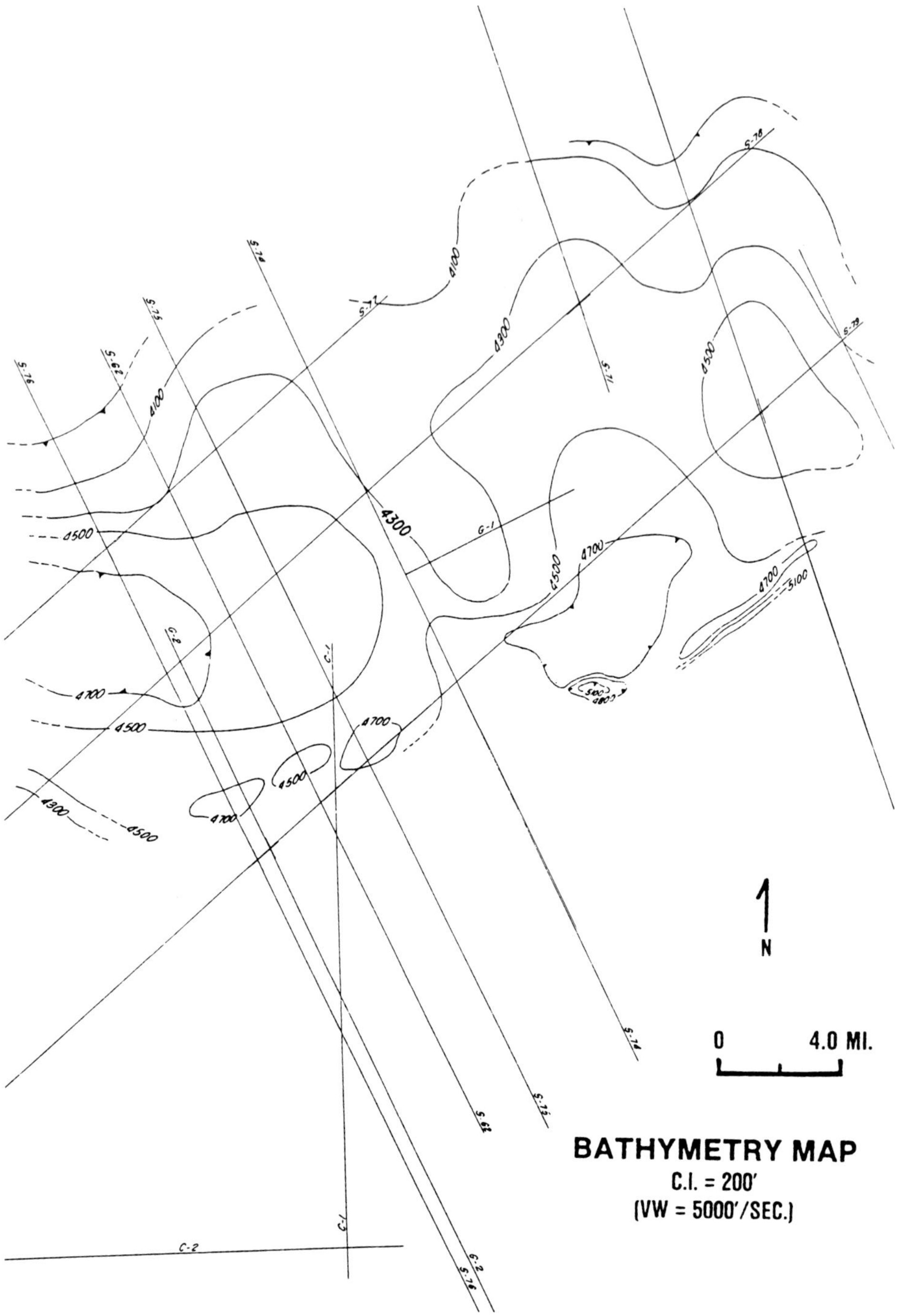

FIGURE 53. Horizon 'e' (bathymetry) map.

the order of 7000 to 7500 ft). In general, water depths decrease from 4900 to 5100 ft near the scarp to approximately 3900 to 4100 ft to the north-northwest regions of the map. One would expect this type of "regional gradient" as one moves from the Lower Slope toward the Upper Slope.

Another interesting feature on the bathymetry map is the northwest to southeast trending,

TABLE 3
General Comparison of Eustatic Sea Level Cycles (Beard et al., 1982) with Fan Sequences (I — VIII) Defined by Feeley et al. (1984), and Depositional Intervals (I — IV) Defined in This Study. (Modified from Feeley et al., 1984)

Feeley et al. (1984)	Beard et al. (1982)	Swiercz (1986)
VIII	Wisconsin-Holocene	
	Q8	
VII	Wisconsin	IV
	Q7	
VI	Wisconsin	
	Q6	
V	Illinoian-Sangamon	III
	Q5	
IV	Illinoian	II
	Q4	
III	Illinoian	
	Q3	
II	Kansan-Yarmouthian	I
	Q2	
I	Nebraskan-Aftonian	
	Q1	

elongate nature of the 4300 ft contour parallel to line S-74. This region is underlain by a salt ridge which indicates that salt does indeed play a role in controlling bathymetry on the Lower Slope. This should not be hard to believe when one considers that the Sigsbee Escarpment (which has approximately 3000 ft of bathymetric relief), is primarily salt controlled.

J. CONCLUSIONS

The central portion of the study area is characterized by northwest to southeast trending contours suggesting that the underlying salt ridges play a major role in controlling deposition in addition to the direction of maximum sedimentation. For example, the complete Interval I unit is found only in the deep western basin. This suggests that the main salt ridge (which parallels line S-74) blocked sediments coming in from the northwest and ponded them in the deep western basin. Not only does this serve to restrict sediment input into the eastern and southeastern regions of the map, but the ponded sediments effectively load the salt and aid in the continued development of the salt ridge. The ridge was obviously present as an active topographic high during Interval II (Illinoian?) time as suggested by the presence of the northwest to southeast trending zone of non-deposition on the Interval II isopach map. The salt ridges found to the northeast of the main salt ridge are somewhat smaller, probably as a result of the limited sediment supply which is needed to facilitate loading and thus mobilize the salt. Thinning across the crests of these salt ridges seen in Intervals III and IV suggests that the ridges were present at least in the form of residual highs at the end of Interval II (Illinoian?) time and then became very active during Interval III (Sangamonian?) and Interval IV (Wisconsinian?) time as the sediment source shifted to the northeast.

It is obvious that the timing of fluctuations in direction and amount of sediment supply played a major role in the geological development of the study area. A comparison of results from this study to work done by Feeley et al., (1984) yields some interesting correlations.

Feeley et al. (1984) mapped eight depositional sequences within the Mississippi Fan. These eight sequences are correlated to the eustatic sea level cycles proposed by Beard et al. (1982) and in turn to Intervals I—IV proposed in this study, as shown in Table 3. Intervals I and II show a general correlation with the Feeley et al. (1984) Sequences II-IV while Intervals III and IV correlate with the Feeley Sequences V-VIII.

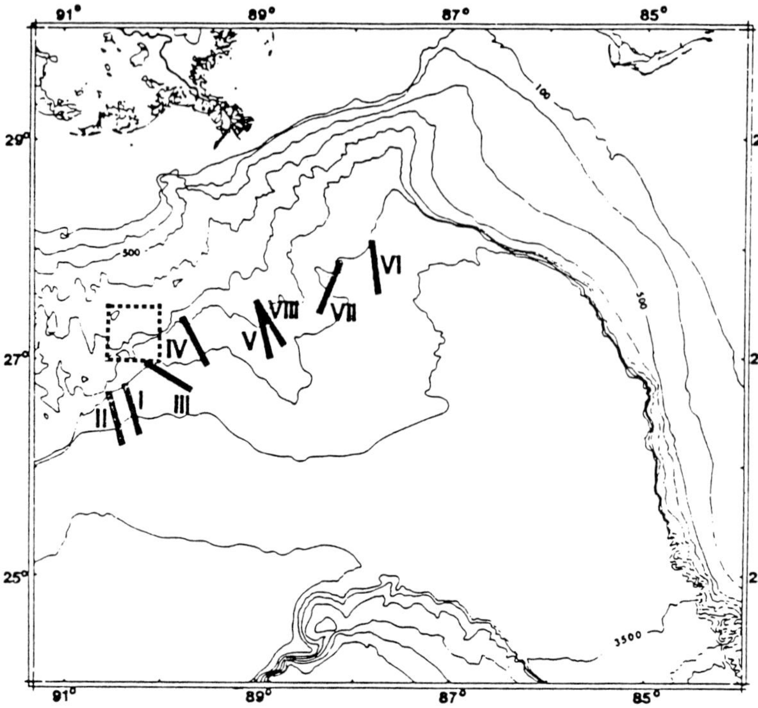

FIGURE 54. Map of the updip position and orientation of the axis of maximum thickness for each of the sequences. Note the gradual shift from west to east of the depocenter through sequence V, then an abrupt shift farther to the east (VI). The fan depocenter then shifted back to the west for sequences VII and VIII. (Modified from Feeley et al., 1984.)

Figure 54 illustrates the position and orientation of the axis of maximum thickness for each of the sequences I—VIII defined by Feeley et al. (1984). Note that for Sequences I and II, the axis is to the far west and is oriented in a NNE direction–the exact conclusion drawn from the isopach map of Interval I in this study. Likewise, Feeley Sequences III and IV also have axes toward the west, and the more westward orientation of especially the Sequence III axis corresponds to the results of the Interval II isopach. The major shift in sediment source from northwest to northeast between Illinoian and Sangamonian time is recorded in both the upper four sequences (V—VIII) defined by Feeley and the upper two intervals (III and IV) defined in this text, including the final slightly westward shift at the end of Wisconsinian time.

VII. SUMMARY DISCUSSION

A. REGIONAL DEPOSITIONAL MODEL

The northern Gulf of Mexico represents a unique geomorphic and geologic province whose continually changing geologic development responds to an interplay of salt tectonics, sediment supply, source direction, and eustatic sea level fluctuations.

The upper slope is characterized predominantly by vertical salt tectonics. These diapirs tend to trap much of the sediments being shed off the shelf within a network of isolated mini-basins. The Lower Slope on the other hand, is composed primarily of lobate salt structures which have been mobilized seaward as a result of a major Pleistocene depocenter along the Texas-Louisiana shelf as described by Paog and Valentine (1976).

The sediments which are supplied to the Green Canyon region along the Lower Slope are most likely to be in the form of turbidites intermixed with pelagic silts and clays, the amount and proportion being directly related to eustatic sea level changes. Periods of sea level high stands are characterized by deposition of fine-grained pelagic material and probably some distal turbidites as well. The rapid lowering of sea level, such as during Upper Illinoian time, resulted in massive erosion and slumping along the shelf thereby dumping huge quantities of material out onto the present-day slope. Undoubtedly, some of this material reached the Lower Slope in the form of slumps, debirs flows, and deep water submarine fan deposits as attested to by the chaotic hummocky nature of Interval II (see Section III). This would suggest that there are reservoir-quality sands within the southern Green Canyon region.

The close correlation between the results of Feeley et al. (1984) and the results of this study (Section VI) relating isopach interval thicknesses to source direction suggests a strong regional control on deposition. Although source direction and sediment supply are controlled on a regional scale, localized effects on deposition are primarily salt controlled. Salt ridges can inhibit sedimentation across the ridge crest, ponding the sediments in the deep basins. The salt ridges can also act as ''funnels'' in which sediments are transported (possibly within channels) out onto the toe of the Scarp.

B. HYDROCARBON POTENTIAL

The sediments within the Gulf of Mexico have a very low (approximately 0.5%) percentage of organic carbon. (In comparison, the North Sea formations are composed of between 10.0 and 22.0% organic carbon.) However, by shear source rock volume alone, the Gulf Coast yields commercial volumes of hydrocarbon and the southern Green Canyon area should be no exception.

The presence of mobile salt within the study area could account for a variety of possible trap types. These include stratigraphic pinchout of sands along flanks of salt, sand bodies juxtaposed directly against salt, domal structures above areas of vertically mobile salt, graben structures above salt, and also growth faults associated with lateral salt withdrawal.

Since there is a large component of lateral salt movement associated with the study area, the withdrawal-related growth faults are probably the most attractive exploration targets. The expanded, down-thrown side of these faults may tend to be more sand rich, as is commonly found updip on the shelf (Larberg, 1986). Specifically, sands associated with the Late Illinoian and Early Wisconsian sea level lowerings would probably represent the best exploration opportunities in terms of reservoir quality sands.

There are two additional possibilities that would be extremely venturesome from an exploration standpoint. First, there is a possibility that large volumes of hydrocarbon may exist and be trapped beneath the lobate salt bodies. However, the thickness of the overlying salt, and the inability to seismically image beneath the salt would be extremely difficult technical obstacles to overcome. A second possibility exists in the large compressional domes found seaward of the scarp (see Section IV). The water depths in excess of 7500 ft found in this region will soon be within production capabilities.

REFERENCES

Amery, G. B., Structure of Sigsbee Scarp, Gulf of Mexico, *Am. Assoc. Pet. Geol. Bull.*, 53, 2480-2482, 1969.

Atwater, G. I. and Forman, M. J. Nature of growth of southern Louisiana salt domes and its effect on petroleum accumulation, *Am. Assoc. Pet. Geol. Bull.*, 43, 2592-2622, 1959.

Beard, J. H., Angree, J. B., and Smith, L. A., Quaternary chronology, paleoclimate, depositional sequences, and eustatic cycles, *Am. Assoc. Pet. Geol. Bull.*, 66, 158-169, 1982.

Buffler, R. T., Worzel, J. L., and Watkins, J. S., Deformation and origin of the Sigsbee Scarp — lower continental slope, northern Gulf of Mexico, in *Proc. 1978 Offshore Technology Conference,* 3, 1425-1433, 1978.

Buffler, R. T., Early history and structure of the deep Gulf of Mexico basin, in Characteristics of Gulf Basin deep-water sediments and their exploration potential, 5th Annu. Gulf Coast SEPM Res. Conf., Proc., 31-34, 1984.

Dobrin, M. B., *Introduction to Geophysical Prospecting,* McGraw-Hill, New York, 1976.

Dupre, W. R., An introduction to the Cenozoic geologic history of the Texas Gulf Coast, prepared for Houston Geological Survey, Department of Geology, University of Houston, Houston, TX, 1985.

Feeley, M. H., Buffler, R. T., and Bryant, W. R., Seismic stratigraphy analysis of the Mississippi Fan, Tech. Report 84-2-T, Texas A&M University, College Station, 1984.

Garrison, L. E., and Martin, R. G., Geologic structures in the Gulf of Mexico basin, U.S. Geological Survey Prof. Paper 773, 1973.

Humphris, C. C., Jr., Salt movement on continental slope, northern Gulf of Mexico, in Framework facies and oil trapping characteristics of the upper continental margin, Bouma, A. H., Moore, G. T., and Colemen, J. M., Eds., *Am. Assoc. Pet. Geol. Studies in Geology,* 7, 69-85, 1978.

Kirk, R. B., A seismic stratigraphic case history in the Eastern Barrow Subbasin, Northwest Shelf, Australia, in Seismic stratigraphy II — an integrated approach, Berg, O. R., and Woolverton, D. G., Eds., *Am. Assoc. Pet. Geol. Memoir 39,* 183-207, 1985.

Larberg, G. M., Contra-regional faulting: salt withdrawal compensation; offshore Louisiana, Gulf of Mexico, in Seismic expression of structural styles, Bally, A. W., Ed., *Am. Assoc. Pet. Geol. Studies in Geol.,* 15, 2, 42-44, 1983.

Larberg, G. M., Personal communication, September 15, 1986, Shell Offshore Inc., New Orleans, LA.

Lehner, P., Salt tectonics and Pleistocene stratigraphy on continental slope of northern Gulf of Mexico, *Am. Assoc. Pet. Geol. Bull.,* 53, 2431-2479, 1969.

Martin, R. G., Distribution of salt structures in the Gulf of Mexico; map and descriptive text, U.S. Geological Survey Map MF-1213, 1980.

Martin, R. G., Diapiric trends in the deep-water Gulf basin, in Characteristics of Gulf Basin deep-water sediments and their exploration potential, 5th Annu. Gulf Coast SEPM Research Conference, Proc. 60-62, 1984.

Mitchum, R. M., Jr., Vail, P. R., and Sangree, J. B., Stratigraphic interpretation of seismic reflection patterns in depositional sequences, *Am. Assoc. Pet. Geol. Memoir 26,* 117-134, 1977.

Peterson, J. J., Personal communication, September 10, 1986, Investment Broker, Shearson Lehman Bros. Inc., Springfield, MA.

Poag, C. W., and Valentine, P. C., Biostratigraphy and ecostratigraphy of the Pleistocene basin, Texas-Louisiana continental shelf, *Gulf Coast Assn. Geol. Soc. Trans.,* 26, 185-256, 1976.

Sangree, J. B., and Widmier, J. M., Seismic interpretation of clastic depositional facies, *Am. Assoc. Pet. Geol. Memoir 26,* 165-184, 1977.

Shih, T., Worzel, J. L., and Watkins, J. S., Northeastern extension of Sigsbee Scarp, Gulf of Mexico, *Am. Assoc. Pet. Geo. Bull.,* 61, pp. 1962-1978, 1977.

Stuart, C. J., and Caghey, C. A., Seismic facies and sedimentology of terrigenous Pleistocene deposits in northwest and central Gulf of Mexico, in Seismic stratigraphy — applications to hydrocarbon exploration, Payton, C. E., Ed., *Am. Assoc. Pet. Geol. Memoir 26,* 249-276, 1977.

Vail, P. R., et al., Seismic stratigraphy and global changes of sea level, in Seismic stratigraphy — applications to hydrocarbon exploration, Payton, C. E., Ed., *Am. Assoc. Pet. Geol. Memoir 26,* 49-212, 1977.

Waters, K. H., *Reflection Seismology: A Tool for Energy Resource Exploration,* John Wiley & Sons, 1981.

Watkins, J. S., et al., Occurrence and evolution of salt in deep Gulf of Mexico, in Framework, facies, and oil trapping characteristics of upper continental margin, Bouma, A. H., et al., Eds., *Am. Assoc. Pet. Geol. Studies in Geology,* No. 7, 1978.

Watkins, J. S., Personal communication, May 4, 1986, Texas A&M University, College Station, TX.

Worzel, J. L., et al., Initial Reports of the Deep Sea Drilling Project, 10, U.S. Government Printing Office, Washington, D.C., 1973, 334 p.

Yorston, H. J. and Fox, J. F., Practical applications of subsurface and seismic methods to prospecting in salt provinces with emphasis on the Gulf Coast tertiary, An exploration workshop, Houston Geological Society, Continuing Education Series, October 23, 1985, Houston, TX.

Chapter 11

GEOACOUSTIC AND GEOLOGICAL CHARACTERIZATION OF SURFICIAL MARINE SEDIMENTS BY *IN SITU* PROBE AND REMOTE SENSING TECHNIQUES

Richard H. Bennett, Huon Li, Michael D. Richardson, Peter Fleischer, Douglas N. Lambert, Donald J. Walter, Kevin B. Briggs, Charles R. Rein, William B. Sawyer, Frank S. Carnaggio, David C. Young, and Samuel G. Tooma

TABLE OF CONTENTS

I. INTRODUCTION

Surficial marine deposits in shallow coastal areas and in deep ocean basins are investigated in detail by *in situ* probe and remote sensing techniques to characterize the geoacoustic and geological properties of surficial sediments. Quantitative static and dynamic data of surficial sediments, three-dimensional structure, morphology, and time-dependent sea floor processes are collected to provide environmental information for basic geological and geophysical research investigations and for naval applications. Geological properties and processes such as bottom roughness, sediment type, sediment transport, physical, mechanical, and acoustic properties, and dynamic wave-current stress on the sea floor sediments are measured, studied and modeled in the field and in the laboratory.

Geosciences research at the Naval Oceanographic and Atmospheric Research Laboratory (NOARL) supports specific naval requirements in mine and Arctic warfare, acoustic bottom and subbottom interaction, coastal engineering, mapping, geoacoustic/geophysical modeling, and antisubmarine warfare. The resulting instrumentation and technology have been applied to other basic research problems in deep-ocean waste disposal, sea floor stability, and sediment transport processes.

Interdisciplinary research is an integral aspect of the surficial sediment investigations carried out by NOARL. Programs are designed to maximize the return on investments in ocean science while making optimum scientific contributions that benefit the U.S. Navy. Studies include site-specific investigations at sea, as well as extensive laboratory studies coupled with theory and modeling.

From an interdisciplinary perspective, the geology input provides the descriptive analysis and background of an investigation (regional or site specific). Geophysical measurements and remote sensing techniques provide qualitative and quantitative data and assessment of the sea floor and subsea-floor structure (stratigraphy) and material properties of the sediment deposits both laterally and vertically. *In situ* geotechnical probes and geoacoustic probes provide the physical, mechanical, and geoacoustic properties "ground truth" data. These data are used with sophisticated geophysical systems data to provide realistic input parameters for geoacoustic, geological, and geotechnical modeling of the dynamic and static behavior of sedimentary deposits at both large and small strains.

The instrumentation was developed for studies of surficial marine sediments. *In situ* probes measure ambient and dynamic pore-water pressure (piezometer), electrical resistivity-conductivity (diver operated and vibrocorer types), hydraulic conductivity (permeameter), and acoustic velocity (shear and compressional waves). Remote sensing measurements include sediment type, grain size, porosity, and bulk density classification by acoustic impedance (acoustic profiler), as well as sea floor morphology and sediment types (sidescan sonar). Microtopography and sea floor roughness are mapped by stereo photogrammetry techniques.

II. DISCUSSION

A. INVESTIGATIONS OF MARINE SEDIMENTS: BASIC AND APPLIED

Understanding the variability of sea floor processes and sedimentary properties in time and space is the focus of research in sea floor geosciences. The studies lead to the formulation of geological and geoacoustic models of the sea floor for the purpose of extending our knowledge of sea floor environments and processes throughout the world's deep-ocean basins, coastal areas, and marginal seas. The studies relate directly to applied problems of interest to the Navy in areas of mine burial, high-frequency acoustics, geoacoustic modeling, coastal engineering, and sea floor mapping.

Basic research investigations provide the Navy with fundamental sea floor environmental data for predictive modeling of remote areas, estimates of bottom and subbottom roughness, scattering, attenuation, estimates of acoustic impedance and reflection coefficients, and other derived geoacoustic parameters. Studies of fine-scale ($<$50 cm) geoacoustic properties and boundary layer acoustics are critical in assessing the subsurface variability, volume scattering and attenuation, and sediment acoustic velocities in shallow-water and deep-ocean basin environments. Field and laboratory testing of marine seiments provides important data for engineering applications and a variety of sea floor environmental requirements. The identification of sediment (soil) types and the determination of their physical and mechanical properties are continuing activities because of the pressing requirements for rapid survey systems to meet Navy requirements. The data obtained from the various *in situ* and laboratory studies are fundamental for evaluating both science and engineering problems in sea floor stability, bearing capacity, mine burial estimates, and the response and behavior of sonar systems.

1. Sediment-Structure Interaction

Sea-floor mapping of bathymetry and morphology, the determination of sediment types and their physical and mechanical properties, bottom roughness, and subbottom layering (stratigraphy) are directly related to the Navy's interest in objects on the sea floor (sediment-structure interaction). Mine burial, cable routing, and permanent or temporary placement of a variety of structures on the sea floor are all important problems that can be solved only by sea floor environmental data and bathymetry studies.

2. Arctic Research

Arctic sea floor environments and subsurface ice characterization are strategically important to the Navy in terms of maneuverability, safety of vessels, and placement of bottom-mounted structures. The NOARL instrumentation adds a new perspective to Arctic environmental research capabilities to address problems of practical consideration, as well as fundamental science issues involving cold-region environments. As a deeper concern develops for understanding world climatic changes, the existing technology will be adapted to unraveling the complex details of the geological and geophysical history of Arctic sea floor environments.

3. Acoustic Bottom Interaction

The development of state-of-the-art instrumentation to measure the acoustic behavior of marine sediments and related properties has been an integral part of the basic research thrust. Theoretical studies and geoacoustic models that predict acoustic wave behavior in the ocean and through the sea floor sediments abound in the literature. These models, however, take into account only a "theoretical" or "simple" environment and often are inadequate for predictive applications. The instrumentation described here is providing the geoacoustic community with real-Earth physical parameters and environmental data for predicting acoustic propagation in marine sediments. The variability of marine sediment properties is the rule rather than the exception, and realistic data and complex geoacoustic models are required to handle the highly variable sea floor materials and environments.

4. High-Frequency Acoustic Bottom Scattering

The scattering of high-frequency acoustic energy at or near the sediment-water interface is controlled by the roughness of the boundary layer (often affected significantly by the benthic community), the impedance mismatch at the sediment-water interface, and volume heterogeneities within the sediment. NOARL often participates in joint acoustic-environmental experiments to understand and model the interaction of acoustic energy with the sea floor. The sea floor is characterized and classified by *in situ* geoacoustic and bottom roughness measurements. The environmental and acoustic measurements are used to validate existing theory and to develop new bottom-scattering and propagation models. Importance is placed on the understanding of the physical, biological, geological, and geochemical processes that cause the high variability in geotechnical, geoacoustic, and roughness, which in turn greatly controls scattering, penetration, and propagation of acoustic energy in sea floor and subbottom sediment.

5. Coastal and Deep-Ocean Engineering

Rapid, more accurate sea floor and subbottom survey techniques are emerging with the development of the NOARL instrumentation. Coastal processes and sea floor properties are important for successful amphibious landings, constructing offshore structures, and placing structures on the bottom. The dynamic environment of the coastal areas are presently being studied by *in situ* and remote sensing techniques to characterize the acoustic, physical, and mechanical properties of marine sediments.

Deep-ocean systems have been developed to operate in extremely high-stress environments for periods of several months. NOARL, in conjunction with Sandia National Laboratories, the Applied Physics Laboratory, the University of Washington, the Woods Hole Oceanographic Institution, and the University of Rhode Island, has developed deep-water (5000 m), high-technology systems to assess the properties of surficial marine sediments. (Percival et al., 1989) Some of this technology is discussed later (see section on piezometer systems).

6. Sea Floor Mapping

An accurate ''picture'' of the sea floor is essential to naval operations and is also a basis for quality geological and geophysical research. Detailed sea floor mapping has been accomplished using sophisticated sonar techniques and is now an integral part of the on-going activities. New systems are also emerging in cooperation with academic institutions that will provide a capability in quantitative sonar imaging of the sea floor. The new systems give the Navy and the scientific community new dimensions in detailed sea floor mapping and acoustic signal processing.

7. Sediment Transport

An ongoing sediment transport project on continental shelves is directed toward the development of reliable models that predict sediment transport processes, including scour and mass transport, as a function of first- and second-order wave–current interactions and wave–sea floor/subbottom interactions and energy transfer. Specific goals are directed toward the understanding of the role of dynamic pore-water pressures in sediment transport and sea floor stability. This NOARL project is part of a larger program managed by the Office of Naval Research. The program, Sediment Transport Events on Shelves and Slopes (STRESS), is focused on understanding the combined effects of waves and currents on sediment transport on continental shelves and slopes. Instrumentation described here is being used in this important basic research program.

III. *IN SITU* PROBES AND MEASUREMENT TECHNIQUES

A. PIEZOMETERS

1. Technical Background

Shallow-water and deep-ocean piezometer probes provide *in situ* data on pore-water pressures and related geotechnical properties of submarine sediments. The shallow-water piezometer investigations were made to determine the feasibility of making long-term pore-pressure measurements of the ambient and dynamic pore pressures and to assess the vertical effective stress (state of stress) and stability of deposits at selected coastal sites. The significance of sediment pore pressures — specifically, excess pressures — has been recognized for decades by soils engineers as a critical property that affects the stability of soil deposits because excess pore pressure reduces the effective stress and strength of a deposit. The effect of surface wave activity during storms is a major factor in triggering slope failures and mass movements in shallow-water submarine sediments of the Mississippi Delta (Bea and Arnold, 1973). Despite the recognized importance of both the role of ambient and dynamic excess pore pressures and the effect of surface wave activity on the ultimate stability of Mississippi Delta sediments, essentially no field data were available to evaluate these aspects before 1975 (Bennett, 1975 and 1982).

The deep-ocean piezometer probe was developed to investigate *in situ* pore-water pressures at nominal ocean depths of 6000 m and at subbottom depths of 1.0 m in support of the Sandia National Laboratories Subseabed Disposal Program. Development of the deep-ocean piezometer probe was an integral part of Sandia National Laboratories' Subseabed Disposal Program (SDP) and their *In Situ* Heat Transfer Experiment (ISHTE). The feasibility of emplacing high-level nuclear waste in fine-grained deep-sea formations of the world's ocean basins was the thrust of the SDP study. ISHTE was designed to provide data for comparing the *in situ* thermal response of marine sediment with numerical predictions. The analyses were used to evaluate the feasibility of subseabed disposal.

The piezometer probe technology can be extended to make pore-pressure measurements for addressing numerous relevant engineering and geological questions and problems. Examples would include problems relating to offshore engineering construction (pipelines, platforms, and settlement), and geological processes (sea floor stability, mass movement, liquefaction potential related to transport and consolidation/compaction [lithification].).

2. Shallow-Water Piezometer Probes

The shallow-water multisensor piezometer probe consists of 0.10-m-diameter seamless steel pipe segments that are 3.05 m in length. The segments are connected with O-ring sealed couplings (Figure 1). Variable-reluctance-type pressure transducers are enclosed in capsules within the probe shell close to a porous filter. This design permits selection of the sensor and porous filter position along the probe length. Filters are coarse corundum and high air entry stones having approximate porosity n and permeability k values of n = 45 to 50%, k = 0.01 to 0.03 cm/sec and n = 35 to 38%, k = 1 to 3 $\times$ 10^{-7} cm/sec, respectively. The reason for the use of different filters was discussed by Bennett and Faris (1979). The probe tip (~5° cone angle) was designed to control sediment disturbance and to produce plane-strain soil deformation during probe insertion. Using this tip design (~5° cone) and an *L/D* (L = length from tip to filter, D = diameter of probe) ratio of ⩾20 from tip to filter, the induced pore pressures (U_1) generated during probe insertion can be related directly to the undrained shear strength of the soil. Also, additional geotechnical parameters can be derived from the pore-pressure data with confidence (Bennett et al., 1985).

The major components of the piezometer system include several subsystems: probe with pressure-sensing transducers, underwater junction box for cables, signal-conditioning electronics, data recorder(s), and power package (Figure 2).

Pressure sensors are of the variable-reluctance-type transducers. Two absolute pressure transducers are used to measure the "hydrostatic" or free water column pressure inside the probe at selected depths. Absolute pressure sensors also are used to measure pore pressure at selected intervals, and excess pore pressure is measured directly with differential pressure sensors, which are placed at selected intervals along the probe length. Excess pore pressure can be derived from the absolute difference between the pore-water measurements and the free water column measurements.

3. Deep-Ocean Piezometer Probes

The piezometer probe design specifications and the types of material required were established on the basis of ISHTE experimental objectives and expected environmental conditions. Limitations were placed on the probe size (that portion of the probe inserted in the sediment) to minimize perturbations in the thermal field induced by the heat source. The required precision of the pressure sensors was 0.34 to 0.69 kPa to achieve experimental goals; however, the precision of the transducers was on the order of ±0.069 kPa. Materials and electronic components were selected on the basis of their ability to withstand severe environmental conditions, such as high ambient hydrostatic pressure (68.9 MPa), high temperature differentials (approximately 296°C with a range of 4° to 300°C) along the 196

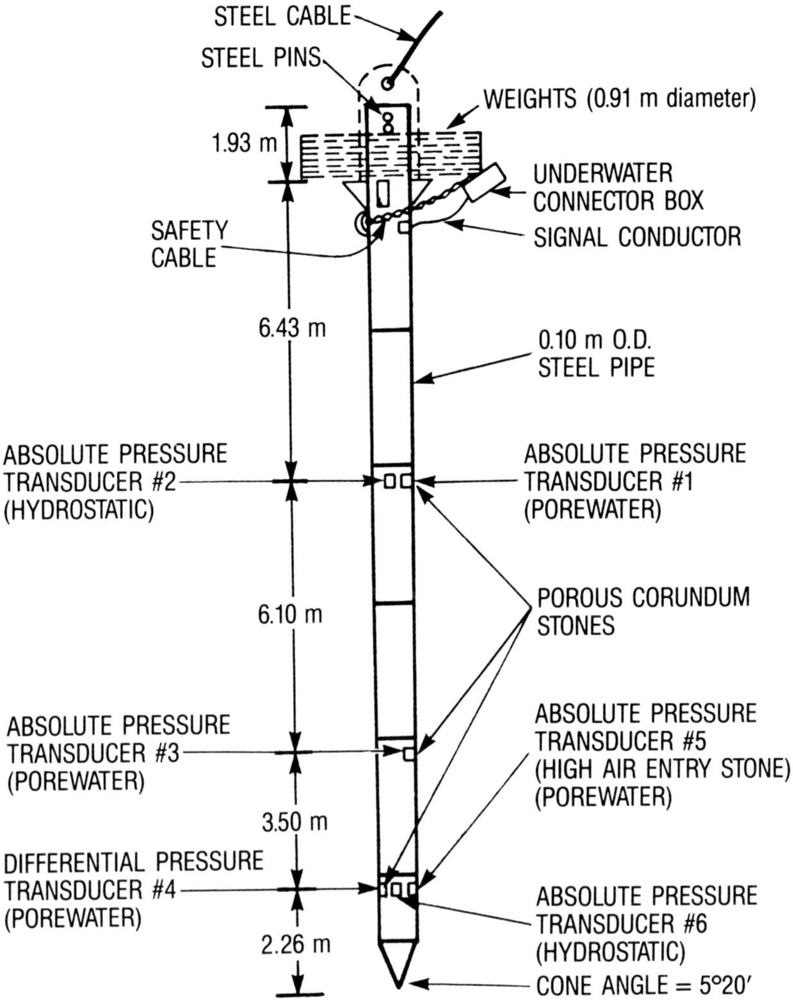

FIGURE 1. Shallow-water piezometer probe (10.2-cm diameter) deployed in East Bay and Main Pass, Mississippi Delta sites. Diagram depicts position of sensors along the probe length.

cm length of the probe, strong electrolyte in contact with materials (approximately 35 ppt seawater salinity), potential for thermogalvanic corrosion, and the predicted small changes in sediment pore-water pressures both spatially and temporally. These factors were the most important in the design of the probe, although other environmental factors were considered: soil strength, biofouling, oxidation/reduction potential, and probe proximity to other instruments.

The piezometer probe(s) consists of an 8-mm-diameter titanium tube that attaches to a tip with a cone angle of 5.3°. A porous stone (filter), which allows pore-water pressure to be transmitted to the pressure sensor, is fastened between the titanium tube and the probe tip (Figure 3). The stone (Figure 3) has a porosity of $n = 45$ to 50% and a permeability of $k = 0.01$ to 0.03 cm/sec, or several orders of magnitude higher than the fine-grained clay tested. Pore pressure is transmitted through the porous stone to an internal tube fastened to the positive side of the pressure sensor. The differential, variable reluctance pressure sensor has a similar internal tube that runs from the negative side of the sensor to the top of the porous-stone retainer but is isolated from the porous stone. The pressure sensor is enclosed

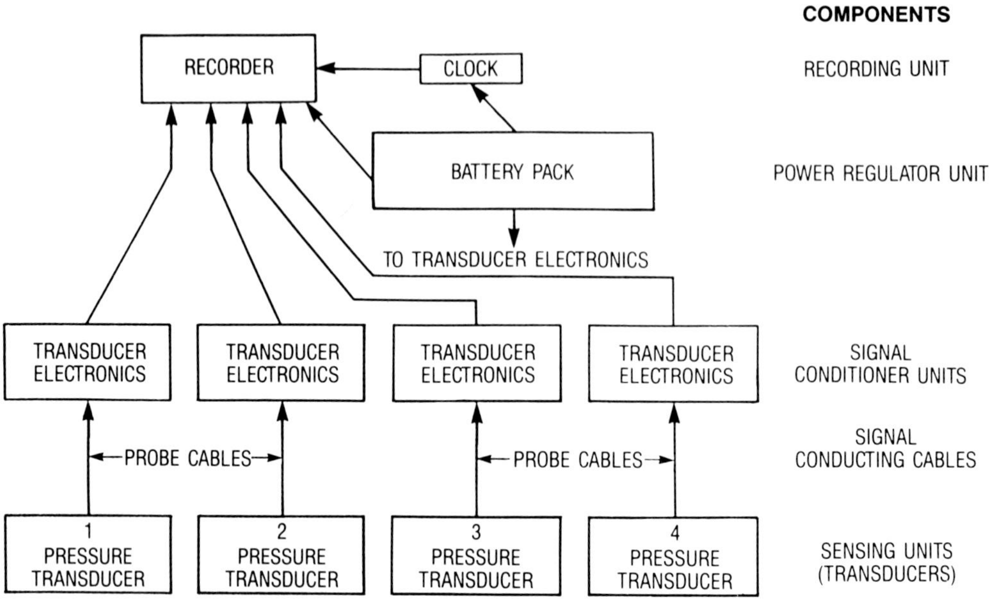

FIGURE 2. Block diagram of major components of the shallow-water piezometer system.

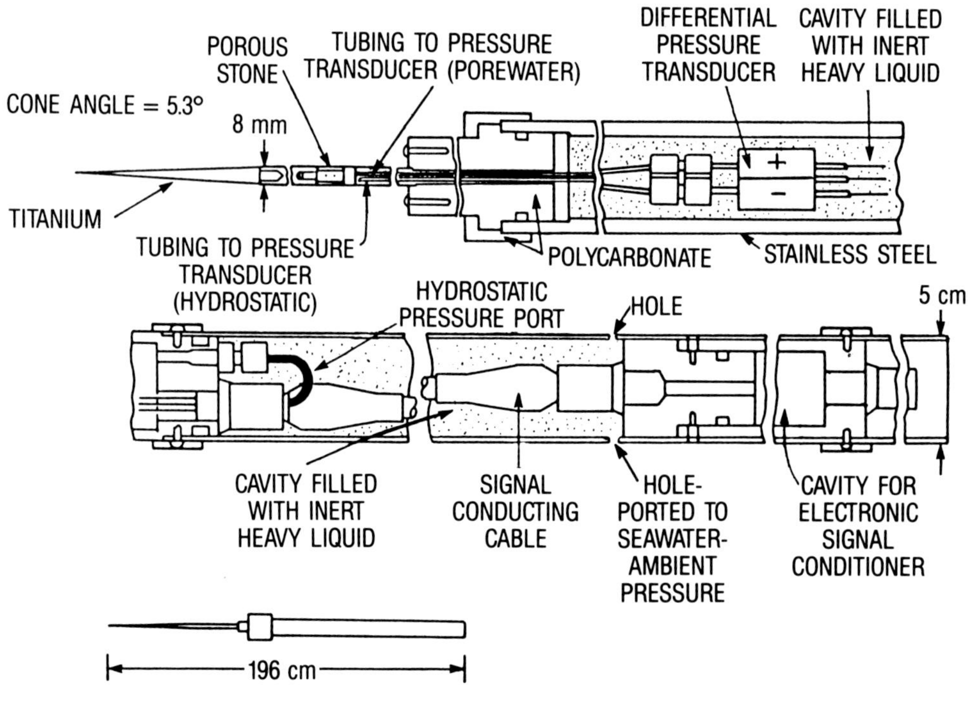

FIGURE 3. General mechanical configuration of the deep-ocean piezometer probe developed for the *In Situ* Heat Transfer Experiment (ISHTE).

in a stainless-steel housing; both are pressure compensated to *in situ* hydrostatic pressures up to ±69 MPa (Figure 3). The transducers have a range of ±68.9 kPa with a pressure cavity volume of 4.26×10^{-2} cm^3 and a full-scale volumetric displacement of only 1.97×10^{-4} cm^3. The size of the transducer is 2.86 by 3.49 by 1.78 cm and weighs 114 g. The

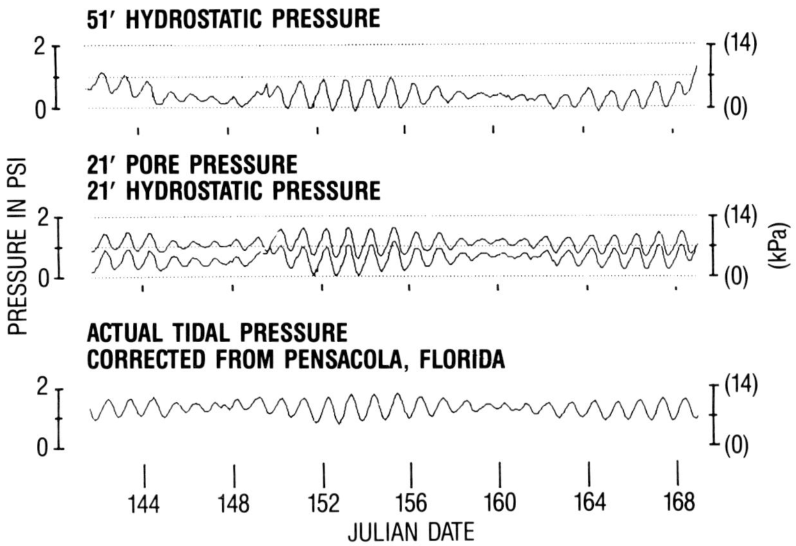

FIGURE 4. Comparison of absolute and differential pressures in response to tidal activity in Block 28, East Bay area. Note "flat" response of differential sensor due to pressure response through sediment and water column. Pressure fluctuation induced by tides in 13.5 m of water.

transducer stainless steel housing is filled with an inert fluid heavier than seawater, and the chamber is exposed directly to seawater through a hydrostatic pressure port when submerged in the ocean (Figure 3). The stainless steel pressure sensor housing is physically separated from the titanium by high dielectric polycarbonate material to minimize galvanic corrosion between the dissimilar metals.

The lengths of the piezometer probes can be changed, depending upon the experimental design objectives and subbottom depth penetration requirements. Only one pore-pressure measurement at a preselected depth below the sediment-water interface (mudline) is possible with each piezometer probe. This limitation is imposed on the probe design for ISHTE because of the requirement to limit the maximum size of the probe diameter to 8 mm in order to minimize perturbation in the induced thermal field generated by the heater.

The piezometer system electronics consist of three main components: pressure sensor, signal conditioner, and interface. Solid-state signal-conditioning electronics are enclosed (at atmospheric pressure) in a stainless steel capsule and located directly above the pressure sensor capsule (Figure 3). Further details of the piezometer system's mechanical and electronic components and performance can be found elsewhere. (Bennett, et al., 1985).

4. Brief Data Example — Piezometer Response to Tidal Activity

The absolute pressure sensors, which monitored pore-water and free water-column pressure over extended periods at the 1977 and 1978 piezometer sites, recorded real changes in tidal activity. From highs and lows obtained at a tide station near Pensacola, FL, tidal pressure fluctuations were computed and corrected to the piezometer site between Southwest Pass and Main Pass based on the *Table of Differences* published by the National Ocean Survey. Small (~1.4 kPa), as well as larger (~7.0 kPa), pressure fluctuations were recorded during days 146 and 160 (small changes; 1977, Figure 4) and days 151 to 156 (large changes; 1977, Figure 4). The response in the 6.5 m (21 ft) pressure sensors showed that sea surface activity of 12 hr or more was felt as real pressure fluctuations through the sediment pore fluid. Similar pore pressure response to tidal activity was recorded in 43.6 m of water to depths of 15.6 m below the mudline at Main Pass (Bennett et al., 1982). The tidal activity

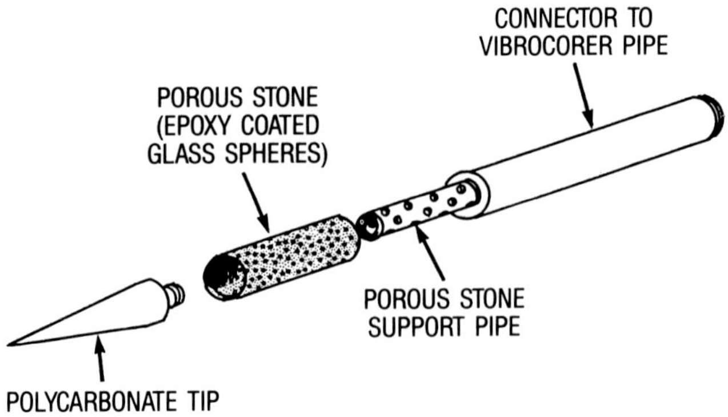

FIGURE 5. Generalized scheme for *in situ* permeability measurement (not to scale).

was measured with absolute sensors at 6.5 and 15.6 m below the mudline. Additional results of the piezometer measurements can be found in the companion paper (Li and others) and elsewhere.

B. *IN SITU* PERMEABILITY

1. Technical Background

Holocene carbonate sediments of the Bahamas and South Florida coastal areas have been studied extensively during the past few decades. Recent studies have focused on carbonate sediment and rock porosity, permeability, and fabric. The depositional microstructure and mass physical properties of both carbonate and terrigenous sediments play critical roles in determining post-depositional properties during burial diagenesis and consolidation processes (Bennett and Hulbert, 1986; Bennett and Nelsen, 1983). A serious lack of reliable geological information and geotechnical data exists for the depositional properties of sediments; obtaining undisturbed samples is difficult and previous technology to measure the required properties at and near the sediment-water interface, i.e., within the upper few centimeters, was inadequate. For these reasons, a permeability probe was developed to obtain *in situ* data of marine sediments (Bennett et al., 1990).

2. Permeability Probe

Permeability measurements were made by using a falling-head permeameter. Before the permeameter can be used, however, it must be calibrated to properly account for the intrinsic head loss associated with flow through the system, which consists of the reservoir, the valve, the tubing, the pipe, and the porous stone. The system is calibrated by establishing flow through the permeameter while it is immersed in water (Bennett et al., 1990). In our application of this technique, a probe that contains a porous segment is emplaced vertically in the sediment. An initial hydrostatic head is established at the porous element via a vertical column of water. The subsequent rate of decay of the pressure head as water flows through the porous element into the surrounding sediment can be related, through Darcy's law, to the permeability of the sediment.

The important components of the permeability probe are illustrated in Figure 5. The probe consists of a cylindrical porous stone positioned between a pipe and a solid conical tip. The porous stone is supported internally by a perforated pipe. The length of the porous stone is 9.9 cm, and the inside and outside diameters are 3.25 and 4.25 cm, respectively (Figure 6). The entire assembly is attached to the vibrocorer driving head. A flexible hose connects the top of the probe assembly with a 25.57-l water reservoir located on the deck of the vessel. The reservoir is calibrated so that the volumetric flow associated with a

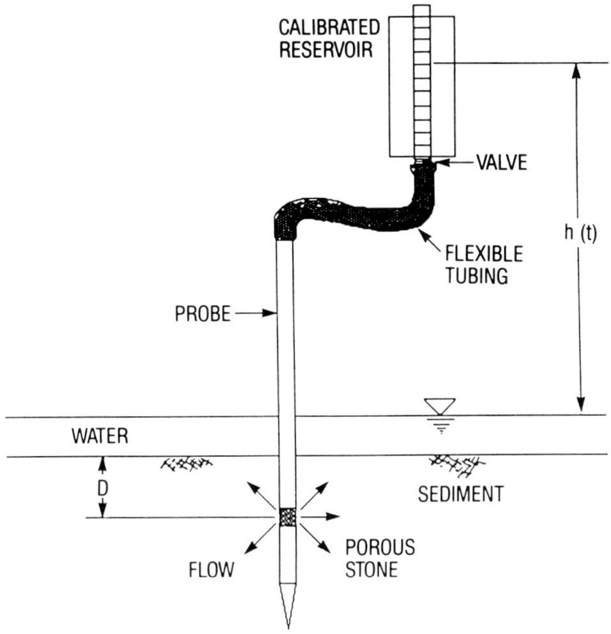

FIGURE 6. Component sketch of *in situ* permeability probe tip.

decrease in pressure head can be determined. The arrangement of the various components is shown schematically in Figure 6.

3. Modeling of *In Situ* Permeability Probe

For analysis purposes, it is assumed that the porous stone can be represented by an isolated, porous sphere of diameter d, with the center located a depth D below the sediment-water interface of a semi-infinite porous medium (Figure 7). If a slowly varying differential pressure head $h(t)$ is maintained at the surface of the sphere, then the instantaneous volumetric flow $Q(t)$ into the sediment is given by (Bennett et al., 1990)

$$Q(t) = \frac{2\pi dK\rho g}{\mu(1 - d/4D)} h(t) \tag{1}$$

where K is the intrinsic, or sediment, permeability; μ is the dynamic viscosity of the saturating fluid; ρ is the fluid density; and g is the acceleration of gravity. The result is obtained by direct analogy with the case of steady heat conduction from an isothermal sphere embedded in a semi-infinite medium with a surface maintained at a temperature different from that of the sphere (Eckert and Drake, 1972). To apply Equation 1, the surface area of the sphere is set equal to the external surface area A of the cylindical porous stone. The sphere diameter is, then,

$$d = \sqrt{\frac{A}{\pi}} \tag{2}$$

Modeling of the permeameter response by Equations 1 and 2 assumes that near-field effects are of minor importance in determining the overall flow rate, so it is not necessary

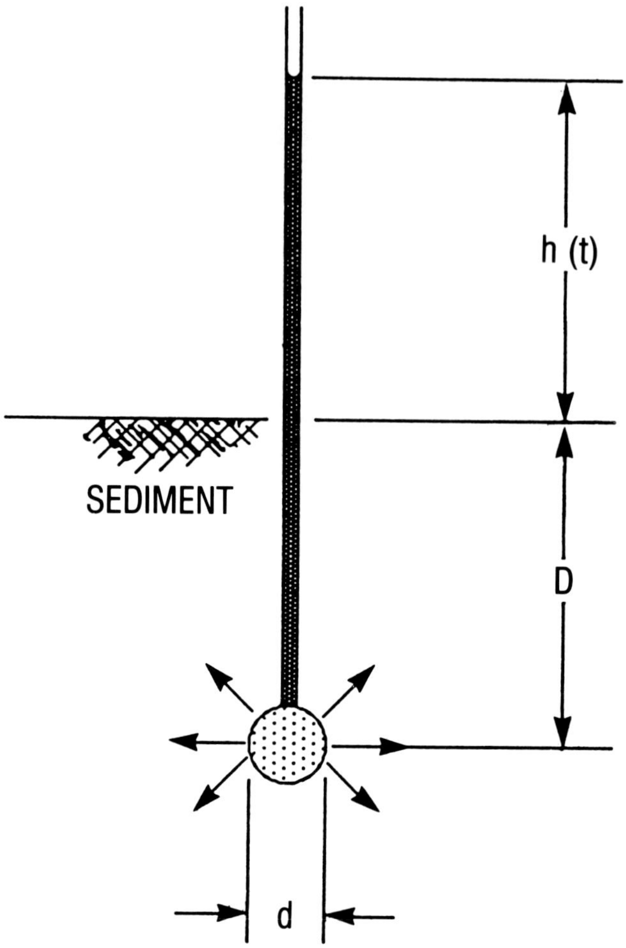

FIGURE 7. Geometry used to model the *in situ* permeability probe.

to use the exact geometry of the porous stone. Conservation of mass requires that the volumeric flow rate *Q(t)* be related to the pressure head *dh/dt* by

$$Q(t) = -b\frac{dh}{dt} \tag{3}$$

where *b* is the cross-sectional area of the reservoir where the head change is measured. Combination of Equations 1 through 3 then provides

$$\frac{dh}{dt} = -Ch(t) \tag{4}$$

where

$$C = \frac{2\pi dK\rho g}{b\mu(1 - d/4D)} \tag{5}$$

In the expression for *C*, the quantity $K\rho g/\mu$ is usually referred to as coefficient of

TABLE 1
Great Bahama Bank *In Situ* Permeabilities

Field sites	Depth feet	Depth cm	Permeability K(D)	Permeability k(cm/s)
Site 1	3	91	69	6.7×10^{-2}
	5	152	60	5.8×10^{-2}
	6	183	44	4.2×10^{-2}
	7.5	229	27	2.6×10^{-2}
	8.5	259	14	1.4×10^{-2}
Site 2	6.7	203	17	1.6×10^{-2}
	9	274	3	0.32×10^{-2}
Site 3	5.5	168	71	6.8×10^{-2}
	7	213	63	6.1×10^{-2}
	8.5	259	23	2.2×10^{-2}

permeability, k, which has units of centimeters per second. The intrinsic permeability, K, which has units of centimeters squared, is a measure of the permeability of a sediment that is independent of the fluid properties. The darcy is a common unit of permeability, and 1 darcy has the value of 9.89×10^{-9} cm^2. For a constant C, integration of Equation 4 provides

$$h(t) = h_0 \exp(-Ct) \tag{6}$$

where h_o is the initial differential head and t is the elapsed time.

The sediment permeability can be inferred using either of two approaches. In the first method, C can be determined from a curve fit of h vs. t based on Equation 6, and the permeability then obtained from Equation 5. A second method is based on integrating Equation 4 over a time interval Δt between elevation heads h_1 and h_2, yielding

$$C = -\frac{1}{\Delta t}\ln\left(\frac{h_2}{h_1}\right) \tag{7}$$

Knowing C, the intrinsic permeability K is, as before, obtained from Equation 5.

4. Measurement Data

The results of *in situ* permeability measurement for Great Bahama Bank are shown in Table 1. The sediments are classified as sand (sites 1 and 2) and silty sand (site 3) and are composed predominantly of oolitic grains (>62.5 μm) supported by aragonitic matrix of silt-size and clay-size particles. It should be noted that the *in situ* measurements are about one order of magnitude higher than the permeability measured in the laboratory. The differences are attributed to sample disturbance during coring, transport, storage, and laboratory testing (Bennett et al. 1990).

C. ELECTRICAL CONDUCTIVITY/RESISTIVITY

1. Technical Background

In situ measurement of sediment electrical conductivity (the reciprocal of resistivity) provides a means of determining the porosity and wet bulk density of sea floor deposits of cohesive materials and noncohesive sands and gravels (Bennett et al., 1983). With present technology, noncohesive sediments are virtually impossible to recover in an undisturbed condition and are generally not suitable for measuring geotechnical and geoacoustic properties with laboratory techniques. For this reason, most of the previous literature (data obtained from cores) for the physical properties of the Great Bahama Bank sediments is highly suspect.

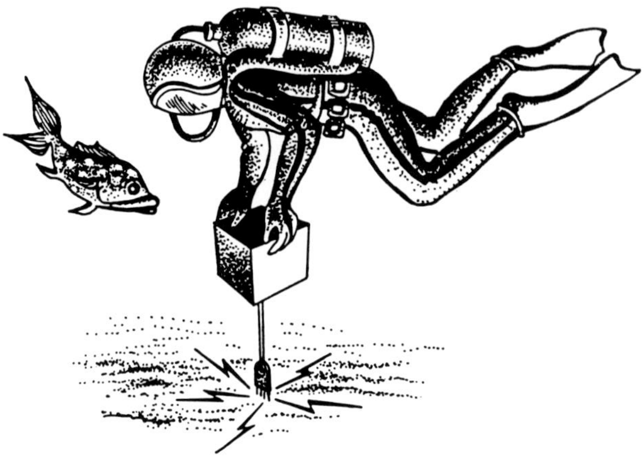

FIGURE 8. Diver-operated conductivity probe.

A new self-contained, diver-operated conductivity probe (Figure 8) was developed to measure *in situ* sea floor sediment conductivity or resistivity, temperature, and depth of sensor embedment (Lambert et al., 1986). This probe is similar to those described in references (Bennett et al., 1983; Hulbert et al., 1981). One major difference, however, is that a probe tip with a considerably smaller diameter (5.5 cm) could penetrate the oolitic sands found in the Great Bahama Bank, compared to the larger tip design (12.5 cm) required to penetrate the coarse carbonates and reefal debris off the Florida Shelf (Hulbert et al., 1981).

2. Resistivity-Porosity

The conductivity probe uses the four-electrode contact technique, which consists of two pairs of electrodes: one pair applies the reference current and the other measures the resulting electrical potential (Figure 9). Details of the systems and *in situ* porosity/bulk density data for coarse-grained carbonates off the Florida Shelf can be found elsewhere (see Bennett, Lambert, and Hulbert references immediately above). The small probe tip design is thoroughly described by Lambert.

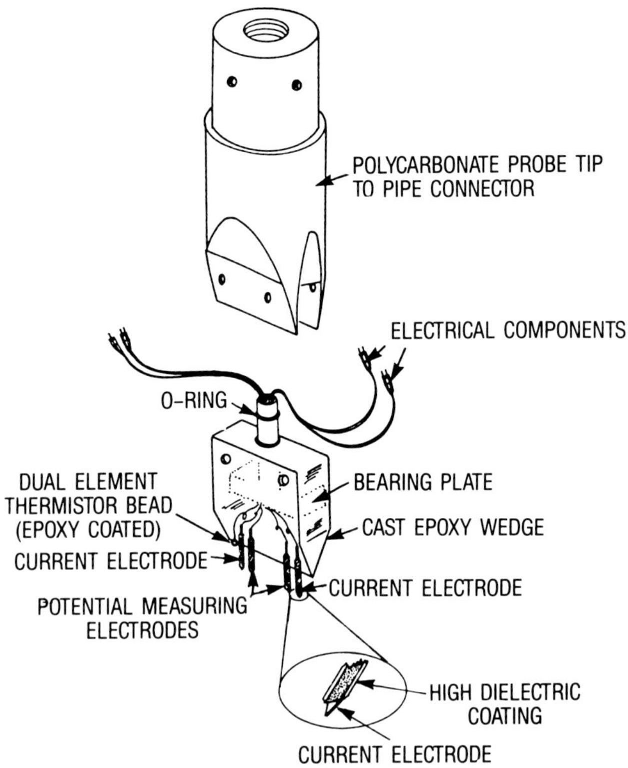

FIGURE 9. Component sketch of *in situ* conductivity probe tip and mounting connector.

The porosity, *n*, of a sedimentary deposit is related to the formation resistivity factor or formation factor, *F*, and was given by Archie (1942).

$$F = n^{-m} \tag{8}$$

This expression was later modified by Dakhnov (1962):

$$F = an^{-m} \tag{9}$$

where *a* is a factor of proportionality that depends upon the ratios of the fine fraction (<62.5 μm) to the coarse fraction (>62.5 μm) of the sediment of interest. The exponent, *m*, is usually called the shape factor (Lambert et al., 1986). Coefficients for Equation 9 have been reported for several sediment types. For sands, Erchul (1974) determined values of 1.4 to 1.6 for the coefficient *a* and 1.0 to 1.1 for the exponent *m*. In the Great Bahama Bank (Figure 10) study, values for *a* and *m* were found to be as follows:

a = 1.27 and m = 1.11 for Sites 1 and 2
(oolitic sand)

a = 1.48 and m = 1.11 for Sites 3 and 4
(muddy oolitic sand)

The relationship between porosity and resistivity is valid for sediments composed of

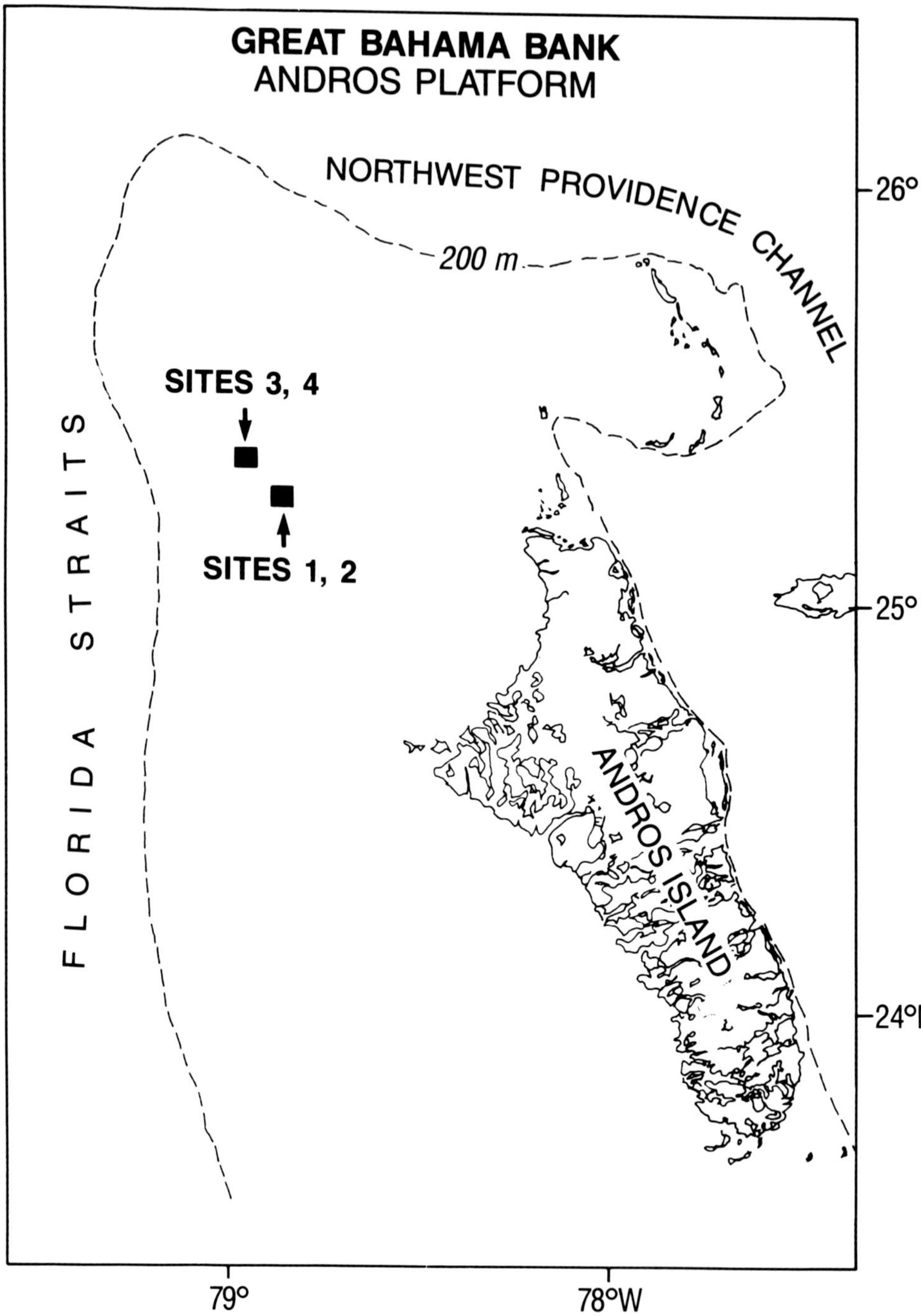

FIGURE 10. Great Bahama Bank, sites 1 through 4.

solid particles. It would not be expected to apply directly to sediments composed of particles with significant intraparticle porosity, such as foram oozes.

3. Measurement Data: Great Bahama Bank

The *in situ* porosity for Grand Bahama Bank sites 1 through 4 is shown in Table 2. The data are plotted as a function of depth of burial in Figure 11. Porosities are somewhat higher

TABLE 2
Great Bahama Bank *In Situ* Porosity

Field sites	Subbottom depth feet	Subbottom depth cm	Formation factor (F)	Porosity from resistivity (n)
Site 1	1	30	3.07	45
	2	61	3.47	40
	3	91	2.92	47
	4	122	2.72	50
	5	152	2.90	47
	6	183	2.82	48
	6.2	191	2.96	46
Site 2A	2	61	3.36	41
	2.5	76	3.02	46
	3	91	3.22	43
Site 2B	3	91	3.12	44
	4.8	145	3.04	45
	5	152	2.77	49
	6.2	191	3.46	40
	6.8	206	3.53	40
	7.8	236	3.50	40
	8.2	251	3.85	36
Site 3A	2.8	84	2.87	55
	8.8	267	5.50 (rock)	30?
Site 3B	1	30	2.55	61
	2.3	69	2.73	57
	3.2	99	2.77	57
	4	122	2.99	53
	5.2	160	3.11	51
	6.2	191	3.01	52
	7	213	3.23	49
	7.2	221	3.79	42
Site 4	1	30	2.73	57
	7	213	3.27	49

and decrease with depth more dramatically at sites 3 and 4 (range n = 61 to 42%; ~30% for rock tested at site 3A) compared to sites 1 and 2 (range n = 50 to 36%). The porosity of the upper 1 m averages 45% at sites 1 and 2 but is considerably higher (58%), at sites 3 and 4. The porosities at sites 1 and 2 are comparable to the initial (depositional) porosities estimated for the oolitic Miami limestone of the lower Florida Keys (Schmoker and Hester, 1986). The high porosities are predominantly a function of the matrix porosities and nature of the microfabric (Bennett, et al., 1990, part I). Further details can be found in Bennett et al. (1990, part II).

D. SEA FLOOR LANDER

A device to transport instrumentation from a surface craft to the sea floor and return is being developed. It is a three-legged aluminum framework with feet 3.1 m apart below vertical legs that are 1.5 m long. A platform at the 1.5-m level will support most of the payload instrumentation. The overall height is 3 m. A 1.5-m vertical guideway is centered above the platform. The top view and a photo of the lander are shown in Figures 12 and 13, respectively.

The device will be deployed with a 500-lb weight in the upper end of the vertical guideway. On command, the weight will be released at the sea floor to drive a 1.4-m piezometer probe into the surface. The weight will then be recovered. The lander and

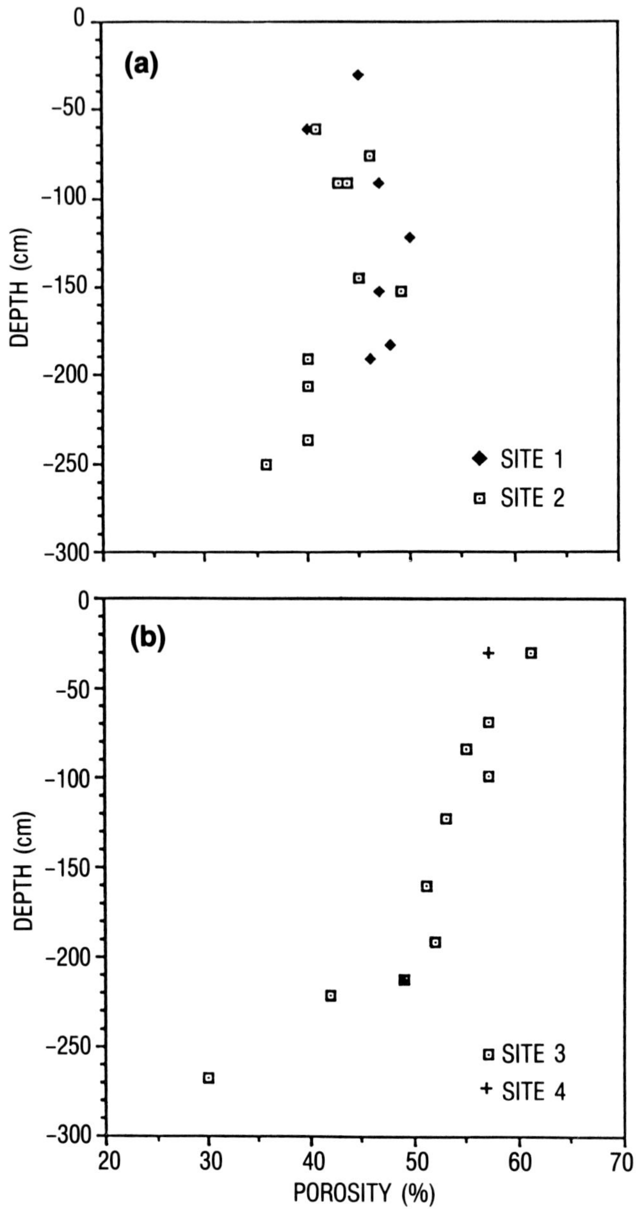

FIGURE 11. *In situ* porosity vs. depth in sediment, sites 1 through 4.

instrumentation will remain unattended to gather data until commanded to release a float to take a recovery line back to the surface. Deployments should span several weeks. Maximum depth is presently limited by the length of the recovery line. Operation to 200 m deep is envisioned.

The instrumentation payload on the platform is limited only by the requirement that control and operation of the lander not be adversely affected. Among the instrument candidates with the piezometer probe are sensors of current, heading, acceleration, strain, and temperature, as well as a camera. Accessories required in addition to the recovery system are a battery pack and control electronics, a data acquisition system, and acoustic releases. Sediment samplers could be added.

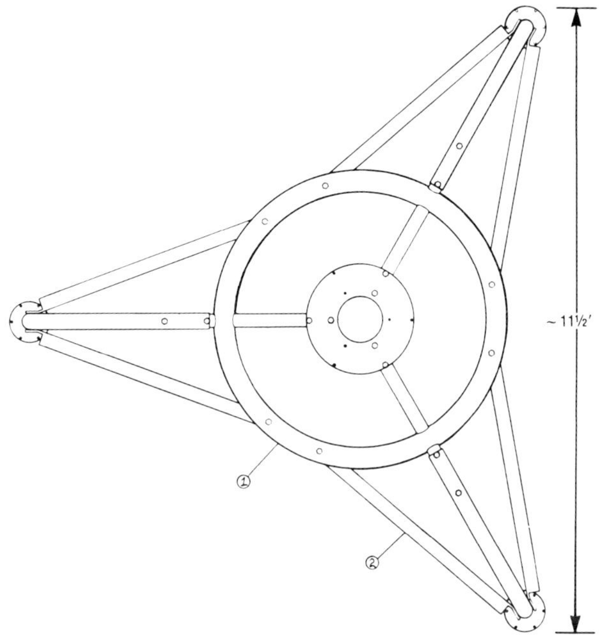

FIGURE 12. Top view of the sea floor lander.

FIGURE 13. Photo of the sea floor lander.

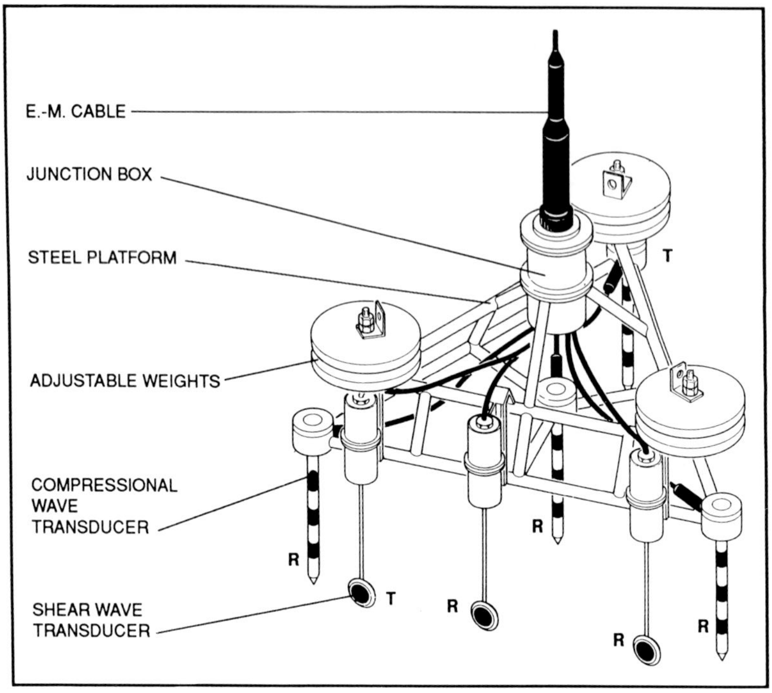

FIGURE 14. *In situ* geoacoustic measurement system.

E. *IN SITU* GEOACOUSTIC PROBES

1. Technical Background

In recent years, scientists from such diverse fields as geophysics, sea floor engineering, sedimentology, soil mechanics, and underwater acoustics have devoted considerable attention to the measurement of compressional and shear wave velocity and attenuation. These fundamental sediment properties are important to predicting the stability of sediment slopes, the consolidation behavior of sediments, the strength of marine foundations, the propagation of underwater sound through water and sediments, and the scattering and penetration of acoustic energy from and into the sediment.

Sediment geoacoustic properties have been measured *in situ* using probes deployed remotely from ships. Deployment techniques vary from scuba divers and research submersibles, to between and down drillholes using various vibratory techniques. Various seismic reflection and refraction techniques, direct path measurements of surface waves using various combinations of ocean bottom seismometers and geophones, and laboratory methods using pulse and other geotechnical techniques related to the stress/strain behavior of sediments also have been employed.

2. Description of the *In Situ* Geoacoustic Measurement System

a. General

A system was designed, built, and tested for measuring near-surface sediment geoacoustic properties, including compressional and shear wave velocity and compressional wave attenuation. The system operates in shallow water with the aid of scuba divers, but can be operated remotely at depths up to 300 m.

Shear and compressional wave probes are attached to a triangular frame (Figure 14). The frame holds adjustable weights to increase penetration into the sediment. Four compressional wave and three shear wave probes are held at constant distances apart while scuba

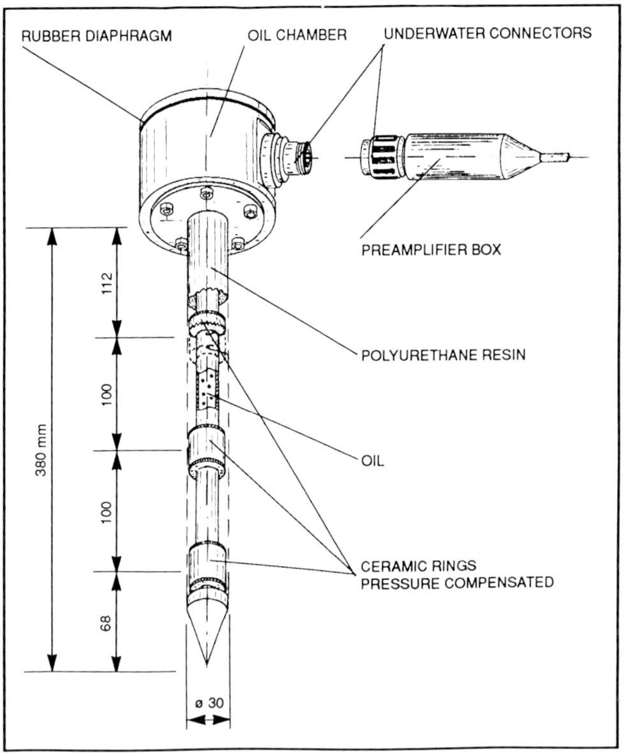

FIGURE 15. Compressional wave probe.

divers guide them into the sediment. Generous amounts of neoprene (compressional wave probes) and silicone rubber (shear wave probes) are used to mechanically and electrically isolate all probes from the frame. The system is hardwired to the surface via an electromechanical cable containing 16 twisted pairs of conductors. All signal transmitting, receiving, and recording equipment is at the surface; only the geoacoustic probes with preamplifiers are deployed with the frame.

b. Compressional Wave Probes

Transmit and receive probes consist of three identical, radially poled ceramic cylinders mounted on a hollow steel pipe (Figure 15). The ceramic cylinders are 1.0 in. (25 mm) outside diameter × 1.0 in. (25 mm) long and have 0.125-in.-thick (3.2 mm) walls. Ceramic elements are attached to stainless steel pipe by epoxy resins. The elements are pressure compensated by connection to an oil-filled reservoir with a neoprene diaphragm (Allman, 1974). The probes are potted in hard polyurethane resin (Hardness = 80 Shore A) for protection. The only difference between transmit and receive probes is the addition of a 40-dB gain amplifier near the head of the receiver probes. The radially poled ceramics have omnidirectional transmitting and receiving characteristics in the plane perpendicular to their axis of deployment. Compressional wave probes are mounted on the frame so that ceramic cylinders will be 10, 20, and 30 cm below the sediment-water interface during deployment. Therefore, compressional wave velocity and attenuation measurements are made over nine paths during each deployment.

c. Shear Wave Probes

Transmitters and receivers are identical 1.25 in.^{2} (31.75 mm^2) × 0.019 in.-thick (0.48 mm) bimorph ceramic benders (Figure 16). The ceramics are potted in a stainless steel ring

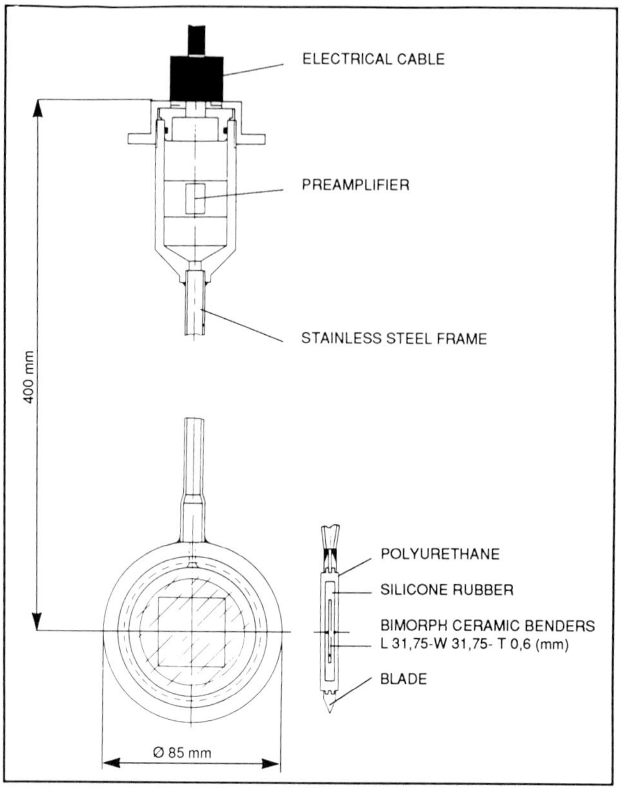

FIGURE 16. Shear wave probes.

with silicone rubber (Hardness = 35 Shore A) to allow relatively unrestricted bender movement. A thin covering of much harder polyurethane resin (Hardness = 80 Shore A) holds the ceramics in place and provides a tough coating to protect the ceramics during insertion into the sediment. The received signals are amplified using a 40 to 60 dB gain amplifier in the head of the receiver probes. Shear waves, polarized in the horizontal plane, travel between the transmitter and the two receivers at a mean depth of 30 cm below the sediment-water interface.

d. Electronics

Basic electronic configuration is the same for all measurements (Figure 17). Both shear and compressional wave ceramic elements are driven with a pulsed sine wave generated with standard pulse and function generators. A power amplifier is used to increase signal level when required. Shear waves are generated as 6-cycle sine waves pulsed every 500 msec. Driving frequency (100 to 2000 Hz) and driving voltage (100 to 230 V p-p) vary, depending on coupling characteristics, sediment shear wave velocity and attenuation, and the path length between receiver and transmitter. Compressional waves are generated as 10-cycle sine waves pulsed every 100 msec. Driving frequency is a constant 58 kHz, and the driving voltage (30 to 100 V p-p) varies, depending on path length, attenuation, and sediment type. The electrical energy is converted to mechanical energy by the transmitter ceramic elements in the geoacoustic probes. Identical ceramic receivers convert the mechanical energy to electrical energy. Preamplifiers, located near the receiver elements, increase signal levels and the signals are filtered, then received and recorded with a digitizing oscilloscope. The travel time for the propagation of acoustic energy between transmitter and receiver is used to calculate acoustic velocities.

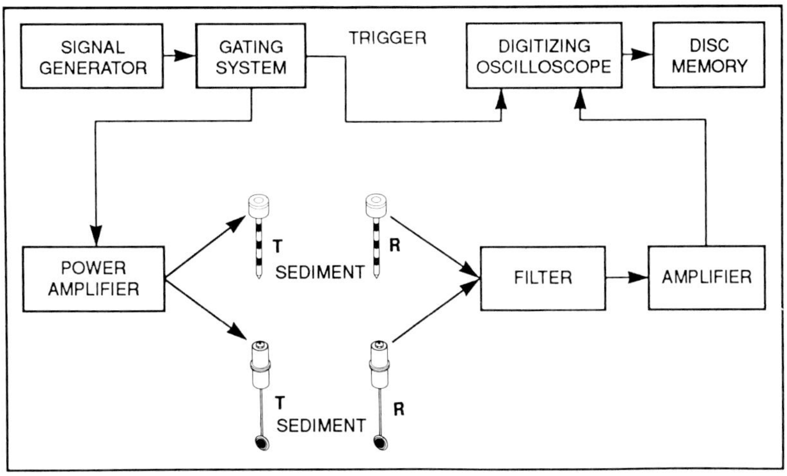

FIGURE 17. Block diagram of *in situ* geoacoustic measurement system electronics.

e. Data Calculations

Compressional wave probes are calibrated in the water column, a few meters above the bottom, before and after insertion into the sediment. Compressional wave velocity is calculated from the difference in time delays in water and sediment between each of the nine possible combinations of ceramic elements. Examples of received signals propagated through the water and the sediment are presented (Figure 18). Sediment compressional wave velocity (*Vp, m/s*) is calculated from the time-arrival differences between water and sediment (Δt):

$$\mathrm{Vp} = \mathrm{Cw}/(1 - \Delta t \times \mathrm{Cw}/d) \qquad (10)$$

where *Cw* (*m/s*) is the measured or calculated acoustic water velocity and *d* is the distance (*m*) between probes. In practice, water velocity (*Cw*) is calculated from temperature, salinity, and depth, and the distance (*d*) between probes is determined from the calculated water velocity and the measured time delay between probes in water. Compressional wave attenuation (dB) is calculated as 20 log of the ratio of received voltage between probes in seawater and sediment. Attenuation is then extrapolated to a 1-m path length and expressed as decibels per meter.

Shear wave velocity is calculated from the time delay and measured distances between the transmit and receive probes. Time delay is measured at the leading edge of the received signal (Figure 19). Distances between transmitter and the near and far shear wave receivers were 30 and 70 cm, respectively. This arrangement does not allow the calculation of shear wave attenuation.

f. Measurement Data

During the development of the *in situ* geoacoustic measurement system, a series of compressional wave velocity and attenuation and shear wave velocity measurements was made at seven sites in the vicinity of the harbor at La Spezia, Italy (Figure 20). A summary of measurement results is presented in Table 3.

Compressional wave velocity ranged from 1464 m/sec in silty-clay sediments at Diga to 1989 m/sec gravels found off Monasteroli (Table 3). Velocities of 1600 to 1700 m/sec were common in sandy sediments. Compressional wave attenuation (at 58 kHz) ranged from the near zero dB/m at muddy sites to 58.5 dB/m in gravels at Monasteroli. Mean attenuation of 4 to 12 dB/m was common in muddy sediments. Higher attenuations (near 30 dB/m)

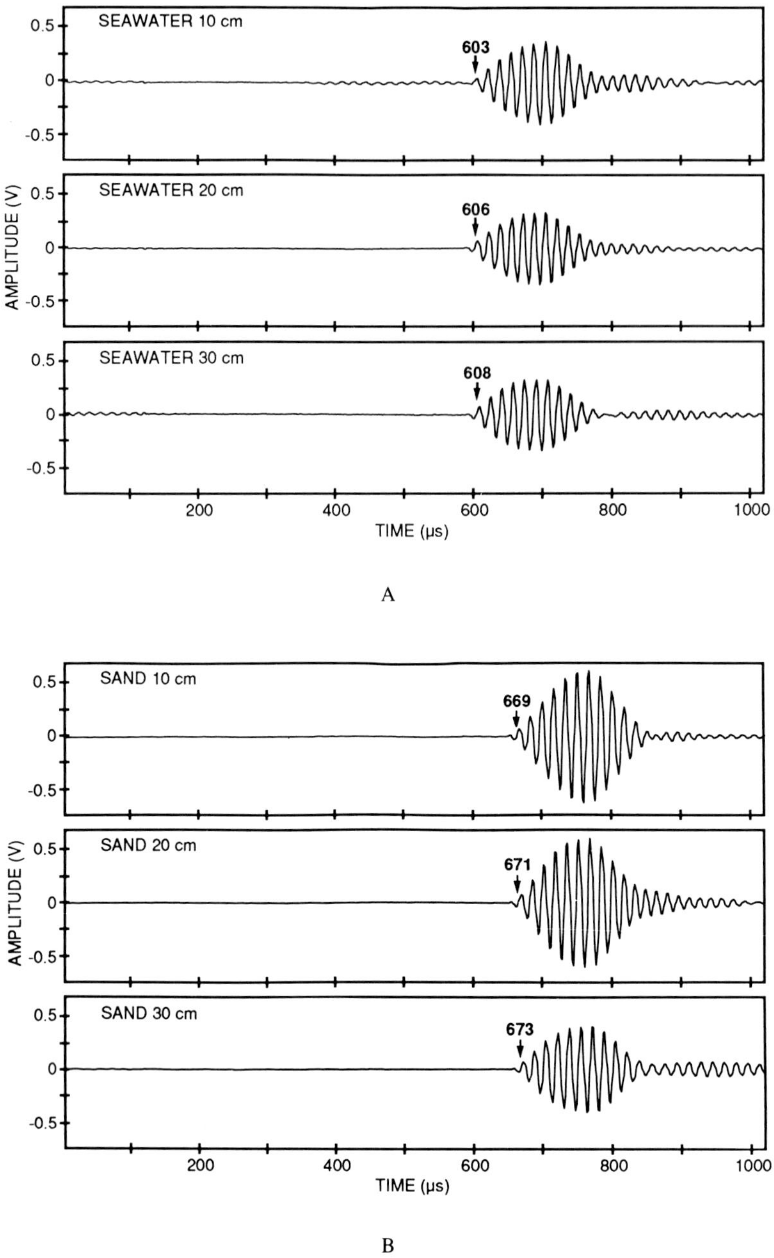

FIGURE 18. Examples of compressional wave signals recorded for a deployment of the geoacoustic measurement system at Monasteroli, a sandy site in the Ligurian Sea. (A) Signals transmitted through 0.9905 to 0.9996 m of seawater; (B) signals transmitted through sediments at 10, 20 and 30 cm below the sediment-water interface. Compressional wave velocity and attenuation were calculated from the differences in received times and voltages between sediment- and water-borne signals.

were found at the sandy sites. Shear wave velocities ranged from 16.4 m/sec in soft silty-clay sediments near Santa Teresa to 89.9 m/sec in the hard-packed sands of Venere Azzura.

g. New-Generation Deployment Systems

The next generation of *in situ* geoacoustic measurement systems was recently tested in

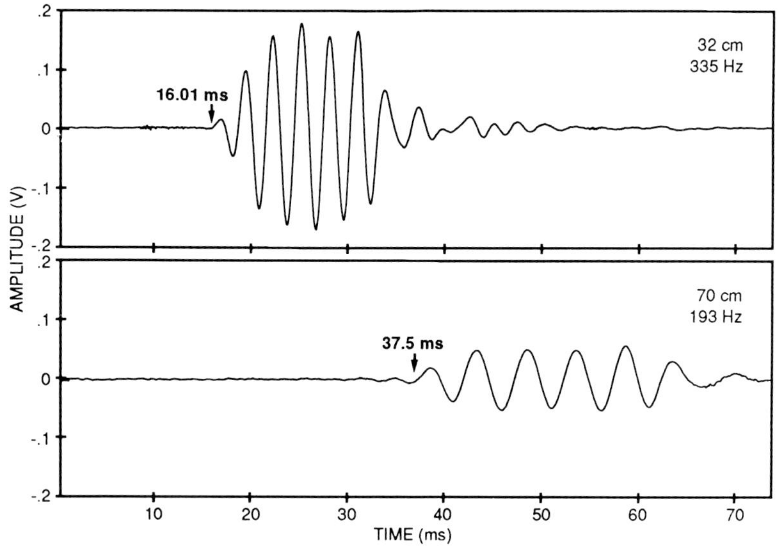

FIGURE 19. Examples of signals recorded from two shear-wave receivers at Santa Teresa, a silty-clay site in the Bay of La Spezia. Calculated shear wave velocities were 20.0 m/sec at 32 cm and 19.7 m/sec at 70 cm distances between probes.

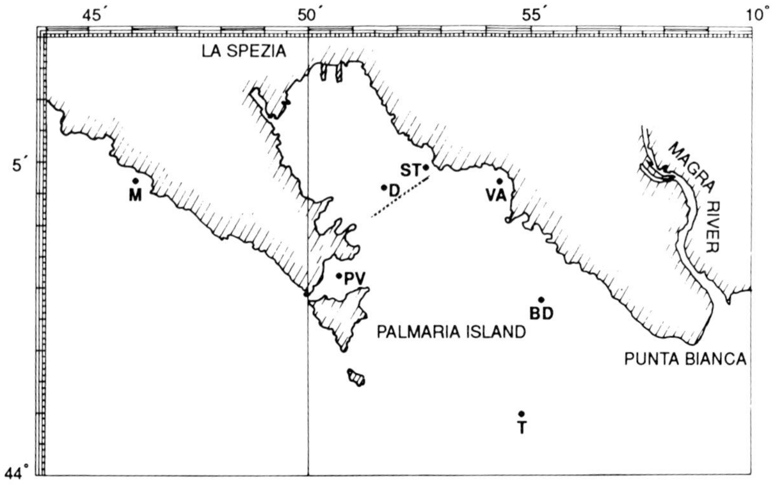

FIGURE 20. Location of test sites for the *in situ* geoacoustic measurement system: Diga (D), Venere Azzura (VA), Santa Teresa (ST), Portovenere (PV), Turf (T), Boa Dragaggio (BD), and Monasteroli (M).

the Adriatic and Norwegian Seas. Compressional and shear wave probes with associated electronics are identical to the old system, but the system is designed to be deployed in water depths up to 300 m without the aid of divers (Figure 21). Future developments include adding probes to measure seawater temperature and sound velocity, sediment temperature, electrical resistivity, and pore pressure (see relevant sections of these papers for probable probe designs). A fourth shear wave transmitter will be added to measure shear wave attenuation. The system will be modified to allow a single coaxial electrical connection to the surface, which will require that acoustic signal generation and receiving electronics are part of the deployment package (Richardson et al., 1990).

TABLE 3
Values of Sediment Geoacoustic Properties Measured *In Situ* in the Vicinity of La Spezia, Italy

		Compressional wave				Shear wave velocity	
		Velocity		Attenuation			
Site	Sediment Type	Mean	Range	Mean	Range	Mean	Range
Diga	Silty-clay	1469	1464—1474	7.0	3.6—14.5	23.6	19.5—28.1
Venere Azzura	Sand	1433	1625—1640	32.8	23.6—43.6	78.8	65.7—89.9
Santa Teresa	Silty-clay	1473	1470—1478	4.9	0.8—8.6	19.7	16.4—23.3
Portovenere	Silty-clay	1479	1470—1485	12.6	3.5—18.8	29.3	24.8—37.4
Turf	Sand	1615	1608—1620	30.9	28.2—34.6	74.0	72.1—75.7
Boa Dragaggio	Sand	1606	1589—1623	31.3	27.2—37.7	40.2	37.0—45.3
Monasteroli	Gravel	1928	1854—1989	57.7	56.1—58.5	52.3	45.3—59.0

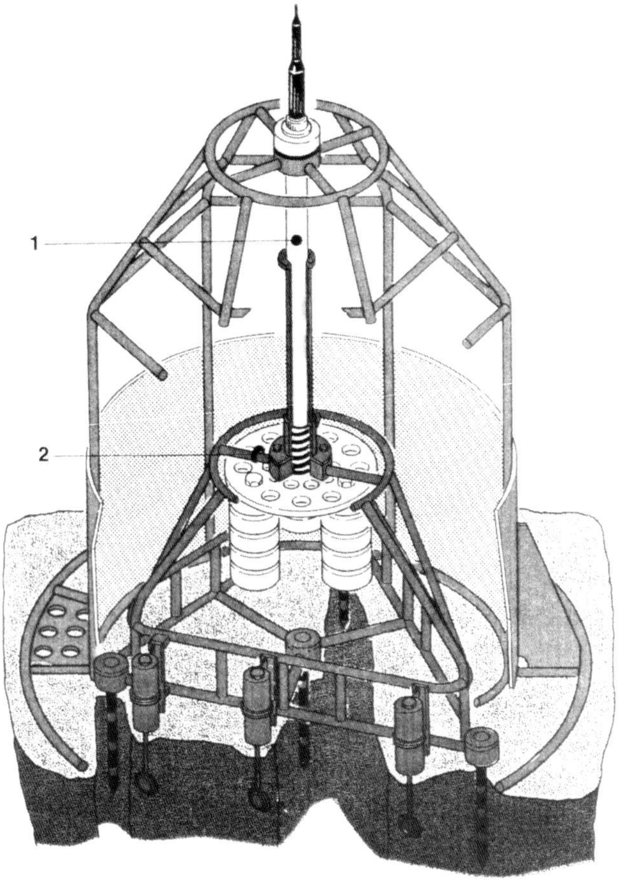

FIGURE 21. A remote deployment platform for the *in situ* sediment geoacoustic measurement system.

IV. REMOTE METHODS OF SEA FLOOR CHARACTERIZATION

A. SIDESCAN SONAR

1. Technical Background

Sidescan sonar has become an indispensable tool with which to characterize the sea floor for geologic interpretations, for site selection and assessment, and for environmental

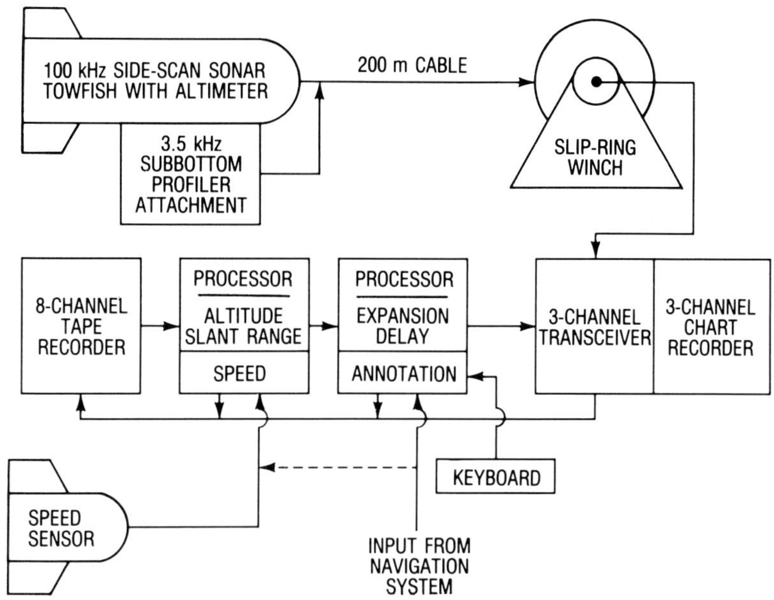

FIGURE 22. Schematic diagram of sidescan sonar system.

and other surveys. The sidescan technique has revealed the true extent of the complexity of surficial sediment relationships, even on ''uniform'' bottoms. It also demonstrated the inadequacy of areal bottom classification based only on spot samples and vertical incidence acoustic profiles. The sidescan sonar system described here, a portable system readily deployable on small vessels, was developed in response to a need for comprehensive bottom characterization required for high-frequency acoustic bottom scattering experiments (Stanic et al., 1988 and 1989) and for other site selection surveys, primarily in continental shelf environments. Topics include a description of the system configuration, operational considerations, and a summary of primary data analysis. Historical background, sidescan sonar theory, and data interpretation have been addressed in a variety of sources (Belderson et al., 1985; Mazel, 1985). Secondary data analysis, which is system independent and includes digital image processing and spectral signal analysis (Reut et al., 1985) for sediment classification, is evolving rapidly as the high data processing rates required by sidescan sonar are becoming readily accessible.

2. Sidescan Sonar Measurement System

The sidescan sonar is a shallow-water system (<200 m) composed mainly of off-the-shelf components. A major factor in its configuration is to achieve maximum flexibility for field operations under varying conditions, for post-operational processing of data, and for interlinking navigation and acoustic data (Figure 22). The main sidescan sonar system comprises a transducer towfish, a tow cable, a winch, a sonar transceiver with paper recorder, a speed sensor, signal and annotation processors, and a magnetic tape recorder. The transmit/receive system is analog, and the signal processing is essentially digital. Nominal frequency of the transducers is 100 kHz. Bandwidth is relatively large. This capacity enhances resolution in the near- to mid-range (to 100 m) over what would be expected at 100 kHz. A narrow horizontal beamwidth of 0.75° also enhances across-track resolution over that of a standard 1° or 1.5° system; towing speed to maintain along-track resolution is only slightly sacrificed.

Maximum swath is 800 m (400-m range per transducer), but in 50-m depths or less, surface noise, refraction effects and limited towfish altitude limit the effective swath to 300 to 400 m. The towfish also carries a 100-kHz altimeter. The altimeter provides a strong initial bottom return, which is used by the signal processor to remove the towfish altitude. A sidescan signal can also be used for this purpose, but since this signal is tuned for uniformity across range, the initial return from soft, watery sediments may be too weak to serve as an effective bottom tracker.

The towfish is deployed with a 600-m armored multiconductor cable or with a 200-m Kevlar® cable from portable electric slip-ring winches, or with a hand-deployed 100-m Kevlar® cable. Depending on tow speed and water depth, either a small or a large depressor is fitted to the towfish. At speeds ranging from 1.0 to 6.5 m/sec the towfish is stable with or without the depressor.

The towfish signals are transmitted and received by the sidescan transceiver recorder unit. Received pulses are adjusted by the time-variable gain amplifiers and are recorded directly on a multichannel paper recorder or on magnetic tape. The transceiver operates at rates up to 30 pulse/sec (50-m swath) for near-bottom, high-resolution work. The left and right sidescan signals, altimeter signal, timing pulse, speed input, annotation, and voice are recorded on an 8-channel analog recorder or a digital audio tape recorder. Recorded signal band widths are 2.5 kHz to 5 kHz, and dynamic range is 43 dB and 72 dB, respectively. The recorded signals are played back through the processor for monitoring during data collection and for reprocessing the collected data in the laboratory.

The signal processor performs geometric corrections on the sidescan signals and creates a plane-view or map-equivalent acoustic image of the sea floor. It employs the altimeter signal to remove the towfish altitude from the sidescan signal and to correct for slant range. It also accepts a speed input to maintain the along-track scale proportional to the across-track scale. Three sources provide speed input: manually adjustable settings, an external speed sensor, or electronic navigational inputs. The external sensor is an impeller mounted on a small towfish that is deployed from the vessel on its own tow cable. It accurately measures speed through the water (a potential problem when strong currents exist) and is responsive to small, short-term speed fluctuations. The electronically computed navigational speed input can be subject to a significant averaging period, and is thus suitable mainly for running long, straight lines.

In addition to event marking and annotation by keyboard, the annotation processor provides signal expansion and delay. The expansion capability allows enlargement of the altimeter profile while operating the sidescan at long ranges and also enlarges items of interest while underway. For post-operational processing, it may be used to enlarge or reduce, a significant aid in sidescan sonar mosaic construction. The annotation processor also accepts data streams from navigation systems for event marking.

A 3.5-kHz subbottom profiler is deployed as an integral attachment to the sidescan sonar towfish. When installed, the profiler occupies the signal path of the altimeter within the sidescan system and also assumes the function of an altimeter. This unit can penetrate 20 m or more into muddy sediments but gives little or no penetration in hard-packed sands. The combined 3.5-kHz sidescan towfish is a somewhat less stable towbody and operates best at low speeds (1.5 to 2.0 m/sec). Its combined weight of over 100 kg necessitates the use of a winch and a davit for deployment and recovery. In shallow-water operation, the 3.5-kHz profiler can also be used to measure true water depth. Although bathymetric measurements are theoretically not feasible without a towfish depth sensor, there is sufficient signal leakage from the profiler that a signal with a towfish-surface-bottom-towfish path can be detected and used for depth measurement.

One of two navigation systems normally supports the sidescan sonar, Loran-C or a C-band microwave transponder system. The advantage of Loran-C is ease of setup and use, whereas that of the transponder system is accuracy. The Loran-C system provides time differences of 0.01 μsec, and under normal conditions provides a short-term operational precision of ± 0.02 μsec (nominally ± 4 m in areas of good Loran-C coverage), although absolute accuracy is considerably less. In practice, the usual limit on navigation accuracy for an uninstrumented sidescan sonar depends on the skill of the operators. Such factors as towing parameters, environmental conditions, and trackline deviations result in errors that are generally larger than the relative accuracy of the Loran receiver. The transponder system, in contrast, provides a positioning accuracy of ± 1 m. But the system is line-of-sight and has an approximate range of 35 km, and considerable effort is required to set up baselines, towers, and power sources.

3. System Operation

Through the experience of numerous field missions, a set of operating conditions has evolved that is generally applicable to shallow-water sidescan sonar surveys. For successful operation, certain minimum vessel characteristics are required. Some compromises are usually involved in a typical charter boat, but any major deficiencies will adversely affect the operation. The vessel should be 12 to 30 m long; smaller vessels do not have sufficient space, and larger vessels tend to be too sluggish to maneuver on small line grids. The vessel requires a stern davit or frame for towfish deployment and retrieval and for centerline towing, stern deckspace, a climate-controlled cabin, and dry storage space. Continuous long-term minimum speed for towing should be 1.5 to 2.0 m/sec, but a high cruising speed is desirable for transit. The vessel should be equipped with an autopilot and have a responsive helm.

In determining adjacent trackline positions for complete sidescan coverage, a swath overlap of 10% to 30% should be included (e.g., roughly 30 m for a 300-m swath; 10 m for a 100-m swath). With Loran-C, greatest trackline accuracy results from running lines of position and maintaining one time difference constant. Over distances of 1 to 5 km, lines of position can usually be considered straight and nearly parallel. Since a straight track is imperative for high-quality sidescan sonar imagery, the real limitation on navigational accuracy is the skill of the helmsman and the handling characteristics of the boat. Waypoints and a trackline deviation indicator also may be used, but navigation by latitude/longitude with a Loran-C receiver should be avoided unless the accuracy of the receiver's latitude/longitude conversion is known and has been tested.

The operating parameters of the sidescan system will depend on the purpose of the survey. The sidescan sonar towfish is optimally flown at one-seventh of the across-track range, but wide variations are acceptable. Towing speeds of 2 to 3 m/sec are best for most situations, depending on the desired along-track resolution. Cable payouts of 30 to 100 m are normal for 20- to 40-m water depths; a cable length-towing depth ratio of three or more is realistic in water depths to several hundred meters. In shallow-water depths the sidescan system is quite sensitive to bubble noise and scattering generated by waves. Generally, the presence of abundant whitecaps or seas over 1.0 to 1.5 m degrades signals to an unacceptable level. Also, the shallow-water column is commonly variable and contributes to refraction of the sidescan signals. Thermoclines, internal waves, and tidal flows may produce refraction effects that will distort, obscure, or in some cases mimic bottom features (Werner, 1982). Navigation data should be recorded on tape or disk at frequent intervals (typically 12 sec for Loran-C) for later processing. If feasible, a depth sounder should be operated during the sidescan survey to collect the complementary bathymetric data needed to interpret the sidescan images or as part of a more extensive bathymetric survey of the area to be sidescanned.

4. Sidescan Sonar Data Analysis

The general processing areas of sidescan sonar data are editing and plotting the navigation track, reprocessing sidescan imagery, and constructing sidescan sonar mosaics. Navigation data are processed to obtain edited and corrected navigation plots of tracklines on a Mercator projection. Large-scale navigation plots are used with the sidescan data for image processing and assembling mosaics, as well as in bathymetric map construction.

Sidescan imagery and 3.5-kHz subbottom profiles can be reprocessed on playback to improve the quality of the display. Enlargement is through the expansion/delay processor. The tape playback speed is reduced for additional enlargement and is useful to subbottom profiles. Some contrast enhancement can be achieved through a combination of adjustments to the tape output levels and the print and threshold settings of the chart recorder. The print and threshold levels also can be used as rudimentary filters; e.g., with a high threshold setting, high-intensity targets can be displayed without bottom backscatter. Speed adjustments on the correction processor are used to make scale adjustments. Geometric correction processing can be bypassed for improved near-field imaging. In general, the processing capabilities of the sidescan sonar system are limited, and external digital image processing makes the best use of all information contained in the signals.

Sidescan mosaic construction is conducted manually by reducing sidescan imagery and matching it to tracklines. Reduction begins with the expansion processor in a reduction mode (by applying a cross-track range value greater than the range of the collected data) with the appropriate speed corrections. After the overlapping trackline strips are assembled and mounted onto a backing, the complete mosaic is photographed. Matching the image strips to the tracklines takes into account towfish layback from the navigation antenna. The overall match requires a compromise between speed adjustment, trackline straightness, and quality of fit to produce a mosaic that is least distorted and capable of being interpreted. It must be kept in mind by the viewer that the straight tracklines so evident of most mosaics are artifacts and convey a greater image veracity with respect to geographic coordinates than actually exists.

5. Measurement Data

Two examples illustrate a range of applications of 100-kHz sidescan sonar to shallow-water site characterization. Figure 23 is a Loran-C navigated mosaic that was constructed to identify and characterize a 0.5-km^2 experimental test site (Stanic et al., 1988). The site is in a 30-m water depth on the northwest Florida shelf in the Gulf of Mexico, and the bottom is uniform, smooth, fine sand. Sidescan swath width is 300 m; areal coverage is about 3.3 km $\times$ 3.8 km. The mosaic shows part of a large sand-wave field. However, the dark (strong acoustic returns), linear features are not caused by morphology, but by coarse, rippled sand located on the steep flanks of the asymmetric sand waves. The areas of low reflectivity are floored by smooth, fine-grained sand. In this example, the actual difference in acoustic backscatter between the two bottom types is small, but sidescan time-variable gain amplifiers were adjusted to be sensitive to small variations in signal strength. Note that the mosaic clearly delineates the test site (center of area) on an extremely variable sea floor. The mosaic was further used to position the experiment and as a basis for an extensive bottom sampling/measurement program.

Figure 24 illustrates the resolution of bottom features achievable with a 100-kHz sidescan sonar. The image shows large 2-m ripples with superimposed 0.5-m ripples in a scour depression of coarse sand substrate; the ripples terminate abruptly at the base of a sand wave of medium-to-fine sand. In this example the towfish was flown 3 m above the bottom, and swath width is 50 m (30 pulses/sec). The image was recorded on magnetic tape at 15 ips and has a signal band width of 5 kHz.

B. ACOUSTIC SEA FLOOR CLASSIFICATION SYSTEM

1. Technical Background

Both the military and civilian communities needed a sea floor classification system that would accurately predict the sediment type and selected geotechnical properties of the sea

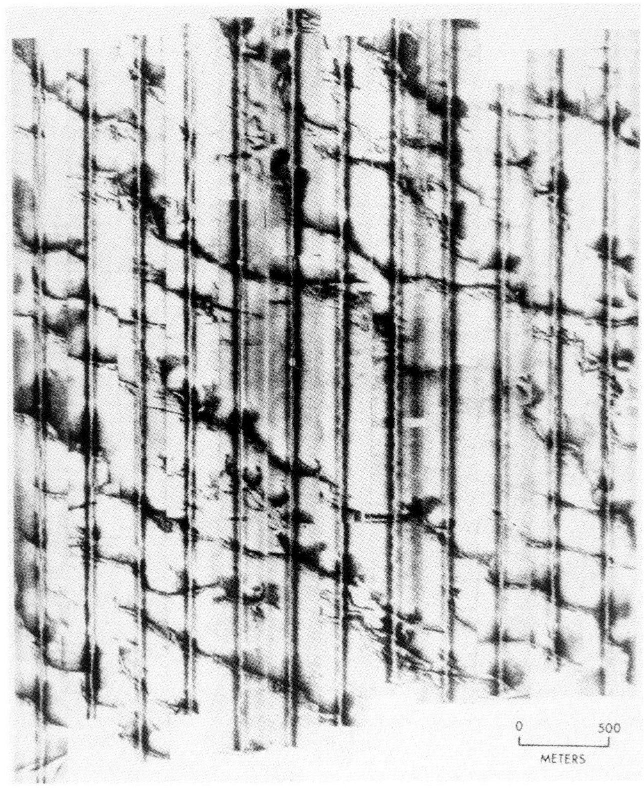

FIGURE 23. Sidescan sonar mosaic of sandwave field on northwest Florida continental shelf.

floor while in an underway, rapid-survey mode. Historically, such properties were determined through extensive, time-consuming laboratory analysis of sediment collected at discrete points using various types of sampling equipment. Analyses provided only information about the sea floor at the site where the sample was collected. The underway rapid-survey, sea floor classification system eliminates the need for approximately 90% of the samples now collected for sediment property delineation. At the same time transitional trends of sediment properties are discriminated and identified areally, since data are collected continuously along tracklines. However, the collection of a few ground-truth samples is still necessary to calibrate a sediment classification system to a particular environment.

Numerous approaches during the past 25 years attempted to develop instruments that would remotely predict sea floor sediment properties with a reasonable degree of precision (Breslav, 1967; Faas, 1967; Akal, 1972; Kogler, 1976; Dodds, 1981; Winn et al., 1980; Winn et al., 1983; Caulfield and Kim, 1983; McLeroy and Shows, 1983; Howard et al., 1983; Caulfield, 1984; Brunson et al., 1985; Corbin, 1986; Satkowiak, 1988; Lambert, 1988 and 1989; and Schock, 1989). Initial tests of these instruments produced varying degrees of success. In 1982, a new remote acoustic sediment classification system, the Echo Strength Measurement System, was developed. The system was demonstrated at the Naval Coastal Systems Center in Panama City, FL, on several test areas in St. Andrews Bay and in the Gulf of Mexico. At the time of the test, the system consisted of a modified high-resolution, echo-sounding recorder (LAZ 72), the microprocessor-controlled echo strength measuring device (EMG2), and a 15-kHz narrow beamwidth (12°) transducer. During the test, the system was configured like a conventional echo sounder with the transducer suspended in the water while the boat transited the test sites along predetermined tracklines.

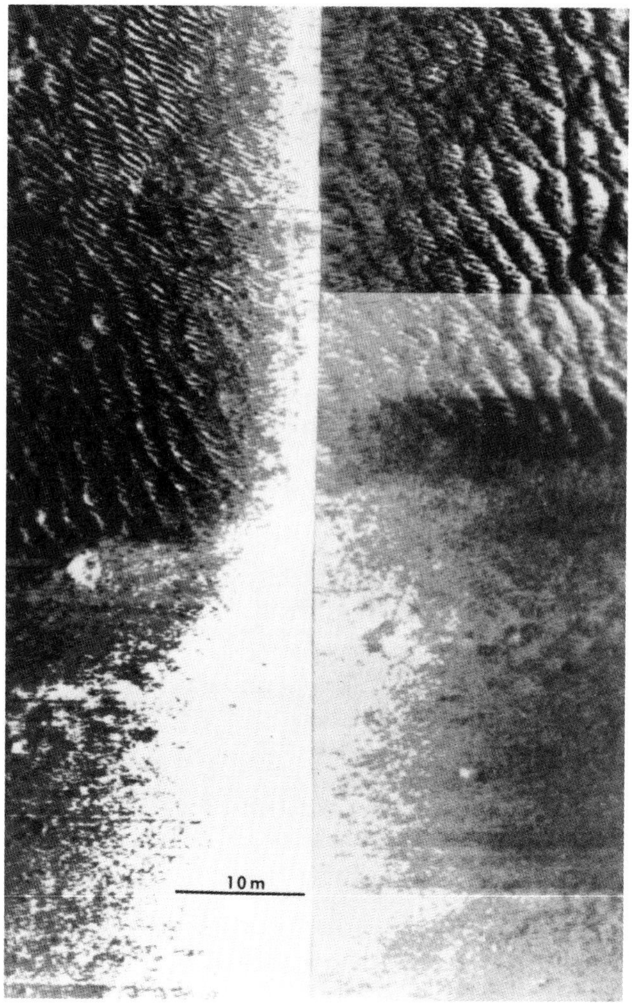

FIGURE 24. A 100-kHz sidescan sonar image, 50-m swath. Superimposed ripple sets are bounded by a sandwave flank, illustrating resolving ability for bottom features.

The LAZ 72 recorder output is an analog paper record of echo amplitude versus travel time. The echo strength is measured by the EMG2 in 10 available adjustable depth intervals in the sea floor. The relative echo strengths are displayed as continuous lines on the analog record (Figure 25). As the echo return from a depth interval increases, the line spread also increases, indicating a more reflective sediment such as a sand. With a soft sediment, the lines displayed are very close together, indicating little reflectance of the acoustic signal. Echo strength values correspond to an average measurement of energy reflected over the time-interval set and not to discrete sediment layers. Minor fluctuations in echo strength due to ship's motion or reflection variations are compensated for by averaging the measured values over an adjustable number of sounding periods.

At the time of the test, no digital recording of the echo strengths was possible, although their output was available at a serial computer interface. This lack of quantitative information left the interpretation of the echo strength lines on the analog record to the operator, which required a subjective decision as to the sediment type and consistency based on operator experience with the system. However, with a minimum of experience, sandy sediments

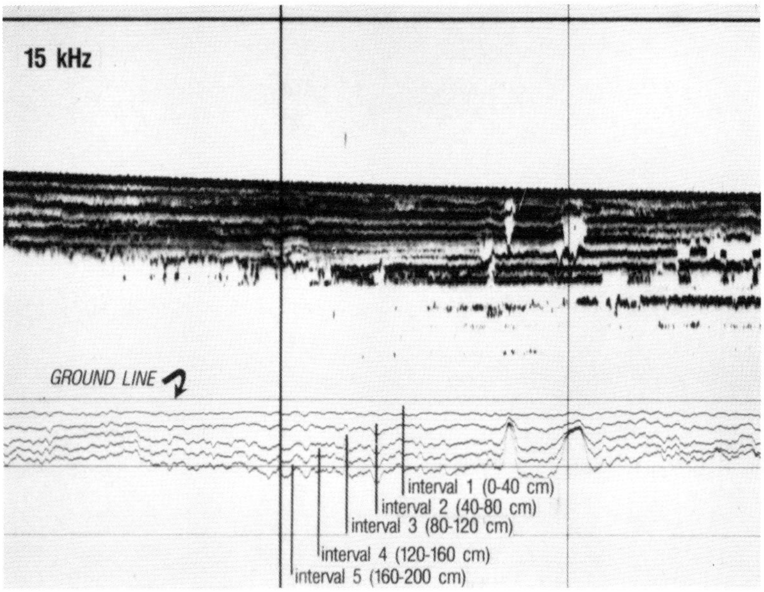

FIGURE 25. Example of ASCS analog record showing echo strength lines.

could be readily differentiated from muds. In their description of the similar tests, Howard et al. (1983) found that sediment porosity estimated off the analog record correlated well with porosity measured on ground-truth samples at the various test sites. This correlation indicated that the system had promise as a sediment classifier: if the echo strength data could be output to a computer where data could be stored; if impedance values of each of the sediment time windows could be computed; and, through empirical relationships, if estimates could be made of the sediment structure, type, and various geotechnical properties.

In 1984, NOARL purchased a system similar to that demonstrated earlier at Panama City. A software package was developed that would allow the recording of the raw echo strengths from each of the ten successive sediment depth intervals in computer memory, while filtering, deconvoluting, and displaying structure, impedance, sediment type, or echo strength in real time as tracklines are run. The first field test of the computerized system was run off Gulfport, MS in June 1985. This system was named the NOARL Acoustic Seafloor Classification System (ASCS).

The ASCS is a high-resolution seismic system that both quantifies and qualifies the acoustic signal returning from various reflectors at the surface and in the sea floor. These measurements are computed by measuring the echo return amplitude and pulse character in ten adjustable time windows that correspond to depth increments in the sediment. Through its microprocessor, the system can then apply algorithms based on multilayer acoustic theory to compute acoustic impedance for each of the depth increments in the sediment. This profile of sediment acoustic impedance is then used in a post-processing mode in combination with the algorithms, primarily based on relationships developed by Hamilton (1980, 1982), to predict sediment structure and type, as well as such geotechnical properties as attenuation, density, porosity, shear strength, sound velocity, and grain size.

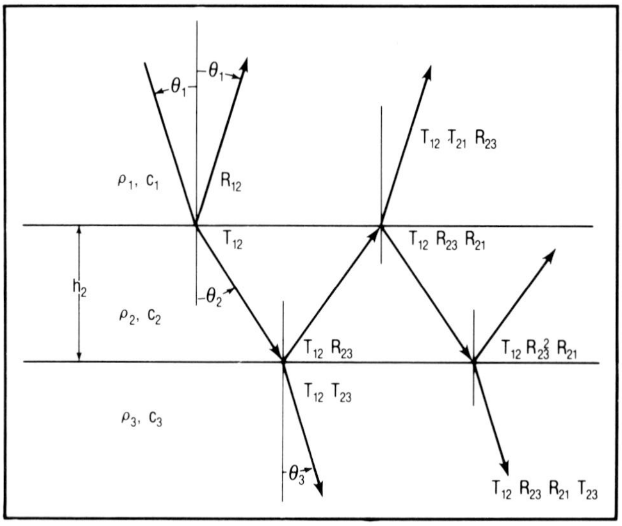

FIGURE 26. Schematic of the reflection and transmission of a plane wave in a layered medium.

2. ASCS Theoretical Considerations and Limits of the Method

The echo strength measuring techniques used with the ASCS are based on conventional multilayer acoustic theory and assume normal incidence with the sea floor. The algorithm used considers transmission, reflection, and attenuation of the signal at each of the boundary surfaces (Figure 26). The impedance (the product of density and sound velocity of the medium) is the property that determines the amount of energy reflected when sound energy passes from a medium of one impedance into another medium of a different impedance. Therefore, the measured acoustic signal reflected from the sea floor can be used to calculate acoustic impedance for each medium. This calculation can be done by time-windowing the return signal, where the amount of energy (echo strength) reflected from various depths in the sea floor can be measured electronically. Each value of echo strength for each time window (corresponding to a particular depth in the sediment) can then be deconvoluted and converted to a reasonable estimate of impedance. The ASCS algorithm can convert ten adjustable-width time windows, assuming an echo of adequate amplitude is received. A more detailed description of the acoustic theory used in the operation of the ASCS can be found in Lambert (1989).

In any system that uses theoretical considerations to apply measurements made of natural phenomena to predict other characteristics, certain compromises must be made. It is often possible to minimize the deleterious effects of such compromises by modeling methods and operational procedures so that the predictions will be satisfactory for the task at hand. During the design and development of the ASCS, several of these compromises were made, but their effect was minimized by the system design; others can be minimized through special operating procedures and "tuning" of the system to a particular environment. Several of the limitations associated with acoustic classification of sea floor sediments and their effect on data output are discussed in the following paragraphs.

The theory assumes that a plane wave strikes the sea floor at normal incidence. Since all transducers must have some directionality, it is virtually impossible to have all of their energy strike the sea floor with an angle of incidence of 0°. In a sediment classification system, it is important for as much energy to intersect the sediment surface in a beamform that is as concentrated as possible. This concentration reduces extraneous scatter and keeps the energy footprint small. With a wide beamwidth there is also a difference in propagation time to a planar surface from the center of the beam to the margin. This difference tends to spread the returning energy from the plane over a wider time interval, thus descreasing the total echo strength at a given instant in time.

Ideally, only very narrow beam-width transducers should be used for sea floor classification. For frequencies used in sediment classification (10 to 30 kHz), however, narrow-focusing transducers are too large and unwieldy. The ASCS compromises here by using transducers (15 and 30 kHz) with a relatively narrow (12°) beamwidth. The problems of keeping the acoustic footprint small and the difference in propagation time from the center of the beam to the margin are fairly well compensated for by towing the transducer near the sea floor and by using time windows over which the echo strength is integrated.

This theory demands an extremely short pulse of a constant amplitude because the pulse length determines spatial resolution in the sea floor. In reality, this function is difficult to produce, since there is a gradual buildup of energy to maximum amplitude followed by a gradual decay of energy. In other words, the sound pulse is never clean with a constant amplitude of energy. This can be compensated for by using the widest bandwidth possible, which will increase the percentage of the pulse that is at maximum amplitude, since the buildup and decay will remain constant. However, the efficiency of the transducer decreases with an increase in bandwidth, and the noise will increase because a wider band is necessary in the receiving channel. In the ASCS, the shortest transmitted pulse length used is about four wavelengths, which corresponds to a pulse length of 40 cm (in water) at a 15-kHz frequency. In the 30-kHz system, the pulse length will be approximately 20 cm. These, then, are the minimum vertical resolutions of the two frequencies.

Some propagation losses in the water column are due to plankton, fish, and various particulates. These incidents can affect echo strength measurements by attenuating only a portion of the transmitted and received energy. Most of this problem can be eliminated in the ASCS by towing the transducer near the sea floor and adjusting the time-variable gain setting.

The maximum depth of penetration into the sea floor is determined by the available dynamic and applied frequencies. A fine sand has an attenuation constant of about 0.8 dB per wavelength (λ). A dynamic output of 40 dB, then, corresponds to 50 λ. Therefore, a frequency of 15 kHz would have a maximum penetration of approximately 2.5 m (two-way travel time). However, in a mud with an attenuation of 0.1 dB/λ the signal would penetrate approximately 20 m. An applied frequency of 30 kHz would have a maximum penetration one-half that of 15 kHz. The ASCS, with the ten time windows at approximately 40-cm depth intervals (15 kHz), will measure echo strength 4 m deep and 2 m deep for the 20-cm increment (30 kHz) time windows in soft sediments.

To simplify the classification algorithms, the ASCS uses a constant sound velocity of 1500 m/sec to compute the water depth and time windows in the sediment, resulting in a small and relatively insignificant error in these computations. This error is most noticeable in such high-velocity sediments as sand, where the error would be greatest. However, because the penetration of the acoustic energy is limited (due to high attenuation), this error is fairly insignificant. Velocities of high penetration sediments (muds) are near 1500 m/sec, which almost negates the problem.

Bottom roughness can affect the ASCS results somewhat due to scatter of the transmitted energy off the sea floor. Much of this problem can be eliminated by towing the transducer near the sediment surface to keep the transmitted energy concentrated over a small area, thus decreasing the chance for scatter away from the receiver. Although the magnitude of thc bottom roughness effect is not known for the present ASCS configuration, it is being investigated.

While towing an acoustic system, the acoustic signal must stay constantly in a normal incidence to the sea floor. Thus, motions due to the pitch and roll of the transducer platform must be eliminated. Otherwise, at least a portion of the transducer's energy will be transmitted obliquely and will be lost to scatter. This loss can be minimized by towing the transducer in a stable towbody, which will keep the transmitted energy perpendicular to the sea floor. Up-and-down motions can often be nearly eliminated by towing the system from an elastic member.

Because the ASCS integrates its output over fixed vertical time windows in the sediment, it is likely that in some of these intervals it will be averaging the echo strength vertically across two media with divergent properties. In an extreme example, such as a sand and a mud interface, the error will be only about 10% of the measured echo strength due to the inherent influence of attenuation on the algorithm (F. Kamkalow, personal communication, 1986). However, the output from the computerized system will give little information about the actual number of sediment layers, particularly if they are small, because of the integration of several small layers over the time windows. This integration is not a serious problem, since any significant reflector will be visible on the analog record.

The previous discussion was primarily concerned with homogeneous sediments that contain distinct layers several wavelengths thick. The analysis and interpretation of these sediments is straightforward, using the theory previously discussed. However, a problem can exist in the interpretation of inhomogeneous sediments, i.e., sediments that contain numerous thin layers with alternating strong and weak acoustic reflectors. This situation most often occurs in mud sediments where sand, silt, or even shell materials can form thin, interbedded reflectors. This type of sediment will produce numerous acoustic reflections that are of short duration and alternate between strong and weak echo strengths. During an acoustic survey, the receiver will encounter a series of echoes of about the same amplitude from all sediment intervals because of the way the ASCS integrates the echo strengths over at least four wavelengths in the individual time windows. If the acoustic impedance is determined from the echo characteristic by using the homogeneous sediment algorithm, then the system will predict a continuous increase in impedance with depth in the sediment. This prediction would not necessarily represent the true sediment impedance profile. Therefore, when applying the algorithm, reflections caused by homogeneous layers with an increase or a decrease in the acoustic impedance must be distinguished from those reflections caused by anisotropically distributed reflectors from inhomogeneous sediment. The ASCS computerized system makes this distinction by first recognizing the sediment as either homogeneous or inhomogeneous from the characteristics of the returning signal. If the system recognizes an inhomogeneous sediment, then the computer will calculate the impedance profile using a slightly modified algorithm.

The theory assumes that the transition between two media happens abruptly. Again, nature does not always cooperate with theory. In this instance, it is fairly common for environmental conditions to change over a long period; thus, there may be a gradual transition in sediment layers from one sediment type to another over a meter or more in depth. An extreme example of this transition can be found in dredged channels where fine materials often settle out and form suspensions of very low and only slightly increasing density. In such cases, even the sediment/water interface is difficult to distinguish. It is therefore possible for the ASCS (particularly the 15 kHz) to miss the upper surface of this suspension because of the very slight increase from the density of the overlying water to the density of the suspension. However, using a transducer of a higher frequency allows a smaller density change per wavelength to be identified.

Thus, it is evident that an acoustic sediment classification system such as the ASCS cannot absolutely predict sediment properties. However, since the ASCS is based on sound acoustic theory, it does have the capability to reliably predict sediment properties while in a rapid survey mode — if certain operational and data processing compromises are made.

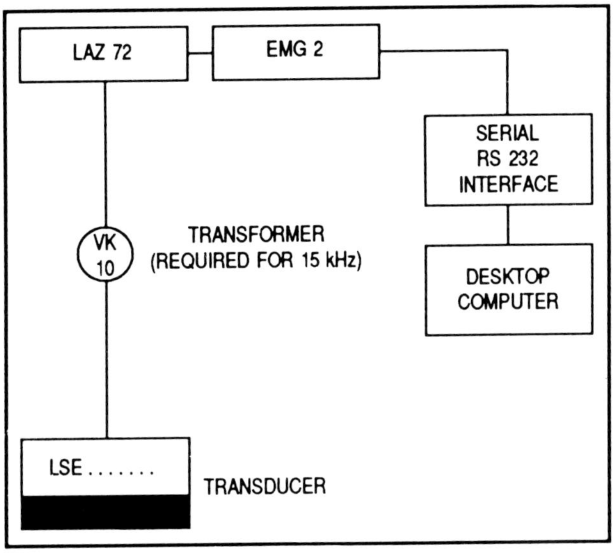

FIGURE 27. Hardware schematic of the computerized ASCS.

3. Acoustic Sea Floor Classification System Description

a. ASCS System Hardware

1. General

The ASCS consists of four hardware components: a central signal processor and system controller, the EMG2; an analog recorder, the LAZ 72; a transducer, either the LSE 179 or LSE 131ZR (15 and 30 kHz, respectively); and a central data storage and processor, which is presently a desktop computer (Figure 27).

2. Echo Strength Measuring Unit

The EMG2 is microprocessor controlled and is the central processor of the echo strength measurements. This unit provides adjustments for the various configurations of the complete system, including digital depth measurements, signal processing, echo strength measurement, and data output display.

The EMG2 was specifically designed to investigate the structure and geological features of the sea floor. It is based on conventional echo-sounding technology but is capable of quantitatively measuring the returning echo intensity. The EMG2 initially functions as a precise digital depth measuring device that can recognize the sediment-water interface, even in extremely soft sediments. In conjunction with the LAZ 72 recorder, a fixed velocity of 1500 m/sec is used to compute water depth below the transducer. Water depth is displayed on a four-digit LED indicator either in decimeters or meters, depending on water depth. The EMG2 automatically suppresses stochastically occurring interferences in the water column by computing a sliding average over several sounding periods; the number of periods is switch-selectable from one to eight. A high-contrast display of objects, such as fish, etc., in the water column is possible, since the averaging mode can be turned off. The system provides for gain adjustment, which also will allow the display of weak mid-water reflectors, such as haloclines and thermoclines, on the echogram. Another gain switch increases the contrast of reflectors in the sea floor by using a depth-controlled amplifier, which is programmable to the transducer being used. This amplifier is capable of up to a 40-dB increase in processed signal contrast. High resolution of the analog output on the recorder is provided by the display of the received signal with a resolution of five pixel elements per mm.

The EMG2 allows the starting depth for echo strength measurement to be varied. Normally it uses automatic bottom-catch circuits to determine the upper surface of the sea floor. This function can be overridden so that the echo strength of water column discontinuities, concentrations of suspended matter, or schools of fishes can be measured.

This system will suppress anomalies in echo strength due to ship motion or reflection conditions by averaging the measured echo strength of each depth interval over an adjustable number of sounding periods from 1 to 64. If this averaging interval is set too wide, then the large number of data points averaged will tend to mask subtle areal variations in the sediment, which would be perceptible with a narrower averaging period. An echo strength amplifier also is provided that allows gain adjustment between −6 and +6 dB in 2-dB increments.

3. Analog Recorder

The output from the EMG2 is displayed on the LAZ 72 recorder similar to a conventional high-resolution subbottom profiler where the echo amplitude vs. travel time is presented graphically. Also displayed is the relative echo strength of either the first or second five of the ten successive switch-selectable time intervals (Figure 25). These intervals are displayed as continuous lines across the chart. The echo strength of the first depth interval is the difference between the ground line and the first line. The echo strength of the second depth interval is the difference between the first and second lines below the ground line. The third, fourth, and fifth intervals are displayed similarly. In this way, no two echo strength lines can cross each other. It is also possible to continuously monitor on an oscilloscope the returning echo strength of each of the ten intervals for each ping. This output is available through BNC jacks on the EMG2 front panel.

The echo strength is also output to the desktop computer via a series RS-232 computer interface. This allows the storage of the raw echo strengths for each of the ten intervals of each ping on the computer hard disk. The computer then will normalize, filter, deconvolute, and display on its color monitor the unfiltered echo strengths, the filtered echo strengths, impedance, or the sediment structure and sediment type in real time while a trackline is being run.

The LAZ 72 recorder is a high-resolution, graphic strip chart recorder that uses standard 9-in.-wide, dry electrosensitive recording paper for display. This unit also contains both the transducer signal transmitting and receiving circuit boards. Controls on the recorder allow the selection of pulse width (0.3, 1.0, 3.0 msec), power output (100 or 1000 W), depth range (15 combinations), and echo amplification (switchable in 11 steps of 6 dB each). The LAZ 72 recorder triggers the output pulse by its styli and operates in either a one- or two-styli mode. Other controls include an amplitude variable/time variable (AV/TV) switch, selection of 20 or 40 log in the TV mode, chart speed (five speeds and off), depth reference lines that divide the paper into five equal increments, and a contrast booster.

4. Transducers

Two transducers have been used with the computerized ASCS system, a 30 kHz (LSE 131ZR) and a 15 kHz (LSE 179). Transducers of several other frequencies are available for use by simply changing the transmitter and receiver boards. The LSE 179 transducer can be operated at either 15 or 30 kHz. In the 15-kHz mode, a transformer (VK10) must be used in line between the transmitter board and the transducer. Both transducers have narrow (12°) cone angles. The LSE 179, if operated in the 30-kHz mode, has a cone angle of only 6°.

Two different methods were used successfully to tow the 15-kHz transducer. Since it is fairly large (0.56-m diameter, 0.15-m height, and 80-kg weight) and since it was expected to be towed from fairly small ships-of-opportunity, we chose to mount it on a catamaran which was designed to be only slightly buoyant and that could be launched at small boat

FIGURE 28. Photograph of the catamaran used to tow the 15-kHz transducer.

FIGURE 29. Photograph of the "tadpole" type towbody used to tow the 30-kHz transducer.

ramps if necessary (Figure 28). The transducer is mounted to a vertical post containing a rack-and-pinion gear driven by an electric motor. The motor allows the transducer to be remotely lowered 1.3 m below the catamaran for towing to keep it well below the turbulence caused by the pontoons. The small positive buoyancy of the catamaran allows it to cut through waves as it is towed. The large transducer mass suspended below the catamaran also provides stability to the sled. Measurements of sled motion using inclinometers indicate this system limits transducer motion to $<1.5°$ from normal incidence to the sea floor in seas up to 5 ft. The 15-kHz transducer also has been mounted in a large, enclosed towbody (Figure 29). To eliminate transducer motion, the transducer was suspended from a two-axis gimbal inside the towbody. Since the towbody is normally hung from an outrigger attached to the ship, it is subject to the ship's vertical motions. Much of the motion can be eliminated by towing the fish from large, elastic members.

The 30-kHz transducer is normally towed inside a "tadpole-type" towbody with excellent results. All three tow vehicles are towed from the side of the ship, outside its wake, to eliminate acoustic signal interferences due to prop wash.

5. Computer System

The desktop computer system consists of a central processing unit (CPU), a keyboard, a printer, and a high-resolution color video monitor. The CPU has 512 kilobytes (kbytes) of main memory (only 32 kbytes are available for computation), a 10-megabyte (Mbyte) hard disk, and dual 400-kbyte floppy-disk drives. An extended bit-map graphics module was added to the CPU to output the color display of impedance and sediment properties to the red-green-blue color monitor. The monitor has a displayable resolution of 960 to 240 pixels. The software that is used for data storage, display, and output uses the operating system (P/OS) and the tool kit extensively.

b. ASCS Software

The software was designed to be a very flexible research tool; as the user's knowledge of the system increases, this knowledge can be input into the software databases to produce improved results. The software is menu driven, similar to P/OS, which makes it user friendly. The P/OS and tool kit were used as the foundation for the software design. Pascal was used as the software language, since it is a structured language and any necessary modifications would be straightforward. The P/OS, the tool kit, and the ASCS software take up 5 Mbytes of the 10-Mbyte hard disk. This is somewhat of a problem, because a relatively small amount of disk space is left for data storage and reprocessing.

The software package provides a fairly simply way to store data (echo strength, water depth, and time) and allows the user to do extensive data reprocessing and interpretation (Figure 30). Once the raw echo strengths are stored on the hard disk, these data can be processed through three separate databases to output sediment structure and type; impedance; and several geotechnical properties, such as wet unit weight, porosity, and sound velocity, attenuation, and mean grain size, for each of the ten depth increments. While the data are being collected, the operator has the option of displaying unfiltered echo strengths, filtered echo strengths, or sediment structure and sediment type in a scrolling tabular format on the CRT. It is possible to output a 16-color scrolling display of sediment impedance with depth of the ten intervals along with time, water depth, and a structure code. The program gives priority to the storage of the raw data during data collection. Because of the large volume of data being collected (usually ten incremental echo strength measurements, water depth and time for each ping) every 0.5 sec, the complexity of the programs, the relatively small operational memory of the computer, and the slowness of the computer, only about 15% of the collected data can be displayed on the CRT in a real-time mode.

The existing software package is made up of four central programs: INITPR, an initiation program; ONLINE, which is used to collect, store, and display data; OFFINI, which is used to reprocess data; and EMGR, which is used to output the processed data to a printer.

INITPR initiates ONLINE (Figure 31) and allows the operator to input information about the components of the system being used (transducer, etc.), the various amplifier gain settings, and LAZ 72 recorder settings. Once these are input, the operator selects the databases to be used during data collection and the display function to be output to the CRT. The program finally goes through a calibration sequence where an initial normalization factor is calculated. This factor is then used throughout the data acquisition process to normalize the data to the system configuration being used, to the various gain settings, and to the environment. Because of the way this "parameter file" is developed, the software is independent of the frequency used for sediment classification. Calibration is usually done over a sand bottom because the reflection coefficient, which is a required input, is usually near 0.5 and

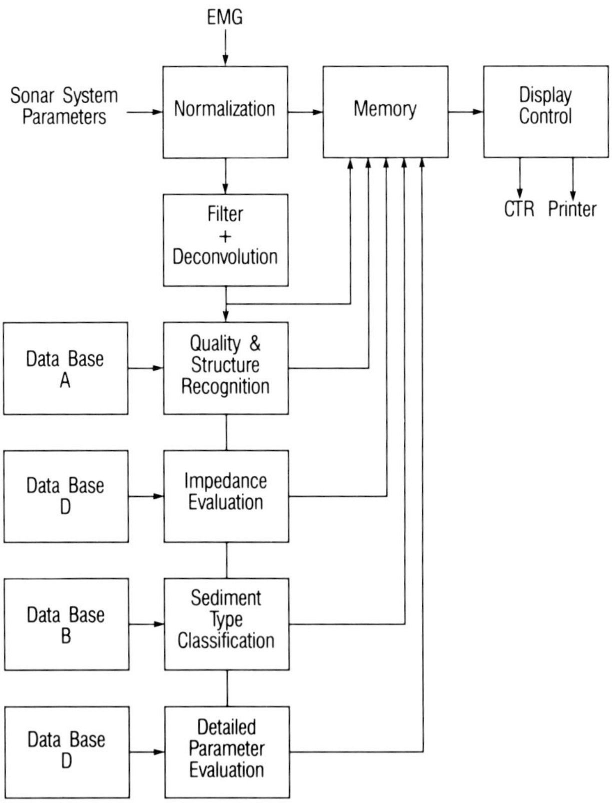

FIGURE 30. Schematic of the ASCS software structure.

is fairly predictable. Once this procedure is completed, the computer stores the information as a parameter file, which is used by the ONLINE program to adjust and output the data to the CRT. The parameter file has no effect on the raw echo strengths being stored on the hard disk, so to produce better results the parameter file can be adjusted later if necessary, or different or modified databases can be selected during the reprocessing stage.

ONLINE stores the incoming raw echo strength measurements on the hard disk as a data file, normalizes the data, filters the data for extraneous values, and deconvolutes it so that the echo strength measurements with depth in the sea floor are accurate (Figure 31). ONLINE then displays as much of the processed data as the available computer time allows on the CRT. This is done according to the databases and display function stored in the parameter file.

OFFINI provides a means to reprocess the stored echo strength measurements in the laboratory. It uses many of the same menus and subprograms as ONLINE but not the data acquisition subprograms. Instead, it has expanded data processing capabilities, including the ability to output values of specific sediment parameters computed from the calculated impedance values via a series of regression equations. During the data reprocessing, OFFINI first provides for the choice of a parameter file to be used for normalization, filtering, and deconvolution of the raw echo strength measurements. It is possible to use the parameter file that was used during data collection, to modify it, or to use a different file for these calculations. Once knowledge of the ASCS system and how it operates under various conditions and in different environments is better understood, it is probable that little or no adjustment of this parameter file will be necessary for reprocessing. Once the parameter file has been selected, the operator can choose the type of data output (structure and sediment

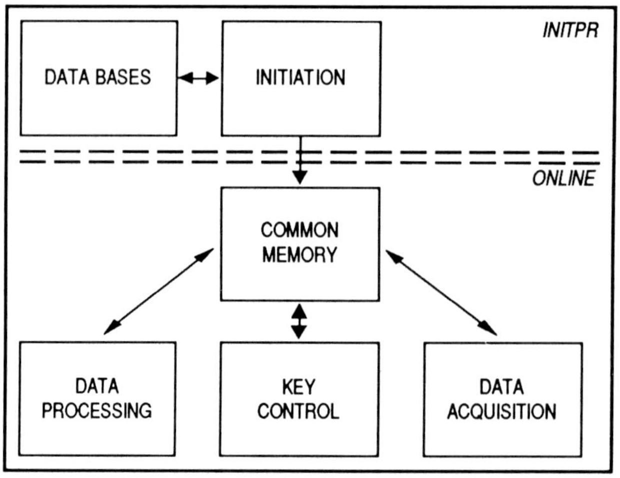

FIGURE 31. Schematic of software structure used during system calibration and data collection.

type, impedances, unfiltered and filtered echo strengths, or the various sediment parameters) and the databases to be used. The choice of various databases is important because they allow the operator to select specific relationships that were developed for a particular environment; that is, the acoustic reflectivity characteristics of a terrigenous sediment will be different from those of carbonate, red clay, siliceous, or Arctic high-silt-content sediment. The other geotechnical properties of these various sediment types also will exhibit wide variation. The reader is referred to Hamilton (1980, 1982) for summarization of much previous research on the geoacoustic properties of various environments. A detailed explanation of the development and use of the various databases can be found in the following section.

Each time OFFINI determines the sediment type or any of the geotechnical properties, it first computes an impedance profile for each acoustic ping using Database A. This value is center averaged over the number of pings set in the parameter file. The impedance profile is then used in conjunction with Database B to predict sediment type and with Database D to estimate any of the several geotechnical properties. The estimate is done by simply comparing the impedance value to a value of the sediment property that has been previously programmed into the database.

The EMGR program simply provides for output to a hard-copy printer the various sediment properties, impedance values, and filtered and unfiltered echo strengths stored as files on the computer hard disk. It provides a way to print either all or only portions of the data file. Alternatives are determined by selecting a time interval over which the data of interest were originally collected.

c. ASCS Databases

Three databases are critical portions of the ASCS software. Database A determines the impedance profile of the sea floor sediment, Database B predicts the sediment type, and Database D estimates the various geotechnical properties. Databases A and B work in conjunction to output impedance, sediment structure, and sediment type. All three databases are used in conjunction to output the geotechnical property estimates.

The three databases were programmed with one standard set of subdatabases called "Continental Terrace". These subdatabases were named after the physical properties relationships developed by Hamilton for a shallow-water terrigenous environment. Although

DATABASE A

1 homogeneous layer
2 strong reverberation
3 inhomogeneous layer over layer
4 layer over inhomogeneous layers
5 several homogeneous layers
6 inhomogeneous layers over several layers
7 several layers over inhomogeneous layers

ORDER OF SEQUENCES

STRUCTURE	1	2	3	4
1	A	A	A	A
2	A	A	A	A
3	A	A	A	A
4	A	B	C	D
5	A	B	C	D
6	A	A	A	A
7	A	A	A	A

POSSIBLE SEQUENCES OF TRANSITIONS

A + + + + + + + + + +
B + − + + + + + + + +
C + + − + + + + + + +
D + − + + + + + + + +
E + + + + + + + + + +
F + + + + + + + + + +
G 0 0 0 0 0 0 0 0 0 0

+ TRANSITION LOW/HIGH IMPEDANCE
− TRANSITION HIGH/LOW IMPEDANCE

LIMIT OF IMPEDANCE : 1.50 - 4.00

SEA AREA : REF15

FIGURE 32. Computer screen display of the current Database A.

these subdatabases were developed primarily from the Hamilton relationships, they were inadequate for predicting impedance profiles and sediment properties during the initial tests of the ASCS system off Gulfport, MS. Therefore, new versions of Databases A, B, and D have evolved, based on considerable ground-truth sampling during ten cruises with the ASCS. These versions were used to collect and process a large amount of ASCS data with a reasonable degree of accuracy. Each of these three databases allows the user to program and store five complete sets of subdatabases specifically developed for different environments, in addition to the Continental Terrace versions. Once stored on the hard disk, these versions can be recalled for data collection or data processing.

In general, the ASCS software uses two basic and critical algorithms based on multilayer acoustic theory to calculate the sediment impedance profile. These algorithms are based on the hypothesis that sediment structure can be considered either homogeneous or inhomogeneous. A homogeneous sediment can be thought of as a vertically uniform mud or sand with little change in sediment properties throughout the given layer or, in this case, the time window set. An inhomogeneous sediment is defined as a time-window interval containing one or more reflectors, such as a mud containing thin sand layers. The ASCS software is designed to recognize these two sediment structures from the return echo and to assign one of the seven structure codes stored in Database A (Figure 32). The sediment impedance profile is calculated using one of these slightly different algorithms or both in combination, depending on the recognized structure code.

Database A is used to describe the sea floor in terms of its generalized structure (Figure 32) and to compute an impedance profile for the ten sediment depth intervals. The computer first recognizes the sediment structure through an algorithm that uses the deconvoluted echo strength values. The computer then calculates an impedance profile using the transitional relationships found in Database A. It must use Database A, since it is not able to discriminate across a reflection boundary of different acoustic impedances, whether this transition is from high to low or low to high impedance. Therefore, Database A has been programmed with seven possible sequences of transition between the boundaries.

The first four (A, B, C, and D) are fixed, but the fifth and sixth (E and F) are adjustable for a particular environment. Transition G indicates to the evaluation program that no further transitions are to be considered. For a particular sediment structure, the computer tests the transition sequence listed under Column 1 in the order-of-sequences table.

```
                         DATABASE B

                  IMPEDANCE

                                 1.  1.  2.  2.  2.  2.  2.  3.  3.  3.  3.  3.  4.
                                 6   8   0   2   4   6   8   0   2   4   6   8   0

TYPE OF homogeneous     G N N H H E E K K J B B B B M M M M M M A A A A A
LAYER inhomogeneous     G N N H H E E K K J B B B B M M M M M M A A I I I

     TYPES OF SEDIMENTS
     A  coarse sand                    H  sand-clay
     B  fine sand                      I  gas
     C  sand silt                      J  silty sand
     D  sil                            K  clayey sand
     E  sand-silt-clay                 L  silty clay sand
     F  clay-silt                      M  medium sand
     G  clay                           N  silty clay

            SEA AREA  REF15
```

FIGURE 33. Computer screen display of the current Database B.

If the calculated impedance profile goes out of range of the impedance limits (1.50 to 4.00) shown in the lower portion of Figure 32, then the computer will test the sequence indicated in Column 2. It will continue to test in Columns 3 and 4 until all of the impedance values stay within range. When they do, the computer accepts this sequence and computes and stores the impedance profile for that particular acoustic ping. Should none of the four sequences lead to a successful result, the first (Column 1) will be used (default sequence) to calculate the impedance profile. The impedance profile is considered by the computer to be valid until the impedance range is exceeded. When this happens, the rest of the depth intervals will be assigned the last reasonable impedance value. The computer then uses this impedance profile to compute sediment type and the geotechnical properties from the relationships found in Databases B and D, respectively.

Database B (Figure 33) compares the impedance profile to the sediment type listed in the table by its corresponding impedance range. Note that two rows of letters (A through N) relate sediment type to the impedance range. One row is for homogeneous sediments, and the other is for sediments that the structure recognition program determined as inhomogeneous. This relation is not particularly important, except possibly in a case where gassy sediments are identified by strong reverberation and give a high impedance value. In the homogeneous row the sediments would classify as a coarse sand instead of a gassy sediment.

Further experience with the ASCS system may indicate other acoustic differences between homogeneous and inhomogeneous sediments and that the sediment-type database should be further modified. The ranges of impedance that correspond to sediment types is shown in Table 4.

Database D relates the ASCS-calculated impedance values to various geotechnical parameters, such as density, porosity, grain size, sound velocity, and attenuation. This database provides the prediction of up to eight different parameters, one at a time. Preprogrammed initially into the database are relationships based, at least partially, on work done by Hamilton for the previously mentioned sediment properties. Extensive testing of these relationships and comparison of the data output to ground truth has led to the system's improvement in sediment property predictive capability.

Three blank files are provided for user-determined relationships. Since Database D is expandable, five separate subdatabases for different environments can be stored and recalled on demand for comparing the impedance values to various sediment properties. Recent NOARL research led to the development of a reasonable empirical relationship between acoustic impedance and sediment shear strength (Lambert, 1989).That relationship is presently programmed into one of the three extra database files.

TABLE 4
Impedance Values vs. Sediment Type Programmed into Database B

Impedance	Homogeneous sediment	Inhomogeneous sediment
1.50—1.60	Clay	Clay
1.61—1.70	Silty clay	Silty clay
1.71—1.80	Silty clay	Silty clay
1.81—1.90	Sandy clay	Sandy clay
1.91—2.00	Sandy clay	Sandy clay
2.01—2.10	Sand-silt-clay	Sand-silt-clay
2.11—2.20	Sand-silt-clay	Sand-silt-clay
2.21—2.30	Clayey sand	Clayey sand
2.31—2.40	Clayey sand	Clayey sand
2.41—2.50	Silty sand	Silty sand
2.51—2.60	Fine sand	Fine sand
2.61—2.70	Fine sand	Fine sand
2.71—2.80	Fine sand	Fine sand
2.81—2.90	Fine sand	Fine sand
2.91—3.00	Medium sand	Medium sand
3.01—3.10	Medium sand	Medium sand
3.11—3.20	Medium sand	Medium sand
3.21—3.30	Medium sand	Medium sand
3.31—3.40	Medium sand	Medium sand
3.41—3.50	Medium sand	Medium sand
3.51—3.60	Coarse sand	Coarse sand
3.61—3.70	Coarse sand	Coarse sand
3.71—3.80	Coarse sand	Gaseous
3.81—3.90	Coarse sand	Gaseous
3.91—4.00	Coarse sand	Gaseous

d. Measurement Data

A data set is described for the types of sea floor classification data and the ASCS can produce, as well as the accuracy and repeatability of the data (in relation to ground truth sediment cores). This data set was collected during an evaluation of the system collected in well-known testbed areas of St. Andrews Bay near Panama City, FL. Although data were collected from six different test areas with varying sediment composition and geotechnical properties, data from only two of these sites are presented here. The S6 site was chosen to demonstrate the ASCS capabilities, since its sediments range from silty clays to medium sands and are typical of the types of sediments found in terrigenous environments. The test results and comparisons to the ground truth sediment cores from the other test sites are comparable to those shown here.

During the Panama City tests,the ASCS 15-kHz transducer was toward on the catamaran described earlier and shown in Figure 28. The catamaran was towed from an outrigger off the stern quarter of the boat. This configuration allowed the transducer to be towed approximately 3 m away from the boat and outside its wake. The 30-kHz transducer was towed in a tadpole-type towbody suspended from the arm of the boat's crane about 3 m off the stern quarter.

Both the 15- and 30-kHz transducers were operated during a total of four transects across the S6 site. A typical 15-kHz high resolution analog record from this site is shown in Figure 34. Site 6 is in an area with transitional sediment types along the trackline. The sediments gradually become finer from the start of the trackline (labeled ''S'' in Figure 34) to the centerpoint (labeled ''C''), where they are classified as silty clays to sandy clays. The ASCS record for this trackline clearly shows this transition in the echo-strength lines displayed at the top of the record. The sediments become coarser from the center to the endpoint (labeled ''E'') as the trackline moves into shallower water. Again, note the transition of these

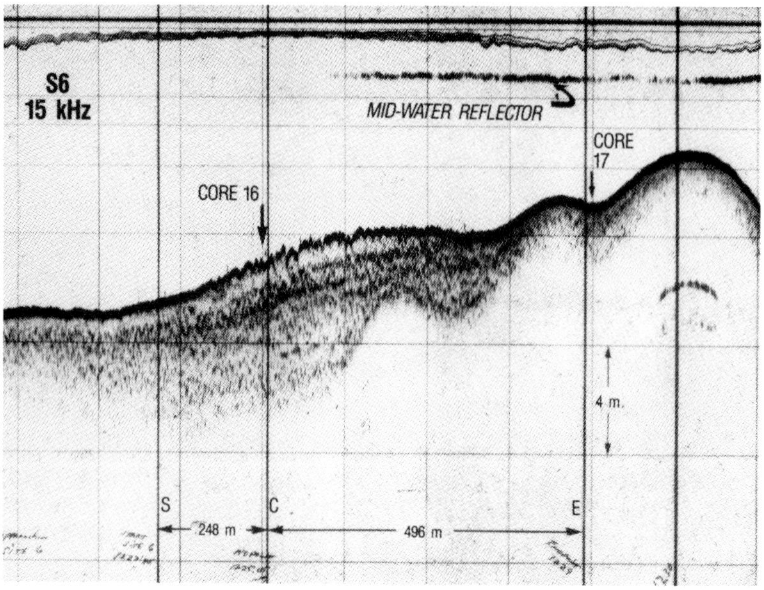

FIGURE 34. A typical 15-kHz analog record from the S6 site.

sediments as shown by the echo-strength lines at the top of the record; as the sediments become more acoustically reflective, their echo-strength lines diverge from the ground line. At the centerpoint the record shows that the sediment surface is quite rough and that maximum height irregularity is 0.5 m. One explanation for this surface roughness is gouging of the soft sediment by shrimp trawls.

Data collected at the centerpoint (Core 16) show a slight increase in sediment coarseness with depth. The ASCS record reflects this coarsening by displaying a transparent layer of soft sediment down to about 1.2 m, where a more reflective sediment is encountered. Maximum penetration of the 15-kHz signal is 6 m at the site of Core 16 (Figure 34). Note the strong midwater reflector shown on the ASCS record from near the centerpoint to past the endpoint. This reflector is probably either a thermocline or a halocline caused by tidal effects.

The sediments at the center of Site 6 (based on Core 16) are composed of muds, which classify as silty clay for the upper 20 cm. From 20 to 120 cm they are sandy clay with sand-silt-clay below. The 15-kHz system predicts silty clay for the upper 80 cm and sandy clay below (Table 5). The 30-kHz system predicts silty clay for the whole sediment column but, like the 15-kHz system, showed a significant impedance change at 80 cm (Table 6). The difference between the core data and the ASCS-predicted data is negligible and is primarily a function of the way Database B was programmed.

In the prediction of sediment structure across this site, the 15-kHz system defines a ''layer over an inhomogeneous layer'', which means it delineates one homogeneous layer for the upper 80 cm and detects several small echo strength returns from below. When this type of prediction occurs, the system switches to the inhomogeneous algorithm and calculates an impedance value for the first inhomogeneous sediment interval (80 to 120 cm). It then records the same value through the remainder of the sediment column because it recognizes the same integrated echo strength throughout the rest of the sediment column — in this case, down to 4 m. The 30-kHz system recognized ''several homogeneous layers'', which in this case means two layers (0 to 80 cm and 80 to 200 cm) at this site. The returning echo

TABLE 5
Comparison of Sediment Type and Structure Predicted by the ASCS to Ground Truth Core Data Collected at the Centerpoint of the 56 Test Area Trackline

		Ground truth			ASCS predicted		
	Core	Interval (cm)	Sediment type	φ Mean	Interval (cm)	Sediment type	Impedance
A. 15 kHz	16	0—40	Sandy clay/silty clay	7.8	0—40	Silty clay	1.72
		40—80	Sandy clay	6.9	40—80	Silty clay	1.72
					— — —		
		80—120	Sandy clay	6.9	80—120	Sandy clay	1.81
		120—160	Sand-silt-clay	6.8	120—160	Sandy clay	1.81
					Structure: Layer over Inhomogeneous		
B. 30 kHz	16	0—20	Silty clay	8.7	0—20	Silty-clay	1.65
		20—40	Sandy clay	7.0	20—40	Silty clay	1.65
		40—60	Sandy clay	7.0	40—60	Silty clay	1.66
		60—80	Sandy clay	6.8	60—80	Silty clay	1.66
				— — —	— — —	— — —	— — —
		80—100	Sandy clay	6.7	80—100	Sandy clay	1.78
		100—120	Sandy clay	6.7	100—120	Sandy clay	1.78
					Structure: Several Homogeneous Layers		

— — — Layer (significant impedance change)

strengths were probably too weak below 120 cm and thus did not exceed the ASCS's minimum threshold. The system then recorded an impedance value of 1.78 for all increments between 120 and 200 cm.

Table 6 compares the core data for wet unit weight, porosity, grain size, and sound velocity to values of these properties estimated by the ASCS (both frequencies) from the calculated impedance data. Both frequencies did an excellent job of predicting the surface values at this site. The 15-kHz system (Table 6) exactly predicted density (1.24 g/cm^3) and porosity (83%) but slightly underestimated grain size and sound velocity by 0.3φ and 5 m/sec respectively. With depth in the sediment, ASCS's ability to predict the core data deteriorated somewhat, but as a core average only underestimated density by 0.08 g/cm^3 and grain size by less than 0.1φ. It slightly overestimated porosity by 2% and sound velocity by 2/msec.

The results with the 30-kHz system (Table 6) are very similar to the 15-kHz system; differences throughout the data set are almost negligible. The greatest variations occurred in the mid-depth values of density and porosity, where maximum divergences of 0.15 g/cm^3 and 10% were found. However, over the first 160 cm of sediment, the 30-kHz system variation averaged only 0.08 g/cm^3 in density, 4% in porosity, 0.1φ in grain size, and 7 m/sec in sound velocity from that measured on Core 16.

Another particularly impressive aspect of the ASCS performance during the processing of the Panama City test data was the repetitiveness with which it predicted the impedance profile at another test site, Battleship Buoy (Table 7). Nine short tracklines were run across the centerpoint of this test area; two were parallel but slightly offset to each other and made with the 15-kHz transducer (Runs 1 and 2). The following day the same two tracklines (Runs 1 and 2) were repeated with the 30-kHz system. After using the system for several hours to survey two other test sites, five crossings were made perpendicular to the original Battleship Buoy tracklines (Crossings 1 through 5). Each trackline was slightly offset from

TABLE 6
Comparison of Sediment Geotechnical Properties Predicted by the ASCS to Ground Truth Core Data Collected at the Centerpoint of the 56 Test Area Trackline

	Interval (cm)	Wet unit weight (g/cm^3)			Porosity (%)			Grain size (mean ϕ)			Sound velocity (m/sec)			Impedance
		Cores	ASCS	Δ	Cores	ASCS	Δ	Cores	ASCS	Δ	Cores	ASCS	Δ	
A. 15 kHz	0—40	1.24	1.24	0	83	83	0	7.8	7.5	−0.3	1530	1525	−5	1.72
	40—80	1.36	1.24	−0.12	76	83	+7	6.9	7.5	+0.6	1530	1525	−5	1.72
	80—120	1.35	1.26	−0.09	77	79	+2	6.9	6.7	−0.2	1531	1535	+4	1.81
	120—160	1.34	1.26	−0.08	79	79	0	6.8	6.7	−0.1	1530	1535	+5	1.81
	160—200	1.38	1.26	−0.12	76	79	+3	7.0	6.7	−0.3	1523	1535	+12	1.81
Average				−0.08			+2			−0.1			+2	
B. 30 kHz	0—20	1.18	1.23	+0.05	88	86	−2	8.7	7.8	−0.9	1530	1515	−15	1.65
	20—40	1.30	1.23	−0.07	79	86	+7	7.0	7.8	+0.8	1530	1515	−15	1.65
	40—60	1.35	1.23	−0.12	77	86	+9	7.0	7.7	+0.7	1530	1516	−14	1.66
	60—80	1.38	1.23	−0.15	76	86	+10	6.8	7.7	+0.9	1530	1516	−14	1.66
	80—100	1.37	1.25	−0.12	76	80	+4	6.7	6.7	0	1530	1531	+1	1.78
	100—120	1.33	1.26	−0.07	78	80	+2	7.2	6.7	−0.5	1531	1531	0	1.78
	120—140	1.34	1.26	−0.08	78	80	+2	6.8	6.7	−0.1	1531	1531	0	1.78
	140—160	1.33	1.26	−0.07	79	80	+1	6.7	6.7	0	1530	1531	+1	1.78
Average				−0.08			+4			+0.1			−7	

TABLE 7
Repetitiveness of the ASCS Acoustic Impedance Profiles During Nine Trackline Surveys Across the Same Midpoint of the Battleship Buoy Test Area

	Interval (cm)	Impedance	
		Run no. 1	Run no. 2
15 kHz	0—40	1.74	1.68
	40—80	1.91	1.84

	Interval (cm)	Impedance						
		Run		Crossing				
		No. 1	No. 2	No. 1	No. 2	No. 3	No. 4	No. 5
30 kHz	0—20	1.64	1.68	1.68	1.66	1.67	1.71	1.66
	20—40	1.65	1.68	1.68	1.66	1.67	1.71	1.66
	40—60	1.82	1.85	1.68	1.66	1.67	1.71	1.66
	60—80	1.83	1.85	1.82	1.82	1.84	1.87	1.82

Layers Indicated from Core 4

0—19: Very soft mud with strong H_2S door
19—39: Soft mud, slightly more sand, strong H_2S odor
39—70: Clayey sand
70—84: Clayey sand, higher sand content (70%)

———Layer (significant impedance change)

the others. Over these nine tracklines across the midpoint, the range of predicted impedances for the first sediment interval (either 0 to 20 cm of 0 to 40 cm) was 0.08. For the second significant layer it was 0.09. The four runs predicted a sediment impedance change at a 40-cm depth, while all of the crossings predicted a change at a 60-cm depth. This difference may be due to the direction of the tracklines and the influence of the 32-ping averaging used to process the data. The runs were started in soft sediment and proceeded past the midpoint to coarser sediments on a ridge. The crossings began on the ridge and proceeded downslope to the midpoint and softer sediments beyond. On the crossing tracklines, the ridge is considerably closer to the midpoint than on the runs. Because of this proximity, the residual effect of the 32-ping averaging possibly influenced the predicted depth of impedance change.

From the preceding test results, it can be concluded that the NOARL ASCS can make reasonable remote predictions of sediment type, structure, and various geotechnical properties while in a rapid survey mode. The ASCS is being transitioned from a research instrument into a routine survey tool for a wide variety of Navy applications. A project is currently underway to convert and upgrade the existing ASCS software to an 80386/80486 microprocessor-based computer, where more real-time output will be available to the operator. The software will also contain its own automated mapping code for producing maps of sediment properties. This project also is conducting a considerable effort to verify and/or enhance the existing ASCS algorithms, which are used to recognize patterns in the acoustic return signal due to sediment structure, type, and composition. Significant improvements in Database A, in particular, are expected from this research. These refinements will enhance the ASCS's capability to remotely classify seafloor sediments while in a rapid survey mode.

C. MICROTOPOGRAPHICAL ROUGHNESS DETERMINED BY STEREO PHOTOGRAMMETRY

1. Technical Background

Using high-resolution sonar mapping systems simultaneously with signal-processing techniques and ground-truth measurements promises significant advances in quantifying small-scale bottom features from remote sensing. Examining bottom roughness on scales of less than 1 m now involves a focused, nonsynoptic approach. Previous measurements of the sea floor microtopography were applied to problems in sediment transport and acoustic bottom interaction. Recent efforts to determine small-scale roughness have involved narrow-beam, high-frequency, close-range sonars, or stereo photogrammetry. Stereo photographic methods have the advantage of higher resolution and two-dimensional representation of the sediment surface. Echo-sounding devices can survey longer scales of roughness but are susceptible to contamination from ship and source-receiver motion. The stereo photographic equipment is not. Thus, accurate measurements of microtopographical roughness can be made, assuming that the cameras are properly calibrated and that proper constraints are realized when compiling photogrammetrical measurements (Briggs, 1989).

2. Methods

Three different types of equipment can profile the sea floor in the course of assessing bottom microtopography. The simplest method for determining the shape of bedforms on the sea floor is to trace the interface. Divers are employed to trace the bottom profile directly onto gridded Mylar sheets attached to a 1.8-m-long acrylic base. The acrylic base is carefully oriented above the bottom with the aid of a carpenter's level and movable metal rods. By carefully following the shape of the bottom with a sliding arm, the interface is traced onto the Mylar with a minimum of sediment disturbance. This method is especially useful when water visibility is poor.

When the optical quality of the water is good, the use of a diver-operated 35 mm underwater stereo camera is preferable. Two 28 mm, water-corrected lenses are separated by 61 mm in the stereo camera, and yield a 55- by 72-cm photographic overlap area at a 91-cm distance from the sea floor. Illumination of the area to be photographed is accomplished with a 100-J underwater strobe mounted in the diver-operated system. Divers control orientation and camera placement and can create a mosaic of stereo photographs to cover a much larger area.

In deeper waters where it is not practical to use scuba divers, stereo photography is conducted remotely with two 70 mm underwater cameras mounted side by side on a balanced steel frame with two 150-J underwater strobes (Figure 35). Various distances between optical axes of the cameras are used, but 40 cm with an inward-directed angle of 4° from the vertical for each camera is optimal for a 122-cm distance from cameras to sea floor. This configuration yields an 80- by 50-cm image overlap area. Cameras and strobes are simultaneously actuated by a weight compass vane, which makes contact with the bottom as the apparatus is alternately raised and lowered while drifting. Despite the inability to create a photographic mosaic of stereo photographs, the wider optical axis and 70 mm format provide superior accuracy and better resolution than the diver-operated stereo camera.

Roughness height values are digitized at regularly spaced intervals from the stereo pair photographs, using a photogrammetric stereocomparator. Orthogonal transects of roughness heights are generated directly with the stereocomparator. The photogrammetric software corrects the measurements for distortion caused by refraction in seawater and lens aberrations. The stereocomparator provides the means for high-frequency sampling of bottom roughness due to its horizonal accuracy of nearly 0.1 mm, as well as the capability to eschew the time-consuming process of contouring the entire image.

Roughness profiles are analyzed for variability in the vertical dimension by use of the

FIGURE 35. The 70 mm underwater cameras.

root-mean-square (rms) roughness statistic (standard deviation of relative height values). Variability in terms of spatial frequency is assessed with the roughness power spectrum. Estimates of the roughness spectrum are filtered by ensemble averaging of spectra from the same orientation. The slope of the roughness spectrum in log-log space signifies the decay of power with increasing spatial frequency and is a convenient parameter to use to describe bottom microtopography.

3. RESULTS

Table 8 displays 11 sites for which microtopographical roughness was determined photogrammetrically. Values of rms height roughness are calculated as the square root of the mean variance of all orientations. Values of rms roughness are a function of hydrodynamic and biological processes. Storm-generated ripples are responsible for roughness expressed off the coasts of California, Washington, and South Carolina. Differences in the magnitude of storms and fetch over which the storms developed are the principal factors affecting the magnitude of rms roughness in hydrodynamically controlled areas. Differences among proximal sites can be due to the sorting of surface sediments. Well-sorted sands are more readily manipulated by oscillatory currents into rippled bedforms than hard-packed or poorly sorted sediments.

Bioturbation of surface sediments may decrease or increase the variability of interface relief. The movement of large populations of the sand dollar *Echinarachnius parma* on the sediment surface contributes to low values of rms height roughness by actually smoothing the microtopographic features. This is the case of the site investigated off Montauk Point.

TABLE 8
Sediment Type and Bottom Roughness Measured at Various Shallow Water Sites

Site	Instrument (camera)	Depth (m)	Sediment type	rms (cm)
Quinault Range, WA	70 mm	49	Fine sand	1.76
Charleston, SC (I)	35 mm	21	Fine sand	0.93
Charleston, SC (II)	35 mm	20	Shell hash	0.29
Charleston, SC (III)	35 mm	20	Medium sand	0.37
Montauk Point, NY	35 mm	35	Fine sand	0.28
Arafura Sea, Australia	70 mm	47	Mud	0.37
Panama City, FL (I)	35 mm	31	Fine sand	0.49
Panama City, FL (II)	35 mm	29	Coarse sand	3.10
Jacksonville, FL	35 mm	27	Shell hash	0.42
Juan De Fuca Canyon, WA	70 mm	280	Mud/till	0.48
Russian River, CA	70 mm	90	Silt	0.32

TABLE 9
Features of Isotropic Roughness Spectra

Site	Slope	Intercept (cm^3)	Pathlength (cm)
Charleston, SC (I)	−2.50	0.09×10^{-3}	31.5
Charleston, SC (II)	−2.05	0.08×10^{-3}	31.5
Montauk Point, NY	−2.72	0.03×10^{-3}	127.5
Arafura Sea, Australia	−2.18	0.69×10^{-3}	35.6
Panama City, FL (I)	−1.92	2.33×10^{-3}	53.3
Panama City, FL (II)	−2.57	0.64×10^{-3}	53.3
Jacksonville, FL	−1.47	5.34×10^{-3}	53.3
Juan De Fuca Canyon, WA	−3.35	0.25×10^{-3}	53.6
Russian River, CA	−2.84	0.07×10^{-4}	50.8

Deeper areas such as the Juan De Fuca Canyon are sometimes biologically controlled areas because of the infrequency or total absence of strong hydrodynamic events. Burrowing infauna, such as worms and crustaceans, create volcano-like cones of reworked sediments when excavating, and this activity may rival the magnitude of the relief generated by storm events.

Concentrations of coarse shell hash, which result from the winnowing of finer sediment by strong storm currents, dominate the bottom of several locations (Table 8). These coarse sediments are a typical feature of continental shelf geology off the coast of the southeastern United States. Microtopographical relief is reduced in this environment due to the winnowing of biogenic structures and the relative resistance to movement of gravel-sized shell hash by hydrodynamic processes.

Slopes of roughness spectra calculated for various locations are displayed in Table 9. The least amount of low-frequency roughness is exhibited by the locations with high concentrations of coarse shell hash. The highest spectrum level in the high frequencies occurs at the Jacksonville location, and this combination of weak low-frequency roughness and strong high-frequency roughness accounts for the smallest value for power-law decay (Table 9). Low spectrum levels in the high spatial frequencies are responsible for the steepest slopes exhibited by Juan De Fuca Canyon, Russian River, Montauk Point, and Panama City locations. The destruction and consequent smoothing of microrelief by sand dollars at Montauk Point may be responsible for the rapid power-law decay and, hence its steep spectral slope. Periodograms of selected spectra are shown in Figure 36.

Determining the behavior of the roughness spectrum over the widest possible range of

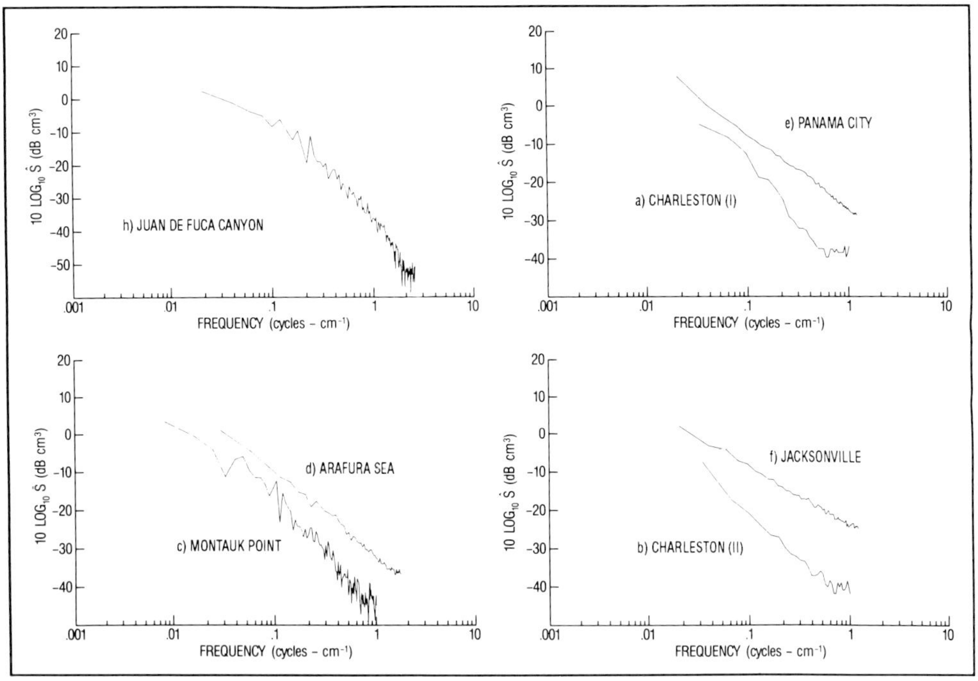

FIGURE 36. Periodograms of selected spectra.

spatial frequencies is essential in developing a valid stochastic model for sea floor microtopography. The extension of the roughness spectrum in the direction of lower spatial frequencies is accomplished by measuring the microtopography over longer pathlengths.

Extension of the roughness spectrum in the direction of higher spatial frequencies requires the digitization of the microtopography at smaller intervals. The practical limitation on the digitizing interval for analysis from stereo photographs is the available data processing time. Although the accuracy of the stereocomparator is better than 1 mm, time constraints reduce the number of profiles that could be digitized at sufficient length with intervals near the resolution limit. Consequently, the trade-off is more roughness profiles at relatively long pathlengths against higher Nyquist frequencies. Bottom roughness spectra analyzed from the Arafura Sea have the highest Nyquist frequency of the tabulated spectra, near 1.8 cycles/cm. The power-law relationship appears to be valid for the entire range of spatial frequencies (Figure 36). Figure 36 shows the averaged spectrum from three roughness profiles obtained from a photograph of the sea floor in Juan de Fuca Canyon. This example extends the Nyquist frequency to near 2.4 cycles/cm.

Photogrammetric measurement of sea floor microtopography is presently the most accurate and precise method of quantifying microroughness parameters. Path lengths of less than 1 m are typically the maximum over which small-scale bedforms can be resolved on an individual basis with underwater photogrammetry.

Photogrammetric alternatives to relying on short roughness profiles for developing stochastic models are increasing the camera-to-bottom distance or overlapping successive stereo photographs. However, each alternative carries its own limitations. Increasing the distance from the sea floor at which stereo photographs are taken results in a loss of resolution, a gain in distortion (unless the distance between cameras is increased accordingly) and, ultimately, insufficient light. In fact, decreasing the camera-to-bottom distance is necessary in locations where water clarity is less than optimal. Overlapping successive stereo photographs necessitates the establishment of a horizontal frame of reference that does not vary from stereo pair to stereo pair. Otherwise, low-frequency roughness is artificially introduced to or removed from the composite profile.

V. SUMMARY

The need for technically sound, problem-solving-oriented, basic research on surficial submarine sediments stemmed from diverse national and international requirements directed at applied problems and applications, in addition to basic scientific curiosity. The applications include military activities, energy resources such as petroleum (hydrocarbon) exploration and recovery, waste disposal and long-term waste containment, offshore construction and sea floor mineral exploration and recovery, and coastal engineering in general. Major science programs, such as the earlier Deep Sea Drilling Project and the present Ocean Drilling Program, have added important dimensions to the understanding of global marine sedimentation in time and space. An important issue concerns the fate of man's changing global environment brought about by the environmental stresses superimposed on the natural long-term variations that occur over time. This issue, as well as those of immediate practical importance, are best addressed through the detailed research and study of surficial marine sediments, which capture and record environmental conditions and changes at the time of deposition (Bennett et al., in press). The *in situ* probes and remote sensing technologies are important tools for addressing not only scientific issues related to marine processes and sediment properties but also to practical problems concerning man's activities in the ocean environment and the long-term quantitative prediction of environmental change. The new technologies will add major dimensions to the required "ground truthing" of emerging sophisticated geophysical mapping systems and seismic profilers.

ACKNOWLEDGMENTS

The authors thank Ms. Sherryl Liddell and staff for graphics production and Ms. Linda H. Jenkins for editorial review, and greatly appreciate the editorial work and proofing of the galleys by Ms. F. Lee Nastav, and for her artwork on figures 1-10. The secretarial help of Ms. Ethel Smith and Ms. E. Fay Ladner are appreciated. The authors gratefully acknowledge support and encouragement of NOARL management during the compilation and writing of this paper. This work was funded by program element 61153N, Mrs. Halcyon E. Morris, program manager, NOARL Contribution SP#032:360:90.

REFERENCES

Akal, T., The relationship between the physical properties of underwater sediments that affect bottom reflection, *Mar. Geol.*, 13(4), 251, 1972.

Allman, R. L., Equipment of *in situ* measurement of sound velocity and attenuation, Naval Undersea Center, San Diego, CA, NUC TP 424, 74, 1974.

Archie, G. E., The electrical resistivity log as an aid in determining some reservoir characteristics *Am. Inst. Mech. Engr.*, 146, 54, 1942.

Bea, R. G. and Arnold, P., Movements and forces developed by wave-induced slides in soft clays, *Proc. 50th Annual Offshore Technol. Conf.*, 731-742, 1973.

Belderson, R. H., Kenyon, N. H., Stride, A. H., and Stubbs, A. R., *Sonographs of the Sea Floor,* Elsevier Publishing, Amsterdam, 185, 1972.

Bennett, R. H., Bryant, W. R., Dunlap, W. A., and Keller, G. H., Initial results and progress of the Mississippi Delta sediment pore water pressure experiment, *Mar. Geotech.*, 1, 327-355, 1975.

Bennett, R. H. and Faris, J. R., Ambient and dynamic pore pressures in fine-grained submarine sediments, *Miss. Delta, Appl. Ocean Res.*, 1, 115-123, 1979.

Bennett, R. H., Burns, J. T., Clark, T. L., Faris, J. R., Forde, E. B., and Richards, A. F., Piezometer probes for assessing effective stress and stability in submarine sediments, *Marine slides and other mass movements,* Saxon, J. and Nieuwenhuis, J. K., Eds., Plenum Press, New York, 1982, 129-161.

Bennett, R. H. and Nelsen, T. A., Seafloor characteristics and dynamics affecting geotechnical properties at shelfbreaks, *The Shelfbreak: Critical Interface on Continental Margins,* Stanley, D. J. and Moore, G. T., Eds., *Soc. Econ. Paleontol. Mineral., Spec. Publ.,* 33, 333, 1983.

Bennett, R. H., Lambert, D. N., Hulbert, M. H., Burns, J. T., Sawyer, W. B., and Freeland, G. L., Electrical resistivity/conductivity in seabed sediments, *CRC Handbook of Geophysical Exploration at Sea,* Geyer, R. A., Ed., CRC Press, Boca Raton, FL 1983, 333-374.

Bennett, R. H., Li, H., Valent, P. J., Lipkin, J., and Esrig, M. I., *In situ* undrained shear strength and permeabilities derived from peizometer measurements, *Strength Testing of Marine Sediments; Laboratory and In Situ Measurements,* Chaney, R. C. and Demars, K. R., Eds., Am. Soc. Test. Mater., Spec. Test. Publ. 883, 1985, 83-100.

Bennett, R. H., Burns, J. T., Nastav, F. L., Lipkin, J., and Percival, C. M., Deep-ocean piezometer probe technology for geotechnical investigations, IEEE J. Oceanic Eng., OE-10, 17-22, 1985.

Bennett, R. H. and Hulbert, M. H., *Clay Microstructure,* IHRDC Press, (presently distributed by Prentice-Hall) 1986.

Bennett, R. H., Li, H., Lambert, D. N., Fischer, K. M., Walter, D. J., Hickox, C. E., Hulbert, M., Yamamoto, T., and Badiey, M., *In situ* pososity and permeability of selected carbonate sediment: Great Bahama Bank — Part I: Measurements, *Mar. Geotech.,* 9(1), 1990.

Bennett, R. H., Fischer, K. M., Li, H., Baerwald, R., Hulbert, M. H., Yamamoto, T., and Badiey, M., *In situ* porosity and permeability of selected carbonate sediment: Great Bahama Bank — Part II: Microfabric, *Mar. Geotech.,* 9(1), 1990.

Bennett, R. H., Hulbert, M. H., and Bryant, W. R., Eds., *Microstructure of Fine-Grained Terrigeneous Marine Sediments from Mud to Shale,* Springer-Verlag, New York, 1991, 582.

Breslau, L. R., Classification of sea-floor sediments with a shipborne acoustical system, WHOI Contribution No. 1678, 1967.

Briggs, K. B., Microtopographical roughness of shallow-water continental shelves, *IEEE J. Ocean Eng.,* 14(4), 360-367, 1989.

Brunson, B. A., McLeron, E. E., Holland, C. W., and Hagg, R. K., Acoustic sea bottom classification: a requirements analysis, Planning Systems, Inc., McLean, VA, PSI Tech. Rep. 335313, 1985.

Caulfield, D. D. and Kim, Y. C., Prediction of shallow sub-bottom sediment acoustic impedance while estimating absorption and other losses, *J. Can. Soc. Explor. Geophys.,* 19(1), 1983.

Caulfield, D. D.,. Shallow seismic derived acoustic core, Proc. Offshore Technology Conference, 4720, 1984.

Corbin, C. A., Comparison of three bottom classifiers, Naval Coastal Systems Center, Panama City, FL, NCSC Tech. Note 840-86, 1986.

Dakhnov, V. N., Geophysical well logging, *Quarterly of the Colorado School of Mines,* 57, 1962.

Dodds, D. J., Surface and volume backscattering of broadband acoustic pulses normally incident on the sea floor: observations and models, Huntec Ltd., Ontario, Canada, Rep. H811-02/5MDJD, 1981.

Eckert, E. R. G. and Drake, R. M., *Analysis of Heat and Mass Transfer,* McGraw-Hill, New York, 1972, 106.

Erchul, R. A., Ocean engineering applications for electrical resistivity techniques, *Offshore Tech. Conf. Paper,* 2012, OTC, 73, 1974.

Faas, R. W., Analysis of the relationship between acoustic reflectivity and sediment porosity, General Dynamics Rep. U417-67-036, 1967.

Hamilton, E. L., Geoacoustic modeling of the sea floor, *J. Acoust. Soc. Am.,* 68(5), 1313-1340, 1980.

Hamilton, E. L. and Bachman, R. T., Sound velocity and related properties of marine sediments, *J. Acoust. Soc. Am.,* 72(6), 1891-1904, 1982.

Howard, W. E., Lott, D. F., and Ingram, C., Evaluation of the Honeywell-ELAC Echo Strength Measurement System for quantitative sea-floor classification measurement, Naval Coastal Systems Center, Panama City, FL, Tech. Rep. 380-83, 1983.

Hulbert, M. H., Lambert, D. N., Bennett, R. H., Freeland, G. L., Burns, J. T., and Sawyer, W. B., Electrical resistivity/conductivity of submarine sediments measured by *in situ* techniques, U.S. Dept. Commerce, 1981, TR ERL 416-AOML 31.

Kogler, F. C., The application of electro-acoustic methods in geological investigations of the near-surface substrate in shelf areas, Geological Institute, Kiel, Germany, Rep. 76/3, 1976.

Lambert, D. N., Carmaggio, F. S., and Valent, P. J., A self-contained diver-operated conductivity/resistivity probe, Current Practices and New Technology in Ocean Engineering, *Am. Soc. Mech. Eng.,* OED-V. II, 385, 1986.

Lambert, D. N., An evaluation of the Honeywell ELAC computerized sediment classification system, Naval Ocean Research and Development Activity, Stennis Space Center, NORDA Rep. 169, 1988.

Lambert, D. N., A new computerized single frequency seafloor classification system, Proc. of Current Practices and New Technology in Ocean Engineering, American Society of Mechanical Engineers, Ocean Engineering Div., 385, 1988.

Lambert, D. N., Progress report: Remote methods of mine impact burial prediction, Naval Ocean Research and Development Activity, Stennis Space Center, MS, Tech. Note 460, 48, 1989.

Li, H., Wang, M. C., and Bennett, R. H., Significance of pore-water pressure in marine sediments: measurements and derived properties, in *Handbook of Geophysical Exploration at Sea, 2nd. Edition: Hydrocarbons,* Geyer, R., Ed., CRC Press, Boca Raton, FL, 1991.

Mazel, C., Side-scan sonar record interpretation, Klein Associates, Salem, NH, 144, 1985.

McLeroy, E. G. and Shows, P.A., Results of the 1982 St. Andrew Bay field tests of the D-83 sediment classifier, Naval Coastal Systems Center, Panama City, FL, NCSC Tech. Note 704-83, 1983.

Percival, C. M., Olson, L. O., Bennett, R. H., Sayles, F. L., and Silva, A. J., Application to subseabed disposal: ISHTE — A model experiment, Bennett, R. H., Ed., Special issue, *Mar. Geotech.,* 8(4), 1989.

Reut, A., Pace, N. G., and Heaton, M. J. P., Computer classification of sea beds by sonar, *Nature,* 314, 426-428, 1985.

Richardson, M. D., Muzi, E., Troiano, L., and Miaschi, B., Sediment shear waves: A comparison of *in situ* and laboratory measurements, in *Microstructure of Fine-Grained Sediments: from Mud to Shale,* Bennett, R. H., Bryant, W. R., and Hulbert, M. H., Eds., Springer-Verlag, New York, 1991, 403.

Satkowiak, L. J., Remote sea bottom classification utilizing the Ulvertech bottom profiler parametric source, Proc. Oceans '88, Marine Technology Society, 1988.

Schmoker, J. W. and Hester, T. C., Porosity of the Miami Limestone (late Pleistocene), lower Florida Keys, *J. Sed. Petrology,* 56, 629, 1986.

Schock, S. G., LeBlanc, L. R., and Mayer, L. A., Chirp subbottom profiler for quantitative sediment analysis, *Geophysics,* 54(4), 445-450, 1989.

Stanic, S., Briggs, K. B., Fleischer, P., Ray, R. I., and Sawyer, W. B., Shallow-water high-frequency bottom scattering off Panama City, FL, *J. Acoust. Soc. Am,* 83, 2134-2144, 1988.

Stanic, S., Briggs, K. B., Fleischer, P., Sawyer, W. B., and Ray, R. I., High-frequency acoustic backscattering from a coarse shell ocean bottom, *J. Acoust. Soc. Am.,* 85, 125-136, 1989.

Werner, F., Nearshore bedform patterns along Rhode Island from side-scan sonar surveys — discussion, *J. Sediment. Petrol.,* 52, 674-677, 1982.

Winn, K., Kogler, F. C., and Werner, F., Simultaneous application of reflection strength recorder, sidescan sonar and sub-bottom profiler in seafloor sediment mapping, *Bottom Interacting Ocean Acoustics,* Plenum Press, New York, 1980, 85.

Winn, K., Becker, G., and Theiten, F., The relationship between sediment parameters and the acoustic reflectivity of the sea-bed, *Proc. Acoustics and the Sea-Bed,* Institute of Acoustics, Underwater Acoustics Group, Bath University Press, 1983, 31-39.

Chapter 12

GEOTECHNICAL PROPERTIES OF MISSISSIPPI RIVER DELTA SEDIMENTS UTILIZING *IN SITU* PRESSURE SAMPLING TECHNIQUES

Mark William Johns

TABLE OF CONTENTS

I. INTRODUCTION

Rapid accumulation of organic-rich sediments and the subsequent anaerobic, biogenic decomposition of organic material on the Mississippi River delta have led to sediment regions with high methane concentrations. The presence of very soft, underconsolidated, gassy sediments creates a variety of problems for the acquisition of seismic data and samples for geotechnical analyses. A pressure core barrel has been developed by Texas A & M University, in conjunction with the U.S. Geological Survey Mississippi Delta Project, to investigate these shallow, gas-charged sediments.

Substantial quantities of biogenic methane gas are found in the surficial sediments (0 to 50 m subsurface) of the Mississippi Delta. The presence of methane gas in the pro-delta sediments is believed to play a major role in sediment instability phenomena. In addition, the acoustic attenuation of high-resolution seismic signals that causes an acoustic wipeout commonly observed over large areas of pro-delta sediments, as well as other regions worldwide, is also believed to be caused by the presence of biogenic and/or thermogenic gas in bubble form.

New techniques developed for geotechnical sampling with a pressure core barrel and pressurized Shelby tubes have provided insight into the effects of methane solubility and the resultant physical characteristics of samples collected and tested at *in situ* and ambient pressures. Sample testing procedures require the use of a manned hyperbaric chamber, pressurized to *in situ* sample pressures. Test results indicate samples that contain high methane concentrations show marked reduction in strength characteristics resulting from evolution of methane gas in bubble form upon release of *in situ* pressures. Gas bubble formation is directly evidenced as increased sample void ratio and porosity. Consolidation test results indicate that this process can be correlated to increased coefficients of consolidation and compressibility.

II. OBJECTIVES

The primary objective of this study is to describe the consolidation characteristics and geotechnical properties of the sediment recovered from seven foundation borings on the Mississippi delta complex utilizing new techniques for maintaining samples under *in situ* pressure conditions. Two new methods were utilized to recover samples of this type. Pressurized core barrel and repressurized Shelby tube samples were obtained at these borings (Denk et al., 1981; Johns et al., 1982). The geotechnical properties studied herein include the following: (1) consolidation characteristics; (2) water content; (3) porosity; (4) void ratio; (5) bulk density; (6) specific gravity of solids; (7) clay mineralogy; (8) particle size distribution; and (9) undrained shear strength. Particular emphasis is placed on the evaluation of the consolidation characteristics obtained by three testing procedures on identical samples. These procedures include testing samples at *in situ* and ambient pressures, using remolded samples as a control test procedure.

III. REVIEW OF PERTINENT LITERATURE

The continental shelf deposits of offshore Louisiana are predominately clayey sediments of late Quaternary age. The nature, distribution and thickness of these deposits reflect the effects of sea level fluctuation during this time and the proximity to the Mississippi River. Near the end of Pleistocene time, sea level stood about 140 m below its present level and what is now the Mississippi delta complex began forming seaward of its present position. As sea level rose at the end of the last glacial epoch, deltaic sedimentation could not keep pace with the rise in sea level; the shoreline receded and a series of retreatal deltas was constructed (Fisk and McClelland, 1959).

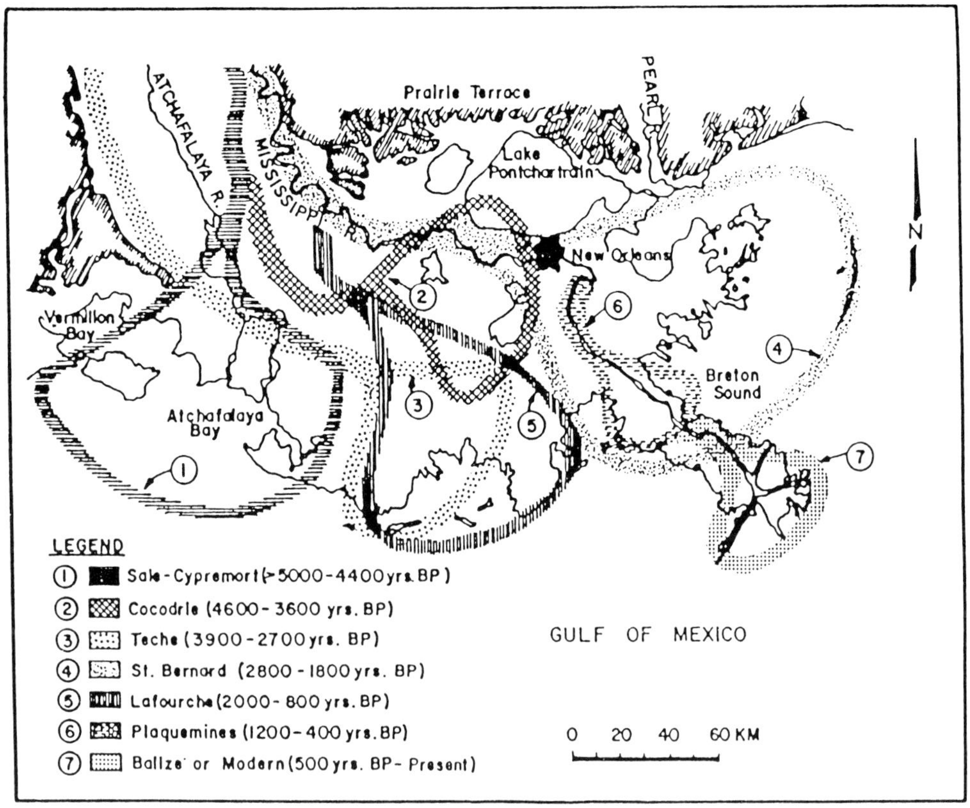

FIGURE 1. Mississippi River deltas (after Kolb and Van Lopik, 1958).

Kolb and Van Lopik (1958) depicted the frontal outlines of these deltas which were formed in response to course changes of the river (Figure 1). The dates shown indicate the duration of significant flow that occurred at each delta. The geologic history and sedimentary patterns of the deltas have been extensively studied (Russell and Russell, 1939; Fisk et al., 1954; Fisk and McFarlan, 1955; Scruton, 1960; Shepard, 1960; Coleman and Gagliano, 1964; Gould, 1965; Morgan, 1965; Kolb and Van Lopik, 1966; Kolb and Kaufman, 1967; Coleman et al., 1974; Coleman, 1976; Roberts et al., 1976; Prior and Coleman, 1981).

The modern Mississippi River has an approximate discharge rate of 15,631 m^3/sec with associated high sediment input to the delta of approximately 2.7×10^7 tons/year (2.4×10^{11} kg) (Coleman, 1981). Preservation of sedimentary organic matter is a common occurrence near river mouths with high depositional rates and is a necessary requirement for the generation of gas (Whelan et al., 1976). The Mississippi river-mouth depositional patterns have been modified by sediment-deforming processes. Several types of deformation resulting from sediment degassing are present, among which mass wasting and flowage induced by wave motion are important processes.

Much time and effort has been spent investigating the results of mass-wasting processes and their effects on bottom-supported structures. A more detailed discussion of processes and the resultant features may be found in Shepard (1955), Morgan et al. (1963), Henkel (1970), Bea (1971), Coleman et al. (1972), Bea and Arnold (1973), Coleman and Wright (1974), Garrison (1974), and Yamamoto (1982). The occurrence of methane gas in recent sediments and its effect on soil stability has been studied to a much lesser degree. Whelan et al. (1975, 1976) found that a notable reduction in undrained shear strength occurs in sediments with high methane concentrations.

The formation of near-surface biogenic gas by bacterial activity occurs as the early stages of diagenesis. Methane is the primary hydrocarbon produced in measurable quantity, but trace quantities of C2-C6 hydrocarbons may also be present (Waples, 1980). Methane production in anoxic sediments occurs only after dissolved sulfate has been depleted by way of biochemical reactions generating sulfide and bicarbonate (Nissenbaum et al., 1972). The presence of sedimentary gas bubbles, primarily methane, distorts acoustical quality (Jones et al., 1958, 1964; Hampton, 1966; Anderson, 1974; Schubel, 1974). Similarly, an apparent reduction in undrained shear strength results from sedimentary gas ebullition (Whelan et al., 1975, 1976; Denk et al., 1981).

Two practical means exist to evaluate the geotechnical strength characteristics of gas-charged sediments as they actually exist under *in situ* conditions: (1) remote *in situ* measurements performed through a borehole with the sediment in place, and (2) measurements performed under controlled pressure conditions (hyperbaric chamber) after collection at *in situ* pressures. Standard *in situ* devices include pressuremeter, remote vane, and remote cone (Doyle et al., 1971). This type of testing has proved useful, but lacks the basic advantage of the latter technique, i.e. the ability to actually perform a basic strength test under variable pressure conditions on samples collected by the same means. Consolidation testing cannot be performed under *in situ* conditions. However, pressurized sampling techniques combined with a pressure-controlled hyperbaric environment and slightly modified consolidometer equipment does allow testing at *in situ* pressure conditions. By this means, consolidation tests can be performed on gas-charged sediments without the release of pressure and the subsequent deleterious effects of gas bubble evolution on the soil structure.

Chiou (1980) performed soil microfabric studies on pressurized and degassed samples collected with the pressure core barrel and found that samples collected and maintained under *in situ* pressures exhibit highly oriented soil microstructure, while degassed samples are typically highly disrupted. Disruption of the microstructure appears to be time-dependent (Chiou, 1981). Similarly, observations of the methane gas concentration studies of pressurized core barrel samples show that at least 98% of the methane is released from the sediment matrix within 3 to 5 hours after pressure reduction to ambient conditions (Whelan et al., 1981).

IV. EXPERIMENTAL PROGRAM

A. BOREHOLE LOCATIONS

Core samples were obtained from several sediment borings located in three different regions of the outer continental shelf lease area of the Mississippi River delta. Locations and water depths of the seven sediment borings are as follows (Figure 2):

1. Boring 1, West Delta, OCS, Block 85, Water Depth 21.0 m, Latitude 28°52.891′N, Longitude 89°28.643′W
2. Boring 1A, West Delta, OCS, Block 85, Water Depth 21.0 m, Latitude 28°52.895′N, Longitude 89°28.650′W
3. Boring 2, West Delta, OCS, Block 85, Water Depth 38.1 m, Latitude 28°54.196′N, Longitude 89°30.133′W
4. Boring 2A, West Delta, OCS, Block 85, Water Depth 45.2 m, Latitude 28°54.214′N, Longitude 89°31.160′W
5. Boring 5, South Pass, OCS, Block 43, Water Depth 39.0 m, Latitude 28°53.057′N, Longitude 89°20.991′W
6. Boring 7, South Pass, OCS, Block 43, Water Depth 45.1 m, Latitude 28°54.115′N, Longitude 89°20.103′W
7. Boring 1, Main Pass, OCS, Block 74, Water Depth 44.2 m, Latitude 29°13.884′N, Longitude 89°54.407′W

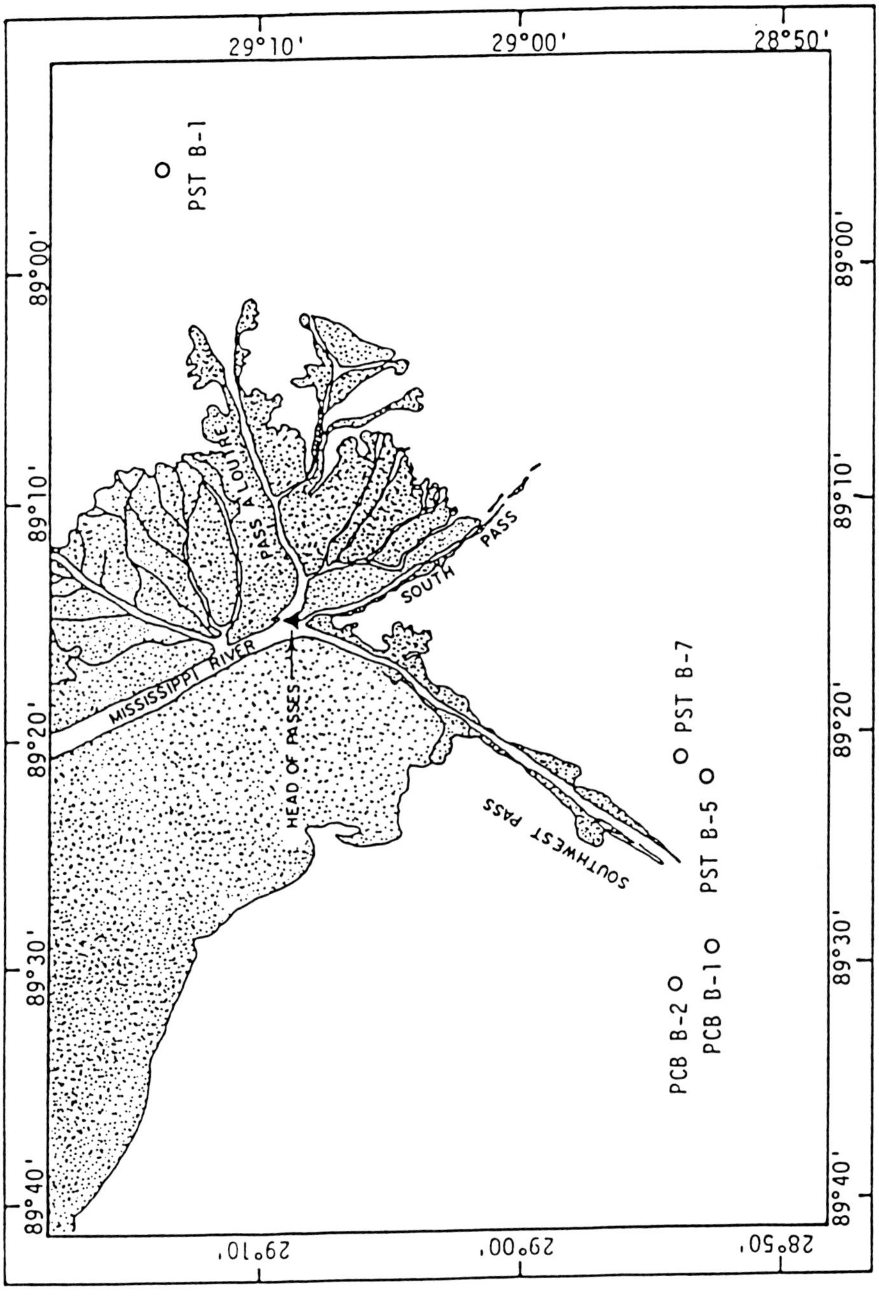

FIGURE 2. Locations of the pressure core barrel (PCB) and pressure Shelby tube (PST) sediment borings.

B. SAMPLE ACQUISITION TECHNIQUES

The samples were obtained on three separate cruises at the sites previously listed. Thirteen pressure core barrel samples were obtained at three drill sites in August 1979. The samples were returned to Texas A & M University and placed in a high-humidity chamber at 5°C where pressures could be maintained until their geotechnical properties could be analyzed.

C. GEOTECHNICAL PROPERTIES

The geotechnical properties investigated in this study include: (1) consolidation characteristics; (2) water content; (3) porosity; (4) void ratio; (5) bulk density (wet unit weight); (6) specific gravity of solids; (7) clay mineralogy; (8) particle size distribution; and (9) cohesive shear strength. Subsampling of the pressure core barrel and pressurized Shelby tube samples was undertaken in a manned hyperbaric chamber at the *in situ* pressure of the given sample. All of the above properties with the exception of cohesive shear strength were determined from sediment subsamples taken before and after the removal of the consolidation sample. Hand-held Torvane shear tests were performed on the open sediment cores within the hyperbaric chamber at *in situ* pressures.

D. CONSOLIDATION TESTING

Consolidation tests were performed on 14 samples from the four foundation borings. These samples were selected by visual inspection at *in situ* pressures in the hyperbaric chamber. Standard X-radiography could not be performed in this situation due to the small size and restrictive environment of the manned hyperbaric chamber.

Pressure core barrel samples showed little, if any, evidence of gas expansion as they were always maintained at *in situ* pressures. The pressure Shelby tube samples often showed signs of degassing within a few centimeters of the sample ends. This characteristically "frothy" or "cottage cheese" void-filled texture was unmistakable (Denk et al., 1981). The use of a hand-held Torvane helped to delineate this degassed section. The consolidation sample was trimmed from the core immediately after the degassed section was delineated and discarded.

Evidence of subaqueous deformation was impossible to determine. Drilling disturbance was visualized in a core as mud intrusions, if present, generally found at the top or bottom of a core. Sampling disturbance was believed to have been held to a minimum. The pressure core barrel sampling tubes have area ratios less than 9% (Denk et al., 1981). Pressure Shelby tube samplers have area ratios between 13.1% and 14.8% (Horslev, 1948).

The collection of truly undisturbed sediment samples is impossible. The basic aim of standard sampling techniques is to minimize mechanical disturbance and changes in water content and void ratio of the soil structure. When utilizing standard sediment sampling techniques in regions of high pore-water methane concentrations, sample recovery to ambient pressure releases this gas in bubble form (Whelan et al., 1975, 1976). This causes disruption of the soil structure, an increase in void ratio, and a decrease in the percent saturation of the sample. Pressurized sampling techniques, additionally, reduce or eliminate the effects of changes in the hydrostatic pressure conditions, thus eliminating or minimizing disturbance to the sediment structure.

Terzaghi's theory of one-dimensional consolidation assumes complete sample saturation. The validity of coefficients derived from standard consolidation testing of gas-charged sediments must be questioned. Lowe et al. (1964) found that gas bubbles within the sediment are highly compressible as compared to the relatively incompressible water. These gas bubbles impede the flow of water in the voids and thus reduce soil permeability (Bjerrum and Huder, 1957).

Specially modified Anteus back pressure consolidometers were used for testing of all samples. The advantages of using the back pressure apparatus is outlined by Lowe et al.

(1964), Bryant et al. (1967), and Shephard et al. (1979). Only modifications to the above-mentioned testing procedures necessary for testing the pressurized samples will be discussed herein.

Each sample was released in a manned hyperbaric chamber which had been pressurized to the sample pressure. The maximum allowable dive depth was 102 psi (225 ft of seawater). When preparing for a dive, the transfer chamber, consolidometer, and the necessary equipment for loading the sample were placed in the hyperbaric chamber. The pressurized transfer chamber and consolidometer were connected through a port to an external gas (helium) supply with an in-line pressure transducer.

At this time, the hyperbaric chamber was manned and pressurized to sample pressure. When sample pressure had been reached, as indicated by the pressure transducer, the transfer chamber was opened and the sample tube removed. Preliminary visual inspection included a sample description and search for evidence of any disturbance caused by degassing or mud intrusion. A hand-held extruder was used to remove the sample from the Shelby tube in approximately 1 cm intervals. Hand-held Torvane measurements were taken between extrusions to delineate any disturbed zones. Generally, a small amount of degassing was found to occur near the ends of the pressurized Shelby tube samples. These gas-disturbed zones were easily distinguished from undisturbed sections as mentioned previously. Subsamples were bagged and labelled for later use in laboratory testing of index properties and for remolded consolidation testing.

If no degassing was present at this point and a consistent undrained shear strength had been maintained for 3 or 4 cm, the consolidation sample was trimmed. The consolidation sample was slowly extruded and trimmed with a stainless steel cutting ring. The pressurized Shelby samples and pressure core barrel samples were trimmed to 4.45 cm and 3.18 cm diameters, respectively. The upper and lower surfaces were then trimmed to plane surfaces using the top and bottom of the ring as a guide. A spacer disk was inserted in the sample, up from the bottom of the ring, after trimming the lower surface and before trimming the upper surface. This permits trimming a sample to a height of approximately 2.54 cm for the pressure Shelby tube samples and approximately 1.91 cm for the pressure core barrel samples.

The sample was then loaded into a teflon-lined confining ring with filter papers and saturated porous stones on top and bottom. Once the sample was in place in the confining ring, the lucite chamber and loading dome were placed in position. The saturating water was added to the water reservoir and the lucite chamber. The oil reservoir was then lowered to the neutral position and the load valve was opened. The loading plate was slowly lowered under its own weight until contact with the upper porous stone was evidenced by lack of movement of the extensiometer. The external gas supply was then regulated to the consolidometer at the sample pressure. The Shelby tube was then resealed in the transfer chamber for later use and hyperbaric chamber decompression was initiated.

During decompression of the hyperbaric chamber, the load valve of the consolidometer was left in the open position with the oil reservoir in a neutral position. Decompression of the divers takes place in stages with rapid pressure reductions occurring at each stage. Decompression of the sample does not occur, as the external gas supply to the consolidometer is maintained at the *in situ* hydrostatic pressure of the sample. Any deflections of the dial gauge due to rapid hyperbaric chamber pressure changes were noted and recorded. When decompression of the divers was complete, the consolidometer was transferred to the laboratory with an external pressure supply, where it was connected to the laboratory air pressure system. Consolidation testing began immediately.

The consolidation test loading sequence began at 3.125 or 6.250 kPa, depending on the soil strength, and continued to 50 kPa at double load increments by raising a sliding reservoir to calibrated heights on a standpipe. Load increments of 100, 200, 400, 800, 1600, and

3200 kPa were applied using the load regulator to control the sliding reservoir pressure. Each load was applied, utilizing double drainage, for a minimum of 24 hr or until linear secondary consolidation was confirmed. Sample height was indicated by an extensiometer which measured the movement of a dial pin resting on the top of the loading plate.

A hysteresis loop was generally performed between 10 and 15% strain. Unloading was accomplished in increments of one-fourth the previous load. The final unloading sequence continued to zero load, where the sample was allowed to swell with no vertical load for 24 hr. A final dial reading was taken and the back pressure was released. The sample was unloaded from the confining ring, oven dried at 105°C for 24 hr and weighed.

The response of consolidation (rate of compression) was determined for each load increment and represented by the coefficient of consolidation, C_v. For each load increment (decrement), compression (swell) vs. logarithm of time was plotted. The end of primary consolidation was determined using the method described by Casagrande (1936). Coefficients of 50% primary consolidation were then determined graphically from these plots.

Compression vs. the square root of time was plotted for each load. The coefficient at 90% primary consolidation was graphically determined for each load using the method outlined by Taylor (1948).

The initial void ratio, e_o, defined as the volume of voids divided by the volume of solids, was calculated for each sample. Decompression of methane-charged sediments to ambient pressures results in degassing and an increase in the void ratio. The cross-sectional area, A, of a consolidation sample remained constant throughout the test; therefore, the initial void ratio can be defined as the height of voids, $2H_o - h_s$, divided by the height of solids, h_s, where H_o is equal to the height of the drainage path. The initial void radio equation becomes:

$$e_o = \frac{(2H_o - h_s)}{h_s} \tag{1}$$

The height of solids is calculated at the end of the consolidation test. The height of solids, h_s, is calculated from:

$$h_s = \frac{W_s}{AG\gamma_w} \tag{2}$$

where W_s is the weight of dried solids (g) in the consolidation sample, A is the cross-sectional area (cm) of the sample, G is the specific gravity of the solids, and γ_w is the unit weight of water (g/cm^3). The void ratio, e_i, at the end of each load was calculated from:

$$e_i = e_o - \frac{\Delta h}{h_s} \tag{3}$$

where Δh is the incremental compression or swell of the sample.

The results of the consolidation test were plotted as void ratio vs. logarithm of vertical effective stress curves on semi-log paper. At 100% primary consolidation for a given load no excess pore pressure exists within the sample; the pore pressure, u, is equal to the back pressure and the vertical effective stress, σ_v', equals the applied pressure.

The effective preconsolidation stress, σ_{vc}', represents the maximum pressure to which the sample had been loaded in nature. The Casagrande method (Casagrande, 1936) was used to graphically determine this stress from the e-log σ_v' curves.

The *in situ* effective overburden stress, σ_{vo}', acting at any given sample depth is the difference between the stress exerted by the overlying sediments and the pore water pressure at that depth. As stated previously, no pore pressure measurements were obtained; therefore, a hydrostatic head of sea water was assumed for the given sample depths. For underconsolidated sediments the pore pressure exceeds the hydrostatic condition; thus, for the sediments studied herein, the effective overburden stress is actually an "apparent" effective overburden stress.

The equation utilized for calculation of effective overburden stress, σ_{vo}' in kPa, was presented by Helwick (1977) for a given interval of known bulk (wet) density γ_b (kg/m^3). The intervals were then summed to the appropriate sample depth. The equation is:

$$\sigma'_{vo} = 9.8065[z_1(\gamma_b) = z_1(1025)] \tag{4}$$

where z_l (m) is the interval depth and 1025 is the unit weight of seawater (kg/m^3).

It is necessary to consider the stress history of the soil to rationally interpret the e-log σ_v' curve. The slope of the e-log σ_v' curve plotted on semi-log paper is a nearly straight line whose slope can be denoted by the compression index, C_c. The compression index is equal to the change in void ratio for a given change in logarithm vertical effective stress. The compression index was calculated for each sample in the region of the virgin slope of the e-log σ_v' curve and can be computed according to the relationship:

$$C_c = \frac{\Delta e}{\Delta \log \sigma'_v} \tag{5}$$

where Δe and $\Delta\sigma_v'$ are the change in void ratio and effective overburden stress over the region of the maximum slope. Standard soil mechanics procedure assumes a positive value for the compression index.

The coefficient of consolidation, C_v, is calculated for each load increment for the compression vs. logarithm of time and square root of time curves. The governing equation is:

$$C_v = \frac{T_v H_o^2}{t} \tag{6}$$

where T_v is the dimensionless time factor, equal to 0.197 for 50% consolidation and 0.848 for 90% consolidation, H_o is the height of drainage path (cm) and t is the time (seconds).

Terzaghi's (1943) equation for the theoretical coefficient of permeability:

$$k = C_v m_v \gamma_w \tag{7}$$

derived from the theory of consolidation, was used to calculate permeabilities using the C_v obtained from the square root of time technique derived by Taylor (1948). The coefficient of volume change, m_v, can be calculated from:

$$m_v = \frac{a_v}{1 + e} \tag{8}$$

The coefficient of compressibility, a_v, equals:

$$a_v = \frac{-\Delta e}{\Delta\sigma'_v} \tag{9}$$

where Δe is the change in void ratio and $\Delta\sigma_v'$ is the change in vertical effective stress (kPa).

E. SHEAR STRENGTH

Calculation of shear strength may be obtained from:

$$\tau = C + \sigma_n \tan\phi \tag{10}$$

where τ is the shear strength, C is the cohesion, σ_n is the normal stress, and ϕ is the angle of internal friction. The shear strength tests were performed directly below that portion of the sediment sample utilized for consolidation, with a Soiltest, Inc. hand-held Torvane. Saturated clayey sediments stressed with no water loss, as these samples were tested, act as purely cohesive materials with ϕ equal to zero. Thus, the undrained shear strength, C_u, is equal to the cohesion of the soil (Moore, 1961), or:

$$\tau = C = C_u \tag{11}$$

F. INDEX PROPERTIES

Index properties investigated include: (1) water content; (2) bulk density (wet unit weight) and specific gravity; (3) grain size; (4) clay mineralogy; and (5) gas concentration. The procedures for determinations of these index properties are similar to those outlined by Lambe (1951) and modified by Cernock (1967).

2. Bulk Density and Specific Gravity

A Beckman air comparison pycnometer with an accuracy of 0.02 cm was utilized to obtain wet and dry volumes of the sediment samples. No direct measurements of pore water salinity were available. However, a salt correction was made assuming a salinity of 35 o/oo (Hamilton, 1971). The correction for pore water salinity affects the computations for water content, specific gravity, void ratio, porosity, and degree of pore water saturation.

2. Grain Size

Approximately 10 g (dry weight) portion of sample was wet sieved through a 62 micron mesh (40) to determine the coarse fraction present. Pipette analysis was run on the less than 62 μm (>40) portion to determine the relative amounts of fine fraction, using the technique described by Krumbein and Pettijohn (1938). This method assumes a specific gravity of 2.65 for spherical particles using Stokes' law for calculation of settling velocities.

3. Clay Mineralogy

Sample preparation techniques for X-ray diffraction of the less than 2 micron clay fraction followed those outlined by Jackson (1956) and modified by Hottman (1975). The Bradley form factor technique described by Scafe and Kunze (1971) was used to obtain relative percentages of clay mineral abundances. The peak height method for estimating relative percentages was chosen for this study to maintain the consistency with previous studies in the Gulf of Mexico.

4. Gas Concentrations

Sections of selected core samples were quickly canned in degassed water on board ship. The canning procedure requires a consistent, rapid transfer (to minimize gas loss) of a sediment section into a 500 ml can filled with 400 ml of degassed water. Sediment is added until the water level is directly at the top of the can, allowing a 100 ml volume of sample to be consistently sealed in the can. This canned sample is subsequently analyzed in the laboratory for methane and other pertinent gasses using gas chromatography (Whelan et al., 1975, 1976).

V. RESULTS

Comparative testing was performed on one pressure core barrel (PCB) sample and 13 pressure Shelby tube (PST) samples. Pressurized, degassed, and remolded specimens were sub-sampled from a given core at intervals of less than 14 cm. All pressurized test specimens were back-pressured to *in situ* pressures, while degassed and remolded samples were tested with no back pressures by standard methods. The remold test for sample B5S28R was performed with a back pressure of 100 psi (689 kPa). Insufficient sample was collected to perform a degassed test on sample B1S2. Therefore, only pressurized and remold test results are available for this specimen.

A. INDEX PROPERTY TEST RESULTS

The index properties for all samples can be found in Table 1. Sample designators contain boring number, sample number, and consolidation test treatment type. Consolidation treatment types are: I, for *in situ* or pressurized test; D, for degassed test; and R, for remold test.

1. Specific Gravity and Wet Bulk Density

Values for wet bulk density (salt-corrected) range between 1.55 and 1.76 mg/m^3 with a mean value of 1.63 mg/m^3 and a standard deviation of 0.06 mg/m^3. Specific gravity of solids varies between 2.54 and 2.77 with a mean value and standard deviation of 2.69 and 0.07, respectively. Bulk density increases while water content and porosity decrease with depth (correlation coefficient $r^2 = 0.90$ and 0.86, respectively).

2. Grain Size

Sand content is low for all samples and ranges between 0 and 3.7%. Silt and clay contents are considerably more variable; however, silt and clay contents show opposite trends with depth. The silt content decreases with depth ($r^2 = 0.86$), while clay content tends to increase with depth ($r^2 = 0.85$). Variation between comparative samples is low, except for samples B7S9 and B5S26, which have variations of 20 and 10% between comparative samples for the silt and clay contents. All samples can be classified as clayey silts, except for B5S15 (all samples) and B7S9I, which contain slightly higher silt contents and are classified as silty clays (Folk, 1974).

3. Water Content and Atterberg Limits

As mentioned previously, all water content determinations include a salt correction of 35 o/oo (Hamilton, 1971). Water contents for all samples vary between 49 and 89%. Comparative samples from Boring 1 are quite similar, variations being 0 to 4%. Boring 5 and 7 samples do not exhibit as close a correlation between comparative samples. Variations between comparative samples range from 0 to 19%, with the samples tested at *in situ* pressure generally having higher values. These samples were tested prior to those for the degassed or remold consolidation tests and, therefore, indicate either partial drying of the samples prior to testing or a loss of porewater resulting from degassing of porewater gasses.

Initial water contents for all samples fall between the liquid limit, w_l, and the plastic limit, w_p. The liquid limits for all samples range between 68 and 107%, while the plastic limiits range between 37 and 63%. The mean value for the liquid limit is 89% with a standard deviation of 12%, while the respective values for the plastic limits are 47 and 7%. Based on the Unified Soil Classification System, these soils can be classified as clays of high plasticity, or as organic clays of high to medium plasticity.

4. Calcium Carbonate Content

Analytical determinations of weight percent calcium carbonate content were made using

TABLE 1
Index Properties for All Samples

Core-section	Test type	Depth z (m)	Water content, w (%)	Initial void ratio, e_o	Initial porosity, n (%)	Wet bulk density, υ_b (mg/m³)	Specific gravity	Undrained shear strength, C_u (kPa)	Liquid limit, w_l (%)	Plastic limit, w_p	Weight percent $CaCO_3$ (%)	Grain Size: Sand (%)	Grain Size: Silt (%)	Grain Size: Clay (%)
B2 S10	I	11.76				2.69								
	D	11.76				2.67								
B1 S2	I	6.72	63	1.626	61.92	1.69	2.68	14.82	77	40	1.21	1.74	41.85	56.41
	R		63	1.630	61.98	1.69	2.68		77	40	2.61			
B1 S4	I	12.41	61	1.511	60.18	1.68	2.60	6.75	74	37	7.05	1.37	47.32	51.31
	D	12.37	61	1.778	64.00	1.66	2.71		74	37	2.66	3.37	44.80	51.83
	R		60	1.628	61.95	1.66	2.71		74	37	2.87			
B1 S6	I	18.78	74	1.995	66.61	1.61	2.71	9.39	97	63	7.44	0.32	33.08	66.60
	D	18.72	75	2.279	69.50	1.61	2.77		97	63	2.64	0.82	34.62	64.56
	R		72	2.163	68.38	1.61	2.77		97	63	2.86			
B1 S8	I	24.32	85	2.283	69.54	1.55	2.75	14.82	103	50	5.37	1.16	46.73	52.11
	D	24.18	89	2.451	71.02	1.55	2.75	14.66	103	50	3.37	0.85	26.82	72.33
	R		85	2.292	69.62	1.55	2.75		103	50	3.31			
B7 S9	I	10.24	62	1.650	62.26	1.69	2.59	15.32	84	48	2.64	0.01	53.49	46.50
	D	10.20	81	2.246	69.19	1.60	2.69	14.33	84	48	2.64	0.03	34.82	65.15
	R		73	1.969	66.32	1.60	2.69		84	48	2.59			
B7 S15	I	16.37	78	2.018	66.87	1.58	2.61	14.89	84	44	2.96	0.00	34.00	66.00
	D	16.31	66	1.823	64.58	1.58	2.61	16.54	84	44	2.96	0.00	34.00	66.00
	R		68	1.708	63.07	1.58	2.61		84	44	3.53	0.06	42.05	57.89
B5 S10	I	13.29	65	1.526	60.41	1.66	2.54	17.48	86	51	4.28	0.00	40.38	59.62
	D	13.21	57	1.414	58.57	1.66	2.54	19.76	86	51	3.26	0.35	44.96	54.69
	R		56	1.325	65.99	1.66	2.54		86	51	2.74	0.35	44.96	54.69
B5 S15	I	19.42	50	1.230	55.16	1.76	2.68	19.15	68	39	3.00	1.42	54.23	44.35
	D	19.47	49	1.395	58.25	1.76	2.68	15.65	68	39	3.84	1.80	52.07	46.13
	R		49	1.283	56.20	1.76	2.68		68	39	3.60	1.80	52.07	46.13
B5 S23	I	36.49	62	1.504	60.06	1.69	2.57	16.47	90	47	3.39	3.68	32.08	64.24
	D	36.44	62	1.728	63.34	1.70	2.68		90	47	3.06	0.43	34.87	64.70
	R		67	1.837	64.75	1.65	2.70		90	47	3.41	0.36	29.65	69.99

B5 S26	I	42.47	85	2.022	66.91	1.56	2.77	21.69	103	52	2.79	0.04	24.39	75.57
	D	42.50	77	2.393	70.53	1.56	2.77	18.94	103	52	2.84	0.82	33.72	65.46
	R		75	2.139	68.14	1.56	2.77		103	52	4.10			
B5 S28	I	45.54	69	1.777	63.99	1.65	2.71	19.77	95	45	4.27	0.10	25.06	74.84
	D	45.49	71	1.941	66.00	1.65	2.71	19.93	95	45	4.72	0.10	25.63	74.27
	R		70	1.821	64.55	1.65	2.71		95	45	4.59	0.10	25.63	74.27
B5 S30	I	51.76	80	2.243	69.16	1.58	2.76	29.69	87	44	2.42	0.03	16.10	83.87
	D	51.80	74	2.245	69.18	1.62	2.76	21.91	87	44	3.72	0.03	19.05	80.92
	R		70	2.128	68.03	1.62	2.76		87	44	4.04	0.12	23.06	76.82
B5 S32	I	57.80	75	2.122	67.97	1.62	2.75	33.52	107	54	11.65	0.57	13.16	86.27
	D	57.74	73	2.091	67.65	1.62	2.75	28.16	107	54	5.56	0.58	15.78	83.64
	R		72	2.050	67.21	1.62	2.75		107	54	4.69	0.58	15.78	83.64

TABLE 2
Semi-Quantitative Analysis (Relative %) of Clay Minerals in Sediments of Borings 1, 5, and 7

Sample	Smectite (%)	Illite (%)	Kaolinite (%)	Chlorite (%)
B1S2	22	58	14	6
B1S4	27	58	11	4
B1S6	33	52	13	2
B1S8	33	58	7	2
B7S9	20	65	11	4
B7S15	20	71	7	2
B5S10	25	65	8	2
B5S15	28	63	7	2
B5S23	29	64	4	3
B5S26	27	64	8	1
B5S28	28	63	7	2
B5S30	18	70	10	2
B5S32	21	64	13	2

a modified carbonate bomb technique (Schink et al., 1979). Values for all samples are low and range between 1.21 and 11.65%. Test results indicate only minor variations exist between comparative samples. Variation appears to be within the standard error (±0.3%) for the test procedure. The mean value is 3.83% with a standard deviation of 1.79%.

5. Clay Mineralogy

One sediment sample was utilized for X-ray diffraction analysis for each set of consolidation samples. These samples were collected from an interval between consolidation samples and, therefore, are assumed to be representative of the relative type of clay minerals present in the sediments of a given consolidation test group. The results of clay mineral analyses indicate a fairly uniform distribution of clay minerals in the sediments studied for pressurized Shelby tube Borings 1, 5, and 7 (Table 2).

The semi-quantitative nature of estimating mineralogical composition and the fact that all calculations are based on 100% clay should be kept in mind. In this study of the less than 2 μm fraction, illite is the dominant mineral (52 to 71%) with smectite (18 to 33%), kaolinite (4 to 14%), and chlorite (1 to 6%) constituting far lesser portions.

6. Gas Concentrations

Methane concentrations for the sampled sections of Boring 1, Main Pass, are quite high (to 47,412 ppm) (Table 3). In addition to the shipboard canned profile, three samples were collected in the manned hyperbaric chamber while consolidation samples were being loaded (Figure 3). These samples were collected as near to the consolidation samples as possible and more closely reflect the sample methane content. However, it must be noted that these values were not measured from the actual consolidation sample. Micro-scale methane variability within a core was not evaluated for this study. The values represent the total gas content including the decompressed or head space gas and matrix values extracted from the pore water and sediment particles at ambient pressure.

Borings 5 and 7 contain relatively lower values of methane concentrations (Table 1). The values plotted in Figure 4 were collected in the hyperbaric chamber, adjacent to the consolidation samples. These values again represent total methane concentration. Three samples contain significant values of methane: B7S9, B5S26, and B5S30. The remainder of the sediments tested have less than 566 ppm.

TABLE 3
Primary Consolidation Characteristics for All Samples

Core-section	Test type	Depth, z (m)	Initial void ratio, e_o	Initial porosity, n (%)	Undrained shear strength, C_u (kPa)	Effective overburden stress, σ'_{vo} (kPa)	Effective preconsolidation stress, σ'_c (kPa)	Compression indices C_c High	Compression indices C_c Low	Methane content (ppm)	Overconsolidation ratio OCR
B2 S10	I	11.76	1.701	62.97			23	0.493	0.493	~60,000	
	D	11.76	1.896	65.47			32	0.429	0.429		
B1 S2	I	6.72	1.626	61.92	14.82	43.56	37	0.656	0.656		0.849
	R		1.630	61.98				0.445	0.445		
B1 S4	I	12.41	1.511	60.18	6.75	79.17	16	0.397	0.397	41,500	0.202
	D	12.37	1.778	64.00		78.91	13	0.518	0.518		0.165
	R		1.628	61.95				0.436	0.436		
B1 S6	I	18.78	1.995	66.61	9.39	115.84	96	0.948	0.640	47,412	0.829
	D	18.72	2.279	69.50		115.49	22	0.806	0.565		0.190
	R		2.163	68.38				1.440	0.497		
B1 S8	I	24.32	2.283	69.54	14.82	154.86	32	0.928	0.589	3,309	0.207
	D	24.18	2.451	71.02	14.66	154.02	37	1.161	0.709		0.240
	R		2.292	69.62				0.765	0.542		
B7 S9	I	10.24	1.650	62.26	15.32	66.48	48	0.824	0.518	16,840	0.722
	D	10.20	2.246	69.19	14.33	66.22	34	0.869	0.585		0.513
	R		1.969	66.32				0.528	0.528		
B7 S15	I	16.37	2.018	66.87	14.89	99.90	41	1.210	0.601	203	0.410
	D	16.31	1.823	64.58	16.54	99.58	35	0.760	0.445		0.351
	R		1.708	63.07				0.474	0.474		
B5 S10	I	13.29	1.526	60.41	17.48	82.37	43	0.468	0.468	566	0.522
	D	13.21	1.414	58.57	19.76	81.87	65	0.518	0.518		0.794
	R		1.325	65.99				0.424	0.424		
B5 S15	I	19.42	1.230	55.16	19.15	126.31	89	0.405	0.405	485	0.705
	D	19.47	1.395	58.25	15.65	126.67	34	0.449	0.369		0.268
	R		1.283	56.20				0.353	0.353		
B5 S23	I	36.49	1.504	60.06	16.47	238.64	51	0.555	0.366	210	0.214
	D	36.44	1.728	63.34		238.31	38	0.581	0.422		0.159
	R		1.837	64.75				0.730	0.730		

B5 S26	I	42.47	2.022	66.91	21.69	270.13	90	0.844	0.844	4,516	0.333
	D	42.50	2.393	70.53	18.94	270.92	60	0.900	0.900		0.222
	R		2.139	68.14				0.604	0.604		
B5 S28	I	45.54	1.777	63.99	19.77	288.83	110	0.956	0.521	120	0.381
	D	45.49	1.941	66.00	19.93	287.72	54	0.732	0.732		0.188
	R		1.821	64.55				0.495	0.495		
B5 S30	I	51.76	2.243	69.16	29.69	322.86	69	0.941	0.702	1,370	0.214
	D	51.80	2.245	69.18	21.91	323.02	74	1.036	0.798		0.229
	R		2.128	68.03				0.605	0.605		
B5 S32	I	57.80	2.122	67.97	33.52	357.87	64	0.717	0.717	170	0.179
	D	57.74	2.091	67.65	28.16	357.52	60	0.741	0.741		0.169
	R		2.050	67.21				0.623	0.623		

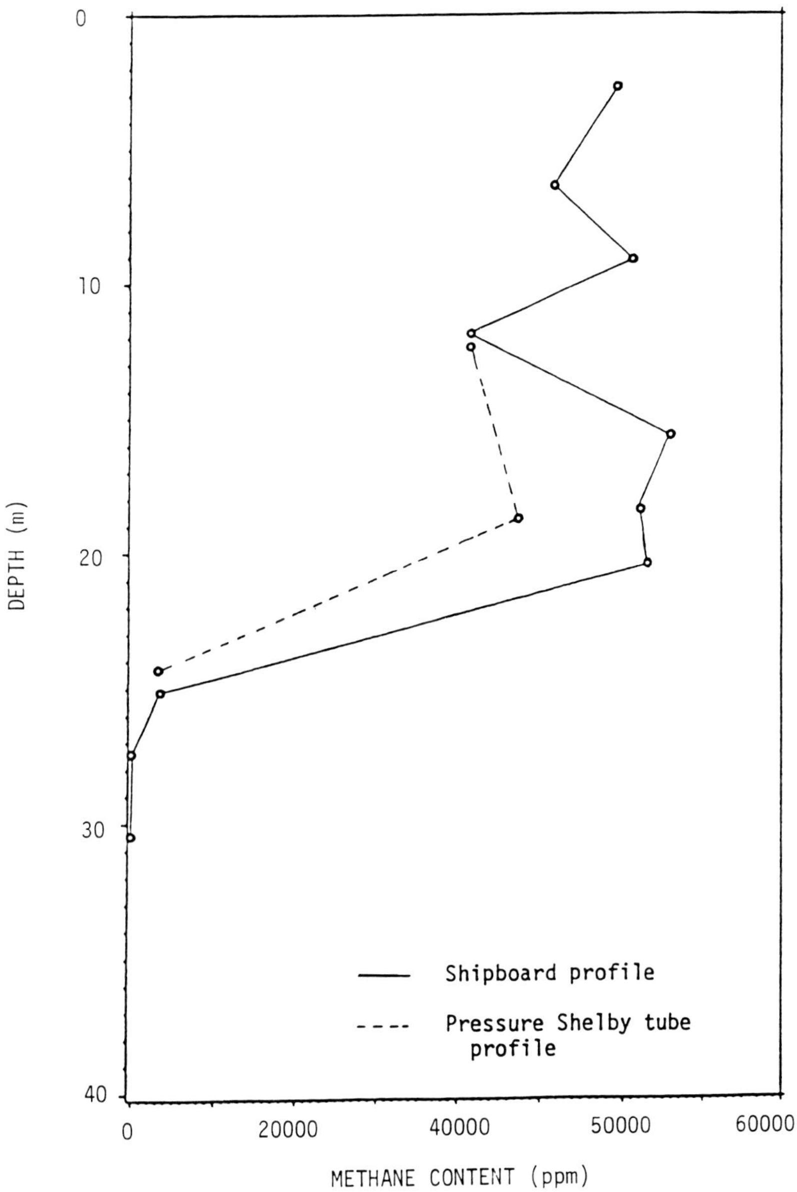

FIGURE 3. Methane concentration vs. depth for PST Boring 1, Main Pass, Block 74.

The interpreted methane concentration vs. depth profile for Boring 2 indicates that sample B2S10 (11.76 m), collected with the pressure core barrel, contains approximately 60,000 ppm (Figure 5). Methane concentrations were analyzed shipboard.

B. CONSOLIDATION TEST RESULTS

The primary consolidation characteristics for all the samples can be found in Table 3. The results of comparative consolidation testing are graphically displayed as e-log σ_v' (Figures 6 through 19). The coefficients of consolidation, C_v, for the respective e-log σ_v' curves are plotted against logarithm of vertical effective stress, C_v-log σ_v', on the same scale as the e-log σ_v' curves to aid in comparison. Because the initial structure of a clay depends on many factors, and the volume changes under pressure are a function of structure, a clay

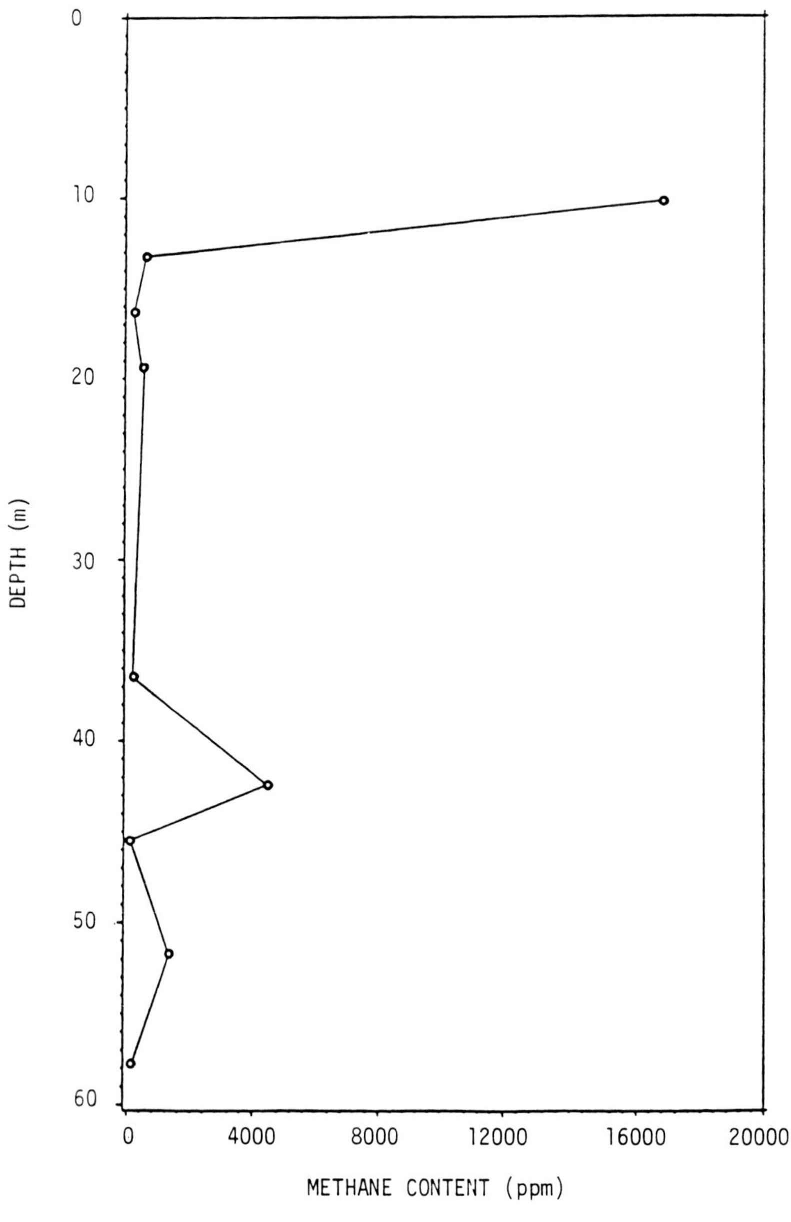

FIGURE 4. Methane concentration vs. depth for PST samples from PST Borings 5 and 7, South Pass, Block 43.

does not have a unique consolidation curve (Mitchell, 1976). It must be noted that sample variability could not be analyzed due to the lack of multiple consolidation analyses for each test procedure.

Sample B7S9 clearly shows the effects of degassing on the soil structure as evidenced in the void ratio and coefficient of consolidation curves vs. logarithm of vertical effective stress (Figure 7). The degassed sample exhibits a drastically different e-log σ_v' curve, clearly indicating greater amounts of compression for a given load than for the *in situ* or remold test results. This is not within the expected normal range of variation for test specimens of the same sample. The compression indices, C_c, for the three specimens are much different (Table 3).

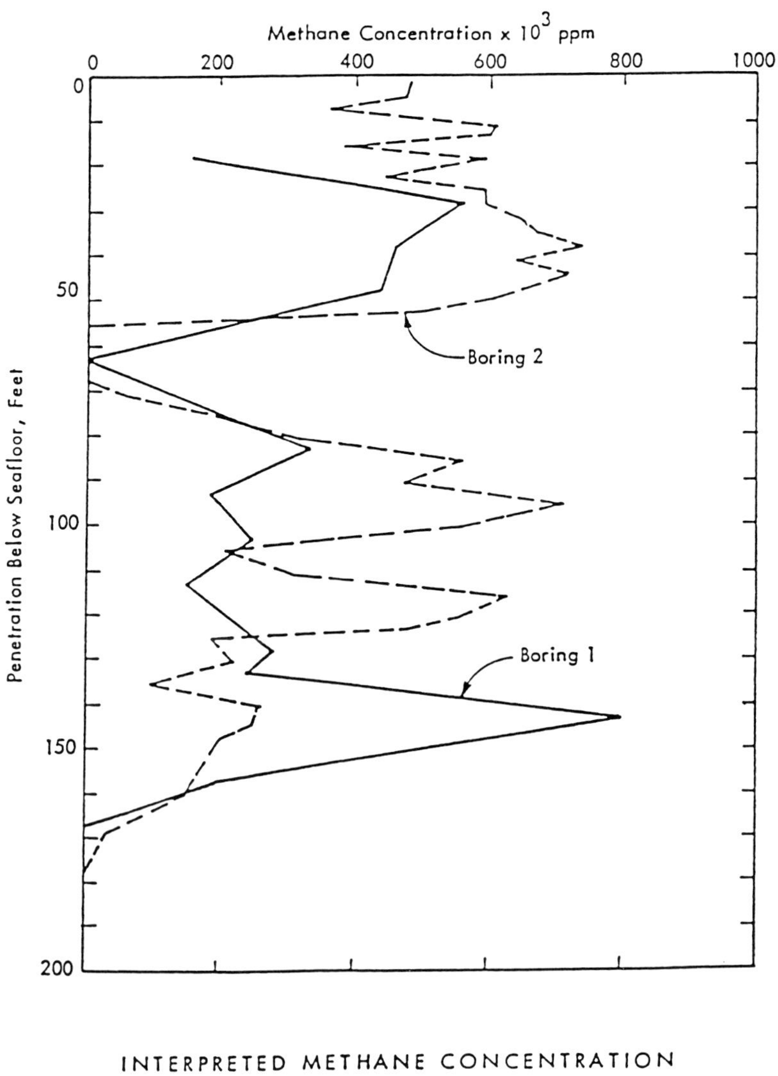

FIGURE 5. Methane concentration vs. depth for PST Boring 2, West Delta, Block 85.

The primary difference between the *in situ* and degassed test results is shown by the time rate of compression curves of Figure 20. This figure presents the comparative square root of time curves of Sample B7S9 for the 50, 100, 200, 400, 800, 1600, and 3200 kPa loadings. For all loads, with the exception of 400 kPa, the t_{90} times are greater for the degassed consolidation curves. For loads less than 400 kPa, the time rate of compression is much larger for the degassed test results. In fact, the differences between the degassed and *in situ* curves increase between 50 and 200 kPa. At the 400 kPa load, the time rate of compression becomes greater for the *in situ* test. This trend continues throughout the remainder of the testing program. The differences in the coefficients of consolidation, C_v, result from these differences.

Compression vs. log time curves for Sample B7S9 show similar results as for the square root of time curves (Figure 21). The time rate of compression is, again, clearly greater for the degassed specimen at loads of 200 kPa and below. For loads of 400 kPa and greater, the time rate of compression for the *in situ* specimen is greater. The t_{50} times are greater for all loadings of the degassed specimen. The resultant C_v-log σ_v' curves at t_{50} are revealing

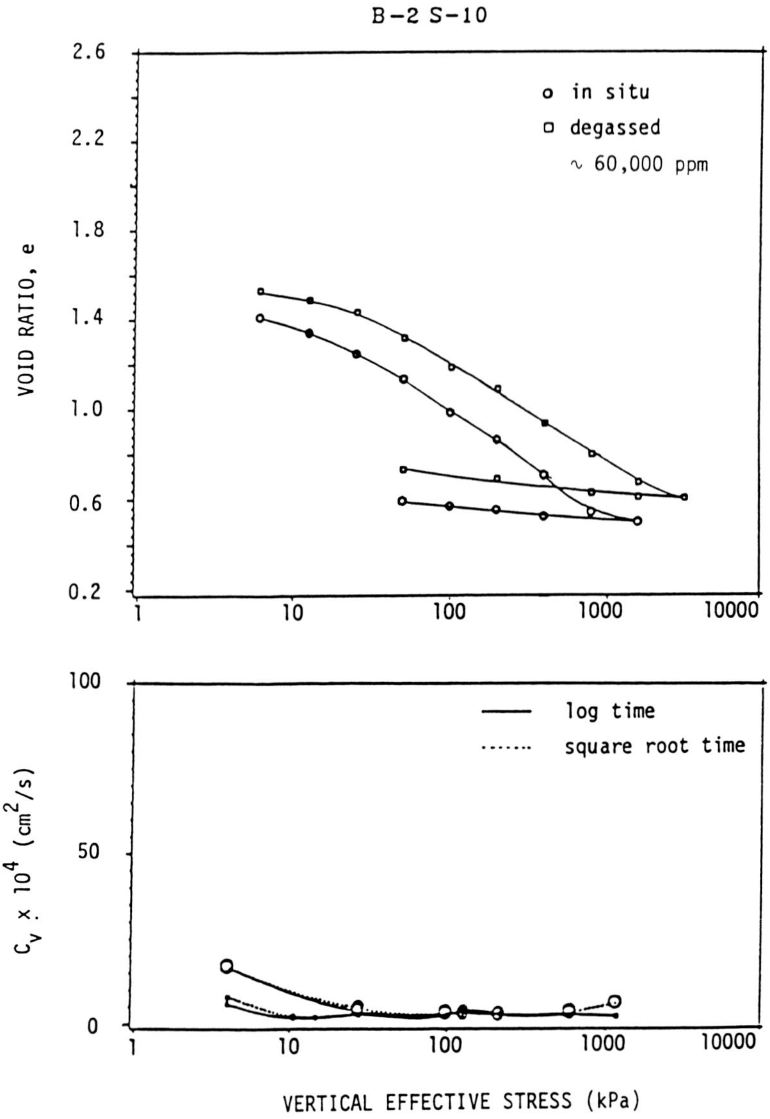

FIGURE 6. Comparative test results for Sample B2S10.

(Figure 7). Initially, the degassed values are larger than the *in situ* values. At approximately 200 kPa, the C_v values for the degassed specimen are slightly less than for the *in situ* specimen. It must be noted that at any given load the void ratio is larger for the degassed sample. However, the magnitude of the difference of the t_{50} values is markedly less (Table 4).

The instantaneous compression with loadings can best be noted by comparing the start height, d_s, of the sample with the calculated height at time zero, d_o, at any given load. Instantaneous compression, Δd, can be defined as d_s-d_o. Initial compression of the degassed samples was found to be considerably larger than for the pressurized samples. This occurs because gas-filled voids are readily compressed in the degassed, partially saturated samples. For samples with 100% pore water saturations, as in the pressurized samples, the addition of an instantaneous normal load results in an equal increase in the pore water pressure. Since permeability and pore water compressibility are low, pore water drainage and, thus, time are required for significant reductions in sample height.

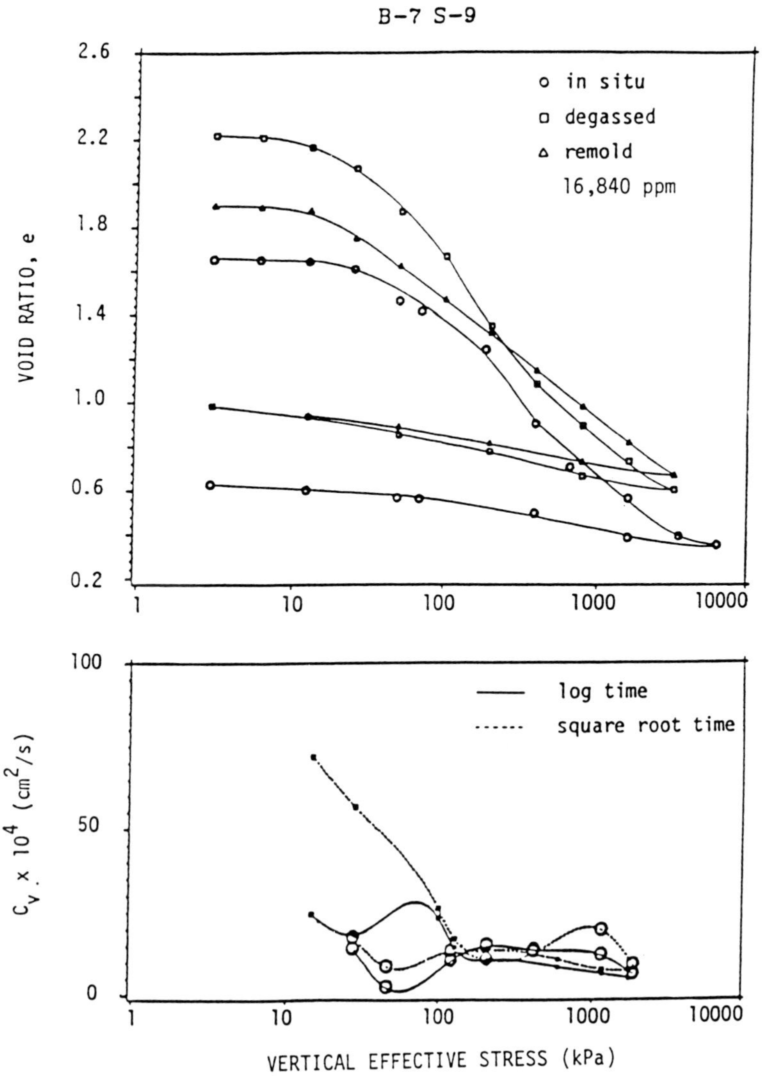

FIGURE 7. Comparative test results for Sample B7S9.

Figure 22 and Table 5 clearly show the greater initial compression for the degassed vs. pressurized samples over the test-loading range. Four points of particular interest to note are: (1) initial compression over the test-loading range is greater for the degassed samples, (2) initial compression increases with increasing methane content, (3) initial compression is greatest over the load range of 25 to 400 kPa, and (4) the total compression for a given degassed specimen is greater than the comparative compression for the *in situ* specimen (Table 6).

Large differences are shown for the coefficients of consolidation, C_v, which are an averaged reflection of the time rate of consolidation. Initially, the degassed values of the coefficient of consolidation, calculated by the square root of time method, are approximately four times greater than the *in situ* test results. Both *in situ* and degassed log and square root of time coefficients of consolidation converge at a vertical effective stress value in excess of the effective preconsolidation stress at approximately 200 kPa. The effective preconsolidation stresses for these samples are 48 kPa and 34 kPa for the *in situ* and degassed tests, respectively.

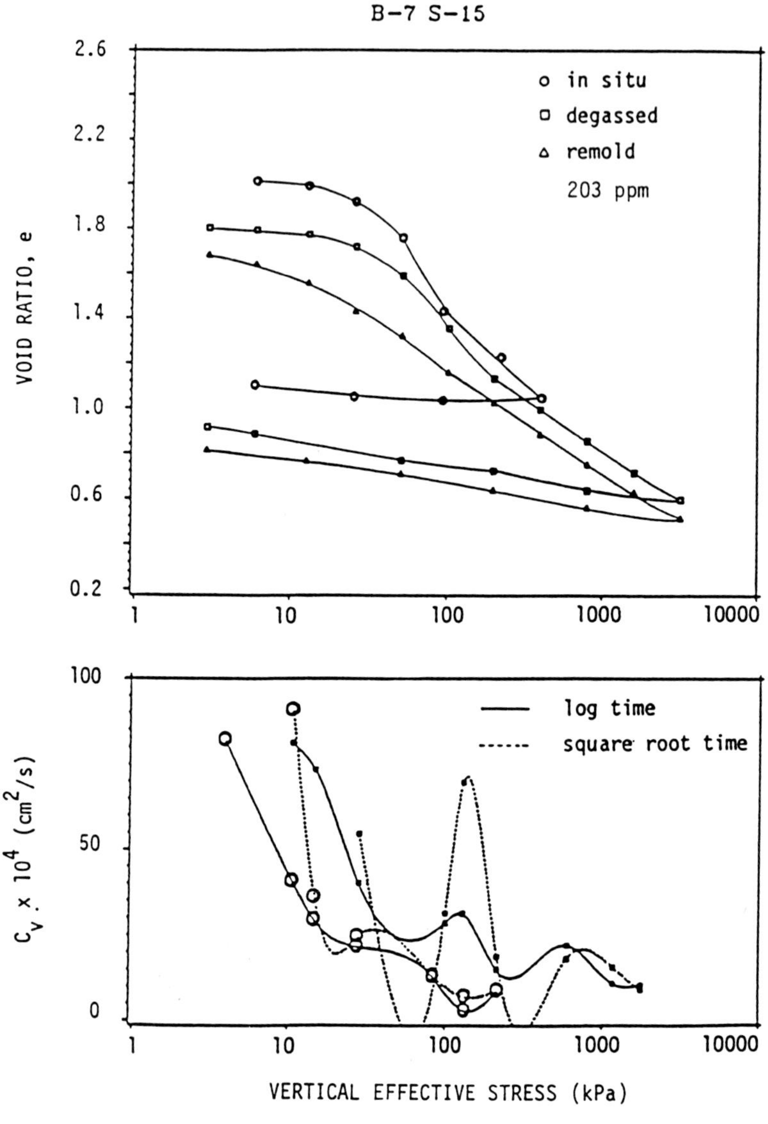

FIGURE 8. Comparative test results for Sample B7S15.

The degassed specimen appears to be reconsolidating to its original state (i.e., that which the *in situ* specimen is at under a given load). But if the inclusion of gas bubbles in the soil matrix tends to reduce effective permeability, then the time rate of consolidation and, thus, the coefficients of consolidation should be less for the degassed test. However, this does not appear to be the case. Lowe et al (1964) noted a similar effect for back-pressured vs. non back-pressured specimens. They presented the possible explanation that the presence of gas bubbles in the soil matrix not only affects the permeability of the soil but also the rate of dissipation of pore pressure.

In this case, it appears that the effect of increased load is much greater than the effect of permeability on compressibility. The highly disturbed matrix of the degassed sample is being readily reoriented under initial load conditions. Both the coefficients of compressibility and consolidation are large for the initial loading conditions. Instantaneous compression of gas-filled voids occurs in all loadings less than the effective preconsolidation stress. Taylor

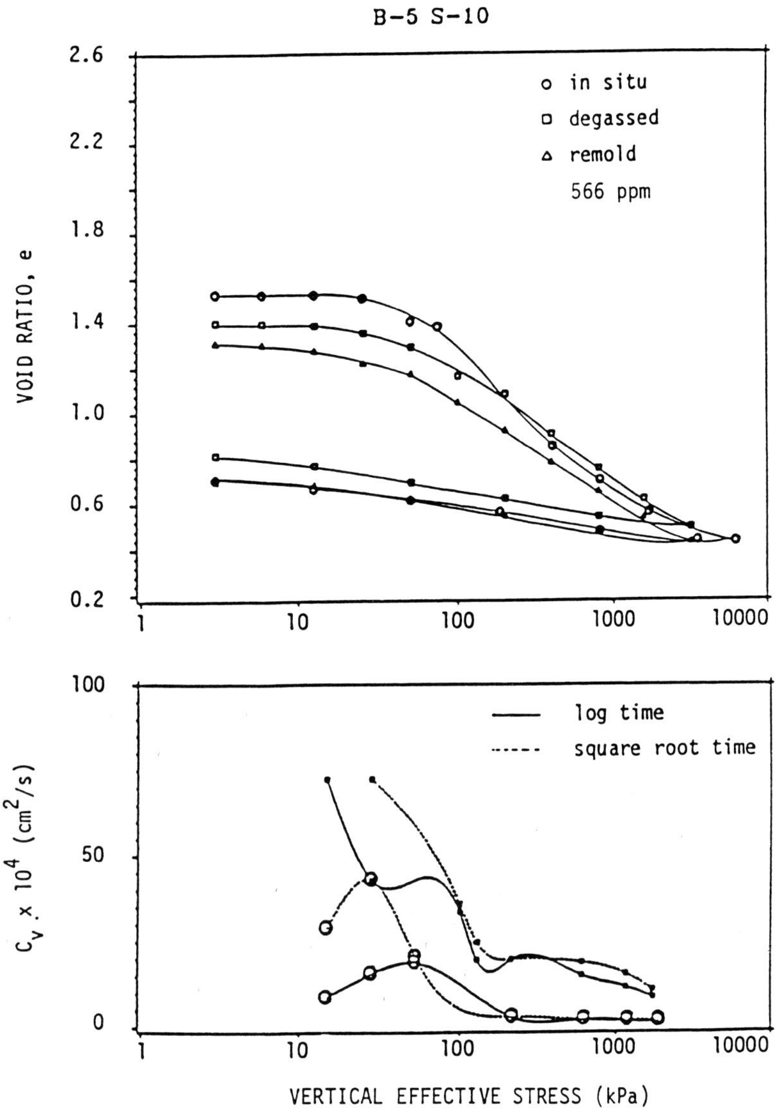

FIGURE 9. Comparative test results for Sample B5S10.

(1948) noted similar rapid initial compression and attributed it, in part, to small amounts of gas in the pores. For loads above the effective preconsolidation stress, compression of gas bubbles no longer constitutes the major portion of compression as the soil matrix is reoriented due to prior compression. At this point, the coefficients of consolidation calculated by the log and square root of time methods are less for the degassed versus the *in situ* test.

Similar results, as noted for the comparative tests of specimen B7S9, are noted for specimens B2S10, B5S23, B5S26, B5S28, B5S30, B5S32, B1S2, B1S4, B1S6, and B1S8. For all the above-mentioned cases, both the coefficients of consolidation calculated by the log and square root of time methods converge. This occurs in the lower portions of the e-log curves, where the vertical effective stress is greater than the effective preconsolidation stress for the respective curves.

The comparative e-log σ_v' and C_v-log σ_v' curves for sample B5S10 are quite dissimilar from the above-mentioned test results (Figure 9). Both the degassed and remold results

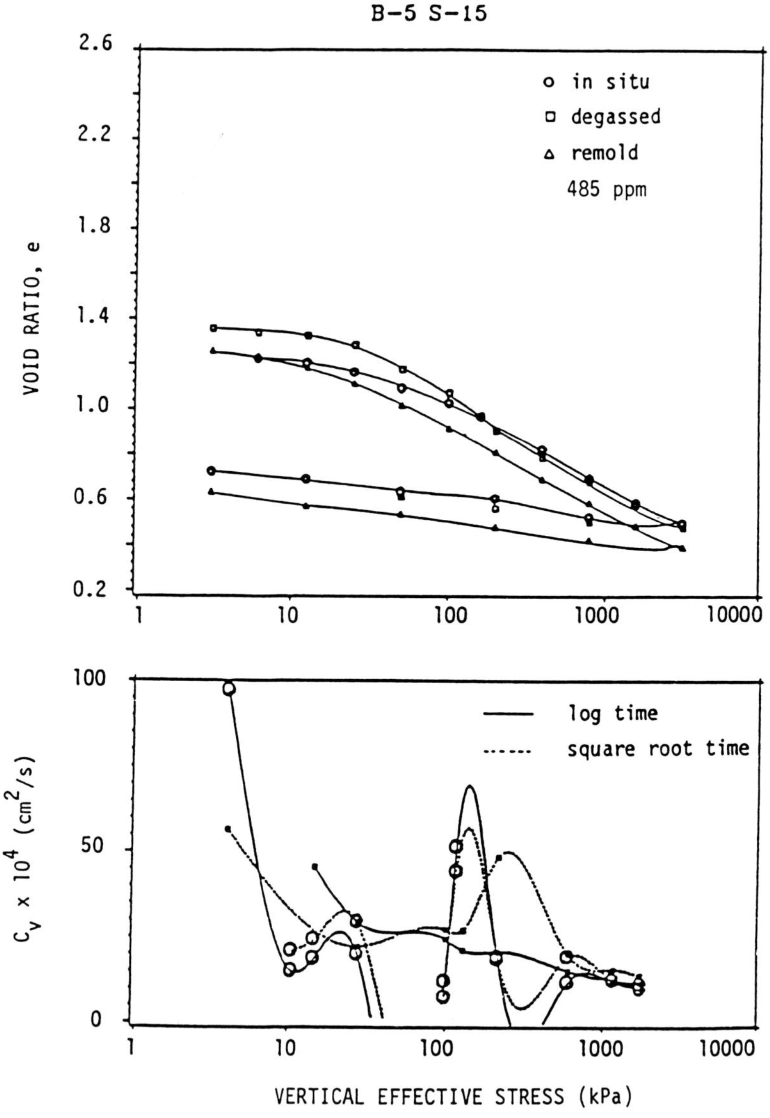

FIGURE 10. Comparative test results for Sample B5S15.

appear quite similar. The e-log σ_v' curves are parallel, indicating similar compression vs. vertical effective stress. The *in situ* test results clearly show a distinct break in curvature and higher void ratios for a given stress. The C_v-log σ_v' curves for the *in situ* and degassed tests are again unique. The degassed results for both the log and square root of time methods are larger by a factor of ten throughout the test. The degassed values converge at approximately 100 kPa, while the *in situ* values converge at approximately 60 kPa. The effective preconsolidation stresses for the degassed and *in situ* samples are 65 and 43 kPa, respectively.

Comparative results of samples B5S15 and B7S15 are similar in many respects. Both degassed e-log σ_v' curves show distinct breaks in curvature at higher vertical effective stresses. However, the *in situ* values for B7S15 have greater void ratios at a given effective stress and show a distinct break in curvature as compared to degassed and remold test results (Figure 8). The *in situ* test results for sample B5S15 have lower values of void ratio for a given effective stress than the degassed test values and do not show a distinct break in curvature (Figure 10).

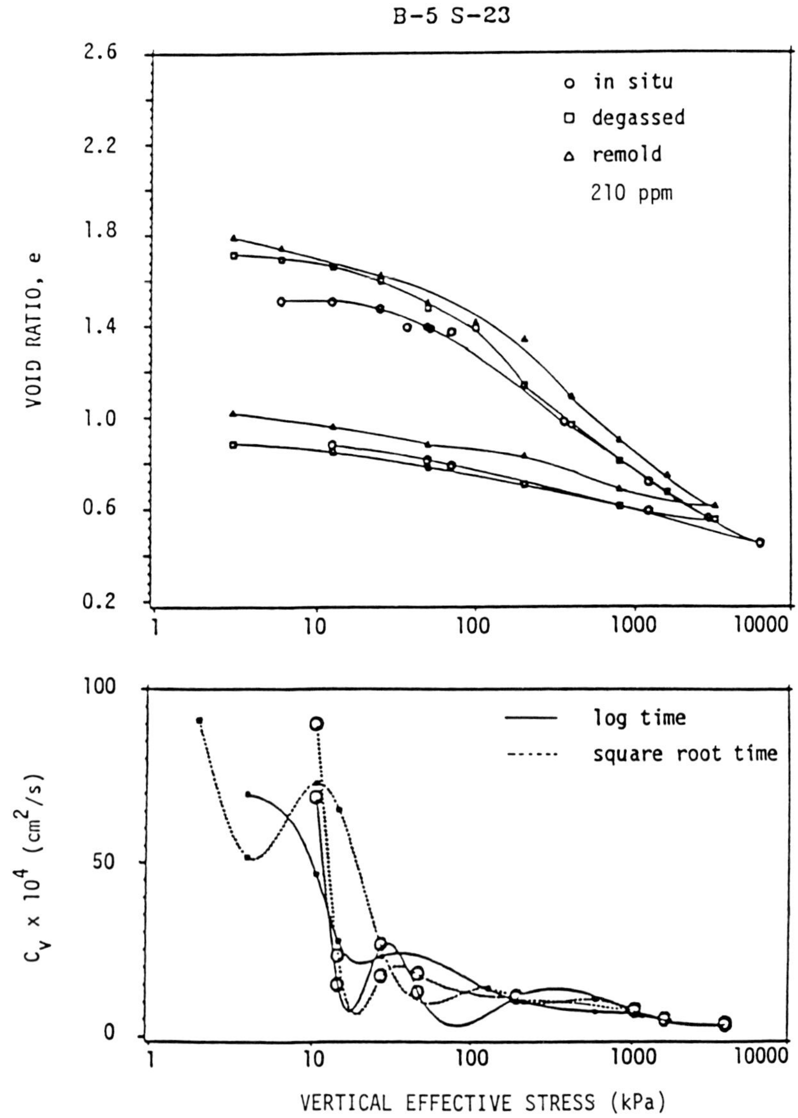

FIGURE 11. Comparative test results for Sample B5S23.

The comparative C_v-log σ_v' curves for both samples B7S15 and B5S15 are quite erratic and only suggest a slight decreasing trend with increasing vertical effective stress.

1. Effective Preconsolidation Stress and Effective Overburden Stress

Effective overburden stress, σ_{vo}', is calculated based on the bulk density profile for the respective borings and assumes no excess pore water pressure (Figure 23). The effective preconsolidation stress, σ_{vc}', is graphically estimated from the e-log σ_v' curves by the Casagrande (1936) technique (Figure 23). The overconsolidation ratio, OCR, is defined as the effective preconsolidation stress, σ_{vc}', divided by the effective overburden stress, σ_{vo}'. The state of consolidation is determined by the evaluation of the OCR value. An OCR value equal to 1 indicates a normally consolidated sediment, while greater than 1 shows overconsolidation, and less than 1 shows underconsolidation. Values for these parameters are listed in Table 3 and graphically displayed vs. depth in Figure 24. Effective preconsolidation stress values falling to the left of the effective overburden stress profile are underconsolidated.

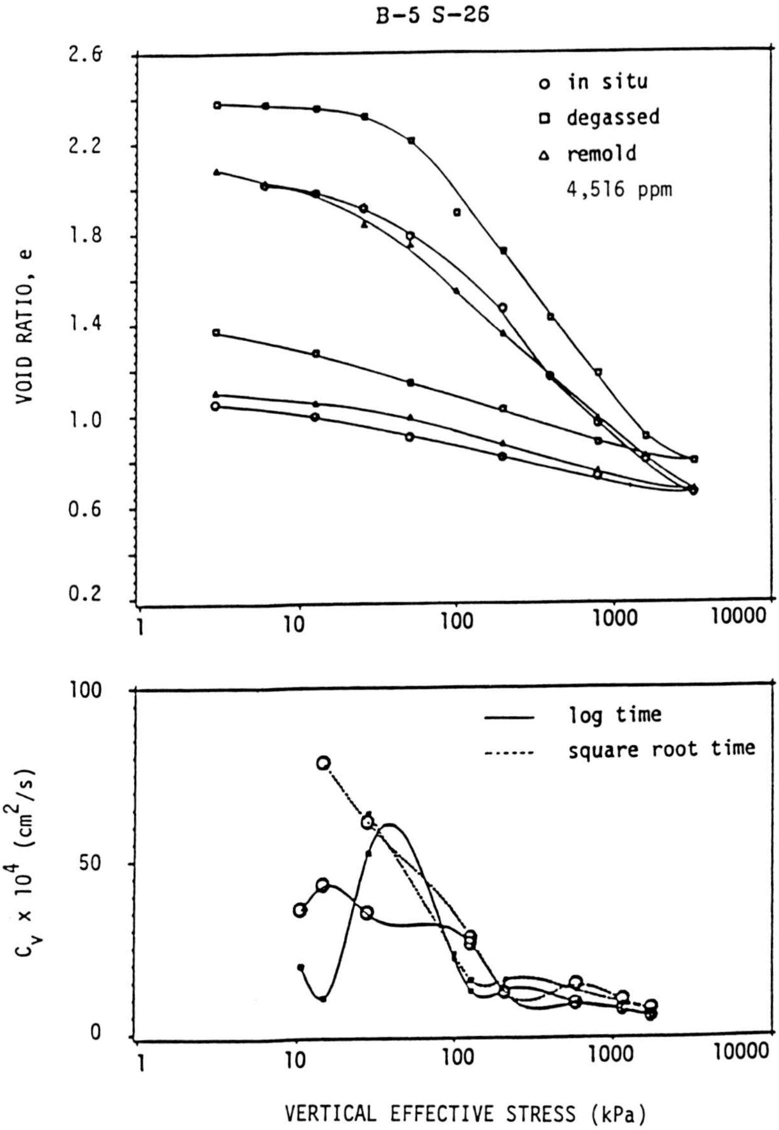

FIGURE 12. Comparative test results for Sample B5S26.

With three exceptions, the values of effective preconsolidation stress are larger for the pressurized tests than those for the degassed tests, and in some cases they are markedly larger. Samples B1S8, B5S10, and B5S30 all exhibit reduced values of effective preconsolidation stress vs. the degassed test results.

The OCR values followed the same trends as the effective preconsolidation values (Figure 24). All samples can be classified as underconsolidated. However, the OCR values for the pressurized tests are greater than for the degassed tests.

2. Compression Indices

Evaluation of the e-log σ_v' curves indicates the presence of double breaks on numerous samples. Therefore, the compression indices, C_c, were calculated for both the high and low portions of these curves (Figures 25 and 26). In 75% of the cases studied, both the high and low compression indices are greater for the degassed versus *in situ* test results. For two

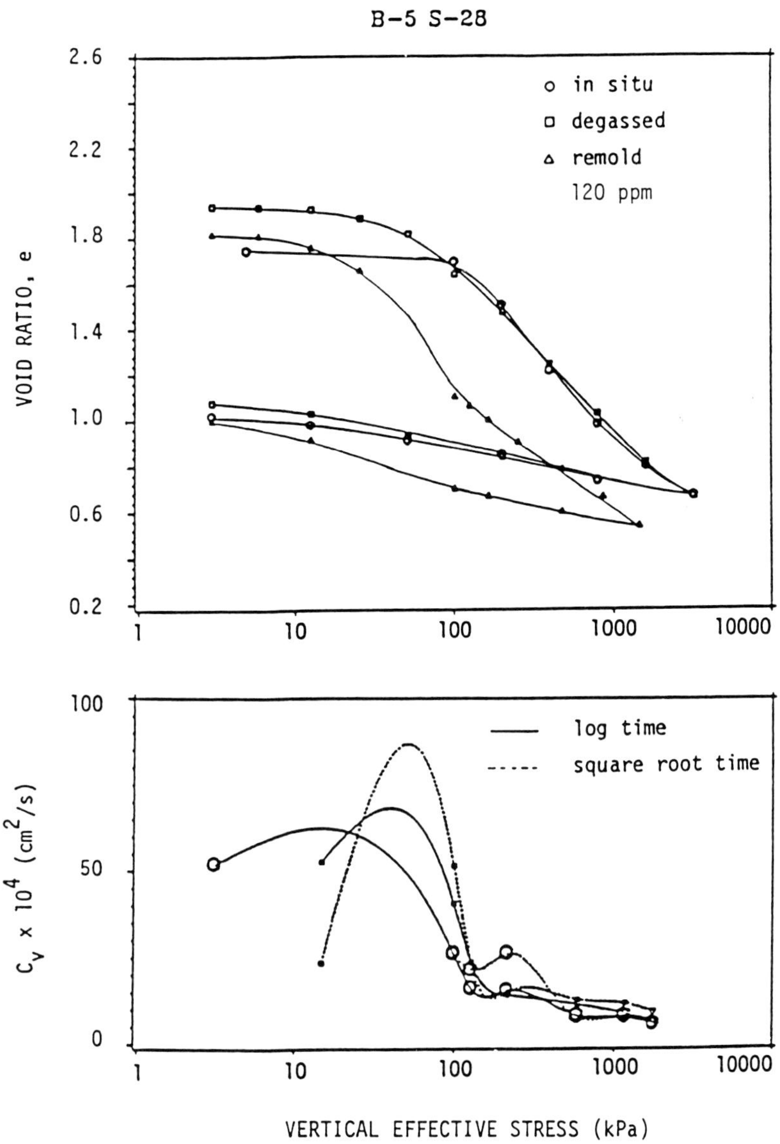

FIGURE 13. Comparative test results for Sample B5S28.

samples that have double curves, B1S6 and B7S15, both C_c values are less for the degassed samples. Additionally, samples B5S15 and B5S28 (both with double curves) have one C_c value less for the pressurized sample. The consolidation curves of the degassed Mississippi delta sediments resemble those of highly sensitive clays, where compressibility is relatively low until the consolidation stress exceeds the preconsolidation stress. It then increases sharply as shown in Figure 7 for sample B7S9. As the void reduces under higher consolidation stresses, the compressibility eventually assumes a lower value.

3. Pore Water Saturation

Pore water saturation values for the consolidation samples provide an interesting insight into the effects of degassing. Pore water saturation values derived by two different methods are shown in Table 7. The calculated values are indicative of theoretical pore water saturation

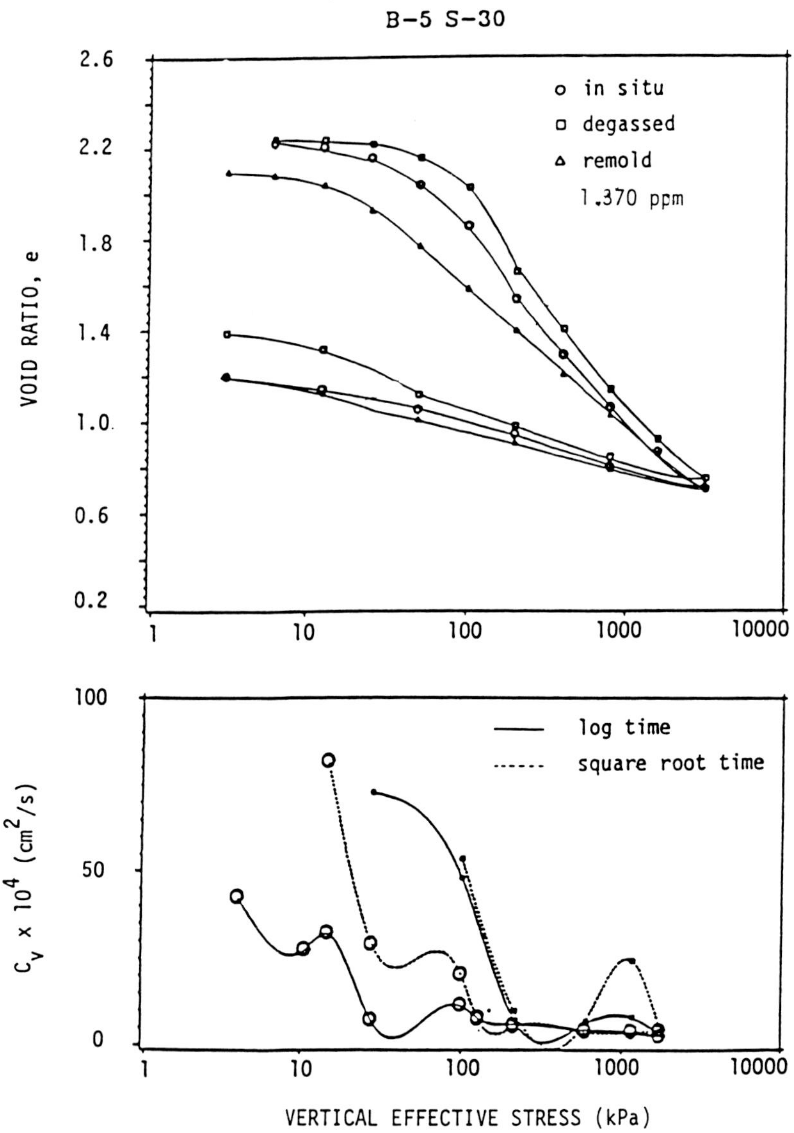

FIGURE 14. Comparative test results for Sample B5S30.

based on the volume of pore water and gas present in the sediment. These values are calculated from:

$$S\ (\%) = \frac{V_w}{V_w + V_g}$$

where S is the degree of saturation, V_w is the volume of water, and V_g is the volume of gas present in a given volume of wet sediment (see Gas Concentrations, this chapter). These values indicate theoretical pore water saturation at standard temperature and pressure (STP) and, therefore, cannot be related to *in situ* conditions. The presence of gasses in the pore water results in pore water saturation values below 100% at STP for all samples.

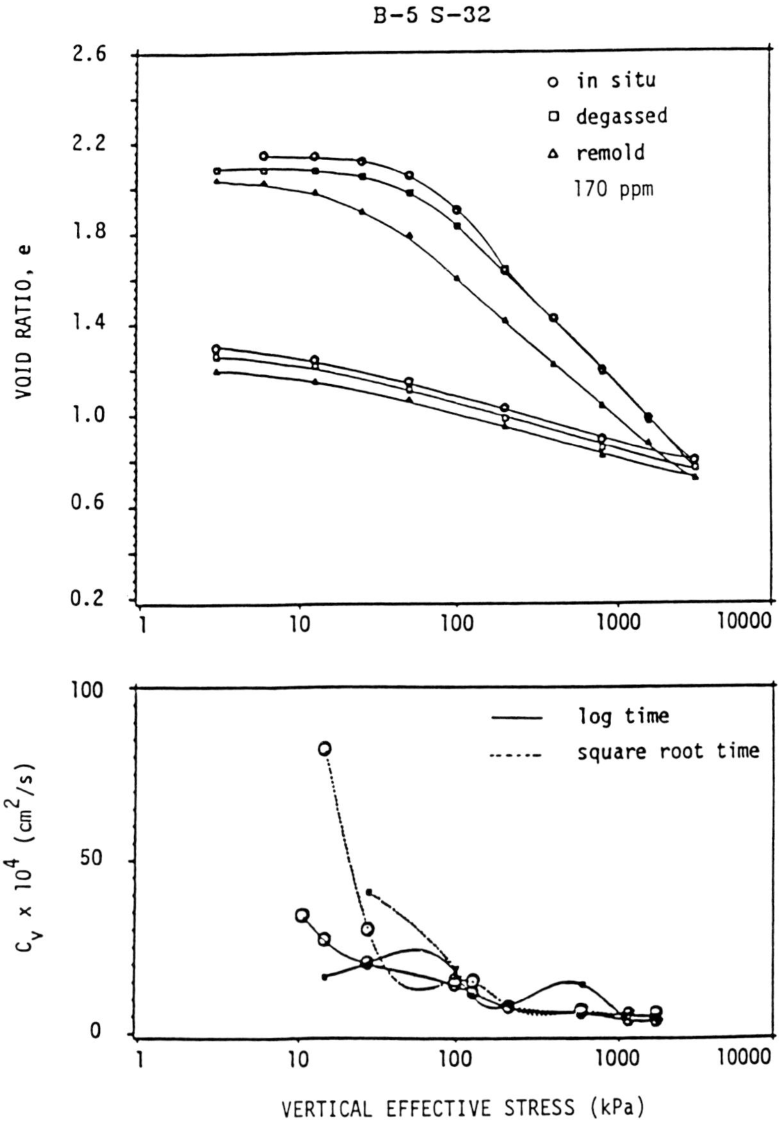

FIGURE 15. Comparative test results for Sample B5S32.

Measured pore water saturation values are based on consolidation test results for the initial conditions and can be calculated from:

$$Gw = se_o \tag{12}$$

where G is the specific gravity of solids, w is the water content, e_o is the initial void ratio, and s is pore water saturation. Nearly all the consolidation samples tested at *in situ* pressures have measured saturation values in excess of 100%. However, there are three exceptions, samples B7S9I, B5S30I, and B5S32I. These samples exhibit saturation values only slightly less than 100%, being undersaturated by a maximum of 2.5%. These low values, in all probability, are within sampling variability.

In marked contrast to the samples tested under *in situ* pressure, the degassed consolidation samples exhibit high degrees of undersaturation. The degree of saturation for the degassed

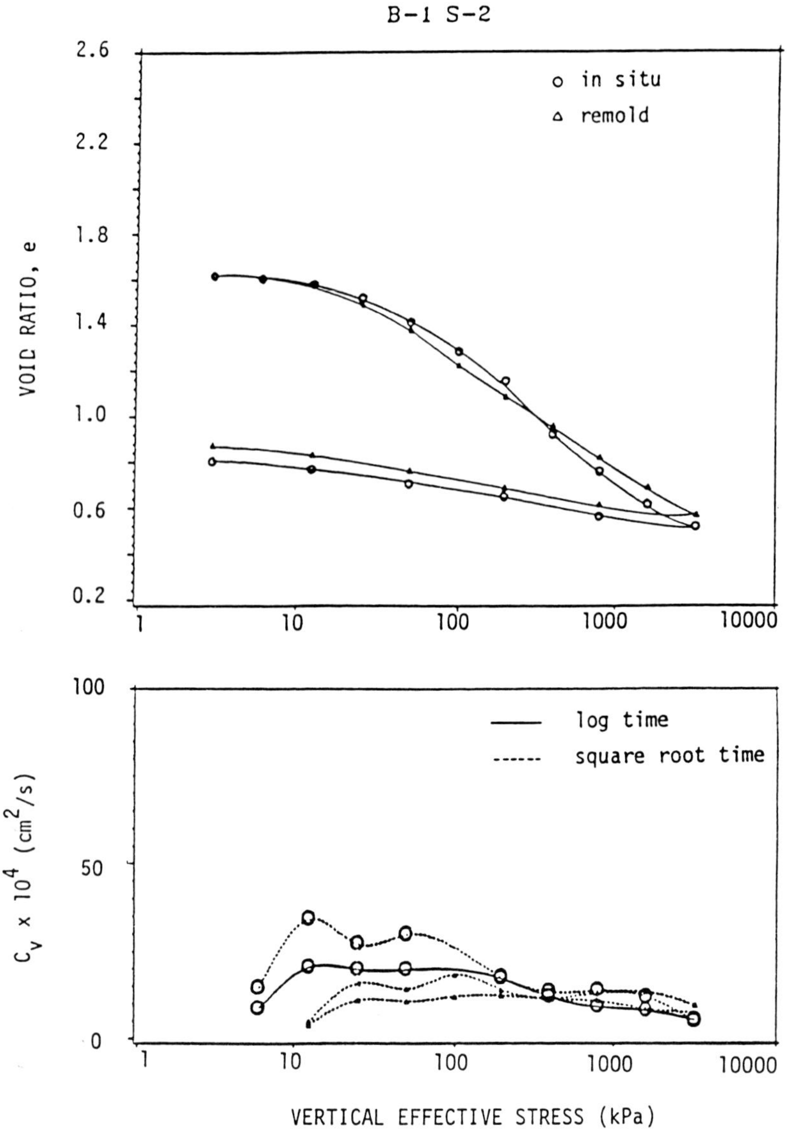

FIGURE 16. Comparative test results for Sample B1S2.

samples ranges between 89 and 99%. The measured pore water saturation values are lower than the calculated theoretical values. A possible explanation is that gas expansion due to a decrease in confining stress results in an increase in pore water pressure and, thus, drainage. Two samples do show measured saturation values of 100%: B1S8D and B5S10D.

4. Void Ratio and Porosity

The effects of porewater degassing are clearly shown as an increase in the initial void ratio or porosity (Figure 27, Table 3). Release of porewater gasses and the relative increase in initial void ratio or porosity is a function of the gas concentration. Three comparative samples exhibit the opposite of this trend, that being a decrease in initial void ratio or porosity upon release of *in situ* pressure. These samples, B7S15, B5S10, and B5S32, all had very low pore water gas concentrations.

Additionally, void ratio was calculated at the effective preconsolidation stress (Table

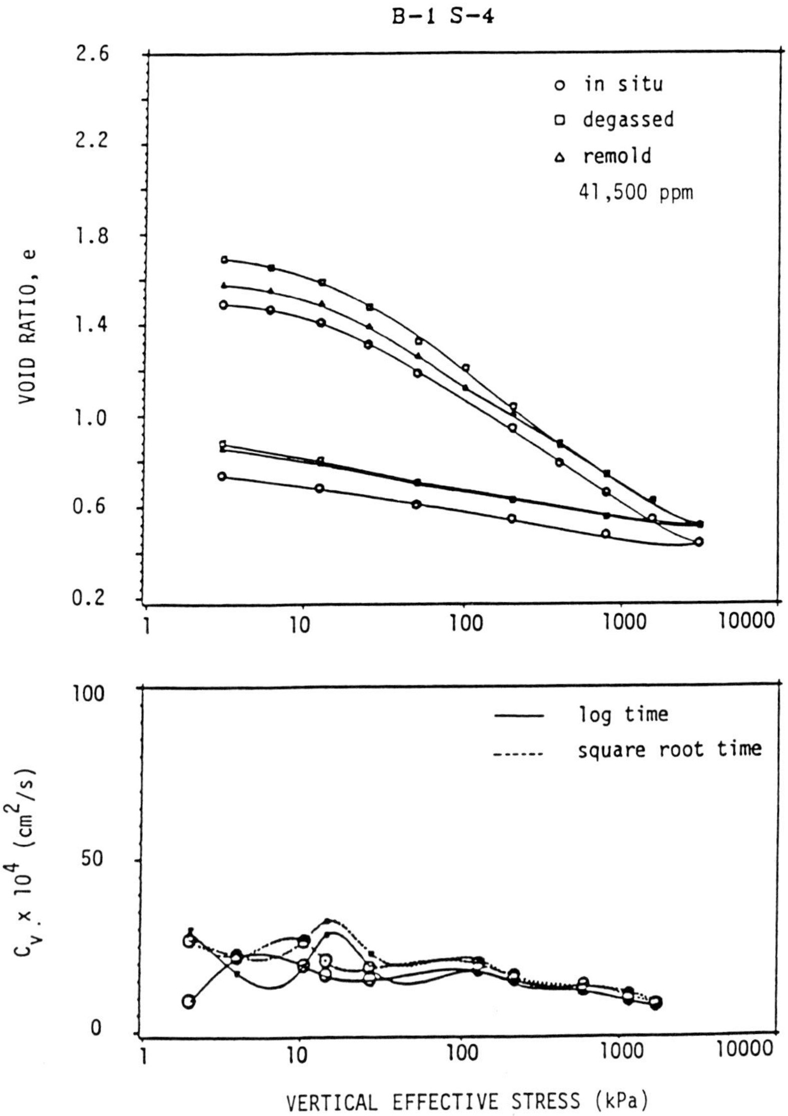

FIGURE 17. Comparative test results for Sample B1S4.

8). The degassed values of void ratio at the preconsolidation stress, e_o, are larger than the comparative *in situ* values (Figure 28). Again, samples B7S15, B5S10, and B5S32 do not follow this trend.

5. Permeability

No direct permeability measurements were performed during consolidation testing; however, permeability was calculated after each load increment. The presence of gas bubbles in a partially saturated soil should act to impede the flow of water in the pores of the soil and thereby reduce the effective permeability of the soil (Bjerrum and Huder, 1957).

Calculated permeability is graphically estimated at the preconsolidation stress for both the *in situ* and degassed test results (Table 8). Figure 29 clearly shows that the calculated permeabilities of the degassed tests are greater than the *in situ* tests. Three samples do not follow this trend: B1S4, B5S30, and B5S32.

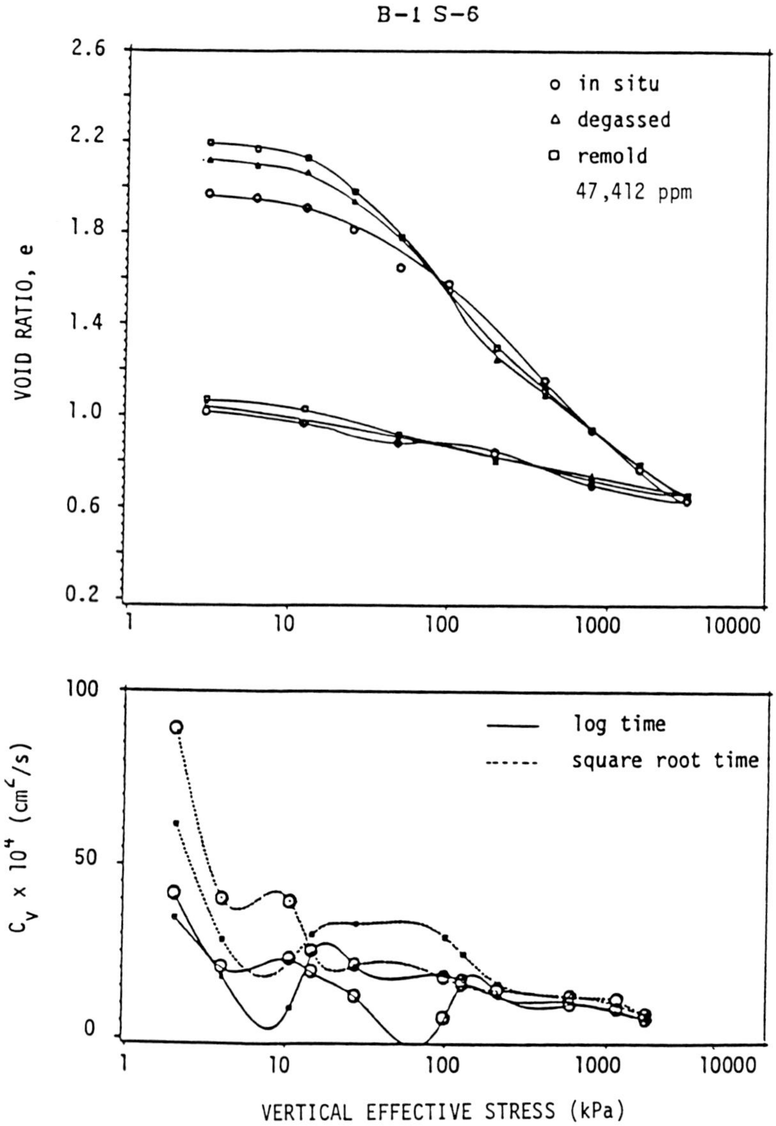

FIGURE 18. Comparative test results for Sample B1S6.

Recalling from Section IV, Experimental Program, the theoretical coefficient of permeability can be calculated from:

$$k = C_v m_v \gamma_w \tag{13}$$

Thus, the coefficients of consolidation and volume change control the value of the theoretical permeability. An increase in either value will result in an increase in calculated permeability, holding γ_w constant.

6. Shear Strength

The basic factor responsible for the strength of a soil is the frictional resistance between soil particles in contact (Mitchell, 1976). As pore water gasses evolve from solution, displacement of soil particles occurs, thereby reducing the intergranular contacts. A reduction

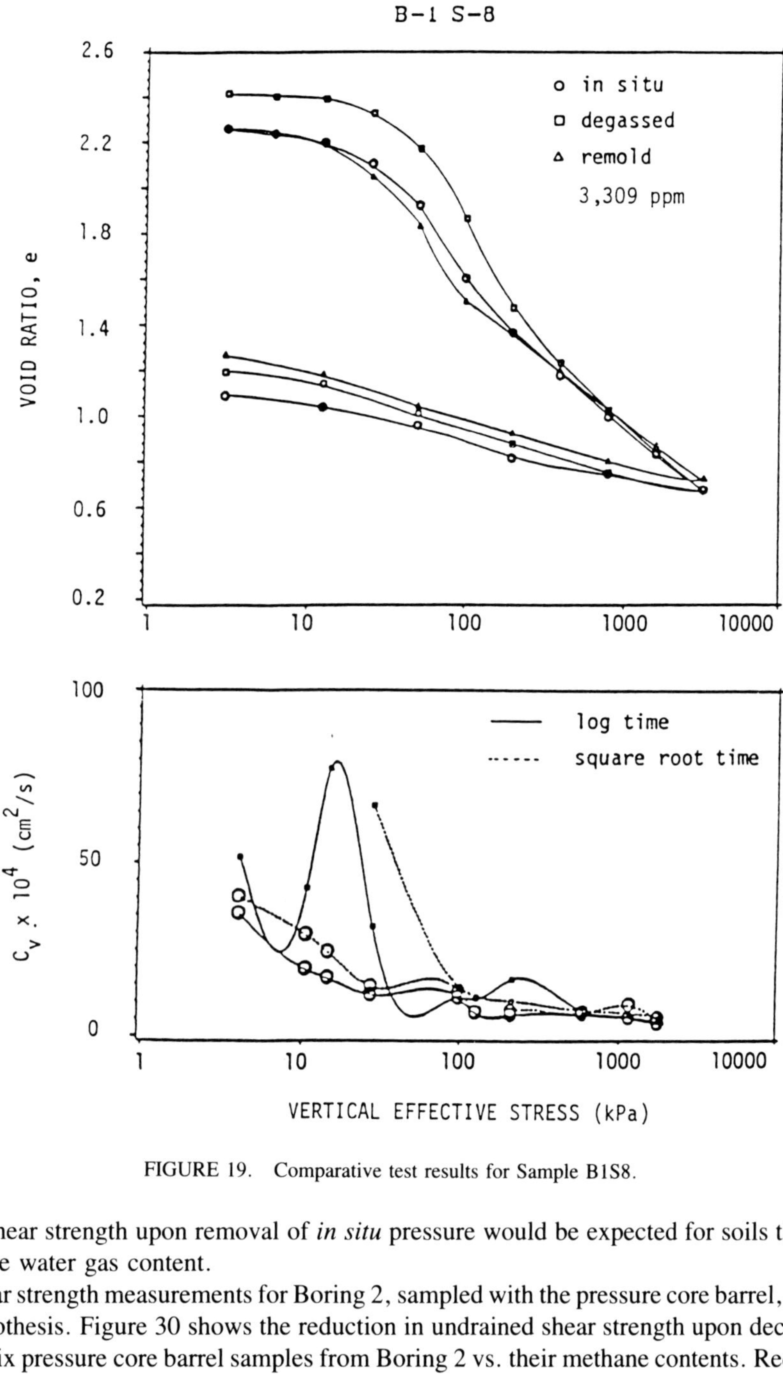

FIGURE 19. Comparative test results for Sample B1S8.

of soil shear strength upon removal of *in situ* pressure would be expected for soils that have high pore water gas content.

Shear strength measurements for Boring 2, sampled with the pressure core barrel, confirm this hypothesis. Figure 30 shows the reduction in undrained shear strength upon decompression of six pressure core barrel samples from Boring 2 vs. their methane contents. Reductions of up to 40% were measured at Boring 2. Whelan et al. (1975) measured similar shear strength reductions of up to 92% in South Pass block 70, where a documented slide occurred. These profiles were obtained from *in situ* remote vane and shipboard degassed measurements (Whelan et al., 1975).

Results of pressure Shelby tube samples tested for strength are not as conclusive as the above-mentioned pressure core barrel tests, nor the remote vane measurements. It must be noted that the pressure Shelby tube results were performed using a hand-held Torvane device (Table 1). The effects of sampling error are more prevalent with the hand-held device because

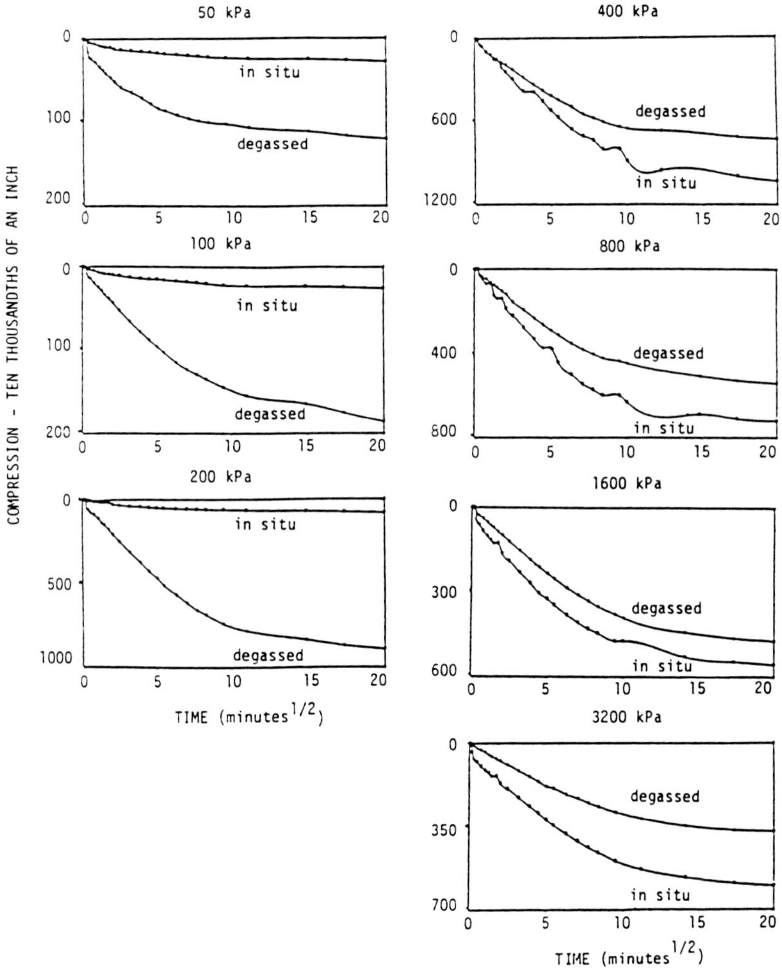

FIGURE 20. Comparative compression vs. square root of time curves for Sample B7S9.

of possible non-uniform rotation rates and improper insertion of the device. However, shear strength test results for the pressure Shelby tube samples show six of nine tests exhibit a reduction of shear strength upon decompression (Figure 31). Samples B7S15, B5S10, and B5S28 show greater values of shear strength for pressurized vs. degassed samples.

VI. DISCUSSION

The conventional theory of consolidation was solved by Terzaghi (1925) for saturated soils. Much work has been carried out to investigate cases which account for the effects of variations of permeability and compressibility during consolidation (Richart, 1957; Lo, 1960; Davis and Raymond, 1965; Janbu, 1965; Barden and Berry, 1965). These factors become important only if the void ratio changes and strains are significant. However, these studies are based on small-strain theory. Additional studies have evaluated the limitation placed on finite-strain theory, non-linear stress-strain relationships, and the effects of self weight of the soil particles (Gibson et al., 1967; Schiffman, 1982; Znidarcic and Schiffman, 1982).

The more advanced of these studies even evaluate the effects of non-constant permeability on soil compressibility (Schiffman, 1982). One of the basic assumptions of consolidation theory which has been largely left alone is that of complete saturation of the soil matrix.

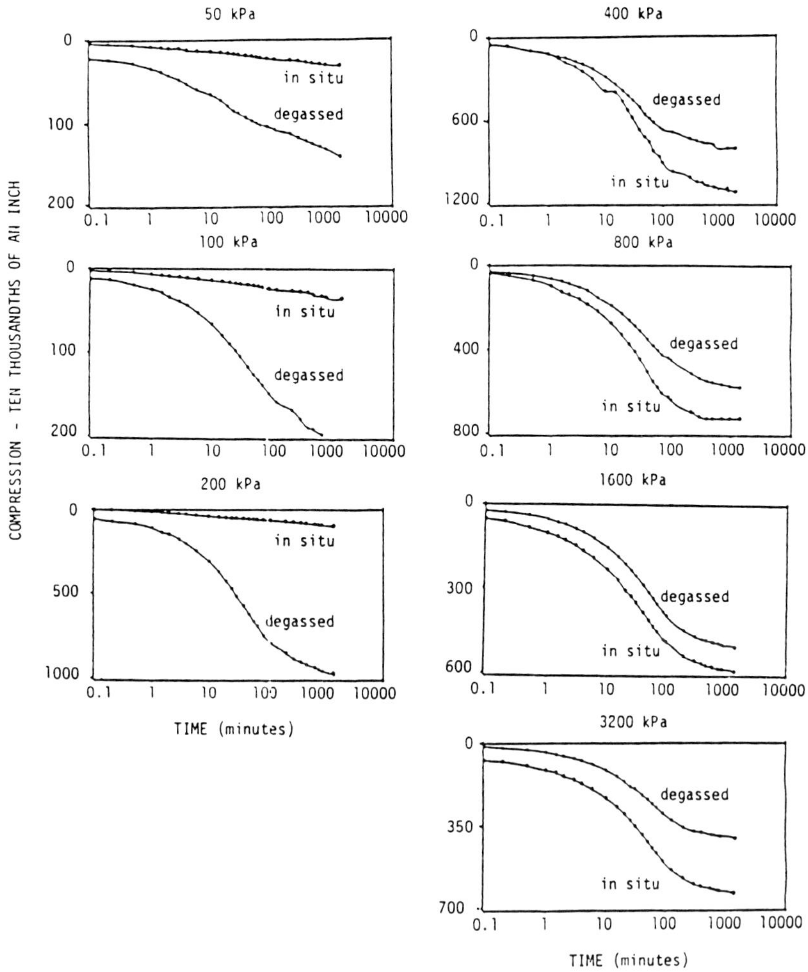

FIGURE 21. Comparative compression vs. log time curves for Sample B7S9.

TABLE 4
Comparative Log and Square Root of Time Values for Sample B7S9

Load	$t_{90 1/2}$ (minutes)		t_{50} (minutes)	
(kPa)	*In situ*	Degassed	*In situ*	Degassed
50	6.25	27.04	3.0	4.1
100	17.64	60.84	2.5	16.0
200	26.01	75.69	2.2	22.0
400	112.36	73.96	21.0	23.0
800	73.96	75.69	17.5	23.0
1600	42.25	86.49	16.5	25.0
3200	72.25	81.0	25.0	27.0

Lowe et al. (1964) investigated the effects of back-pressure consolidation testing on soils. Test results clearly showed that the time rate of consolidation of partially saturated soils under load is significantly different from the time rate of consolidation of a fully saturated soil under the same load. Lowe et al. (1964) cited two basic reasons for this phenomena:

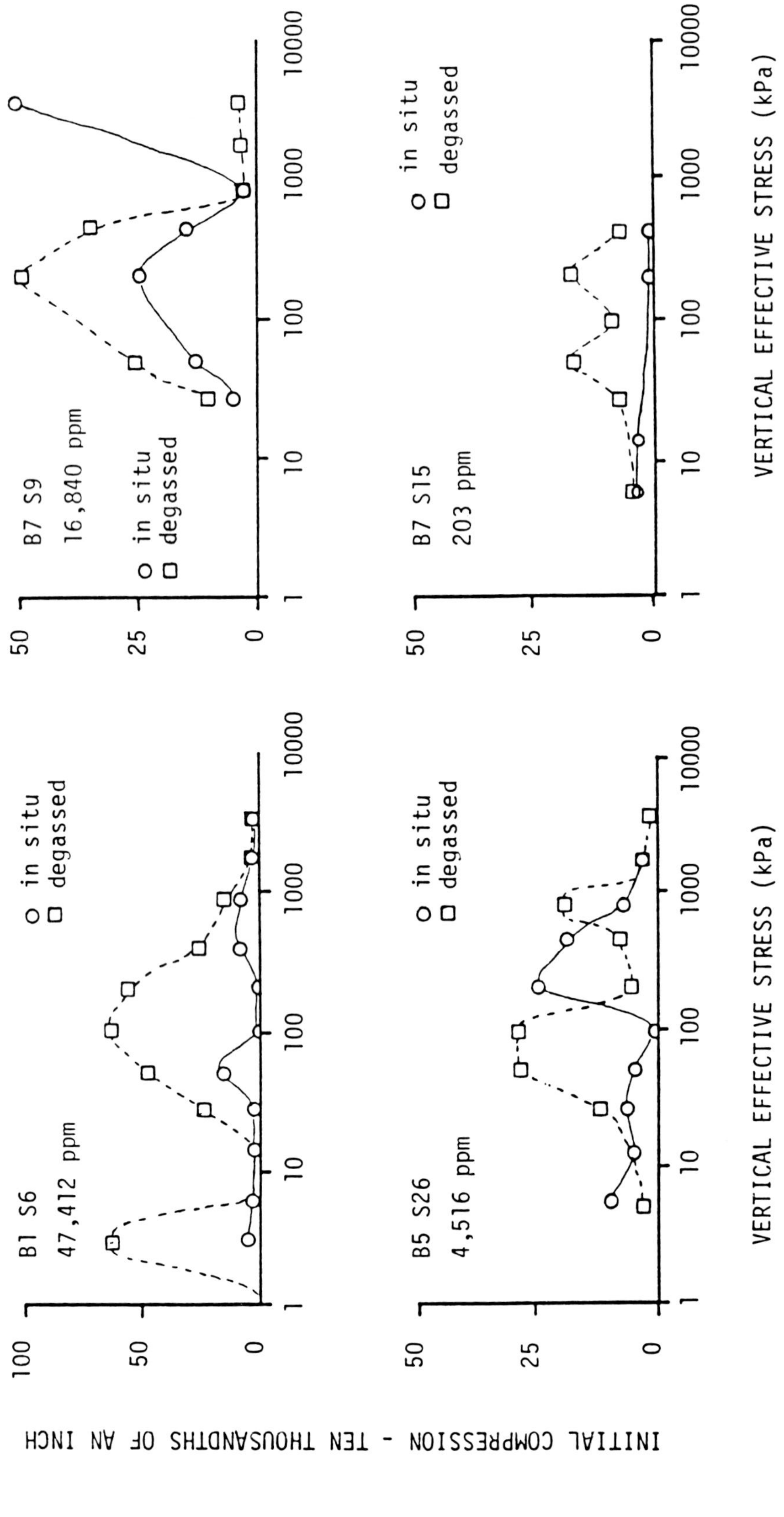

FIGURE 22. Initial compression ($d_s - d_o$) vs. vertical effective stress for four consolidation tests of varying methane concentration.

TABLE 5
Total Initial Compression ($d_s - d_o$) for Loads to 3200 kPa for *In Situ* and Degassed Consolidation Samples from PST Borings 1, 5, and 7

Sample	Total initial compression (10^4 in.) *In situ*	Degassed
B1S2	63	
B1S4	49	162
B1S6	33	244
B1S8	41	97
B7S9	33	123
B7S15[a]	−9	50
B5S10	30	141
B5S15	67	187
B5S23	34	261
B5S26	75	114
B5S28	4	122
B5S30	−7	87
B5S32	16	133

[a] Values indicate the range 3 to 800 kPa.

TABLE 6
Total Compression for Loads to 3200 kPa for *In Situ* Degassed, and Remold Consolidation Samples

Sample	Total compression (inches) *In situ*	Degassed	Remold	Methane conc., (ppm)
B1S2	0.4210		0.4038	
B1S4	0.4271	0.4522	0.4202	41,500
B1S6	0.4530	0.5144	0.5007	47,412
B1S8	0.4868	0.5139	0.4702	3,309
B7S9	0.4861	0.5064	0.4367	16,840
B7S15[a]	0.3306	0.3463	0.3521	203
B5S10	0.4395	0.3794	0.3756	566
B5S15	0.3323	0.3868	0.3805	485
B5S23	0.3748	0.4326	0.4315	210
B5S26	0.4674	0.4741	0.4666	4,516
B5S28	0.3940	0.4377	0.4119	120
B5S30	0.4767	0.4611	0.4098	1,370
B5S32	0.4265	0.4316	0.4356	170
B2S10	0.4469	0.4472		

[a] Values indicate the range 3 to 800 kPa.

(1) The air bubbles in a partially saturated soil are highly compressible, compared to the relatively incompressible water occupying the corresponding pore space in a saturated soil, and (2) the gas bubbles in a partially saturated soil impede the flow of water in the pores of the soil and thereby reduce the effective permeability of the soil (Bjerrum and Huder, 1957).

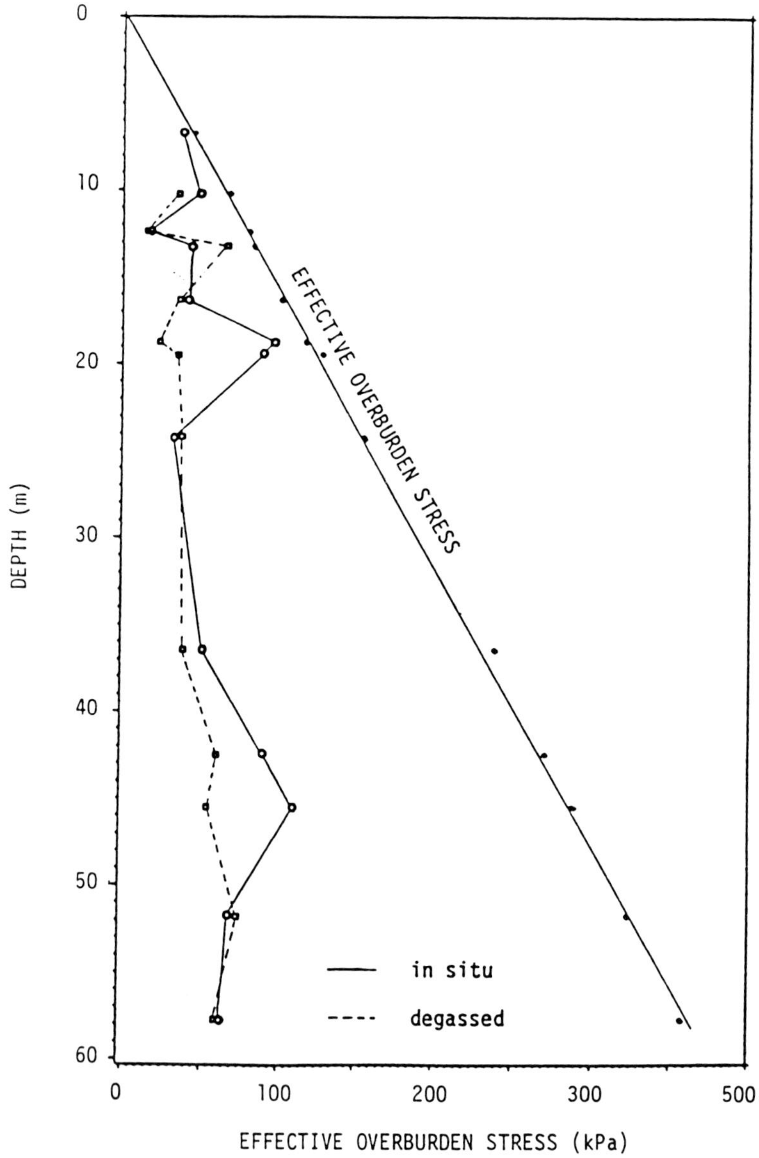

FIGURE 23. Effective overburden stress vs. depth with effective preconsolidation stress for *in situ* and degassed samples from PST Borings 1, 5, and 7.

All of the above-mentioned studies have significantly increased our knowledge of the various aspects of consolidation theory, yet a significant factor has remained unstudied. The effects of disturbance due to ebullition of saturated pore water gasses and its relationship to soil saturation is intuitively easy to understand. However, no studies have attempted to tackle this problem. As sample pressure is released to ambient conditions when a core is recovered, significant stress release can occur. The relative amount of stress release is a function of water depth, effective overburden stress, and the presence of excess pore water pressures at depth.

When no significant levels of pore water gas exist, the soil skeleton will respond to the release of overburden stress, producing negative pore pressures resulting from elastic expansion (Skempton and Sowa, 1963). If mechanical disturbance and moisture loss are

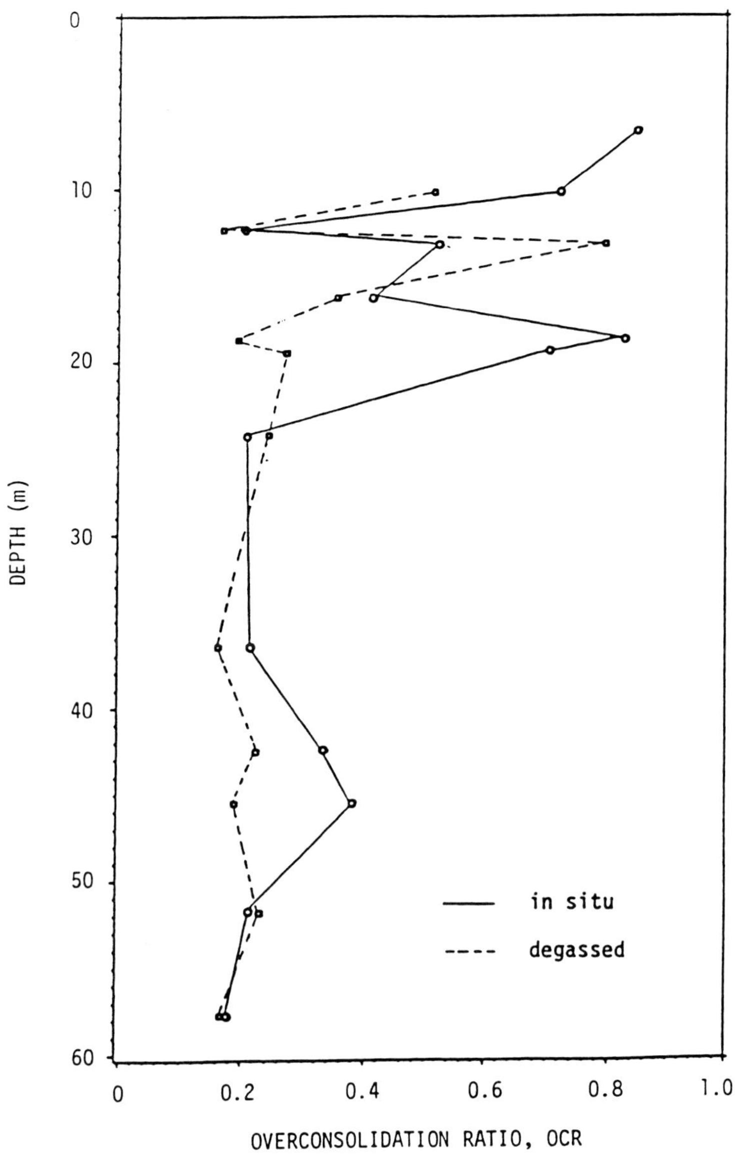

FIGURE 24. Overconsolidation ratio vs. depth for *in situ* and degassed samples from PST Borings 1, 5, and 7.

prevented, this type of disturbance will be minimized, although a strength reduction would be expected.

With the presence of pore water gasses, release of *in situ* stresses results in significant expansion due to the high compressibility of gas as compared with pore water or the soil structure. Expansion to the bubble phase will result in the development of shear stresses within the soil structure. Shear deformations break down particle and aggregate assemblages, resulting in rupture of the interparticle bonds which inhibit swelling (Mitchell, 1976). Gas expansion acts immediately to increase the void ratio and reduce the value of pore water saturation. This is a result of the realignment of clay particles to an altered condition caused by forces exerted by gas expansion (Chiou, 1981).

Consolidation characteristics are greatly affected by the expansion of pore water gasses.

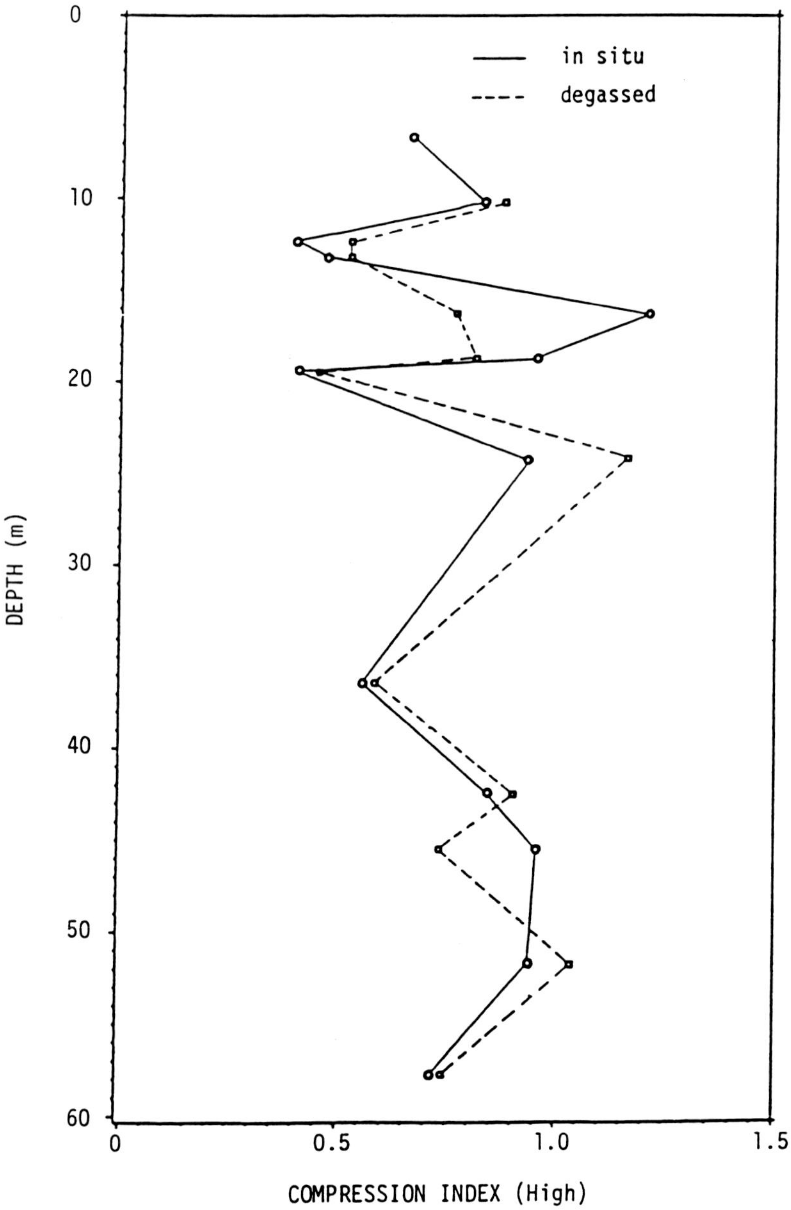

FIGURE 25. High compression indices vs. depth for *in situ* and degassed samples from PST Borings 1, 5, and 7.

Compressibility of a degassed sediment during consolidation testing is initially controlled by the rapid collapse of gas-filled voids, resulting in large coefficients of consolidation and compressibility. At a stress in excess of the effective preconsolidation stress, the coefficients of consolidation and compressibility become essentially identical to the results of a comparative specimen tested under pressurized conditions. So, for the initial consolidation loads, soil permeability does not affect compressibility; rather, compressibility is controlled by the presence of macro-voids.

The volume change that occurs during consolidation is due to the reduction of the volume of voids (laterally constrained sample). For saturated soils the rate of consolidation is gov-

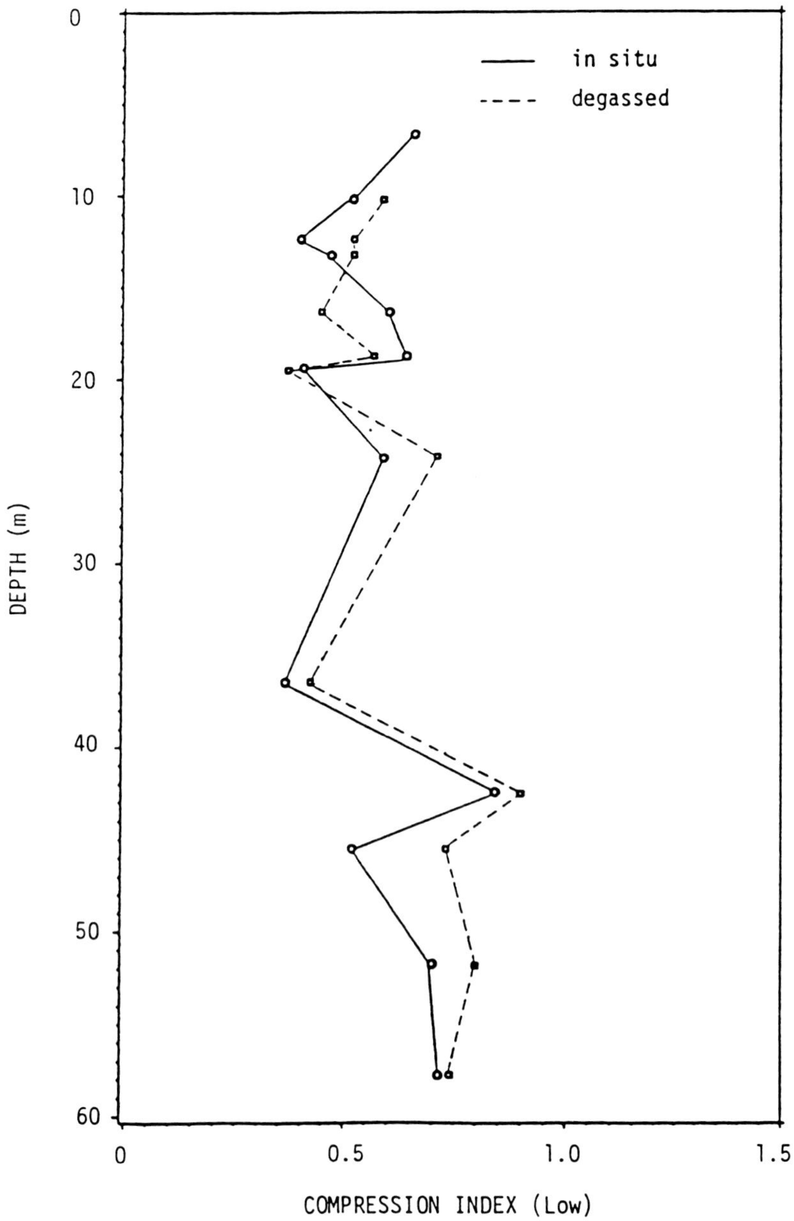

FIGURE 26. Low compression indices vs. depth for *in situ* and degassed samples from PST Borings 1, 5, and 7.

erned by the rate at which pore water can escape from the soil. During the consolidation test instantaneous loads are applied to the specimen. Immediately after application of a normal load, $\Delta\sigma_v$, the entire load is resisted by an increase in pore water pressure equal to $\Delta\sigma_v$ in a saturated soil. With the passage of time, volume changes occur as pore water drains from the soil and the effective stress increases. Eventually, pore water pressure returns to hydrostatic and the applied load and the effective stress are in equilibrium.

For partially saturated soils the instantaneous application of a load results in instantaneous compressions of gas-filled voids. This is evident upon comparison of the initial specimen heights on either the compression vs. square root of time or compression vs. logarithm of

TABLE 7
Pore Water Saturation Values for Consolidation Test Samples of PST Borings 1, 5, and 7 (All Values Calculated at STP)

	Saturation %			
	Calculated		Measured	
Sample	*In situ*	Degas	*In situ*	Degas
B1S2				
B1S4	93.88	93.81	100^+	93.01
B1S6	93.52	93.57	100^+	90.61
B1S8	99.54	99.55	100^+	100^+
B7S9	97.46	97.70	97.89	97.41
B7S15	99.97	99.97	100^+	94.86
B5S10	99.91	99.91	100^+	100^+
B5S15	99.92	99.92	100^+	94.46
B5S23	99.97	99.97	100^+	96.11
B5S26	99.37	99.34	100^+	89.14
B5S28	99.98	99.98	100^+	99.38
B5S30	99.81	99.80	98.50	90.51
B5S32	99.98	99.98	97.60	95.36

time curves. Initial and total compression for a given normal load was shown to be considerably larger for the degassed samples vs. the pressurized samples. Furthermore, compression is a function of methane concentration.

Based on the grain size, calcium carbonate content, Atterberg limits, and mineralogy, the constituents of the comparative consolidation samples tested here are basically identical throughout the three cored sections. Variable gas concentrations, pore water saturation, undrained shear strength, and initial void ratios clearly indicate the effects of degassing between comparative *in situ* and degassed consolidation samples. The resultant effects of degassing on the soil microstructure drastically altered the consolidation characteristics of a given set of samples.

Samples B7S15, B5S10, and B5S32 exhibit consolidation characteristics which can be accounted for by sample variability. Although methane gas concentration was measured for all pressure Shelby tube samples, the exact localization of methane in the sediment matrix cannot be clearly delineated. Methane gas concentration analyses were performed on samples collected as close as possible to the consolidation sample. Although these values may more closely represent those of the consolidation sample, gas concentration measurements were not performed on the actual consolidation sample. The initial void ratio was found to be greater for the *in situ* tests. However, these samples contained low methane concentrations.

The pressure core barrel and pressure Shelby tube sampling techniques discussed previously are a significant advancement to promote the acquisition of relatively undisturbed samples. However, significant problems exist with both techniques.

Developmental problems with the pressure core barrel have led to the acquisition of only a single set of comparative consolidation test results (B2S10). Difficulties with maintaining sample pressures during post-sampling periods resulted in the loss of numerous samples. Yet, prior to this, sample acquisition was a serious problem due to mechanical difficulties with the pressure core barrel.

The pressurized Shelby tube method of sample acquisition does allow the sample to reach ambient pressures momentarily. This period is minimized to reduce the possibility of gas expansion; however, in a few cases it is felt that disturbance due to gas expansion did occur. It must be noted that disturbance was minimized. Yet minor amounts of gas expansion can produce dramatically different consolidation test results.

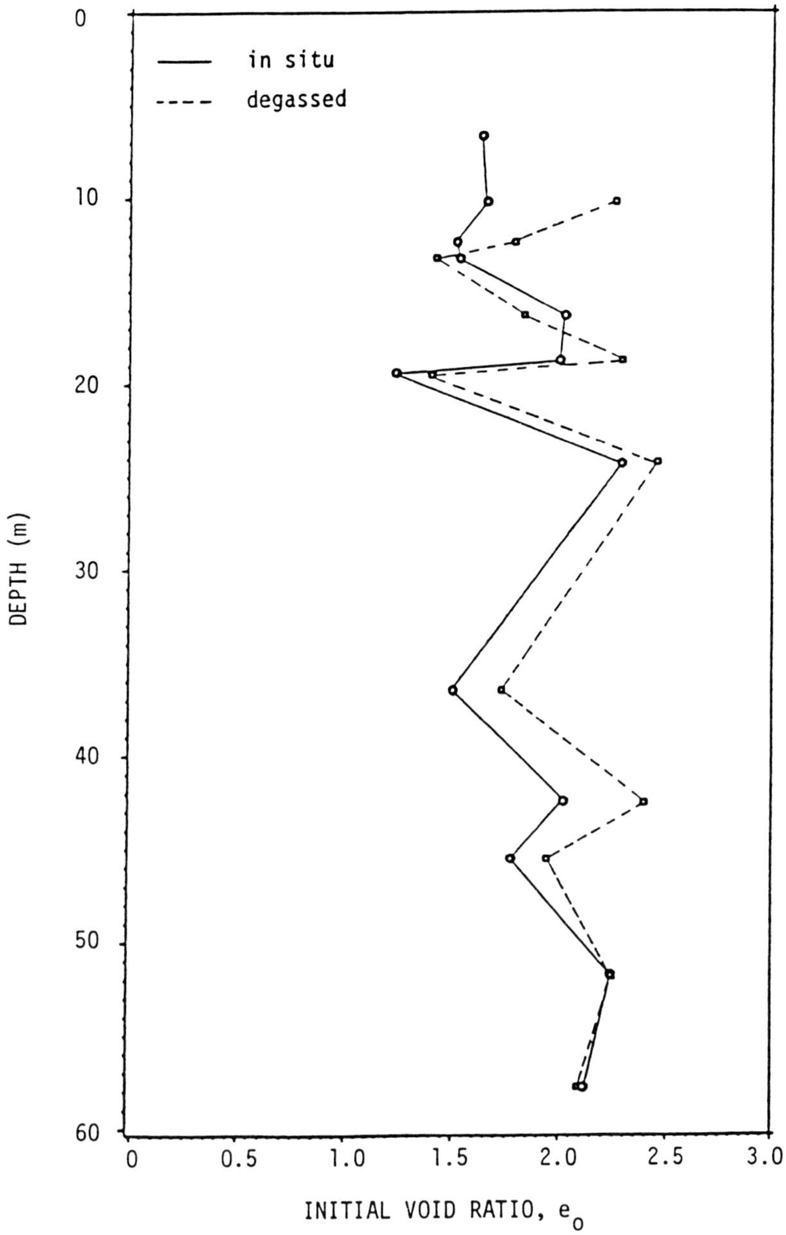

FIGURE 27. Initial void ratio vs. depth for *in situ* and degassed samples from PST Borings 1, 5, and 7.

In addition to sampling problems, the use of a manned hyperbaric chamber is extremely limiting. Space and time limitations are rigid. Ideally, when loading an oedometer, careful sample trimming and loading practices are exercised. Yet, in the limited environment of the hyperbaric chamber it becomes necessary to trim and load the oedometer within a time limit of 55 min. As can be expected, sample disturbances may occur.

To alleviate the problems associated with a manned hyperbaric chamber, an unmanned mini-hyperbaric chamber is being developed and tested, as discussed earlier. This tool was tested during a cruise in August 1983, at which time the pressure core barrel performed

TABLE 8
Consolidation Characteristics Calculated at the Effective Preconsolidation Stress for All Samples

Sample	Depth (m)		Void ratio, e_c		Permeability, k_c (cm/s)	
	In situ	Degas	*In situ*	Degas	*In situ*	Degas
B1S2	6.72		1.4641		2.515×10^{-8}	
B1S4	12.41	12.37	1.3790	1.5840	7.312×10^{-8}	4.269×10^{-8}
B1S6	18.78	18.72	1.5776	2.0170	2.386×10^{-9}	8.129×10^{-8}
B1S8	24.32	24.18	2.0509	2.2540	4.969×10^{-8}	1.615×10^{-7}
B7S9	10.24	10.20	1.5354	1.8868	2.078×10^{-9}	5.630×10^{-8}
B7S15	16.37	16.31	1.8137	1.6676	7.149×10^{-8}	1.572×10^{-7}
B5S10	13.29	13.21	1.4324	1.2555	2.612×10^{-8}	6.121×10^{-8}
B5S15	19.42	19.47	1.1157	1.2461	2.316×10^{-8}	2.013×10^{-7}
B5S23	36.49	36.44	1.3852	1.5337	3.985×10^{-8}	6.716×10^{-8}
B5S26	42.47	42.50	1.6980	2.1464	9.720×10^{-9}	1.047×10^{-7}
B5S28	45.54	45.49	1.6753	1.8053	1.796×10^{-8}	2.817×10^{-7}
B5S30	51.76	51.80	1.9633	2.0927	2.474×10^{-7}	7.459×10^{-8}
B5S32	57.80	57.74	2.0916	1.9479	5.425×10^{-8}	3.489×10^{-8}
B2S10	11.76	11.76	1.2589	1.4011	4.915×10^{-8}	3.645×10^{-7}

flawlessly. However, mechanical problems associated with the internal mechanisms of the mini-hyperbaric chamber resulted in no sample data.

The consolidation test data acquired during this study are the first of their kind. Actual *in situ* consolidation testing is at present physically impossible; therefore, simulated pressure maintenance has resulted in a unique set of test data. The results of the consolidation testing program, although not 100% positive for all cases, clearly indicates unique relationships between the consolidation characteristics and the method of sample acquisition.

VII. CONCLUSIONS

As a result of this research the following conclusions can be drawn:

1. Degassed samples exhibit sharper breaks in the consolidation e-log σ_v' curves. The compressibility of these specimens is relatively low until the consolidation stress exceeds the preconsolidation stress. It then increases sharply as the void ratio reduces under higher consolidation stresses. The compressibility eventually assumes a lower value, acting much like a sensitive clay. At stress values less than the preconsolidation stress the corresponding coefficients of compressibility are shown to be greater.
2. Degassing of pore water gasses alters the soil microstructure, thus resulting in:

   - Reduced undrained shear strength,
   - Increased void ratio,
   - Increased compression index,
   - Decreased preconsolidation pressure,
   - Decreased overconsolidation ratio,
   - Decreased porewater saturation, and
   - Microfabric disruption.

3. An increase in calculated permeability results from sediment degassing when relative permeability should decrease due to the presence of gas-filled void space. Large values of calculated permeability result from large values of the coefficients of volume change and consolidation for the degassed test results.

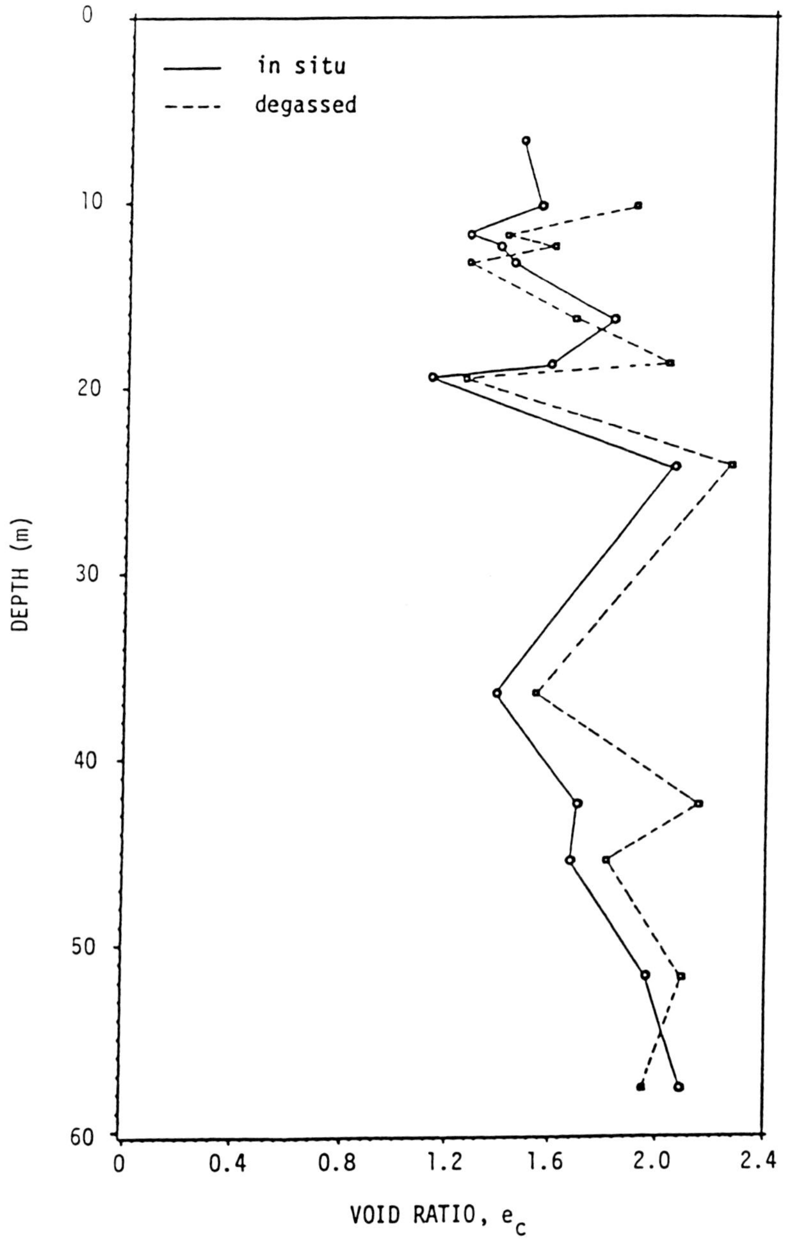

FIGURE 28. Void ratio calculated at the effective preconsolidation stress vs. depth for *in situ* and degassed samples from Borings 1, 2, 5, and 7.

4. The coefficients of consolidation calculated by log time and square root of time methods converge for the *in situ* and degassed tests at a stress beyond the preconsolidation stress for the given curves. The lower portion of the virgin curve is a region in which the soil compressibiity decreases and the degassed and *in situ* values converge.
5. Due to the variability of methane gas concentrations, a more accurate measurement of the actual sample gas content must be made for a more precise evaluation of test results.

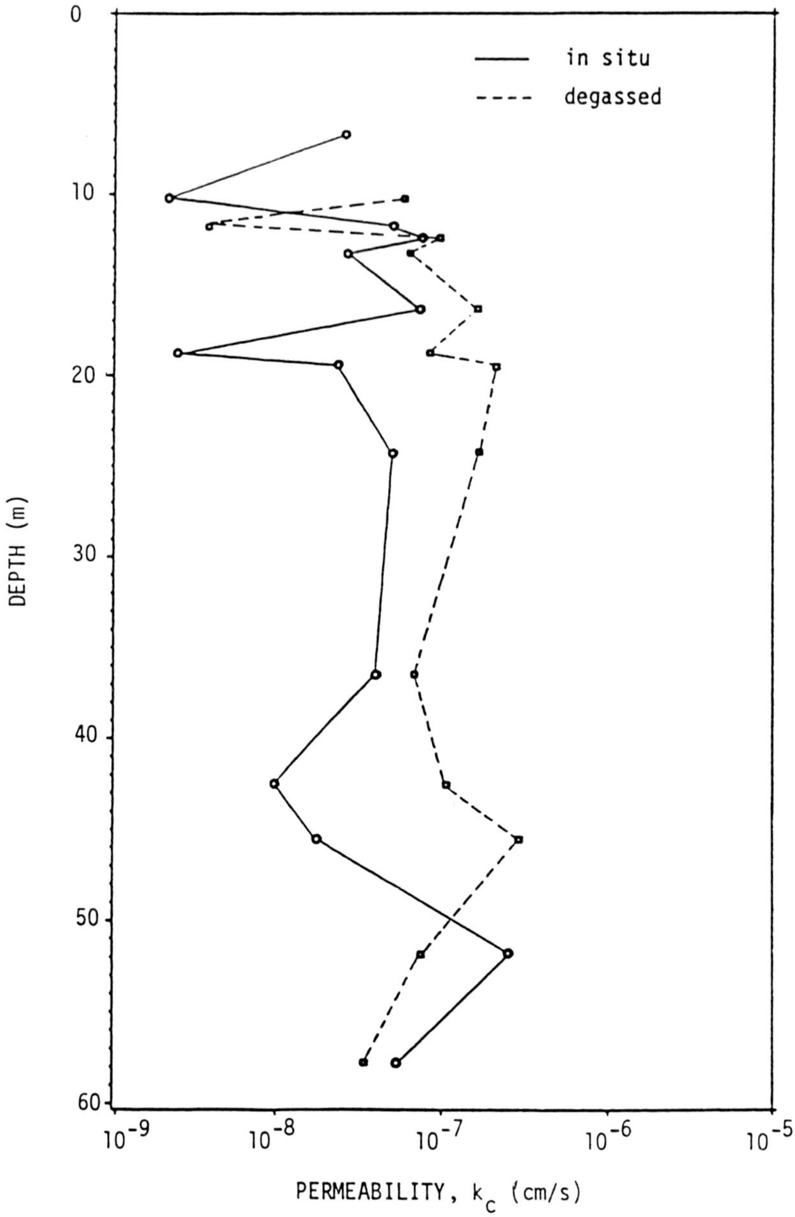

FIGURE 29. Permeability calculated at the effective preconsolidation stress vs. depth for *in situ* and degassed samples from Borings 1, 2, 5, and 7.

6. Instantaneous compression of gas-filled voids during initial loading of consolidation testing was found to be significantly larger for the degassed, partially saturated samples. Similarly, total compression for a given load was greater for degassed vs. *in situ* specimens. In addition, compression is a function of methane gas concentration.

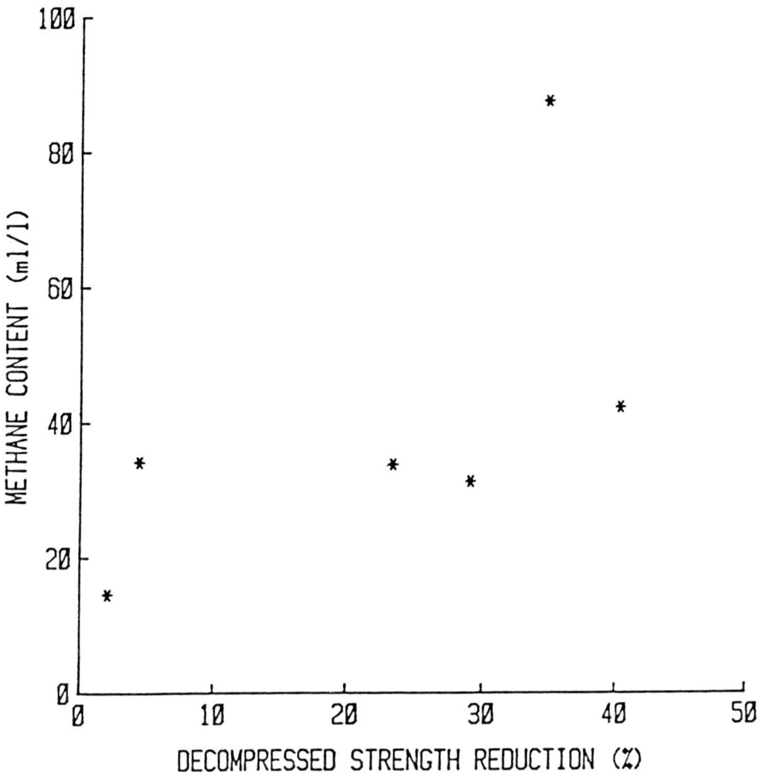

FIGURE 30. Methane content vs. decompressed strength reduction for PCB Boring 2.

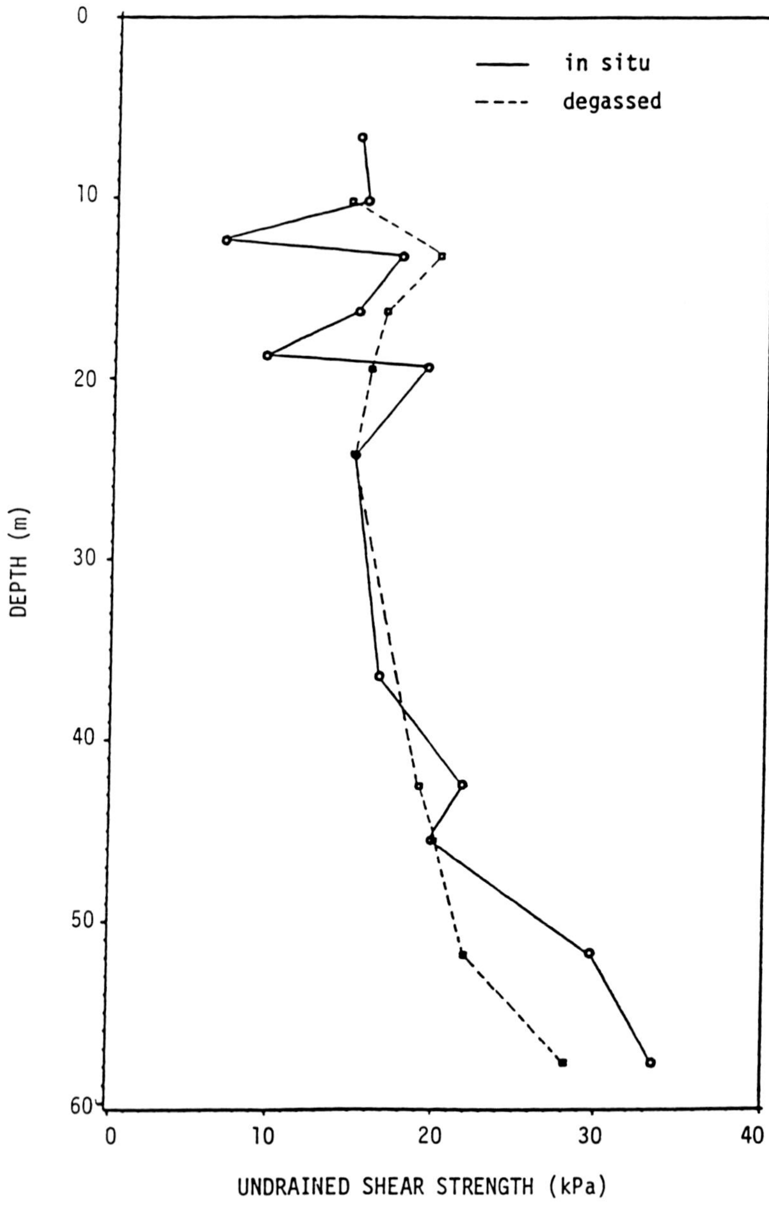

FIGURE 31. Undrained shear strength vs. depth for *in situ* and degassed samples from PST Borings 1, 5, and 7.

REFERENCES

Anderson, A. L., Acoustics of gas bearing sediments, National Technical Information Service, U.S. Department of Commerce, Springfield, VA, 1974, 162 p.

Barden, L. and Berry, P. L., Consolidation of normally consolidated clay, *Proc. Am. Soc. Civ. Eng.*, SM5, 4881, 15-35, 1965.

Bea, R. G., How sea floor slides affect offshore structures, *Oil Gas J.*, 69, 88-92, 1971.

Bea, R. G. and Arnold, P., Movements and forces developed by wave induced slides in soft clays, Eighth Offshore Technol. Conf., 1, 731-742, 1973.

Bjerrum, L. and Huder, J., Measurement of the permeability of compacted clays, Proc. 4th Conf. Soil Mech. Foundation Eng., London, I, 6-8, 1957.

Bryant, W. R., Cernock, P., and Morelock, J., Shear strength and consolidation characteristics of marine sediments, Gulf of Mexico, *Mar. Geotechnol.*, 1, 1-14, 1967.

Casagrande, A., The determination of the preconsolidation load and its practical significance: *Proc. 1st Int. Conf. Soil Mech. and Found. Eng.*, Cambridge, Mass., p. 60, 1936.

Cernock, P. J., Consolidation characteristics and related physical properties of selected sediments from the Gulf of Mexico, Master's thesis, Texas A&M University, 137 p., 1967.

Chiou, W.-A., A new technique for preparing *in situ* marine sediments for clay fabric study, in 38th Annu. *Proc. Elec. Micro. Soc. Am.*, G. W. Bailey, Ed., San Francisco, CA, 1980.

Chiou, W.-A., Clay fabric of gassy submarine sediments, Ph.D. Dissertation, Texas A&M University, 248 p., 1981.

Coleman, J. M., *Deltas: Processes of Deposition and Models for Exploration,* Continuing Education Publication Co., Champaign, IL, 101 p. 1976.

Coleman, J. M., *Deltas: Processes of Deposition and Models for Exploration,* Continuing Education Publication Co., Champaign, IL, 101 p. 1981.

Coleman, J. M., Ferm, J. C., and Saxena, R. S., Deltas Recent and Ancient, New Orleans Geological Society, Continuing Education Seminar, November 169 p, 1972.

Coleman, J. M. and Gagliano, S. M., Cyclic sedimentation in Mississippi River deltaic plain deposits, *Trans. Gulf Coast Assoc. Geol. Soc.*, 14, 67-80, 1964.

Coleman, J. M., Suhayda, J. N., Whelan, T., and Wright, L. D., Mass movement of Mississippi River delta sediments: *Trans. Gulf Coast Assoc. Geol. Soc.*, 24, 49-68, 1974.

Coleman, J. M. and Wright, L. D., Formative mechanisms in a modern depocenter, in *Stratigraphy and Petroleum Potential of the Northern Gulf of Mexico, Part II,* New Orleans Geol. Soc. Sem., January 1974, p. 90-139.

Davis, E. H., and Raymond, G. P., A non-linear theory of consolidation, *Geotechnique,* 15, 161-173, 1965.

Denk, E. W., Dunlap, W. A., Bryant, W. R., Milberger, L. J., and Whelan, T. J., III, A pressurized core barrel for sampling gas-charged marine sediments, *Proc. 13th Offshore Technol. Conf.*, 4, 43-52, 1981.

Doyle, E. H., McClelland, B., and Ferguson, G. H., Wire-line vane probe for deep penetration measurements of ocean sediments, presented at Third Offshore Technol. Conf., April 1971, Houston, TX, (preprint OTC 1327).

Fisk, H. N. and McClelland, B., Geology of continental shelf off Louisiana: Its influence on offshore foundation design, *Bull. Geol. Soc. Am.*, 70, 1369-1394, 1959.

Fisk, H. N. and McFarlan, E., Jr., Late Quaternary deltaic deposits of the Mississippi River, in *The Crust of the Earth, Geol. Soc. Am. Spec. Pap.* 62, p. 279-302.

Fisk, H. N. McFarlan, E., Jr., Kolb, C. R., and Wilbert, L. J., Jr., Sedimentary framework of the modern Mississippi delta, *J. Sediment. Petrol.*, 24, 76-99, 1954.

Folk, R. L., *Petrology of Sedimentary Rocks,* Hemphills, Austin, TX, 1974, 159 p.

Garrison, L. E., The instability of surface sediments on parts of the Mississippi delta front, U.S. Geological Survey Open File Report, Corpus Christi, TX, 18, 1974, 661-673.

Gibson, R. E., England, G. L., and Hussey, M. J. L., The theory of one-dimensional consolidation of saturated clays, *Geotechnique,* 17, 261-273, 1967.

Gould, H. R., The Mississippi delta complex, in *Deltaic Sedimentation, Modern and Ancient,* Morgan, J. P., Ed., Spec. Publ. Soc. Econ. Paleont. Miner., 15, 1965, 3-30.

Hamilton, E. L., Prediction of *in situ* acoustic and elastic properties of marine sediments, *Geophysics,* 36, 266-284, 1971.

Hampton, L. D., Acoustic properties of sediments, Defense Research Laboratory, University of Texas, Acoustic Report 254 (DRL-A-254), 1966.

Helwick, S. J., Jr., Engineering properties of shallow sediments in West Delta and South Pass, outer continental shelf lease areas, offshore Louisiana, Master's thesis, Texas A&M University, 82 p., 1977.

Henkel, D. J., The role of waves causing submarine landslides, *Geotechnique,* 20, 75-80, 1970.

Horslev, M. J., Subsurface Exploration and Sampling of Soils for Civil Engineering Purposes, Waterways, Experiment Station, Vicksburg, MS, 1948.

Hottman, W. E., Areal distribution of clay minerals and their relationship to physical properties, Gulf Of Mexico, Master's thesis, Texas A&M University, 53 p., 1975.

Jackson, M. L., *Soil Chemical Analysis — Advanced Course,* Department of Soil Science, University of Wisconsin, Madison, 1956, 991 p.

Janbu, N., Consolidation of clay layers based on non-linear stress-strain, *6th Int. Conf. Soil Mech.*, 2, 83-87, 1965.

Johns, M. W., Taylor, E., and Bryant, W. R., Geotechnical sampling and testing of gas-charged marine sediments at *in situ* pressures, *Geo-Marine Letters,* 2, 231-236, 1982.

Johns, W. D., and Grim, R. E., Clay mineral composition of Recent sediments from the Mississippi River delta, *J. Sediment. Petrol.*, 48, 186-199, 1958.

Jones, J. L., Leslie, C. B., and Barton, L. E., Acoustic characteristics of a lake bottom, *J. Acoust. Soc. Am.,* 30, 142, 1958.

Jones, J. L., Leslie, C. B., and Barton, L. E., Acoustic characteristics of underwater bottoms. *J. Acoust. Soc. Am.,* 36, 154, 1964.

Kolb, C. R. and Kaufman, R. I., Prodelta clays of southeast Louisiana, in *Marine Geotechnique,* Richards, A. F., Ed., University of Illinois Press, Chicago, 1967, 3-21.

Kolb, C. R. and Van Lopik, J. R., Geology of the Mississippi River deltaic plain, U.S. Corps of Engineers, Waterways Experiment Station, Technical Reports 3-483 and 3-484, 1958.

Kolb, C. R. and Van Lopik, J. R., Depositional environments of the Mississippi River deltaic plain, southeastern Louisiana, in *Deltas in Their Geologic Framework,* Shirley, M. L., Ed., Houston Geologic Society, 1966, 17-61.

Krumbein, W. C. and Pettijohn, F. J., *Manual of Sedimentary Petrography,* Appleton-Century-Crofts, New York, 1938, 549 p.

Lambe, T. W., *Soil Testing for Engineers,* John Wiley & Sons, New York, 1951, 165 p.

Lambe, T. W. and Whitman, R. V., *Soil Mechanics,* John Wiley & Sons, New York, 1969, 553 p.

Lo, K. Y., Discussion on Rowe, measurement of the coefficient of consolidation of lacustrine clay, *Geotechnique,* 10, 36-39, 1960.

Lowe, J., III, Zaccheo, P. F., and Feldman, H. S., Consolidation testing with back pressure, *J. Soil Mech. and Found. Div. Proc. Am. Soc. Civil Engrs.* 90, 69-86, 1964.

Mitchell, J. K., *Fundamentals of Soil Behavior,* John Wiley & Sons, New York, 1976, 422 p.

Moore, D. G., Submarine slumps, *J. Sediment. Petrol.* 31, 343-357, 1961.

Morgan, J. P., Depositional processes and products in the deltaic environments, in *Deltaic Sedimentation, Modern and Ancient,* Morgan, J. P., Ed., Soc. Econ. Paleontol. Mineral, Spec. Publ., 15, 31-47, 1965.

Morgan, J. P., Coleman, J. M., and Gagliano, S. M., Mudlumps at the mouth of South Pass, Mississippi River: Sedimentology, paleontology, structure, origin and relation to deltaic processes, Louisiana State University, Baton Rouge, Coastal Studies Series 10, 1963, 190 p.

Nissenbaum, A., Presley, B. J., and Kaplan, I. R., Early diagenesis in a reducing fjord, Saanich Inlet, British Columbia. I. Chemical and isotopic changes in major components of interstitial water, *Geochim. Cosmochim. Acta,* 36, 1007-1027, 1972.

Prior, D. B. and Coleman, J. M., High-resolution deep-tow survey imagery, Gulf of Mexico, *Geo-Marine Letters,* 1, 261-266, 1981.

Richart, F. E., A review of theories for sand drains, *J. Soil Mech. and Found. Div. Proc. Am. Soc. Civil Engrs.,* 83, SM3, 1957.

Roberts, H. H., Cratsley, D. W., and Whelan, T., III, Stability of Mississippi delta sediment as evaluated by analysis of structural features in sediment borings, *Proc., 8th Offshore Technol. Conf.,* 1, 9-28, 1976.

Russell, R. J. and Russell, R. D., Mississippi River delta sedimentation, in Trask, P. D., Ed., *Recent Marine Sediments,* AAPG, 1939, 153-157.

Scafe, D. W. and Kunze, G. W., A clay mineral investigation of six cores from the Gulf of Mexico, *Mar. Geol.,* 10, 69-85, 1971.

Schiffman, R. L., The consolidation of soft marine sediments, *Geo-Marine Letters,* 2, 199-203, 1982.

Schink, J. C., Stockwell, J. H., and Ellis, R. A., An improved device for gasometric determination of carbonate in sediment, *J. Sediment. Petrol.,* 49. 651-653, 1979.

Schubel, J. R., Gas bubbles and the acoustically impenetrable or turbid character of some estuarine sediments, in *Natural Gases in Marine Sediments,* Kaplan, I. R. Ed., Plenum Press, New York, 1974, 99-135.

Scruton, P. C., Delta building and the deltaic sequence, in *Recent Sediments, Northwest Gulf of Mexico,* Shepard, F. P., Phleger, F. B., and van Andel, T. H., Eds., AAPG, 1960, 82-102.

Shepard, F. P., Delta front valleys bordering the Mississippi distributions, *Geol. Soc. Am. Bull.* 66, 1489-1498, 1955.

Shepard, F. P., Mississippi delta: marginal environments, sediments, and growth, in *Recent Sediments, Northwest Gulf of Mexico,* Shepard, F. P., Phleger, F. B., and van Andel, T. H., Eds., AAPG Bull., 56-81, 1960.

Shephard, L. E., Bryant, W. R., and Dunlap, W. A., Geotechnical properties and their relation to geological processes in South Pass outer continental lease area Blocks 28, 47 and 48, offshore Louisiana: Technical Report 79-5-T, Texas A&M University 1979, 125 p.

Skempton, A. W. and Sowa, V. A., The behavior of saturated clays during sampling and testing, *Geotechnique,* 13, 269-290, 1963.

Taylor, D. W., *Fundamentals of Soil Mechanics,* John Wiley & Sons, New York, 1948, 700 p.

Terzaghi, K., Principles of soil mechanics. III. Determination of permeability of clay, *Eng. News-Rec.,* 95, 832, 1925.

Terzaghi, K., *Theoretical Soil Mechanics,* Chapman and Hall, London, 1943, 510 p.

Waples, D., Organic Geochemistry for Exploration Geologists, Burgess Publishing, Minneapolis, 1980, 151 p.

Whelan, T., III, Bryant, W. R., and Dunlap, W. A., Methane concentration and distribution in pressure core samples from Mississippi delta sediments (abstract), *Am. Assoc. Pet. Geol.,* 65, 1007 p. 1981.

Whelan, T., III, Coleman, J. M., Roberts, H. H., and Suhayda, J. N., The occurrence of methane in Recent deltaic sediments and its effect on soil stability, *Int. Assoc. Eng. Geol. Bull.* 14, 55-64, 1976.

Whelan, T., III, Coleman, J. M., Suhayda, J. N., and Garrison, L. E., The geochemistry of Recent Mississippi River delta sediments: Gas concentration and sediment stability, *Proc. 8th Offshore Technol. Conf.*, 3, 71-84, 1975.

Yamamoto, T., Experiments of wave-driven soil transport in clay beds, *Geo-Marine Letters,* 2, 205-208, 1982.

Znidarcic, D., and Schiffman, R. L., On Terzaghi's concept of consolidation, *Geotechnique,* 32, 387-389, 1982.

Chapter 13

OCEANIC SEDIMENTATION AND GEOTECHNICAL STRATIGRAPHY: HEMIPELAGIC CARBONATES AND RED CLAYS

Elliott Taylor

TABLE OF CONTENTS

I. INTRODUCTION

Developing an understanding of sediment depositional processes and the relationship between successive deposits is the purpose of stratigraphic research. Stratigraphic descriptions have enabled scientists to gain insight into geologic history through events recorded by the formation of sedimentary sequences. Studies in biostratigraphy, for instance, have developed and refined a time scale for correlating sedimentary events. Lithostratigraphic studies define contrasting sediments types within a deposit, thus reflecting historical changes of sediment source, depositional environment or diagenetic processes. The common stratigraphic approaches to studying sedimentary sequences have, however, been relegated to descriptive aspects of primarily static properties. A sedimentary sequence is not a static feature. The physical and mechanical characteristics of a deposit represent sediment attributes that are closely related to the dynamics of a sediment column. The description of these aspects of the sediment column, therefore, addresses the dynamic characteristics of sediments. This approach, coupled with knowledge of common static properties, greatly enhances our understanding of geologic history obtained through stratigraphic analysis.

The descriptive analysis of geotechnical properties in a vertical sequence represents the basic concept of stratigraphic analysis. Stratigraphy is the descriptive study of the relationship, in terms of time and space, of a defined set of characteristics in a sedimentary or rock column. The qualifier "geotechnical" describes those characteristics utilized in the descriptive analysis of sedimentary units. Thus, geotechnical stratigraphy is defined as the study of vertical, successive characteristics such as index properties (bulk density, porosity, water content, etc.), state and behavior of consolidation, strength parameters, and other inherent physical and mechanical properties of a sediment that define a distinct geotechnical aspect of a given sedimentary unit. The purpose of this research is to establish the geotechnical stratigraphy of sedimentary columns from two distinct marine sedimentary environments. This stratigraphy is then related to other conventional stratigraphies such as bio-, litho-, and seismic stratigraphies. This study also identifies the interaction of sedimentary processes in the geotechnical stratigraphic record for both depositional environments. A final summary is made addressing the common effects between two different sediment types regarding these interactions.

A. STUDY AREAS

The specific study sites for the samples used in this analysis are, to a degree, sites of opportunity. However, the sedimentary conditions and environments represented by these sites characterize two diverse marine settings. It is appropriate that the new concept of geotechnical stratigraphy be applied to widely different sedimentary conditions. Description of the geotechnical stratigraphy of two basically diverse environments fulfills two important aspects of this research. First, it tests the applicability of this qualitative scheme of analysis for sediments of different types and histories. Secondly, it creates an initial basis for comparison with subsequent studies into more diverse and select sedimentary conditions.

The two sites of this study fit into the basic subdivision of marine sedimentary envi-

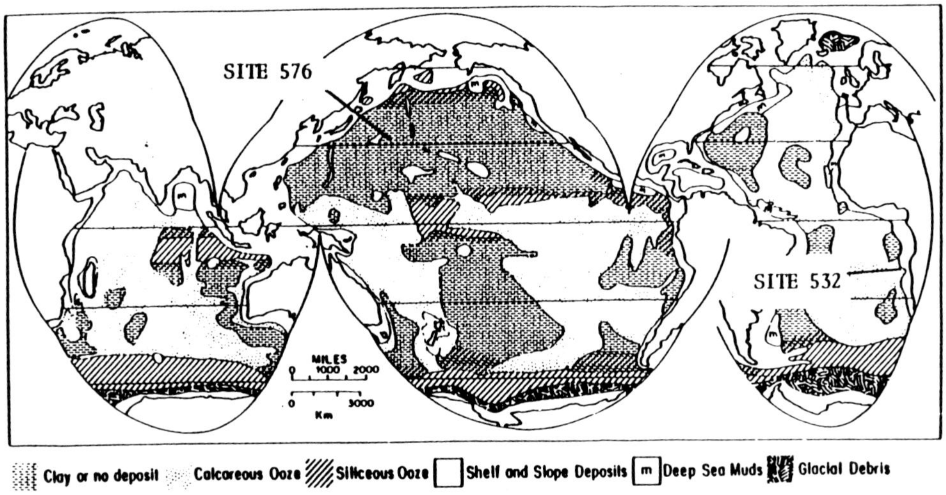

FIGURE 1. Distribution of recent sediments on the ocean floor and general location of the red clay (Site 576) and hemipelagic carbonate (Site 532) study areas. (Map from Berger, *The Geology of Continental Margins*, Springer-Verlag, New York, 1974, 213. With permission.)

ronments described by Gorsline (1980) consisting of margin affected and oceanic. This subdivision distinguishes sediment environments primarily by the mode of sediment emplacement. The oceanic environment is identified with a predominantly pelagic deposition, whereas margin affected zones contain deposits emplaced by a variety of mechanisms ranging from downslope transport to pelagic rain. The sites in this study are predominantly oceanic; however, as Gorsline (1980) indicated, even small influxes of shelf derived material can strongly alter the characteristics of deposited sediments and therefore are considered margin affected.

The two sites of this study were initially identified by the Sedimentation and Physical Properties Panel of the Joint Oceanographic Institutions for Deep Earth Sampling (JOIDES) group. These sites were to be drilled using the newly developed hydraulic piston corer (HPC) in the non-indurated sedimentary section. The cores were to be preserved intact and shipped back to shore for geotechnical analyses. These recommendations were adopted by the Deep Sea Drilling Project (DSDP) and accomplished on Legs 75 and 86 of this project. The selected sites represent diverse sedimentary systems, and are located in areas of good paleontologic control. Their general geographic location and sedimentary affinities can be seen in Figure 1.

The margin affected sedimentary environment is found on the flank of the Walvis Ridge off the southwestern coast of Africa. The area has a mixed sediment source consisting of terrigenous material supplied from offshelf and slope transport and from high biogenic input derived from production of tests by planktonic organisms in the overlying water column. This sediment mixture, referred to as hemipelagic, has a variable contribution from its two sources which is related to the intensity of upwelling and corresponding productivity, and/or the offshore supply of land-derived material.

The oceanic sedimentary environment for this study is located on the abyssal plain of the northwest Pacific Ocean (Figure 1). The sedimentary deposit here is composed of a very fine, pelagic red clay. The sediments lie below the carbonate compensation depth, hence keeping currently deposited material free of carbonates. The main source of material forming these deposits are wind transported silts and clays probably derived from the Asian continent (Leg 86 Shipboard Party, 1982).

B. PREVIOUS WORK

1. Geotechnical Properties and Sedimentation

Two sedimentary processes are intimately tied to geotechnical properties of a sediment; depositional processes emplacing the unit and processes active in post-depositional alterations of the original sediment. The relationship between these processes and the physical-mechanical (geotechnical) properties has been the subject of numerous studies in the field of marine geotechnique. Silva (1974) pointed out the importance of understanding sediment processes such as erosion, transport, deposition and slumping as they relate to geotechnical properties. He also indicated the need for understanding effects of diagenesis and consolidation on geotechnical behavior.

The variety of depositional processes emplacing sediments on the sea floor is extremely large. These processes range from an oceanic pelagic rain of fine clay or skeletal remains to catastrophic slumps or slides on continental shelves and slopes. The geotechnical properties of mass movements are important to understand in areas where instabilities of the sediment column may result in significant downslope transport. It is the strength of the sediment in question that resists this movement. Relationships between slumping and geotechnical properties were presented by Bennett et al., (1977, 1980), Keller et al., (1979) and Booth (1979). Physical and mechanical properties of deposits related to mass flow were presented by Hein and Gorsline (1981) and Baltuck et al., (1985). Sedimentary processes related to deltaic deposition and progradation, and their geotechnical properties were described by Shephard et al., (1979) and Helwick and Bryant (1980).

A review of marine geotechnical literature clearly points out that the majority of the effort involved has been relegated to nearshore and continental margin environments. Nevertheless, some work has been done for areas located in environments without high terrigenous input. Keller and Bennett (1968) described differences between geotechnical characteristics of both Atlantic and Pacific basin sediments. They found that most variations of these properties were related to sediment type and site of deposition. Horn et al. (1974) compared the textural and physical properties of sediments from the Atlantic and Pacific basins having similar compositions. They described a coarser texture for Pacific red clays than North Atlantic and they found a wider range of properties for carbonates of the Atlantic. Hamilton (1970) described physical and acoustic properties of ocean sediment types and found these could be studied relative to their composition and marine physiographic setting.

The second sedimentary process dictating the observed physical properties of a sedimentary sequence are the post-depositional changes occurring in the sediment. Deposited sediment is subject to several processes dominated by stress conditions and diagenesis. Application of stresses arise from processes ranging from current action or bioturbation, through intermediate stresses resulting from self weight, to those related to tectonic deformation. Current action on the surficial sediment may lead to erosion and removal of sediments, or it may simply winnow the sediment components leaving behind a lag type deposit. Effects of erosion and current scour are believed to be reflected in the distribution of physical properties with depth at areas along the Middle America Trench (Shephard et al., 1982; Taylor and Bryant, 1985). Another process significantly altering primary sedimentary features as well as geotechnical properties is bioturbation (Richardson and Young, 1980; Rowe, 1974). Bioturbation leads to reworking of sediments, often providing a disturbance effect resulting in a particle rearrangement, consolidation and enhancement of the strength of surficial sediments.

The previous two processes, erosion and bioturbation, are generally active in the upper portions of sedimentary columns. However, the sediment columns themselves are undergoing a slower but generally steady deformation. This deformation is known as compaction, whereby the sediment void space (volume occupied by non-solid material) decreases as a function of time. The compaction, or consolidation, is effective in driving off the pore water and rearranging process particles (sediment fabric). This is a critical aspect of sedimentary processes since it involves the interaction of stress history and geotechnical properties. Consolidation of marine sediments has been the subject of numerous studies. The reader is referred to the following for an overview of the subject: Bryant et al. (1981); Hamilton (1976); Rieke and Chilingarian (1974); Skempton (1970); and Richards (1967).

Finally, compressional and shear stresses associated with tectonic settings can greatly alter an original deposit. Imbricate thrust sheets, deformation ridges, sediment subduction or simple injection of pore water fluids into adjacent sediments in the vicinity of convergent margins probably represents the most significant mechanical alteration affecting sediments. The depositional settings in this study, however, represent sediment drape conditions that have not been subject to tectonic events such as described above.

Diagenetic alterations of sediments are of two types: mechanical and chemical. In essence, mechanical diagenesis is the type of change occurring in sediments as a response to applied stresses. Chemical diagenesis, on the other hand, depends on lithology, temperature, time, pressure, and pore water chemistry. The degree to which chemical diagenesis alters sediments, however, is quite variable and the consequent effects range from undetectable to a complete alteration of the original deposit. Processes such as contact solution, recrystallization, mineral alteration and cementation represent some chemical diagenetic actions which have been identified as the dominant factors in both chemical and mechanical post-depositional alteration of a deposit. The effects of chemical diagenesis in soft, unconsolidated sediments are generally minimal and the dominant aspect of post-depositional changes of a sediment are mechanical. However, when sediments become somewhat lithified the chemical diagenetic aspect begins to dominate these alterations.

C. GEOTECHNICAL PROPERTIES OF THE STUDY AREAS

1. Southwest African Margin

The continental margin off southwest Africa was drilled on two DSDP legs. Two sites (362 and 362A) were drilled on the Abutment Plateau portion of the Frio Ridge segment of the Walvis Ridge in January 1975, DSDP Leg 40 (Bolli, Ryan et al., 1978). A series of cores from a sediment-filled trough on the eastern flank of the Walvis Ridge were obtained in August 1980, DSDP Leg 75. These latter core sections represent a thick sequence of Holocene to upper Miocene sediments adjacent to the previously drilled Site 362.

Hottman (1978) studied the physical properties of sediments recovered by conventional rotary drilling during DSDP Leg 40. The upper section of the drilled holes were severely disturbed which created difficulties in interpreting the data. Hottman conducted his research in an effort to determine the possibility of utilizing physical properties as a tool to interpret portions of the geologic history of the West African continental margin. The sediments from this region are principally biogenic carbonates and silicates with variable degrees of terrigenous clays and high organic carbon content (Hay, Sibuet et al., 1984). Hottman concluded from his work that the physical properties of sediments from the Walvis Ridge area are dominated by the calcium carbonate content. Lithology of this area reflects variations of the amount of input and preservation of pelagic and terrigenous derived material. Hence, Hottman's conclusion reaffirms the conclusion reached by Gorsline (1980) indicating the importance of characterizing sedimentary environments by the presence of terrigenous input, since this component has a dominant effect on sediment composition and behavior. Hottman found that the sediments containing greater than 50% carbonate are apparently underconsolidated. However, he indicated that this underconsolidation may indeed be an artifact of the manner in which we define normal consolidation, and that perhaps this concept needs to be reconsidered in light of sediment composition and interparticle nature. Hottman also found these carbonate sediments to be less compressible than equivalent fine grained sediments, corroborating the results of Bryant et al. (1974).

The extent of geotechnical work off the southwest coast of Africa is obviously limited. Nevertheless, considerable work has been done regarding geotechnical characteristics of pelagic to hemipelagic calcareous and siliceous sediments (Bryant et al., 1974; Richards and Hamilton, 1967). A unique aspect of the setting for this site is that it is situated under a region of intense upwelling. Under these conditions we may expect to see some of the aspects of physical properties described by Busch and Keller (1981) and Reimers (1982) for sediments underlying the upwelling regime off Peru and Chile. Such properties include higher plasticity indices and shear strengths when compared to sediments of lower organic contents at similar water contents.

2. Pacific Deep Sea Pelagic Clays

A widespread sediment type of the Pacific Ocean sea floor is a very fine, well sorted pelagic clay commonly referred to as red clays. Red clays have been identified throughout the world oceans, being deposited in water depths in excess of the carbonate compensation depth. Their textural and compositional aspects have been described in detail, as has their distribution throughout the oceans (Lisitzin, 1972).

The geotechnical nature of red clays has been the subject of various studies. Hamilton (1964) and Richards and Hamilton (1967) described the basic consolidation characteristics of red clays from the MOHOLE site. They pointed out that these clays, and other marine sediments, are typically overconsolidated near surface and tend to approach similar void ratios at stresses of approximately 10,000 kPa. Keller and Bennett (1968) described physical properties of red clay deposits from both the Atlantic and Pacific basins. They concluded that red clays have a generally high water content and porosity (up to 260% and 90%, respectively) with low shear strengths and low bulk densities. Keller (1969) described the variations in physical properties of a red clay deposit associated with a seamount. He found that a relationship exists between the topography of the seamount (2960 m) and the mass properties of the surrounding clay deposit. Keller believed this relationship may result from a winnowing effect of bottom currents leaving coarser material at the seamount and increasing fines away from the topographic high. The radial distribution of properties away from the seamount included increases of cohesion and bulk density, and associated decreases of water content and porosity. Keller also pointed out that no systematic variation of these properties was observed in relation to smaller sea floor relief (40 to 100 m). Bennett et al., (1981) performed a clay fabric analysis on a red clay sample and described the fabric as a highly complex structure of linked clay flocs with large intravoid spaces which explains the high water contents and porosities of these particular clays.

Recent work on red clay samples includes a series of studies carried out for the sub-seabed disposal program sponsored by the Department of Energy and Sandia National Laboratories. Some results of these studies are presented in Walker (1981) and Dadey (1983). Walker (1981) studied sediments obtained by several piston core sampling programs, and found the same feature of overconsolidation in the upper section of the red clay units although this overconsolidated state seemed to extend to depths greater than in other sediment columns. He listed several possible reasons for this state of consolidation including assumed incorrect depths for overburden calculations and possible cementation between particles.

He also described the distribution of physical properties of these sediments with depth, however his samples were from less than 14 m subbottom. Dadey (1983) described the consolidation aspect of sediments from the West Pacific, summarizing results from previous studies and from efforts of the work performed on the DSDP Leg 86 dedicated geotechnical core. She interpreted the overconsolidated state of the upper red clay sections to be a result of an increased internal strength of the sediment formed by interparticle bonding.

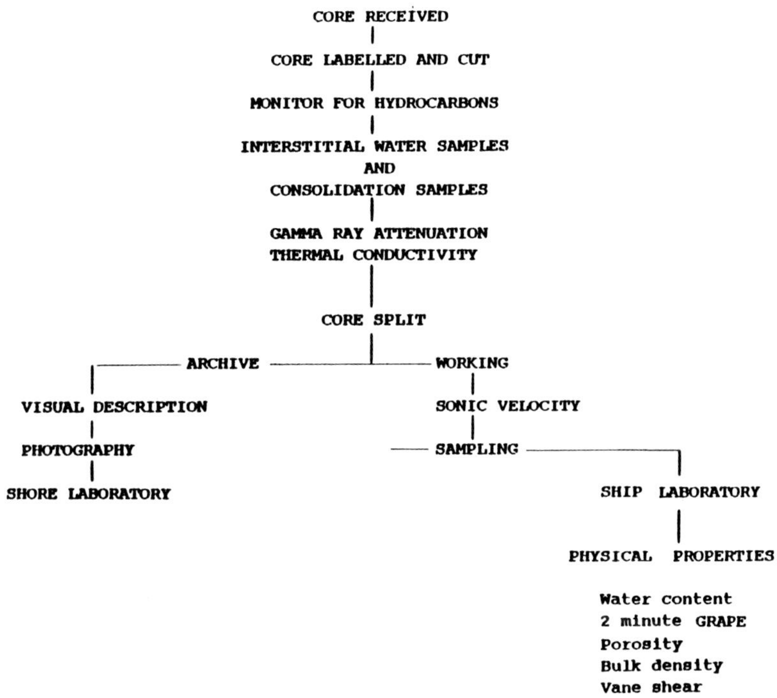

FIGURE 2. Block diagram of standard DSDP procedures for core handling and physical properties testing.

II. METHODS AND MATERIALS

A. SHIPBOARD PROCEDURES

The two sites of this study were each sampled by three holes. At each site, one of the holes was dedicated to geotechnical sampling. Dedicated geotechnical cores were obtained by DSDP using the hydraulic piston core (HPC). The advantage of utilizing the HPC over conventional drilling is that it results in less sampling disturbance. This was pointed out by Mayer (1982) in a comparative analysis of physical properties of sediments obtained using standard rotary drilling and HPC sampling. Sampling disturbance resulting from the penetration of the HPC is generally minimal, but has been noticed to deteriorate when the sediments being sampled have shear strengths on the order of 80 to 100 kPa. Walton et al. (1983) performed a series of measurements on cores obtained on DSDP Leg 72 and geotechnically characterized the HPC sampled sediments. They found that the tool retrieved relatively undisturbed terrigenous muds (clays). However, HPC coring of nannofossil and foraminifer, calcareous oozes resulted in severe disturbance generally associated with liquefaction of the sediment during core punch-in. X-radiographic analysis of the geotechnical cores retrieved from DSDP Hole 532A, however, showed limited signs of disturbance.

The cores obtained at Holes 532A and 576A were sectioned into 1.5 m-long segments immediately after the arrival of the core on deck. They were subsequently labelled, sealed, and stored upright in a refrigerator at 4.5°C until later processing onshore. Adjacent holes were sampled routinely, generally following the flowchart shown in Figure 2 for DSDP shipboard analysis of its cores.

Physical properties measured on the Glomar Challenger consisted of compressional wave velocity, water content, GRAPE (gamma ray attenuation porosity evaluator) bulk density and porosity, and undrained vane shear strength. Analytical procedures for these tests were described in detail by Boyce (1976, 1977).

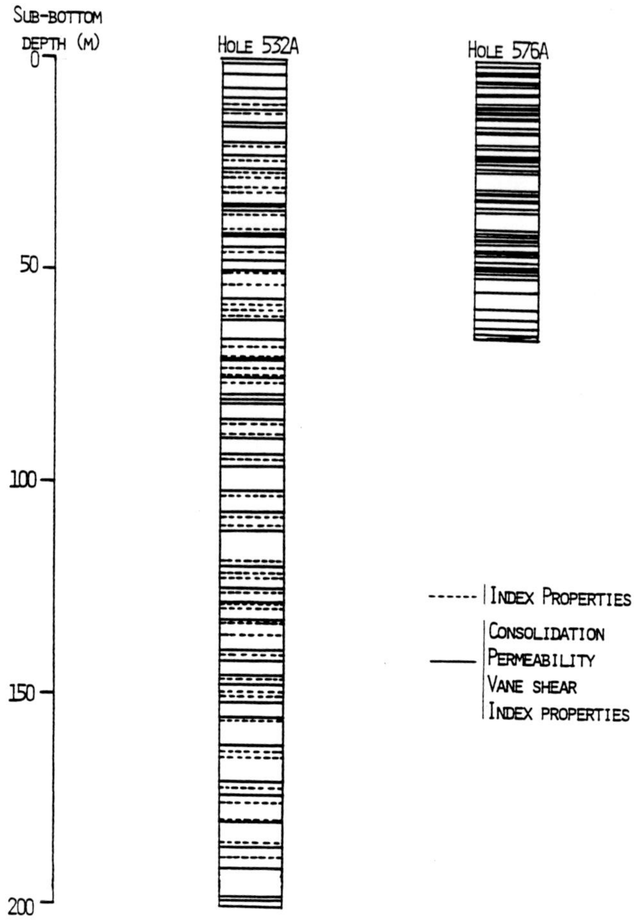

FIGURE 3. Geotechnical sampling coverage of cores from Holes 532A and 576A.

B. SHOREBASED PROCEDURES

1. Sampling Program

The activity involved for obtaining, sampling and studying the geotechnical cores was directed by a joint effort of members from several institutions forming the Marine Geotechnical Consortium (MGC). Members of the MGC devised the sampling and testing program for cores from these two holes. The individuals and institutions involved in the detailed analysis and sampling varied from one core to the next but the dominant activity was provided by Texas A&M University, University of Colorado, University of Rhode Island, San Diego State University, and Humboldt State University.

The shorebased sampling program was carried out at Lamont-Doherty Geological Observatory for Hole 532A cores and at Scripps Institution of Oceanography for Hole 576A. Sets of 6.5 cm diameter full-round samples were selected to provide full coverage of the changing geotechnical character of the sediments at each site. All full-round samples were capped and sealed with wax to prevent dessication. Samples taken in sections adjacent to the preserved samples were obtained at the same time of core processing for water content, bulk density and vane shear analyses. The frequency and distribution of sampled material for both sites is shown in Figure 3 which also indicates the type of test performed for each sample.

2. Testing Program

The testing program for these cores is outlined in reports of the geotechnical properties for these DSDP Legs (Marine Geotechnical Consortium, 1984 and 1985). The testing and analysis programs varied between the different groups of MGC, however all groups did perform routine analyses of index properties (water content, bulk density, porosity) on subsamples of the material used for other studies.

The testing program at Texas A&M University (TAMU) included studies of index properties, grain size and calcium carbonate content analyses, consolidation and permeability testing, and vane shear strength determinations. Some samples were also obtained for fabric and clay mineralogy studies. All samples were X-radiographed in the full-round on a Picker Andrex A/5 Model 1541 unit using Kodak Industrex AA film. X-radiographs were used to distinguish disturbance in the samples by using cracks, edge-drag traces and mixing features associated with coring as evidence of core induced disturbance. X-radiographs were also used to assess the presence or absence of natural sedimentary structures, and when present these were in turn used to correlate the sedimentary sequence with material from the adjacent holes.

All samples used for geotechnical testing were selected from sediments showing no signs of disturbance. In general, most of the samples from both sites were of high quality and zones of disturbance were easily avoided. The majority of disturbance in cores was frequently limited to the upper portion of each HPC core. This is probably a result of material falling downhole between successive cores. Another probable cause of this disturbance is that the top sediments in a core travel the full length of the liner during penetration, thus undergoing more deformation than the lower sediments. The overall testing scheme followed at TAMU is described sequentially in Figure 4 showing the procedure used for all samples processed. Samples were first tested for undrained shear strength measured parallel to bedding using a Wykeham-Farrance vane shear rotating at a rate of 33 or 60 degrees per minute. Occasional vane measurements were also made using a hand-held torque watch.

Subsequently, the sediment was slowly extruded from the butrate liner in the same direction of coring and a 6.35 cm diameter subsample was cut and trimmed to a thickness of approximately 2.54 cm for consolidation testing and permeability measurements. Material from above the consolidation sample was used to obtain water content and in some cases sampled for microfabric studies. Trimmings from the sampling procedure were used for calcium carbonate, grain size, and pycnometer analyses. Clay mineralogy and Atterberg limits for Hole 576A samples were performed on material from below the consolidation sample.

Measured index properties include water content, porosity, bulk density, grain density, void ratio, liquid limit and plasticity index. These measurements represent gravimetric and volumetric relationships and are defined by Lambe and Whitman (1969). Volumetric measurements were made using a Beckman Model 930 Air Comparison Pycnometer and helium as the purging gas. All index properties were corrected for salt content, except Atterberg limits, following the method described by Hamilton (1971) and assuming a pore water salinity of 35 parts per mil.

Percent calcium carbonate was obtained for samples using the Scheibler method described by Bouma et al. (1969) for Hole 532A samples and using a modified carbonate bomb technique for sediment from Hole 576 (Schink et al., 1979). Grain size analyses were performed using sieve techniques for the coarse fraction (greater than 62 microns) and the pipette method for silts and clays (Folk, 1974). Atterberg limits were done using ASTM methods D423 and D424.

Consolidation testing was performed in Anteus back-pressure consolidometers which were modified to allow for falling head permeability tests. The basic design and use of the Anteus back-pressured consolidometer was presented by Lowe et al. (1964) and Shephard

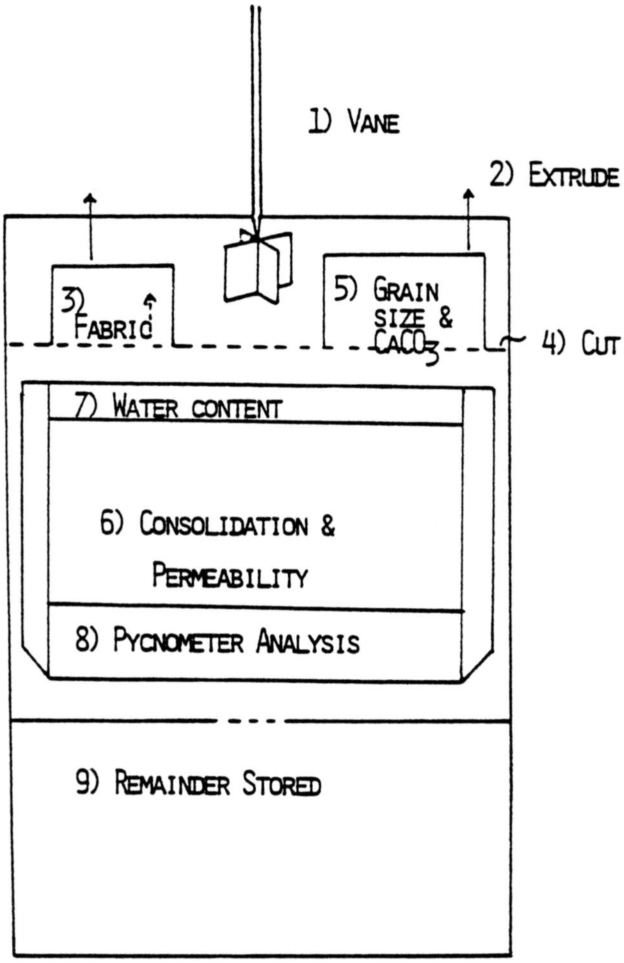

FIGURE 4. Sub-sampling procedure for 10 cm full-round cores for geotechnical analysis and testing.

et al. (1979). Modification of the units, as described by Shephard, for falling head permeability tests consisted of attaching a back-pressured column of seawater to the lower stone under the consolidation sample. This lower stone sits in a cup fitted with an O-ring seal creating an impermeable barrier between the stone cup and the consolidation sample confining ring. A three-way valve is then used to direct flow in or out of the sample. Flow out of the sample is obtained during the consolidation process. This flow is two-way, through upper and lower stones, and is routed to a reservoir. Flow into the sample is from one direction during the permeability testing phase. Seawater is directed from the graduated column, through the lower stone and sample, and ultimately to the water reservoir. The rate of flow of seawater through the sample is then used to calculate permeability following the falling head test computation outlined by Lambe (1951). These permeability tests were run after the completion of each load increment during the consolidation tests.

Clay mineralogy was determined for a limited number of samples using a Philips X-ray diffractometer with a copper-potassium nickel filtered source and scanning at 2° to 2Θ from 2 to 45°. The sample preparation and analytical procedures used were described by Hottman (1975). Relative percentages of smectite, illite, chlorite and kaolinite were estimated using the techniques described by Biscaye (1964, 1965).

TABLE 1
Logistics of DSDP Leg 75 Geotechnical Hydraulic Piston Core Hole and Adjacents at Site 532: Leg 40 Holes Near this Site are Also Listed

Hole 532A	
Drilled August 21-23, 1980	19° 44.64' S 10° 31.13' E
Water depth 1331 m	Penetration 199.6m
Hole 532	
Drilled August 20-21, 1980	19° 44.61' S 10° 31.13' E
Water depth 1331 m	Penetration 250.8 m
Hole 532B	
Drilled August 23-25, 1980	19° 44.66' S 10° 31.13' E
Water depth 1331 m	Penetration 291.3 m
Holes 362 and 362A	
Drilled January 17-23, 1975	19° 45.45' S 10° 31.95' E
Water depth 1325 m	Penetration 805.0 m Hole 362
	1081.0 m Hole 362

Transmission electron microscope (TEM) photographs were taken for select samples. These photographs were done using the instrumentation at the Texas A&M University Electron Microscopy Center. The method used for TEM sample preparation was described by Chiou (1981) and photomicroscopy performed using a Zeiss-IOC transmission electron microscope at 80 kV.

Details of the geotechnical tests performed by other members of MGC can be found in Marine Geotechnical Consortium (1984, 1985).

III. RESULTS

A. HEMIPELAGIC CARBONATES

1. Location

The hemipelagic carbonates used in this study were obtained by HPC drilling during DSDP Leg 75. Hole 532A represents the geotechnical site, with Holes 532 and 532B as the adjacent holes offset a distance of 50 and 30 m, respectively. Exact locations of holes drilled by DSDP in this area off southwest Africa are listed in Table 1, along with the total penetration and dates of drilling. Hole 532A was drilled in a sediment filled trough found on the eastern flank of the Walvis Ridge (Figures 5 and 6).

2. Sedimentology

The principal oceanographic conditions, sediment sources, accumulation rates, and lithological composition of sediments found in the area of Site 532 are presented to provide a perspective of sedimentological conditions prevailing at this locale. The southwestern continental margin of Africa underlies a distinct oceanographic setting. The sites shown in Figures 5 and 6 are overlain by an upwelling regime in which nutrient rich Antarctic Intermediate Water reaches the upper water column, creating an environment of high biological productivity. This upwelling regime is a result of a seasonally developed westward divergence of the Benguela Current which has been active in this region since the Miocene (Siesser, 1980). This upwelling regime results in rapid accumulation of biogenic material in the sediments and is frequently accompanied by relatively high organic carbon contents.

The summary data for the recovered sediments at Holes 532 and 532B was described by Hay, Sibuet et al. (1984). Sediments at Site 532 consist of Holocene to upper Miocene material with biogenic and clay components. The mixture of biogenic remains and terrestrial-derived clays represents a typical hemipelagic sedimentary setting, and also reflects the

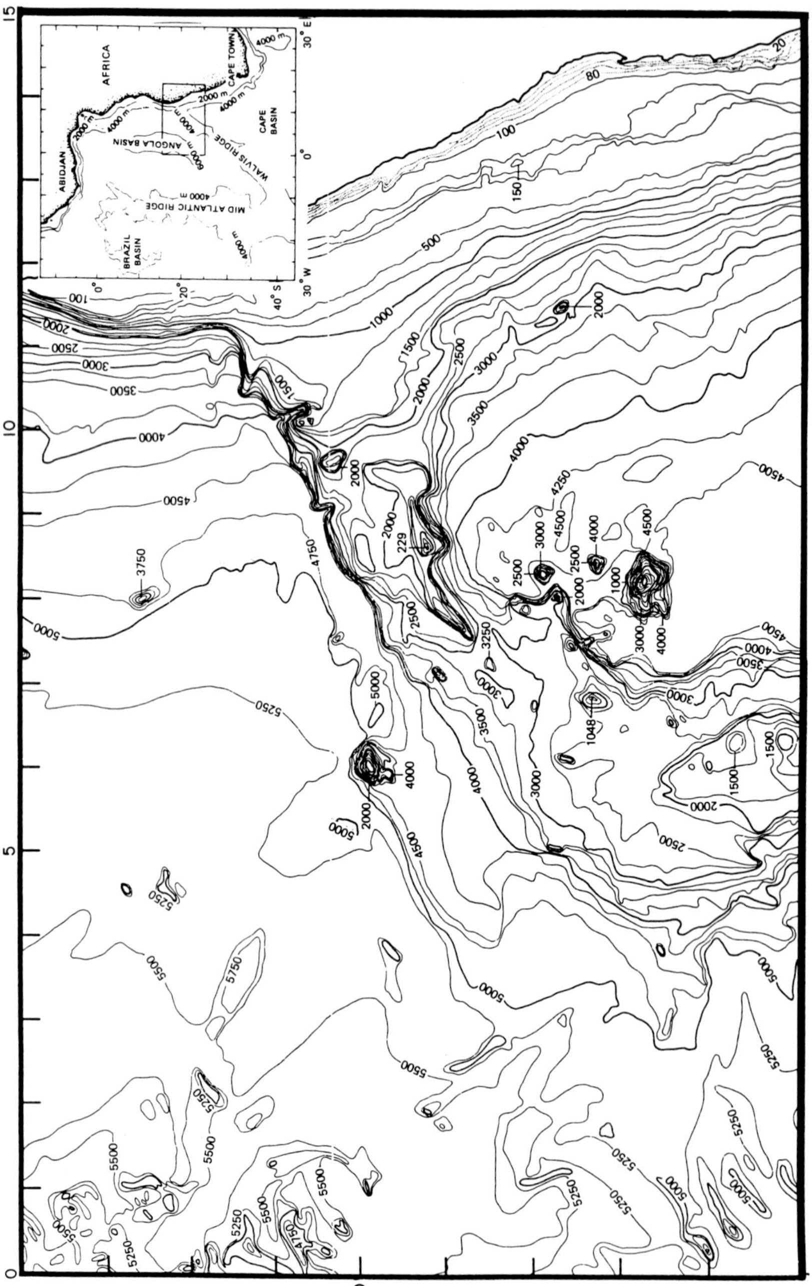

FIGURE 5. Location map of DSDP Sites 532 and 362. (Bathymetric map from Rabinowitz and Simpson, *Initial Reports of DSDP*, U.S. Govt. Printing Office, Washington, D.C., 74, 795, 1984.)

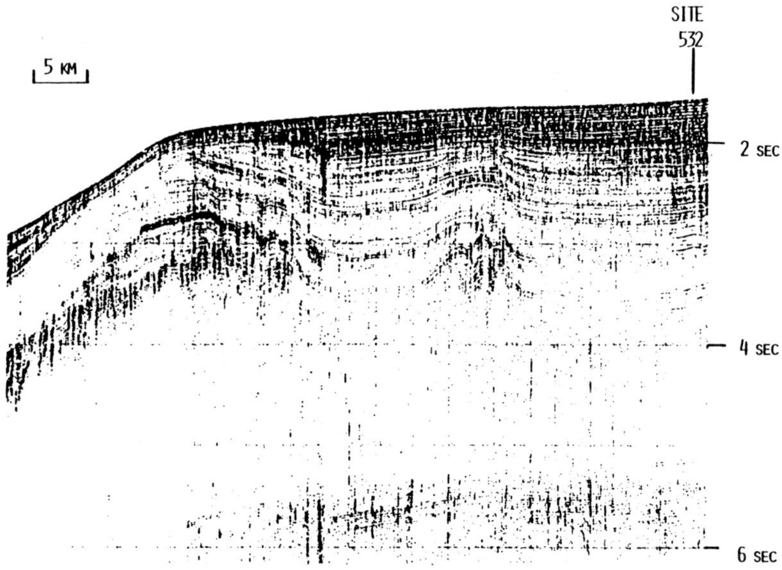

FIGURE 6. Glomar Challenger seismic profile at Site 532 (from Hay, Sibuet et al., U.S. Govt. Printing Office, Washington, D.C., 1984.)

margin-affected marine sedimentary environment mentioned previously. The sediments in cores from Site 532 were divided into three principal units based on the amount of biogenic and terrigenous components: Unit 1: calcareous, siliceous and clayey biogenic sediments of Pleistocene age (0 to 49.5 m sub-bottom), Unit 2: diatom-nannofossil marl, dominantly siliceous, of late Pliocene age (49.5 to 64.5 m sub-bottom), Unit 3: nannofossil marl, of late Pliocene to late Miocene (64.5 to 117.3 m sub-bottom). Superimposed on this stratification are alternating dark and light bands with dark bands representing zones rich in organic carbon, clay, and pyrite. The cyclicity of these bands may be indicative of alternating periods of enhanced productivity in the overlying water column.

The overall distribution of grain size is shown in Figure 7. Clay and silt-sized material constitutes the major portion of the sediment. Clay size material increases with depth from roughly 25% of the total size distribution near surface to 70% at 200 m sub-bottom. Sand size material forms a surprisingly high portion of sediment near surface (up to 35%) but rapidly decreases below 36 m sub-bottom to remain near 1 to 2% of the total. This changing character of grain size reflects compositional changes in the sediment section. Increasing nannofossil components in the lower portions of the core are reflected by increases of clay and silt size material. In addition, breakage of foraminiferal tests which form the bulk of the sand size fraction results in an increase of silt and clay size proportions. Three silt-rich lenses (more than 55% silt) occur at 19, 50, and 71 m sub-bottom.

X-ray diffraction analysis, performed on the Glomar Challenger, for sediments from this site show the presence of quartz, apatite, pyrite and feldspar although the core is clearly dominated by the carbonates. Clay minerals found here include illite as the predominant fraction, with minor amounts of kaolinite/chlorite, a mixed-layer mineral (smectite), and clinoptilolite. The percentage of calcium carbonate relative to the weight of dry sediment ranges from 11 to 76% (Figure 7). In general, the percentage of carbonate decreases from approximately 40% at the mudline to a minimum of 15% at approximately 50 m sub-bottom. Below 50 m, the carbonate content increases to about 70% at 90 m and then decreases to near 20% at the base of the core. This general trend in carbonate reflects, to a degree, the major lithological units identified by the Leg 75 shipboard party.

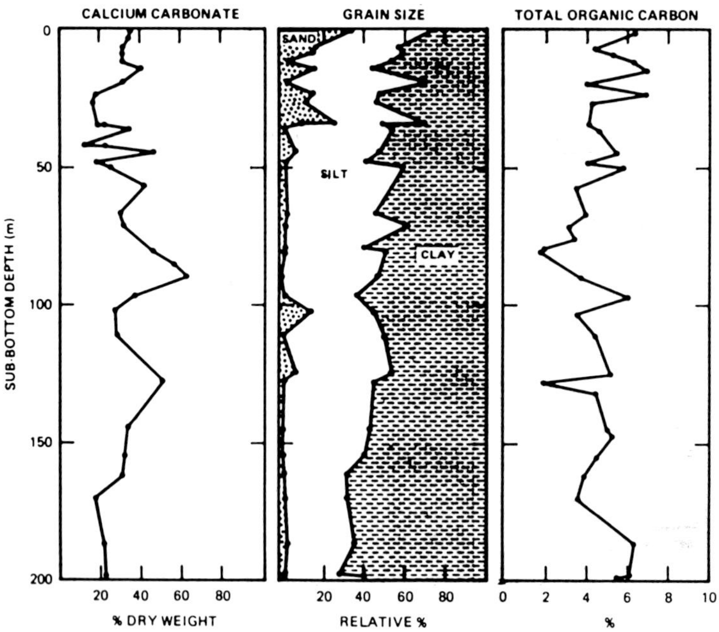

FIGURE 7. Compositional characteristics of the hemipelagic carbonate sequence from DSDP Site 532.

The organic carbon contents of sediments from Holes 532 and 532B were determined by shipboard scientists. Percentages of organic carbon relative to the dry, carbonate-free weight of sediments vary from lows of approximately 1% to a high of 6% (Figure 7). High organic carbon content reflects either, or both, a strong upwelling environment with increased productivity and deposition of biogenic debris, or reduced oxidation rates of bottom sediments. Composition of the organic matter suggests that it is predominantly of marine origin (Hay, Sibuet et al. 1984). Sediment accumulation rates for Site 532A sediments were computed utilizing the relationship described by van Andel et al. (1975). This technique uses porosity and bulk density measurements to adjust the sediments accumulated to a measure of the volume of accumulating solids, thus the resulting values are not readily affected by changes in the sediment column arising from consolidation.

Accumulation rates at Site 532 vary from a high of 43.95* 10^2 g/cm^{2}* my to 31.91* 10^2 g/cm^{2}* my at the surface (Figure 8). These sediment accumulation rates show a consistent increase with depth in the sediment section which may be an artifact of using porosities and bulk densities measured on board the ship that had not been corrected for rebound. Percentage accumulations of the different biogenic fractions and of the nonbiogenic material varies more with depth than the overall rates. The nonbiogenic, or terrigenous, fraction increases with depth from approximately 10* 10^2 to 20* 10^2 g/cm^{2}* my while biogenic fractions differ in relative proportions to each other, yet overall remain fairly constant throughout the cored interval.

3. Geotechnical Properties

Vertical distributions of bulk density, porosity, water content, and void ratio are shown in Figure 9. Water content, porosity and void ratio decrease rapidly downhole in the upper 15 m of the sediment column at Site 532A. Water contents, for instance, decrease from values near 140% to approximately 90% in the uppermost interval. A rapid decrease of water content in the uppermost sediments is characteristic of most calcareous marine sediments (Hamilton, 1976). At Hole 532A, however, a certain contribution to this behavior of index properties may result from the breakdown of foraminiferal tests. The index properties

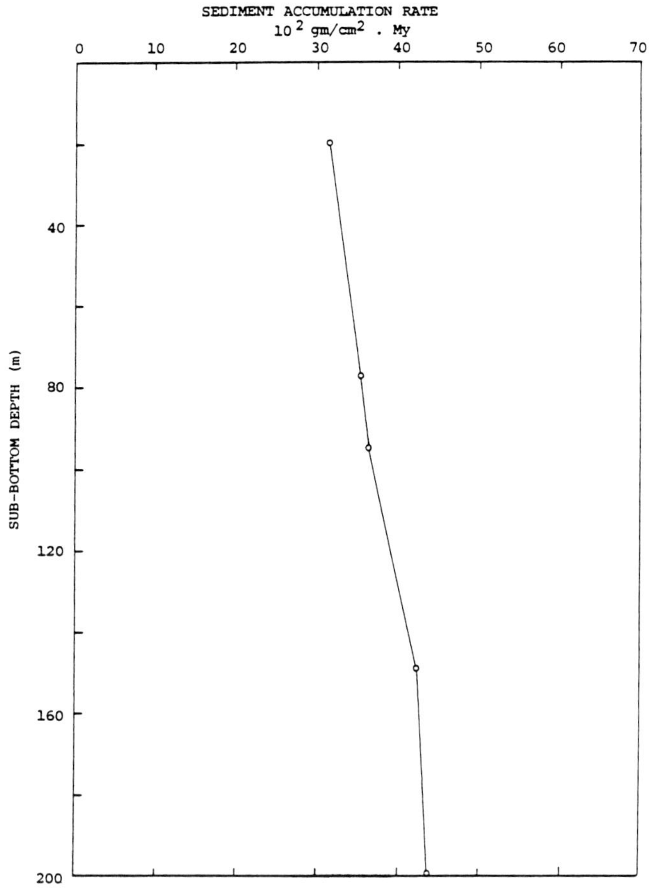

FIGURE 8. Calculated sediment accumulation rates at DSDP Site 532.

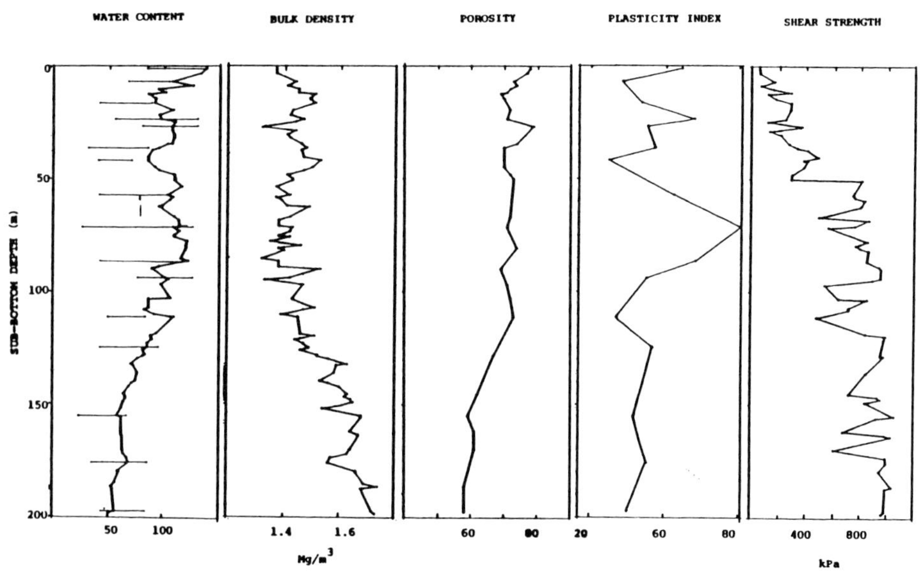

FIGURE 9. Index properties of DSDP Hole 532A sediments. Water content profile shows natural water content and bars indicate the plastic limit (left end) and liquid limit (right end).

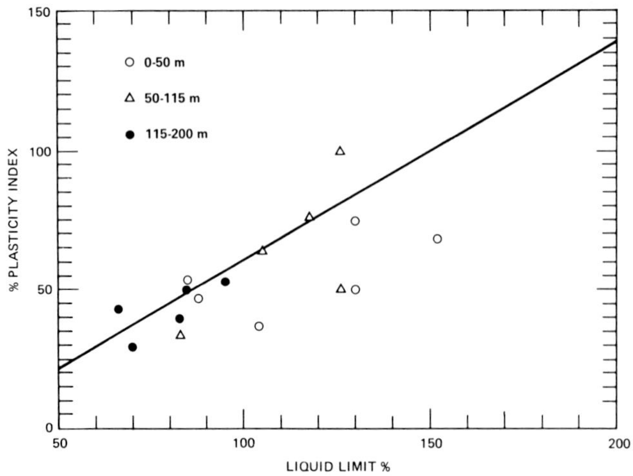

FIGURE 10. Plasticity chart for DSDP Hole 532A sediments. Most sediments fall within the zone typified by inorganic, elastic silts and organic clays of medium to high plasticity.

below 15 m sub-bottom show no distinct increase or decrease, but generally fluctuate around a common value until reaching a depth near 120 m. At this depth water contents decrease from about 100% to 72% at 150 m, representing a change of 28% over a 30-m interval. Below 175 m, the water content decreases to approximately 50% at the bottom of the recovered sediment section. Little difference of index properties is exhibited relative to the lithological units.

Results of Atterberg limits are shown in Figures 9 and 10. The plotted data, grouped according to lithological subunits, shows that no distinctive difference occurs between these units relative to their limits. Figure 9 illustrates the relationship of liquid and plastic limits to the natural water content of the sample. Most sediments tested have a natural water content that lies between the plastic and liquid limit.

Results of shear strength tests are given in detail in Marine Geotechnical Consortium (1984). A summary of the strength character of the hemipelagic oozes in question is presented in Figure 9, showing the variation of vane shear test results with depth. The shear strength increases steadily to a depth of approximately 90 m sub-bottom. Below this depth the values of shear strength are variable and show little subsequent increase. The trend of shear strength in the lower half of the cores from Hole 532A is most likely an artifact resulting from failure of the sediment by cracking. Therefore, the subsequent calculated vane shear strength is not valid since failure is assumed to occur as a cylinder. The upper 100 m of sediment at this site have an average Su/Po′ ratio, or shear strength to effective overburden pressure (pressure resulting from the overlying sediment column) of 0.19 (Figure 11). Comparison of this value with ratios obtained from triaxial and direct shear tests of strength indicates that 0.19 may indeed be a low estimate. Values of 0.5 to 0.7 were obtained using strengths determined from these latter tests, agreeing closely with data obtained earlier by Nacci et al. (1974) for other deep sea carbonate sediments. Vane strength data is tabulated in Taylor (1984).

Consolidation results are listed in Taylor (1984), providing detailed information regarding overburden pressures, overconsolidation ratios, void ratios at various overburdens, compression indices, and expansion indices. The means and ranges of several consolidation parameters are presented in Table 2. Examples of decreasing void ratio of some samples as a function of increased vertical pressure (e-log P′ curves) are illustrated in Figure 12. The zone of sharp curvature in these plots corresponds to the points at which vertical pressures exerted during the consolidation test were proceeding from a recompression stage to a virgin compression of the sediment. The compression index, Cc, of these samples is the slope of

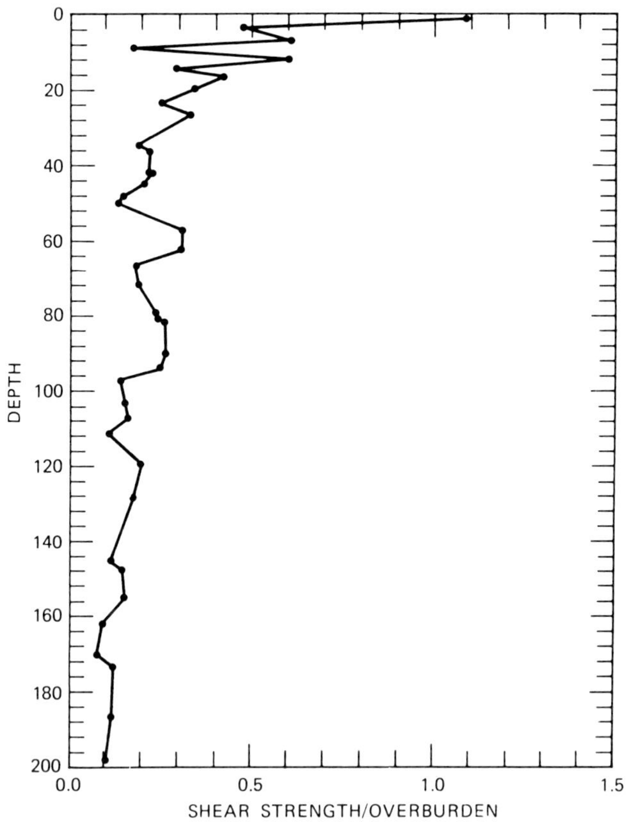

FIGURE 11. Vane shear strength — effective overburden pressure ratios (Su/Po′) for sediments from DSDP Hole 532.

TABLE 2
Means and Ranges of Consolidation and Permeability Characteristics of Hemipelagic Sediments from DSDP Hole 532A

Variable	Mean	Minimum	Maximum
OCR	3.18	0.28	7.33
Cc	0.90	0.42	1.55
Rc	0.10	0.03	0.17
Permeability (cm/sec)			
Direct	1.60E-6	2.19E-8	2.29E-5
Indirect	2.11E-7	1.23E-8	6.92E-7

the virgin portion of this curve. The variation of the compression index with depth (Figure 13) shows a general increase from approximately 0.50 near the sediment surface to about 1.35 near 85 m sub-bottom. Values of Cc show a subsequent decrease below this depth back to 0.50 at the base of this hole.

The rebound indices, Rc, for Hole 532A sediments are plotted against depth in Figure 13. Surface values of approximately 0.06 increase to an average of 0.14 at 40 to 50 m sub-bottom. The value of Rc below 50 m fluctuates around a mean of 0.10 to a sub-bottom depth of 135 m. Below 135 m, the rebound coefficients decrease steadily to a minimum of

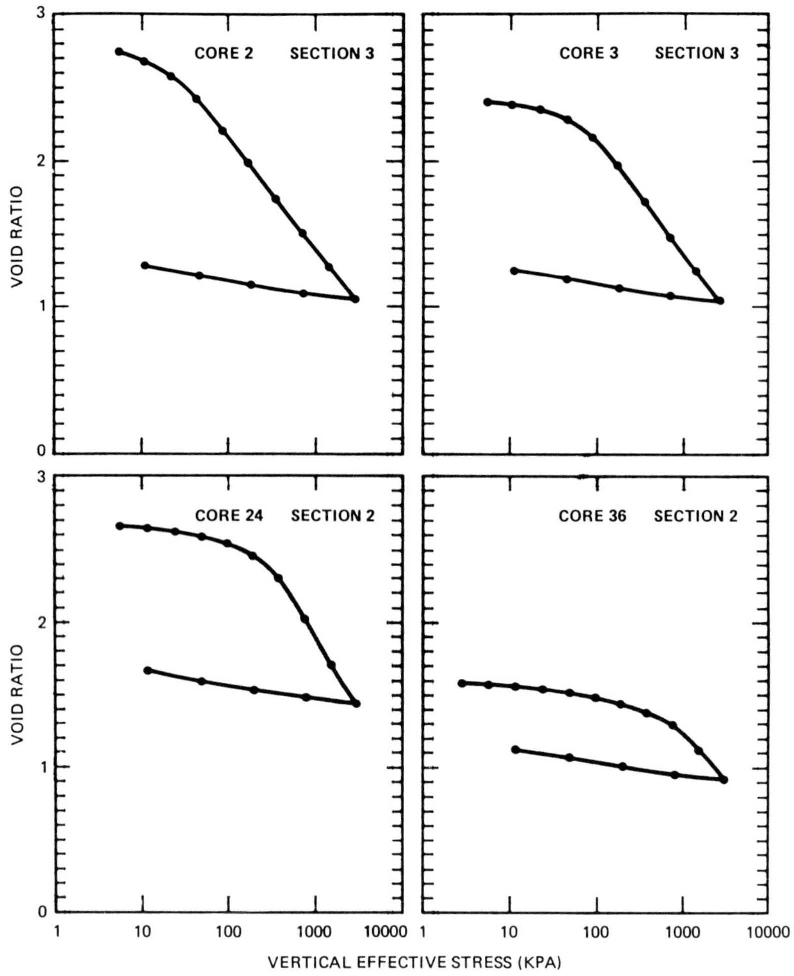

FIGURE 12. Characteristic void ratio vs. log of effective vertical pressure (e-log P′) curves for sediments from DSDP Hole 532A. Samples shown are from sub-bottom depths of 6.90, 11.75, 102.95, and 155.10 (clockwise from top left).

0.03 at the base of the hole. Low Rc values in the uppermost section are probably due to breakage of foraminiferal tests with consolidation. This breakage results in the destruction of sediment framework and in turn reduces the ability of the sediment to rebound elastically. The increase of Rc in the underlying sediments is most likely due to the increased proportion of clay in the sediment matrix, which contributes elasticity to these sediments. Low Rc values in the lowermost section probably reflect incipient cementation and/or diagenesis of particles. Bonding may become more important than the clay fraction as the controlling factor in rebound.

Consolidation test results can also be used as indicators of the degree of disturbance a sample has undergone relative to a naturally consolidated sediment. A commonly utilized indicator of disturbance is the degree of curvature of the break from recompression to virgin compression of the e-log P′ curve. A remolded sample quite often shows a large radius of curvature, whereas intact samples usually have a smaller radius of curvature (Lambe and Whitman, 1969; Casagrande, 1936). The e-log P′ curves shown in Figure 12 have a sharp curvature after the initial reloading, as do other samples from this site, which suggests that the samples tested for this project were not very disturbed.

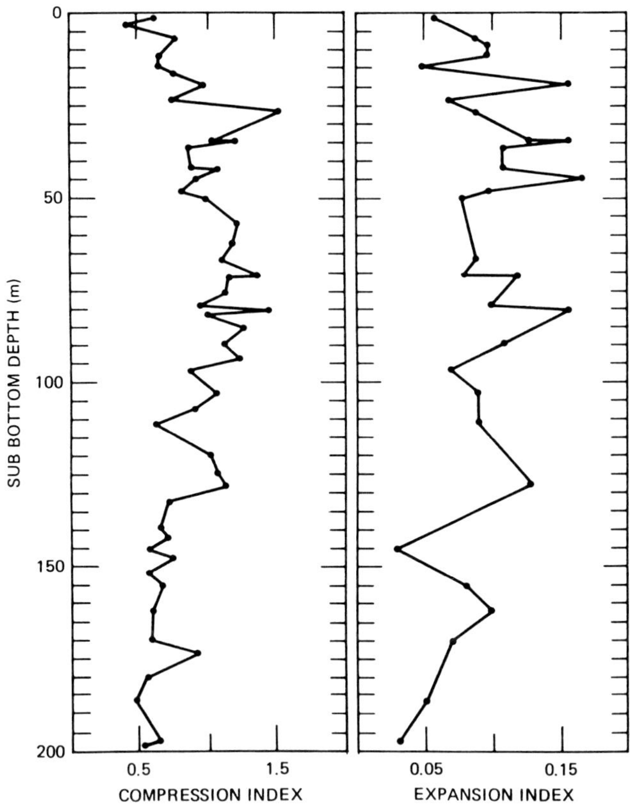

FIGURE 13. Elastic characteristics of the calcareous hemipelagic sediments from DSDP Hole 532A.

The overconsolidation ratio, OCR, is the ratio of the maximum past vertical pressure (Pc′) the sediment had equilibrated with relative to its *in situ* vertical pressure (Po′). The value of Pc′ is determined graphically following the method described by Casagrande (1936). This point generally falls within the area of maximum curvature of the e-log P′ curve. Standard geotechnical practice defines an overconsolidated sediment as having an OCR greater than 1, or a past maximum overburden pressure in excess of that calculated as existing *in situ*. An underconsolidated sediment has an OCR less than 1. The changing character of OCR vs. depth for the Hole 532A sediments is shown in Figure 14.

These sediments are generally very close to being normally consolidated. Overconsolidated sediments are found very near the mudline, however this is a result of comparing the very low, in situ vertical overburden pressure with an inherent strength of the deposited material. It can be seen from the data (Figure 14) that this discrepancy is pronounced. However, the resulting difference between Pc′ and Po′ is small, as seen in Figure 15. Comparison of Figures 14 and 15 to the profile shown in Figure 11 indicates that Su/Po ratios of 0.12 to 0.33 represent the range found for normally consolidated hemipelagic carbonate sediments. Some instances of underconsolidation were observed at a sub-bottom depth of approximately 50 m and between 150 to 185 m (Figures 14 and 15). The underconsolidated state of these sediments may indicate disturbance induced during coring or disturbance resulting from significant degassing of the sediment during recovery.

Permeability for these sediments, expressed as hydraulic conductivity (Rieke and Chilingarian, 1974), is tabulated in Taylor (1984). Permeabilities discussed in this study are interpolated to correspond with void ratios at their *in situ* effective vertical overburden

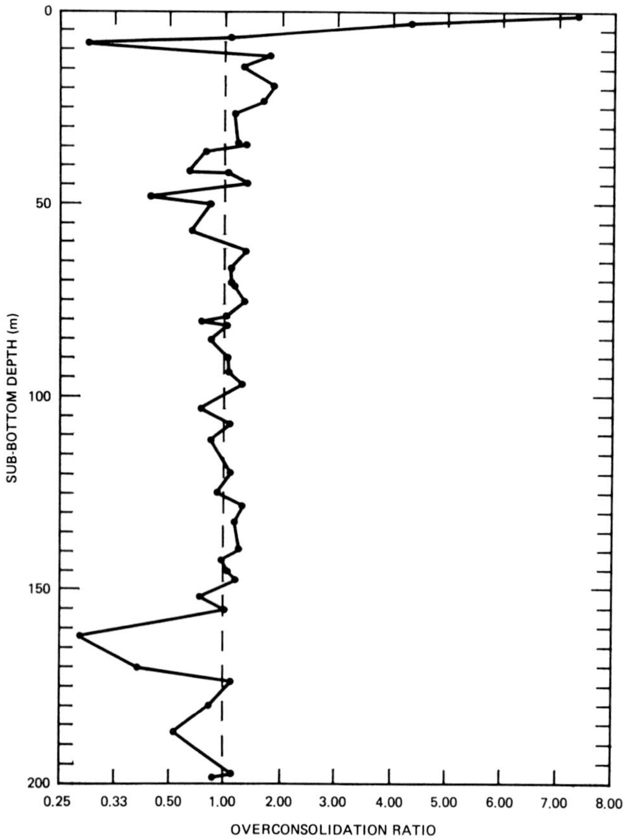

FIGURE 14. Profile of overconsolidation ratios (OCR) for sediments of DSDP Hole 532A.

pressures. These values represent the closest approximation to the true, downhole state. Permeabilities are computed from direct falling-head measurements or from consolidation test results and are plotted with respect to depth in Figure 16. A fair amount of spread between direct and indirect measurements of permeability can be seen from this figure. Direct permeability measurements were performed using different techniques and high hydraulic gradients. High hydraulic gradients may, in turn, produce seepage induced consolidation resulting in faster flow rates of fluid from and through the sample. Eliminating the direct measurements of permeability results in the distribution shown in Figure 17. The scatter in the permabilities is greatly reduced as is the downhole gradient. Surface values near 10^{-6} cm/sec decrease gradually to approximately $2*10^{-7}$ cm/sec at a depth of 125 m. A shift of the permeability occurs at this depth, with values near $7*10^{-8}$ cm/sec decreasing gradually to near $3*10^{-8}$ cm/sec at 200 m.

B. RED CLAYS

1. Location

In May 1982, a transect of sites in the northwestern Pacific was drilled on DSDP Leg 86 to answer basic questions concerning the paleo-oceanographic history of the Kuroshio current system. Another objective of drilling was to sample an undisturbed representative or "type" section of pelagic clay for geotechnical testing. Hole 576A, located approximately 300 km east of the Shatsky Rise (Figure 18) was sampled using the HPC and preserved for dedicated geotechnical studies. Adjacent holes 576 and 576B were also drilled with the HPC

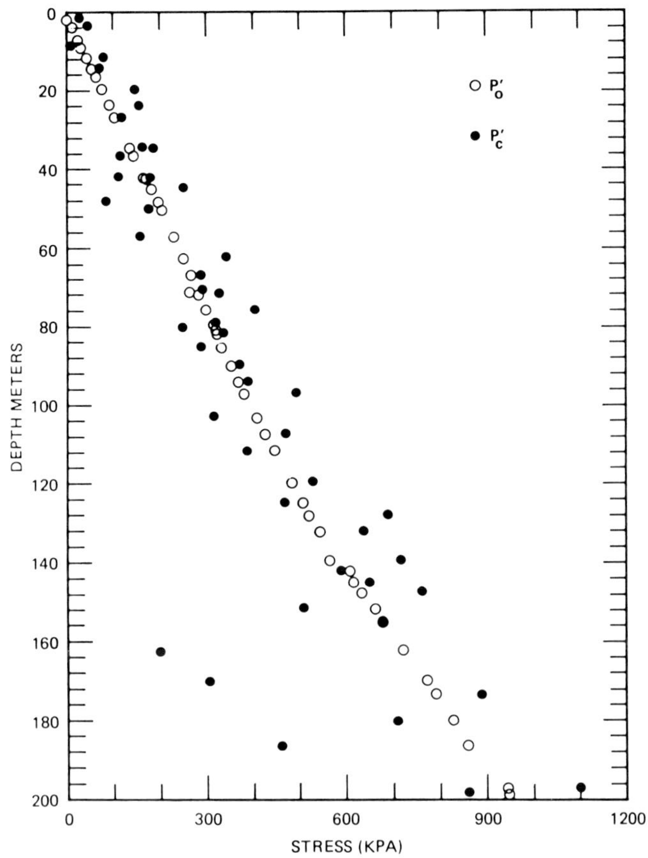

FIGURE 15. Profiles of effective overburden pressure (Po′, Δ) and preconsolidation pressure (Pc′, *) calculated by the Casagrande reconstruction technique for DSDP Hole 532 sediments.

and offset by approximately 80 and 30 m from 576A, respectively. Exact locations of these holes, dates of drilling, and total penetration are listed in Table 3.

2. Sedimentology

The principal factors defining the sedimentological features at Site 576 consist of ocean circulation and sediment source, composition, and accumulation rates. The location of Site 576 is on the Pacific abyssal plain at a water depth of 6217 m. The site is located within the oceanographic setting described as the mid-Pacific gyre, a region with very slow currents. Sediments accumulating in this region originate from eolian transported material, current transported sediment, and from volcanic sources. The majority of sediments are derived from the Asian continent. They are transported easterly, and slowly settle through the water column. The considerable distance between the main sediment sources and Site 576 results in accumulations of very fine material, or pelagic clay. Most of this material had probably remained in suspension for considerable amounts of time. Sediment input from biogenic sources is minimal, probably due to the relatively barren environment of mid-ocean gyres for productivity. Also, the depth of the sea floor below the carbonate compensation depth accounts for the lack of calcareous material in these clays. Authigenic sediment also forms some of the material found at Site 576, and may even become the predominant material of these sediments at depth (Leinen, 1985).

Coring at Site 576 resulted in the complete recovery of the sedimentary section from

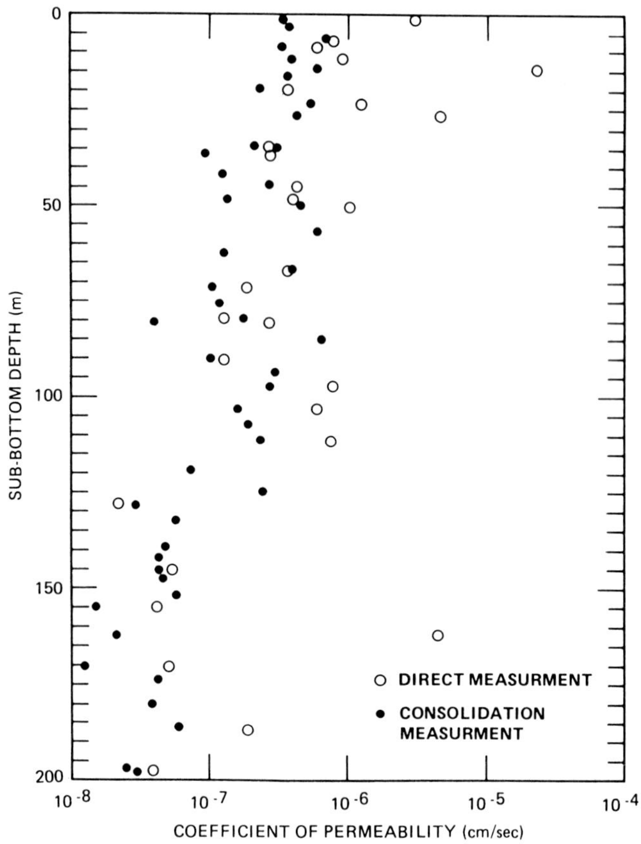

FIGURE 16. Vertical profile of the coefficients of permeability for sediments of DSDP Hole 532A. This figure shows both direct (Δ) test results and those calculated from consolidation tests (*) interpolated for Pc′ conditions.

Holocene to Upper Cretaceous. The sediments, as described by the shipboard party, consist of two basic units. The upper unit extends from the mudline to 55 m sub-bottom and is composed of the typical red clay material. This unit overlies a mixed unit of red clay and carbonate ooze. The overall lithological stratigraphy at this site is given below as described by Heath, Burkle et al. (1985): Sub-unit I-1: brownish pelagic clay of Pliocene and Quaternary age, with a gradational lower contact (0 to 28 m sub-bottom), Sub-unit I-2: homogeneous, dark brown, slick pelagic clay, fine grained, magnesium rich, and of Miocene to Paleocene age (28 to 55 m sub-bottom), Unit II; interbedded material similar to sub-unit I-2 with pale brown, nannofossil ooze of Campanian to Early Maastrichtian age. Some carbonate layers are gradational, with sharp lower basal contacts, resembling turbidites.

Grain size analyses obtained for samples from this site yielded the results shown in Figure 19 and tabulated in Taylor (1984). The grain size distribution with depth shows a clear predominance of fine grain material. Silt size material constitutes nearly 30% of the total sample in the upper 20 m, but below this depth it decreases to form only 10 to 15% of the sediment. The lowest sediment, a calcareous ooze, shows an increase of the silt-size fraction. The percentage of sand size material in the red clays mimics the silt fraction, with maximum values in the upper 20 m and then decreasing to almost insignificant amounts below this depth. Sand size material observed under the microscope consists primarily of quartz and micro-nodules, probably manganese. Silt size material, as seen through SEM

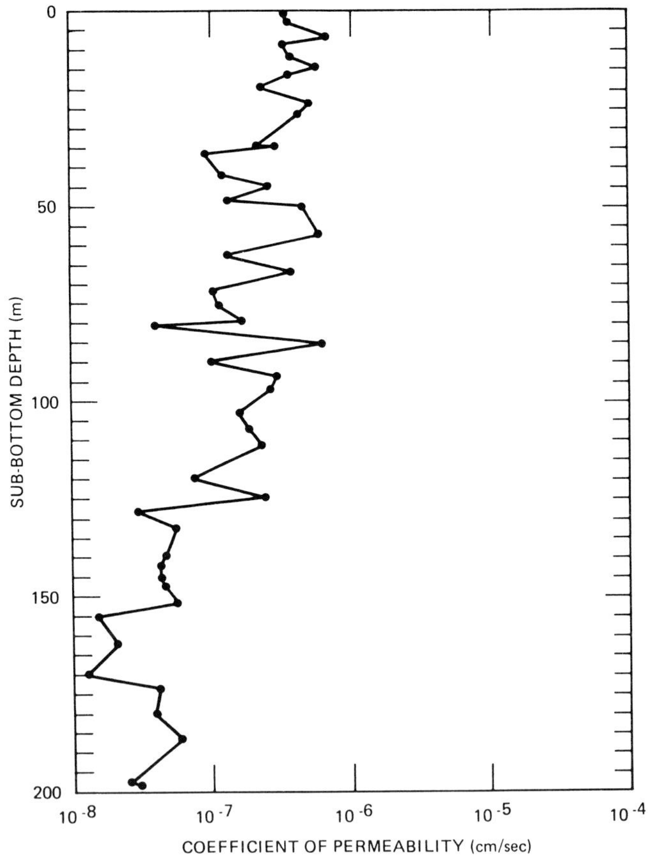

FIGURE 17. Calculated coefficients of permeability from consolidation tests for sediments of DSDP Hole 532A. Permeabilities are interpolated to the corresponding pre-consolidation pressure.

and TEM photography, appears composed of two principal types; a crystalline group of particles formed mostly of quartz and feldspar, and a second group composed of bonded clays which form silt-size flocs (Figure 20).

These clay flocs resisted the deflocculation process used for pipette analysis, and, hence, the actual proportions of clay to silt-size material in these sediments are obviously not quite correct. These flocs, also observed and described by Bennett et al. (1981), are believed responsible for some of the peculiar geotechnical characteristics of the upper 20 m of sediment at this site. Taylor (1984) also lists relative percentages of illite, chlorite, smectite, and kaolinite for material finer than 2 μm. These analyses indicate a downhole increase of smectite and decrease of illite. The variations in clay mineralogy may, however, be overwhelmed by the large increase of amorphous material that occurs in the lower half of these cores.

Carbonate measurements in the sediments were only of significance in the calcareous ooze horizons. Within these horizons, calcium carbonate was found to range between 80 and 92% of the total sample. As indicated, the principal component of the carbonate in the lower layer is formed of nannofossils.

Sedimentation rates were estimated on the Glomar Challenger using several approaches. A variety of methods were used for determining the age of sediments. The need for several dating techniques was necessary because of the paucity of fossils in the sediment. The

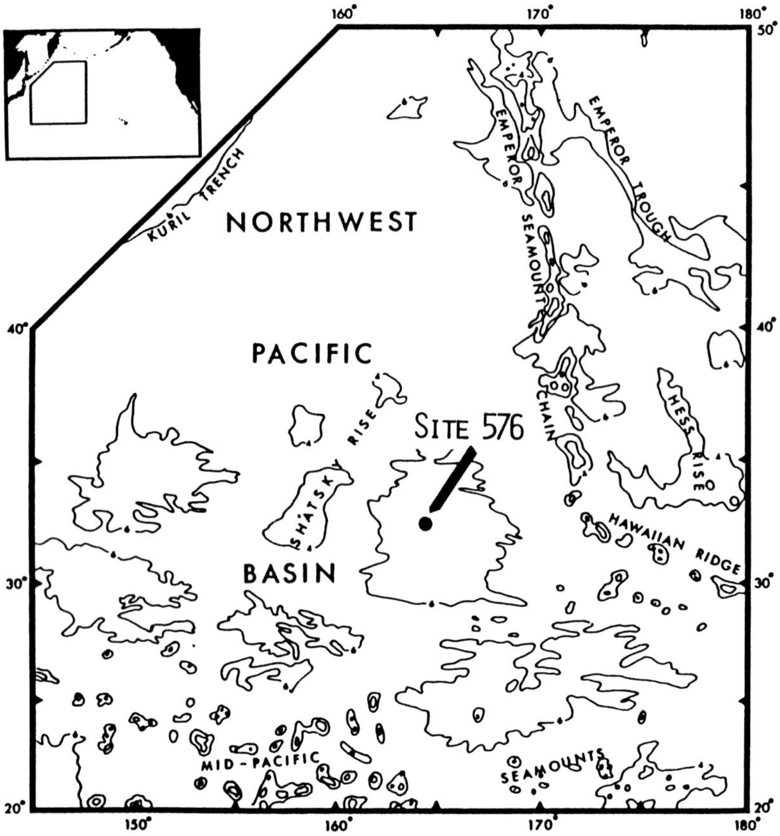

FIGURE 18. Location map of DSDP Site 576. (Base map from Damuth et al., *Geol. Soc. Am. Bull.*, 94, 381, 1983.)

TABLE 3
Logistics of DSDP Leg 86 Geotechnical Hydraulic Piston Core Hole and Adjacents at Site 576

Hole 576A	
Drilled May 18-19, 1982	32° 21.38' N 164° 16.52' E
Water depth 6217 m	Penetration 65.7 m
Hole 576	
Drilled May 16-17, 1982	32° 21.36' N 164° 16.54' E
Water depth 6217 m	Penetration 69.2 m
Hole 576B	
Drilled May 19-20, 1982	32° 21.37' N 164° 16.52' E
Water depth 6217 m	Penetration 74.8 m

methods used for dating include magnetostratigraphy, icthyolithic stratigraphy, and age determinations from the few radiolarian remains found (Doyle and Riedel, 1985; Heath et al., 1985; Morely, 1985). Estimates of sedimentation rates vary from 10 to 0.2 m/my based on all dating information (Heath et al., 1985). The generalized record of sediment accumulation is shown in Figure 21. Rates decrease from 10 to 3 m/my in sub-unit I-2, and are in general lower than any of the other sites cored during DSDP Leg 86.

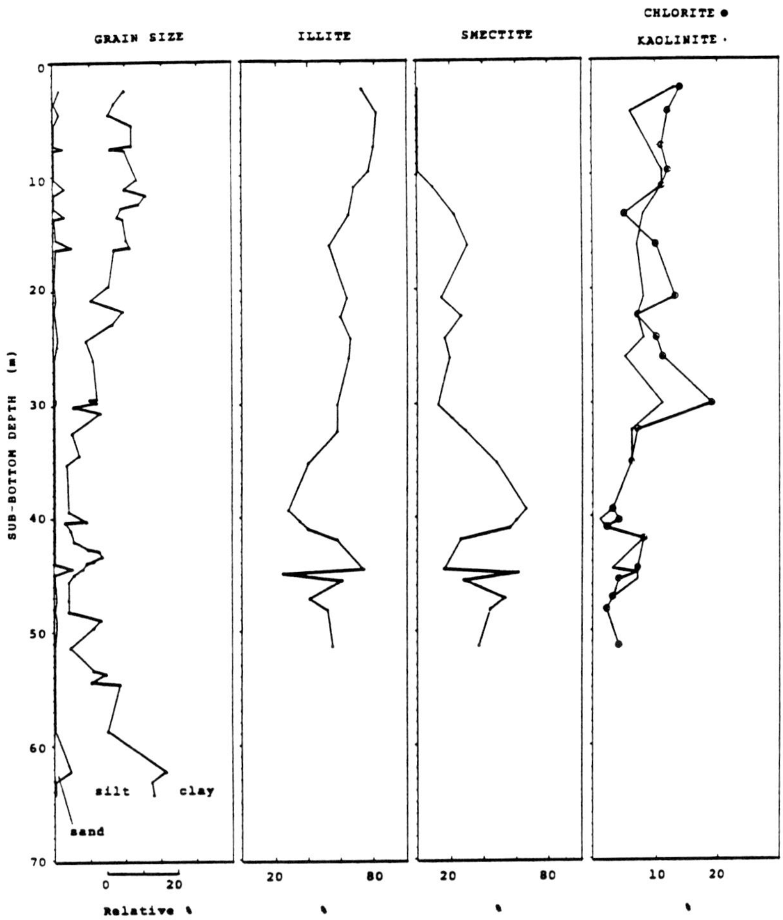

FIGURE 19. Compositional profiles of sediments from DSDP Hole 576A. Mineralogies are based on the 2 micron fraction only and assume that these minerals form 100% of the clay fraction.

3. Geotechnical Properties

Measurements of index properties for Hole 576A sediments are tabulated in Taylor (1984) and shown as depth profiles in Figure 22. These data represent the most intensive geotechnical testing program undertaken on DSDP cores. The water content, bulk density, and void ratio profiles change little in the upper 18 m. Water content decreases from roughly 180% at 18 m to 118% at 20 m sub-bottom. Water content increases below 28 m to approximately 145% at 34 m and then gradually decreases to near 115% at the base of subunit I-2. By contrast, the nannofossil ooze at the base of the hole has a water content of only 40 to 50%. Other index properties possess similar trends for these sediments. Liquid and plastic limits are listed in Taylor (1984) along with the plasticity index. A vertical profile of Atterberg limits is shown in Figure 22. Liquid and plastic limits are practically constant in the upper 28 to 30 m. At 30 m, the limits show a marked increase, with the liquid limits well in excess of the natural water content. Finally, the nannofossil ooze contact is readily observable as an inflection in the profile at 54 m sub-bottom. A plot of the liquid limits and plasticity indices on the Casagrande plasticity chart shows a fairly good clustering of pelagic clays distinct from the ooze which centers around a classification for medium plasticity silty clay (Figure 23). Two clusters for pelagic clay are also shown, pointing out the distinct character of sediments from above and below 28 m sub-bottom.

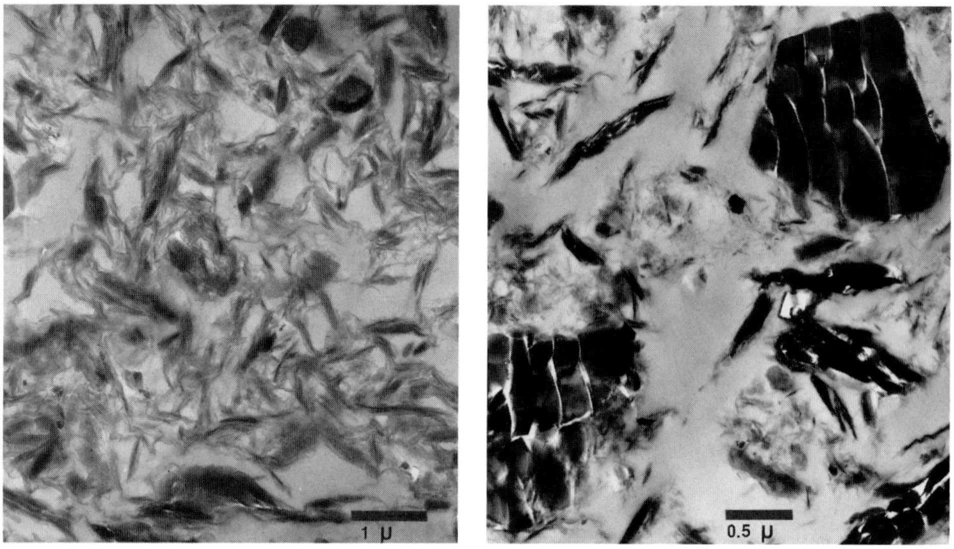

FIGURE 20. TEM photomicrographs of red clay floc fabric.

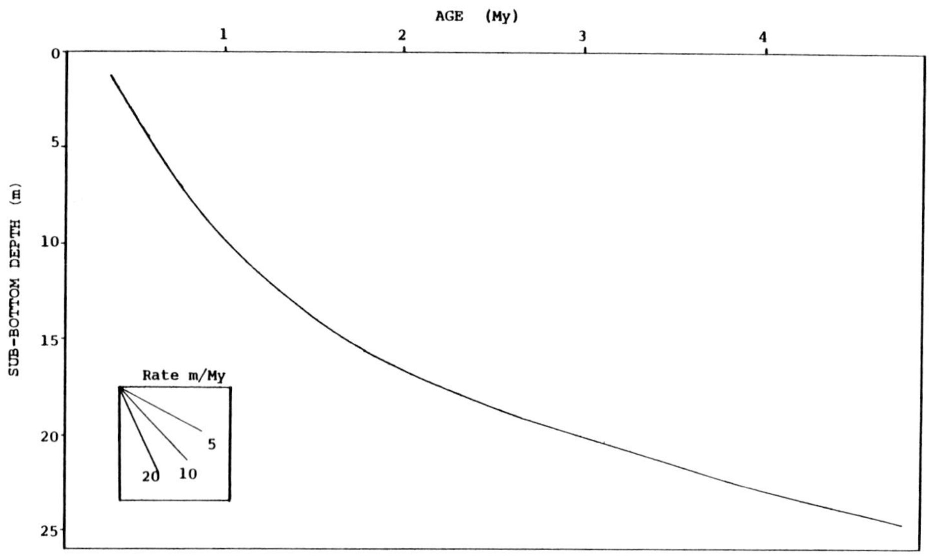

FIGURE 21. Sedimentation rates at Site 576 in the Northwest Pacific based on the time scale from magnetic stratigraphy. (Modified from Heath et al., 1985.)

Numerous strength measurements were performed down the core yielding undrained vane shear strength. A plot of this strength measurement vs. depth is shown in Figure 22. Vane shear strengths increase monotonically from a low of approximately 6 kPa near the mudline to highs on the order of 50 kPa at 54 m sub-bottom. The boundary between sub-units I-1 and I-2 is distinguished by a break in the trend of strengths at 28 m. The lower sub-unit has an initially lower strength than the overlying sediments of sub-unit I-1. The carbonate ooze found below 54 m has very low strength which may reflect coring-induced disturbance. The shear strength vs. in-situ effective overburden pressure (Su/Po′) ratio for the red clays at Site 576A decreases rapidly in the uppermost 12 m of sediment (Figure 24).

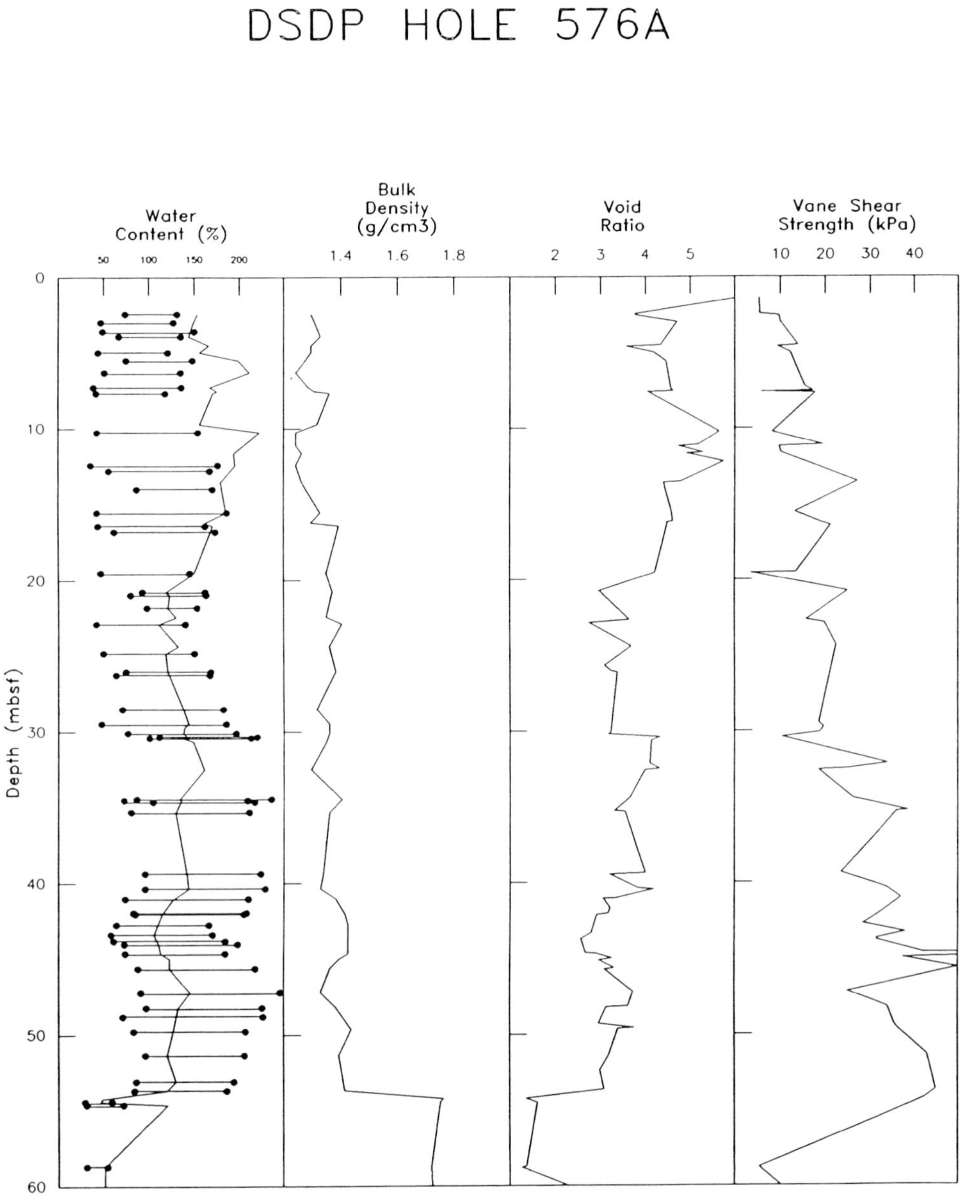

FIGURE 22. Index properties profiles for sediments from DSDP Hole 576A. Bars in the water content column illustrate the plastic limit (left end) and liquid limit (right end).

The Su/Po′ ratios range from highs of 1.20 near the mudline to lows of 0.25. This ratio then remains nearly constant at 0.25 from 12 m to the contact with Unit II where it decreases to less than 0.1.

Skempton (1957) defined a relationship of Su/Po′ ratios and plasticity indices for normally consolidated sediments. Using this relationship, red clay sediments at Hole 576A have a Su/Po′ range of 0.31 to 0.55, which would suggest that all sediments below 20 m are underconsolidated. Using this same argument, the sediments above approximately 7 m are overconsolidated, as indicated by the results of consolidation and other strength testing. All sediments from below 20 m, however, are not found to be in the underconsolidated state as suggested by the consolidation tests. This discrepancy in apparent states of consolidation

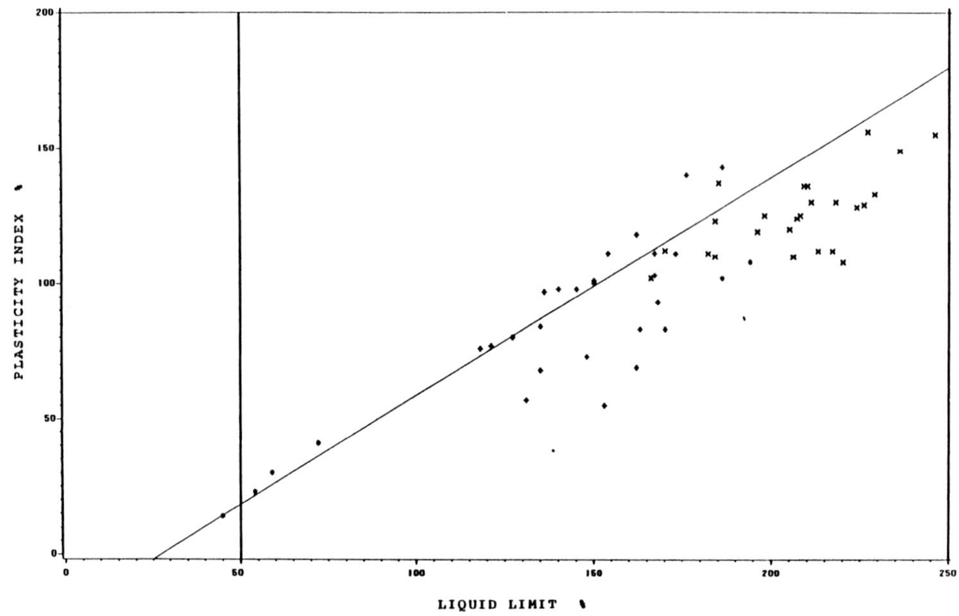

FIGURE 23. Plasticity chart for sediments of DSDP Hole 576A. Red clay sediments from zero to 28 m (+) are distinct from the lower (28 to 53 m) group (x). The carbonate ooze from below 53 m is very distinct (*) from both red clay groups.

is indicative of the need for using caution when trying to apply standard geotechnical techniques to marine sediments in general.

Consolidation and permeability tests were performed on 80 samples from this site. The results of these tests are listed in Taylor (1984). Typical e-log P′ curves are shown in Figure 25. The mean values and ranges of several consolidation parameters and permeabilities are presented in Table 4.

The shape of the e-log P′ curves is often used as an indicator of the amount of disturbance the sediment may have undergone. From the shape of the curves shown in Figure 25 we can deduce that these samples have not been noticeably disturbed. However, a consolidation test performed on an intentionally remoulded sample shows a well developed break from reloading to the virgin portions of the e-log P′ curve (Figure 26). Although this argument has had widespread use and is generally a good indicator of disturbance, the applicability of this data as a measure of disturbance in red clays is questionable.

A plot of OCR vs. depth is presented in Figure 27. The computed state of consolidation, given by OCR, indicates an overconsolidated upper unit of sediment extending to a depth of 27 m sub-bottom. Another representation of the state of consolidation is illustrated in Figure 28 in which values of both *in situ* effective overburden pressure, Po′, and the preconsolidated pressure, Pc′, are shown. This representation minimizes the distortion of overconsolidation ratios obtained when considering very small pressures, such as those existing near the mudline. Despite this, some degree of overconsolidation is apparent for the uppermost sediments. Comparison of these latter measures of the state of consolidation to the Su/Po′ ratios shown in Figure 24 suggests that a range of 0.15 to 0.32 defines a normally consolidated red clay sediment. Considerable debate still remains regarding the standard methods of defining the state of consolidation from oedometer tests. Three different techniques were used in this study to estimate the state of consolidation of the red clays. Each technique yielded different results (Marine Geotechnical Consortium, 1985).

Measures of the elastic character of the red clay sediments are given by the compression

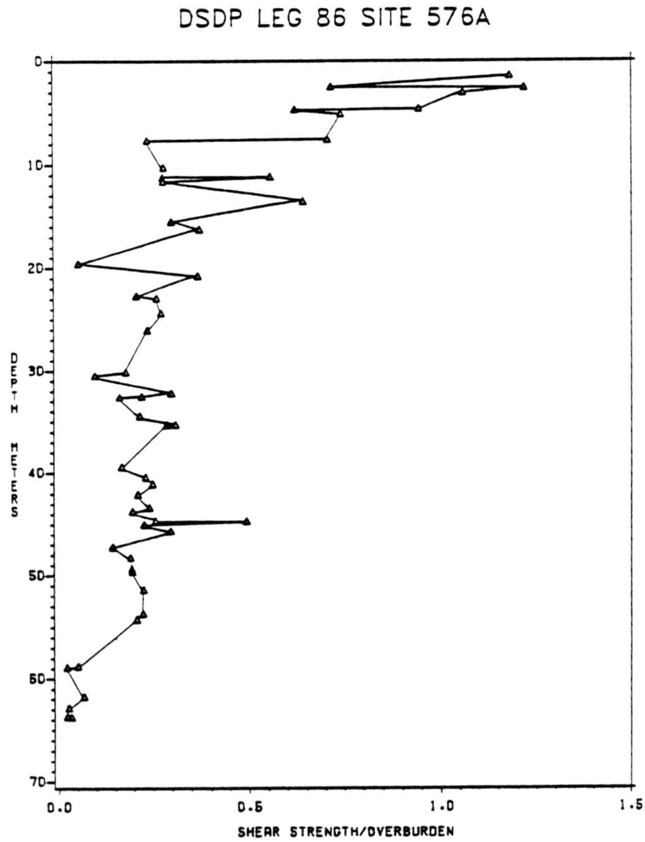

FIGURE 24. Vane shear strength-natural effective overburden pressure ratios (Su/Po′) for sediments of DSDP Hole 576A.

index, Cc, and the expansion index, Rc. As illustrated in Figure 29, the compression index increases from approximately 1.5 at the mudline to a maximum of 3.0 at a depth of 13 m. Values of Cc decrease below 13 m to approximately 1.6 at 21 m and then vary slightly from this value until the nannofossil ooze is reached at 54 m sub-bottom. Measures of the expansion index show a close relationship to depth, increasing from near surface values of 0.08 to values of 0.33 at 53 m sub-bottom.

The value of Rc drops abruptly in the nannofossil ooze to values of 0.03. The Rc profile indicates an increasing elasticity of sediments with depth, and shows the practically nonelastic behavior of the calcareous ooze. Consolidation tests having several episodes of rebound, such as during hysteresis loops, show that the values of Rc during these rebounds are generally similar. The fact that an e-log P′ curve having a hysteresis loop exhibits similar rebound characteristics during the unloading portion of the loop and the unloading at the end of a test suggests that the Rc values represent an inherent characteristic of the sediment with minor response to overburden. This being the case, the variation of Rc with depth at Hole 576A represents a changing character of sediment behavior that is not directly related to its depth of burial.

The mean and the range of permeabilities for the red clay section are shown in Table 4. These represent values consistent with the expected low permeabilities of very fine grained sediments. The permeability shows a continously decreasing gradient with depth (Figure 30). The major decrease in permeability occurs in the top 22 m, and becomes less pronounced with greater depth. The occurrence of calcareous ooze is clearly identified by the abrupt

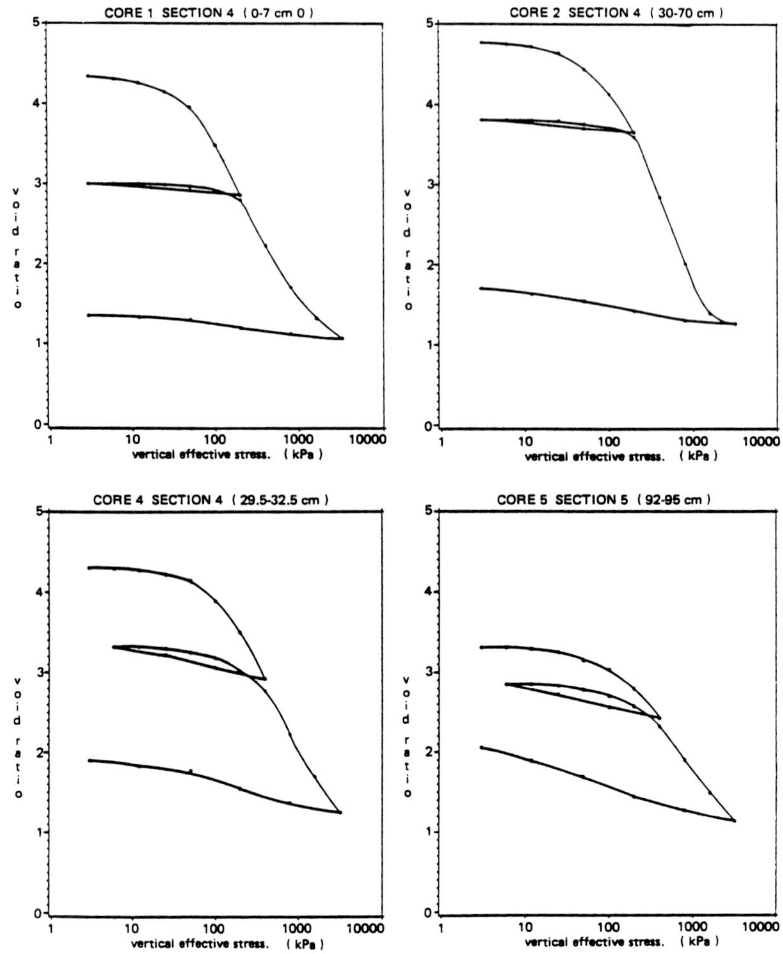

FIGURE 25. Typical void ratios vs. effective vertical pressure (e-log P′) curves for red clay sediments from DSDP Hole 576A. Samples are from sub-bottom depths of 4.52, 13.52, 32.47, and 45.60 m (clockwise from top left).

TABLE 4
Means and Ranges of Consolidation and Permeability Characteristics of Red Clays From DSDP Hole 576A

Variable	Mean	Minimum	Maximum
OCR	2.02	0.36	10.11
Cc	1.74	0.30	3.08
Rc	0.16	0.01	0.34
Permeability (cm/sec)			
Direct	3.76E-7	6.60E-8	1.30E-5
Indirect	2.34E-7	6.20E-9	6.48E-5

increase of permeability at 53 m sub-bottom. Two data sets of permeability are illustrated in Figure 30; one computed from consolidation testing and the other from direct measurement. The direct measurement of permeability shows a consistently higher value than that of the indirect measurement. This difference may be a result of the limitation of Terzaghi's (1943)

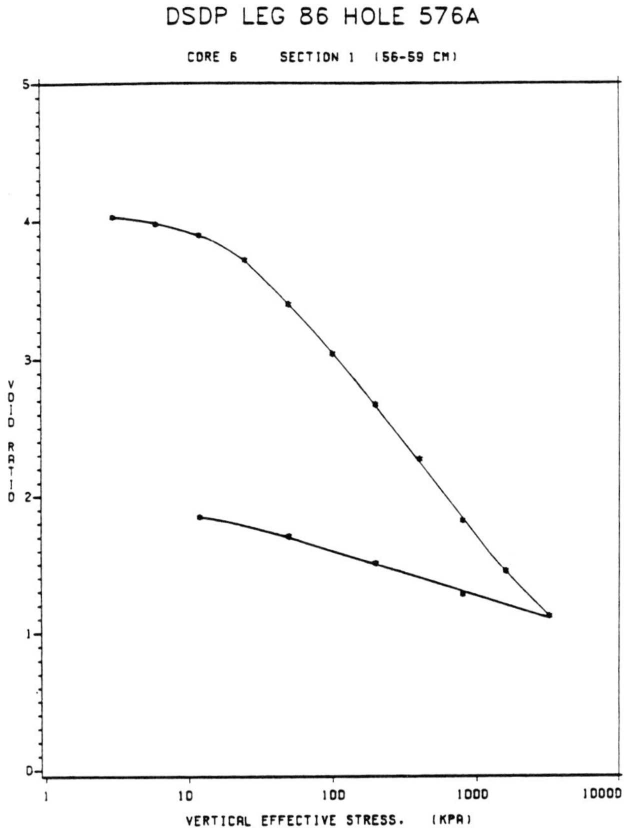

FIGURE 26. Void ratios vs. effective vertical pressure (e-log P′) curve for an intentionally remolded red clay sample from DSDP Hole 576A.

theory for consolidation, an increased flow rate caused by seepage-induced consolidation during the direct test, or a combination of both. Though the absolute values differ, the profiles obtained from the two methods are parallel.

IV. DISCUSSION

A. GEOTECHNICAL STRATIGRAPHY

The fundamental hypothesis of this study is that geotechnical properties of a sedimentary sequence can be used to define that sequence's stratigraphy in a new way. As a corollary, the new method produces a stratigraphy in agreement with standard techniques such as lithostratigraphy or biostratigraphy, and also adds unique information which aids in the geologic interpretation of the sequence.

The term "geotechnical stratigraphy" was introduced by Taylor (1984) as the logical name for this new concept. The idea behind the concept is to define those measures that define sedimentary units having similar geotechnical characteristics and to establish the position of such units.

One distinctive point regarding the data presented for the study sites is that most of these are dependent on burial depth, or consolidation. Parameters with this dependency are listed in Table 5. These are some of the tools used for defining the geotechnical stratigraphy. In essence, these data reflect sediment reactions to the applied overburden resulting from the overlying sedimentary column. Changes in the gradients of the properties with depth

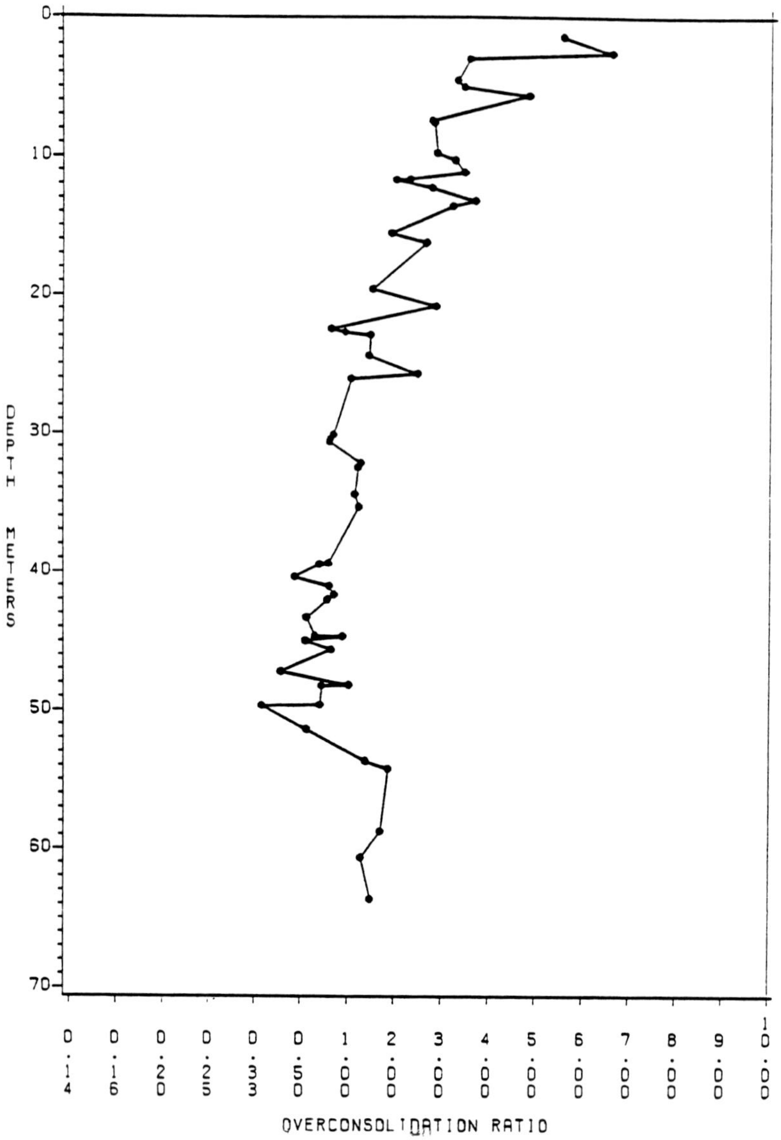

FIGURE 27. Overconsolidation ratio (OCR) profile for sediments of DSDP Hole 576A.

may be indicative of sediments with different mechanical behavior and/or with a different sedimentary history. Geotechnical stratigraphic boundaries are defined by offsets or inflections in the gradients of geotechnical properties vs. depth functions.

A second technique used to locate geotechnical boundaries is to analyze parameters that are independent of depth or overburden. The data employed for this approach consist of overconsolidation ratios, Su/Po′ ratios, Atterberg limits, expansion indices, and compression indices. Data with similar values, or clusters, reflect sediments having distinct characteristics. Sedimentary units with similar depth-independent characteristics represent common inherent properties of the sediment. The stratigraphic division defined using depth independent measurements does not detract from the geotechnical stratigraphic classification defined by depth-dependent parameters. However, using only the depth-independent measurements can present difficulties for defining successive or sequential geotechnical units.

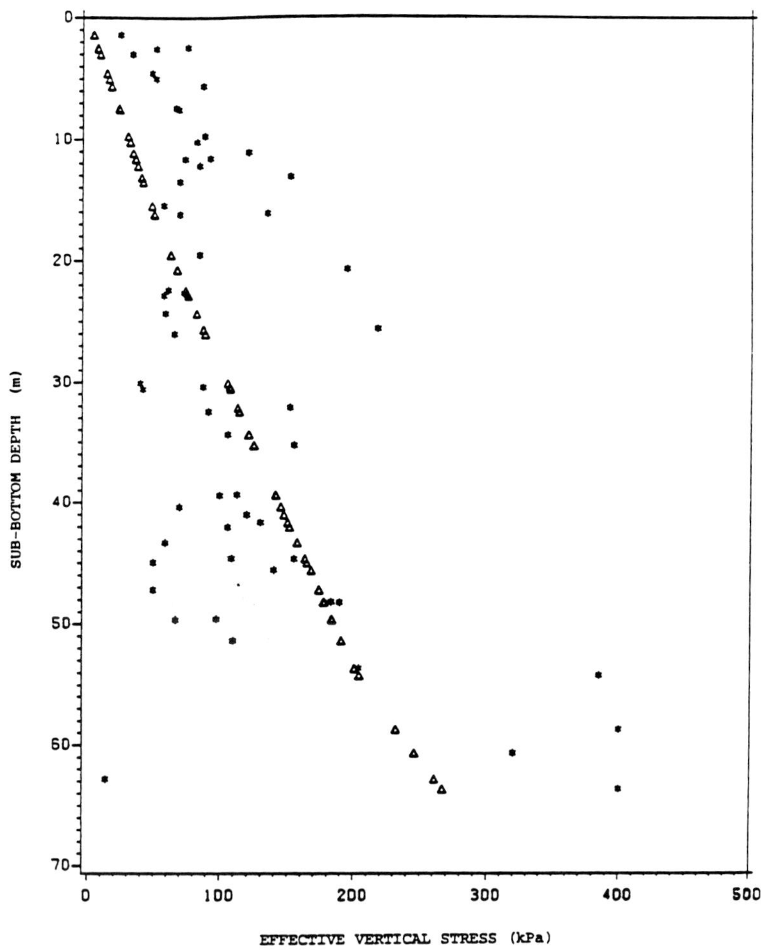

FIGURE 28. Vertical profile of effective overburden (Po′, Δ) and pre-consolidation pressures (Pc′, *) for sediments from DSDP Hole 576A.

B. GEOTECHNICAL STRATIGRAPHY OF SITE 532

The vertical zonation of the geotechnical properties for sediments from Hole 532A was applied to depth curves for several geotechnical parameters. The first division of stratigraphic units was obtained using changes in trends. Boundaries for distinct zones are separated by reversals of trends, inflections of a particular trend or by an apparent discontinuity of properties (Figure 31).

Perhaps the most sensitive and, at the same time, abundant measurement of geotechnical properties for these sediments is water content. The profile of water content at Hole 532A shows numerous trends having well defined breaks. Figure 32 depicts this variation of water content with depth and also shows the boundaries of selected zonations. Zones 1, 2, 3, 5, 8, and 9 in this figure show strong decreases of water content with depth. Their contacts are marked either by a reversal in trend of water content (between zones 1-2, 3-4, 5-6, 9-10), an inflection, or an offset in the values of water content (between zones 8-9). Zone 7 is characterized by a highly variable water content possessing no discernible trend.

The same technique used to define the zonation of the sedimentary sequence on the basis of water contents for Hole 532A sediments was applied to a number of geotechnical parameters. These include both depth dependent and independent parameters. The fact that different geotechnical properties present diverse sensitivities to sedimentological conditions

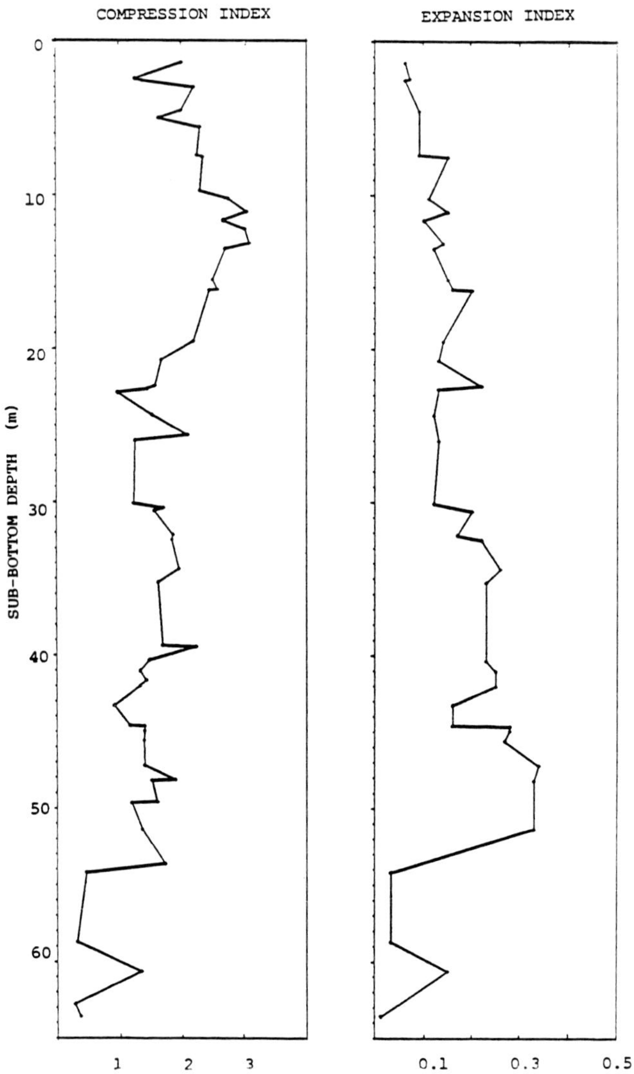

FIGURE 29. Elastic characteristics of sediments from DSDP Hole 576A.

results in distinct zonations for each parameter. On the basis of water content there were 11 discernible units in Hole 532A. Only four units were distinguished for OCR values. In addition to variations in the sensitivity of the variables to changes in the sedimentary sequence, there are also variations in the sampling density with respect to depth among the various measurements. These differences inevitably lead to variations in the number of units defined for each variable.

Following the determination of the zonation of these hemipelagic sediments for each geotechnical parameter, correlations were made between the distinct properties to determine common geotechnical stratigraphic units. Boundaries identified from the various geotechnical measurements are listed in Table 6. The correlation developed from all of these zonations was then used as the source for constructing stratigraphic units. The degree to which particular boundaries are significant was estimated from the number of parameters defining the same horizons. For instance, five parameters define the boundaries at 35, 90, 115, 135 and 155 m sub-bottom, whereas the boundaries at 50 and 145 m were only distinguished by one

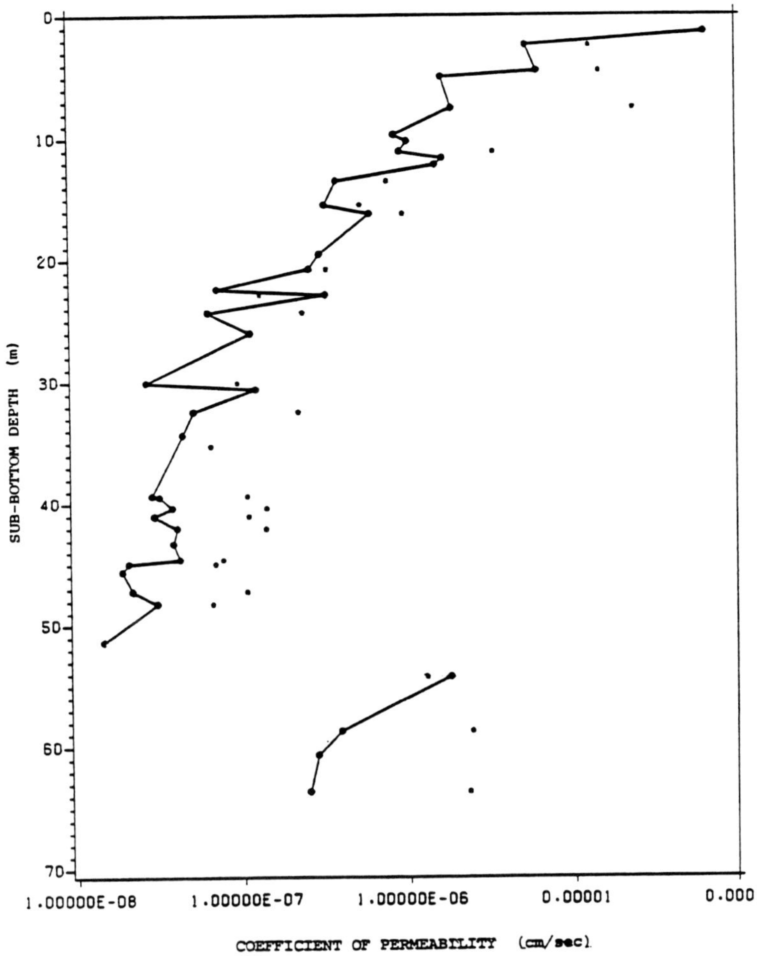

FIGURE 30. Vertical profiles of coefficients of permeability for sediments from DSDP Hole 576A. Direct measures of permeability (unconnected) and those calculated from consolidation test results (connected) are shown.

TABLE 5
Depth Dependent Parameters

Parameter	Group
Water content	Index properties
Bulk density	Index properties
Porosity	Index properties
Void ratio	Index properties
Coefficient of permeability	
Shear strength	

variable, shear strength. Those boundaries identified by five or more parameters are given greater geotechnical significance than others.

Primary boundaries may be defined as those horizons having the greatest number of independently identified geotechnical contrasts. Successively less confidence in the significance of a boundary is given as the number of identifiers diminishes. Figure 33 shows the zonations resulting from this analysis. The number of geotechnical units increases from six, identified by five of nine parameters, to ten units identified by three out of nine measures. The maximum number of units distinguished at this hole is 11 based on water content alone.

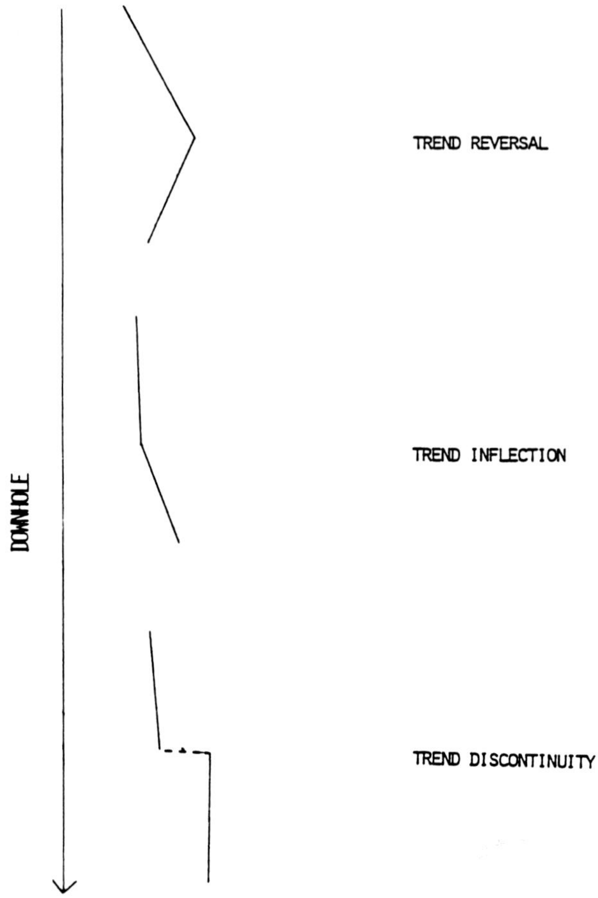

FIGURE 31. Criteria used in defining zones of geotechnical properties with distinct downhole trends.

The critical point is, of course, the coincidence of boundary position for these units as identified independently by the different geotechnical measures.

C. GEOTECHNICAL STRATIGRAPHY OF SITE 576

Ten geotechnical parameters were used to determine the geotechnical stratigraphy of the sediment sequence of Site 576. The zones were selected for each property independently of the others. The depths for some boundaries may have been loosely confined but are listed, within their uncertainty, to the same point defined by other properties having greater coverage. Zonations for Hole 576A sediments varied from a high of ten, using water contents as a single parameter, to a low of four using either OCR values, preconsolidation stresses, or permeabilities alone.

The boundaries selected for the possible geotechnical stratigraphic units are listed in Table 7. The boundary between the red clay and carbonate ooze, at 53 m sub-bottom, was identified by all geotechnical parameters. Coincidence of this lithologic boundary with the geotechnical boundary demonstrates the complimentary nature of the methods. Other strong boundaries were found at 28 and 37 m sub-bottom, as indicated by seven of ten possible parameters. Figure 34 shows the hierarchy and zonation of geotechnical stratigraphic units for Hole 576A. The number of geotechnical units increases from two, defined by all geotechnical properties, to ten identified by two only variables.

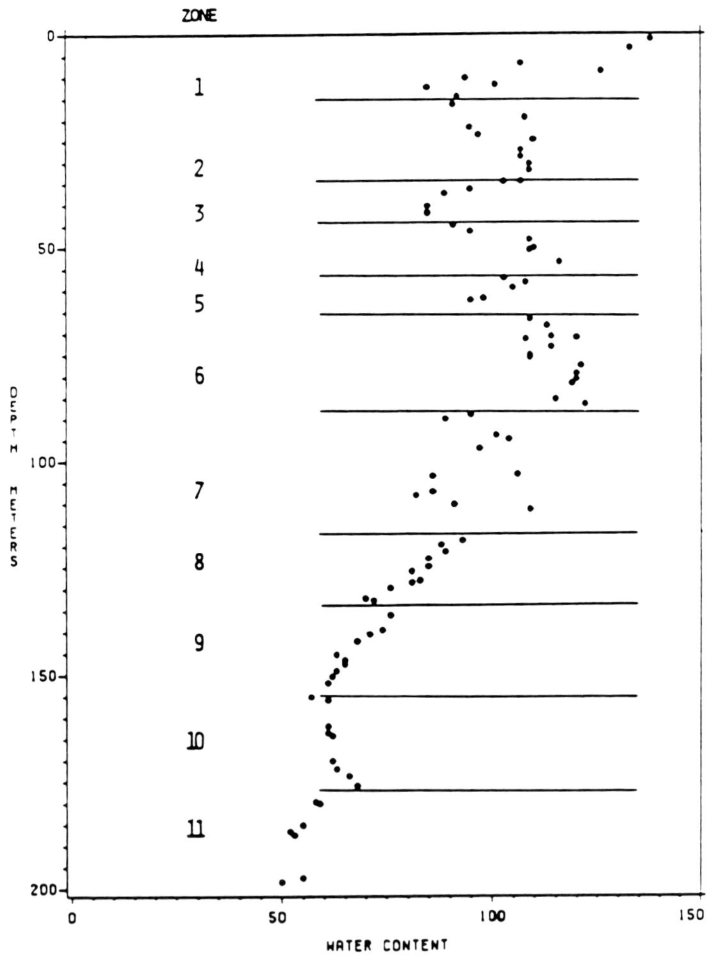

FIGURE 32. Zonation of water content data for DSDP Hole 532A sediments using changes in downhole trends to define boundaries.

D. STRATIGRAPHIC RELATIONSHIPS

A test of the use of geotechnical stratigraphy as an interpretive tool for understanding sedimentary sequences can be made by comparing this stratigraphy with conventional stratigraphic descriptions. The stratigraphic comparisons tested include litho-, bio-, magneto-, and seismic stratigraphies. Most details of these latter stratigraphies are results of analysis performed on cores from adjacent holes because the cores used in this study were dedicated for geotechnical studies and not available for standard analysis.

1. Geotechnical Stratigraphic Interrelationships

The data used for between-hole comparisons of geotechnical stratigraphies is that obtained from standard DSDP procedures from abroad the Glomar Challenger (Boyce, 1984; Schultheiss, 1985). Boyce (1984) performed the physical properties measurements on sediments from the Walvis Ridge area. Using these data, the geotechnical stratigraphic zonation of Hole 532 was defined. Two boundaries, indicated by decreases of porosity, occur at 49 to 50 and 175 m sub-bottom. Boundaries marked by increases of porosity occur at the 90 and 115 m levels. These distinctive contacts are correlated to the two principal boundaries between geotechnical units II-III and III-IV in Hole 532A. The decreases of porosity mark

TABLE 6
Geotechnical Contacts Marking Zones for Different Parameters Measured From DSDP Hole 532A Samples

Depth to contact (m sub-bottom)	Presence (●) or Absence (○) of Contact*									
	w%	n%	rho	Su	Su/Po′	Cc	OCR	k	Rc	Total
10	○	○	○	○	○	○	●	○	○	1
15	●	●	●	○	●	○	○	○	○	4
35	●	●	●	○	●	○	○	●	○	5
45	●	○	●	○	○	○	○	○	●	3
50	○	○	○	●	○	○	○	○	○	1
55	●	○	●	○	●	●	○	○	○	4
58	○	○	○	○	○	○	●	●	○	2
65	●	○	○	○	●	○	○	○	○	2
90	●	●	●	○	○	○	○	●	●	5
95	○	○	○	●	●	○	○	○	○	2
115	●	●	●	●	○	●	○	○	○	5
135	●	○	●	○	●	●	○	○	●	5
145	○	○	○	●	○	○	○	○	○	1
155	●	●	○	○	●	○	●	●	○	5
162	○	○	○	○	○	○	○	○	●	1
175	●	○	●	○	●	○	○	○	○	3
Resolved	10	5	8	4	8	3	3	4	4	16

*w% = water content; n% = porosity; rho = bulk density; su = undrained shear strength; others as discussed in text.

the levels described between units IIA-1 and 2 and between units VI-1 and-2 at Hole 532A. The further sub-division of units at Hole 532 may show additional correlations with Hole 532A; however, data produced from shipboard analysis were scarce and lacked sufficient definition to enable more contacts to be identified. The existence and intercorrelation of the identified horizons strongly supports the idea of their reality. That is, the geotechnical properties are not induced by disturbance involved in coring. The equal depth of boundaries also confirms the lateral continuity of these properties, enabling direct lateral correlation of the geotechnical stratigraphy of Hole 532A to the stratigraphic analyses from the sister holes.

The physical properties of the red clay section were determined aboard the ship by Schultheiss (1985). Profiles of vane shear strength, bulk density, water content, porosity, and GRAPE bulk density were used to define the geotechnical stratigraphy of sediments from Holes 576 and 576B. Significant changes in these trends occur at depths of 3, 12, 30, 44, 51, and 55 m sub-bottom. The very sharp contact at 55 m corresponds to the occurrence of carbonate ooze and is the equivalent of the 53 m boundary between geotechnical units I and II identified in Hole 576A. The geotechnical stratigraphy determined by using shipboard data is shown in Figure 35. Using the same criteria as that used for Hole 576A, zonations were defined and listed. Four major geotechnical units are identified at Holes 576 and 576B, as opposed to the two defined for Hole 576A. The division between units III and IV at Hole 576 corresponds to the unit I-II break at Hole 576A.

The approximate 2 m difference between the sub-bottom position of this latter boundary and the one at approximately 53 m is probably a result of difficulties encountered during coring operations on Holes 576 and 576B. These two holes had some flow-in intervals making it difficult to accurately determine the true depth of penetration. Comparison of the trends of water content profiles between Holes 576A and 576 show common boundaries occurring in Hole 576A at 28, 41, 49, and 53 m.

The geotechnical stratigraphic boundaries identified by both data sets are listed in Table 8. Some matches are made for zones in Hole 576A defined using the two parameter criteria as shown in the right-hand column of Figure 34. Again, the vertical density of data coverage

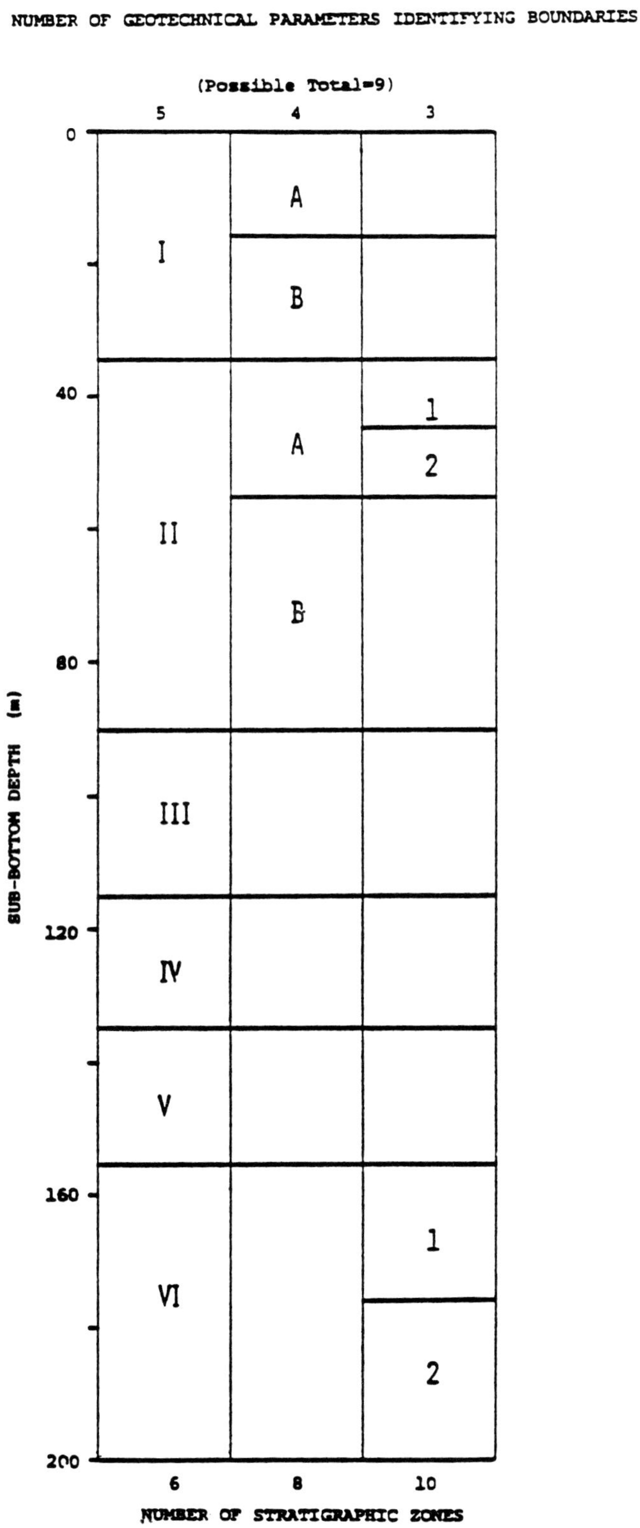

FIGURE 33. Geotechnical stratigraphic sub-division of DSDP Hole 532A sediments.

TABLE 7
Geotechnical Contacts Marking Zones for Different Parameters Measured From DSDP Hole 576A Samples

Depth to contact (m sub-bottom)	Presence (●) or absence (○) of contact*										
	w%	rho	LL	Su	Su/Po′	Pc′	OCR	Cc	Rc	k	Total
10	●	●	●	●	○	○	●	○	○	○	5
14	○	○	○	○	●	○	○	●	○	○	2
17	○	○	○	○	○	○	○	○	●	○	1
20	●	○	●	○	○	●	○	○	○	○	3
28	●	●	○	●	●	○	○	●	●	●	7
30	○	○	●	○	○	○	○	○	○	○	1
35	●	●	○	○	○	○	○	○	○	○	2
37	●	●	○	●	●	●	●	○	●	○	7
41	●	○	○	○	○	○	○	●	○	○	2
44	●	●	○	○	○	○	○	○	○	○	2
49	●	○	●	○	○	○	○	○	○	○	2
53	●	●	●	●	●	●	●	●	●	●	10
Resolved	9	6	5	4	4	3	3	4	4	2	12

*see footnote to Table 6 for abbreviations; LL = liquid limit.

at a particular site and the sensitivity of the measurements create differences in the stratigraphic definition. Nevertheless, comparison of the total data set from the red clay section strongly supports the presence of at least four stratigraphic units and suggests a consistent 2 m offset between adjacent holes.

E. CONVENTIONAL STRATIGRAPHIC SUMMARIES

1. Site 532

Analyses of planktonic foraminifera and nannofossils were used to construct the bio- and chronostratigraphy of sediments from the Walvis Ridge area. The resultant biostratigraphic description was obtained from the site report for Leg 75 (Hay, Sibuet et al., 1984). The section shown in Figure 36 illustrates the geochronostratigraphy of these sediments. This sections spans from Holocene to Pliocene with relative contacts between epochs shown as intervals of uncertainty rather than discrete lines.

Lithostratigraphic zonation was defined by shipboard visual analysis of sediment composition. As indicated earlier, three units were identified, as shown in Figure 36. Unit I consists of a foram-nannofossil marl and ooze extending from the mudline to 50 m. Unit II extends from 50 to 115 m sub-bottom and is composed of diatom-nannofossil marl. The marl is underlain by Unit III which was identified as a nannofossil marl.

A seismic reflection profile from the Site 532 area was obtained aboard the Glomar Challenger using a 15-in.^{3} water gun. This provided high resolution of the upper sedimentary sequence which correlated well with one obtained by the Bundesanstalt fur Geowisenschaften und Rohstoffe shown in Figure 37. Typical seismic stratigraphic units are not present on this scale. However, two reflectors fall within the depth range of Hole 532A. The depths to reflectors were calculated using shipboard velocity measurements and are shown in Figure 36 occurring at sub-bottom depths of 90 and 160 m.

2. Site 576

The summary of stratigraphic sub-divisions of sediments from the red clay study area is shown in Figure 38. Biostratigraphic analyses were performed using a few radiolarian remains (Morely, 1985) and ichthyolith, or fish debris, data (Doyle and Riedel, 1985). The resultant geochronography indicates sediments of Quaternary age near the mudline and

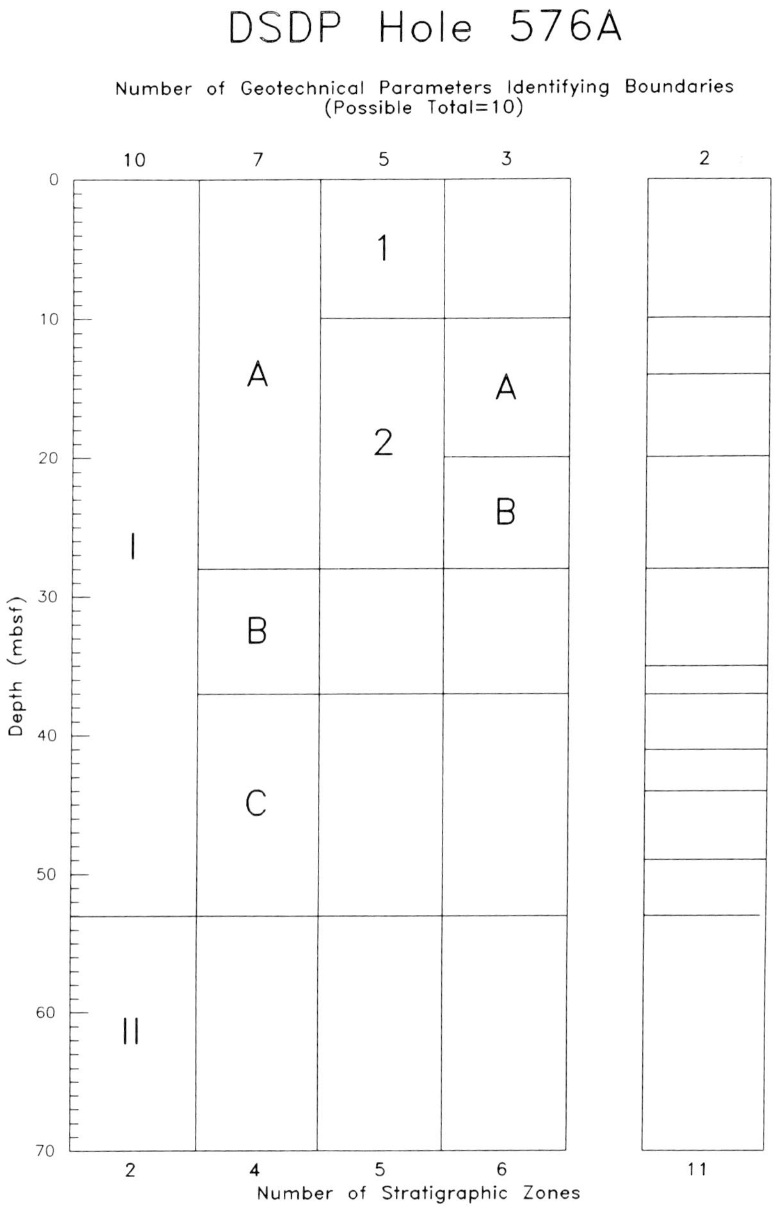

FIGURE 34. Geotechnical stratigraphic sub-division of DSDP Hole 576A sediments.

sediments of Cretaceous age (Maastrichtian-Upper Campanian) at the base of the hole. Doyle and Riedel (1985) suggest that a better method for establishing the chronostratigraphy is to define probablistic biostratigraphic boundaries, such as shown in the second column of Figure 38. The zones shown here represent the depth interval in which a particular boundary is likely to be encountered. Currently, lack of a sufficiently large ichthyolith data set precludes refining these boundaries.

A comprehensive magnetic study was performed on sediments from Site 576 by Heath et al. (1985). Lack of sufficient magnetic signature in sediments from below 27 m resulted in limiting the magnetostratigraphy to the upper portion of the sediment column. Column 3 in Figure 38 shows boundaries between the principal magnetic zones with the lower end of the Gilbert epoch measured at approximately 25 m sub-bottom.

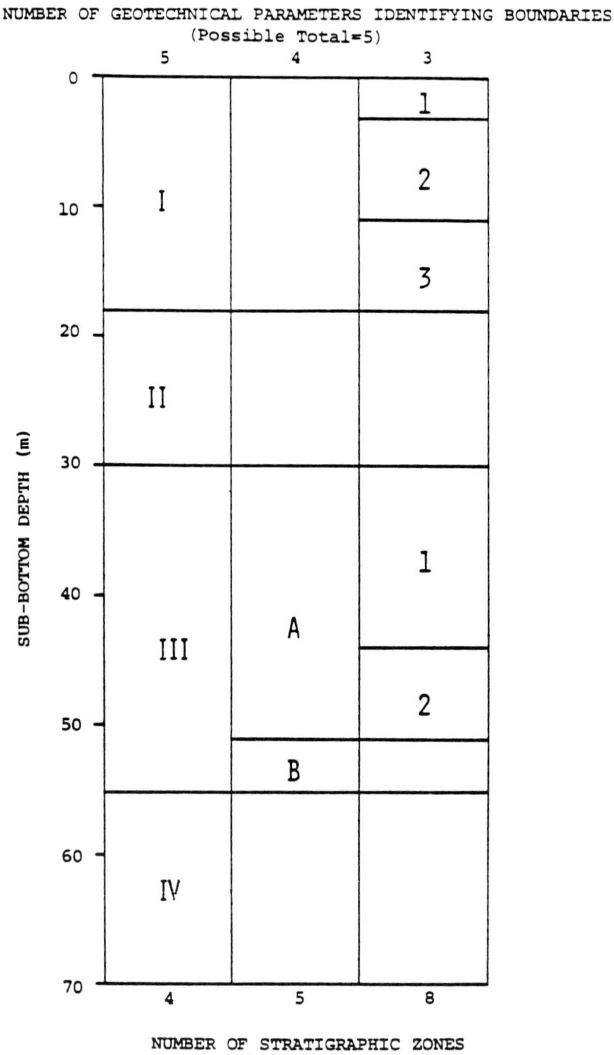

FIGURE 35. Geotechnical stratigraphic sub-division of DSDP Holes 576 and 576B sediments.

Seismic stratigraphic analysis for a wide area around Site 576 was performed by Jacobi et al. (1985). Using 3.5 kHz echograms, they define three principal seismic units (Figure 38). Seismic unit I consists of continuous, multiple, sharp and parallel reflectors. This unit extends from the sea floor to R2a (at 26 m sub-bottom). An intermediate reflector, R1, believed to correspond to an ash layer occurs at a depth of 12 m. Other weak reflectors are found in this unit at depths near 16 m. Unit II extends from R2 to R3 (top of carbonate ooze) and is characterized as a thick, transparent section with discontinuous, sharp to slightly prolonged, transparent reflectors. One reflector occurs about two thirds of the way down Unit II and is discontinuous. Finally, Unit III has its upper limit at reflector R3, corresponding to the appearance of carbonate ooze. This unit has a composite echo type depending on the location of seafloor relative to abyssal hill region. Reflectors within this unit are smooth and strong.

The lithostratigraphic sub-division of sediments of the northwest Pacific site was based on visual classification of sediments, microscopic analysis and some mineralogical studies.

TABLE 8
Equivalent Geotechnical Boundaries of all Site 576 Measurements

Hole 576A		Holes 576 and 576B	
Depth (mbsf)	Units	Depth (mbsf)	Units
10	IA—1 and 2	11	I—2 and 3
28	IA and B	30	II and III
41	unspecified	44	IIIA—1 and 2
49	unspecified	51	IIIA and B
53	I and II	55	III and IV

Two principal units were identified by the DSDP Leg 86 shipboard scientific party (Figure 38). The upper unit, extending from the mudline to approximately 55 m sub-bottom, represents the red clay section. Unit II underlies this previous unit and is composed of interbedded red clay and nannofossil ooze becoming strictly ooze with depth. Unit I was further subdivided into parts A and B. Subunit IA is a pelagic, yellowish brown, burrowed clay whereas subunit IB is composed of a dark brown, slick homogeneous pelagic clay. The contact between the two subunits is gradational, occurring around a sub-bottom depth of 28 m.

F. CORRELATION OF STRATIGRAPHIC UNITS

Definition of the geotechnical stratigraphy for the two sedimentary columns in this study is based on interpretation of the changing character of geotechnical properties with depth. The distribution of geotechnical properties with depth can be variable, and the degree to which this variation is a function of measurement error or a trait of the sedimentary units is difficult to establish. Geotechnical measurements generally result in the destruction in the original characteristics of the sediment which makes repetitive measurements impossible. Some instances of repeated measures were made for water contents and variability arising from these was limited to less than 5%. The distinction of sediment units based on geotechnical properties must be sufficiently strong so that measurement variability is not the cause for selecting different boundaries. Several approaches to testing the validity of the hypothesis of geotechnical stratigraphy can be made in an effort to distinguish real geotechnical contacts.

One test of the boundaries selected for the stratigraphic division of the study site is the duplication of boundaries identified by independent geotechnical measures. As discussed previously, numerous contacts were identified by separate variables. Two other approaches to testing the validity of the hypothesis are to first test the consistency of geotechnical stratigraphic definition with standard techniques and, second, to discern whether this analysis adds knowledge to the interpretation of a sedimentary sequence. These two tests are presented in continuation.

1. Site 532

Four stratigraphic schemes measuring distinct properties of sediments from the Walvis Ridge hemipelagic sequence are shown in Figure 39. The correlation of boundaries for the different stratigraphies is also noted in the figure, illustrating those stratigraphies whose measured parameters are sensitive to common processes controlling the sediment deposit. Four geotechnical stratigraphic boundaries are found to correlate with other contacts. The lithostratigraphic contact between Unit I and II and the biostratigraphic contact marking the lower-upper Pleistocene bear a prominent correlation with the IIA-1 and -2 boundary of

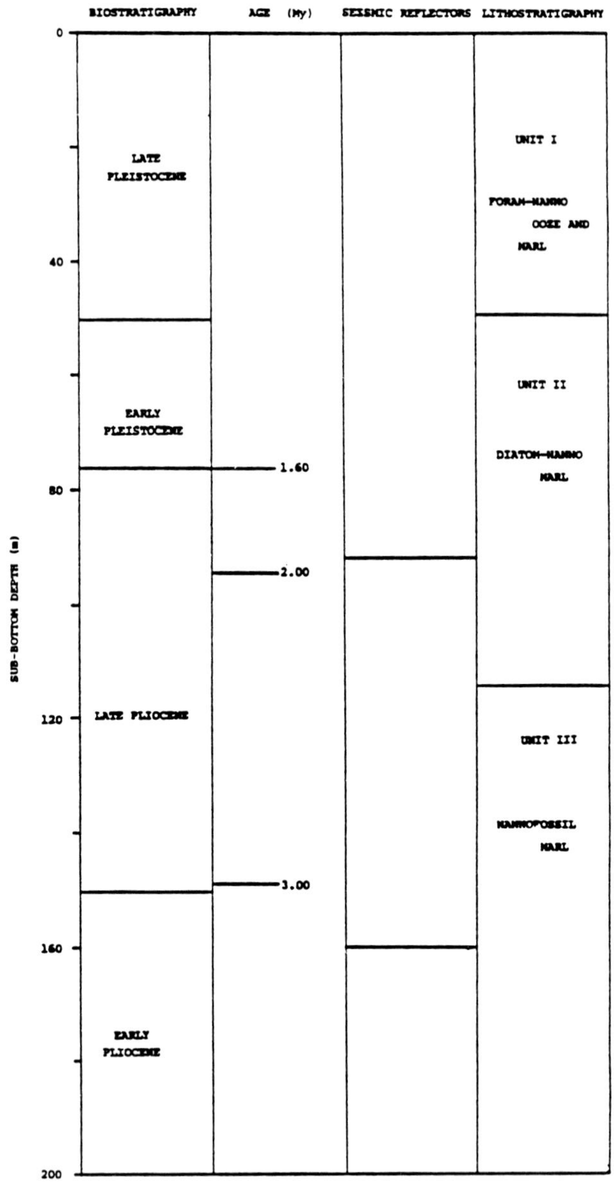

FIGURE 36. Stratigraphic methods and sub-divisions of sediments recovered from DSDP Site 532.

geotechnical parameters. A correlation is made between the contact of geotechnical units II and III with seismic reflector R1.

Lithostratigraphic units II and III are marked by a contact reflected in the geotechnical record by the boundary between units III and IV. Another strong correlation is found between geotechnical, bio-chronographic, and seismic stratigraphies at approximately 150 m sub-bottom. Here reflector R2, the lower-upper Pliocene boundary, and the contact of geotechnical units V and VI appear to have responded to a common sedimentary condition.

The correlation of four out of nine contacts in the geotechnical stratigraphy with other stratigraphies suggests that the former is indeed sensitive to primary sedimentary conditions. A more interesting point, however, is that all other stratigraphic boundaries, with the ex-

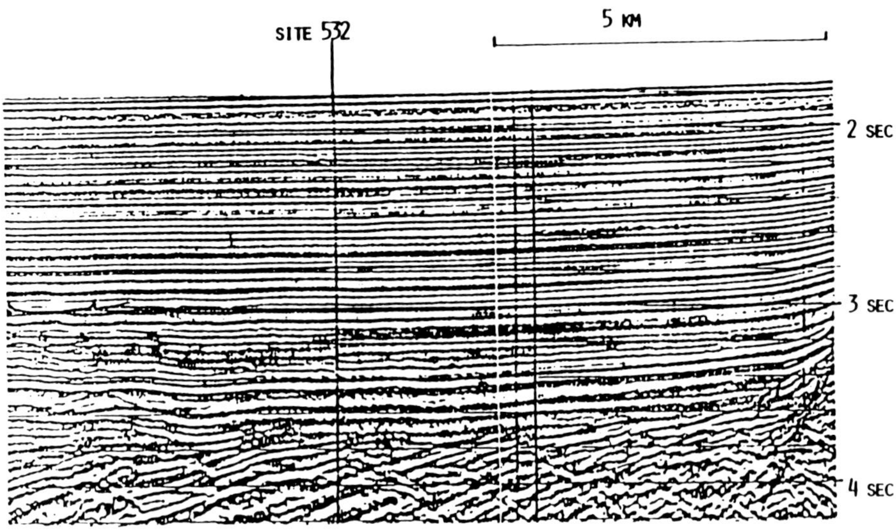

FIGURE 37. Processed seismic line BGR-36 and location of Site 532 (from Hay, Sibuet et al., *Initial Reports of DSDM,* U.S. Govt. Printing Office, Washington, D.C., 75(1), 1984.)

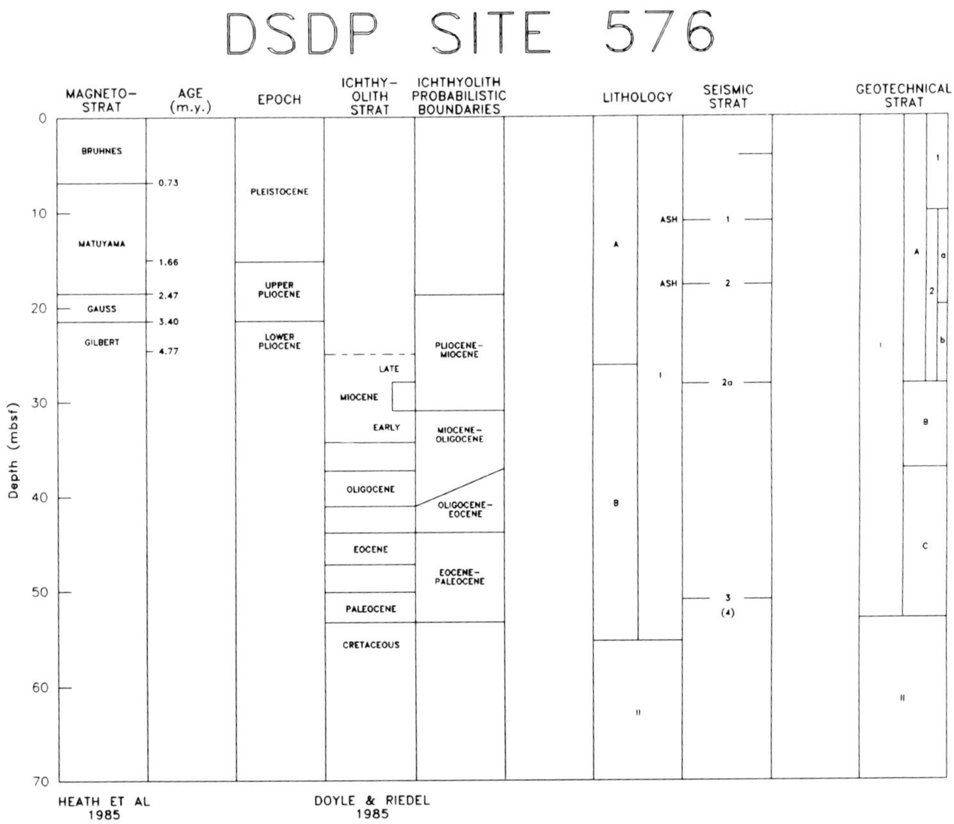

FIGURE 38. Stratigraphic methods and sub-divisions of sediments recovered from DSDP Site 576. Age markers are estimated from both magnetic and ichthyolith stratigraphic boundaries.

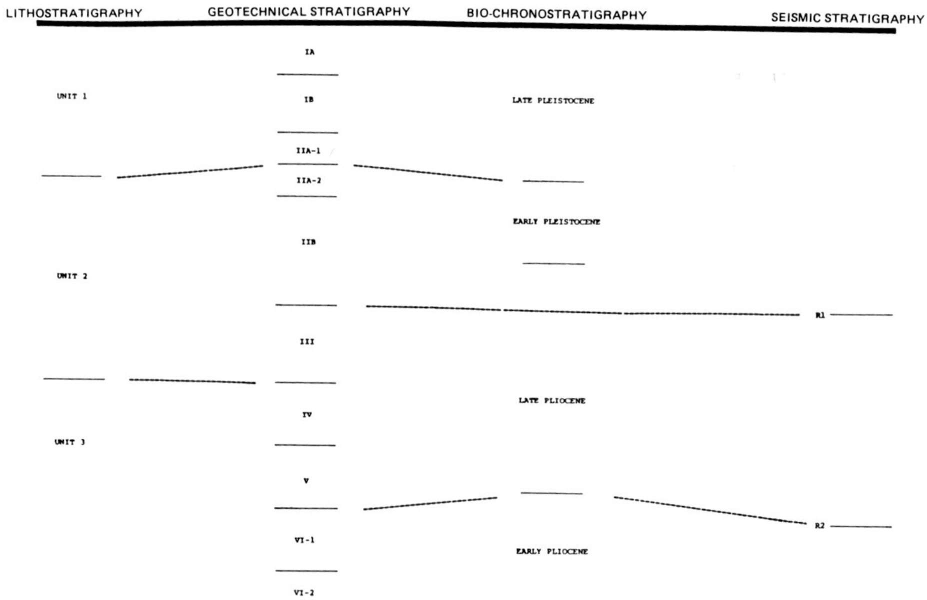

FIGURE 39. Correlations of conventional and geotechnical stratigraphies defined for the hemipelagic carbonate sequence at DSDP Site 532.

ception of the lower Pleistocene-upper Pliocene contact, are related to events in the geotechnical record. This raises the question of how sensitive other stratigraphic schemes are relative to the geotechnical data. For instance, one lithostratigraphic contact does not match at all with the other conventional stratigraphies, while one of two seismic reflectors matches nicely with a chronostratigraphic event. The geotechnical stratigraphy appears to be far more sensitive to fluctuations of sedimentary parameters forming and altering a deposit than those stratigraphies identifying either compositional or seismic properties.

2. Site 576

Five stratigraphic descriptions of Pacific red clay sediments obtained by DSDP Leg 86 and their intercorrelations are shown in Figure 40. Using the geotechnical stratigraphic scheme defined by three parameters, correlations are found with several other stratigraphic boundaries. The boundary between geotechnical units IA-1 and -2 matches closely with seismic reflector R1. The division of geotechnical units IA-2b and IB marks a dominant contact identified by reflector R2a which limits the base of seismic unit I. This contact probably corresponds to the Miocene-Pliocene boundary defined at the base of the Gilbert event in the magnetostratigraphic column, and delimited to a certain depth range by ichthyolith results.

The boundary between geotechnical units IA and IB also marks the division between lithostratigraphic units IA and IB. The IB to IC contact in the geotechnical column may be indicative of the Eocene-Oligocene boundary as suggested by the limiting depth range described in the biostratigraphic results. Finally, the strongest single boundary prevalent throughout all stratigraphic schemes is the red clay to carbonate ooze transition, forming the contact of geotechnical units I and II, seismic units II and III (below reflector R3), the Cretaceous-Paleocene boundary, and lithologic units I and II.

Incorporation of a finer scale geotechnical stratigraphic sub-division was made using two parameters to distinguish contacts. As shown in Figure 40, these boundaries do have their counterparts from other stratigraphic techniques. This means that stratigraphic changes

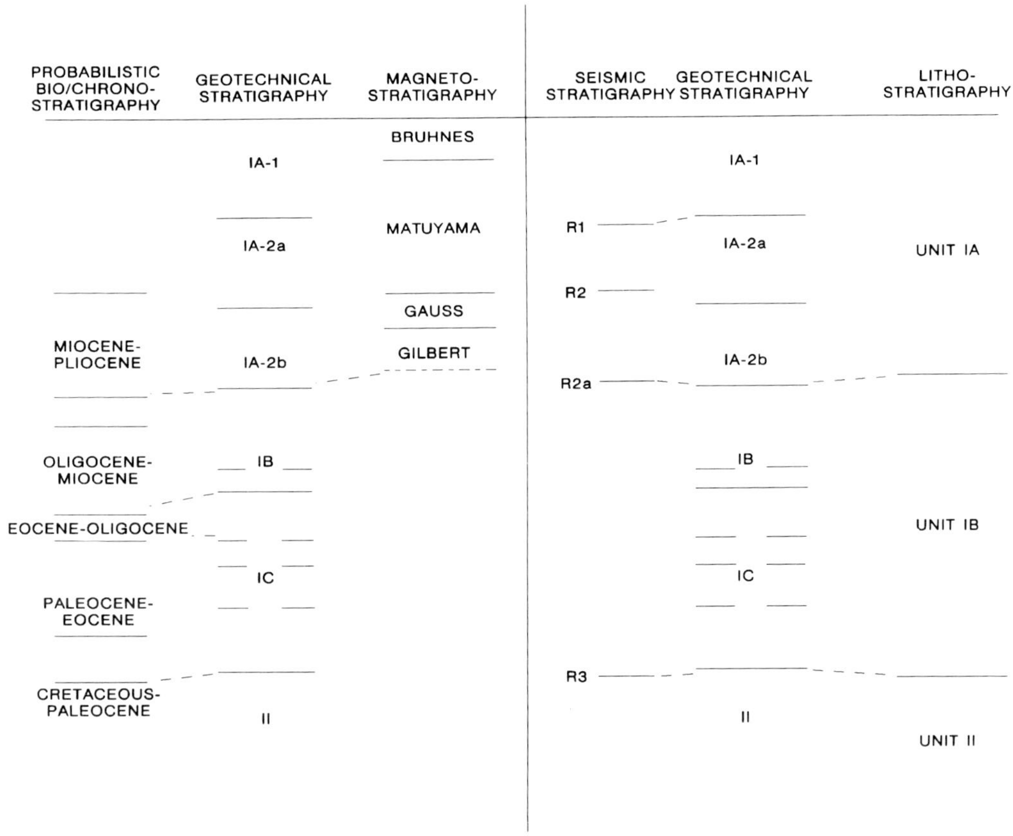

FIGURE 40. Correlations of conventional and geotechnical stratigraphies defined for the red clay sequence at DSDP Site 576.

indicated by the sensitive geotechnical measures should be carefully analyzed for their information content. For instance, seismic reflectors are correlated with geotechnical boundaries. Furthermore, the depth ranges of bio- and chronostratigraphic contacts between Oligocene and Miocene, and the Paleocene-Eocene are correlated with breaks in the geotechnical stratigraphy using this finer scale of resolution. The only geotechnical stratigraphic boundary uncorrelated with conventional stratigraphies is the one at 49 m sub-bottom. This contact is present in both geotechnical stratigraphies for Site 576 (Hole 576A and Holes 576-576B) which suggests that it represents a real event. However, it is not accounted for by other stratigraphic techniques. As for Site 532, the geotechnical stratigraphy is remarkably consistent with that derived from standard stratigraphic techniques.

G. GEOTECHNICAL STRATIGRAPHY AND SEDIMENTATION

Sediments as studied through stratigraphic analysis provided us with a sequence of events forming the geologic record. Depositional and post-depositional processes form and alter a sedimentary deposit. A change in a sedimentary process will often result in a change in sediment characteristics, although the effects range from the imperceptible to a drastically altered sediment section. Thus, the stratigraphic description of any qualitative or quantitative sediment characteristic reflects the history of the sediment as recorded by that particular aspect. It is important to note, though, that not all sediment properties are sensitive to the full spectrum of variations occurring in the sedimentary processes. Hence this results in the diversity of different stratigraphic units and lack of perfect correlation between different stratigraphies.

Geotechnical properties often allow discrimination between the depositional and post-depositional processes to which a sedimentary unit is subjected. Depositional fluctuations can include changes of sediment source and/or the process of sediment emplacement or rate of accumulation. Post-depositional effects are those such as compaction, deformation, erosion, or diagenesis. The combination of these diverse conditions altering the sediment record is generally registered in sediment composition, structure, geotechnical and other related properties. An approach to understanding the interaction of sedimentary processes and the geotechnical record is to study changes of sediment characteristics across the defined stratigraphic contacts and to identify the processes causing those changes.

H. SITE 532: HEMIPELAGIC CARBONATES

Sedimentation at Site 532 on the flank of the Walvis Ridge off Southwest Africa is dominated by the biogenic supply of mainly calcareous and some siliceous remains of plankton. A supplemental input is provided by a terrigenous component derived from shelf material which reaches the flank of this ridge either as pelagic detrital rain or, to a lesser degree, from downslope transport mechanisms such as turbidity currents. The terrigenous input to the sediments varies and is responsible for the major portion of fluctuation of carbonate with depth (Gardiner et al. 1984). Variation in sediment composition at Site 532 is due to variation of the dominating planktonic organism (i.e. coccolithophore, foraminifera, or diatom), dissolution rates of biogenic material, variability of terrigenous supply, and variability in preservation of organic matter. The principal variations of these factors are reflected in the lithostratigraphic units and their correlated geotechnical units. It is noticeable, however, that not all geotechnical stratigraphic boundaries are correlated with contacts of lithologic units.

The scale at which distinction of lithologic units were defined is somewhat broad when compared to the manner in which the geotechnical stratigraphy was defined. Finer scale variations of lithologic composition may indeed contribute to the behavior of geotechnical properties. With this in mind, geotechnical stratigraphic contacts were matched with variations of lithologic components including sand/silt, carbonate, and organic carbon contents. The resulting comparison showed that all contacts of geotechnical units are related to compositional shifts, though not necessarily to all aspects of composition. The exception to this correlation was the contact at 135 m sub-bottom. The sensitivity of geotechnical properties to lithology has been pointed out by numerous researchers in the field of marine geotechnique, and this study simply confirms the importance of this aspect of sedimentation on geotechnical properties.

The relationship of lithologic variations to the change in geotechnical properties is not simple. Both the 35 and 55 m boundaries show decreases of water contents (25 and 15%, respectively) (Figure 9). The 35 m boundary is marked by a 16% decrease of sand and silt, and the 55 m boundary is marked by a 20% decrease of carbonate and a minor (4%) decrease of silt and sand (Figure 7). Comparing these two events to the 175 m change of properties illustrates the probable interplay of lithology and sedimentary processes that affect these boundaries. At 175 m, water contents decrease 10%, yet carbonate content increases 28% and sand/silt 2%. Here, water content shows the opposite response to compositional changes as that seen at 55 m. Another example of the diverse relationship of this geotechnical and lithologic characteristic is the increase of OCR values associated with the 55 and 175 m horizons (Figure 14) even though these two levels are marked by the lithologic changes already noted. The OCR values suggest that the boundary may be due to an event such as erosion, or a sedimentation hiatus, leaving a contact with a relatively overconsolidated section of underlying material compared to the non-eroded cover.

Reflectors R1 and R2 are believed to correspond to erosional surfaces or depositional hiatuses at Site 532 at 90 and 155 m sub-bottom. The 90 m event was identified as corresponding to a depositional hiatus indicated by breaks in the biostratigraphic record. Relatively abrupt downhole decreases in the percent of silt correspond to these hiatuses, as well as a sharp decrease of organic carbon content at the 90 m break (Figure 7). Water content exhibits a sharp decrease at 90 m, yet no distinct shift is associated with the 155 m event (Figure 9). Similar comparisons of other geotechnical properties at these two contacts indicate a stronger response to the 90 m event than to the one at 155 m. This may suggest that the hiatus at 90 m is indeed erosional, having resulted in removal of overburden and leaving the distinct shift in physical properties. The apparent hiatus at 155 m may be either a result of a very mild erosional event, erosion followed by slow accumulation and/or winnowing, or simply an episode of non-deposition. Such processes would result in the shift to lower OCR values as opposed to the increased OCR values of an eroded section such as at the 90 meter contact. The lack of significant inflections in geotechnical properties across the 155 m boundary coupled with the decrease of OCR can be created by a period of non-deposition followed by a period of relatively rapid sediment accumulation. Another possible reason for the observed swing in OCR values at this depth is that they are a result of coring induced disturbance, or disturbance related to degassing of sediments during the retrieval process (Marine Geotechnical Consortium, 1984). As mentioned earlier, however, samples were selected in an attempt to keep effects of disturbance to a minimum and disturbance was not detected in X-radiographs.

The boundary at 90 m is also associated with a probable change in surface water biota, evident in the change of sediment accumulation rates of the different biogenic components (Hay, Sibuet et al., 1984). At approximately 2 Ma, both terrigenous and calcareous accumulations dropped significantly, with an accompanied influx of opaline material. This change of sediment supply does not necessarily reflect a change of productivity, but it does represent an exchange of diatoms for coccolithophores as primary producers in the upwelling system. This may be a result of intrusion of colder water into the Benguela Current or increased upwelling of Antarctic Intermediate Water (Siesser, 1978; Gardiner et al., 1984). Sediment properties such as composition and geotechnical aspects of the material on either side of the 90 m event are a result of the basic shift in sediment source and components.

I. SITE 576: RED CLAY

The pelagic clay environment of the Northwest Pacific has not always been a site of red clay accumulation. The presence of nannofossil ooze in the lowermost sections of the DSDP holes represents a period in time in which abundant coccolithophore production was taking place in the overlying water column. It also is indicative of the position of that portion of the seafloor relative to oceanic zones of productivity and to the depth of the carbonate compensation depth (CCD).

Crustal cooling of the oceanic lithosphere, with a resultant increase in depth, and the plate motion away from zones of high productivity placed the site in less advantageous positions for carbonate accumulation. The terminal Cretaceous event marked a major decrease in coccolithophore productivity and coincides with an extremely significant sedimentary hiatus (Kennett, 1982). This hiatus may represent an erosional event, increased bottom water circulation, increased dissolution rates of carbonates, or a combination of all of these.

The dramatic changes that took place at this time are documented by all stratigraphic studies of sediments from this site. In this sense, the geotechnical stratigraphic break at 53 m in Hole 576A represents changing sedimentary processes resulting in changing lithology as well as probable active post-depositional modification of the previously deposited material in the form of erosion. The different lithologies have such contrasting mechanical characteristics that the consolidation of these sediments with time simply enhances the other geotechnical differences between the units.

The overlying red clay section at Site 576 was sampled in an attempt to retrieve what was expected to be a very typical and normally consolidated section of fine-grained marine sediment. As indicated in the results of this and associated research, the geotechnical behavior of this sediment section does not follow this pre-conceived notion. Reasons for the discrepancy between the expected trend of geotechnical properties with depth and those found from laboratory work can be associated with the historical development of the section and processes active in this region of the Pacific.

Geotechnical characteristics of red clay sediments are a function of sediment composition and texture, and of sedimentary processes modifying the previously deposited material. The predominant origin of these sediments is eolian transported silts and clays from the Asiatic continent. The clays are primarily illite. The fluctuation of wind patterns during the Cenozoic and the migration of the Pacific plate during this same time have been the major causes of changing sediment composition, both texturally and mineralogically (Hollister et al., 1979; Leinen, 1985). The increased concentration of X-ray amorphous material in the lower red clay section has been attributed to a lower level of detrital input and concurrent authigenic sediment formation in the form of iron and manganese oxides. The upper section, however, represents an increased pulse of detrital material which Leinen and Heath (1985) match with an increase of detrital quartz incorporated in the red clays.

Hollister et al. (1979) studied geotechnical properties of some shorter cores of red clay from the Mid-Pacific Gyre region. They found a lack of variation of geotechnical properties in the red clays which they attributed to the lack of mineralogic variations within the sedimentary section. This is contradictory to the results of DSDP Site 576. The most significant mineralogic change in Site 576 red clays, based on analyses performed by Leinen (personal communication), occurs at 28 m sub-bottom, corresponding to the contact between lithostratigraphic units IA and B (Figure 38). Above this contact are the detrital, quartz rich, lighter brown clays, and below it lie the dark red clays. The geotechnical stratigraphic contact between IA and IB reflects the change in physical-mechanical properties of the sediment units above and below this depth.

There are other compositionally defined boundaries at this site. The contact at 10 m sub-bottom has been associated with seismic reflector R1 (Figure 38), and is believed to be an ash layer. Shifts of geotechnical properties across this contact probably correspond to a decrease in silt and an increase of smectite (Figure 19). Reflector R1 may simply be a return from the ash layer; however, it may also be a response to a contrast of acoustic impedence at this same level caused by an increase of bulk density. Similarly, the 44 m sub-bottom geotechnical contact can be attributed to a compositional change indicated by an abrupt decrease in smectite and increases of illite and silt (Figure 19).

The origin of acoustic reflectors in the red clay area can be linked to changes in composition or fluctuations of geotechnical properties. Variability of the geotechnical properties in siliceous oozes located to the north of Site 576A has been correlated with reflectors described by Damuth et al. (1983). The strength of these reflectors decreases to the south, however. This attenuation has been attributed to the decrease in the contrast of physical properties. The number of ash layers is also believed to decrease southward.

Biostratigraphic work in red clays is often hampered by the lack of sufficient fossiliferous material to use. The sediments at Site 576 are no exception with fish scale and teeth debris forming the only significant fossiliferous fraction. To compound the dating problem, ichthyolith research and its stratigraphy are relatively new sciences, and biostratigraphic chronology of these remains has not been refined to the scale that other biochronologies have. Time resolution is on the order of 4 to 5 million years in most of the section, and ambiguity of dated horizons is readily apparent when depth intervals for probabilistic chronostratigraphic breaks are given. The result of this gross time scale is that it becomes difficult to define boundaries marking geochronostratigraphic units. Because of this difficulty, geotechnical stratigraphic boundaries may serve as a tool to locate these contacts if the sedimentary processes creating contacts have sufficiently influenced the geotechnical record.

Damuth et al. (1983) and Jacobi et al. (1985) described sediment morphological characteristics, namely bedforms, erosional surfaces, and sediment isopachs, that clearly indicate the influence of bottom currents on sediment accumulation. The predominant oceanographic condition believed to be influencing part of this bottom flow is channelling of Pacific Bottom Water flow around topographic highs in the form of contour currents. This flow can also be augmented by tidal action, which in some areas of the Mid-Pacific abyssal plain have been measured at 5 to 10 cm/sec (Hollister et al., 1979). Keller (1969) found a pronounced association between geotechnical properties of a red clay deposit and seamount topography. He decided that the radial distribution of these properties around the seamount was a result of sediment winnowing by current action. The changes of bottom water flow in the abyssal hill area of the Northwest Pacific may also be recorded in the historical development of the section at Site 576.

Site 576 lies on a smoother seafloor than the area described by Keller (1969) or the area of modern bedforms and erosional surfaces described by Damuth et al. (1983). The setting here is not appropriate for channelized flow, and scour by bottom currents is probably not significant under present oceanographic conditions. However, historical variations in oceanographic conditions associated with tectonic and sea level variations have resulted in numerous widespread depositional hiatuses throughout the Pacific and may also have been recorded at Site 576.

Moore et al. (1978) list a number of sedimentary hiatuses occurring in cores recovered by DSDP from around the world. The timing of these hiatuses can be somewhat constrained within the red clay sequence by utilizing the geochronostratigraphic scale constructed from ichthyolith and magnetic data. Within the depth range defined by these stratigraphies, numerous occurrences of hiatuses as described by Moore et al. (1978) match with contacts of geotechnical stratigraphic intervals. Thus, the geotechnical stratigraphic breaks can refine the chronology of this sediment section and can be used to correlate local and worldwide hiatuses. The Cretaceous-Tertiary boundary, well defined at Site 576, represents one of the most important boundaries in stratigraphic records, and was addressed earlier. At approximately 54 Ma, near the Paleocene-Eocene boundary, numerous hiatuses occur within Pacific sedimentary sequences. The origin of these hiatuses is uncertain; however, its effect at Site 576 is indicated by the geotechnical contact at 44 m. It is at this depth that major fluctuations in composition also occur (Figure 19). Thus, the interplay of these various factors may all contribute to the geotechnical boundary defined at this level. A mid-Eocene maxima of hiatuses is found at 42 to 43 Ma, probably associated with the opening of the Drake Passage (Kennett, 1982). Increased bottom water circulation may have resulted in non-deposition or erosion of sediments in most ocean basins. At Site 576 this event may be correlated with the geotechnical contact at 41 m. Geotechnical properties at this depth show slight downhole changes such as decreased water contents and increased Su/Po′ ratios (Figures 22 and 24) that may be indicative of erosional processes. The Eocene-Oligocene boundary of 38 Ma is marked by another hiatus maxima. Kennett (1982) describes this hiatus maxima, which occurred into early Oligocene, as a response to major oceanic cooling, formation of Antarctic sea ice and increased production of Antarctic Bottom Water. The result of these events led to increased northward bottom water circulation and development of hiatuses throughout the Western Pacific.

The contact of geotechnical units IB and IC matches the chronostratigraphic interval defined by ichthyolith stratigraphy for this boundary. Most geotechnical properties exhibit a change in character across this boundary; however, the manner in which these changes occur is not the expected response to removed overburden. Again, the aspect in which different factors interact to form the measured geotechnical properties of the sediment creates difficulties in applying a simple model to account for the observed changes. In addition, the lack of complete knowledge of sedimentary processes that were active at this site precludes forming definitive statements concerning the effects of these on physical properties. The contact at 41 m suggests a compositional shift takes place across this boundary, or that sediment composition and loading after the event responsible for the hiatuses was such that it restricted normal consolidation of the underlying section. The time between approximately 37 and 18 Ma represents a period of hiatus minima in the Northwest Pacific. Kennett (1982) described a major pulse of volcanism occurred between 30 to 12 Ma associated with activity in the Bonin-Mariana-Yap and Northwest Japan arcs. The increase of volcanic ash input to the red clay sediment would result in an increase of smectite, which may be the source of the smectite peak found at approximately 39 m sub-bottom at Site 576 (Figure 19). The pulse may also be reflected in the geotechnical properties indicated as a downhole increase of water content at 37 m (Figure 22), with related shifts of other physical properties. Inadequate sampling coverage in the depth range near 37 m sub-bottom makes it difficult to distinguish between factors operative at 37 or 41 m. Measurements of finer scale fluctuations would probably resolve the difference between these two events.

The middle Miocene contains another widespread event correlated with the formation of North Atlantic deep water and the formation of the major icecap of East Antarctica. The event led to numerous sedimentary hiatuses and is dated between 17 and 18 Ma. The event may correspond to the 35 m event signalled in the geotechnical stratigraphy (Figure 40). This hiatus is reflected by a sharp drop of water content across the contact with a corresponding increase in bulk density (Figure 22). This geotechnical marker also correlates with a shift in sediment texture and mineralogy indicated by an increase in the percent of clay, in general, and of smectite in particular (Figure 19). Increases of either clay or smectite, however, generally result in increased water content which is just the opposite of the observed trend. Once again, an erosional or depositional event is a more likely cause for the observed changes in physical properties.

The lithologic contact at 28 m is probably related to the onset of increased volcanic and terrigenous input to the Northwest Pacific during the Miocene. This increase of terrigenous material continued until reaching its culmination in Pliocene and Pleistocene times. The late Miocene is also marked by another occurrence of hiatuses related to shifts in spreading directions and restricted equatorial flow. At that time both the Indo-Pacific and Panama passages were restricted (Moore et al., 1978). The record of events creating these hiatuses may be included or masked by the lithologic changes occurring at the 28 m level. It is at this depth that this event is expected to occur based on correlation with the biostratigraphic record.

Between 6.5 and 5 Ma, the latest Miocene had intensified deep water circulation and continued cooling (Kennett, 1982). A major regression took place during this time identified in seismic sequences worldwide (Vail et al., 1977). Reflector R2a and the geotechnical stratigraphic boundary between units IA and IB can be matched to this event. Abrupt decreases in water content, the change in position of natural water content relative to its liquid limit, and a shift in silt content mark the downhole changes taking place across this contact (Figures 22 and 19).

The uppermost two geotechnical stratigraphic boundaries at Site 576 are correlated with seismic reflectors. Reflector R2, at approximately 18 m sub-bottom, was correlated to the widespread occurrence of an ash layer (Jacobi et al., 1985). Pliocene-Pleistocene volcanic activity was greatly increased over previous volcanism and larger amounts of terrigenous, eolian, and detrital material were supplied to the Northwest Pacific Basin. The geotechnical contact at 16 m corresponds closely to the geochronometric age datum of approximately 1.6 to 1.7 Ma. Vail et al. (1977) described a global sea level regression event occurred at this time although existence of widespread hiatuses were not documented for this age by Moore et al. (1978). The boundary marks a weak geotechnical contrast and correlates with the set of weak, discontinous reflectors described by Jacobi et al. (1985). Both stratigraphic markers at this depth may correspond to a downhole decrease of silt content and to a peak in smectite (Figure 19).

The uppermost geotechnical contact and reflector R1 occur at a sub-bottom depth of approximately 10 m. The occurrence of an ash layer at this depth may be the main source of the strong, continuous reflector found extensively throughout this region. At this same depth, though, a major reversal of the geotechnical characteristics of the red clays takes place indicated by a change of gradient directions (Figure 22). For example, water contents increase from 0 to 10 m, then shift and decrease below 10 m. Associated with this boundary is the first measurable amount of smectite, with a concommitant decrease of illite in the clay fraction (Figure 19). The overconsolidated aspect of the uppermost section approaches normal consolidation at this depth as seen in both the OCR and Po′-Pc′ profiles (Figures 26 and 27). The change of geotechnical character at this depth is most likely a function of the compositional and mechanical properties of the sediment. An abrupt change in clay mineralogy and the increasing importance of overburden relative to the internal strength of the sediment becomes critical at this depth, marking a shift toward a normally consolidating sedimentary unit.

V. CONCLUSIONS

Typical geotechnical studies of oceanic sediments have often focused on the direct relationship of physical and mechanical properties to sedimentary characteristics such as composition, mode of emplacement, or depositional environment. Such studies have led to increased understanding of the interrelationships of these different aspects. However, a systematic approach to distinguishing geotechnically distinct sedimentary horizons within a given deposit has been lacking.

DSDP coring of two relatively deep sedimentary sections allowed this concept to be developed and tested for its correlation to standard geological studies. The hypothesis tested in this study is that geotechnical properties can be used to define sediment units from a new approach and at the same time complement other conventional stratigraphies. Discussion of the techniques used for defining boundaries of geotechnical stratigraphic units, the correlation of these boundaries with conventional stratigraphies, and the interpretive value of geotechnical properties relative to geological processes all form part of a positive test of this hypothesis.

The primary results and advantages of studying the geotechnical stratigraphy of sedimentary sections are: (1) sediments are progressively sub-divided in a vertical succession according to distinct geotechnical trends and values, (2) the different geotechnical units represent sections of sediment having either, or both, distinct mechanical behavior and/or different stress histories, and (3) modelling of sediment compaction or decompaction, such as in subsidence problems or geometric reconstructions of sedimentary deposits, has a qualitative base for defining the analytical, quantitative treatment of mechanically, or historically distinct sediments.

The use of geotechnical stratigraphy offers a means not only to analyze physical and mechanical behavior of sediment sections, but also used to refine and correlate other stratigraphic studies. This research project has demonstrated that: (1) distinct geotechnical stratigraphic units reflect both lithologic variations and distinct historical changes in sedimentary processes, (2) strong correlations are found between the contacts identifying geotechnical stratigraphic units with geochronostratigraphic, biostratigraphic, lithostratigraphic and seismic stratigraphic contacts, despite lack of complete intercorrelations between the conventional stratigraphies, (3) presence of sedimentary hiatuses, either as non-depositional or erosional, are readily identified by geotechnical stratigraphic boundaries, (4) chronostratigraphic definition of sediment sections poor in fossiliferous remains can be further constrained through association with geotechnical stratigraphic definition of the sediment column, and (5) the geotechnical characteristics of oceanic, pelagic sediments are strongly dominated by mechanical consolidation, sediment source (lithology), and oceanographic processes leading to the development of sedimentary hiatuses.

REFERENCES

Baltuck, M., Taylor, E., and McDougall, K., Mass movement along the inner wall of the mid-America Trench off Costa Rica, in *Initial Reports of DSDP,* U.S. Government Printing Office, Washington, D.C., 84, 745-766, 1985.

Bennett, R., Lambert, D. N., and Hulbert, M. H., Geotechnical properties of a submarine slide area on the U.S. continental slope northeast of Wilmington Canyon, *Mar. Geotechnol.,* 2, 245-261, 1977.

Bennett, R., Freeland, G. L., Lambert, D. N., Sawyer, W. B., and Keller, G. H., Geotechnical properties of surificial sediments in a mega-corridor. U.S. Atlantic continental slope, rise and deep sea basin, *Mar. Geol.,* 38, 123-140, 1980.

Bennett, R., Bryant, W. R., and Keller, G., Clay fabric of selected submarine sediments: fundamental properties and models, *J. Sediment. Petrol.,* 51, 217-232, 1981.

Berger, W. H., Deep-sea sedimentation; in *The Geology of Continental Margins,* Burke, C. A., and Drake, C. L., Eds. Springer-Verlag, New York, 1974, 213-241.

Biscayne, P. E., Distinction between kaolinite and chlorite in Recent sediments by X-ray diffraction, *Am. Mineral.,* 49, 1281-1289, 1964.

Biscayne, P. E., Mineralogy and sedimentation of recent deep-sea clay in the Atlantic Ocean and adjacent seas and oceans, *Geol. Soc. Am. Bull.,* 76, 803-832, 1965.

Bolli, H. M., Ryan, W. B. F., et al. *Initial Reports of the Deep Sea Drilling Project,* No. 40, U.S. Government Printing Office Washington, D.C., 1978.

Booth, J. S., Recent history of mass wasting on the upper continental slope, Northern Gulf of Mexico, as interpreted from the consolidation states of the sediment, *Soc. Econ. Paleontol. Mineral, Spec. Publ.* 27, 153-164, 1979.

Bouma, A., Bryant, W. R., and Davies, D. R., Topics and techniques, Gulf of Mexico, Texas A&M Univ. Res. Pound. Ref. 69-5T, 148p., 1969.

Boyce, R. E., Definitions and laboratory techniques of compressional sound velocity parameters and wet water content, wet bulk density and porosity parameters by gravimetric and gamma ray attenuation techniques, in *Initial Reports of DSDP,* U.S. Government Printing Office, Washington, D.C., 33, 931-958, 1976.

Boyce, R. E., Deep Sea Drilling Project procedures for shear strength measurement of clayey sediment using modified Wykeham Farrence laboratory vane apparatus, in *Initial Reports of DSDP,* U.S. Government Printing Office, Washington, D.C., 36, 1059-1068, 1977.

Boyce, R. E., Deep Sea Drilling Project drill Sites 530 and 532 in the Angola Basin and on the Walvis Ridge: interpretation of induction log data, sonic log data, and laboratory sound velocity, density, porosity-derived reflection coefficients, and vane shear strength, in *Initial Reports of DSDP,* U.S. Government Printing Office, Washington, D.C., 75 (part 2), 1137-1177, 1984.

Bryant, W. R., Deflache, A. P., and Trabant, P. K., Consolidation of marine clays and carbonates, in *Deep Sea Sediments — Physical and Mechanical Properties,* Inderbitzen, A. L., Ed., Plenum Press, New York, 1974, 209-214.

Bryant, W. R., Bennett, R. H., and Katherman, C. E., Shear strength, consolidation, porosity, and permeability of oceanic sediments, in *The Sea,* vol. 7, Emiliani, C., Ed., John Wiley & Sons, New York, 1981, 1555-1616.

Busch, W. H. and Keller, G. H., The physical properties of Peru-Chile continental margin sediments — the influence of coastal upwelling on sediment properties, *J. Sediment. Petrol.* 51, 705-719, 1981.

Casagrande, A., The determination of the pre-consolidation load and its practical significance, *Proc. Int. Conf. Soil Mech. Found. Eng.,* 3, 60-64, 1936.

Chiou, W. A., Clay Fabric of Gassy Submarine Sediments, Ph.D. dissertation, Texas A&M University, 248p., 1981.

Dadey, K. A., Stress history of unlithified marine sediments in the Northwest Pacific, M.Sc. thesis, University of Rhode Island, 148p., 1983.

Damuth, J. E., Jacobi, R. D., and Hayes, D. E., Sedimentation processes in the Northwest Pacific Basin revealed by echo-character mapping studies, *Geol. Soc. Am. Bull.,* 94, 381-395, 1983.

Doyle, P. S. and Riedel, W. R., Ichthyolith biostratigraphy of Western North Pacific pelagic clays, DSDP Leg 86, in *Initial Reports of DSDP,* U.S. Government Printing Office, Washington, D.C., 86, 349-366, 1985.

Folk, R. L., *Petrology of Sedimentary Rocks,* Hemphill Publishing, Austin, TX, 1974, 182p.

Gardiner, J. V., Dean, W. E., and Wilson, C. R., Carbonate and organic carbon cycles and the history of upswelling at Deep Sea Drilling Project Site 532, Walvis Ridge, South Atlantic ocean, in *Initial Reports of DSDP,* U.S. Government Printing Office, Washington, D.C., 75(part 2), 905-921, 1984.

Gorsline, D. S., Deep water sedimentologic conditions and models, *Mar. Geol.,* 38, 1-22, 1980.

Hamilton, E. L., Consolidation characteristics and related properties of sediments from the experimental Mohole (Guadalupa Site), *J. Geophys. Res.* 69, 4257, 1964.

Hamilton, E. L., Prediction of deep-sea sediment properties: state of the art, in *Deep-sea Sediments, Physical and Mechanical Properties,* Inderbitzen, A. L., Ed. Marine Science, 2, Plenum Press, New York, 1970, 1-43.

Hamilton, E. L., Prediction of in-situ acoustic and elastic properties of marine sediments, *Geophysics,* 36, 266-284, 1971.

Hamilton, E. L., Variations of density and porosity with depth in deep sea sediments, *Sediment. Petrol.,* 46, 280-300, 1976.

Hay, W. W., Sibuet, J. C., et al., *Initial Reports of the Deep Sea Drilling Project,* No. 75 (part 1), U.S. Government Printing Office, Washington, D.C., 1984.

Heath, G. R., Burkle, L. H., et al., *Initial Reports of the Deep Sea Drilling Project,* No. 86, U.S. Government Printing Office, Washington, D.C., 1985, 804 p.

Heath, G. R., Rea, D. H., and Levi, S., Paleomagnetism and accumulation rates of sediments at Site 576 and 578, DSDP Leg 86, Western North Pacific, in *Initial Reports of DSDP,* U.S. Government Printing Office, Washington, D.C., 86, 459-502, 1985.

Hein, F. J. and Gorsline, D. S., Geotechnical aspects of fine-grained mass flow deposits; California Continental borderland, *Geo-Marine Letters,* 1, 1-5, 1981.

Helwick, S. J. and Bryant, W. R., Prediction of the geotechnical properties of Late Quaternary Mississippi Delta deposits, Tech. Report 80-12-T, Texas A&M University, 243p., 1908.

Hollister, C. D. et al, Geological assessment of the MPG-1 region, Central North Pacific, in *Sub-Seabed Disposal-Program, Annual Report,* Jan.-Dec., 1978, 2, Talbert, D. M., Ed., SAND 79, 1678, 9-120, 1979.

Horn, D. R., Delach, M. N., and Horn, B. M., Physical properties of sedimentary provinces, North Pacific and North Atlantic Ocean, in *Deep-Sea Sediments, Physical and Mechanical Properties,* Inderbitzen, A. L., Ed., Marine Science, 2, Plenum Press, New York, 1974, 417-441.

Hottman, W. E., Physical properties of sediments from the Continental Margin of western Africa, Ph.D. dissertation, Texas A&M University, 126p., 1978.

Hottman, W. E., Areal distribution of clay minerals and their relation to physical properties, M.S. Thesis, Texas A&M Univ., 53p. 1975.

Jacobi, R. D., Hayes, D. E., and Damuth, J. P., High frequency seismic studies near DSDP Sites 576 and 578, in *Initial Reports of DSDP,* U.S. Government Printing Office, Washington, D.C., 86, 23-50, 1985.

Keller, G. R., Mass properties of the seafloor in a selected depositional environment, Proc. Civil Eng. in the Oceans, II, Miami Beach, FL, 1969, 857-877.

Keller, G. R. and Bennett, R. H. Mass physical properties of submarine sediments in the Atlantic and Pacific Basins, in Proc. XXIII Internat. Geol. Congress, 8, 1968, 33-50.

Keller, G. R., Lambert, D. N., and Bennett, R. H., Geotechnical properties of continental slope deposits — Cape Hatteras to Hydrographic Canyon, in *Geology of Continental Slopes,* Doyle, L. J. and Pilkey, O. H., Eds., *Soc. Econ. Paleon. Mech. Spec. Publ.* 27, 1979, 131-151.

Kennett, J. P., *Marine Geology,* Prentice-Hall, Englewood Cliffs, NJ, 1982, 813p.

Lambe, T. W., 1951, *Soil Testing for Engineers,* John Wiley & Sons, New York, 1951, 165p.

Lambe, T. W. and Whitman, R. V., *Soil Mechanics,* John Wiley & Sons, New York, 1969, 553p.

Leg 86 Shipboard Party, Challenger investigates NW Pacific, *Geotimes,* 27(11), 20-22, 1982.

Leinen, M., Quartz content of Northwest Pacific Site 576A and implications for Cenozoic eolian transport, in *Initial Reports of DSDP,* U.S. Government Printing Office, Washington, D.C., 86, 581-588, 1985.

Leinen, M. and Heath, G. R., Paleochemical signatures in Cenozoic sediments from the Central North Pacific, *Geochem. Cosmochim. Acta.,* 1985.

Lisitzin, A. P., Sedimentation in the World Ocean, *Soc. Econ. Paleon. Mech. Spec. Publ.* 17, 218p., 1972.

Lowe, J., Zaccheo, P. P., and Feldman, H. S., Consolidation testing with back-pressure, *J. Soil Mech. Found. Eng. Div., Am. Soc. Civil Eng.,* 90, 69-86, 1964.

Marine Geotechnical Consortium, Geotechnical properties of sediments from Walvis Ridge, Deep Sea Drilling Project, Leg 75, Hole 532A, in *Initial Reports of DSDP,* U.S. Government Printing Office, Washington, D.C., 75, 1109-1127, 1984.

Marine Geotechnical Consortium, Geotechnical properties of Northwest Pacific pelagic clays, Deep Sea Drilling Project, Leg 86, Hole 576A, in *Initial Reports of DSDP,* U.S. Government Printing Office, Washington, D.C., 86, 723-758, 1985.

Mayer, L. A., Physical properties of sediment recovered on Deep Sea Drilling Project Leg 68 with the hydraulic piston corer, in *Initial Reports of DSDP,* U.S. Government Printing Office, Washington, D.C., 68, 403-408, 1982.

Moore, T. C., van Andel, Tj. H., Sancetta, C., and Pisias, N., Cenozoic hiatuses in pelagic sediments, *Micropaleontology,* 24, 113-138, 1978.

Morely, J. J., Radiolarian biostratigraphy of sediments from the Northwest Pacific, DSDP Leg 86, in *Initial Reports of DSDP,* U.S. Government Printing Office, Washington, D.C., 86, 399-422, 1985.

Nacci, V. A., Kelly, W. E., Wang, M. C., and Demars, K. R., Shear and stress-strain characteristics of cemented deep-sea sediments, in *Deep-Sea Sediments; Physical and Mechanical Properties,* Inderbitzen, A. L., Ed., Plenum Press, New York, 129-150, 1974.

Rabinowitz, P. D. and Simpson, E. S. W., 1984. Geophysical site survey results on the Walvis Ridge: in *Initial Reports of DSDP,* U.S. Government Printing Office, Washington, D.C., 74, 795-826, 1984.

Reimers, C. E., Organic matter in anoxic sediments off Central Peru: Relations of porosity, microbial decomposition and deformation properties, *Mar. Geol.,* 46, 175-197, 1982.

Richards, A. F., *Marine Geotechnique,* University of Illinois Press, Urbana, 326p. 1967.

Richards, A. F. and Hamilton, E. L., Investigations of deep sea sediment cores, III. Consolidation, in *Marine Geotechnique,* Richards, A. F., Ed., University of Illinois Press, Urbana, 93-117, 1967.

Richardson, M. D. and Young, D. K., Geoacoustic models and bioturbation, *Mar. Geol.* 38, 205-218, 1980.

Rieke, H. H., III, and Chilingarian, G. V., *Compaction of Argillaceous Sediments, Developments in Sedimentology 16,* Elsevier Scientific Publishing, New York, 1974, 424p.

Rowe, G. T., The effects of benthic fauna on the physical properties of deep-sea sediments, in *Deep Sea Sediments, Physical and Mechanical Properties,* Inderbitzen, A. L., Ed., Marine Science, 2, Plenum Press, New York, 1974, 381-400.

Schink, J. C., Stockwell, J. H., and Ellis, R. A., An improved device for gasometric determination of carbonate in sediment: *J. Sediment. Petrol.,* 49, 651-653, 1979.

Schultheiss, P. J., Physical and geotechnical properties of sediments from the Northwest Pacific; DSDP Leg 86, in *Initial Reports of DSDP,* U.S. Government Printing Office, Washington, D.C., 86, 701-722, 1985.

Shephard, L. E., Bryant, W. R., and Dunlap, W. A., Geotechnical properties and their relation to geologic processes in South Pass Outer Continental Shelf Lease Area Blocks 28, 47, and 48, Offshore Louisiana, Tech. Report 79-5-T, Texas A&M University, 1979, 125p.

Shephard, L. E., Bryant, W. R., and Chiou, W. A., Geotechnical properties of Middle America Trench sediments, Deep Sea Drilling Project Leg 66, in *Initial Reports of DSDP,* U.S. Government Printing Office, Washington, D.C., 66, 475-504, 1982.

Siesser, W. G., Leg 40 results in relation to continental shelf and onshore geology, in *Initial Reports of DSDP,* U.S. Government Printing Office, Washington, D.C., 40, 965-979, 1978.

Siesser, W. G., Late Miocene origin of the Benguela upwelling system off Northern Nambia, *Science,* 208, 283-285, 1980.

Silva, A. J., Marine geomechanics: Overview and projections, in *Deep-Sea Sediments; Physical and Mechanical Properties,* Inderbitzen, A. L., Ed., Marine Science, 2, Plenum Press, New York, 1974, 45-76.

Skempton, A. W., Discussion on the planning and design of the New Hong Kong Airport, by Grace and Henry, *Proc. Inst. Civ. Eng.,* 7, 305-307, 1957.

Skempton, A. W., The consolidation of clays by gravitational compaction, *Q. J. Geol. Soc. London,* 125, 373-411, 1970.

Taylor, E. and Bryant, W. R., Geotechnical properties of sediments from the Mid-America Trench and Slope, in *Initial Reports of DSDP,* U.S. Government Printing Office, Washington, D.C., 84, 745-766, 1985.

Taylor, E., Oceanic sedimentation and geotechnical stratigraphy: hemipelagic carbonates and red clays. Ph.D. dissertation, Texas A&M University, 1984, 144p.

Terzaghi, K., *Theoretical Soil Mechanics,* John Wiley & Sons, New York, 1943.

Vail, P. R., Mitchum, R. M., Jr., Todd, R. G., Widmier, J. M., Thompson, S., III, Sangree, J. N., Bubb, J. N., and Hatlelid, W. G., Seismic stratigraphy and global changes of sea level, in *Seismic Stratigraphy — Applications to Hydrocarbon Exploration,* Payton, C. E., Ed., Am. Assoc. Petrol. Geol. Mem. 26, 1977, 49-212.

van Andel, Tj. H., Heath, G. R., and Moore, T. C., Cenozoic history and paleoceanography of the central equatorial Pacific ocean, *Geol. Soc. Am. Mem.,* 143, 1975.

Walker, G. R., Geotechnical properties of deep sea sediments in the Northwestern Pacific Ocean, M.Sc. thesis, University of Rhode Island, 1981, 131p.

Walton, W. H., Sangrey, D. A., and Miller, S. A., 1983. Geotechnical engineering characterization of hydraulically piston-cored deep ocean sediments, in *Initial Reports of DSDP* U.S. Government Printing Office, Washington, D.C., 72, 537-549, 1983.

INDEX

A

B

C

D

E

F

G

H

M

N

Q

R

T

U

V

W

Y

Z